STRUCTURAL HEALTH MONITORING

STRUCTURAL HEALTH MONITORING
A MACHINE LEARNING PERSPECTIVE

Charles R. Farrar
Los Alamos National Laboratory, USA

Keith Worden
University of Sheffield, UK

A John Wiley & Sons, Ltd., Publication

This edition first published 2013
© 2013 John Wiley & Sons, Ltd

Registered office
John Wiley & Sons Ltd, The Atrium, Southern Gate, Chichester, West Sussex, PO19 8SQ, United Kingdom

For details of our global editorial offices, for customer services and for information about how to apply for permission to reuse the copyright material in this book please see our website at www.wiley.com.

Library of Congress Cataloging-in-Publication Data

Farrar, C. R. (Charles R.)
  Structural health monitoring : a machine learning perspective / Charles R. Farrar, Keith Worden.
    p. cm.
  Includes bibliographical references and index.
  ISBN 978-1-119-99433-6 (cloth)
  1. Structural health monitoring.   I. Worden, K.   II. Title.
  TA656.6.F37 2012
  624.1′71–dc23

                                                          2012018036

A catalogue record for this book is available from the British Library and the Library of Congress.

ISBN: 978-1-119-99433-6

Typeset in 9/11pt Times by Aptara Inc., New Delhi, India

Contents

Preface

This book is the result of the two authors' collaborations on the subject of Structural Health Monitoring (SHM) dating back to the mid-1990s. It is interesting to point out that the two authors, somewhat independently and at approximately the same time, decided that the statistical pattern recognition approach to SHM was the best framework to use for these problems. With this realisation, it was then possible to tap into the extensive set of pattern recognition algorithms developed by the statistics and machine learning communities and effectively apply them to the damage detection problem even though in almost all cases they were not originally developed with damage detection in mind. This viewpoint on SHM has developed to a major extent within the literature and it is therefore a primary goal for this book to provide the readers with a summary of the structured statistical pattern recognition approach and introduce them to the tools of machine learning as they are applied in the context of the SHM problem. In addition, the authors hope that the readers will appreciate the very general nature of this approach and see that it is well suited to deal with all the sources of variability encountered in any real-world damage detection problem.

The authors believe that the material is presented at a level suitable for upper-level undergraduates, postgraduate students and practising engineers. In fact, much of the material in this book is based on industry short courses that have been taught more than 20 times since 1997, a graduate class on SHM that has been taught numerous times at the University of California, San Diego and a graduate class taught since 2002 at the University of Sheffield. An attempt has been made to produce a book that is as self-contained as possible and the two appendices on signal processing and linear structural dynamics were added with that goal in mind. As such, readers at the levels listed above with an appropriate numerate background should find that they do not need too many prerequisites to understand the majority of the material presented in this book. However, readers should keep in mind that SHM is a very multidisciplinary topic and the materials presented herein would 'traditionally' be covered to some degree in courses on structural dynamics, nondestructive evaluation, signal processing, detection theory, machine learning, probability and statistics, and sensor networks. When trying to cover such a broad range of technologies, the authors here were faced with the formidable challenge of striking a balance between a complete and detailed treatment of these subjects while maintaining the book at a reasonable length. This has led to difficult decisions regarding the content; clearly, many chapters could themselves be expanded into books (e.g. Chapter 4, Sensing and Data Acquisition Issues). The tack taken here is to explain the material in sufficient detail that the reader can understand the concepts and key issues. The authors then make use of citations to point the reader to more detailed summaries of the various topics. Finally, it should be pointed out that the material is presented serially, following the statistical pattern recognition paradigm introduced in the first chapter, but the authors recommend that this statistical pattern recognition paradigm should actually be thought of and implemented in a much more integrated manner.

In addition to presenting the theory for the various aspects of the SHM process, the authors have tried to demonstrate the concepts in a variety of manners. These demonstrations include numerical simulations, tests on well-controlled laboratory experiments specifically designed for SHM and, when possible, a wide variety of examples on 'real-world' structures ranging from civil infrastructure to telescopes to aircraft. These demonstrations are not only supposed to show how a particular method works but they are also intended to highlight issues and challenges with these various methods.

When one states that they are going to take a statistical pattern recognition approach to the SHM problem and make use of machine learning algorithms to implement this approach, there is often the misconception that this precludes the use of physical models. The development of such physical models has traditionally been the mainstay of engineering research. However, because of the widely varying length and time scales associated with damage initiation and evolution, the geometric complexity of most real-world systems and the inevitable operational and environmental variability encountered in most SHM problems, such physical modelling for SHM becomes challenging. It should be made clear that the statistical pattern recognition approach, which is the foundation for this book, in no way precludes the use of such physical models. If such models are available, and have been validated, the SHM process can only be improved based on insights gained from these models. However, by taking a pattern recognition approach, we are not constraining ourselves to a particular model form and, in a sense, we are allowing the structure to 'talk to us' more directly.

Finally, it must be acknowledged that for most applications SHM is still primarily a research topic. The authors believe this book gives an up-to-date summary of the field and provides the most general approach to addressing the SHM problem that has been proposed to date. However, it is anticipated that this technology will continue to evolve with new methodologies continually being proposed. Even for the topics presented herein, the discussions are by no means complete. As an example, in Chapter 7 there are many more features that have been proposed in the literature that are not summarised in this chapter. Therefore, the readers are encouraged to seek out the other books that have been published on SHM, most of which have appeared in the last ten years, for a broader treatment of this subject. Because of the life-safety and economic advantages that SHM solutions can provide, the future is very bright for this technology. The applications of SHM are expanding and diversifying. As such, the authors hope this book contributes to the 'health' of this technical field.

Acknowledgements

The work summarised in this book represents contributions from numerous students and colleagues who both authors have been interacting with over the last 20 years. An attempt is made to identify the many people whose work is in some way highlighted in this book. The difficulty with such acknowledgements is that inevitably someone is inadvertently left out and the authors would like to apologise for such omissions up front. These acknowledgements are spelled out chapter by chapter for each author

Chuck Farrar's Acknowledgements

A great deal of thanks must go to the UK Royal Academy of Engineering for the award of a Distinguished Visitor Fellowship for this author; that fellowship was used to complete large portions of this book. Next, Mike Todd from the University of California, San Diego (UCSD) and Gyuhae Park from Los Alamos National Laboratory (LANL) are gratefully acknowledged. They have helped develop, evolve and run the LANL-UCSD Engineering Institute for the last 10 years where it has grown from a dynamics summer school into an international education and research collaboration. Through their endeavors, this institute has been able to maintain a focus on SHM research and keep this research thrust active and growing throughout this time. Furthermore, this author has enjoyed the support of all his LANL managers over the years he has pursued SHM research. In particular, he would like to acknowledge the support and friendship of Steve Girrens, who is currently the Associate Director for Engineering at LANL. This author has worked with Steve for 29 years and his backing and encouragement have been extremely valuable from so many different perspectives.

Chapter 1. For this author the initial ideas for the statistical pattern recognition paradigm, which is the foundation of this book, came primarily from discussions in the mid-1990s with two computer scientists at LANL, Vance Faber and Dave Nix, who used a similar approach to address many of their pattern classification problems.

Chapter 2. Much of the material for this overview has come from two extensive literature reviews that were the effort of many contributing authors. The lead authors on these reviews were Scott Doebling, when he was a postdoctoral fellow at LANL, and Hoon Sohn, when he was a technical staff member at LANL. Tom Duffey is also acknowledged for contributing to the portion of this chapter related to rotating machinery.

Chapter 4. The wireless sensor node work highlighted in this chapter has been the work of a group of graduate research assistants at LANL starting with Neal Tanner and followed by David Allen, Jarrod Dove, David Mascarenas, Tim Overly, and currently led by Stuart Taylor. The helicopter wireless energy delivery system was the focus of David Mascarenas' PhD dissertation. LANL technical staff members Gyuhae Park and Kevin Farinholt have been the primary mentors for these students. Gyuhae Park has also led the impedance-based sensing work briefly summarised in this chapter.

Chapter 5. Many people from New Mexico State University (NMSU), Sandia National Laboratory (SNL) and LANL took part in the I-40 Bridge test. First, Ken White from NMSU must be acknowledged as the tests were his idea and he secured the funding for this project. Albert Migliori is also acknowledged for inviting this author to participate in this project, which is how he got started in the field of SHM. Two students, Kerry Cone from the University of New Mexico (UNM) and Wayne McCabe from NMSU, along with Bill Baker from UNM and Randy Mayes from SNL were key to the success of this project.

Gerard Pardoen from the University of California–Irvine must be thanked for inviting this author to participate in his concrete column testing program. This work was carried out with the help of Phil Cornwell from the Rose-Hulman Institute of Technology and Erik Strasser and Hoon Sohn, who at that time were graduate students at Stanford University.

Bill Baker from UNM is again acknowledged for his work in designing the eight-degree-of-freedom test structure. Similarly, David Mascarenas designed the simulated building structure and was responsible for its fabrication while he was a graduate research assistant at LANL.

As with the I-40 Bride tests, numerous people have participated in the many tests performed on the Alamosa Canyon Bridge since 1996. Again, Ken White at NMSU is credited with getting this bridge designated as a test structure by the New Mexico Highway and Transportation Department. Scott Doebling and Phil Cornwell led many of the tests whose results are summarized in this book.

Chapter 6. Brett Nadler and Jeni Wait, who were technical staff members at LANL, are thanked for their impact testing of the composite plates and subsequent ultrasonic scanning of these plates.

Chapter 7. Francois Hemez from LANL is gratefully acknowledged for his help with the section on temporal moments and guidance on the model updating sections of this chapter. Scott Doebling was also a significant contributor to the model updating sections. Gyuhae Park and Eric Flynn from LANL have provided considerable input to the guided wave and impedance measurements sections. Kevin Farinholt is thanked for his development of the test structure shown in Section 7.2 and Chris Stull from LANL is thanked for the numerical simulations summarised in Section 7.3. The COMAC values for the I-40 Bridge were computed by Eloi Figuereido as part of his graduate studies carried out at LANL. The load-dependent Ritz vectors are taken from work done by Hoon Sohn when he was a graduate student at Stanford University. Results from the I-40 Bridge shown throughout this chapter are taken from work done by David Jauregui from NMSU when he was a graduate research assistant at LANL. Dave Nix was the person who introduced this author to the use of time series models for damage detection. The mutual information results are taken from the work that Tim Edwards at SNL did as part of a project for this author's SHM class.

Chapter 8. Amy Robertson is acknowledged for her work on the Holder exponent that she performed when she was a technical staff member at LANL.

Chapter 10. The control charts shown in this chapter were generated by Eloi Figuereido as part of his graduate studies carried out at LANL. Chris Stull, Jim Wren and Stuart Taylor from LANL are acknowledged for their experimental and analytical work on the Raptor telescope project.

Chapter 11. The example on support vector regression is based on the work performed by Luke Bornn from the University of British Columbia when he was a LANL graduate research assistant.

Chapter 12. This author would like to thank Gyuhae Park for impedance measurement material presented in the sensor system design section. Mike Todd is acknowledged for providing the data from the composite-hull ship and Hoon Sohn is thanked for his extensive analysis of these data. Again, Eloi Figuereido must be thanked for the comparative study of the various machine learning algorithms that was part of his graduate studies carried out at LANL. Dustin Harvey did much of the work to define the look-up table example while he was a graduate research assistant at LANL.

Chapter 13. Once more, Gyuhae Park is thanked for the impedance measurement example used in this chapter.

Chapter 14. This chapter is taken extensively from an article that appeared in the *Philosophical Transactions of the Royal Society* and this author wants to thank the co-author of that article, Nick Lieven, currently a pro-vice chancellor at the University of Bristol, for allowing us to use this material in this book.

Keith Worden's Acknowledgements

In Sheffield, the main acknowledgements are to Wieslaw Staszewski (now in AGH University Krakow) and Graeme Manson, who have been working in the field of SHM with the author since he started and have been an unfailing source of ideas and support throughout. Keith would also like to thank Geof Tomlinson for providing the post-doctoral position that distracted him into SHM from the (then) comfort zone of nonlinear system identification. In the research climate prevalent now, it is almost impossible to make progress without the support of talented and dedicated PhD students and Research Associates; Keith considers himself blessed to have had the opportunity to collaborate with a stream of outstanding researchers (in very rough chronological order): Janice Dulieu-Barton, Cecilia Surace, Julian Chance, Andreas Kyprianou, Hoon Sohn, Karen Holford, Rhys Pullin, Mark Eaton, David Allman, Gaetan Kerschen, Tze Ling Lew, Daley Chetwynd, Gareth Pierce, Faizal Mustapha, Jose Zapico, Luis Mujica, Mike Todd, Gyuhae Park, Frank Stolze, Jyrki Kullaa, Arnaud Deraemaeker, Evangelos (Vaggelis) Papatheou, Rob Barthorpe, Thariq Hameed bin Sultan, Anees Ur Rehman and Lizzy Cross. Many of these people are still valued colleagues, although some inevitably escaped. Keith would also like to thank a stream of excellent and committed undergraduate and masters project students who carried out work to such a high standard that it was published. Many of the results in the current book were at least influenced by the work of such students and in some cases their actual results are included; if a direct attribution is absent in any case, this is an omission and is subject to apology. On a chapter-by-chapter basis, Keith would like to thank a number of people whom Chuck hasn't already acknowledged.

Chapter 9. The acoustic emission work was carried out using experimental data provided by Professor Karen Holford and Dr Rhys Pullin of the University of Wales Cardiff – some of the detailed analysis is the work of Steve Rippengill as part of his MEng project work. (AE data provided by Karen and Rhys also appears in the brief illustration of PCA in Chapter 6.)

Chapter 10. The results in Section 10.3 pertaining to operational variations are the result of joint work with Cecilia Surace of the Politecnico di Torino, Italy. It was through work with Cecilia that the importance of operational and environmental variations first impinged on Keith. The work on the Gnat aircraft presented in Chapter 10, like all work presented throughout on the Gnat, only happened because of financial support from the late Dr David Allman of DERA (now QinetiQ) and the collaboration (and saint-like patience) of Dr Graeme Manson (Sheffield). By supporting our SHM work, David allowed a much more thorough access to experimental validation than would have been possible for us otherwise; as a friend and colleague he is sadly missed. The section on control charts in Chapter 10 owes much to material Keith has learned from Jyrki Kullaa (Aalto University, Helsinki, Finland). Finally, the material on extreme value statistics owes a great deal to numerous discussions over the years with Professor Hoon Sohn of KAIST, South Korea.

Chapter 11. Again, Keith would like to thank Graeme Manson for his collaboration on the Gnat aircraft. The analysis of the Gnat data based on support vector machines was largely carried out by Alex Lane as part of his undergraduate project. In terms of genetic optimisation for feature seclection, Keith thanks Vaggelis Papatheou and Graeme Manson again; much of the detailed analysis in the final section was carried out as part of the final year project of Gabrielle Hilson.

Chapter 12. The material on cointegration in this chapter was provided by Dr Lizzy Cross, to whom sincere thanks are due.

Appendix B. The material on the composite beam experiment was provided by Mr Nikolaos (Nikos) Dervilis (Sheffield) and is appreciated.

Throughout, any data presented on Lamb wave propagation as part of a Sheffield collaboration almost always arose from work carried out with Dr Gareth Pierce and Professor Brian Culshaw of the University of Strathclyde, Glasgow, Scotland, or with Professor Wieslaw Staszewski (now AGH University, Krakow, Poland). Also, Keith would like to thank Dr Rob Barthorpe (Sheffield) for a number of interesting and enlightening conversations on SHM over recent years, particularly pertaining to model-based approaches and issues of validation and verification.

Finally, Keith would like to apologise to his children Anna and George for all the time spent writing this book (among other things) that was not spent with them. Thanks for your patience and understanding guys, I'll try and be better.

Overall from both authors, the foremost acknowledgement here must go to Dr Lizzy Cross for her careful proofreading of, and comments on, almost all of the manuscript. Her efforts ensured that (sometimes rough) drafts were converted into chapters and that the minimum of mistakes survived into the final versions. It should be said that any mistakes that do remain are the fault of the authors and if any readers become aware of such mistakes the authors would like to be informed (but politely ☺). Finally, the authors would like to thank a number of the Wiley editors, latterly Ms Liz Wingett, for their polite but firm attempts to stop them missing an infinite sequence of deadlines.

1

Introduction

Modern societies are heavily dependent upon structural and mechanical systems such as aircraft, bridges, power generation systems, rotating machinery, offshore oil platforms, buildings and defence systems. Many of these existing systems are currently nearing the end of their original design life. Because these systems cannot be economically replaced, techniques for damage detection are being developed and implemented so that these systems can continue to be safely used if or when their operation is extended beyond the design basis service life. Also, in terms of the design and introduction of new engineering systems, these often incorporate novel materials whose long-term degradation processes are not well understood. In the effort to develop more cost-effective designs, these new systems may be built with lower safety margins. These circumstances demand that the onset of damage in new systems can be detected at the earliest possible time in an effort to prevent failures that can have grave life-safety and economic consequences.

Damage detection is usually carried out in the context of one or more closely related disciplines that include: structural health monitoring (SHM), condition monitoring (CM), nondestructive evaluation (NDE) – also commonly called nondestructive testing, or (NDT), health and usage monitoring system (HUMS), statistical process control (SPC) and damage prognosis (DP).

The term *structural health monitoring* (SHM) usually refers to the process of implementing a damage detection strategy for aerospace, civil or mechanical engineering infrastructure. This process involves the observation of a structure or mechanical system over time using periodically spaced dynamic response measurements, the extraction of damage-sensitive features from these measurements and the statistical analysis of these features to determine the current state of system health. For long-term SHM, the output of this process is periodically updated information regarding the ability of the structure to continue to perform its intended function in light of the inevitable ageing and degradation resulting from the operational environments. Under an extreme event, such as an earthquake or unanticipated blast loading, SHM could be used for rapid condition screening, to provide, in near real time, reliable information about the performance of the system during the event and about the subsequent integrity of the system.

Condition monitoring is analogous to SHM, but specifically addresses damage detection in rotating and reciprocating machinery, such as that used in manufacturing and power generation (Worden and Dulieu-Barton, 2004).

Both SHM and CM have the potential to be applied on-line, that is during operation of the system or structure of interest. In contrast, *nondestructive evaluation* (NDE) is usually carried out off-line after the site of the potential damage has been located. There are exceptions to this rule, as NDE is also used as a monitoring tool for *in situ* structures such as pressure vessels and rails. NDE is therefore primarily

Structural Health Monitoring: A Machine Learning Perspective, First Edition. Charles R. Farrar and Keith Worden.
© 2013 John Wiley & Sons, Ltd. Published 2013 by John Wiley & Sons, Ltd.

used for damage characterisation and as a severity check when there is a priori knowledge of the damage location (Shull, 2002).

*Health and usage monitoring system*s (HUMSs) are closely related to CM systems, but the term has largely been adopted for the specific application to damage detection in rotorcraft drive trains (Samual and Pines, 2005). In that context, the health monitoring portion of the process attempts to identify damage, while the *usage monitoring* records the number of load cycles that the system experiences for the purposes of calculating fatigue life consumption.

Statistical process control (SPC) is process-based rather than structure-based and uses a variety of sensors to monitor changes in a process, with one possible cause of a change being structural damage (Montgomery, 2009).

Once damage has been detected, the term *damage prognosis* (DP) describes the attempt to predict the remaining useful life of a system (Farrar *et al.*, 2003).

Condition monitoring, NDE and SPC are without doubt the most mature damage detection disciplines as they have made the transition from a research topic to actual engineering practice for a wide variety of applications. However, it is a widely held belief that SHM is in the process of making the transition into the application domain. This book will focus primarily on SHM as the authors believe that the time has come for a comprehensive and fundamental exposition of the basic principles of this branch of damage detection.

1.1 How Engineers and Scientists Study Damage

Materials scientists and engineers are the primary classes of technologists that study damage; in this, they commonly approach the problem by asking one or more of the following questions (in no particular order):

1. What is the cause of damage?
2. What can be done to prevent damage?
3. Once present, how are the effects of damage mitigated?
4. Is damage present?
5. How fast will the damage grow and exceed some critical level?

The answers to these questions will depend on whether one takes a material science point of view or an engineering point of view. As an example, the materials scientist may address question 1 by studying the initial imperfections at the grain boundary scale as shown in Figure 1.1 and attempt to develop tools that predict how these imperfections coalesce and grow under various loading conditions. They might also study properties such as surface finish that result from the manufacturing process or develop an understanding of material ageing and degradation processes at the micro-scale. In contrast, the engineer may attempt to establish allowable strength, deformation or stability criteria associated with the onset of damage. A materials scientist might approach the second question by designing new materials that are less susceptible to a particular type of damage (e.g. use of stainless steel in corrosive environments) while the engineer might incorporate alternate design strategies for manufacturability and reliability or prescribe operational and environmental limits for system use. Damage mitigation strategies might be accomplished with the development of self-healing materials, which is currently a focus of materials science research (Zwaag, 2007). Alternatively, engineers will prescribe maintenance and repair or limit operations (e.g. slow the speed of a vehicle) as a damage mitigation strategy.

Questions 4 and 5 are the focus of SHM and here the difference between how the material scientists and engineers address the problem is related to the length scale on which they study the problem. Additionally, a distinction arises based on the ability to do the damage assessment with the system in operation or if the assessment needs to be performed with the system in or out of service. More

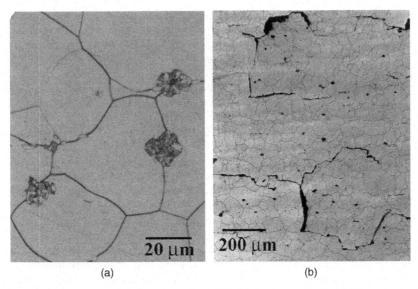

(a)   (b)

Figure 1.1 (a) Inclusions at the grain boundaries in U-6Nb. (b) A micrograph of a U-6Nb plate showing crack propagation along inclusion lines after shock loading (source: D. Thoma, Los Alamos National Laboratory).

drastically, the assessment may be carried out in a destructive manner. Materials scientists will often perform damage detection at the microscopic level using thin sectioning of the material to recreate a three-dimensional image of the microstructure. As previously mentioned, traditional NDE methods are applied to assess incipient macroscopic damage at the material and component level, typically with the system out of service. Wave propagation approaches to SHM, which can be used to assess damage with the system in operation, are also being used to assess incipient damage at the macroscopic material and component scale. Finally, other forms of SHM, like vibration-based approaches, can also be used to assess damage from the component to full system scale, as can CM, HUMS and SPC.

1.2   Motivation for Developing SHM Technology

Almost all private and government industries want to detect damage in their products as well as in their manufacturing infrastructure at the earliest possible time. Such detection requires these industries to perform some form of SHM and is motivated by the potential life-safety and economic impact of this technology. As an example, the semiconductor manufacturing industry is adopting this technology to help minimise the need for redundant machinery necessary to prevent inadvertent downtime in their fabrication plants. Such downtime can cost these companies on the order of millions of dollars per hour. Aerospace companies, along with government agencies in the United States, are investigating SHM technology for detection of damage to space shuttle control surfaces hidden by heat shields. Clearly, such damage detection has significant life-safety implications. Also, as an example from the civil engineering context, there are currently no quantifiable methods to determine if buildings are safe for reoccupation after a significant earthquake. SHM technology may one day provide a means of minimising the uncertainty associated with current visual post-earthquake damage assessments. The prompt reoccupation of buildings, particularly those associated with manufacturing, can significantly mitigate economic losses associated with major seismic events. Finally, many portions of our technical infrastructure are approaching or exceeding their initial design life. As a result of economic issues, these civil, mechanical and aerospace structures are being used in spite of ageing and the associated damage accumulation. Therefore, the ability to monitor the health of these structures is becoming increasingly important.

Maintenance philosophies have evolved to minimise the potential negative life-safety and economic impacts of unforeseen system failures. Initially, *run-to-failure* approaches to engineering system maintenance were used. With this approach the system is operated until some critical component fails and then that component is replaced. This procedure requires no investment in monitoring systems, but it can be extremely costly as failure can occur without warning. Clearly, this approach to maintenance is unacceptable when life-safety is a concern.

A more sophisticated maintenance approach that is used extensively today is referred to as *time-based maintenance*. This maintenance approach requires that critical components are serviced or replaced at predefined times or use intervals regardless of the condition of the component. A typical example is the recommendation that one changes the oil in their car after it has been driven a certain distance or at some prescribed time interval. This maintenance is done regardless of the condition of the oil. Another example is the requirement that a missile be retired after a certain number of captive-carry flight hours on the wing of an aircraft. Time-based maintenance is a more proactive approach than run-to-failure and it has made complex engineering systems such as commercial aircraft extremely safe. In some cases usage monitoring systems are deployed in conjunction with the time-based maintenance approach. Such a system might record the number of manoeuvres performed by a high-performance aircraft that exceed a certain threshold acceleration level. Maintenance would then be performed after the aircraft has accumulated some predefined number of these peak acceleration readings.

SHM is the technology that will allow the current time-based maintenance approaches to evolve into *condition-based maintenance* philosophies. The concept of condition-based maintenance is that a sensing system on the structure will monitor the system response and notify the operator that damage or degradation has been detected. Life-safety and economic benefits associated with such a philosophy will only be realised if the monitoring system provides sufficient warning such that corrective action can be taken before the damage or degradation evolves to some critical level. The trade-off associated with implementing such a philosophy is that it potentially requires more sophisticated monitoring hardware to be deployed on the system and more sophisticated data analysis procedures to interrogate the measured data.

Defence agencies are particularly motivated to develop SHM capabilities and to move to a condition-based maintenance philosophy in an effort to increase *combat asset readiness*. Military hardware is only effective if it is deployed for its combat mission. Minimising the maintenance intervals for the equipment maximises its availability for combat missions. Also, when such equipment is subjected to noncatastrophic damage from hostile fire, as shown in Figure 1.2, there is a need to rapidly assess the extent of this damage in an effort to make informed decisions about completing the current mission, about repair requirements and about subsequent use of the hardware.

Finally, many companies that produce high-capital-expenditure products such as airframes, jet engines and large construction equipment would like to move to a business model where they lease equipment as opposed to selling it. With these models the company that manufactures the equipment would take on the responsibilities for its maintenance. SHM has the potential to extend the maintenance cycles and, hence, keep the equipment out in the field where it can continue to generate revenues for the owner. Furthermore, the equipment owners would like to base their lease fees on the amount of system life used up during the lease time rather than on the current simple time-based lease fee arrangements. Such a business model will not be realised without the ability to monitor the damage initiation and evolution in the rental hardware.

1.3 Definition of Damage

In the most general terms, damage can be defined as changes introduced into a system, either intentionally or unintentionally, that adversely affect the current or future performance of that system. These

Figure 1.2 Damage sustained by an A-10 Thunderbolt during a 2003 Iraq War mission (source: US Air Force).

systems can be either natural or man-made. As an example, an anti-aircraft missile is typically fired to intentionally introduce damage that will immediately alter the flight characteristics of the target aircraft. Biological systems can be unintentionally subjected to the damaging effects of ionising radiation. However, depending on the levels of exposure, these systems may not show the adverse effects of this damaging event for many years or even future generations.

This book is focused on the study of damage identification in structural and mechanical systems. Therefore, *damage* will be defined as *intentional or unintentional changes to the material and/or geometric properties of these systems, including changes to the boundary conditions and system connectivity, which adversely affect the current or future performance of these systems.*

Thinking in terms of length scales, all damage begins at the material level as shown in Figure 1.1 and such material-level damage is present to some extent in all systems. Materials scientists and condensed-matter physicists commonly refer to such damage as *defects*; the term encompasses voids, inclusions and dislocations. Under appropriate loading scenarios, the material-level damage progresses to component- and system-level damage at various rates. *Failure* occurs when the damage progresses to a point where the system can no longer perform its intended function. Often failure is defined in terms of exceeding some strength, stability or deformation-related performance criterion.

Clearly, even though damage is present in all engineered systems at some level, modern design practices can account for this low-level damage and the systems perform as intended. The structure can often continue to perform its intended function when damage has progressed beyond the levels considered in design, but usually this performance is at some reduced level. As an example, the aircraft shown in Figure 1.2 was able to return to its base despite the severe damage it sustained. However, it is doubtful if it could perform at its original design levels during that return flight. Also, it may be some time before the structure experiences the appropriate loading conditions for the damage to cause a reduced level of performance. An extreme example of this situation occurred during the last flight of the space shuttle *Columbia*. Insulating foam impact during the launch caused damage to the shuttle. However, it was not

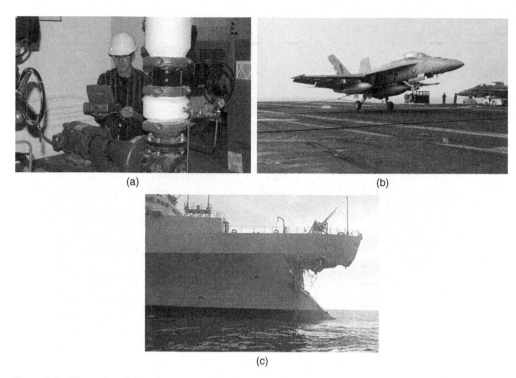

Figure 1.3   Illustration of three damage accumulation time scales. (a) Monitoring incremental damage accumulation in rotating machinery, (b) scheduled discrete damage accumulation resulting from a carrier landing (source: US Navy), (c) unscheduled discrete damage accumulation resulting from a ship-to-ship collision (source: US Navy).

until re-entry into the atmosphere when thermal environments were experienced that caused this damage to rapidly progress to catastrophic failure.

In terms of time scales, damage can accumulate incrementally over long periods of time, as in the case of damage associated with fatigue or corrosion. Damage can also progress very quickly, as in the case of critical fracture. Finally, scheduled discrete events such as aircraft landings and unscheduled discrete events such as birdstrike on an aircraft or transient natural phenomena such as earthquakes can lead to damage. Examples of damage developed over various time scales are shown in Figure 1.3.

A great deal of this book will be concerned with vibration-based approaches to SHM; this is amply justified by the fact that, in most practical scenarios, changes to a structural system caused by damage manifest themselves as changes to the mass, stiffness and energy dissipation characteristics of the system. Damage can also manifest itself as changes to the boundary conditions of a structure that reveal themselves as changes to the structure's dynamic response characteristics. As discussed earlier, the effects of the damage may become apparent on different time scales. The following examples illustrate situations where damage induces changes in one or more dynamical characteristics:

- A crack that forms in a mechanical part produces a change in geometry that alters the stiffness characteristics of that part while having almost no influence on the material characteristics or boundary conditions of the structure. Depending on the size and location of the crack, the adverse effects to the system can be either immediate or may take some time before they alter the system's performance.

Figure 1.4   Scour of bridge piers caused by increased flow rates that erode the supporting soil resulting in changes to the bridge's boundary conditions (source: US Geological Survey).

- Scour of a bridge pier is the process whereby increased flow rates around a pier erode the surrounding soil, as shown in Figure 1.4. This can be viewed as a change to the boundary conditions of the bridge that can compromise its structural integrity. However, this form of damage does not alter the local mass or stiffness properties of the structure itself.
- The loss of a lead balancing weight on a car wheel, the subsequent excessive wear of the tyre, loss of handling and loss of ride comfort is an example where the change of mass of the mechanical system can be viewed as the damaging event. In this case the stiffness and boundary conditions of the system are not altered by the damaging event.
- Finally, the loosening of a bolted connection in a structure is damage that alters the connectivity between elements of the structure while the stiffness and mass characteristics of the structural elements are not altered. Often this form of damage adds additional energy dissipation mechanisms to the structure, which would be reflected as an increase in measured vibrational damping properties.

1.4   A Statistical Pattern Recognition Paradigm for SHM

Implicit in the previous definition of damage is that the concept of damage is not meaningful without a comparison between two different states of the system, one of which is assumed to represent the initial, and often undamaged, state. This point is illustrated by Figure 1.5, which shows an apparently damaged highway bridge even though a close examination shows that pedestrians are still using this bridge to cross the river. Almost all readers, even if they have no background in damage assessment or bridge engineering, would affirm that Figure 1.5 shows a damaged bridge although there is no documentation indicating the initial state of this structure for comparison. This observation would appear to contradict the previous statement that a comparison with an undamaged state is needed to definitively conclude that the current observation represents a damaged condition. However, the readers' conclusion that this bridge is damaged is based on a mental comparison with the hundreds or thousands of examples of undamaged bridges that they have observed in their daily lives. Therefore, even for this very extreme case of damage some form of an initial condition comparison is needed before it can be stated that the pictured condition represents a damaged state for that structure. The other point to be made by this

Figure 1.5   A bridge located in Dagupan, Philippines, that was damaged by an earthquake in 1990.

example is that the reader employed pattern recognition in the process of mentally comparing the bridge image shown in Figure 1.5 to their internal database of previously observed healthy bridges. It will be argued in this book that pattern recognition provides a fundamental framework for carrying out SHM, although in most SHM applications this pattern recognition will need to be applied to mechanical or electrical sensor data, such as time-history readings as opposed to images. This book will also attempt to formalise this pattern recognition process by using the principles of machine learning.

Pattern recognition implemented through machine learning algorithms is a mature discipline. In abstract terms the theory provides mathematical means of associating measured data with given class labels. In the context of SHM, one wishes to associate the measured data with some damage state, the simplest – and arguably most important – problem being that of distinguishing between the states 'healthy' and 'damaged' for a structure. In mathematical terms there are a number of distinct approaches to pattern recognition, the main ones being the *statistical*, *neural* and *syntactic* approaches (Schalkoff, 1992). As all engineering problems are subject to various degrees of uncertainty, the statistical approach to pattern recognition appears to stand out as a natural approach for SHM purposes. As it will be seen in later chapters, neural network approaches can also be interpreted in statistical terms and also offer a robust means of dealing with SHM problems. In the example of the bridge earlier, the reader assigned the label *damaged* to the bridge by making a comparison of the structure with an internal database of healthy bridges representations. This database will have been accumulated, or *learned*, over an earlier period of time. The concept of learning representations from *training* data will be exploited throughout this book as the means of accomplishing pattern recognition. The mathematical framework needed for the problem is well established as the field of *machine learning* (Cherkassky and Mulier, 2007).

A general statistical pattern recognition (SPR) paradigm for an SHM system can be defined through the integration of four procedures (Farrar, Doebling and Nix, 2001):

1. Operational evaluation,
2. Data acquisition,
3. Feature selection and
4. Statistical modelling for feature discrimination.

Data normalisation, cleansing, compression and fusion are processes inherent in steps 2 to 4 of this paradigm. These processes can be implemented in either hardware or software and typically some combination of the two is used. The concept of machine learning enters into this paradigm primarily in steps 3 and 4.

Here, the idea of machine learning can be simply stated; it is to 'learn' the relationship between some features derived from the measured data (step 3) and the damaged state of the structure. If such a relationship between these two quantities exists, but is unknown, the learning problem is to estimate the function that describes this relationship using data acquired from the test structure – the *training* data. This estimation process is the focus of step 4. Learning problems naturally fall into two classes. If the training data comes from multiple classes and the labels for the data are known, the problem is one of *supervised learning*. If the training data do not have class labels, one can only attempt to learn intrinsic relationships within the data, and this is called *unsupervised learning*. Unsupervised learning can also be used to construct a model for a given single class that can then be used to test new data for consistency with that class; when used in such a manner, the process leads to novelty detection algorithms.

When one mentions the use of machine learning, there is often the misconception that this is an entirely data-driven process that makes no use of physics-based modelling. In fact, this need not be the case. In order to elaborate on this point it is useful now to discuss competing approaches to SHM. It is generally accepted that there are two main approaches, the 'inverse-problem' or 'model-based' approach and the 'data-based' approach.

The inverse-problem approach is usually implemented by building a physics-based or law-based model of the structure of interest; this is commonly a finite element (FE) model, although other modelling methods are used. Once the model is built, based on a detailed physical description of the system, it is usually updated on the basis of measured data from the real structure. This updating brings up an important point; it is very difficult to build an accurate model of a structure from first physical principles. Information or insight will be lacking in many areas, for example, and the exact nature of bonds, joints and so on can be difficult to specify. Another issue is that material properties may not be known with great accuracy; this is a common problem for civil engineers who will typically work with concrete. The updating step, then, adjusts the built model in such a way as to make it conform better with data from the real structure. The mathematical framework for this procedure is dominated by linear algebraic methods (Friswell and Mottershead, 1995). After updating, one has an accurate model of the structure of interest in its normal condition. When data from a subsequent monitoring phase become available, if any deviations from the normal condition are observed (e.g. the natural frequencies of the structure change), a further update of the model will indicate the location and extent of where structural changes have occurred, and this provides a damage diagnosis.

The data-based approach, as the name suggests, does not proceed from a law-based model. One establishes training data from all the possible healthy and damage states of interest for the structure and then uses pattern recognition to assign measured data from the monitoring phase to the relevant diagnostic class label. In order to carry out the pattern recognition, one needs to build a statistical model of the data, for example, to characterise their probability density function. This approach depends on the use of machine learning algorithms. In the data-based approach one can still make effective use of law-based models as a means of establishing good features for damage identification; this is discussed extensively in Chapters 7 and 8 later in this book.

There are pros and cons for both approaches; the reader can find a detailed discussion of these in Barthorpe (2011). In any case, the distinction between the two philosophies is not as clear-cut as one might wish. The model-based approach depends critically on the availability of training data for the initial update step; the data-based approach also establishes a model, but a statistical one. For various reasons, which will be elaborated later, the authors of this book believe firmly in the data-based approach.

The four steps of the SPR paradigm for SHM advocated in this book are briefly described below; the rest of the book is organised around this paradigm.

1.4.1   Operational Evaluation

The process of *operational evaluation* attempts to provide answers to four questions regarding the implementation of a damage identification investigation:

1. What is the life-safety and/or economic justification for performing the structural health monitoring?
2. How is damage defined for the system being investigated and, for multiple damage possibilities, which cases are of the most concern?
3. What are the conditions, both operational and environmental, under which the system to be monitored functions?
4. What are the limitations on acquiring data in the operational environment?

Operational evaluation begins to set limitations on what will be monitored and how the monitoring will be accomplished; it tries to tailor the damage identification process to features that are unique to the system being monitored and attempts to exploit unique features of the damage that is to be detected. Operational evaluation is discussed in more detail in Chapter 3.

1.4.2   Data Acquisition

The data acquisition portion of the SHM process involves selecting the excitation methods, the sensor types, number and locations, and the data acquisition/storage/transmittal hardware (see Chapter 4). This portion of the process will be application-specific. Economic considerations will play a major role in making decisions regarding the data acquisition hardware to be used for the SHM system. The interval at which data should be collected is another consideration that must be addressed. For earthquake applications it may be prudent to collect data immediately before and at periodic intervals after a large event. If fatigue crack growth is the failure mode of concern, it may be necessary to collect data almost continuously at relatively short time intervals once some critical crack has been identified.

1.4.3   Data Normalisation

The process of separating changes in the measured system response caused by benign operational and environmental variability from changes caused by damage is referred to as *data normalisation* (see Chapter 12). Examples of such variability are:

- An aircraft will change its mass during flight. A continuous change is caused by the burning of fuel; abrupt changes can be caused by the dropping of stores. Both of these effects are operational issues. If an in-flight SHM system were based on resonance frequencies, one would not wish to infer damage when a change occurred for benign reasons.

- The stiffness properties of a bridge can and do change with temperature. This variation can be quite complex; for example, the behaviour of the Z24 Bridge in Switzerland was observed to change when the ambient temperature dipped below the freezing point of the deck asphalt (Peeters, Maeck and Roeck, 2001). The variation described is a result of an environmental change; bridges are also susceptible to operational changes like variations in traffic loading.

Because system response data will often be measured under varying operational and environmental conditions, the ability to normalise the data becomes very important to the damage detection process; without this, changes in the measured response caused by changing operational and environmental conditions may be mistaken as an effect of damage. Additional measurements may be required to provide the information necessary to normalise the measured data and the need for this should be considered in the operational evaluation stage. When environmental or operational variability is an issue, the need can arise to normalise the data in some temporal fashion to facilitate the comparison of data measured at similar times of an environmental or operational cycle. Often the data normalisation issues will be key challenges to the field deployment of a robust SHM system.

1.4.4   Data Cleansing

Data cleansing is the process of selectively choosing data to pass on to or reject from the feature selection process. The data cleansing process is usually based on knowledge gained by individuals directly involved with the data acquisition. As an example, an inspection of the test setup may reveal that a sensor was loosely mounted and, hence, based on the judgement of the individuals performing the measurement; this set of data or the data from that particular sensor may be selectively deleted from the feature selection process. Signal processing techniques such as filtering and resampling can also be thought of as data cleansing procedures.

1.4.5   Data Compression

Data compression is the process of reducing the dimension of the measured data. The concept of data or feature dimensionality is discussed in more detail in Chapter 7. The operational implementation of the measurement technologies needed to perform SHM inherently produces large amounts of data. A condensation of the data is advantageous and necessary when comparisons of many feature sets obtained over the lifetime of the structure are envisioned. Also, because data will be acquired from a structure over an extended period of time and in an operational environment, robust data reduction techniques must be developed to retain feature sensitivity to the structural changes of interest in the presence of environmental and operational variability. To give further aid in the extraction and recording of the high-quality data needed to perform SHM, the statistical significance of the features will need to be characterised and used in the condensing process.

1.4.6   Data Fusion

Data fusion is the process of combining information from multiple sources in an effort to enhance the fidelity of the damage detection process. The fusion process may combine data from spatially distributed sensors of the same type such as an array of strain gauges mounted on a structure. Alternatively, heterogeneous data types including kinematic response measurements (e.g. acceleration) along with environmental parameter measurements (e.g. temperature) and measures of operational parameters (e.g. traffic volume on a bridge) can be combined to determine more easily if damage is present. Clearly, data fusion is closely related to the data normalisation, cleansing and compression processes.

1.4.7   Feature Extraction

The part of the SHM process that arguably receives the most attention in the technical literature is the identification of data features that allows one to distinguish between undamaged and damaged states of the structure of interest (Doebling *et al.*, 1996; Sohn *et al.*, 2004) (see Chapters 7 and 8). A damage-sensitive feature is some quantity extracted from the measured system response data that indicates the presence (or not) of damage in a structure. Features vary considerably in their complexity; the ideal is a low-dimensional feature set that is highly sensitive to the condition of the structure. Generally, a degree of signal processing is required in order to extract effective features. For example, if one wished to monitor the condition of a gearbox, one might start by attaching an accelerometer to the outer casing. This sensor would yield a stream of acceleration–time data. To reduce the dimension of the data without compromising the information content, one might use the time series to compute a spectrum. Once the spectrum is available, one can then extract only those spectral lines centred around the meshing harmonics, as these are known to carry information about the health of the gears. This specific feature extraction process is quite typical in that it involves both mathematical operations or transformations and the use of a priori engineering judgement. Another useful source of diagnostic features is to build (or learn) a physical or data-based parametric model of the system or structure; the parameters of these models or the predictive errors associated with these models then become the damage-sensitive features. Inherent in many feature selection processes is the fusing of data from multiple sensors (see Chapter 4) and subsequent condensation of these data. Also, various forms of data normalisation are employed in the feature extraction process in an effort to separate changes in the measured response caused by varying operational and environmental conditions from changes caused by damage (Sohn, Worden and Farrar, 2003).

1.4.8   Statistical Modelling for Feature Discrimination

The portion of the SHM process that has arguably received the least attention in the technical literature is the development of statistical models for discrimination between features from the undamaged and damaged structures. Statistical model development is concerned with the implementation of algorithms that operate on the extracted features to quantify the damage state of the structure; they are the basis of the SPR approach. The functional relationship between the selected features and the damage state of the structure is often difficult to define based on physics-based engineering analysis procedures. Therefore, the statistical models are derived using machine learning techniques. The machine learning algorithms used in statistical model development usually fall into two categories, as alluded to earlier. When training data are available from both the undamaged and damaged structure, *supervised learning* algorithms can be used; *group classification* and *regression analysis* are primary examples of such algorithms. In the context of SHM, *unsupervised learning* problems arise when only data from the undamaged structure are available for training. *Outlier* or *novelty detection* methods are the primary class of algorithms used in this situation. All of the algorithms use the statistical distributions of the measured or derived features to enhance the damage detection process. Chapters 9, 10 and 11 discuss in more detail the statistical modelling portions of the SHM process as implemented using machine learning principles.

The damage state of a system can in principle be arrived at via a five-step process organised along the lines of the hierarchy discussed in Rytter (1993). This process attempts to answer the following questions:

1. Is there damage in the system (existence)?
2. Where is the damage in the system (location)?
3. What kind of damage is present (type)?
4. How severe is the damage (extent)?
5. How much useful (safe) life remains (prognosis)?

Answers to these questions in the order presented represent increasing knowledge of the damage state. When applied in an unsupervised learning mode, statistical models can typically be used to answer questions regarding the existence (and sometimes, but not always, the location) of damage. When applied in a supervised learning mode and coupled with analytical models, the statistical procedures can, in theory, be used to determine the type of damage, the extent of damage and the remaining useful life of the structure. The statistical models are constructed in such as way as to minimise false diagnoses. False diagnoses fall into two categories: (1) *false-positive* damage indication (indication of damage when none is present) and (2) *false-negative* damage indication (no indication of damage when damage is present). If one wishes, one can design diagnostic systems that weight the costs of the two error types differently.

Statistical models are used to implement two types of SHM. *Protective monitoring* refers to the case when damage-sensitive features are used to identify impending failure and shut the system down or alter its use in some other manner before catastrophic failure results. In this case the statistical models are used to establish absolute values or thresholds on acceptable levels of feature change. *Predictive monitoring* refers to the case where one identifies trends in data features that are then used to predict when the damage will reach a critical level. This type of monitoring is necessary to develop cost-effective maintenance planning. In this case statistical modelling is used to quantify uncertainty in estimates of the feature's time rate of change.

1.5   Local versus Global Damage Detection

Interest in the ability to monitor a structure and detect damage at the earliest possible stage is pervasive throughout the civil, mechanical and aerospace engineering communities. Most current damage-detection methods are NDE-based using visual or localised experimental methods such as acoustic or ultrasonic methods, magnetic field methods, radiography, eddy-current methods and thermal field methods (Hellier, 2001; Shull, 2002). All of these experimental techniques require that the vicinity of the damage is known a priori and that the portion of the structure being inspected is readily accessible. Subject to these limitations, such experimental methods can detect damage on or near the surface of the structure. However, surface measurements performed by most standard NDE procedures cannot provide information about the health of the internal members without costly dismantling of the structure. As an example, micro-cracks were found in numerous welded connections of steel moment-resisting frame structures after the 1994 Northridge earthquake (Darwin, 2000). These connections are typically covered by fire-retardant and nonstructural architectural material. Costs associated with inspecting a single joint and then reinstalling the fire-retardant and architectural cladding can be on the order of thousands of dollars per joint. A typical twenty-storey building may have hundreds of such joints. Clearly, there is a tremendous economical advantage to be gained if the damage assessment can be made in a nonintrusive and more cost-effective manner.

In addition to the local inspection methods, there has been a perceived need for quantitative global damage detection methods that can be applied to complex structures. Among other things, this has led to the development of, and continued research into, methods that examine changes in the vibration characteristics of the structure. As discussed earlier, the basic premise of vibration-based damage detection is that damage will alter the stiffness, mass or energy dissipation properties of a system, which, in turn, alter the measured global dynamic response properties of the system. Although the basis for vibration-based damage detection appears intuitive, its actual application poses many significant technical challenges. The most fundamental challenge is the fact that damage is typically a local phenomenon and may not significantly influence the lower-frequency global response of a structure that is normally measured during vibration tests, particularly those where the response to ambient excitation is measured. Stated another way, this fundamental challenge is similar to that found in many engineering fields where there is a need to capture the system response *on widely varying length scales*, and such system modelling and measurement has proven difficult.

More recently researchers have been studying hybrid multiscale sensing approaches to SHM (Park *et al.*, 2003). Such approaches rely on active sensing systems for local damage detection and the same sensor/actuators are used in a passive mode to measure the influence of damage on the global system response. Here the term *active* refers to systems where actuators are incorporated with the sensing system to provide a known input to the structure that is designed to enhance the damage detection process.

Fundamentally, there will always be a trade-off between the cost associated with deploying a local sensing system over a large area of the structure and the lack of fidelity associated with more global sensing systems. For most applications some hybrid system will most likely be employed that is based on a priori knowledge of specific areas on the structure that are most likely to experience damage.

1.6   Fundamental Axioms of Structural Health Monitoring

Because of the economic and safety implications associated with accurate damage identification, many new SHM studies and resulting technical advances have been seen in recent years. The advances that have been made in these more global approaches to damage detection, herein referred to simply as damage detection, are the result of coupling recent advances in various technologies such as computer hardware, sensors, computational mechanics, experimental structural dynamics, machine learning/SPR and signal analysis. Correspondingly, the number of papers dealing with this subject that have appeared in the technical literature has been rapidly growing for the past twenty years.

Based on information published in this extensive literature, the authors feel that the field has matured to the point where several fundamental axioms, or accepted general principles, have emerged (Worden *et al.*, 2007; Farrar, Worden and Park, 2010). The axioms are first presented here, without further justification, for the reader to keep in mind while viewing the rest of this book. The authors believe that evidence will be presented to justify each axiom and some of the subtleties associated with each will be directly addressed in the subsequent chapters. These axioms will be revisited in Chapter 13 with more detailed justification as a means of summarising the material presented throughout this book. The axioms are:

Axiom I. All materials have inherent flaws or defects.

Axiom II. Damage assessment requires a comparison between two system states.

Axiom III. Identifying the existence and location of damage can be done in an unsupervised learning mode, but identifying the type of damage present and the damage severity can generally only be done in a supervised learning mode.

Axiom IVa. Sensors cannot measure damage. Feature extraction through signal processing and statistical classification are necessary to convert sensor data into damage information.

Axiom IVb. Without intelligent feature extraction, the more sensitive a measurement is to damage, the more sensitive it is to changing operational and environmental conditions.

Axiom V. The length and time scales associated with damage initiation and evolution dictate the required properties of the SHM sensing system.

Axiom VI. There is a trade-off between the sensitivity to damage of an algorithm and its noise rejection capability.

Axiom VII. The size of damage that can be detected from changes in system dynamics is inversely proportional to the frequency range of excitation.

Axiom VIII. Damage increases the complexity of a structure.

1.7   The Approach Taken in This Book

This book is designed to provide an overview of the current state of the art in SHM for those just entering the field and for those who want to begin applying this technology to real-world problems. More specifically, the authors believe this book will provide a systematic approach to addressing damage detection problems through the machine learning/SPR paradigm for SHM that forms the primary theme for this book. The second theme for this book is that successful implementation of a structural health monitoring process requires a synergistic, multidisciplinary approach. Finally, an important conclusion will be that there is no one damage detection method that is applicable to all structural and mechanical systems. These themes will be emphasised throughout this book and will be reinforced through numerous examples.

Depending on the application, SHM can be thought of as either a new emerging field, as in cases when applied to civil engineering and aerospace infrastructure, or as a fairly mature technology when it is viewed in the context of CM for rotating machinery. To further enhance the reader's understanding of this technology, descriptions of many methods that have been reported and applied beyond a laboratory setting will be presented along with summaries of methods currently under development at various research institutes. In addition, the current limitations of this technology and assumptions associated with a particular feature or statistical classification procedure will be continually emphasised throughout the book. Application perspectives from aerospace, civil and mechanical engineering communities will be used as examples. Also, practical issues related to implementation of damage identification technologies will be presented. In summary, the authors hope that this book will provide the reader with a general background in SHM technology, an understanding of the limitations of this technology, the areas of current research in damage detection and the appropriate reference material such that the reader can further study this subject in the context of their particular application.

References

Barthorpe, R. (2011) On model- and data-based approaches to structural health monitoring. PhD Thesis, Department of Mechanical Engineering, University of Sheffield.

Cherkassky, V and Mulier, F.M. (2007) *Learning from Data: Concepts, Theory and Methods*, Wiley-Blackwell.

Darwin, D. (ed.) (2000) Steel moment frames after Northridge. Special Issue of *Journal of Structural Engineering*, **126**(1).

Doebling, S., Farrar, C., Prime, M. and Shevitz, D. (1996) Damage Identification and Health Monitoring of Structural and Mechanical Systems from Changes in Their Vibration Characteristics: A Literature Review. Los Alamos National Laboratory Report LA-13070-MS, available online from Los Alamos National Laboratory, http://library.lanl.gov/.

Farrar, C., Doebling, S. and Nix, D. (2001) Vibration-based structural damage identification. *Philosophical Transactions A*, **359**(1778), 131.

Farrar, C., Worden, K. and Park, G. (2010) Complexity: a new axiom for structural health monitoring? 5th European Workshop on Structural Health Monitoring, Sorrento, Italy.

Farrar, C., Sohn, H., Hemez, F.M. *et al.* (2003) Damage Prognosis: Current Status and Future Needs. Los Alamos National Laboratory Report LA-14051-MS, available online from Los Alamos National Laboratory, http://library.lanl.gov/.

Friswell, M.I. and Mottershead, J.E. (1995) *Finite Element Model Updating in Structural Dynamics*, Springer.

Hellier, C.J. (2001) *Handbook of Non-destructive Evaluation*, McGraw-Hill, New York.

Montgomery, D.C. (2009) *Introduction to Statistical Quality Control*, John Wiley & Sons, Inc., Hoboken, NJ.

Park, G., Sohn, H., Farrar, C.R. and Inman, D.J. (2003) Overview of piezoelectric impedance-based health monitoring and path forward. *Shock and Vibration Digest*, **35**(6), 451–463.

Peeters, B., Maeck, J. and Roeck, G.D. (2001) Vibration-based damage detection in civil engineering: excitation sources and temperature effects. *Smart Materials and Structures*, **10**, 518–527.

Rytter, A. (1993) Vibration Based Inspection of Civil Engineering Structures. Building Technology and Structural Engineering. Aalborg University, Aalborg, Denmark.

Samual, P.D. and Pines, D.J. (2005) A review of vibration-based techniques for helicopter transmission diagnostics. *Journal of Sound and Vibration*, **282**, 475–508.

Schalkoff, R.J. (1992) *Pattern Recognition: Statistical, Structural and Neural Approaches*, John Wiley & Sons, Ltd, Chichester, Sussex.

Shull, P. (ed.) (2002) *Nondestructive Evaluation: Theory, Techniques and Applications*, Marcel Dekker, New York.

Sohn, H., Worden, K. and Farrar, C.R. (2003) Statistical damage classification under changing environmental and operational conditions. *Journal of Intelligent Materials Systems and Structures*, **13**(9), 561–574.

Sohn, H., Farrar, C., Hemez, F. *et al.* (2004) A Review of Structural Health Monitoring Literature from 1996–2001. Los Alamos National Laboratory Report LA-13976-MS, available online from Los Alamos National Laboratory, http://library.lanl.gov/.

Worden, K. and Dulieu-Barton, J. (2004) An overview of intelligent fault detection in systems and structures. *Structural Health Monitoring*, **3**(1), 85.

Worden, K., Farrar, C., Manson, G. and Park, G. (2007) The fundamental axioms of structural health monitoring. *Proceedings of the Royal Society A: Mathematical, Physical and Engineering Science*, **463**(2082), 1639.

Zwaag, S.v.d. (2007) *Self Healing Materials: An Alternative Approach to 20 Centuries of Materials Science*, Springer.

2

Historical Overview

It is the authors' speculation that damage or fault detection, as determined by changes in the dynamic system response, has been practised in a qualitative manner, using acoustic- or vibration-based techniques, since man has been using tools. One of the earliest references to SHM that the authors have found is the tap testing that was performed to detect cracks in railroad wheels in the 1800s (Higgins, 1895; Stanley, 1995). In the last thirty years there has been a significant increase in the number of studies that attempted to evolve SHM from these early qualitative approaches to more sensitive and quantifiable methods for damage detection. During this time SHM has evolved to the point where it is beginning to make the transition from a research topic to actual application on a variety of structural systems. The widespread and growing interest in this technology is attested to by the dedicated annual and biannual conferences focusing specifically on this subject that have emerged in the last fifteen years.[1,2,3,4]

This chapter will provide an overview of the evolution in SHM technology by summarising the developments that have been made in four distinct application areas: rotating machinery, offshore oil platforms, aerospace structures and bridges. The state of SHM that resulted from technology developments associated with these various applications is then summarised by comparing and contrasting the rotating machinery and bridge applications. The reader is directed to two extensive reviews of the SHM literature for more detailed discussions of the various applications summarised in this chapter (Doebling *et al.*, 1996; Sohn *et al.*, 2004).

2.1 Rotating Machinery Applications

Although this book will focus on applications of SHM to structural systems, a brief summary of rotating machinery applications is included because this application has made the transition from a research topic to actual practice. As previously stated, when applied to rotating machinery, SHM is referred to as *condition monitoring* and this field has its own dedicated conferences and technical societies.[5,6] Condition monitoring began with somewhat crude qualitative procedures such as holding a screw driver on the machinery housing and listening for changes in the acoustic signal or feeling changes in the

---

[1] The 8th International Workshop on Structural Health Monitoring, Palo Alto, CA, 2011.

[2] SPIE Symposium on Smart Structures and Materials and Nondestructive Evaluation and Health Monitoring, San Diego, CA, 2011.

[3] The 8th International Conference on Damage Assessment of Structures, Beijing, China, 2009.

[4] The Fifth European Structural Health Monitoring Workshop, Sorrento, Italy, 2010.

[5] The 62nd Meeting of the Society for Machinery Failure Prevention Technology, Virginia Beach, VA, 2008.

[6] COMADEM International.

Structural Health Monitoring: A Machine Learning Perspective, First Edition. Charles R. Farrar and Keith Worden.
© 2013 John Wiley & Sons, Ltd. Published 2013 by John Wiley & Sons, Ltd.

vibration response of the screw driver, as described in a historical overview of the subject provided by Mitchell (2007). The monitoring procedures have evolved to the point where measurements are made with modern data acquisitions systems, but qualitative interpretation of vibration signatures both in the frequency and (to a lesser extent) in the time domain is still the primary data interrogation procedure. Numerous summaries and reviews of these approaches are available in textbook form, including detailed charts relating various machine failures to specific characteristics of the measured dynamic response (Braun, 1986; Hewlett-Packard, 1991; Wouk, 1991; Crawford, 1992; Mitchell, 1993; Taylor, 1994; Eisenmann and Eisenmann, 1997; Bently and Hatch, 2003; Randall, 2011). As an example, one such chart (Mitchell, 1993) shows that gear misalignment will result in higher harmonics of the gear meshing frequency and that these harmonics will have sidebands associated with the shaft rotation frequencies. For a concise overview of this technology the reader is referred to the following articles (Randall, 2004a, 2004b; Jardine, Lin and Banjevic, 2006). The approach taken has generally been to consider the detection of damage qualitatively on a fault-by-fault basis by examining vibration signatures measured on the equipment housing for the presence and growth of peaks in spectra at certain frequencies, such as integer multiples of shaft speed. A primary reason for this approach has been the inherent nonlinearity associated with damage in rotating machinery. Recently, more general approaches to damage detection in rotating machinery have been developed. These approaches utilise formal analytical methods to assess both the presence and level of damage on a statistical basis (Li *et al.*, 1991; Roth and Pandit, 1999).

2.1.1 Operational Evaluation for Rotating Machinery

The definition of damage is typically straightforward for rotating machinery and includes chipped gear teeth, shaft misalignment and damage to bearing race surfaces. Economic arguments for performing condition monitoring will vary depending on the application, but can generally be quantified (Wouk, 1991). As with other SHM applications, life-safety benefits associated with condition monitoring are again application specific and harder to quantify. Often, there are a limited number of well-defined damage scenarios that are being monitored and the possible locations of the damage are known a priori. The primary operational limitation on acquiring data is that the machine will typically be in operation and performing its normal function or will be in a transient start-up or shut-down mode. In its *in situ* environment many other machines will most likely produce additional dynamic excitation sources that must be accounted for in the damage detection process. Limitations to acquiring vibration data can vary widely. For many applications the limitations will be based on administrative criteria such as the availability of personnel to make the necessary measurements. In other applications the machine may be located in hazardous environments allowing for only limited access time.

2.1.2 Data Acquisition for Rotating Machinery

Sensors and data acquisition equipment used to monitor rotating machinery are discussed in the previously cited reference books. The selection and placement of appropriate transducers depends upon the type of machinery and its construction, and these issues are discussed in detail in various references (Eisenmann and Eisenmann, 1997). The primary vibration transducer used for condition monitoring is the piezoelectric accelerometer. These accelerometers have a broad operating frequency range and are well suited for monitoring roller bearings and gear trains. Accelerometers are typically used in conjunction with single-channel signal analysers so that the machinery vibration output signal can be viewed in the frequency domain as well as a function of time, that is amplitude-frequency, amplitude-time and waterfall plots (three-dimensional plots showing multiple spectra acquired at different times). An engineer using a single-channel, hand-held signal analyser is shown monitoring a piece of rotating machinery in Figure 1.3(a). An example of a waterfall plot obtained from such measurements is shown in Figure 2.1. Velocity transducers and noncontact displacement transducers are also widely used. Noncontact eddy

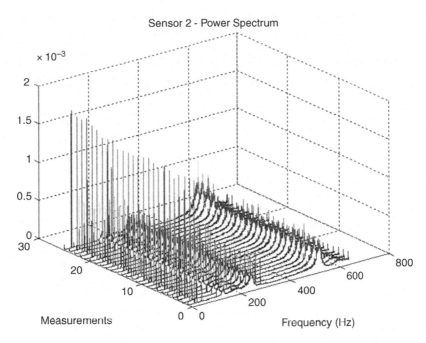

Figure 2.1   A waterfall plot from an accelerometer mounted on a bearing housing.

current displacement transducers find application in the monitoring of shaft motion and position relative to fluid-film bearings. A set of two noncontact transducers, mounted at right angles, is often used to determine the orbit of the shaft in its bearing. In general, all data acquisition equipment for condition monitoring is commercially available, off-the-shelf technology provided by a number of different companies.

2.1.3   *Feature Extraction for Rotating Machinery*

As previously mentioned, there exist numerous detailed charts of anticipated characteristic faults and their associated features for a variety of machines and machine elements (Crawford, 1992; Mitchell, 1993).

These charts show that many commonly used features for rotating machinery are based on the presence of peaks in acceleration spectra at certain multiples of shaft rotational frequency and their growth or change with time. These types of features are quite distinct to the type of machine element, the specific fault and in some cases the level of damage. Therefore, it may be possible to locate the defective machine element (e.g. bearing or gears), isolate the specific fault in the element and determine the level of damage based purely on these features. Commercial software packages for this type of feature extraction and associated classification, including automated expert diagnostic systems specifically designed for the isolation of faults based on vibration signatures, which are integrated with the data acquisition systems, are commercially available (Watts and Van Dyke, 1993).

Feature extraction methods can be further broken down into time-domain methods, transformed-domain methods and time-frequency methods. Time-domain methods have particular application to roller bearings, as these elements typically fail by fatigue cracking and the associated material spalling on one of the bearing or race contact surfaces. Time-domain methods for roller bearing analysis include peak amplitude, root-mean-squared amplitude, crest factor analysis, kurtosis analysis and shock pulse counting (Ma and Li, 1993). Some of these features are discussed in more detail in Chapters 7 and 8 of this book. As an example, kurtosis is the normalised fourth statistical moment of some measured data (see Chapter 6). If surface roughness of bearing races is used as an indicator of damage, then for an undamaged surface the profile is random, corresponding to a Gaussian irregularity distribution producing a Gaussian vibration response with a theoretical kurtosis value of three. A kurtosis value significantly different from three indicates that the irregularities are no longer Gaussian in depth and/or location distribution, which can be taken as an indication of damage (Martin, 1986; Volker and Martin, 1986). Proprietary time-domain condition monitoring methods and associated instrumentation are commercially available for the detection of defects involving repetitive mechanical impacts, primarily associated with roller bearings (Le Bleu and Xu, 1995).

Frequency-domain methods typically characterise changes in machine vibration spectra that result from damage. Frequency-domain approaches for roller bearings include Fourier spectra of synchronised-averaged time histories, 'cepstrum' analysis (specifically the inverse Fourier transform of the logarithm of the Fourier spectra magnitude squared (Randall, 2011)), sum and difference frequencies analysis, the high-frequency resonance technique and short-time signal processing (Ma and Li, 1993). Quantitative evaluation of faults in gears can also be detected using peaks in the cepstrum as indicators of harmonics (Tang *et al.*, 1991). Cepstral approaches have also been used for spectral-based fault detection in helicopter gearboxes (Kemerait, 1987).

Time-frequency methods have their application in the investigation of rotating machinery faults exhibiting nonstationary effects; that is, the vibration response signal's properties (e.g. frequency content, root-mean-square amplitude and other statistics) vary with time, as discussed in Appendix A. Nonstationary effects are associated with machinery in which the dynamic response differs in the various phases of the operating cycle. Examples include reciprocating machines, localised faults in gears and damage in cam mechanisms. The wavelet transform (see Appendix A) has been applied to fault detection and diagnosis of cam mechanisms and to helicopter gearboxes (Wang and McFadden, 1996; Dalpiaz and Rivola, 1997; Staszewski and Worden, 1997). Additional references provide comparative studies of frequency-domain and time-frequency methods to fault detection in various types of rotating machinery (Elbestawi and Tait, 1986; Petrilli *et al.*, 1995).

2.1.4 *Statistical Modelling for Damage Detection in Rotating Machinery*

Once features have been selected and extracted from the data recorded on the rotating machinery, the next step is to infer whether or not damage is present, the type of damage and possibly the level of that damage. As introduced in Chapter 1 and discussed throughout this book, this process can generally be described as a problem in pattern classification. Informally, skilled individuals can use their experience with previous undamaged and damaged systems and the changes in the features associated with previously observed damage cases to deduce the presence, type and level of damage. This approach is an example of the application of *informal supervised learning*.

More formal supervised learning methods founded in machine learning have been introduced for damage detection in rotating machinery. These methods place the system of interest, as represented by one or more features, into either an undamaged category or one or more damaged categories (Chin and Danai, 1991). The classification techniques reported in the rotating machinery literature include: Bayesian classification, Kth nearest neighbour rules, artificial neural network classifiers (Lin and Wang, 1993) and more recently support vector machines (Widodo and Yang, 2007), some of which will be discussed in more detail later in this book. A particularly powerful technique is that of artificial neural networks for statistical pattern classification. As an illustration, artificial neural networks were used in conjunction with various preprocessing algorithms to detect a controlled tooth fault in a pair of meshing spur gears based on changes in a measured vibration response (Staszewski and Worden, 1997). The problem of rotor imbalance of a multidisc shaft has been investigated with neural networks. The 'input' to the neural network consists of conditions of imbalance; the 'output' is represented by measured bearing reactions. A learning algorithm is then used to 'train' the network to relate bearing reactions with presence (and possibly level) of imbalance (Stevenson *et al.*, 1991). A pattern recognition analysis scheme as applied to roller bearing condition monitoring makes use of features relating to the sum-frequency components of bearing defect frequencies and their harmonics. A linear discriminatory operator is then developed to detect localised damage to bearing components (Li *et al.*, 1991). Unsupervised learning approaches to fault detection in rotating machinery have received less attention in the technical literature. A sequential hypothesis test is one approach to unsupervised learning or novelty detection that has been applied to rotating machinery (Singer *et al.*, 1990). Jardine, Lin and Banjevic (2006) provide a summary of statistical classification and machine learning techniques that have been applied to condition monitoring of rotating machinery.

2.1.5   Concluding Comments about Condition Monitoring of Rotating Machinery

As previously mentioned, condition monitoring of rotating machinery has matured to the point where this technology is used as part of predictive maintenance procedures for a wide variety of machinery, some of which perform very high consequence functions in terms of life-safety or economic impact. As an example, condition monitoring is being developed to monitor components in nuclear reactors (Bond *et al.*, 2003). The US Navy has adopted this technology for monitoring propulsion systems on more than nine classes of ships. Their monitoring system interfaces with supply management in an effort to coordinate the availability of replacement parts with maintenance needs identified by the monitoring systems (DiUlio *et al.*, 2003). Finally, another indication of the maturity of condition monitoring is the extensive set of standards that are available for this application. Table 2.1 summarises some of the standards available for vibration measurements and condition monitoring of rotating machinery.

2.2   Offshore Oil Platforms

During the 1970s and 1980s the oil industry made considerable efforts to develop vibration-based damage detection methods for offshore platforms. This damage detection problem is fundamentally different from that of rotating machinery because the damage location is unknown and because the majority of the structure is not readily accessible for measurement. To circumvent these difficulties, a common methodology adopted by this industry was to simulate candidate damage scenarios with numerical models, examine the changes in resonance frequencies that were produced by these simulated changes and correlate these changes with those measured on a platform (Vandiver, 1975, 1977). A number of very practical problems were encountered including measurement difficulties caused by platform machine noise, instrumentation difficulties in hostile environments, changing mass caused by marine growth and varying fluid storage levels, temporal variability of foundation conditions and the inability of wave motion to excite higher vibration modes.

2.2.1   Operational Evaluation for Offshore Platforms

Structural health monitoring research for offshore oil platforms has been driven by both economic and life-safety issues. As an example, deep water platforms can represent over a billion US dollar capital investment before any revenues are generated from the platform. In addition, these structures can have many people working and living on the platform at any time. There have been numerous cases where

Table 2.1   Standards for vibration measurements and condition monitoring of rotating machinery.

Standard designation	Subject
Mil-STD 167-1	Mechanical Vibrations of Shipboard Equipment (Type 1 – Environmental and Type II – Internally Excited)
Mil-STD 167-2	Mechanical Vibrations of Shipboard Equipment (Reciprocating Machinery and Propulsion System and Shafting) Types III, IV and V
American Petroleum Institute Standard 670	Machinery Protection Systems
American Society of Nondestructive Testing ASNT-217	Corrosion: Machine System Condition Monitoring
ANSI S2.17	Machinery Vibration Measurements
ISO 2372	Mechanical Vibration of Machines with Operating Speeds from 10 to 200 rps – Basis for Specifying Evaluation Standards
ISO 3945	In-Place Evaluation of Larger Machinery
ISO 7919	Mechanical Vibration of Non-Reciprocating Machines – Measurement on Rotating Shafts and Evaluation Criteria
ISO 10816	Mechanical Vibration – Evaluation of Machine Vibration by Measurement on Non-Rotating Parts
ISO 13372	Condition Monitoring and Diagnostics of Machines – Vocabulary
ISO 13373-1	Condition Monitoring and Diagnostics of Machines – Vibration Condition Monitoring General Procedures
ISO 13374	Condition Monitoring and Diagnostics of Machines – Data Processing, Communication and Presentation
ISO 13379	Condition Monitoring and Diagnostics of Machines – General Guidelines for Data Interpretation and Diagnostic Techniques
ISO 13380	Condition Monitoring and Diagnostics of Machines – General Guidelines on Using Performance Parameters
ISO 13381	Condition Monitoring and Diagnostics of Machines – Prognostics
ISO 14830	Condition Monitoring and Diagnostics of Machines – Tribology-Based Monitoring and Diagnostics
ISO 14839	Mechanical Vibration – Vibration of Rotating Machinery Equipped with Active Magnetic Bearings
ISO 15242	Rolling Bearings – Measuring Methods for Vibration
ISO 17359	Condition Monitoring and Diagnostics of Machines – General Guidelines
ISO 18436	Condition Monitoring and Diagnostics of Machines – Requirements for Training and Certification of Personnel
Hydraulic Institute M122 (ANSI/HI 9.6.5)	Rotodynamic (Centrifugal and Vertical) Pump Guidelines for Condition Monitoring
General Motors Corporation	Vibration Standard for Machinery and Equipment

collapse of these platforms have resulted in the loss of life of all those on the platform.[7] Fires resulting from damage to an offshore platform can be seen in Figure 2.2.

There are a wide variety of damage scenarios that are of interest to the offshore oil industry. These include ship impact on the structural elements, corrosion damage and fatigue damage accumulation. However, most of the damage detection studies that were performed numerically simulate either a crack in a structural member or complete removal of the member. Many studies were performed on *in situ* platforms (Duggan, Wallace and Caldwell, 1980; Kenley and Dodds, 1980; Crohas and Lepert, 1982; Nataraja, 1983; Whittome and Dodds, 1983). In some cases damage was actually introduced

---

[7] See http://home.versatel.nl/the_sims/rig/i-fatal.htm.

Figure 2.2   Fires resulting from damage on an offshore oil platform in the Gulf of Mexico (courtesy of the US Coast Guard).

in decommissioned platforms (Kenley and Dodds, 1980). Also, tests were performed on scale-model platforms tested in laboratory settings (Begg *et al.*, 1976; Swamidas and Chen, 1992) including tests in water tanks (Yang, Dagalakis and Hirt, 1980; Osegueda, Dsouza and Qiang, 1992). In some of these studies the data were given to researchers without knowledge of the specific damage scenario in an effort to determine if damage could be detected in a blind manner.

Limitations on performing the *in situ* monitoring focus on the fact that only a small fraction of the structure is located above the water line. Even these locations are often difficult and dangerous to access. If one attempts to make below-water-line measurements, it becomes very expensive and much more dangerous if divers are used for these measurements. Alternatively, one can obtain below-water-line measurements by mounting the sensors and their wiring into the structural elements, which are generally tubular truss elements, during construction. For long-term monitoring the corrosive saltwater environment can adversely affect the sensors and their wiring systems.

There are numerous sources of variability that must be accounted for when applying SHM to *in situ* offshore platforms. These include varying mass from things such as marine growth, fluid storage and equipment. Water ingress to structural elements can also cause mass changes and alter the energy dissipation characteristics of the structure. Another source of variability is the fact that the inputs used to induce dynamic system response are often nonstationary. In addition, there are usually multiple input sources occurring simultaneously (e.g. wave input simultaneously occurring with drilling machinery in operation). In some cases a measure of the input excitation may be available, while in other cases, such as with wave motion, there is no direct measure of the input. Finally, it was noted that when a new platform is put in place, the pressure on the foundation that is transferred to the ocean floor soil will slowly change the soil properties over time, which in turn will influence the global dynamic response characteristics of the platform (Brincker *et al.*, 1995).

As a result of such studies, researchers have defined requirements for an oil platform monitoring system (Loland and Dodds, 1976). These requirements include:

- Ambient (sea and wind) excitation must be used to extract the resonance frequencies.
- Vibration spectra must remain stable over long periods of time.
- Instruments must withstand environmental challenges.
- Mode shapes must be identified from above-water measurements.
- The system must offer financial advantages over the use of divers.

The requirements were developed based on Loland and Dodds' practical experiences learned by monitoring three North Sea platforms for six to nine months and inspection requirements that came into effect in 1974 regarding the structural integrity of UK offshore oil platforms.

2.2.2   Data Acquisition for Offshore Platforms

This industry has focused on making acceleration measurements using commercially available piezo-electric accelerometers mounted on the above-water portion of the platform. This measurement scenario is analogous to inferring the global dynamic response characteristics of a cantilevered structure from a limited number of measurements made near the free end of the structure. These sensors are monitored with standard commercial data acquisition systems. Vibration-based SHM studies for offshore platforms have primarily used ambient inputs from wave motion and platform operations as the excitation source (Martinez and Quijada, 1991).

There are a limited number of studies where the researchers have developed data acquisition systems designed specifically for offshore platform applications. As an example, engineers have developed a vibration excitation and data acquisition system that can be mounted on a truss element below the water surface. This system, referred to as a 'vibro-detection device,' provided a local excitation capability and a local system response measurement capability and was mounted on a test brace of the Total-ABK living quarters platform in the Arabian Gulf (Crohas and Lepert, 1982). The vibro-detection device applied a 44 kN (5 ton) input to the truss elements over a frequency range of 0.5 to 45 Hz and measured the below-water response with submersible triaxial accelerometers. These one-of-a-kind, application-specific, data acquisition systems are expensive. In general, it is very expensive to obtain below-water-line measurements on offshore platforms.

2.2.3   Feature Extraction for Offshore Platforms

Most of the previously cited damage detection studies for offshore oil platforms examine changes in basic modal properties (resonance frequencies and mode shapes) that are extracted from measured acceleration response time histories. Such features are also discussed in additional studies (Wojnarowski *et al.*, 1977; Coppolino and Rubin, 1980; Yang, Chen and Dagalakis, 1984). Other investigators have used parameters of time-series models developed from these same acceleration–time histories as indicators of system change in the platforms (Brincker *et al.*, 1995). In many cases numerical modelling approaches were used where changes in modal properties predicted by simulating postulated damage with finite element models were compared to changes in modal properties that were estimated from a measured system response. In theory, this approach allows one to detect, locate and estimate the extent of damage. These damage-sensitive features including those based on modelling approaches will be discussed in more detail in Chapter 7.

2.2.4   Statistical Modelling for Offshore Platforms

In most of the cited references formal statistical classification procedures have not been applied to the damage-sensitive features obtained from offshore oil platform SHM studies. However, there have been numerous sensitivity studies performed in conjunction with offshore oil platform damage detection studies (Vandiver, 1975, 1977; Begg *et al.*, 1976; Loland and Dodds, 1976; Duggan, Wallace and Caldwell, 1980; Kenley and Dodds, 1980). With these studies the authors attempted to estimate how much change a phenomena such as marine growth can produce on the measured modal properties. They then made an estimate of the amount of damage that would have to be present to produce changes in the modal properties larger than those changes resulting from the nondamage phenomena.

2.2.5   Lessons Learned from Offshore Oil Platform Structural Health Monitoring Studies

Developing damage detection strategies for offshore platforms was one of the first large-scale research efforts in the SHM field. Although the oil industry largely abandoned this technology, some very important lessons were learned that apply to many other SHM applications. The most enduring lesson learned is the importance of data normalisation (see Chapter 12). Environmental variability was shown to produce significant shifts in the damage-sensitive features (modal properties in this case) from causes such as marine growth, water ingress and wave motion. These changes have to be separated from the changes in dynamic response caused by damage and this problem is pervasive in most other SHM applications that are not performed in well-controlled settings. These issues prevented adaptation of this technology, and efforts at further developing this SHM technology for offshore platforms were largely abandoned in the early 1980s.

2.3   Aerospace Structures

The aerospace community began to study the use of damage detection technology during the late 1970s and early 1980s for a variety of civilian and defence applications (Ikegami, 1999; Kudva, Grage and Roberts, 1999). Early work focused on loads monitoring where a limited number of sensors are tracked to count load cycles and/or to count the number of times certain threshold response levels are exceeded. Loads monitoring continues to be one primary structural health assessment tool used in practice (Goranson, 1997). Such approaches are often referred to as *usage monitoring*. The development of SHM for aerospace applications has continued and increased considerably in technical sophistication with current applications being investigated for commercial and military aircraft, the National Aeronautics and Space Administration's (NASA) space station and the next generation of reusable launch vehicles. One unique aspect of aerospace SHM applications is that regulatory agencies have been involved with the certification of the health and usage monitoring systems (HUMS) that are deployed on rotorcraft (Samual and Pines, 2005; Pawar and Ganguli, 2007).

The application of SHM to aerospace structures has yielded several systems that have made the transition from research to practice. The most notable are rotorcraft HUMS and the space shuttle modal inspection system (SMIS) programme. Other systems are currently being tested and this industry continues to expend considerable resources on the development of new SHM technology.

Perhaps the most refined forms of SHM performed in the aerospace industry are the HUMS used by the rotorcraft industry. These systems evolved from condition monitoring of rotating machinery and were developed for commercial rotorcraft in response to the significant number of crashes experienced by helicopters servicing North Sea oil platforms. In parallel, these systems were developed for military rotorcraft. Figure 2.3 shows US Army personnel checking the output of a HUMS deployed on a UH-60M Blackhawk helicopter. With the introduction of a HUMS for main rotor and gearbox components on

Figure 2.3 US Army personnel checking the output from a helicopter HUMS (courtesy of Spc. J. Andersson US Army).

large rotorcraft, these systems have been shown to reduce 'the fatal hull loss within the UK to half what could have otherwise been expected had HUMS not been installed' (McColl, 2005). The essential features of this success are that the rotor speed – although not the torque – is maintained typically within 2% of nominal for all flight regimes and that there is a single load path through the power transmission system with no redundancy. These constraints provide a basis for a stable vibration spectrum from which changes in measured parameters can be attributed to component deterioration. As such, the use of vibration data trending for predictive maintenance can be shown to increase rotor component life by 15% (Silverman, 2005). The development of a HUMS is summarised by Carlson, Kershne and Sewersky (1996) and Cleveland and Trammel (1996) with early concepts described by Astridge (1985). These systems are used to diagnose faults in helicopter drive trains, engines, oil systems and the rotor system. Most significant is the fact that HUMS have been endorsed by the Federal Aviation Administration (FAA) and the Civil Aviation Authority (CAA) as part of an acceptable maintenance strategy, with the first certified HUMS being flown in the United Kingdom in 1991. Other references that discuss HUMS in more detail, including practical implementation issues, are Chronkite (1993), Hess and Hardman (1999), Robeson and Thompson (1999) and White (1999).

The space shuttle was designed as the first reusable space vehicle and as such required inspections to assess structural integrity after each flight where it experiences launch, spaceflight, re-entry and landing loading environments. In response to this need, the SMIS was developed to identify fatigue damage in components such as control surfaces, fuselage panels and lifting surfaces. These areas are covered with a thermal protection system making them inaccessible and, hence, impractical for conventional local nondestructive evaluation methods. The SMIS has been successful in locating damaged components that are covered by the thermal protection system and all orbiter vehicles have been periodically subjected to SMIS testing since 1987. Early shuttle inspections applied modal test techniques for nondestructive evaluation of the orbiter structure. As an example, testing was performed on the orbiter body flap, which is used to shield the main engines from heat and to provide pitch control during atmospheric re-entry (West, 1982). Single-point random excitation was used to acquire frequency response functions from the flap. Between modal tests, the flap was exposed to an acoustic environment similar to operating conditions. It was observed that the frequencies of the first three modes decreased following the acoustic exposure. Upon disassembly and inspection of the test article, indications of galling in the spherical bearings at the actuator–rib interfaces were discovered. Additionally, shear clips in the interface between the trailing edge wedge and the flap ribs were found to contain significant cracking. It was noted that the conventional visual, X-ray and ultrasonic inspection techniques had failed to locate this damage. Also, the conventional techniques require the removal of at least some orbiter thermal protective system tiles, whereas the modal inspection technique did not. Further development and refinement of the SMIS has been documented in several articles (West, 1984; Hunt *et al.*, 1990; Grygier, 1994; Grygier *et al.*, 1994; Pappa, James and Zimmerman, 1998; Sirkis *et al.*, 1999).

The development of SHM systems for reusable launch vehicles continued as NASA began to design a next-generation launch vehicle. Strategies were proposed for rapid damage diagnosis and decision making that focused on a distributed sensor system (Melvin, 1997). In addition to the launch vehicle itself, the composite fuel tanks were surfacing as one of the critical items for long-term health monitoring. One such monitoring system for a next-generation launch vehicle was successfully demonstrated during 1996 flight tests at White Sands Missile Range (Baumann *et al.*, 1997). This flight test is shown in Figure 2.4.

Space station applications have primarily driven the development of experimental/analytical methods aimed at identifying damage to truss elements caused by space debris impact. These methods are very similar in concept to those developed and applied to the previously discussed offshore platforms. They use inverse modelling approaches where analytical models of the undamaged structure are correlated with measured modal properties from both the undamaged and damaged structure. Changes in stiffness indices as assessed from the two model updates are used to locate and quantify the damage.

Since the mid-1990s, studies of damage identification for composite materials have been motivated by the development of a composite fuel tank for a reusable launch vehicle as well as the increasing use

Figure 2.4   DC-XA flight tests with an onboard SHM system at White Sands Missile Range (courtesy of NASA).

of composite materials in all types of commercial and military aircraft. The failure mechanisms, such as delamination caused by debris impacts, debonding of glued joints and corresponding material response for composite fuel tanks are significantly different from those associated with metallic structures. Often, such damage is located below the surface of the structure, thus increasing the challenges of the damage detection. Also, the composite fuel tank problem presents challenges because the sensing systems must not provide a spark source. This challenge has led to the development of SHM based on fibre optic sensing systems. Active pulse-echo and pitch-catch wave propagation-based damage detection approaches, acoustic emission passive wave propagation methods and active thermography methods (Maldague and Moore, 2001) have been developed for damage detection in composite materials.

There are many other examples of more specialised aircraft SHM systems such as monitoring nitrogen pressure in welded chrome-moly fuselage tubing to detect crack and corrosion damage (Kounis, 2007) and the development of SHM to monitor rapid satellite assembly and deployment for the US Air Force's Operationally Responsive Space capability, where the goal is to assemble and launch a satellite within one week of mission definition (Arritt *et al.*, 2007). Aircraft engine condition monitoring systems are described in Tumer and Bajwa (1999) and Jaw (2005). Finally, Staszewski, Boller and Tomlinson (2003) and Boller and Buderath (2007) both provide more detailed discussions of SHM applied to aerospace structures.

2.3.1   Operational Evaluation for Aerospace Structures

The development of SHM for aerospace applications has been driven by both life-safety and economic concerns. Clearly, any technology that prevents a manned aircraft crash will provide a life-safety benefit. In the year 2000, 30% of the commercial aircraft fleet had reached its design life. As with other infrastructures, there is a desire to operate these assets beyond their original design life and, hence, more attention must be paid to identifying damage before it reaches a critical level. Military operators are interested in this technology because of the potential to reduce lifecycle costs, knowing that over the lifetime of a military aircraft system, the maintenance costs will exceed the purchase price of that system. Several articles directly address the impact of HUMS on the lifecycle cost of rotorcraft (Land and Weitzman, 1995; Forsyth and Sutton, 2001). Operationally, military users also want to maximise combat asset readiness of their aircraft, which necessitates minimising the amount of maintenance required to keep the aircraft in an effective operating mode. Furthermore, additional economic advantages can be gained if the SHM system can reliably prevent unnecessary dismantling and the associated reassembly of mechanical and structural components. The basis for operational evaluation is already established within the aerospace sector through the widespread implementation of reliability centred maintenance (Moubray, 1997). A detailed discussion of operational evaluation concepts applied to unmanned aerial vehicles can be found in MacConnell (2007).

There are a wide variety of damage types that are of interest to the aerospace industry, with corrosion and cracking in metallic components and delamination, debonding, fibre breakage and matrix microcracking in composite components being some of the most prevalent. Moisture intrusion into sandwich constructions is another form of damage associated with composite components. Low-velocity impact damage that can result in nonvisible or barely visible damage is of particular concern for composite aerospace components. Additionally, overloading can cause yielding, fastener failure, failure of adhesive repair bonds and damage to protective coatings. Overloads can occur from foreign object impact including birdstrikes, hail, lightning, battle damage and unintentional damage from inspection activities and impact with ground service equipment. In some cases critical flaw sizes and likely damage locations have been defined for various damage types. One of the most prevalent damage concerns in commercial aircraft is fatigue cracks that form around rivets in the fuselage. Damage sustained by the Aloha Airlines flight in 1988, shown in Figure 2.5, was attributed to such cracks. As a result of this accident, the US Federal Aviation Administration set up an Airworthiness Assurance Non-Destructive Inspection Validation Center in Albuquerque, New Mexico, that is managed by Sandia National Laboratory. This centre

Figure 2.5 Damage caused by fatigue cracking around rivets that connect the fuselage skin to the airframe (courtesy of the National Transportation Safety Board).

develops a variety of damage detection technologies for aircraft inspection. It also maintains specimens with known damage that can be used to validate new damage detection technologies.

Constraints on performing the damage assessment include: weight limitations for the sensing system, the need for sensors that do not pose a spark hazard when monitoring near fuel, widely varying operational and environmental conditions, lack of accessibility to critical structural components, and the influence of nonstructural components such as cables and insulation on the dynamic response of the structure.

Environmental variability is particularly acute for space structures where the required operating temperature range for sensors can vary from −250 °C to 120 °C. A thorough overview of on-orbit damage identification issues for trusses is presented by Kashangaki (1991). In jet engines temperatures exceeding 1000 °C make in-flight monitoring very challenging. Operational variability can include: changing mass associated with fuel consumption and varying payloads, changing aerodynamic loading caused by various in-flight manoeuvres, changes in air speed and turbulence, and the number of take-offs and landings. All of these sources of variability must be accounted for when performing SHM on aerospace structures.

2.3.2 Data Acquisition for Aerospace Structures

Sensing and data acquisition systems for aerospace SHM vary widely depending on the specific application. Many of these systems are commercially available and deployed onboard the structure while it is in flight, such as the rotorcraft HUMS.[8] In-flight SHM systems have only become practical with the evolution of microelectronics that allow such systems to be deployed onboard aircraft with minimal weight penalties. Other systems like the SMIS used commercial ground-based dynamic data acquisition systems for tests performed with the shuttle out of service. Generally, large amounts of data can be acquired in-flight, necessitating a robust data management strategy (Hall, 1999).

---

[8] See www.goodrich.com.

Quantities such as strain and acceleration are the most common dynamic response parameters that are monitored using conventional electrical resistance strain gauges and piezoelectric accelerometers, respectively. In addition, the aerospace industry has studied the use of fibre optic strain sensors for SHM applications because they are lightweight, they do not produce a spark source and because their size makes them somewhat nonintrusive (Kabashima, Ozaki and Takeda, 2000). The small size of optical fibres has motivated several studies where researchers have embedded the fibres into composite materials for more direct measures of the composite material's response characteristics. The fibre optic sensors have also been used to measure temperature. Other types of sensors have been deployed for corrosion monitoring as part of a US military demonstration of health monitoring systems for large structural components that investigated different sensing modalities (Van Way *et al.*, 1995). Researchers at Sandia National Laboratory have used scanning laser vibrometers to measure the dynamic response of a DC-9 fuselage in an attempt to identify damage to structural elements beneath the skin (James, 1996).

Most commonly, when tests are performed in-flight, the aerospace industry measures the response of the structure to operational and environmental loading conditions such as those caused by aerodynamic forces and engine vibration. In cases with the MIR space station the astronauts themselves have excited the structure by jumping, and impacts associated with shuttle docking procedures have also been used as an unmeasured excitation source (Kammer and Steltzner, 2001). More recently, the aerospace industry has started to use active sensing systems, where a small actuator is mounted on the structure to produce a local excitation signal tailored to enhance the damage detection process (Crawley, 1994). Acoustic emissions studies rely on measuring the response of the system that results when strain energy is released during damage initiation or growth (Staszewski, Boller and Tomlinson, 2003). During ground vibration tests performed for damage assessment purposes, electrodynamic shakers are used to excite the structure, as is the case with the SMIS that is shown in Figure 2.6. Finally, a variety of impact devices such as

Figure 2.6 An electrodynamics shaker is attached to the shuttle vertical stabiliser in preparation for an experimental modal analysis (courtesy of NASA).

a hammer or even a coin are used for tap tests that detect delaminations or debonding in composite materials (Cawley and Adams, 1989).

2.3.3   Feature Extraction and Statistical Modelling for Aerospace Structures

Early studies from the 1980s first report the use of modal properties such as mode shapes and resonance frequencies as the features used in aerospace structure damage detection studies (NASA, 1988). Modal parameters have continued to be the most common features used in aerospace SHM studies. Rather than simply looking at changes in these parameters as indicators of damage and its location, the aerospace industry has done extensive work on developing model updating procedures for damage assessment; these methods are reviewed in detail in Doebling *et al.* (1996). With these procedures a finite element model is used analytically to generate mode shapes and resonance frequencies for the undamaged system. Similar quantities are obtained from experimental modal analyses. The finite element model is then systematically updated by modifying the stiffness, mass and/or damping matrices using a constrained optimisation procedure in an effort to minimise the difference between the analytical and experimental modal parameters. After a potentially damaging event, this process is repeated using the previously updated finite element model as the baseline condition. Changes to the stiffness indices that are based on a subsequent optimisation process using the newly measured modal parameters are indicative of damage and can potentially be used to locate and quantify the extent of damage. Feature extraction based on model updating is discussed further in Chapter 7. To validate model updating procedures, numerous experiments have been carried out on trusses where damage is assumed to be caused by micrometeorites and results in the severing of a truss element (Letchworth and McGowan, 1988; Doebling *et al.*, 1997). Figure 2.7 shows one such experiment being conducted on a planar truss in a zero-*g* environment.

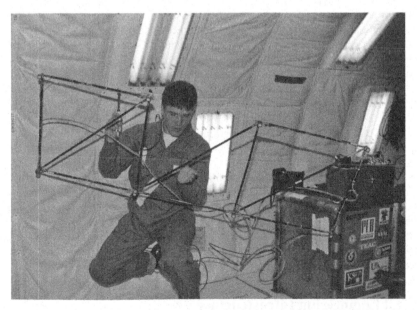

Figure 2.7   An experimental modal analysis being performed on a truss with one element removed to simulate damage. The tests are being conducted in a zero-*g* environment to investigate difficulties with performing such experiments in space (courtesy of Professor David Zimmerman).

More recently, wave propagation-based damage detection procedures have been applied to aerospace structures (Park *et al.*, 2003; Cesnik, 2007; Mal, Banerjee and Ricci, 2007). In particular, these tests have been used to identify delaminations and debonding in composite materials. These methods can be simple tap tests on composite components, acoustic emission tests or they can make use of more sophisticated active pitch-catch or pulse-echo sensing techniques. Examining distortions in waveforms that result when elastic waves are reflected off the damage boundary are the primary means of detecting damage. In the case of acoustic emissions the generation of elastic waves that result when the damage initiates or propagates in the system is used to indicate damage.

2.3.4 Statistical Models Used for Aerospace SHM Applications

Most aerospace SHM studies reported in the open literature make limited use of rigorous statistical classification procedures or machine learning algorithms in the damage detection process. One notable exception is the use of the sequential probability ratio test (previously mentioned regarding condition monitoring of rotating machinery) for outlier detection associated with jet engine damage (Herzog *et al.*, 2005). Other examples of statistical classification procedures and machine learning algorithms applied to aerospace SHM include Ganguli, Chopra and Haas (1998), Hayton *et al.* (2007) and Worden and Manson (2007).

2.3.5 Concluding Comments about Aerospace SHM Applications

In summary, the aerospace application of SHM has technical challenges that are, in general, similar to the challenges faced by those for offshore oil platforms. Both types of structures experience widely varying environmental and operational conditions that can affect sensor readings. In the case of aerospace structures these changes include widely varying thermal, vibratory and acoustic environments, changing mass that results from fuel consumption and changing aerodynamic load caused by varying atmospheric conditions and/or variations in how the aircraft is operated. Both types of structures also have the common issue that many portions of the structure are difficult to access when trying to retrofit an existing structure with a sensing system. Some unique challenges faced by the aerospace industry are the restrictions on the weight of the SHM system that can be deployed during flight. Also, the sensing system cannot be a spark source when monitoring fuel tanks, which are a structure of considerable interest for reusable launch vehicle applications. The aerospace industry has been very proactive in exploring advanced sensing technology for SHM, including the use of fibre optic strain sensors, active wave propagation methods using piezoelectric patches, acoustic emission sensors and scanning laser vibrometers. However, the majority of the aerospace SHM studies reported to date use features based on parameters of linear models that are fit to measure data even though many damage phenomena of interest are inherently nonlinear. Statistical classification of the damage-sensitive features is in its initial phase of development for most of these aerospace structure studies.

Finally, it should be noted that for many applications such as commercial aircraft, the aerospace industry has a long record of safely maintaining their assets. This record is based on accepted and well-validated inspection methods that are understood and trusted by practitioners and regulatory agencies. When one considers the costs associated with current inspection procedures and large inspection infrastructure that is already in place, there will most likely be a very slow and conservative evolution to SHM where the SHM system must be shown to provide improved, cost-effective and reliable damage assessment with minimal false indications of damage.

2.4 Civil Engineering Infrastructure

The civil engineering community has studied vibration-based damage assessment of bridge structures since the early 1980s. The physical size of these structures presents many practical challenges for SHM.

Regulatory requirements in Asian countries, which mandate the companies that construct the bridges to certify their structural health periodically, are driving current research and commercial development of vibration-based bridge monitoring systems. One such example is that found on the Tsing Ma Bridge in Hong Kong where over 1000 channels of data acquisition are deployed (Ko, Ni and Chan, 1999; Lau *et al.*, 1999).

In the United States an extensive survey of bridge failures since 1950 points out that responses of engineers to these failures have been reactive (Shirole and Holt, 1991). Bridge design modifications and inspection programme changes are often made only in response to catastrophic failures. The collapse of the Tacoma Narrows Bridge more than a half century ago is a classic example of this reactive attitude because it led to the inspection and modification of other suspension bridges. The widespread introduction of a nationwide systematic bridge inspection programme was directly attributed to the catastrophic bridge collapse at Point Pleasant, West Virginia, in 1967 (White, Minor and Derucher, 1992). Design modifications for the seismic response of bridges have been made as a direct consequence of the damage suffered by these structures during the 1971 San Fernando earthquake (Gates, 1976).

At present, bridges in the United States are generally rated and monitored during federally mandated biennial inspections, largely with the use of visual inspection techniques. This procedure is slow, not quantifiable and portions of the bridge are inaccessible for such visual inspection (Federal Highway Administration, 2001). There is the possibility that damage can go undetected at inspection or that cracks in load-carrying members can grow to critical levels between inspection intervals (Gorlov, 1984), with the most widely publicised of such failures being the I-35 Bridge that collapsed in Minneapolis during the summer of 2007 (National Transportation Safety Board, 2008); see Figure 2.8. In addition to these more gradual damage accumulation mechanisms, sudden damage leading to bridge collapse also occurs as a result of collisions. For example, the AMTRAK railroad bridge collapse in Southeastern United States in 1993 involved the collision of a barge with the bridge pier. More than 13% of identified failures of US bridges since 1950 are attributed to collisions (Shirole and Holt, 1991).

Figure 2.8   The 2007 I-35 Bridge collapse in Minneapolis, Minnesota (courtesy of the US Navy).

Based on these concerns, quantitative, possibly continuous, approaches to bridge damage monitoring are being studied as a means to augment and/or replace the current visual inspection methods. Additionally, the use of an active damage detection system may be appropriate in some cases. For example, such a system could detect sudden significant damage to the bridge structure resulting from collision and trigger a traffic control system to close the bridge. Commercial systems for bridge health monitoring have been available since the late 1990s (Nigbor and Diehl, 1997), but in most cases they provide only minimal instrumentation without the spatial resolution to detect the early onset of local damage. Finally, based on the growing interest in SHM technology for bridge structures, the US Federal Highway Administration has developed a centre to validate nondestructive evaluation methods for bridges.[9]

2.4.1 Operational Evaluation for Bridge Structures

The application of SHM to bridge structures is motivated by several unexpected bridge collapses that have resulted in loss of life, as well as the desire to reduce lifecycle costs associated with these structures through the use of condition-based maintenance. The collapse of the I-35 Bridge in Minneapolis during the summer of 2007 is an example of such a catastrophic failure that for a short time focused a great deal of attention on the need for robust damage detection systems. However, there is no study that has definitively shown that SHM can perform better damage assessment than the currently mandated biennial visual inspections, or that SHM can provide an economic advantage over current maintenance procedures.

The definition of damage for bridges can vary widely as many bridges are one-of-a-kind structures. Typical damage concerns include corrosion, fatigue cracks, loss of prestressing forces and scour at the bridge pier. In many cases the damage accumulates in a gradual manner over long periods of time. In other cases, such as collisions, the damage can occur over a very short time span. Further difficulties arise when one tries specifically to define the damage to be monitored because there has not been enough experience with failures of the particular structure being studied or with similar structures. For large bridges it is imperative that the damage definition be restricted to the greatest extent possible so as to make the monitoring practical with a cost-effective number of sensors.

The need to perform the monitoring in a manner that does not impede traffic flow dictates that the monitoring occurs when the bridge is subjected to changing traffic patterns coupled with typical environmental variability. These operational and environmental constraints pose a formidable challenge to bridge health monitoring. Traffic flow can vary on a 24-hour cycle and a weekly cycle. Environmental variability can occur over a 24-hour cycle (Farrar *et al.*, 1997), over a seasonal cycle (Askegaard and Mossing, 1988) and intermittent cycles caused by varying rainfall conditions. Limitations on making dynamic response measurements on the bridge include accessibility issues for portions of the bridge during normal operating conditions. As an example, traffic usually prevents the topside of the deck from being instrumented. For large bridges many of the structural elements are difficult to instrument because it is impractical to access them in a safe and economic manner. Finally, the sensors and associated data transmission hardware are subjected to harsh environments that make equipment reliability a serious issue (Nigbor and Diehl, 1997).

2.4.2 Data Acquisition for Bridge Structures

As with most other SHM applications, the primary sensors used for bridge health monitoring are piezo-electric accelerometers. Force-balance accelerometers and electric-resistance and vibrating-wire strain gauges are also widely used. More recently, fibre optic sensors utilising Bragg gratings have been studied as a means of increasing channel counts for bridge monitoring in a cost-effective manner (Todd *et al.*, 1999). In addition to the motion-measuring devices, anemometers, thermocouples, weigh-in-motion

---

[9]See http://www.fhwa.dot.gov/research/tfhrc/labs/nde/.

Figure 2.9   A commercial dynamic data acquisition system used to perform an experimental modal analysis on the Alamosa Canyon Bridge.

sensors, sensors specifically designed to measure tension in support cables (Wang, 2005) and electro-chemical sensors for corrosion monitoring have been used in various bridge monitoring studies.

The damage to be monitored and the allowable sensor budget typically dictate sensor placement. Most bridge health monitoring studies reported to date utilise less than 100 sensors. At the extreme is the previously mentioned Tsing Ma Bridge in Hong Kong that has been instrumented with over 1000 sensors.

Excitation for dynamic response measurements is typically associated with ambient inputs caused by vehicular or pedestrian traffic, wind or wave motion. Several research studies have directly measured the inputs to the structure. These inputs are applied with either electrodynamic or hydraulic shakers, an impacting device or the release of a tensioned cable (Farrar *et al.*, 1999; Peeters, Maeck and De Roeck, 2000b).

Data transmission and recording are done primarily with commercial data acquisition systems similar to those used for most mechanical vibration applications (McConnell, 1995). Figure 2.9 shows one such system being deployed for an experimental modal analysis that was part of a damage detection study on the Alamosa Canyon Bridge in southern New Mexico (see Chapter 5). Research into wireless data acquisition systems for large civil engineering structures began in the mid-1990s (Straser, 1998) and is being studied extensively by the civil engineering SHM community (Spencer, Ruiz-Sandova and Kurata, 2004; Lynch, 2007). Data acquisition intervals are not well defined for bridges as most studies have been conducted in a research mode where long-term monitoring has not been addressed. However, as with any continuous monitoring system, when a bridge monitoring system is deployed, it will generate large amounts of data that will require effective archiving and retrieval strategies so that trends in the response can be tracked over long periods of time.

2.4.3   Features Based on Modal Properties

Over the past 30 years numerous studies have been made that examine changes in the vibration response of the bridge as a global indicator of damage. Initial studies examined changes in conventional modal properties such as resonance frequencies and mode shapes, but the lower-frequency global modes of the structure, which are the ones that tend to be measured in either measured-input or ambient modal tests, were found to be insensitive to local damage. In addition, these low-frequency global modal properties have been shown to be sensitive to changing operational and environmental conditions, which further confound their use as damage indicators. Often these parameters have to be estimated from ambient excitation tests with no direct measurement of the system input, requiring the development of alternate system identification approaches from those used in standard measured-input experimental modal analyses (Farrar and James, 1997).

During the last 20 years more sophisticated methods of damage detection that extract features from the modal parameters such as modal strain energy, uniform load flexibility shapes, and finite element model updating similar to that developed for aerospace applications have been developed and applied to *in situ* bridges with mixed success (Biswas, Pandey and Samman, 1989; Farrar and Jauregui, 1996; Pandey, Biswas and Samman, 1991; Raghavendrachar and Aktan, 1992; Zhang and Atkan, 1995; Simmermacher, 1996). More recently, investigators have started to develop local active inspection techniques based on pattern recognition applied to wave propagation data that monitor very specific portions of the bridge structure such as loss of preload in a bolted connection (Park *et al.*, 2008).

2.4.4   Statistical Classification of Features for Civil Engineering Infrastructure

Studies of bridge structures that apply statistical classification algorithms to the damage detection process are somewhat limited. However, several studies have recognised that environmental and operational variability can cause significant changes in the dynamic response characteristics of the bridge and these changes must be separated from the changes in dynamic response caused by damage (Cornwell *et al.*, 1999; Farrar *et al.*, 2000). In an effort to quantify this variability, several researchers have developed statistical procedures to assess the accuracy of the modal parameter estimates obtained from measured frequency response function data (Doebling and Farrar, 2001; Doebling, Farrar and Goodman, 1997). Recently, studies are beginning to examine environmental and testing variability of the measured modal properties in detail (Peeters, Maeck and De Roeck, 2000a) and others have developed models that predict the influence of these sources of variability on the model properties (Sohn *et al.*, 1999).

Two studies on *in situ* bridge structures have attempted to propagate uncertainty on the damage-sensitive features through a damage classification process (Peeters and De Roeck, 2000; Doebling, Farrar and Goodman, 1997). In a test of the Alamosa Canyon Bridge in New Mexico, the variability in the estimates of modal parameters was based on Monte Carlo and bootstrap approaches to sampling of the measured frequency response functions. A finite element model was then used to simulate damage in the bridge. Changes in modal properties caused by the simulated damage as predicted by the numerical model were then compared to the 95% confidence intervals on the estimates of the modal parameters derived from the measured data. Some modal parameters were found to change by a statistically significant amount as a result of damage. A conclusion of this study was that resonance frequency changes caused by damage are small relative to changes in mode shapes. However, the resonance frequencies can be estimated with less variability and, hence, were shown in some cases to be more sensitive indicators of damage.

2.4.5   Applications to Bridge Structures

There are many *in situ* bridge damage detection studies reported in the literature (Peeters and De Roeck, 2000; Farrar *et al.*, 1994; Sohn *et al.*, 2004). In many cases these studies are destructive with permanent damage being introduced into the structure. In general, the destructive studies are performed over a relatively short time span so that long-term environmental and operational variability have not been considered. Often the damage that is introduced does not represent actual damage that will occur in the structure. As an example, torch cuts are performed on steel girders to simulate fatigue cracks (Farrar *et al.*, 1994), but in this study the torch cut produced cracks that were wide and that did not open and close under dynamic loading in the same manner that fatigue cracks would respond to such loading. These studies are augmented by even more studies where monitoring systems are deployed on bridge structures that are in use, which prevents the intentional introduction of controlled damage scenarios. In these cases the monitoring takes place over longer periods and the effects of changing environmental and operational conditions on the measured system response can be assessed (Wentzel, 2009).

2.5 Summary

This chapter has summarised applications of SHM to various types of infrastructure that have driven the development of this technology. These applications identified various issues and challenges for SHM. Note there are many other applications of SHM to different types of structures not discussed in this chapter. The bridge and rotating machinery applications represent two extremes in successful applications of the damage detection technology. In general, the application of SHM to rotating machinery (i.e. condition monitoring) has made the transition from a research topic to successful implementation by practising engineers. In contrast, SHM for larger structures, such as bridges, has been studied for many years, but this application has, in most cases, not progressed beyond the research phase. To conclude this discussion of the applications that have driven SHM technology development, a summary directly comparing these two applications is presented. This comparison further emphasises some of the research directions that must be followed if SHM for large structural systems is to gain the same acceptance that condition monitoring has in the rotating machinery industry:

1. Motivation. Damage detection in bridges has been primarily motivated by the prevention of loss of life; damage detection in rotating machinery is motivated largely by economic considerations often related to minimising production downtime. Clearly, there are exceptions where bridges are being monitored to facilitate timely and cost-effective maintenance and where failure of rotating machinery can have life-safety implications, as, for example, in the fracture of jet engine turbine blades.
2. Availability. Highway bridges are generally one-of-a-kind items with little or no data available from the damaged structure. Rotating machines are often available in large inventories with data available from both undamaged and damaged systems. It is much easier to build databases of damage-sensitive features from these inventories and, hence, supervised machine learning (see Chapter 11) can be much more readily accomplished for rotating machinery.
3. Definition of damage. For rotating machinery there are a finite number of well-defined damage scenarios and the possible locations of that damage are limited to fairly small spatial regions. Many bridge damage detection studies do not specifically define either the damage type or location, and the areas where damage can occur are relatively large, making local damage detection a challenge.
4. Operational evaluation. In practical health-monitoring applications, measured vibration inputs are not applied to either type of system. Rotating machinery typically exhibits responses to harmonic-like inputs that are stationary, while traffic tends to produce inputs that are typically assumed to be random in nature and are often nonstationary.
5. Data acquisition. Because the approximate location of the damage is generally known, vibration test equipment for rotating machinery can consist of a single sensor and a single-channel FFT analyser. Monitoring of bridges is normally performed with few channels distributed over a relatively large spatial region. For damage detection on a highway bridge, 100 data acquisition channels, which in most cases would be a very large bridge monitoring system, currently represent a sparsely instrumented bridge. Also, a permanent *in situ* data acquisition system for bridge structures can represent a significant capital outlay and further funds would be needed to maintain such a system over extended

periods of time. Such bridge monitoring systems will also require a sophisticated telemetry system that is typically not needed for rotating machinery applications.

6. Feature selection. A well-developed database of features corresponding to various types of damage has been developed by the rotating machinery community. Many of these features are qualitative in nature and have been developed by comparing vibration signatures from undamaged systems to signatures from systems with known types, locations and levels of damage. Many of the features observed in the vibration signatures of rotating machinery result from nonlinear behaviour exhibited by the damaged system. Features used to identify damage in bridge structures are most often derived from linear modal properties such as resonance frequencies and mode shapes. These features are identified before and after damage and require a distributed system of sensors. Few studies report the development of damage-sensitive features for bridge structures based on nonlinear response characteristics.

7. Statistical model building. The rotating machinery literature reports many more studies that investigate the application of statistical pattern classifiers to the damage detection process than have been reported for civil engineering infrastructure applications. Rotating machinery is often located in a relatively protected environment and operates under relatively consistent conditions. The primary sources of extraneous vibration inputs are other rotating machinery in the vicinity. Changes in damage-sensitive features caused by environmental and operational variability are significant and must be accounted for in bridge applications through statistical pattern classifiers. However, the literature shows little application of this technology to bridge damage detection studies.

Clearly, the application of vibration-based damage detection to rotating machinery is a much more mature technology than that associated with large civil engineering infrastructure. This comparison shows that a pressing need for the large system applications is to define a limited number of damage scenarios to be monitored. Such a limitation will reduce the requirement for an expensive and difficult to maintain distributed sensing system. Advances in low-cost, wireless instrumentation and data acquisition systems can make a major contribution to the large structures applications. Further developments of sensitive analogue-to-digital converter technology will reduce the noise floor in measurements, allowing for better assessment of the high-frequency structural response. It is postulated that this high-frequency response is more sensitive to local damage. Also, to account for variability in ambient loading conditions and environmental variability, it is imperative that the statistical pattern classifiers and associated data normalisation procedures must be adopted for these SHM applications. Without this technology it will be difficult to determine if changes in the identified features are caused by damage or by varying operational/environmental conditions. Identifying new damage-sensitive features, particularly those that are based on a nonlinear, time-varying response, will always be a focus of research efforts for all SHM applications. Finally, there is a pressing need to make measurements on large one-of-a-kind structures such as bridges. Experience gained by analysing data from *in situ* structures will be instrumental in developing new damage-sensitive features as well as defining new and improved hardware for the vibration measurements and robust approaches to data normalisation.

References

Arritt, B.J., Buckley, S.J., Ganley, J.M. *et al.* (2007) *Responsive Satellites and the Need for Structural health Monitoring. International workshop on Structural Health Monitoring*, Technomic Publishing, Palo Alto, CA.

Askegaard, V. and Mossing, P. (1988) Long term observation of RC-bridge using changes in natural frequencies. *Nordic Concrete Research*, **7**, 20–27.

Astridge, D.G. (1985) The health and usage monitoring of helicopter systems – the next generation. Proceedings of the 41st Forum of the American Helicopter Society, Alexandria, VA, American Helicopter Society.

Baumann, E.W., Becker, R.S., Ellerbrock, P.J. and Jacobs, S.W. (1997) DC-XA Structural Health Monitoring System. Smart Structures and Materials: Industrial and Commercial Applications of Smart Structures Technologies, SPIE 3044, SPIE.

Begg, R.D., Mackenzie, A.C., Dodds, D.J. and Loland, O. (1976) Structural integrity monitoring using digital processing of vibration signals. Proceedings of the Offshore Technology Conference, Houston, TX.

Bently, D.E. and Hatch, C.T. (2003) *Fundamentals of Rotating Machinery Diagnostics*, ASME Press.

Biswas, M., Pandey, A.K. and Samman, M.M. (1989) Diagnostic experimental spectral/modal analysis of a highway bridge. *The International Journal of Analytical and Experimental Modal Analysis*, **5**, 33–42.

Boller, C. and Buderath, M. (2007) Fatigue in aerostructures – where structural health monitoring can contribute to a complex subject. *Philosophical Transactions of the Royal Society: Mathematical, Physical and Engineering Sciences*, **365**(1851), 561–588.

Bond, L.J., Jarell, D.B., Koehler, T.M. *et al.* (2003) Pacific Northwest National Report, Richland, WA, PNNL-14304, available online from US Department of Energy Office of Scientific and Technical Information, www.osti.gov.

Braun, S. (1986) *Mechanical Signature Analysis – Theory and Applications*, Academic Press, Inc., London.

Brincker, R., Kirkegaard, P.H., Anderson, P. and Martinez, M.E. (1995) Damage detection in an offshore structure. Proceedings of the 13th International Modal Analysis Conference, Nashville, TN.

Carlson, R.G., Kershne, S.D. and Sewersky, R.A. (1996) Sikorsky health and usage monitoring system (HUMS) program. Proceedings of the 52nd Forum of the American Helicopter Society, Alexandria, VA, American Helicopter Society.

Cawley, P. and Adams, R.D. (1989) Sensitivity of the coin-tap method of nondestructive testing. *Materials Evaluation*, **47**, 558–563.

Cesnik, C. (2007) Review of guided-wave structural health monitoring. *The Shock and Vibration Digest*, **39**(2), 91–114.

Chin, H. and Danai, K. (1991) A method of fault signature extraction for improved diagnosis. *Journal of Dynamic Systems, Measurement, and Control*, **113**, 634–638.

Chronkite, J.D. (1993) Practical application of health and usage monitoring (HUMS) to helicopter rotor, engine and drive system. Proceedings of the 49th Forum of the American Helicopter Society, Alexandria, VA, American Helicopter Society.

Cleveland, G.P. and Trammel, C. (1996) An integrated health and usage monitoring system for the SH-60B helicopter. Proceedings of the 52nd Forum of the American Helicopter Society, Alexandria, VA, American Helicopter Society.

Coppolino, R.N. and Rubin, S. (1980) Detectability of structural failures in offshore platforms by ambient vibration monitoring. Proceedings of the 12th Annual Offshore Technology Conference, Houston, TX.

Cornwell, P.J., Farrar, C.R., Doebling, S.W. and Sohn, H. (1999) Environmental variability of modal properties. *Experimental Techniques*, **23**(6): 45–48.

Crawford, A.R. (1992) *The Simplified Handbook of Vibration Analysis*, Computational Systems, Inc., Knoxville, TN.

Crawley, E.F. (1994) Intelligent structures for aerospace: a technology overview and assessment. *AIAA Journal*, **32**(8), 1689–1699.

Crohas, H. and Lepert, P. (1982) Damage-detection monitoring method for offshore platforms is field-tested. *Oil and Gas Journal*, **80**(8), 94–103.

Dalpiaz, G. and Rivola, A. (1997) Condition monitoring and diagnostics in automatic machines: comparison of vibration analysis techniques. *Mechanical Systems and Signal Processing*, **11**(1), 53–73.

DiUlio, M., Savage, C., Finley, B. and Schneider, E. (2003) Taking the integrated condition assessment system to the year 2010. 13th International Ship Control Systems Symposium, Orlando, FL.

Doebling, S., Farrar, C., Prime, M. and Shevitz, D. (1996) Damage Identification and Health Monitoring of Structural and Mechanical Systems from Changes in Their Vibration Characteristics: A Literature Review. Los Alamos National Laboratory Report LA-13074-MS, available online from Los Alamos National Laboratory, http://library.lanl.gov/.

Doebling, S.W. and Farrar, C.R. (2001) Estimation of Statistical distributions for modal parameters identified from averaged frequency response function data. *Journal of Vibration and Control*, **7**(4), 603–624.

Doebling, S.W., Farrar, C.R. and Goodman, R. (1997) Effects of measurement statistics on the detection of damage in the Alamosa Canyon Bridge. Proceedings of the 15th International Modal Analysis Conference, Orlando, FL, Society for Experimental Mechanics.

Doebling, S.W., Hemez, F.M., Peterson, L.D. and Farhat, C. (1997) Improved damage location accuracy using strain energy-based mode selection criteria. *AIAA Journal*, **35**(4), 693–699.

Duggan, D.M., Wallace, E.R. and Caldwell, S.R. (1980) Measured and predicted vibrational behavior of Gulf of Mexico platforms. Proceedings of the 12th Annual Offshore Technology Conference, Houston, TX.

Eisenmann, S.R.C. and Eisenmann, J.R.C. (1997) *Machinery Malfunction Diagnosis and Correction: Vibration Analysis and Troubleshooting for the Process Industries*, Prentice-Hall, Upper Saddle River, NJ.

Elbestawi, M.A. and Tait, H.J. (1986) A comparative study of vibration monitoring techniques for rolling element bearings. Proceedings of the XXth International Modal Analysis Conference, Society for Experimental Mechanics.

Farrar, C.R. and James, G.H. (1997) System identification from ambient vibration measurements on bridges. *Journal of Sound and Vibration*, **205**(1), 1–18.

Farrar, C.R. and Jauregui, D. (1996) Damage Detection Algorithms Applied to Experimental and Numerical Modal Data from the I-40 Bridge. Los Alamos National Laboratory Report LA-13074-MS, available online from Los Alamos National Laboratory, http://library.lanl.gov/.

Farrar, C.R., Baker, W.E., Bell, T.M. *et al.* (1994) Dynamic Characterization and Damage Detection in the I-40 Bridge over the Rio Grande.

Farrar, C.R., Doebling, S.W., Cornwell, P.J. and Strasser, E.G. (1997) Variability of modal parameters measured on the Alamosa Canyon Bridge. Proceedings of the 15th International Modal Analysis Conference, Orlando, FL, Society for Experimental Mechanics.

Farrar, C.R., Duffey, T.A., Cornwell, P.J. and Doebling, S.W. (1999) Excitation methods for bridge structures. Proceedings of the 17th International Modal Analysis Conf, Orlando, FL, Society for Experimental Mechanics.

Farrar, C.R., Cornwell, P.J., Doebling, S.W. and Prime, M.B. (2000) Structural Health Monitoring Studies of the Alamosa Canyon and I-40 Bridges. Los Alamos National Laboratory Report LA-13635-MS, available online from Los Alamos National Laboratory, http://library.lanl.gov/.

Federal Highway Administration (2001) Reliability of Visual Inspection for Highway Bridges. 1. Final Report FHWA-RD-01-020.

Forsyth, G.F. and Sutton, S.A. (2001) Using economic modeling to determine and demonstrate affordability. Tiltrotor/Runway Aircraft Technology and Applications Specialists Meeting of American Helicopter Society, Arlington, VA.

Ganguli, R., Chopra, I. and Haas, D.J. (1998) Helicopter rotor system fault detection using physics-based model and neural networks. *AIAA Journal*, **36**(6), 1078–1086.

Gates, J.H. (1976) California's seismic design criteria for bridges. *ASCE Journal of Structural Engineering*, **102**(12), 2301–2313.

Goranson, U.G. (1997) *Jet Transport Structures Performance Monitoring. Structural Health Monitoring Current Status and Perspectives*, Technomic Publishing Company, Palo Alto, CA.

Gorlov, A.M. (1984) Disaster of the I-95 Mianus River Bridge – where could lateral vibration come from? *ASME Journal of Applied Mechanics*, **51**, 694–696.

Grygier, M.S. (1994) Modal Test Technology as Non-destructive Evaluation of Space Shuttle Structures. NASA Conference Publication 3263, NASA Technical Reports Server, http://ntrs.nasa.gov/search.jsp.

Grygier, M.S., Gaspar, J., West, W. and Wilson, B. (1994) SMIS Analysis Report: OV-102 Control Surfaces (Post STS-65). Johnson Space Center Test Branch Report JSC-26819, Houston, TX.

Hall, S.R. (1999) *The Effective Management and Use of Structural Health Data. Structural Health Monitoring 2000*, Technomic Publishing, Palo Alto, CA.

Hayton, P., Utete, S., King, D. *et al.* (2007) Static and dynamic novelty detection methods for jet engine health monitoring. *Philosophical Transactions of the Royal Society: Mathematical, Physical and Engineering Sciences*, **365**(1861), 493–514.

Herzog, J.P., Wegerich, S.W., Hanlin, J. and Wilks, A.D. (2005) High performance condition monitoring of aircraft engines. Proceedings of CT2005 ASME Turbo Expo 2005: Power for Land Sea and Air, Reno-Tahoe, NV.

Hess, A.J. and Hardman, W. (1999) SH-60 helicopter integrated design system (HIDS) program experience and results of seeded fault testing. DSTO Workshop on Helicopter Health and Usage Monitoring Systems, Melbourne, Australia.

Hewlett-Packard (1991) Effective Machinery Measurements using Dynamic Signal Analyzers, Application Note 243-1.

Higgins, S. (1895) Inspection of steel-tired wheels. *Proceedings of New York Railroad Club*, **5**, 988–989.

Hunt, D.L., Weiss, S.P., West, W.M. *et al.* (1990) Development and implementation of a shuttle modal inspection system. *Sound and Vibration*, **24**(9), 34–42.

Ikegami, R. (1999) *Structural Health Monitoring: Assessment of Aircraft Customer Needs. Structural Health Monitoring 2000*, Technomic Publishing, Palo Alto, CA.

James, G.H. (1996) Development of Structural Health Monitoring Techniques using Dynamics Testing. Sandia National Laboratory Report SAND96-0810, Albuquerque, NM, available online from US Department of Energy Office of Scientific and Technical Information, www.osti.gov.

Jardine, A.K.S., Lin, D. and Banjevic, D. (2006) A review of machinery diagnostics and prognostics implementing condition-based maintenance. *Mechanical Systems and Signal Processing*, **20**, 1483–1510.

Jaw, L.C. (2005) Recent advancements in aircraft engine health management (EHM) technologies and recommendations for the next step. Proceedings of Turbo Expo 2005: 50th ASME International Gas Turbine and Aeroengine Technical Congress, Reno-Tahoe, NV.

Kabashima, S., Ozaki, T. and Takeda, N. (2000) Damage detection of satellite structures by optical fiber with small diameter. Proceedings of the Conference on Smart Structures and Materials 2000: Smart Structures and Integrated Systems, Newport Beach, CA, SPIE.

Kammer, D.C. and Steltzner, A.D. (2001) Structural identification of Mir using inverse system dynamics and Mir/Shuttle docking data. *Journal of Vibration and Acoustics*, **123**(2), 230–237.

Kashangaki, T.A.-L. (1991) On-Orbit Damage Detection and Health Monitoring of Large Space Trusses – Status and Critical Issues. NASA Technical Memorandum 104045, available online from NASA Technical Reports Server, http://ntrs.nasa.gov/search.jsp.

Kemerait, R.C. (1987) A new Cepstral approach for prognostic maintenance of cyclic machinery. Proceedings of IEEE Southeastcon 87, Tampa, FL.

Kenley, R.M. and Dodds, C.J. (1980) West Sole WE Platform: detection of damage by structural response measurements. Proceedings of the 12th Annual Offshore Technology Conference, Houston, TX.

Ko, J., Ni, Y. and Chan, T. (1999) Dynamic monitoring of structural health in cable-supported bridges. Proceedings of the Conference on Smart Structures and Materials 1999: Smart Systems for Bridges, Structures, and Highways, Newport Beach, CA, SPIE.

Kounis, J.T. (2007) Expedition 350 tastefully rugged. *Pilot Getaways*, 2–8.

Kudva, J.N., Grage, M.J. and Roberts, M.M. (1999) *Aircraft Structural Health Monitoring and Other Smart Structures Technologies – Perspectives on Development of Future Smart Aircraft. Structural Health Monitoring 2000*, Technomic Publishing Company, Palo Alto, CA.

Land, J. and Weitzman, C. (1995) How HUMS systems have the potential of significantly reducing the direct operating cost for modern helicopters through monitoring. Proceedings of the 51st Forum of the American Helicopter Society, Alexandria, VA, American Helicopter Society.

Lau, C.K., Mak, W.P.N., Wong, K.Y. *et al.* (1999) *Structural Health Monitoring of Three Cable-Supported Bridges in Hong Kong. Structural Health Monitoring: 2000*, Technomic Publishing, Palo Alto, CA.

Le Bleu, J.J. and Xu, M. (1995) Vibration monitoring of sealess pumps using spike energy. *Journal of Sound and Vibration*, **28**, 10–16.

Letchworth, R. and McGowan, P.E. (1988) Space Station: A Focus for the Development of Structural Dynamics Scale Model Technology for Large Flexible Space Structures. AIAA SDM Issues of the International Space Station, A Collection of Technical Papers, Williamsburg, VA.

Li, C.J., Ma, J., Hwang, B. and Nickerson, G.W. (1991) Pattern recognition based bicoherence analysis of vibrations for bearing condition monitoring. Sensors, Controls, and Quality Issues in Manufacturing, American Society of Mechanical Engineers.

Lin, C.-C. and Wang, H.-P. (1993) Classification of autoregressive spectral estimated signal patterns using an adaptive resonance theory neural network. *Computers in Industry*, **22**, 143–157.

Loland, O. and Dodds, J.C. (1976) Experience in developing and operating integrity monitoring system in North Sea. Proceedings of the Offshore Technology Conference, Houston, TX.

Lynch, J.P. (2007) An overview of wireless structural health monitoring for civil structures. *Philosophical Transactions of the Royal Society A*, **365**(1851), 345–372.

Ma, J. and Li, C.J. (1993) Detection of localized defects in rolling element bearings via composite hypothesis test. Symposium on Mechatronics, American Society of Mechanical Engineers.

MacConnell, J. (2007) ISHM & Design: a review of the benefits of the ideal ISHM system. IEEE Aerospace Conference Big Sky, MT, IEEE.

Mal, A., Banerjee, S. and Ricci, F. (2007) An automated damage identification technique based on vibration and wave propagation data. *Philosophical Transactions of the Royal Society: Mathematical, Physical and Engineering Sciences*, **365**(1851), 479–492.

Maldague, X.P.V. and Moore, P.O. (2001) *Infrared and Thermal Testing Nondestructive Testing Handbook*, American Society for Nondestructive Testing.

Martin, H.R. (1986) Statistical moment analysis as a means of surface damage detection. Proceedings of the XXth International Modal Analysis Conference.

Martinez, M.E. and Quijada, P. (1991) Experimental modal analysis in offshore platforms. Proceedings of the 9th International Modal Analysis.

McColl, J. (2005) HUMS in the era of CAA, JAA, EASA and ICAO. Proceedings of the Eleventh Australian International Aerospace Congress, Melbourne.

McConnell, K.G. (1995) *Vibration Testing Theory and Practice*, John Wiley & Sons, Inc., New York, NY.

Melvin, L. (1997) Integrated vehicle health monitoring (IVHM) for aerospace vehicles, in *Structural Health Monitoring: Current Status and Perspectives* (ed. F. Chang), Technomic Publishing, Palo Alto, CA, pp. 705–714.

Mitchell, J.S. (1993) *Introduction to Machinery Analysis and Monitoring*, PenWel Books, Tulsa, OK.

Mitchell, J.S. (2007) From vibration measurements to condition based maintenance seventy years of continuous progress. *Journal of Sound and Vibration*, **41**(1), 62–75.

Moubray, J. (1997) *Reliability-Centered Maintenance*, Industrial Press, New York, NY.

NASA (1988) Structural Fault Detection of a Light Aircraft Structure Using Modal Technology.

Nataraja, R. (1983) Structural integrity monitoring in real seas. Proceedings of the 15th Annual Offshore Technology Conference, Houston, TX.

National Transportation Safety Board (2008) Collapse of I-35W Highway Bridge, Minneapolis, Minnesota, August 1, 2007. Highway Accident Report NTSB/HAR-08/03.

Nigbor, R.L. and Diehl, J.G. (1997) *Two Years' Experience Using OASIS Real-Time Remote Condition Monitoring System on Two Large Bridges. Structural Health Monitoring: Current Status and Perspectives*, Technomic Publishing, Palo Alto, CA.

Osegueda, R.A., Dsouza, P.D. and Qiang, Y. (1992) Damage Evaluation of Offshore Structures Using Resonant Frequency Shifts. Serviceability of Petroleum, Process, and Power Equipment, ASME PVP239/MPC33.

Pandey, A.K., Biswas, M. and Samman, M.M. (1991) Damage detection from changes in curvature mode shapes. *Journal of Sound and Vibration*, **145**(2), 321–332.

Pappa, R.S., James, G.H. and Zimmerman, D.C. (1998) Autonomous modal identification of the space shuttle tail rudder. *Journal of Spacecraft and Rockets*, **35**(2), 163–169.

Park, G., Sohn, H., Farrar, C.R. and Inman, D.J. (2003) Overview of piezoelectric impedance-based health monitoring and path forward. *Shock and Vibration Digest*, **35**(6), 451–463.

Park, G., Overly, T.G., Farinholt, K.M. *et al.* (2008) Experimental investigation of wireless active-sensor nodes using impedance-based structural health monitoring. Proceedings of 15th SPIE Conference on Smart Structures and Non-destructive Evaluation, San Diego, CA, SPIE.

Pawar, P.M. and Ganguli, R. (2007) Helicopter rotor health monitoring – a review. *Proceedings of the Institution of Mechanical Engineers, Part G: Journal of Aerospace Engineering*, **221**(5): 631–647.

Peeters, J.M.B. and De Roeck, D. (2000) Damage identification on the Z24 Bridge using vibration monitoring. European COST F3 Conference on System Identification and Structural Health Monitoring, Madrid, Spain.

Peeters, J.M.B., Maeck, J. and De Roeck, G. (2000a) Dynamic monitoring of the Z24 Bridge: separating temperature effects from damage. European COST F3 Conference on System Identification and Structural Health Monitoring, Madrid, Spain.

Peeters, J.M.B., Maeck, J. and De Roeck, G. (2000b) Excitation sources and dynamic system identification in civil engineering. European COST F3 Conference on System Identification and Structural Health Monitoring, Madrid, Spain.

Petrilli, O., Paya, B., Esat, I.I. and Badi, M.N.M. (1995) Neural network based fault detection using different signal processing techniques as pre-processor. Structural Dynamics and Vibration, New York, American Society of Mechanical Engineers, pp. 97–101.

Raghavendrachar, M. and Aktan, A.E. (1992) Flexibility by multireference impact testing for bridge diagnostics. *ASCE Journal of Structural Engineering*, **118**, 2186–2203.

Randall, B.R. (2004a) State of the art in monitoring rotating machinery – Part 1. *Journal of Sound and Vibration*, **38**(3), 14–21.

Randall, B.R. (2004b) State of the art in monitoring rotating machinery – Part 2. *Journal of Sound and Vibration*, **38**(5), 10–17.

Randall, R.B. (2011) *Vibration-Based Condition Monitoring – Industrial, Aerospace and Automotive Applications*, John Wiley & Sons, Ltd, Chichester, UK.

Robeson, E. and Thompson, B. (1999) Tools for the 21st century: MH-47E SUMS, in *Structural Health Monitoring: 2000* (ed. F.K. Chang), Technomic Publishing, Palo Alto, CA, pp. 179–189.

Roth, J.T. and Pandit, S.M. (1999) Condition monitoring and failure prediction for various rotating equipment components. Proceedings of the 17th International Modal Analysis Conference, Kissimmee, FL, Society for Experimental Mechanics.

Samual, P.D. and Pines, D.J. (2005) A review of vibration-based techniques for helicopter transmission diagnostics. *Journal of Sound and Vibration*, **282**, 475–508.

Shirole, A.M. and Holt, R.C. (1991) Planning for a comprehensive bridge safety assurance program. *Transportation Research Record*, **1290**, 39–50.

Silverman, H. (2005) T-HUMS – AH64 lead the fleet (LTF) summary and glimpse at Hermes 450 MT-HUMS. Proceedings of the Eleventh Australian International Aerospace Congress, Melbourne.

Simmermacher, T.W. (1996) Damage detection and model refinement of coupled structural systems. PhD Dissertation.

Singer, R.M., Gross, K.C., Walsh, M. and Humenik, K.E. (1990) Reactor coolant pump monitoring and diagnostic system. Proceedings of the 2nd International Machinery Monitoring and Diagnostic Conference, Los Angeles, CA.

Sirkis, J., Childers, B., Melvin, L. *et al.* (1999) Integrated Vehicle Health Monitoring (IVHM) on Space Vehicles: A Space Shuttle Flight Experiment. Damage Assessment of Structures, Dublin.

Sohn, H., Dzwonczyk, M., Straser, E.G. *et al.* (1999) An experimental study of temperature effects on modal parameters of the Alamosa Canyon Bridge. *Earthquake Engineering and Structural Dynamics*, **28**, 879–897.

Sohn, H., Farrar, C., Hemez, F. *et al.* (2004) A Review of Structural Health Monitoring Literature from 1996–2001. Los Alamos National Laboratory Report LA-13976-MS, available online from Los Alamos National Laboratory, http://library.lanl.gov/.

Spencer, B.F., Ruiz-Sandova, L.M.E. and Kurata, N. (2004) Smart sensing technology: opportunities and challenges. *Structural Control and Health Monitoring*, **11**(4), 349–368.

Stanley, R.K. (tech. ed.) (1995) *Nondestructive Testing Handbook 9: Special Nondestructive Testing Methods*, American Society of Nondestructive Testing.

Staszewski, W.J. and Worden, K. (1997) Classification of faults in gear boxes – pre-processing algorithms and nueral networks. *Neural Computing and Applications*, **5**(3), 160–183.

Staszewski, W., Boller, C. and Tomlinson, G.R. (2003) *Health Monitoring of Aerospace Structures*, John Wiley & Sons, Ltd.

Stevenson, W.J., Brown, D.L., Rost, R.W. and Grogan, T.A. (1991) The use of neural nets in signature analysis – rotor imbalance. Proceedings of the 9th International Modal Analysis Conference, Florence, Italy, Society for Experimental Mechanics.

Straser, E.G. (1998) A wireless health monitoring system for civil structures. Doctoral Dissertation, Department of Civil and Environmental Engineering, Stanford University.

Swamidas, A.S.J. and Chen, Y. (1992) Damage Detection in a Tripod Tower Platform (TTP) Using Modal Analysis. ASME Offshore Technology.

Tang, H., Cha, J.-Z., Wang, Y. and Zhang, C. (1991) The principle of cepstrum and its application in quantitative fault diagnostics of gears. Modal Analysis, Modeling, Diagnostics, and Control – Analytical and Experimental, American Society of Mechanical Engineering.

Taylor, J.I. (1994) *Back to the Basics of Rotating Machinery Vibration Analysis*, Vibration Consultants, Inc., Tampa Bay, FL.

Todd, M.D., Johnson, G., Vohra, S. *et al.* (1999) *Civil Infrastructure Monitoring with Fiber Optic Bragg Grating Sensor Arrays. Structural Health Monitoring 2000*, Technomic Publishing, Palo Alto, CA.

Tumer, A.Y. and Bajwa, A. (1999) A survey of aircraft engine health monitoring systems. 35th AIAA/ASME/SAE/ASEE Joint Propulsion Conference and Exhibit, Los Angeles, CA.

Van Way, C.B., Kudva, J.N., Schoess, J.N. *et al.* (1995) Aircraft Structural Health Monitoring System Development – Overview of the Air Force/Navy Smart Metallic Structures Program. Proceedings of the Conference on Smart Structures and Materials: Smart Structures and Integrated Systems, San Diego, CA, SPIE.

Vandiver, J.K. (1975) Detection of structural failure on fixed platforms by measurement of dynamic response. Proceedings of the 7th Annual Offshore Technology Conference, Houston, TX.

Vandiver, J.K. (1977) Detection of structural failure on fixed platforms by measurement of dynamic response. *Journal of Petroleum Technology*, **29**, 305–310.

Volker, E. and Martin, H.R. (1986) Application of kurtosis to damage mapping. Proceedings of the XXth International Modal Analysis Conference.

Wang, M.L. (2005) *Damage Assessment and Monitoring of Long-Span Bridges. Structural Health Monitoring 2005 – Advances and Challenges for Implementation*, Technomic Publishing, Palo Alto, CA.

Wang, W.J. and McFadden, P.D. (1996) Application of wavelets to gearbox vibration signals for fault detection. *Journal of Sound and Vibration*, **192**(5), 927–939.

Watts, B. and Van Dyke, S.J. (1993) An automated vibration-based expert diagnostic system. *Journal of Sound and Vibration*, **26**, 14–20.

Wentzel, H. (2009) *Health Monitoring of Bridges*, John Wiley & Sons, Ltd, Chichester, UK.

West, W.M.J. (1982) Single point random modal test technology application to failure detection. *The Shock and Vibration Bulletin*, **52**(4), 25–31.

West, W. (1984) Fault Detection in Orbiter OV-101 Structure and Related Structural Test Specimens. Loads and Structural Dynamics Branch Report, NASA-Johnson Space Center, Houston, TX.

White, D. (1999) Helicopter usage monitoring using MaxLife System. DSTO Workshop on Helicopter Health and Usage Monitoring Systems, Melbourne, Australia.

White, K.R., Minor, J. and Derucher, K.N. (1992) *Bridge Maintenance, Inspection and Evaluation*, Marcel Dekker, New York, NY.

Whittome, T.R. and Dodds, C.J. (1983) Monitoring offshore structures by vibration techniques. Proceedings of the Design in Offshore Structures Conference.

Widodo, A. and Yang, B.-S. (2007) Support vector machine in machine condition monitoring and fault diagnosis. *Mechanical Systems and Signal Processing*, **21**(6), 2560–2574.

Wojnarowski, M.E., Stiansen, S.G. and Reddy, N.E. (1977) Structural integrity evaluation of a fixed platform using vibration criteria. Proceedings of the Offshore Technology Conference, Houston, TX.

Worden, K. and Manson, G. (2007) The application of machine learning to structural health monitoring. *Philosophical Transactions of the Royal Society: Mathematical, Physical and Engineering Sciences*, **365**(1851), 515–538.

Wouk, V. (1991) *Machinery Vibration Measurement and Analysis*, McGraw-Hill, New York.

Yang, J.C.S., Chen, J. and Dagalakis, N.G. (1984) Damage detection in offshore platforms by the random decrement technique. *ASME Journal of Energy Resources Technology*, **106**(1), 38–42.

Yang, J.C.S., Dagalakis, N. and Hirt, M. (1980) Application of the Random Decrement Technique in the Detection of Induced Cracks on an Off-shore Platform Model. Computational Methods for Offshore Structures, ASME Publication AMD-37.

Zhang, Z. and Atkan, A.E. (1995) The damage indices for constructed facilities. Proceedings of the 13th International Modal Analysis Conference, Orlando, FL, Society for Experimental Mechanics.

3

Operational Evaluation

This chapter will describe in more detail the issues associated with the four questions that must be answered as part of the operational evaluation step in the SHM process. As stated in Chapter 1, these four questions are:

1. What are the life-safety and/or economic justifications for performing the structural health monitoring?
2. How is damage defined for the system being investigated and, for multiple damage possibilities, which cases are of the most concern?
3. What are the conditions, both operational and environmental, under which the system to be monitored functions?
4. What are the limitations on acquiring data in the operational environment?

In some cases, a significant effort will have to be expended to provide quantified answers to these questions. However, without well-defined answers to these questions, it is unlikely that SHM systems will be developed or be used for anything but research programmes.

A variety of individuals' input will be required to answer these questions. In general, different input will be required for SHM systems that are being developed for existing structures as opposed to a system that is being developed for a structure still in the planning or design stages. People who may be called upon to provide this input will include, but are not limited to, system designers, system operators, maintenance personnel, financial analysts and regulatory officials.

3.1 Economic and Life-Safety Justifications for Structural Health Monitoring

To date, numerous SHM studies have been carried out with little concern for the economic issues associated with field implementation of the technology. This situation has arisen because many of these studies have been carried out as part of a more basic, one-of-a-kind, research effort aimed at developing proof-of-concept technology demonstrations. Prime examples of such systems are the ones that have been put in place on large suspension bridges such as the Tsing Ma Bridge near Hong Kong (Lau *et al.*, 1999) and the Commodore Perry Bridge in Delaware (Barrish, Grimmelsman and Aktan, 2000). However, it is unlikely that any private company or government institution will be willing to fund the development of an SHM capability for field deployment unless it can be shown that this capability can provide a quantifiable and enhanced life-safety and/or economic benefit relative to the currently employed damage detection strategy. Economic benefits can be realised in a variety of manners including reduced maintenance cycles, reduced warranty obligations, increased manufacturing capacity and increased system availability.

Life-safety advantages provided by SHM are often harder to quantify than economic advantages. Also, life-safety issues are not independent from economic issues. Litigations associated with injuries or fatalities resulting from damage to a structure add a significant economic impact to the human tragedy. Furthermore, after many structural failures resulting in fatalities, such as the space shuttle accidents and most accidents associated with large commercial aircraft, there is a considerable expense associated with the accident investigation.

The process of demonstrating the economic benefits of a damage detection capability will set limits on all other aspects of the SHM paradigm. In particular, the data acquisition hardware budget and the budget for time that can be spent developing numerical simulations of the damage scenarios will be limited by the need to show a positive rate-of-return on investment in SHM technology. Therefore, if one is to take SHM beyond the proof-of-concept level, at a minimum, the following questions related to the life-safety/economic benefits issue must be presented to authorities making the decision to invest in the system:

1. What are the limitations of the currently employed damage detection methodology?
2. What are the advantages provided by the proposed SHM system?
3. How much will the proposed SHM system cost?
   (a) How long will it take to develop and validate the proposed SHM system?
   (b) How much will it cost to maintain the proposed SHM system?

A final economic issue that should be considered is the portion of the budget that is funding the SHM system development. Typically, for most high-expenditure aerospace, civil and mechanical systems, the design and construction budget is much larger than the annual maintenance budget. Therefore, the capital expenditures associated with the hardware requirements of the SHM system will not seem as extreme if they are included in the design and construction portion of the budget. In contrast, this same hardware can be a significant fraction of the system's annual maintenance budget, which poses challenging economic issues when trying to incorporate an SHM capability in a retrofit mode.

3.2   Defining the Damage to Be Detected

The success of any damage detection technique will be directly related to the ability to define that damage that is to be detected in as much detail as possible and in as quantifiable terms as possible. Here the definition of damage can include issues such as the type of damage to be detected (e.g. fatigue crack), the threshold level of damage that must be detected (e.g. a 2-mm-long, through-thickness crack), the critical level of damage that produces failure or that will no longer allow for a planned safe shut down of the system (a 5-cm-long crack), locations where the particular type of damage accumulates in the structure (e.g. welded beam-to-column connections) and the tolerable or anticipated rate of damage growth.

In the examples cited above, the damage is quantified in terms directly related to the type of failure. There will also be cases where the damage will be quantified in an indirect manner. As an example, damage may be defined as delamination of a composite aircraft wing component. The critical level may be defined as an amount of damage that produces a certain change in the flutter characteristics of the aircraft.

Because of the costs, system-level testing to failure for the purpose of defining the damage to be detected is rarely done. Instead, these definitions are often based on prior observed behaviour of damaged systems. Component-level testing is a more cost-effective experimental approach that can also be used to develop such damage definitions. At other times numerical simulations of the damaged system will be used to establish these definitions. Finally, there will be many cases where these definitions will be developed in an ad hoc manner based on the experience and intuition of people familiar with the equipment.

Large complex structural systems are usually made up of numerous components. Typically, there will be multiple damage mechanisms that will be of concern. However, an SHM system that is optimal for detecting one type of damage may not be useful for detecting an alternate damage condition. Also, there will be cost limitations on the development and deployment of the SHM system. Therefore, it is imperative to define the various possible damage scenarios and to establish a priority for detecting damage associated with these scenarios. Such prioritisation will be based on some study of the relative probability of occurrence associated with the various damage scenarios versus the consequences of the respective damage scenarios. With such a prioritisation established, the SHM system can be designed to address the most critical damage concerns, which will make optimal use of a given SHM budget.

The definition of damage is the first step in identifying the required data acquisition system capabilities, the candidate features to be extracted for damage detection and the statistical models that will be employed for the feature discrimination. To date, many SHM research studies suffer from taking the approach where the damage detection process is first developed and then the procedure is studied in an effort to define the damage that it can detect.

3.3 The Operational and Environmental Conditions

When deployed on a structure outside a controlled laboratory setting, the damage detection process will have to deal with structures that experience changing operational and environmental conditions. These changing operational and environmental conditions will produce changes in the measured system response and it is imperative that these changes are not interpreted as indications of damage. Varying temperature and moisture levels are two common environmental conditions that must be accounted for during the damage detection process. Changing mass is a common operational variable that results from fuel usage and varying payloads. Running equipment or vehicles at varying speeds and through varying manoeuvres are other operational parameters that can significantly influence dynamic quantities that are being measured as part of the damage detection process. As an example, the offshore oil platform shown in Figure 3.1 poses many of these issues for an SHM system. Inputs to this platform will vary depending on the sea states and the operating equipment. Mass of this system will vary depending on the fluid stored on the platform and the amount of marine growth. The corrosive environment associated with salt water can also pose limitations for equipment that may be used in the monitoring process.

Defining and quantifying the range of operational and environmental conditions under which the monitoring is to be performed will help to define the required data acquisition system capabilities better. Often sensors will have to be added to monitor the changing operational and environmental conditions in an effort to develop a procedure that normalises the data to remove trends caused by these effects. Also, the sensors themselves will have environmental and operational limitations under which they can properly function. As with the damage definition portion of the operational evaluation step, definition of the operational and environmental conditions under which the monitoring is to be performed will also impact other portions of the SHM process including the data normalisation procedures, feature selection and statistical modelling.

3.4 Data Acquisition Limitations

Finally, one must assess the limitations associated with making measurements on the *in situ* structure. These limitations may result from a variety of considerations including those of an economic or environmental nature, structure size, physical access, adverse effects of the sensors on the system operation and even nontechnical issues. An example of a nontechnical issue that will put limitations on the measurement system arises when the owner does not want the public to see the sensing hardware for fear

Figure 3.1 The operational and environmental conditions associated with offshore oil platforms pose many challenges for a structural health monitoring system (courtesy of US National Oceanic and Atmospheric Administration).

they will lose confidence in the safety of the system being monitored. As previously mentioned, all data acquisition will be somewhat limited by economic considerations and the desire to demonstrate that the SHM activity can provide a positive rate-of-return on investment.

Data acquisition limitations can manifest themselves in a variety of ways as the following examples illustrate. Certain types of sensors will not perform well in extreme thermal or radiation environments. Sensors that are used to monitor equipment in electric power substations must be able to withstand the electromagnetic fields in these plants. Physical access to critical portions of the structure is a particular issue when the SHM system is being developed as a retrofit as opposed to being planned during the initial system design. Portions of an aircraft wing are difficult to access after the wing is built. Access to the foundation of an offshore oil platform is extremely difficult once the structure is put in place. A hip prosthesis is impossible to instrument directly without invasive surgery. Sensors that require an electric signal for operation, such as piezoelectric accelerometers, can potentially have an adverse effect on weapon systems or fuel tanks where detonation or deflagration is a concern. For aerospace structures weight restrictions will place severe limitations on the sensor system to be deployed.

In summary, it is imperative that issues such as these be addressed before the SHM process is developed further. If these limitations are not defined early on in the SHM system development process, one runs the risk of defining a system that will not perform its intended function.

3.5 Operational Evaluation Example: Bridge Monitoring

Monitoring a highway bridge such as the one shown in Figure 3.2 for fatigue crack growth is used as an example to demonstrate, in general terms, how one might formulate the response to the four questions that

Figure 3.2   A steel girder bridge that requires periodic inspection for fatigue crack damage.

make up the operational evaluation portion of the SHM paradigm. Fatigue crack growth in steel bridge structures is a concern to highway departments from a life-safety perspective and from the perspective of economically efficient maintenance. Currently, the US government requires bridges to be visually inspected every two years (White, Minor and Derucher, 1992), as shown in Figure 3.3. Therefore, it must be demonstrated that the SHM system can provide higher-fidelity damage detection at a cost equal to or less than that associated with the required visual inspections.

For certain steel girder bridge designs (e.g. the I-40 Bridge discussed in Chapter 5), it is known from past experience that fatigue cracks form at locations where the seats supporting the lateral floor beams are welded to the web of the longitudinal plate girders. These cracks result from out-of-plane bending of the plate girder web caused by traffic loading that is transferred from the floor beam to the seat. Therefore, information is available regarding the type of damage that is to be detected and its location. Note that there will be other types of damage that are of concern, but crack formation in the web of the longitudinal plate girders is of primary concern because there is no redundancy for these structural elements. Damage parameters that are harder to quantify will be the level of damage to be detected and a tolerable rate of damage growth. Bridge engineers and maintenance personnel will have to work together to establish these SHM system parameters.

This particular type of damage may manifest itself in terms of changes to the vibration signature caused by a fatigue crack opening and closing under the normal traffic loading. Generation of resonance frequency harmonics is a common feature of a vibration signature associated with cracks opening and closing (see Chapter 8). Therefore, the instrumentation and data acquisition system will need to be designed such that it has the highest potential to detect this feature, which may necessitate the ability to measure a higher-frequency portion of the response spectrum.

Figure 3.3 An engineer performs a visual inspection of a steel girder bridge (courtesy of US Federal Highway Administration).

Ideally, a sensing system for bridge fatigue crack growth will monitor the bridge during its normal operations. During such operation there can also be considerable changes in the environmental conditions including temperature changes, rain and wind. Traffic, wind and ground motion will provide the primary inputs to the structure that can cause the fatigue cracks to open and close. These excitation sources will not be stationary and their variability will have to be quantified. Certain changes associated with operational and environmental conditions will occur at regular intervals. These intervals can occur within a day (e.g. traffic flow changes associated with rush hours, night-to-day temperature fluctuations), within a week (e.g. traffic flow changes during the week versus the weekend) or seasonally (e.g. rainy versus dry seasons). Other changes may be more random, such as traffic slow-downs caused by an accident. All these scenarios must be considered in the development of the hardware and software portions of an SHM system that monitors a bridge for fatigue crack growth.

If fatigue crack growth in highway bridges is to be monitored on an operating bridge, then there will be limitations on where the sensors can be placed such that they do not interfere with traffic flow. The changing weather conditions under which the bridge operates place further constraints on the data acquisition system that can be deployed, as does the physical size of the bridge. Also, many portions of the bridge will be difficult to access, which will place limitations on where the sensors can be located.

This example shows that many aspects of the SHM problem are first addressed in the operational evaluation portion of the SHM process. It should be noted that in actual practice this qualitative discussion would have to be supplemented by some more rigorous engineering analysis to better quantify the critical size of damage that must be detected. An assessment would also have to quantify what a tolerable damage growth rate is for the fatigue loading that the bridge will experience. Such issues tend to be overlooked in many research studies.

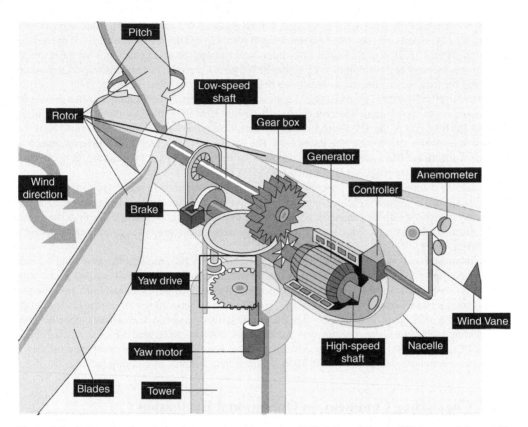

Figure 3.4 Schematic of a wind turbine generator (courtesy of US Office of Energy Efficiency and Renewable Energy).

3.6 Operational Evaluation Example: Wind Turbines

The use of wind turbines is increasing as societies look for alternative clean energy sources. A three-blade horizontal axis wind turbine is typical of many utility-scale turbine designs[1] (Barkley, Jacobs and Rutherford, 2006). The wind energy is captured by the turbine blades, which then turns a drive train connected to a generator as depicted in Figure 3.4. Wind turbine profitability depends on the amount of electricity produced, which in turn depends on both the size of the turbine, the wind speed and the reliability of the rotor and generator. Utility-scale rotor diameters range from 50 to 100 metres. Typical turbines can produce between 700 kW to 2.5 MW of electricity, with some designs producing up to 5 MW.

Because turbines are typically located in remote areas, away from people, there are no life-safety justifications for implementing an SHM system. Motivation for a wind turbine SHM is purely economic. Wind turbine economics can be broken down into the categories of initial investment, installation cost, income from electricity sales and operation and maintenance (O&M) costs. One can assume that the initial investment is in the range of US$1–1.5 million per megawatt. Furthermore, current annual O&M costs might be assumed to be approximately 1.5–3% of the original investment price. If a 5-megawatt

---

[1] This example is taken from a 2006 report entitled 'Structural Health Monitoring of Wind Turbine Blades', by W. Barkley, L. Jacobs and A. Rutherford, submitted as part of a project for a graduate class on SHM offered by the Structural Engineering Department at the University of California, San Diego.

turbine is considered, then the initial investment will be about $5–7.5 million/turbine and annual O&M costs using a 2% figure are $100–150 K/year/turbine. It is anticipated that a major overhaul of the turbine, which might cost 15–20% of the initial investment, will be required at 20 years and for this example corresponds to a cost of $1–1.5 million using the 20% figure. Currently, technicians inspect turbines to determine the necessary maintenance. Ideally, an SHM system would replace or augment this inspection process in an effort to minimise annual O&M costs and allow for a major overhaul only when the turbine condition dictates such a need. The cost defined above for O&M as well as for the overhaul set limits on what the total cost of the SHM system (development, deployment and maintenance) must be if it is to provide a positive rate-of-return on investment. The development cost can be divided over numerous turbines if identical SHM systems are to be deployed on multiple turbines. In addition, this evaluation also places limits on the required service life of the SHM system.

Damage in a turbine blade is typically caused by fatigue resulting in some type of cracking or delamination of the composite blade, loosening of torque in the blade root or damage to the rotating machinery associated with the generator. Therefore, one must define a critical crack size or delamination area, critical levels of torque reduction and/or generator damage concerns the system must be capable of detecting.

Environmental and operating conditions for wind turbines can be quite severe, depending on their location. Not only do these turbines have to withstand vibration and bending loads caused by the wind and the resulting cyclic stresses, but they also experience diurnal changes in temperature, precipitation and lightning strikes. Because the turbine rotates, communication between sensors on the blades and the data collection point can be a challenge. Additionally, the turbines generate an electromagnetic field that can interfere with wireless data transmission. Finally, some turbines are located offshore, making access more difficult if the SHM system itself needs maintenance. For either land-based or offshore turbines, a telemetry system is needed to transfer the SHM information to a central monitoring facility.

3.7   Concluding Comment on Operational Evaluation

These examples are intended to emphasise that the efforts expended for the operational evaluation during the SHM system conception are essential to successful SHM. However, other documented studies of the operational evaluation portion of the SHM process that have been reported in the literature are somewhat limited as such studies are rarely conducted for research programmes. Industries will, in general, not undertake the development of SHM systems without performing some form of operational evaluation and some of these studies are starting to be reported in the literature (MacConnell, 2007). Once the operational evaluation has been completed, the other portions of the SHM process can be developed with an increased probability that the end system will be able to serve its intended function in a cost-effective manner and identify the damage that is of concern.

References

Barkley, W., Jacobs, L. and Rutherford, A. (2006) Structural Health Monitoring of Wind Turbine Blades. Report submitted to Structural Engineering Department at the University of California San Diego.

Barrish, R.A., Grimmelsman, K.A. and Aktan, A.E. (2000) Instrumented monitoring of the Commodore Barry Bridge. Proceedings of the SPIE Nondestructive Evaluation of Highways, Utilities, and Pipelines IV.

Lau, C.K., Mak, W.P.N., Wong, K.Y. *et al.* (1999) *Structural Health Monitoring of Three Cable-Supported Bridges in Hong Kong. Structural Health Monitoring 2000*, Technomic Publishing, Lancaster, PA.

MacConnell, J. (2007) ISHM & Design: A review of the benefits of the ideal ISHM system. IEEE Aerospace Conference, Big Sky, MT, IEEE.

White, K.R., Minor, J. and Derucher, K.N. (1992) *Bridge Maintenance, Inspection and Evaluation*, Marcel Dekker, New York.

4

Sensing and Data Acquisition

4.1 Introduction

Obtaining accurate measurements of a system's dynamic response is essential to SHM. There are many different sensors and data acquisition systems that can be applied to the SHM problem and the one employed will be application specific. As discussed in Chapter 3, the design of the appropriate sensor system begins during the operational evaluation stage of the ideal SHM process. A detailed summary of all of these systems is beyond the scope of this book. Instead, the approach taken in this chapter will be to summarise general sensor network paradigms employed for SHM and to identify issues that must be considered when developing the SHM sensing and data acquisition system. This chapter will begin to address these issues by presenting two different general instrumentation strategies for damage detection. These strategies are presented in increasing order of sophistication. These strategies are then discussed in terms of the fundamental issues to achieving a viable damage detection capability. Next, this discussion addresses more specific design issues associated with the sensors and data acquisition systems that must be addressed.

It is acknowledged that sensing technology is one of the most rapidly developing fields related to SHM and, therefore, one must always be looking for new technologies that are applicable to the SHM problem. Throughout this chapter it will be emphasised that the most fundamental issue for SHM sensor system design is the need to capture the structural response on widely varying length and time scales. Also, the sensing system development must be done in an integrated manner as the selection of one component (e.g. the transducer) often places requirements on other system components such as the analogue-to-digital (A/D) converter as well as limitations on the feature extraction and statistical modelling portions of the process. Finally, a reality that must be kept in mind as the sensing and data acquisition system is developed is that there will be a finite budget for this hardware, which will always place limitations on the system that is eventually deployed.

4.2 Sensing and Data Acquisition Strategies for SHM

Almost all sensor systems are deployed for one or more of the following applications:

1. Detection and tracking problems.
2. Model development, validation and uncertainty quantification.
3. Control systems.

Structural Health Monitoring: A Machine Learning Perspective, First Edition. Charles R. Farrar and Keith Worden.
© 2013 John Wiley & Sons, Ltd. Published 2013 by John Wiley & Sons, Ltd.

Structural health monitoring is essentially a detection and tracking problem. The first goal of any SHM sensor system development is to make the sensor reading as directly correlated with, and as sensitive to, damage as possible (*detection*). Then one wishes the sensor readings and associated damage-sensitive features extracted from these data to change in a monotonic fashion with increasing damage levels (*tracking*). At the same time one also strives to make the sensors as independent as possible from all possible sources of environmental and operational variability. There are a variety of approaches to developing a sensing and data acquisition strategy that adequately meets these goals. Two general approaches will be discussed here in an attempt to highlight the effort required to develop a robust SHM sensor network and the practical limitations that must be considered in such a design process.

4.2.1 Strategy I

An often-used approach to deploying a sensing system for damage detection, particularly associated with earlier SHM studies, consists of adopting a sparse sensor array. This array is installed on the structure after fabrication, possibly following an extended period of service. Sensors are typically chosen based on previous experience of the SHM system developer and commercial availability. The selected sensing systems often have been commercially available for some time and the technology may be more than twenty years old. Excitation is often limited to that provided by the ambient operational environment. The physical quantities that are measured are often selected in an ad hoc manner without an a priori quantified definition of the damage that is to be detected or any a priori analysis that would indicate that these measured quantities are sensitive to the damage of interest. This approach dictates that damage-sensitive data features are selected 'after the fact' using archived sensor data and ad hoc algorithms. This scenario represents many real-world systems, particularly those deployed on civil engineering infrastructure (Farrar *et al.*, 1994).

The most common damage detection approach associated with Strategy I is based on the assumption that the undamaged and damaged structure are subjected to nominally similar excitations. Features determined using trial-and-error approaches, or physical intuition of the SHM system developers, equipment operators or maintenance personnel, are extracted from the measured response and correlated to damage using a variety of methods that vary in their level of mathematical sophistication. When data are not available from the damaged structure, the damage detection process reverts to some form of outlier or novelty detection (as discussed in Chapter 1, the problem is one of unsupervised learning). In this case, trial-and-error procedures or physical intuition are again used to define the features that will have enhanced sensitivity to damage and to set thresholds that define when these features can be identified as outliers to some preset level of statistical confidence. Despite the ad hoc nature of this process, Strategy I is sometimes effective for damage detection. This approach is often enhanced by a comprehensive historical database of measured system response and associated damage observations. For example, the availability of information regarding the damage state and corresponding measurement results for large numbers of nominally identical units will significantly improve the ability to detect damage in subsequent units when Strategy I is employed. Typically, such sensing systems are not designed to measure the parameters necessary to allow one to separate operationally and environmentally induced changes in the measured dynamics response from changes caused by damage. In general, if one is 'retrofitting' an SHM system to some existing structure, it is likely that some close variant of Strategy I will result.

4.2.2 Strategy II

Strategy II is a more coupled analytical/experimental approach to the sensor system definition that incorporates some significant improvements over Strategy I. First, damage is well defined and to some extent quantified through the operational evaluation process before the sensing system is designed. Next, the sensing system properties, and any relevant actuator properties, are defined based on the results of numerical simulations of the damaged system's dynamic response or physical experiments where simulated damage is inflicted on the structure. Note that the inclusion of an actuation capability implies that an active sensing approach will be used as opposed to the passive approach associated with Strategy I. The data analysis procedures (e.g. feature extraction and statistical discrimination) that will be employed in the damage detection application are also considered when developing the sensing and data acquisition system. This process of defining the sensor system properties will often be iterative. Sensor types and locations are chosen because the numerical simulations or physical tests and associated optimisation procedures show that the expected type of damage produces known, observable and statistically significant effects in features derived from the measurements at these locations. Additional sensing requirements are then defined based on how changing operational and environmental conditions affect the damage detection process. Methods for data archiving and telemetry are considered in the sensor system design along with long-term environmental ruggedness and maintenance issues. However, all sensors and data acquisition hardware are still chosen from the commercially available products that best match the defined sensing system requirements.

Strategy II incorporates several enhancements that would typically improve the probability of damage detection:

1. It has well-defined and quantified damage information that is based on initial system design information, numerical simulation of the postulated damage process, qualification test results, maintenance records, and/or system autopsies.
2. It uses sensors that are shown to be sensitive enough to provide data that can be used to identify the predefined damage when the measured data are coupled with the feature extraction and statistical modelling procedures.
3. Active sensing is incorporated into the process where a known input is used to excite the structure with a waveform tailored to the damage detection process.
4. Sensors are placed at locations where responses are known from analysis, experiments and past experience to be sensitive to damage.
5. Additional measurements are made that can be used to quantify changing operational and environmental conditions.

Only recently have SHM studies reported in the technical literature started to take the Strategy II approach to developing a sensing system (Flynn, 2010). In actuality, most sensing systems used to detect damage take an approach somewhere in between Strategy I and Strategy II. Even if one does adopt Strategy II, a major limitation is reliance on off-the-shelf sensing technology, most of which is not designed with SHM in mind. This situation is likely to persist until appropriately large markets can be developed for SHM.

4.3 Conceptual Challenges for Sensing and Data Acquisition Systems

The two strategies described above suggest general conceptual challenges to effective damage detection from a sensing and data acquisition system perspective. These challenges include the following:

1. The need to develop a quantified definition of the damage that is to be detected before the sensor system is designed.
2. The ability to capture local and system-level responses; that is, the need to capture responses on widely varying length and time scales.
3. The need for a well-defined sensing system design methodology.
4. The need to separate environmental and operation effects on sensor readings from the effects caused by damage.

5. The need to integrate the feature extraction and statistical modelling algorithms with the sensing system design process.
6. The ability to archive data in a consistent, retrievable manner for long-term monitoring.
7. The ability to transmit information regarding the system condition to maintenance personnel or a control system.
8. The ability of the sensing system to function with minimal maintenance over long periods of time.
9. The need to minimise the cost of the sensing and data acquisition system.

The feature extraction, data normalisation and statistical modelling portions of the process can greatly influence the definition of the sensing system properties. Before such issues can be confronted, two important questions that were first discussed in Chapter 3 must be addressed.

First, one must answer the question, 'What is the damage to be detected?' As discussed in Chapter 3, the answer to this question must be provided in as quantifiable a manner as possible and address issues such as (1) type of damage (e.g. crack, loose connection, corrosion), (2) threshold damage size that must be detected, (3) probable damage locations and (4) anticipated damage growth rates. The more specific and quantifiable this definition, the more likely it is that one will optimise the sensor budget to produce a system that has the greatest possible fidelity for damage detection. Second, an answer must be provided to the question, 'What are the environmental and operational variabilities that must be accounted for?' To answer this question, one will not only have to have some ideas about the sources of such variability but will also have to think about how to accomplish data normalisation. Typically, data normalisation will require some combination of sensing system hardware and data interrogation software. However, these hardware and software approaches will not be optimal if they are not developed in a coupled manner. The data normalisation topic is discussed in much more detail in Chapter 12.

In summary, from the discussion in this section it becomes clear that there are a lot of issues to address when designing an SHM sensor system. Also, the ability to convert sensor data into structural health information is directly related to the coupling of the sensor system hardware development with the data interrogation procedures. The subsequent portions of this chapter will address specific sensing system issues associated with damage detection.

4.4   What Types of Data Should Be Acquired?

Instrumentation, which includes sensors and data acquisition hardware, first translates the system's dynamic response into an analogue voltage signal that is proportional to the response quantity of interest. This process is typically accomplished though the transduction properties of the sensing material that allows for the conversion of the response variable (e.g. acceleration) to some other field (most often an electric signal). Next, the analogue signal is discretely sampled to produce digital data. To begin defining a sensing system for damage detection, one must first define the types of data to be acquired. The data types fall into four general categories of:

1. Dynamic input and response quantities (e.g. input force, strain or acceleration),
2. Other damage-sensitive physical quantities (e.g. electromagnetic fields, chemicals),
3. Environmental quantities (e.g. temperature or wind speed) and
4. Operational quantities (e.g. traffic volume or vehicle air speed).

There are many commercially available sensors that can be used to measure these various physical quantities and there are also emerging sensor technologies that will have a tremendous impact on the future of SHM.

4.4.1   Dynamic Input and Response Quantities

The dynamic response of a structure is a function of its mass, stiffness and damping properties as well as the applied input. Damage is typically assumed to change the stiffness and damping properties of a structure; hence it is intuitive that one should monitor the response of and the input to the structure as part of a damage detection strategy.

The 'traditional' sensors used to measure dynamic response include electric-resistance strain gauges, displacement transducers such as linear variable differential transducers (LVDTs) and piezoelectric accelerometers. In general, the accelerometers provide an absolute measurement at a point on the structure while the displacement and strain sensors provide relative measurements over typically short gauge lengths. These sensors are used extensively for aerospace, civil and mechanical engineering applications. Conditioning electronics for these sensors have evolved from bulky vacuum tube systems to small, sophisticated, solid-state devices integrated directly with the sensor. A wide variety of these types of sensors, which are appropriate for many different applications, are commercially available.

Current commercially available piezoelectric accelerometers have proven to be reliable and stable, and to date are probably the most commonly used sensor for SIIM. These accelerometers incorporate onboard signal conditioning and may soon have onboard A/D conversion. However, these traditional sensors are relatively expensive (hundreds of dollars for a conventional piezoelectric accelerometer) and to date they are typically not integrated with microprocessors. In contrast, high-quality piezoelectric discs suitable for direct strain measurement can be obtained for a few dollars.

The principal emerging sensing technologies for measurement of dynamic response quantities that are being applied to SHM include microelectromechanical systems (MEMS), piezoelectric actuator/sensors and fibre optic strain sensors. Commercially available MEMS devices can measure strain, as well as angular and linear acceleration. A MEMS accelerometer that measures three angular and three linear accelerations is shown in Figure 4.1. Once fully developed, MEMS sensors have the potential to impact a variety of sensing activities based on their versatility, small size and low cost when manufactured in large numbers. These properties will allow the sensor density on a structural system to increase significantly, which is essential to improve damage detection fidelity. MEMS can be integrated with onboard computing to make these sensors self-calibrating and self-diagnosing. This integration of the sensor with microprocessors defines the 'smart sensor' concept. Inhibiting MEMS use today are issues such as mounting schemes, traceable calibration, demonstrated long-term survivability and the lack of sensors specifically designed for structural monitoring activities.

The most common input quantity measured for SHM purposes is the input force from some type of electrodynamic or hydraulic actuator. This quantity is most commonly measured with a piezoelectric force transducer or pressure sensor. Although the sensing element is configured differently, these sensors work in a similar manner to the piezoelectric accelerometer. These force and pressure sensors are commercially available in an array of different designs that meet the measurement needs for a wide range of applications.

The electrical resistance strain gauge is the most common strain sensor used for SHM. Piezoelectric materials are also widely used to measure the 'dynamic' strain, as opposed to 'static' strain, which is measured by traditional strain gauges (Sirohi and Chopra, 2000). Strain can be related to the electric charge generated in the piezoelectric material when it is deformed by the structure to which it is mounted. Piezoelectric materials are capable of interacting over a wide range of frequencies, which makes them particularly useful for sensing on varying length and time scales. Both piezoceramics and flexible piezofilms are commercially available and come in a variety of forms ranging from thin rectangular patches to complex shapes obtained with MEMS fabrication. More importantly, piezoceramic materials can also serve as an actuator because they can apply local excitations in response to an applied electric field (see Section 4.11).

Recently, there have been numerous SHM studies that are making use of fibre optic strain gauges. Here a selectable gauge length of a single long fibre with multiple Bragg gratings (Todd, Johnson and Althouse, 2001) is queried to obtain the strain (with up to picostrain accuracy). This technology

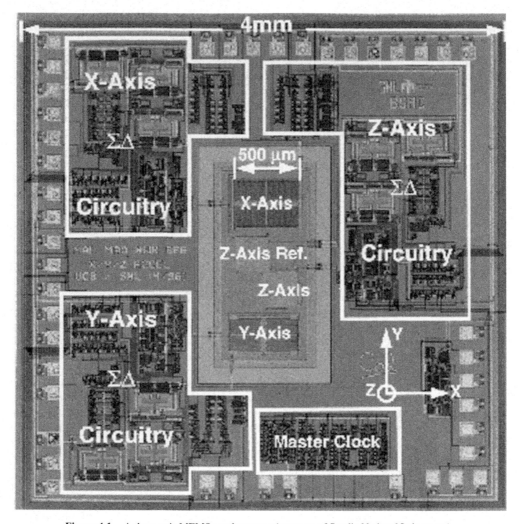

Figure 4.1 A three-axis MEMS accelerometer (courtesy of Sandia National Laboratory).

can allow a single fibre with the length of a bridge girder to monitor strain at numerous locations along the girder. Furthermore, the fibre can be embedded in manufactured parts, such as composite aerospace components. Features of fibre optic sensors that are advantageous for SHM applications are their immunity to electromagnetic and radio-frequency (RF) interference. Also, these sensors are not a spark source, which is a key issue if monitoring is to be done near combustible materials such as rocket fuel tanks. These sensors are nonintrusive (on the order of a human hair in thickness), extremely lightweight (a key advantage for aerospace applications) and have proven to be very rugged (Johnson, 2000). Furthermore, they can be formed into rosettes to yield a more accurate estimate of the strain field at a particular location (Betz *et al.*, 2003). As with all strain measuring devices, fibre optic strain gauges are sensitive to changing thermal fields, but temperature compensation strategies have been developed to mitigate this problem.

The previous discussion of measuring dynamic response quantities has focused on local measurements. In addition, there are some more global sensing technologies that are commercially available. More mature global sensing technologies include scanning laser Doppler velocimeters, digital image correlation, active thermography systems and acoustic field detectors. Global sensors can scan a surface of a structure and in some cases, with proper signal processing, they can identify damaged areas. The disadvantages associated with global sensors include a fairly high procurement cost, the need for a visual access to the measured part and the need to remove the structure from service to carry out the test. An emerging technology in this area of global sensing is chemical coatings that emit a particular signature when cracked. This technology has already been demonstrated through the application of pressure-sensitive paints for wind tunnel testing (Engler, Klein and Trinks, 2000).

4.4.2   Other Damage-Sensitive Physical Quantities

One is not limited to sensing only the quantities discussed above for SHM studies. In fact, there are many other physical quantities that are related to various types of damage and sensors have been developed to detect these quantities. Acoustic emission associated with damage initiation and progression is one example of a mature sensing technology that is employed for different SHM applications (Paget, Atherton and O'Brien, 2004). The measurement of changes in electric impedance across a piezoelectric sensor/actuator is another measurement technique that is discussed extensively in the SHM literature. In these studies it has been shown that changes in the mechanical impedance of the structure that results from damage will influence the electrical impedance measured across the sensor (Park et al., 2003). This type of measurement process has been advanced by the recent commercial availability of low-cost microchip impedance analyzers.[1] Changes in electromagnetic properties have been measured and correlated with damage in bridge cables (Wang, 2005). Corrosion detection is of particular interest for all types of aerospace, civil and mechanical infrastructure SHM. Some corrosion sensors detect chemical by-products of the corrosion process (Elster et al., 1999). Clearly, there are potentially many other physical quantities that can be shown to be correlated with damage, and new sensors for measuring these quantities are continually being reported in the technical literature.[2]

4.4.3   Environmental Quantities

If changes in environmental quantities produce changes in the measured response similar to those produced by damage, a measure of the environmental quantity will be necessary to separate the environmental effects from the damage effects. Such a case will necessitate that environmental quantities such as temperature, pressure and moisture content must be measured. Note that if damage changes the features in a manner that is in some way orthogonal to the changes produced by environmental effects, then a measure of the environmental parameters may not be necessary (see Chapter 12). There are a wide variety of well-developed sensing technologies to measure environmental quantities that can be readily adapted to the SHM problem (Webster, 1999). Not only must these environmental quantities be measured to assess their influence on the structure but they must also be measured to assess their impact on the sensors. In many cases it is difficult to assess and quantify the impact of environmental effects such as electromagnetic fields and radiation fields on the long-term stability of sensors (Holbert et al., 2003) (see Section 4.8.4 below).

---

[1] www.analogue.com/static/imported-files/data_sheets/AD5933.pdf.
[2] www.sensorsmag.com.

4.4.4 Operational Quantities

Similar to environmental quantities, operational quantities may also produce changes in data used to derive damage-sensitive features that necessitate the measurement of the operational parameters. Operational quantities include such things as traffic volume for a bridge, mass loading on an offshore oil platform or the amount of fuel in an aeroplane wing. Operation speeds and manoeuvres of vehicles are other examples of quantities that can influence SHM sensor readings. Operational quantities for mechanical equipment that can help identify damage include equipment and cooling fluid temperatures, fluid pressures and flow rates, oil debris and power consumption. As with environmental quantities there are many well-developed sensing technologies for measurement of operational variables that can be readily adapted to SHM applications.

4.5 Current SHM Sensing Systems

Sensing systems for SHM consist of some or all of the following components:

1. Transducers that convert changes in the field variable of interest (e.g. acceleration, strain, temperature) to changes in an electrical signal (e.g. voltage, impedance, resistance).
2. Actuators that can be used to apply a prescribed input to the system (e.g. a piezoelectric transducer bonded to the surface of a structure).
3. A/D converters that transform the analogue electrical signal into a digital signal that can subsequently be processed on digital hardware. For the case where actuators are used, a digital-to-analogue (D/A) converter will also be needed to change a prescribed digital excitation signal to an analogue voltage that can be used to control the actuator.
4. Signal conditioning.
5. Power.
6. Telemetry.
7. Processing.
8. Memory for data storage.

The number of sensing systems available for SHM even now is enormous and these systems vary quite a bit depending upon the specific SHM activity. Two general types of SHM sensing systems are described below.

4.5.1 Wired Systems

Here wired SHM systems are defined as ones that transfer data and power to or from the sensor over a direct wired connection from the transducer to the central data analysis facility, as shown schematically in Figure 4.2. In some cases the central data analysis facility is then connected to the Internet such that the processed information can be monitored at a remote location. Digitising, cleansing, recording and data storage all occur at the one central location. Subsequent data analysis also occurs at this same location.

There are a wide variety of such systems. At one extreme is peak-strain or peak-acceleration sensing devices that notify the user when a certain threshold in the measured quantity has been exceeded. A more sophisticated system often used for condition monitoring of rotating machinery is a piezoelectric accelerometer with a built-in charge amplifier connected directly to a hand-held, single-channel fast Fourier transform (FFT) analyser. Here the central data storage and analysis facility is the hand-held FFT analyser. At the other extreme are custom designed systems with hundreds of data channels containing numerous types of sensors that cost on the order of multiple millions of dollars, such as the sensing system deployed on the Tsing Ma Bridge in China (Ni, Wang and Ko, 2001).

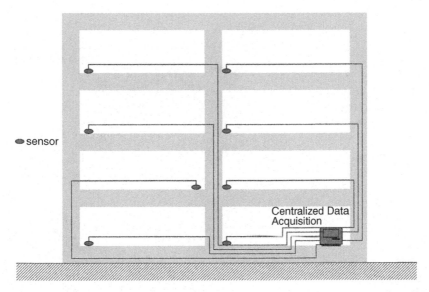

Figure 4.2 Paradigm I: a wired sensor network connected to a central data acquisition system running off AC power.

There are a wide range of commercially available wired systems, some of which have been developed for general purpose data acquisition and others that have been specifically developed for SHM applications. Those designed for general purpose data acquisition typically can interface with a wide variety of transducers and also have the capability to drive actuators. The majority of these systems have integrated signal conditioning, data processing and data storage capabilities and operate on alternating current (AC) power. Those designed to operate on batteries typically have a limited number of channels and are limited in their ability to operate for long periods of time.

One wired system that has been specifically designed for SHM applications consists of an array of piezoelectric patches made of lead zirconate titanate (PZT) embedded in a Mylar sheet that is bonded to a structure (Lin *et al.*, 2001). The PZT patches can be used as either an actuator or sensor. Damage is detected, located and in some cases quantified by examining the attenuation of signals propagated between different sensor–actuator pairs or by examining the characteristics of waves reflected from the damage. An accompanying computer is used for signal conditioning, A/D and D/A conversion, data analysis and display of final results. The system, which runs on AC power, is shown in Figure 4.3.

4.5.2 Wireless Systems

Since about the year 2000, researchers have been adapting general purpose wireless sensor nodes to SHM applications. In one of the first studies published on the application of wireless embedded systems to SHM, a damage detection algorithm was modified to the limitations of commercial off-the-shelf wireless sensing and data processing hardware (Tanner *et al.*, 2003). A wireless sensing system of 'Motes' running the TinyOS operating system developed at The University of California, Berkeley, was chosen because of their commercial availability and their built-in wireless communication capabilities. A Mote consists of modular circuit boards integrating a sensor (in this study a MEMS accelerometer), microprocessor, A/D converter and wireless transmitter, all powered by two AA batteries. A significant reduction in power consumption can be achieved by processing the data locally and only transmitting the results. The system was demonstrated using a small portal structure with damage induced by loss of preload in a bolted joint. The tested Mote system is shown in Figure 4.4. However, several problems

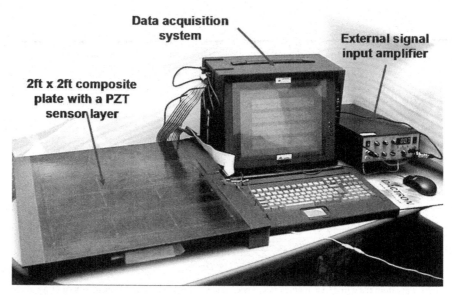

Figure 4.3   An example of a wired data acquisition system designed specifically for SHM applications. This system consists of 16 piezoelectric patches in a Mylar sheet. The sensors are connected to a data acquisition system through the ribbon wire.

Figure 4.4   A 'Mote' sensor node that includes a microprocessor, sensor, A/D converter and radio. A penny has been placed on the node for scale.

were encountered including significant degradation in telemetry capability as the battery discharged, very limited processing capability allowing only the most rudimentary data interrogation algorithms to be implemented and difficulties coupling the accelerometer to the structure.

Lynch *et al.* (2002) presented hardware for a wireless peer-to-peer SHM system. Using off-the-shelf components, the authors coupled sensing circuits and a wireless transmitter with a computational core allowing the decentralised collection, analysis and broadcast of parameters indicating the structure's health. The final hardware platform included two microcontrollers for data collection and computation connected to a spread-spectrum wireless modem. The software was tightly integrated with the hardware and included the wireless transmission module, sensing module and application module. The application module implemented a time-series-based SHM algorithm. This integrated data interrogation process required communication with a centralised server to retrieve model coefficients.

More recent studies have provided state-of-the-art reviews of current 'smart sensing' technologies that include compiled summaries of wireless work in the SHM field using small, integrated sensor and processor systems (Spencer, Ruiz-Sandova and Kurata, 2004; Lynch, 2007). A *smart sensor* is defined as a sensing system with an embedded microprocessor and wireless communication. Many smart sensors discussed in these articles are still at the stage where they simply sense and transmit data. The Mote platform is discussed as an impetus for development of the next generation of SHM systems and a new generation of Motes is also outlined. The authors also raise the issue that current smart sensing approaches scale poorly to the types of systems with densely instrumented arrays of sensors that will be required for future SHM.

In order to develop a truly integrated SHM system, the data interrogation processes must be transferred to embedded software and hardware that incorporate sensing, processing and the ability to return a result either locally or remotely. Most off-the-shelf solutions currently available, or in development, have a deficit in processing power that limits the complexity of the software and SHM processes that can be implemented. Also, many integrated systems are inflexible because of tight integration between the embedded software, the hardware and sensing. More recently, researchers have implemented distributed data interrogation algorithms where processing is done across the sensor network to enhance the computational capabilities of those sensor systems (Swartz and Lynch, 2006; Zimmerman *et al.*, 2008).

To implement computationally intensive SHM processes, one research group selected a single-board computer as a compact form for increased processing power (Allen, 2004; Farrar *et al.*, 2006). Also included in the integrated system was a digital signal processing board with six A/D converters providing the interface to a variety of sensing modalities. Finally, a wireless network board was incorporated into the node to provide the ability for the system to relay structural information to a central host, across a network, or through local hardware. Figure 4.5 shows the prototype of this sensing system. Each of the hardware parts was built in a modular fashion and loosely coupled through the transmission control protocol or Internet protocols. By implementing a common interface, changing or replacing a single component does not require a redesign of the entire system. By allowing processes developed in the graphical linking and assembly of syntax structure (GLASS) client to be downloaded and run directly in the GLASS node software, this system became one of the first SHM hardware solutions where new algorithms could be created and loaded dynamically. This modular nature did not lead to the most power-optimised design, but instead achieved a flexible development platform that could be used to find the most effective combination of algorithms and hardware for a specific SHM problem.

4.6   Sensor Network Paradigms

The sensor systems discussed in the previous section have led to three types of sensor network paradigms that are either currently being used for SHM or are the focus of current research efforts in this field. These paradigms are described below. Note that the illustrations of these systems show them applied to

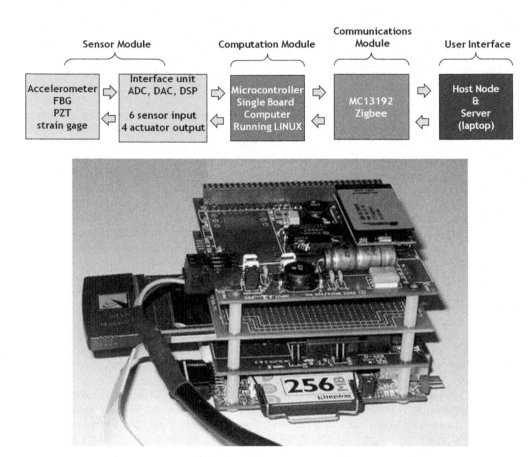

Figure 4.5   A sensor node incorporating a single-board computer to increase processing power.

a building structure; however, these paradigms can be applied to a wide variety of aerospace, civil and mechanical systems and the building structure is simply used for illustration purposes.

4.6.1   Sensor Arrays Directly Connected to Central Processing Hardware

Figure 4.2 shows a sensor network directly connected to the central processing hardware. Such a system is arguably the most commonly used for SHM studies. The advantage of this approach is the wide variety of commercially available off-the-shelf systems that can be used for this type of monitoring and the wide variety of transducers that can typically be interfaced with such a system. Also, such systems have the advantage that recordings from multiple channels are more easily time-synchronised, which is particularly important when the damage sensitive features are based on relative information between sensors. For SHM applications, these systems have been used in both a passive and active sensing manner. A significant limitation of such systems is that they are difficult to deploy in a retrofit mode because they usually require AC power, which is not always available. In addition, the deployment of such a system can be challenging, with potentially over 75% of the setup time attributed to the installation of system wires and cables for larger-scale structures such as those used for long-span bridges (Lynch *et al.*, 2003). Furthermore, experience with field-deployed systems has shown that the wires can be costly to maintain

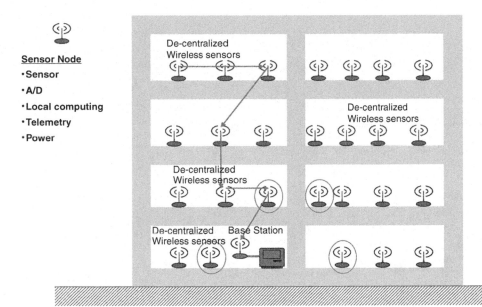

Figure 4.6 Paradigm II: decentralised processing with each sensor node running off battery power and utilising a 'hopping' telemetry protocol. The 'Mote' shown in Figure 4.4 is one such sensor node that can be deployed to form this type of sensor network.

because of general environmental degradation and damage caused by agents external to the system, like rodents and vandals (Nigbor and Diehl, 1997).

4.6.2 Decentralised Processing with Hopping Connection

The integration of wireless communication technologies into SHM methods has been widely investigated in order to overcome the limitations of wired sensing networks (Lynch, 2002). Wireless communication can remedy the cabling problem of the traditional monitoring system and significantly reduce the maintenance cost. The schematic of the decentralised wireless monitoring system, which is summarised in detail by Spencer, Ruis-Sandova and Kurata (2004) is shown in Figure 4.6.

For large-scale SHM, however, several very serious issues arise with the current design and deployment scheme for decentralised wireless sensing networks (Lynch *et al.*, 2002; Spencer, Ruiz-Sandova and Kurata, 2004). First, the current wireless sensing design usually adopts ad hoc networking and hopping that results in a problem referred to as data collision. Data collision is a phenomenon that results from a network device receiving several simultaneous requests to store or retrieve data from other devices on the network. With increasing numbers of sensors, a sensor node located close to the base station will experience more data transmission, possibly resulting in a significant bottleneck. Because the workload of each sensor node cannot be evenly distributed, the chances of data collision increase with expansion of the sensing networks. Another issue is that, with this approach, time synchronisation of sensors at different nodes is more difficult than for the wired system. Finally, the decentralised wireless sensing network scales poorly in an active-sensing system deployment. Because active sensors can serve as actuators as well as sensors, the time synchronisation between multiple sensor/actuator units is again a challenging task. Because of the processor scheduling or sharing, the use of multiple channels on one

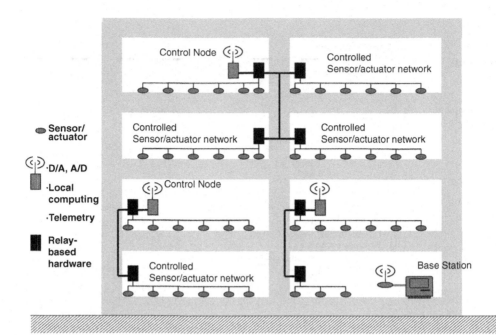

Figure 4.7   Paradigm III: a hybrid decentralised sensor system with local multiplexing and a hopping telemetry protocol.

sensor node with a single A/D converter will reduce the sampling rate, which provides neither a practical nor equitable solution for active-sensing techniques that typically interrogate higher-frequency ranges.

4.6.3   Decentralised Processing with Hybrid Connection

The hybrid connection network advantageously combines the desirable characteristics of the previous two networks, as illustrated in Figure 4.7. At the first level, several sensors are connected to a relay-based piece of hardware, which can serve as both a multiplexer and general purpose signal router, shown in Figure 4.7 as a black box. This device will manage the distributed sensing network, control the modes of sensing and actuation, and multiplex the measured signals. The device can also be expanded by means of daisy-chaining. At the next level, replicates of this hardware are linked to a decentralised data control and processing station. This control station is equipped with data acquisition boards, onboard computer processors and wireless telemetry, which is similar to the architecture of current decentralised wireless sensors. This device will perform the duties of a relay-based hardware control, data acquisition, local computing and transmission of the necessary computed results to the central base station. At the highest level, multiple data processing stations are linked to a central base station that delivers a damage report back to the user. Hierarchical in nature, this sensing network can efficiently interrogate large numbers of distributed sensors and active sensors while maintaining an excellent sensor–cost ratio because only a small number of data acquisition and telemetry units is necessary. This hierarchical sensing network is especially suitable for active-sensing SHM techniques (Dove, Park and Farrar, 2006). Researchers have shown that expandability of the sensing network is of the most importance for significantly larger numbers of active sensors, as the number of channels on a decentralised wireless sensor is limited because

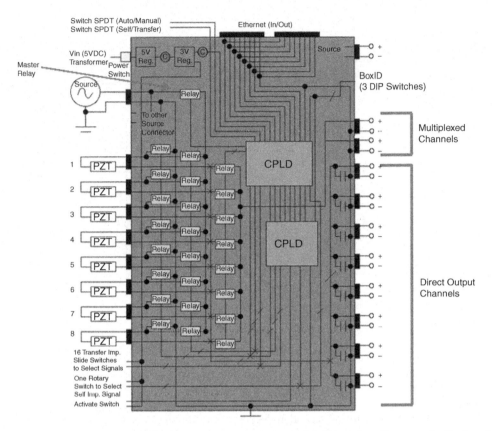

Figure 4.8 The relay-based hardware shown as a black box in Figure 4.7. This device will manage the distributed sensing network, control the modes of sensing and actuation, and multiplex the measured signals.

of the processor sharing and scheduling. The prototype of the relay-based hardware (the 'black box' shown in Figure 4.7) is illustrated in Figure 4.8.

4.7 Future Sensing Network Paradigms

The sensing network paradigms described in the previous section have one characteristic in common. The sensing system and associated power sources are installed at fixed locations on the structural system. As previously stated, the deployment of such sensing systems can be costly and the power source may not always be available. A new, energy-efficient future sensing network is currently being investigated and is shown in Figure 4.9. This system couples energy-efficient embedded sensing technology and remote interrogation platforms based on either robots or unmanned aerial vehicles (UAVs) to assess damage in structural systems (Mascarenas, 2008). This approach involves using a mobile host node (delivered via a UAV or robot) to generate a radio-frequency (RF) signal near the receiving antennas connected to sensor nodes that have been embedded on the structure. Once a capacitor on the sensor node is charged by the RF energy emitted from the host node on the UAV, the sensors measure the desired response (e.g. impedance, strain, wave attenuation) at critical areas on the structure and transmit the signals back to a processor on the mobile host.

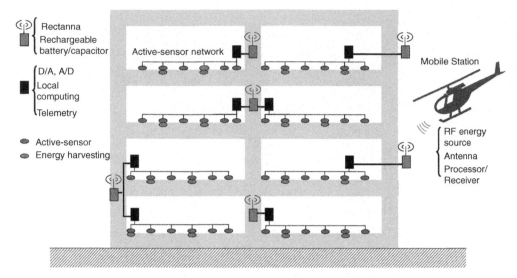

Figure 4.9 A new sensor network strategy where power and processing are brought to the sensor nodes using a robotic device. The sensor nodes are powered on demand by means of wireless energy transmission.

Figure 4.10 shows a sensor node that has been developed for this mode of remote powering and telemetry. This sensor node uses a low-power integrated circuit that can measure, control and record an impedance measurement across a piezoelectric transducer. The sensor node integrates several components, including a microcontroller for local computing and sensor node control, a radio for wireless data transmission, multiplexers for managing up to seven piezoelectric transducers per node, energy storage media and several triggering options including wireless triggering. In one package is realised a comprehensive, self-contained wireless active-sensor node for SHM applications. It was estimated that this sensor node requires less than 70 mW of total power to operate, measure, compute and transmit results to the mobile host. Considering this amount of power consumption, the sensor node is within the range of the wireless energy transmission capabilities provided by the host node on the UAV as well as energy harvesting devices such as small solar arrays (Taylor *et al.*, 2009b).

One UAV with its power source, telemetry and computing can be used to interrogate an entire sensor array placed on the structure and then can be used for other structures that have similar embedded sensor arrays. A recent field demonstration of this sensor network strategy is shown in Figures 4.11 and 4.12 where a remotely controlled helicopter and ground-based robotic vehicle (Taylor *et al.*, 2009a) were used to deliver power to sensor nodes mounted on a bridge structure. This technology will be directly applicable to rapid structural condition assessment of buildings and bridges after an earthquake where the sensor nodes may need to be deployed for decades during which conventional battery power will be depleted. Furthermore, this technology may be adapted and applied to damage detection in a variety of other civilian and defence-related structures such as nuclear power plants, where it is advantageous to minimise human exposure to hazardous environments during the inspection process. A review of other applications of robotic devices for SHM sensing can be found in Huston (2005).

4.8 Defining the Sensor System Properties

After one has defined the quantities to be measured and the general types of sensors to be used, the next step is to define the requisite sensor system properties. One of the major challenges in defining sensor

Figure 4.10   The sensor node that is designed to receive power wirelessly from a remote host as depicted in Figure 4.9. This sensor node can measure impedance across up to seven piezoelectric sensors.

Figure 4.11   Field demonstration of wireless power delivery to a sensor node embedded on a bridge structure. The receiving antenna can be seen suspended from the bottom flange of the bridge girder.

Figure 4.12   Wireless power delivery to a series of sensor nodes embedded on a bridge structure using a ground-based robot.

properties is that these properties need to be defined a priori and typically cannot be changed easily once a sensor system is in place. These sensor properties include type, bandwidth, sensitivity (dynamic range), number, location, stability, reliability, cost, power and telemetry. To address this challenge, a coupled analytical and experimental approach to the sensor system design (previously discussed in Section 4.2) should be used in contrast to the ad hoc procedures used for many current damage detection studies. The outcome of such a design process may indicate that additional sensors are needed to quantify the effects of varying operational and environmental conditions on the damage detection process. As an example, it has been shown that the dynamic properties of a bridge structure vary significantly with temperature (Farrar *et al.*, 2002); however a measure of ambient air temperature does not correlate with the change in dynamic properties. Instead, it was found that the change in dynamic properties was correlated with the temperature differential across the bridge, which implies the need for multiple temperature sensors in the sensing system.

4.8.1   Required Sensitivity and Range

Sensitivity is defined as the smallest change in the measured parameter that can be resolved by the sensing system. Range defines the largest value of the parameter that can be measured by the system. Sensitivity and range are a function of both the sensor and the data acquisition system. The resolution of the system's A/D converter (i.e. the number of bits used to digitise the analogue signal) will be the key parameter that defines the sensitivity. Adequate sensitivity and dynamic range are required to separate a low-level local response caused by damage (e.g. cracks opening and closing) from a large-amplitude global response such as that caused by aerodynamic loads on aircraft or earthquake loading on buildings. Range and sensitivity are not independent and, in general, as the range of the sensor increases, the sensitivity decreases. This dependence typically results in sensing systems that have to compromise uneasily between these two related properties. An alternative will be to develop multiple sensing systems, each of which is designed with different sensitivity and range requirements. For single sensing systems designed to capture response over a wide range and to have high sensitivity (as is becoming possible with 32-bit systems), a concern is the ability to calibrate the sensors over the entire range of measurements.

4.8.2   Required Bandwidth and Frequency Resolution

Bandwidth refers to the range of the system response frequencies that can be captured by the sensing system. Frequency resolution is the smallest change in frequency that can be resolved by the sensing system. Sensing system bandwidth and frequency resolution are also a function of both the sensor and the data acquisition hardware. Local response characteristics are required to identify the onset of damage (e.g. onset of delamination in the skin of a composite aircraft wing), which tends to manifest itself in the higher-frequency portions of the response spectrum. Global response characteristics are required to capture the influence of damage on the system-level performance (e.g. changes in the flutter characteristics of the aircraft resulting from damage to the wing). The global system response is typically characterised by the lower-frequency portion of the response spectrum. Therefore, SHM applications will often require sensors with a large bandwidth if they are to be able to capture both local and global responses. However, a confounding issue is that as the bandwidth goes up, the frequency resolution of the sensing system typically goes down. A strategy that might be employed to address this issue, similar to that discussed in Section 4.12 below, is the use of multiple sensing systems that are designed to capture various portions of the frequency response spectrum with various frequency resolutions.

4.8.3   Sensor Number and Locations

Decisions regarding the number and location of sensors must balance the economic advantages of an optimal sensing system with the enhanced reliability of a redundant sensing system. Traditional sensing for SHM has almost exclusively taken an approach that utilises relatively few sensors often distributed in a somewhat uniform grid on the structure. The primary concern here is that if one sensor fails the system might no longer be adequate for the damage detection task. With advances in MEMS technology it may be possible to provide sensing redundancy in an economic manner and to reduce the need for an optimal sensing system.

Intuitively, sensors should be located near expected damage locations. However, depending on the response quantity being monitored, there are cases where the damage may be more observable at other locations. For example, if a crack occurs at the fixed end of a cantilever beam, an accelerometer mounted at the free end of the beam measuring acceleration normal to the axis of the beam will be more sensitive to the damage than one mounted in the same orientation at the fixed end. In contrast, a strain gauge located near the crack at the fixed end and measuring deformation in the longitudinal direction will be more sensitive to this damage than a strain gauge mounted in the same orientation at the free end. Therefore, it is critical to determine that the expected type of damage produces known, observable and statistically significant effects in features derived from the measured quantities at the chosen transducer locations. It is well known from control theory that the observability of a system depends critically on the location of the sensors and the desired feature to be extracted (Kailath, 1979). For instance, if one would like to measure the second resonance frequency of a structure and use this value as a metric for damage, mounting the sensor at the node of the second mode will doom such an algorithm to failure. The issues associated with integrating observability calculations for local damage and its influence on global system behaviours into the sensor system design have received limited attention in SHM studies undertaken to date. However, there are studies that have developed genetic algorithms or neural networks to optimise a given sensor budget based on a damage observability criterion (Staszewski *et al.*, 2000; Worden and Burrows, 2001). The former of these two references discusses how a 'fail-safe' sensor network may be optimised in such a way that if a sensor fails the remaining sensors will provide a network with appropriate capability.

4.8.4   Sensor Calibration, Stability and Reliability

When discussing calibration, stability and reliability, it must be clear that these concepts are applied to the entire sensor system and not just the sensor itself. Calibration is the process of determining the relationship between the field variable to be measured and the signal generated that is eventually passed on to the feature extraction process. Stability refers to how the calibration is varying with time. Because SHM often involves a comparison of measured responses before and after a damaging event, stability is generally the more critical property for SHM sensor systems. Well-defined sensor calibration procedures exist, but approaches for establishing sensor stability are less well defined.

Most sensors are calibrated at a specialised calibration facility with well-established protocols and standards. These standards mandate that sensors are calibrated at regular intervals. However, in many SHM applications it may not be practical to remove a sensor for calibration at these regular intervals. This current approach to calibration is expected to endure, but for embedded sensor systems it is desirable to supplement these procedures by incorporating a self-diagnosing and self-calibrating capability directly into the sensors.

Stability raises several important issues. In some cases measurements are acceptable with significant error resulting from faulty calibration, as long as this error remains constant from one measurement to another one made at some future time. If relative information is desired over short time spans, such as before and after an earthquake, slow drift in the calibration may be acceptable when examining data from a single sensor. Stability becomes a more significant issue as the data from larger numbers of sensors are fused. In this case the issue is not only the stability of the individual sensors but the relative rates of change in calibration of all the sensor pairs. Clearly, the time scales over which the structure will be monitored and the time scales associated with the damage evolution will be key parameters when establishing acceptable levels of stability for the sensing system.

Reliability refers to the ability of the sensor to continue functioning over extended periods of time and stability is one component of reliability. Confidence in the sensing system is a prime consideration for SHM. If the sensing system is compromised, then the overall confidence in the SHM process output is undermined. For SHM sensing systems several reliability considerations emerge:

1. The required sensor life.
2. The ability of the sensor to endure extreme environments. This issue encompasses the nontrivial problem of sensor selection for extreme environments. Examples of such environments include in-service jet engine turbine blades exposed to extreme temperatures, high-temperature components in oil refineries and extreme temperature fluctuations in space environments and nuclear power plant fluid systems that are exposed to radiation fields.
3. The ability of the sensor to survive the structure's loading conditions. Sensors may fail through outright destruction from excessive loading while the component being monitored endures.
4. The reliability of the sensor relative to a structure or component that it is monitoring. For example, reliable parts may have failure rates of 1 in 100 000 over several years in time. Sensors are often small, complex assemblies with built-in microelectronics, so sensors subjected to the same operational and environmental loading conditions may fail because of inherent flaws more often than the component being monitored.
5. The ability to detect a faulty sensor. Loss of the sensor signal may be falsely interpreted as component failure, not sensor failure. False indications of damage or damage precursors that might result from a faulty sensor are extremely undesirable. If this occurs often the sensor will subsequently be overtly or covertly ignored.

Recently, several studies have focused on issues of sensor validation (Park et al., 2006b). Here, sensor validation refers to the capability of detecting and isolating a faulty sensor in a sensing network. It has been pointed out that the field of sensor validation has received very little attention in the structural dynamics community compared to the process control community in the chemical engineering field

(Friswell and Inman, 1999). These latter authors proposed a sensor validation method based on the comparison between the subspace of the response and the subspace generated by the lower structural modes. Their method was further extended by generating new residuals using modal filtering approaches (Abdelghani and Friswell, 2007). Another study has relied on an auto-associative neural network, which is known to implement nonlinear principal component analysis (see Chapter 10), for the detection of sensor failures (Worden, 2003). Linear principal component analysis (see Chapter 6) has also been used to perform detection, isolation and reconstruction of a faulty sensor (Kerschen *et al.*, 2005). However, these studies are usually limited to those sensors used for measuring lower-frequency global vibration modes.

Several studies have been conducted on the validation of surface-bonded PZT sensor-actuator transducers for higher frequency local damage detection applications. The goal of one ongoing study is to improve the durability and survivability of PZT active sensors in a typical aerospace environment (Blackshire, Martin and Cooney, 2006). Debonding identification algorithms have been proposed that monitor the resonance of a PZT sensor using an electrical impedance measurement (Saint-Pierre *et al.*, 1996; Giurgiutiu, Zagarai and Bao, 2002). As the debonding area between the PZT wafer and the host increases, the shape of the PZT wafer's resonance becomes sharper and more distinctive, and the magnitudes of the host resonances are reduced. However, this method is not able to account for the sensor fracture that may simultaneously occur with debonding, as the sensor breakage will also change the resonances of a PZT sensor. Another PZT sensor debonding identification scheme makes use of a sensitive closed-loop control system that can be destabilised by a slight frequency shift caused by small edge debonding (Sun and Tong, 2003). Although the method shows great sensitivity – 0.1% debonding identification was achieved in a simulation study – the issues associated with how to differentiate the frequency shift caused by structural damage from the actuator debonding was not fully addressed.

Shear deformation across the bond layer can also influence the electromechanical impedance measurements made with PZT sensor actuators. It was found that the bond layer can significantly modify the measured admittance signatures. To mitigate this problem, the use of adhesives with high shear modulus, the smallest practicable bond thickness and small-sized PZT transducers were recommended in order to minimise the influence of the bond layer on the PZT measurements (Bhalla and Soh, 2004). These authors also suggested that the imaginary part of the electrical admittance of PZT transducers may play a meaningful role in detecting deterioration of the bond layer. Further investigations of the electrical admittance as a PZT diagnostic have been conducted where it was shown that this property can be used to assess both sensor fracture as well as the bond condition (Park *et al.*, 2006b; Park *et al.*, 2009). The influence of these failure mechanisms on wave propagation and impedance-based SHM procedures using PZTs was addressed in a follow-on study (Park *et al.*, 2006a). More recently, an automated approach to PZT sensor-actuator diagnostics was proposed that also assesses the influence of temperature fluctuations on these devices (Overly *et al.*, 2009).

In general, a completely broken PZT active sensor can be easily identified if the sensor does not produce any meaningful output or an actuator does not reasonably respond to applied voltage signals. However, if only a small fracture or debonding occurs within the materials, the sensors/actuators are still able to produce sufficient performance (with distorted signals after the sensor fracture), potentially leading to a false indication of the structural condition.

In summary, to achieve the most reliable SHM system, it is imperative that effective sensor validation and diagnostic procedures are adopted in the SHM process. Although the majority of research into sensor validation for SHM applications has focused on active sensing systems based on PZT sensor/actuators, these procedures must also be developed for passive sensing systems.

4.9   Define the Data Sampling Parameters

After the sensors have been selected, a subsequent step in developing an SHM sensing system is to define the sampling parameters for the data acquisition system. These parameters include the sampling rate, the sampling duration and when to sample the data. Appropriate values for these parameters will depend on the structure, the expected loading rates, the expected type of damage and the expected rate of damage growth. Also, these parameters will be dependent on the data storage, processing and transmission strategies developed for the SHM process. The sampling rate and duration will influence the bandwidth and frequency resolution that can be obtained with the data, which will have a direct impact on the feature extraction process. If it is important to characterise the environmental or operational variability, then many samples of these parameters may be required over long time periods in an effort to quantify these sources of variability. Once a baseline has been established, data may be obtained periodically with the period based on loading rates and assumed damage growth rates to some critical level. Alternatively, data may be obtained only after extreme or anomalous events, such as when a projectile impact triggers the data acquisition system or when new environmental or operational conditions not previously experienced are encountered.

4.10 Define the Data Acquisition System

The data acquisition system digitises the analogue sensor signals, applies some form of data cleansing (typically filtering or decimation), transmits the data, records the data and stores the data for further analysis. Often the host computer that controls the hardware performing these functions will also be the computer that is used in the feature extraction and statistical modelling portions of the SHM process. Issues that must be considered in designing this hardware include:

1. All the previously discussed issues associated with the sensors and the data sampling.
2. The types of data cleansing to be employed prior to feature extraction.
3. The amount of memory that is needed.
4. The required telemetry including issues such as maximum range, amount of bandwidth available and susceptibility to electromagnetic interference.
5. The user interface.
6. Power requirements.
7. The required environmental and operational ruggedness of the system.
8. The speed at which feature extraction and statistical modelling must be done.

An example of an integrated sensing and processing system is the high-explosives radio telemetry (HERT) system (Bracht, Pasquale and Petersen, 2000) (see Figure 4.13) developed by a Los Alamos

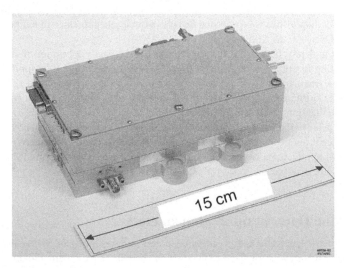

Figure 4.13   High-explosives radio telemetry system (courtesy of T. Petersen, Los Alamos National Laboratory).

National Laboratory–Honeywell, Inc. team for weapons flight test monitoring. The HERT system can measure, record, process and transmit data from 32 fibre optic sensor channels. A field-programmable gate array (FPGA) is used for local data processing and sensor diagnostics. In addition, the system has been developed to survive intercontinental ballistic missile flight environments.

4.11   Active versus Passive Sensing

Most field-deployed structural health monitoring strategies examine changes in quantities such as strain or acceleration to detect and locate damage. These methods typically rely on the ambient loading environment as an excitation source and, hence, are referred to as passive sensing systems. The difficulty with using such excitation sources is that they are often not stationary (see Appendix A). The nonstationary nature of these signals requires robust data normalisation procedures to be employed in an effort to determine that the change in the dynamic response quantity is the result of damage as opposed to changing operational and environmental conditions. Also, there is no control over the excitation source and it may not excite the type of system response useful for identifying damage at an early stage. However, for large structures, particularly most civil engineering infrastructures, ambient excitation is the only practical way to excite the global dynamic response of the structure.

As an alternative, a sensing system can be designed to provide a local excitation tailored to the damage detection process. Piezoelectric materials such as PZT are being used for such active sensing systems. Because PZT produces an electrical charge when deformed, PZT patches can be used as dynamic strain gauges. Conversely, the same PZT patches can also be used as actuators because a mechanical strain is produced when an electrical field is applied to the patch. This material can exert predefined excitation forces into the structure. The use of a known and repeatable input makes it much easier to process the response signals for damage detection. For instance, by exciting the structure in an ultrasonic frequency range, the sensing system can focus on monitoring changes of structural properties with minimum interference from operational and environmental variability, which tend to be low-frequency in nature. These sensor/actuators are inexpensive, generally require low power, and are relatively nonintrusive (as shown in Figure 4.14).

Examples of documented successes in active local sensing for damage detection using PZT are the impedance-based methods (Park *et al.*, 2003) and Lamb wave-propagation methods (Cesnik, 2007). The impedance method monitors the variations in mechanical impedance resulting from damage and the mechanical impedance is coupled with the electrical impedance of the PZT sensor/actuator. For this method, the PZT acts simultaneously as a discrete sensor and actuator. A schematic of the impedance method is shown in Figure 4.15.

For the Lamb wave-propagation method, one PZT is activated as an actuator to launch elastic waves through the structure and responses are measured by an array of the other PZT patches acting as sensors. The structure can be systematically surveyed by sequentially using each of the PZT patches as an actuator and the remaining PZT patches as sensors. The technique looks for possible damage by tracking changes in transmission velocity and wave attenuation/reflections.

A composite plate with a PZT sensor layer is shown in Figure 4.3. The Lamb wave-propagation active sensing method described above was applied to this plate after damage was introduced by a projectile impact. For this method excitations are in the high-frequency range (typically above 30 kHz) where there are measurable changes in structural response for even incipient damage associated with crack formation, debonding, delamination and loose connections.

4.12   Multiscale Sensing

Depending on the size and location of the structural damage and the loads applied to the system, the adverse effects of the damage can be either immediate or may take some time before they alter the system performance. In terms of length scales and as noted in Chapter 1, all damage begins at the material level

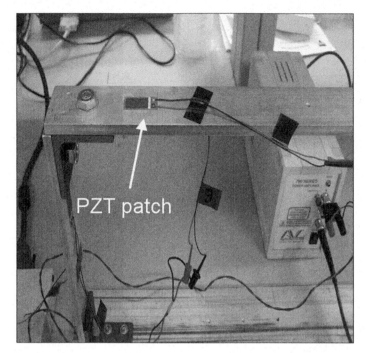

Figure 4.14   A PZT sensor/actuator being used to monitor a bolted connection.

and then under appropriate loading conditions progresses to component and system-level damage at various rates. Sensing systems that are able to capture the responses over widely varying length and time scales have not been substantially investigated by researchers, although it is quite possible to use the same piezoelectric patches discussed in the previous section in both an active, high-frequency mode to assess changes in a local system response and in a passive mode to capture the lower-frequency global response of the system. As an example, in the active mode, the piezoelectric sensors can be used to detect and find damage on a local level using relatively higher frequency excitation and response measurements in conjunction with Lamb wave-propagation damage detection methods. This type of active sensing can be used, for example, to detect delaminations in the composite skin on the wing of an unmanned aerial vehicle. In addition, these same sensors can be used in a passive mode to monitor the low-frequency global modal response of the wing when it is subjected to aerodynamic loading. This global response data can be used to assess the effect of the delamination on the flutter characteristics

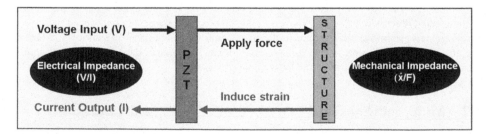

Figure 4.15   Schematic of the impedance damage detection method.

of that aircraft as determined by analysis of the coupling between the first bending and torsion mode of the wing.

4.13   Powering the Sensing System

Provision for power is a major issue that must be addressed when developing any SHM sensing system. There are four possible approaches to powering the system:

1. Direct connection to AC power.
2. Battery power.
3. A combination of AC power with battery backup.
4. Energy harvesting.

The location where the sensors are to be placed, the anticipated sensing system life and the required sensing duty cycle will dictate which power source or combinations of power sources will be used. The availability of existing power and the ease with which batteries can be replaced will be a function of the sensor locations. In most cases the power options will be more limited if the sensors are being installed on an existing structure as opposed to being incorporated into the design of a new structure. Wireless sensing systems will typically rely on battery power. This constraint significantly influences many aspects of wireless sensing and data acquisition system design. As an example, the coupling of a local processing capability with a wireless sensing module is dictated by the fact that it takes much less power to process data than to transmit data. Therefore, the design philosophy associated with these systems is to process the data locally to the greatest extent possible and then transmit only the essential information. With a two-way communication capability, the local sensing and processing units can also revert to a 'sleep mode' for energy conservation and they can be activated when a 'wake-up' signal is broadcast from the central monitoring facility (Park *et al.*, 2008).

A major consideration in using a dense sensor array is the problem of providing power to the sensors. This demand leads to the concept of 'information as a form of energy'. Obtaining information requires expenditure in terms of energy. If the only way to provide power is by direct connections, then the need for wireless communications protocols is eliminated, as the cabled power link can also be used for the data transmission. However, if a wireless communication protocol is used, the development of micropower generators will provide significant advantages over battery power sources as the concept of autonomous embedded sensing cannot be realised if one has to periodically replace batteries. A possible solution to the problem of localised power generation is to use technologies that enable harvesting ambient energy to power the sensor nodes. Forms of energy that may be harvested include thermal, vibration, acoustic and solar. Energy harvesting from mechanical vibration sources is new technology and the focus of many current research efforts while harvesting solar energy is a more mature technology. Commercial products specifically designed for harvesting energy from mechanical vibrations are now coming on to the market.[3,4,5]

4.14   Signal Conditioning

As stated in Chapter 1, data cleansing, normalisation, compression and fusion can be part of the data acquisition process as well as the feature selection and statistical modelling portions of the SHM process. Here, those processes that are implemented in the data acquisition system electronics are briefly

---

[3] www.kcftech.com.

[4] www.mide.com.

[5] www.microstrain.com.

discussed. One of the most common forms of data cleansing integrated into the data acquisition hardware is the analogue low-pass filter – referred to as an anti-aliasing filter – that is designed to eliminate high-frequency data based on the maximum system sampling rate. An AC coupling filter, which is a high-pass filter (e.g. 3 dB/octave roll-off from 2 Hz to 0 Hz), performs a data normalisation function that is used to remove direct current (DC) offsets from a signal. Strain gauge bridge circuits can be used to normalise the sensor reading for changing thermal environments. Amplifiers are also used to scale the reading from a variety of different sensors. Certain sensors, such as peak strain sensors, perform a data compression function (Mascarenas, 2008). Other sensor designs can fuse data from two translational measurements to estimate the rotational response.

4.15 Sensor and Actuator Optimisation

Few researchers have addressed the issue of developing a systematic approach to the design of a sensor system for SHM. In very general terms, one approach is to consider the sensor system design as an unconstrained or constrained optimisation problem. As an example, one study has employed machine learning to optimise sensor number and location (Worden and Burrows, 2001). In terms of a constrained optimisation problem, the designer would like to maximise the 'damage observability' subjected to a wide variety of possible constraints such as cost, weight, power (when active sensing is used) and allowable locations. A challenge to actually implement this approach is establishing accurate mathematical definitions for damage observability and its relation to the various sensor system properties. This challenge is confounded by the fact that quite a few sensor system parameters may influence observability and the interactions between these various parameters may not be well understood.

One approach to solve the optimisation problem is to determine (or assume) that a particular sensor to be employed has a certain damage detection resolution (e.g. can detect a 1-mm crack through the thickness of a plate within a 15-cm radius of the sensor). Then assume that there are an infinite number of sensors, which in turn maximises observability. Next, optimisation procedures such as genetic algorithms or gradient descent methods are used to maximise the observability while retaining some fraction of the infinite sensor array. This process produces a sensor layout with a minimum number of sensors placed at locations that maximise damage observability. Note that this optimisation problem will become much more complicated when 'real-world' issues such as operational and environmental variability have to be addressed. Also, one must consider the trade-offs between an optimal sensing system and a redundant sensing system in terms of reliability. As previously stated, if one sensor or senor node fails in an optimal system, it is most likely no longer optimal. An interesting and pragmatic approach to estimating the number of sensors needed per unit area of structure for Lamb wave SHM is discussed in Croxford *et al.* (2007).

With the advent of active sensing approaches there can be SHM applications where the excitation is selectable, and this excitation should be chosen to maximise damage observability. As a simple example, consider a beam or column with a crack that is nominally closed because of a preload. If the provided excitation is not sufficient to open and close the crack, the detectability of the crack in the measured output will be severely limited. Thus, if possible, it is important to answer the question: 'Given ever-present physical limits on the level of excitation, and limited outputs that can be measured, what excitation should be provided to a system to make damage most detectable?' When one considers that an excitation may be viewed as a time series with hundreds or thousands of free parameters, optimisation in this high-dimensional space appears to be a daunting task. However, as demonstrated in a recent study, a gradient-based technique may be used to address this high-dimensional optimisation problem in which the gradient can be calculated very efficiently (Bement and Bewley, 2009a, 2009b). The method does require a model of the system and the accuracy of that model will influence the results. The cited study showed that the observability of damage can be increased several orders of magnitude through the use of an optimised input in an active sensing system. Other investigators have shown similar improved damage detection results using input waveforms that have been optimised using evolutionary algorithms (Olson *et al.*, 2007; Olson, Overbey and Todd, 2009).

More recently, a Bayesian experimental design approach to sensor optimisation has been proposed where a cost function is defined both in terms of the sensor system costs as well as the costs of making decisions based on the particular sensor design and the prior probabilities of damage (Flynn, 2010). With this cost function defined one can employ various optimisation procedures to search candidate sensor system designs to find the one that minimises this total cost of decision (the system is damaged or undamaged). It should be noted that this formulation provides a framework for both optimising the sensing system and the detection algorithm. Although this formulation is quite general, the cited reference shows applications of this procedure to the optimal sensor system and detector algorithm design for wave propagation-based damage detection in plate structures.

4.16   Sensor Fusion

Sensor fusion is the process of integrating data from a multitude of sensors with the objective of making a more robust and confident decision than is possible with any one sensor alone. There are numerous reasons why multisensor systems are desirable (Esteban and Starr, 1999):

1. Higher signal-to-noise ratio.
2. Robustness and reliability. Enough information may be available to form a decision even if a subset of the sensors fails. Note that in order to have fault tolerance, it is necessary to design in redundancy and thus increase the number of measurements. This redundancy should always be removable in the signal processing stage or the complexity of the pattern recognition and decision problem will increase.
3. Information regarding independent features in the system can be obtained.
4. Extended coverage gives a more complete picture of the system.
5. Improved resolution.
6. Increased confidence in the results. This confidence may be achieved simply by observing the agreement between several independent measurements or could go as far as providing statistics for uncertainty assessment.
7. Improved hypothesis discrimination.
8. Reduced measurement times (under certain circumstances).

There are many ways of implementing sensor fusion strategies. Common to all is that the single-sensor processing chain of Figure 4.16(A) is replicated a number of times and the chains are fused together. An example strategy, which is usually called *central-level* or *centralised* fusion is shown in Figure 4.16(B) (Klein, 1999). (Figure 4.2 shows a central-level fusion strategy for a sensor network.) Each of the sensors provides information to a central feature extraction unit and thus attempts to remove both *inter-* and *intra-*sensor redundancy. Another approach could be a pattern-level fusion architecture as in Figure 4.16(C), where the feature extraction is carried out for each sensor independently. There is no reason why individual sensor's chains cannot be fused at different points, which means that there are as many fusion strategies as there are ways of connecting the chains (e.g. the one shown in Figure 4.16(D)). Clearly from these examples, sensor fusion and data fusion can be combined to form a more general fusion process.

A sensor/data fusion architecture can therefore be thought of as a directed graph with input nodes corresponding to the sensors and output nodes corresponding to the decisions. The information flowing through the graph is initiated as the sensor values together with estimates of the sensor confidence, and

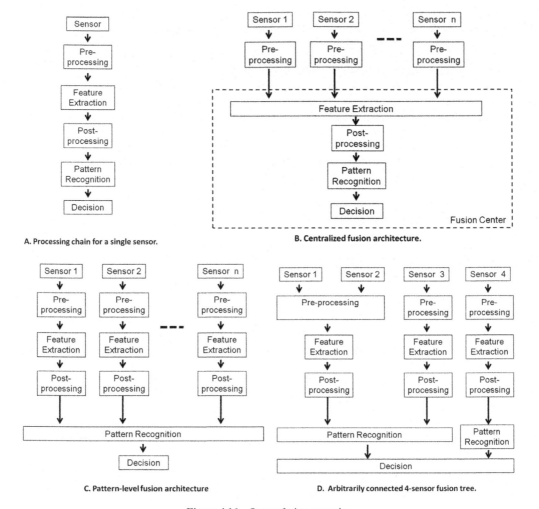

Figure 4.16   Sensor fusion strategies.

is condensed and refined at each stage of its passage and therefore continually changes its nature. Fusion occurs at the vertices and will require different techniques depending on the position or level of the vertex. The different fusion levels are:

1. Raw sensor-level fusion. The data from two or more of the sensors is directly combined before any preprocessing has been performed.
2. Feature-level fusion. Two or more preprocessed sensor signals are combined to produce a single feature vector for classification. This may be as simple as concatenation or may be as complex as a nonlinear mapping, such as nonlinear principal component analysis.
3. Pattern-level fusion. Two or more feature vectors are combined and passed to the pattern recognition algorithm.
4. Decision-level fusion. Two or more decisions or classifications are combined in such a way as to produce a decision with higher confidence.

The first attempts at formalising the discipline of data fusion were made in the military domain of *Command, Control, Communications and Intelligence* or C3I. From these efforts various data fusion models such as the Joint Director of Laboratories Model (JDL) (Waltz and Llinas, 1990), Observe, Orientate, Decide and Act Model (OODA), also known as the Boyd Model (Boyd, 1987), and Waterfall models (Bedworth, 1994) were developed. A survey was made of the fusion models described in these references, as well as some others, and it was concluded that each has limitations (Bedworth and O'Brien, 1991). In an effort to design a model that incorporated the desirable features of the standard fusion models and overcame their limitations, the authors drew up the following 'wish list' for the ideal fusion model:

1. It should define the order of processing.
2. It should make the cyclic nature of the system explicit.
3. It should admit representation from multiple viewpoints.
4. It should identify the advantages and limitations of various fusion approaches.
5. It should facilitate the clarification of task-level measures of performance and system-level measures of effectiveness.
6. It should use a general terminology that is widely accessible.
7. It should not assume that applications are defence orientated.

The solution proposed is the *Omnibus model* (Bedworth and O'Brien, 1991). The model uses the fine levels of definition of the Waterfall model inside a cyclic structure similar to the OODA loop. The model is shown in Figure 4.17.

The fact that the Omnibus model provides a unification of the previously discussed models is clear from the diagram. Because the Omnibus model is freed from the defence orientation of previous models and does not use a terminology specific to military applications, and because it incorporates the best features of the influential JDL, OODA and Waterfall models, it is suggested that it is the most appropriate (currently available) framework for discussing applications to SHM. Because of the explicit inclusion of the *control* level, the model actually provides an instance of sensor/actuator fusion. In general, not all

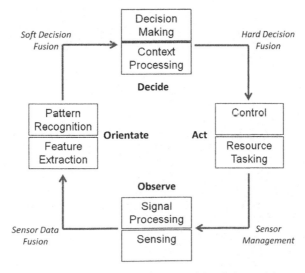

Figure 4.17 The Omnibus sensor and data fusion model.

features of the Omnibus model will be relevant for an SHM application. However, the generality of the model makes it an excellent model to use when one is planning a sensor and data fusion process.

4.17   Summary of Sensing and Data Acquisition Issues for Structural Health Monitoring

In this chapter, the current sensor system design research that is being conducted to address the data acquisition portion of the SHM problem has been summarised. Several sensor systems that have been developed specifically for SHM were discussed. These sensor systems led to the definition of several general SHM sensor network paradigms. All of these paradigms have relative advantages and disadvantages. Also, the paradigms described are not at the same level of maturity and, hence, some may require more development to obtain a field-deployable system while others are readily available with commercial off-the-shelf solutions.

Currently, there is no well-established procedure to design a sensing system for SHM. To address this issue, an integrated analytical and experimental approach to the sensor system development has been suggested in contrast to the ad hoc procedures often used for damage detection studies. This coupled strategy should yield considerable improvements over current approaches to SHM sensing system design, but at the cost of more up-front engineering. First, critical failure modes of the system must be defined and, to the greatest extent possible, quantified. This quantification is accomplished from either past experience or by using information obtained from high-fidelity numerical simulations. This information is essential to defining the sensor system properties. Additional sensing requirements can also be ascertained if the effects of changing operational and environmental conditions on the measured quantities and measurement system have been characterised. As an alternative, changing operational and environmental conditions can be included in the numerical models in an effort to determine how these conditions affect the damage detection process. The feature extraction and statistical modelling portions of the SHM process can also dictate certain parameters of the data acquisition system. Therefore, an integrated sensor system design approach must consider these aspects of the SHM process as well.

This chapter has emphasised that the key issue for developing an SHM sensor system is the ability to capture system response on widely varying length and time scales at a reasonable cost. It is necessary for any damage detection sensing system to acquire data that encapsulates any change in system properties that may affect the system's ability to perform its intended function in a safe and economical manner. Although sensor configurations with a limited number of sensors will provide an indicator of change to the global properties, higher-density sensor arrays are required, not only to provide localised information relating to damage but also to provide for redundancy. Perhaps the most important aspect of the sensing system is that it must be more reliable than the system being monitored.

This chapter has also summarised practical implementation issues associated with the SHM sensor system in an effort to suggest the need for a more mathematically and physically rigorous approach to future SHM sensing system design. It should be noted that the fundamental axioms for SHM that were introduced in Chapter 1 and discussed in more detail in Chapter 13 are based on the information published in the extensive amount of literature on SHM over the last 20 years. Of the eight axioms proposed, seven are closely related to sensing aspects of the SHM problem and, therefore, should be considered when designing any SHM sensor network.

Finally, it is important to emphasise that there is no sensor that can measure damage (see Axiom IVa, Chapter 13). Rather, the sensors measure the response of a system to its operational and environmental loading or the response to inputs from actuators embedded with the sensing system. Depending on the sensing technology deployed and the type of damage to be identified, the sensor readings may be more or less directly correlated to the presence and location of damage. Data interrogation procedures (feature extraction and statistical modelling for feature classification) are the necessary components of an

SHM process that convert the sensor data into information about the structural condition. Furthermore, it is reiterated that to achieve successful SHM the sensors and data acquisition system will have to be developed in conjunction with these data interrogation procedures.

References

Abdelghani, M. and Friswell, M.I. (2007) Sensor validation for structural systems with multiplicative sensor faults. *Mechanical Systems and Signal Processing*, **21**(1), 270–279.

Allen, D.W. (2004) Software for manipulating and embedding data interrogation algorithms into integrated systems – special application to structural health monitoring. Masters Thesis, Department of Mechanical Engineering, Virginia Polytechnic and State University, Blacksburg, VA.

Bedworth, M. (1994) *Probability Moderation for Multilevel Insformation Processing*, Defence Evaluation and Research Agency, Malvern, UK.

Bedworth, M. and O'Brien, J. (1991) *The Omnibus Model: A New Model of Data Fusion* (preprint), Defence Evaluation and Research Agency, Malvern, UK.

Bement, M.T. and Bewley, T. (2009a) Optimal excitation design for damage detection using adjoint based optimization Part 1. Theoretical development. *Mechanical Systems and Signal Processing*, **23**(3), 783–793.

Bement, M.T. and Bewley, T. (2009b) Optimal excitation design for damage detection using adjoint based optimization Part 2. Experimental verification. *Mechanical Systems and Signal Processing*, **23**(3), 794–803.

Betz, D.C., Thursby, G., Culshaw, B. and Staszewski, W.J. (2003) Acousto-ultrasonic sensing using fiber Bragg gratings. *Smart Materials and Structures*, **12**, 122.

Bhalla, S. and Soh, C.K. (2004) Electromechanical impedance modeling for adhesively bonded piezo-transducers. *Journal of Intelligent Material Systems and Structures*, **15**(12), 955.

Blackshire, J.L., Martin, S. and Cooney, A. (2006) Characterization and modeling of bonded piezoelectric sensor performance and durability in simulated aircraft environments. *3rd European Workshop on Structural Health Monitoring*, Destech Publications Inc., Granada, Spain.

Boyd, J. (1987) A discourse on winning and losing. Montgomery, AL, Maxwell Air Force Base Lecture.

Bracht, R., Pasquale, R.V. and Petersen, T. (2000) High explosive radio telemetry on exoatmospheric re-entry flight vehicle. International Telemetry Conference, San Diego, CA, United States.

Cesnik, C. (2007) Review of guided-wave structural health monitoring. *The Shock and Vibration Digest*, **39**(2), 91–114.

Croxford, A., Wilcox, P., Drinkwater, B. and Konstantinidis, G. (2007) Strategies for guided-wave structural health monitoring. *Proceedings of the Royal Society A: Mathematical, Physical and Engineering Science*, **463**(2087), 2961.

Dove, J.R., Park, G. and Farrar, C.R. (2006) Hardware design of hierarchal active-sensing networks for structural health monitoring. *Smart Materials and Structures*, **15**, 139–146.

Lister, J.L., Greene, J.A., Jones, M.E. et al. (1999) Optical-fiber-based chemical sensors for detection of corrosion precursors and by-products. Proceedings of the SPIE Chemical, Biochemical and Environmental Fiber Sensors X, SPIE.

Engler, R.H., Klein, C. and Trinks, O. (2000) Pressure sensitive paint systems for pressure distribution measurements in wind tunnels and turbomachines. *Measurement Science and Technology*, **11**(7), 1077.

Esteban, J. and Starr, A. (1999) Building a data fusion model. Proceedings of the EuroFusion 99 International Conference on Data Fusion, Stratford-upon-Avon, UK.

Farrar, C.R., Baker, W.E., Bell, T.M. et al. (1994) Dynamic Characterization and Damage Detection in the I-40 Bridge over the Rio Grande. Los Alamos, NM, Los Alamos National Laboratory Report LA-1276.

Farrar, C.R., Cornwell, P.J., Doebling, S.W. and Prime, M.B. (2002) Structural Health Monitoring Studies of the Alamosa Canyon and I-40 Bridges (LA-13635-MS). Los Alamos, NM, Los Alamos National Laboratory.

Farrar, C.R., Allen, D.W., Park, G. et al. (2006) Coupling sensing hardware with data interrogation software for structural health monitoring. *Shock and Vibration*, **13**(4), 519–530.

Flynn, E.B. (2010) A Bayesian experimental design approach to structural health monitoring with application to ultrasonic guided waves. Doctorate, Department of Structural Engineering, University of California, San Diego.

Friswell, M.I. and Inman, D. (1999) Sensor validation for smart structures. *Journal of Intelligent Material Systems and Structures*, **10**(12), 973–982.

Giurgiutiu, V., Zagarai, A. and Bao, J.J. (2002) Piezoelectric wafer embedded active sensors for aging aircraft structural health monitoring. *International Journal of Structural Health Monitoring*, **1**(1), 41–61.

Holbert, K.E., Sankaranarayanan, S., McCready, S. et al. (2003) Response of piezoelectric acoustic emission sensors to gamma radiation. Proceedings of the 7th European Conference on Radiation and Its Effects on Components and Systems, IEEE.

Huston, D. (2005) *Robotic Surveillance Approaches for SHM. Structural Health Monitoring 2005, Advancement and Challenges for Implementation*, DEStech Publication, Inc.

Johnson, G.A. (2000) Surface effect ship vibro-impact monitoring with distributed arrays of fiber Bragg gratings. Proceedings of the 18th International Modal Analysis Conference, San Antonio, TX.

Kailath, T. (1979) *Linear Systems*, Prentice Hall, Inc., Englewood Cliffs, NJ.

Kerschen, G., Boe, P.D., Golinval, J. and Worden, K. (2005) Sensor validation using principal component analysis. *Smart Materials and Structures*, **14**(1), 36–42.

Klein, L.A. (1999) *Sensor and Data Fusion Concepts and Applications*, Society of Photo-Optical Instrumentation Engineers (SPIE).

Lin, M., Qing, X., Kumar, A. and Beard, S. (2001) SMART layer and SMART suitcase for structural health monitoring applications. Proceedings of the SPIE Smart Structures and Materials Conference, Newport Beach, CA, SPIE.

Lynch, J.P. (2002) Decentralization of wireless monitoring and control technologies for smart civil structures. Doctorate, Department of Civil and Environmental Engineering, Stanford University, Palo Alto, CA.

Lynch, J.P. (2007) An overview of wireless structural health monitoring for civil structures. *Philosophical Transactions of the Royal Society A*, **365**(1851), 345–372.

Lynch, J., Law, K., Kiremidjian, A. *et al.* (2002) Validation of a wireless modular monitoring system for structures. SPIE 9th International Symposium on Smart Structures and Materials, San Diego, CA, SPIE.

Lynch, J.P., Partridge, A., Law, K.H. *et al.* (2003) Design of a Piezoresistive MEMS-based accelerometer for integration with a wireless sensing unit for structural monitoring. *ASCE Journal of Aerospace Engineering*, **16**(3), 108–114.

Mascarenas, D.D.L. (2008) "Mobile Host" Wireless Sensor Networksss – A New Sensor Network Paradigm for Structural Health Monitoring Applications. PhD Dissertation, Department of Structural Engineering, University of California, San Diego, CA.

Ni, Y.Q., Wang, B.S. and Ko, J.M. (2001) Simulation studies of damage location in Tsing Ma Bridge deck. Proceedings of the Nondestructive Evaluation of Highways, Utilities, and Pipelines IV, Bellingham, WA, SPIE.

Nigbor, R.L. and Diehl, J.G. (1997) *Two Years' Experience Using OASIS Real-Time Remote Condition Monitoring System on Two Large Bridges. Structural Health Monitoring: Current Status and Perspectives*, Technomic Publishing, Palo Alto, CA.

Olson, C.C., Overbey, L.A.L.A. and Todd, M.D. (2009) An experimental demonstration of tailored excitations for improved damage detection in the presence of operational variability. *Mechanical Systems and Signal Processing*, **23**(2), 344–357.

Olson, C.C., Todd, M.D., Worden, K. and Farrar, C.R. (2007) Improving excitations for active sensing in structural health monitoring via evolutionary programming. *ASME Journal of Vibration and Acoustics*, **129**(6), 784–802.

Overly, T.G., Park, G., Farinholt, K.M. and Farrar, C.R. (2009) Piezoelectric active-sensor diagnostics and validation using instantaneous baseline data. *Sensors Journal, IEEE*, **9**(11), 1414–1421.

Paget, C.A., Atherton, K. and O'Brien, E. (2004) Modified acoustic emission generated in a full-scale aircraft wing subjected to simulated flight loading. *Composites Technologies for 2020: Proceedings of the Fourth Asian–Australian Conference on Composite Materials*, Woodhead Publishing Ltd, Cambridge, England.

Park, G., Sohn, H., Farrar, C.R. and Inman, D.J. (2003) Overview of piezoelectric impedance-based health monitoring and path forward. *Shock and Vibration Digest*, **35**(6), 451–463.

Park, G., Farrar, C.R., di Scalea, F.L. and Coccia, S. (2006a) performance assessment and validation of piezoelectric active-sensors in structural health monitoring. *Journal of Smart Material and Structures*, **15**, 1673–1683.

Park, G., Farrar, C.R., Rutherford, C.A. and Robertson, A.N. (2006b) Piezoelectric active sensor self-diagnostics using electrical admittance measurements. *ASME Journal of Vibrations and Acoustics*, **128**, 469–476.

Park, G., Rosing, T., Todd, M.D. *et al.* (2008) Energy harvesting for structural health monitoring sensor networks. *ASCE Journal of Infrastructure Systems*, **14**(1), 64.

Park, S., Park, G., Yun, C.B. and Farrar, C.R. (2009) Sensor self-diagnosis using a modified impedance model for active-sensing structural health monitoring. *International Journal of Structural Health Monitoring*, **8**(1), 71–82.

Saint-Pierre, N., Jayet, Y., Perrissin-Fabert, I. and Baboux, J.C. (1996) The influence of bonding defects on the electric impedance of piezoelectric embedded element. *Journal of Physics D (Applied Physics)*, 2976–2982.

Sirohi, J. and Chopra, I. (2000) Fundamental understanding of piezoelectric strain sensors. *Journal of Intelligent Material Systems and Structures*, **11**(4), 246.

Spencer, B.F., Ruiz-Sandova, l.M.E. and Kurata, N. (2004) Smart sensing technology: opportunities and challenges. *Structural Control and Health Monitoring*, **11**(4), 349–368.

Staszewski, W., Worden, K., Wardle, R. and Tomlinson, G. (2000) Fail-safe sensor distributions for impact detection in composite materials. *Smart Materials and Structures*, **9**(3), 298.

Sun, D. and Tong, L. (2003) Closed-loop based detection of debonding of piezoelectric actuator patches in controlled beams. *International Journal of Solids and Structures*, **40**(10), 2449–2471.

Swartz, R. and Lynch, J. (2006) A multirate recursive arx algorithm for energy efficient wireless structural monitoring. 4th World Conference on Structural Control and Monitoring, San Diego, CA.

Tanner, N.A., Wait, J.R., Farrar, C.R. and Sohn, H. (2003) Structural health monitoring using modular wireless sensors. *Journal of Intelligent Material Systems and Structures*, **14**(1), 43.

Taylor, S.G., Farinholt, K.M., Flynn, E.B. *et al.* (2009a) Mobile-agent based wireless sensing network for structural monitoring applications. *Measurement Science and Technology*, **20**(2), 1–14.

Taylor, S.G., Farinholt, K.M., Park, G. *et al.* (2009b) Wireless Impedance Device for Electromechanical Impedance Sensing and Low-Frequency Vibration Data Acquisition. Sensors and Smart Structures Technologies for Civil, Mechanical, and Aerospace Systems, San Diego, CA, SPIE.

Todd, M., Johnson, G. and Althouse, B. (2001) A novel Bragg grating sensor interrogation system utilising a scanning filter, a Mach-Zehnder interferometer and a 3 × 3 coupler. *Measurement Science and Technology*, **12**(7), 771.

Waltz, E. and Llinas, J. (1990) *Multisensor Data Fusion*, Artech House, Norwood, MA.

Wang, M.L. (2005) *Damage Assessment and Monitoring of Long-Span Bridges. Structural Health Monitoring 2005 Advances and Challenges for Implementation*, Technomic Publishing, Palo Alto, CA.

Webster, J.G. (1999) *The Measurement, Instrumentation, and Sensors Handbook*, CRC Press.

Worden, K. (2003) Sensor validation and correction using auto-associative neural networks and principal component analysis. Proceedings of the IMAC-XXI, Orlando, FL.

Worden, K. and Burrows, A. (2001) Optimal sensor placement for fault detection. *Engineering Structures*, **23**(8), 885–901.

Zimmerman, A.T., Shiraishi, M., Swartz, A. and Lynch, J.P. (2008) Automated modal parameter estimation by parallel processing within wireless monitoring systems. *ASCE Journal of Infrastructure Systems*, **14**(8), 102–113.

5

Case Studies

Numerous case studies on *in situ* structures and laboratory test structures designed specifically for SHM research will be used to illustrate different concepts throughout this book. As these structures will reoccur regularly, it is convenient to provide here a summary of the structure descriptions, the testing strategies employed with them and the associated data acquisition procedures. Subsequent chapters dealing with feature extraction and pattern recognition will discuss the analyses of data obtained from these structures. When available, detailed reports summarising the tests are cited. If the readers are unfamiliar with any of the basic signal processing terms used throughout this chapter, they are advised to consult Appendix A, which provides all of the necessary background.

5.1 The I-40 Bridge

The I-40 Bridge over the Rio Grande consisted of twin spans (there were separate bridges for each traffic direction) made up of a concrete deck supported by two welded-steel plate girders and three steel stringers (Farrar *et al.*, 1994). Prior to its demolition this bridge was destructively tested for the purpose of developing and validating SHM procedures on an *in situ* structure. These tests were carried out by a team of faculty and students from New Mexico State University, staff members and students from Los Alamos National Laboratory and staff members from Sandia National Laboratory.

Loads from the stringers were transferred to the plate girders by floor beams located at 6.1-m (20-ft) intervals. Cross-bracing was provided between the floor beams. Figure 5.1 shows an elevation view of the portion of the bridge that was tested. The cross-sectional geometries of each of the two bridges comprising the I-40 crossing are shown in Figure 5.2, while Figure 5.3 shows the actual substructure of the bridge.

Each bridge was made up of three identical sections. Except for the common pier located at the end of each section the sections were structurally independent. A section had three spans; the end spans were of equal length, approximately 39.9 m (131 ft) and the centre span was approximately 49.7 m (163 ft) long. Five plate girders were connected with four bolted splices to form a continuous beam over the three spans. The portions of the plate girders over the piers had increased flange dimensions, compared with the mid-span portions, to resist the higher bending stresses at these locations. Connections that allowed for longitudinal thermal expansion were located where the plate girders attach to the abutment and where the plate girders attach to piers 2 and 3 of the section that was tested (Figure 5.1). A connection

Structural Health Monitoring: A Machine Learning Perspective, First Edition. Charles R. Farrar and Keith Worden.
© 2013 John Wiley & Sons, Ltd. Published 2013 by John Wiley & Sons, Ltd.

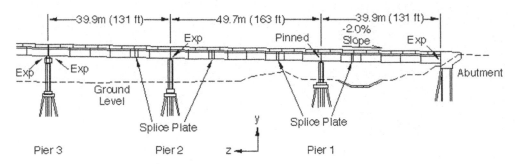

Figure 5.1   Elevation view of the portion of the eastbound bridge that was tested.

that prevented longitudinal translation was located at the base of each plate girder where they attached to pier 1.

The damage that was introduced was intended to simulate fatigue cracking that has been observed in plate girder bridges. This type of cracking results from out-of-plane bending of the plate girder web and usually begins at welded attachments to the web such as the seats supporting the floor beams (shown in Figure 5.2). Four levels of damage were introduced to the middle span of the north plate girder close to the seat supporting the floor beam at mid-span. Damage was introduced by making various torch cuts in the web and flange of the girder. The first level of damage consisted of a 0.61-m-long (2-ft-long) cut through the web approximately 9.5 mm wide (3/8 in wide) centred at mid-height of the web. Next, this cut was continued to the bottom of the web. The flange was then cut halfway in from either side directly below the cut in the web. Finally, the flange was cut completely through, leaving the top 1.2 m (4 ft) of the web and the top flange to carry the load at this location. The various levels of damage designated E-1 to E-4 are shown in Figure 5.4. The testing was carried out in a number of separate campaigns, the details of which are given in the following subsections.

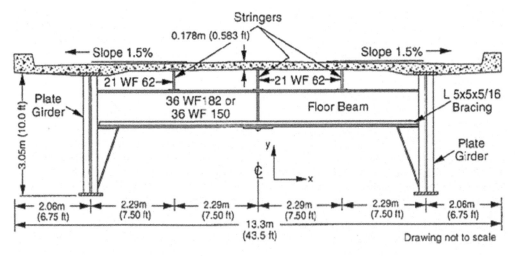

Figure 5.2   Typical cross-section geometry of the I-40 Bridge.

Figure 5.3   I-40 Bridge substructure.

5.1.1   Preliminary Testing and Data Acquisition

The preliminary vibration measurements were made on the I-40 Bridge over a period when temperatures ranged from morning lows of 1.7–3.9 °C (35–39 °F) to afternoon highs around 14 °C (58 °F). Measurements were made on the middle and eastern spans of the eastern-most section of the bridge carrying eastbound traffic. Traffic, which had been funnelled on to the two northern-most lanes, provided an

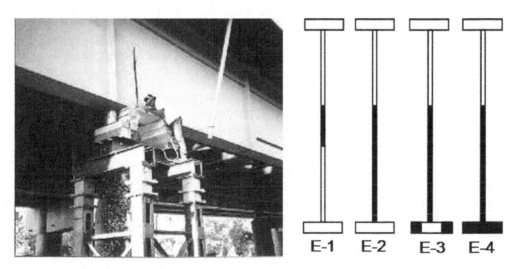

Figure 5.4   Damage that was introduced into the I-40 Bridge in four increments.

ambient vibration source for these measurements. Acceleration responses of the bridge were measured at the various locations with two types of commercial off-the-shelf (COTS) integral-circuit piezoelectric accelerometers, one set with a nominal sensitivity of 10 mV/g and a specified minimum frequency of 1 Hz, and another set with a nominal sensitivity of 1 V/g and a lower-frequency bound of 0.01 Hz. Power was supplied to the accelerometers by a two-channel spectrum analyser. Sampling parameters were established that allowed frequency ranges of 6.25, 12 and 50 Hz to be displayed with 400 spectral lines. Estimates of spectral quantities (power spectral density (PSD), cross-power spectra (CPS), transmissibility function (TF) and coherence functions) were determined from 10 to 30 averages, with 20% overlap. Transmissibility functions have proved useful for SHM in a number of studies; whereas the standard frequency response function (FRF) is essentially the ratio of a response spectrum to an input spectrum, a TF is the ratio of two response spectra from different points on the structure. To eliminate problems associated with direct current (DC) offsets, analogue alternating current (AC)-coupling high-pass filters were applied to the signals.

5.1.2 Undamaged Ambient Vibration Tests

Following the preliminary measurements, additional ambient (traffic) vibration tests were conducted three months later on the undamaged structure. These tests were intended to identify the structure's resonance frequencies, modal damping values and the corresponding mode shapes. Weather conditions were considerably different from those during the preliminary tests as temperatures would range from morning lows around 18 °C (65 °F) to afternoon highs of around 38 °C (100 °F). The thermal expansion associated with these higher temperatures produced noticeable changes in the angles of the rocker bearings. It was also noted that the east end of the top flange on the south girder was in contact with the top of the concrete abutment. Although it is assumed that this contact was a result of thermal expansion, similar observations had not been made during the earlier tests; hence, the state of this boundary condition corresponding to cooler temperatures was unknown. Wind was very light during all ambient vibration tests and was not considered a significant input source.

As during the preliminary tests, traffic had been funnelled on to the two northern lanes of the crossing. Significantly different traffic flow could be observed at various times when data were being acquired. During morning and afternoon rush hours the traffic would slow down considerably, thus producing lower-level excitations in the bridge. At midday, the trucks crossing the bridge at high speeds would cause higher-level excitations that resulted in some of the sensor signals exceeding the maximum values specified for their respective data acquisition channels. The final ambient vibration test was done just prior to the forced vibration tests when all traffic had been removed from the eastbound bridge. For this test the ambient vibration source was provided by the traffic on the adjacent new eastbound bridge and the existing westbound bridge that was transmitted through the ground to the piers and abutment.

The data acquisition system used in these ambient vibration tests consisted of a computer workstation, 29 data acquisition modules that provided power to the accelerometers and performed analogue-to-digital (A/D) conversion of the accelerometer signals, a signal processing module that performed the required fast Fourier transform (FFT) calculations and a commercial data acquisition/signal analysis software package. A 3500-watt AC generator was used to power this system in the field.

Integral circuit piezoelectric accelerometers were used for the ambient vibration measurements. These accelerometers had a nominal sensitivity of 1 V/g, a specified frequency range of 1–2000 Hz and an amplitude range of ±4g's. Thirty centimetre-long (12-in-long) 50-ohm cables were connected to the accelerometers. These cables were then connected to various lengths of two-conductor, polyvinyl-chloride-jacketed, 20-gauge cable ranging from 21 m to 89 m (70 ft to 291 ft) that were, in turn,

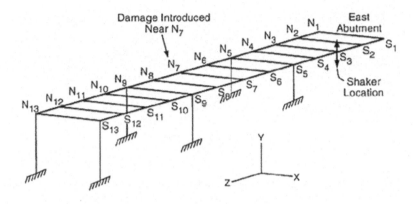

Figure 5.5 I 40 Bridge sensor locations.

connected to the input modules. The sensors were attached to the inside web of the plate girder at mid-height to measure response in the vertical direction at the locations shown in Figure 5.5.

To avoid overloads, simple mechanical filters consisting of double-sided mounting tape placed between the accelerometers and mounting blocks were used to isolate the accelerometers from the high-frequency inputs. Below 1 kHz, laboratory tests showed that the isolation system had no effect on the measured signal.

The data acquisition system was set up to measure acceleration–time histories and calculate PSDs, CPS and TFs. The CPS and TFs were calculated using either the sensor S-2 or S-6, as shown in Figure 5.5, specified as the reference channel (i.e. the denominator of the TF ratio). Sampling parameters were specified that calculated the TF from 64-s, 32-s or 16-s time windows discretized with 1024 samples. Therefore, the spectral quantities were calculated for frequency ranges of 0–6.25 Hz, 0–12.5 Hz and 0–25 Hz. Typically, 100 averages were used to calculate the 0–6.25 Hz TFs, 30 averages were used to calculate the 0–12.5 Hz TFs and 75 averages were used to calculate the 0–25 Hz TFs. Frequency resolutions of 0.015 625 Hz, 0.031 25 Hz and 0.0625 Hz were obtained for the 0–6.25 Hz TFs, the 0–12.5 Hz TFs and the 0–25 Hz TFs, respectively. Hanning windows were applied to the time signals to minimize leakage and AC coupling was specified to minimize DC offsets.

Additional data were acquired using eleven additional accelerometers placed in the global Y direction (see Figure 5.5) at a nominal spacing of 4.9 m (16 ft) along the mid-span of the North plated girder. All accelerometers in this group were located at mid-height of the girder and had a nominal sensitivity of 500 mV/g. The actual spacing of these accelerometers is shown in Figure 5.6. The same data acquisition system, similar wiring and identical sampling parameters as those used for the ambient vibration tests were again used with these sensors.

5.1.3   Forced Vibration Tests

The main portion of the testing was a series of forced vibration tests conducted on the undamaged bridge followed by a series of forced vibration tests of the structure with the various levels of damage shown in Figure 5.4. Eastbound traffic had been transferred to a new bridge just south of the one being tested. The westbound traffic continued on the original westbound bridge. Sandia National Laboratory provided a hydraulic shaker that generated the measured force input for these tests. Excitation from traffic on the adjacent bridges could be felt when the shaker was not running. The load cell located between the

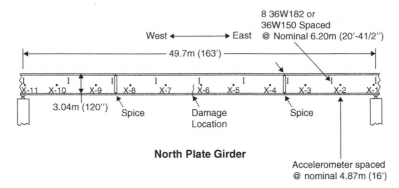

Figure 5.6 Locations of additional accelerometers mounted on the girder that was damaged.

hydraulic actuator and reaction mass showed that the vibration from traffic on the adjacent bridges, transferred through the ground to the piers and abutment of the bridge being tested, caused the bridge deck to put a peak force of 68 kg (150 lb) into the reaction mass. Temperatures ranged from morning lows of 14 °C (56 °F) to afternoon highs of 27 °C (80 °F). The east end of the south girder was observed to no longer be in contact with the concrete on the top of the abutment. Wind, although not measured, was not considered significant during these tests.

The shaker used in this test campaign consisted of a 9840-kg (21 700-lb) reaction mass supported by three air springs resting on top of drums filled with sand. A 998-kg (2200-lb) hydraulic actuator bolted under the centre of the mass and anchored to the top of the bridge deck provided the input force to the bridge. A random signal generator was used to produce an input with uniform frequency content between 2 and 12 Hz, providing an approximately 907-kg (2000-lb) peak force random input. An accelerometer mounted on the reaction mass was used to measure the total force transferred to the bridge through the drums and actuator. The shaker, shown in Figure 5.7, was located over the south plate girder directly above point S-3 shown in Figure 5.5.

The data acquisition system, mounting blocks, cabling, accelerometers and generator used for the forced vibration tests were identical to those used for the ambient vibration tests. An additional input module was used to monitor the accelerometer located on the reaction mass. Sampling parameters were specified so that responses with a frequency content in the range of 0–12.5 Hz could be measured. All computed frequency-domain quantities (PSDs, CPS, FRFs and coherence functions) were based on 30 averages with no overlap. A Hanning window was applied to all time samples used in these calculations.

Analyses of data recorded from this test structure are discussed in Sections 7.8, 7.9, 7.10, 8.2 and 12.2. Some data from these tests can be downloaded from http://institutes.lanl.gov.ei/software-and-data/.

5.2 The Concrete Column

Students and faculty in the University of California, Irvine's (UCI) Civil Engineering Department performed quasi-static, cyclic load tests on seismically retrofitted, reinforced-concrete bridge columns. The primary purpose of these tests was to study the relative strength and ductility provided by two retrofit construction procedures. The first retrofit procedure extended the diameter of the existing column with cast-in-place concrete. The second procedure extended the diameter of the existing column using shotcrete, which was sprayed on to the exterior of the existing column. Experimental modal analyses

Figure 5.7 Hydraulic actuator and reaction mass used to excite the I-40 Bridge during forced vibration tests.

were performed on the columns. These modal tests were performed at different stages during the quasi-static load cycle testing when various levels of damage had been introduced into the structure. These data were subsequently used to develop and validate damage detection algorithms.

The test structures consisted of two 0.61-m-diameter (24-in-diameter) concrete bridge columns that were subsequently retrofitted to 0.91-m-diameter (36-in-diameter) columns. The first column tested, labelled column 3, was retrofitted by placing forms around an existing column and placing additional concrete within the form. The second column, labelled column 2, was extended to 0.91 m diameter by spraying concrete in a process referred to as *shotcreting*. The shotcreted column was then finished with a trowel to obtain a circular cross-section.

The 0.91-m-diameter portions of both columns were 3.45 m (136 in) in length. The columns were cast on top of a 1.42-m-square (56-in-square) concrete foundation that was 0.635 m high (25 in high). A 0.61-m-square concrete block that had been cast integrally with the column extended 0.46 m (18 in) above the top of the 0.91-m-diameter portion of the column. This block was used to attach a hydraulic actuator to the columns for static cyclic testing and to attach the electrodynamic shaker used for the experimental modal analyses. As is typical of actual retrofits in the field, a 0.038-m gap (1.5-in gap) was left between the top of the foundation and the bottom of the retrofit jacket. This gap meant that the longitudinal reinforcement in the retrofitted portion of the column did not extend into the foundation. The concrete foundation was bolted to the 0.61-m-thick (2-ft-thick) testing floor in the UCI laboratory during both the quasi-static cyclic tests and the experimental modal analyses. Figure 5.8 shows the test structure geometry.

The only measured material property for the columns was the 28-day ultimate strength of the concrete and the test-day ultimate strength. The 28-day ultimate strength of foundations was 32 MPa (4600 psi).

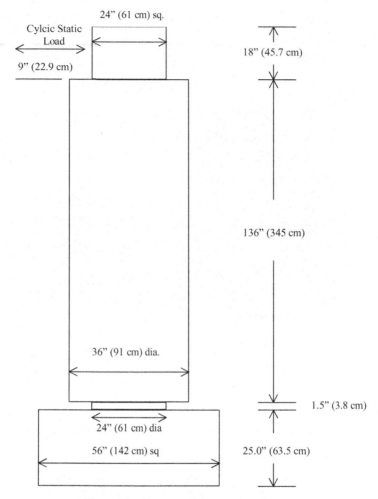

Figure 5.8 Geometry of the concrete columns.

The test-day ultimate strength was not measured for the foundations. The 0.61-m-diameter column 28-day ultimate strength was 30 MPa (4300 psi) and the test-day ultimate strength was 33 MPa (4800 psi). The 28-day ultimate strength of the retrofit portion of the structures was 36 MPa (5200 psi). On the test day the strength of the retrofit concrete was found to be 34 MPa (4900 psi).

The 0.61-m-diameter initial column reinforcement consisted of an inner circle of 10 #6 (19-mm diameter, 3/4-in diameter) longitudinal rebars with a yield strength of 516 MPa (74 900 psi). These bars were enclosed by a spiral cage of #2 (13.5-mm diameter, 1/4-in diameter) rebar having a yield strength of 207 MPa (30 000 psi) and spaced at a 0.18-m pitch (7-in pitch). 50 mm of concrete cover (2-in cover) was provided for the spiral reinforcement. The retrofit jacket had 16 #8 (25-mm diameter, 1-in diameter) longitudinal rebars with a yield strength of 414 MPa (60 000 psi). These bars were enclosed by a spiral cage of #6 rebar spaced at a 150-mm pitch (6-in pitch). The spiral steel also had a yield strength of 414 MPa. Again, 50 mm cover was provided for this reinforcement. Lap-splices 0.43 m (17 in) in length were used to connect the longitudinal reinforcement of the existing 0.61-m column to the foundation.

5.2.1   Quasi-Static Loading

Prior to applying lateral loads, an axial load of 400 kN (90 000 lb) was applied to simulate the dead loads that an actual column would experience. A steel beam was placed on top of the column. Vertical steel rods, fastened to the laboratory floor, were tensioned by jacking against the steel beam, which, in turn, applied a compressive load to the column.

A hydraulic actuator was used to apply a lateral load to the top of the column in a cyclic manner. The loads were first applied in a force-controlled manner to produce lateral deformations at the top of the column corresponding to $0.25\Delta y_T$, $0.5\Delta y_T$, $0.75\Delta y_T$ and Δy_T. Here Δy_T is the lateral deformation at the top of the column corresponding to the theoretical first yield of the longitudinal reinforcement. The structure was cycled three times at each of these load levels.

Based on the observed response, a lateral deformation corresponding to the actual first yield, Δy, was calculated and the structure was cycled three times in a displacement-controlled manner to that deformation level. Next, the loading was applied in a displacement-controlled manner, again in sets of three cycles, at displacements corresponding to $1.5\Delta y$, $2.0\Delta y$, $2.5\Delta y$, and so on until the ultimate capacity of the column was reached. This manner of loading puts incremental and quantifiable damage into the structures. The axial load was applied during all static tests.

5.2.2   Dynamic Excitation

For the experimental modal analyses the excitation was provided by a COTS electrodynamic shaker mounted off-axis at the top of the structure. The shaker, which rested on a steel plate attached to the concrete column, is shown in Figure 5.9. A horizontal load was transferred from the shaker to the structure through a friction connection between the supports of the shaker and the steel plate. This force was measured with an accelerometer mounted on the sliding mass of the shaker. This mass was measured to be 31 kg (0.18 lb s^2/in). A 0–400 Hz bandwidth uniform random signal was sent from a source module in the data acquisition system to the shaker, but feedback from the column and the dynamics of the mounting plate produced an input signal that was not uniform over the specified frequency range. The same level of excitation was used in all tests except for one at twice this nominal level, which was performed as a linearity check.

5.2.3   Data Acquisition

Forty COTS piezoelectric accelerometers were mounted on the structure as shown in Figure 5.10. Note that locations 2, 39 and 40 had accelerometers with a nominal sensitivity of 10 mV/g and were not sensitive enough for the measurements being made. At locations 33, 34, 35, 36 and 37 were accelerometers that had a nominal sensitivity of 100 mV/g. All other channels were accelerometers with a nominal sensitivity of 1 V/g. During the test on the shotcrete column (column 2) the accelerometer at location 23 had to be replaced with an accelerometer that had a sensitivity of 1 V/g.

Data were sampled and processed with a COTS dynamic data acquisition system. This system included a source module used to drive the shaker, five eight-channel input modules that provided power for the accelerometers and performed the A/D conversion of accelerometer signals, and a signal processing module that performed the necessary FFT calculations. A laptop computer was used for data storage and as a platform for the software that controlled the data acquisition system.

Data acquisition parameters were specified such that FRFs, input and response PSDs, CPS and coherence functions in the range of 0–400 Hz could be calculated. Each spectrum was calculated from 30 averages of 2-s-duration time histories discretised with 2048 points. These sampling parameters produced a frequency resolution of 0.5 Hz. Hanning windows were applied to all measured time histories prior to the calculation of spectral quantities.

Figure 5.9 Shaker used to provide lateral excitation to the columns.

A second set of measurements was acquired from 8-s-long time histories discretised with 8192 points. Only one average was measured. A uniform window was specified for these data, as the intent was to measure only a time history.

Analyses of data recorded from this test structures are discussed in Sections 7.1, 7.2, 7.8, 7.11, 8.2.5 and 8.2.8. Data from these tests can be downloaded from http://institutes.lanl.gov.ei/software-and-data/.

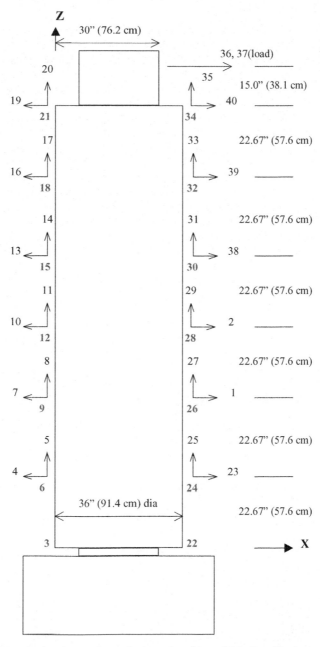

Figure 5.10 Accelerometer locations and coordinate system for modal testing. Numbers not positioned at an arrowhead indicate accelerometers mounted in the *y* direction.

5.3 The 8-DOF System

An eight-degree-of-freedom (DOF) system was designed and constructed to study the effectiveness of various vibration-based damage identification techniques in a controlled laboratory setting. The system was formed with eight translating masses connected by springs. Damage was simulated by changing the stiffness characteristics of the spring joining two of the masses and by placing an impacting mechanism between two adjacent masses. A schematic of the system is shown in Figure 5.11.

Each mass is a disc of aluminium 25.4-mm thick (1-in thick) and 76.2 mm in diameter (3 in in diameter) with a centre hole. The hole is lined with a Teflon bushing. There are small steel collars on the end of each disc. The masses all slide on a highly polished steel rod that supports the masses and constrains them to translate along the rod. The masses are fastened together with coil-springs epoxied to the collars, which are, in turn, bolted to the masses as shown in Figures 5.12 and 5.13. The undamaged configuration of the system is the state for which all springs are identical and have a linear spring constant. Two types of damage may be simulated, linear and nonlinear. Either type of damage may be located between any adjacent masses in the system. These damage types are described below.

Linear damage is defined as a change in the stiffness characteristics of the system such that the system can still be modelled by the standard linear differential equations of motion for a vibrating system after the damage. Linear damage in the model is simulated by replacing an original spring with another linear spring that has a spring constant less than that of the original. The replacement spring may be located between any adjacent masses and thus simulate different locations of damage. The replacement spring may have different degrees of stiffness reduction to simulate different levels of damage.

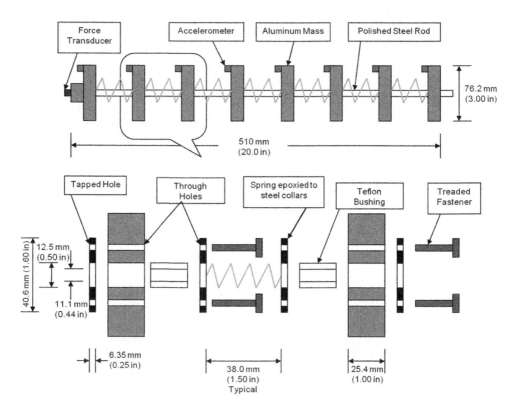

Figure 5.11   Schematic diagram of the eight degree-of-freedom system.

Figure 5.12 Eight degree-of-freedom system attached to an electrodynamic shaker with accelerometers mounted on each mass.

Figure 5.13 Impactors and bumpers (in boxes) used to simulate damage that produces a nonlinear response.

Nonlinear damage is defined as the occurrence of impacts between two adjacent masses. It is simulated by placing rods (impactors) on a given single mass that limits the amount of motion that the mass may move relative to the adjacent mass. Figure 5.13 shows the hardware used to produce nonlinear damage. When the distance between the mass and the ends of the rods is equal to the initial clearance, impact occurs. This impact simulates damage caused by spring deterioration to a degree that permits contact between adjacent masses or, in a simplified manner, the impact from the closing of a crack during vibration. The degree of damage is controlled by changing the amount of relative motion permitted before contact and changing the hardness of the bumpers on the impactors.

5.3.1 Physical Parameters

The nominal values of the 8-DOF system parameters are as follows:

> **Mass 1**: 0.5593 kg (0.003 19 lb s^2/in). This mass is located at the end where the shaker was attached. It is greater than the others because of the hardware needed to attach the shaker.
>
> **Masses 2 through 8**: 0.4194 kg (0.002 39 lb s^2/in)
>
> **Spring constants**: 56.7 kN/m (322 lb/in) (undamaged)
> 43.0 kN/m (244 lb/in) (24% stiffness reduction)
> 49.0 kN/m (278 lb/in) (14% stiffness reduction)
> 52.6 kN/m (299 lb/in) (7 % stiffness reduction)

Spring locations are designated by a sequential number with the spring closest to the end of the system where the excitation is applied designated as '1'. The 'damaged' spring location is given by a number, counting from the excitation end. Damping in the system is caused primarily by Coulomb friction. Every effort was made to minimise the friction through careful alignment of the masses and springs and by applying a lubricant between the Teflon bushings and the support rod.

5.3.2 Data Acquisition

Measurements made during the damage identification tests were of the excitation force applied to mass 1 and the acceleration response of all masses. Excitation was accomplished with either an impact hammer or a 215-N (50-lb) peak-force electrodynamic shaker shown in Figure 5.12. The data acquisition used in this study was a COTS data acquisition system. This system was composed of a source module, four eight-channel input modules (which provided power for the accelerometers and performed A/D conversion of the transducer signals). A signal-processing module performed the necessary FFT calculations. A laptop computer was used for data storage and as the platform for the software that controlled the data acquisition system. The force transducer used had a nominal sensitivity of 22 mV/N (100 mV/lb) and the accelerometers had a nominal sensitivity of 10 mV/G. Each data file was recorded in universal data file format (type 58) containing the FRFs, input PSD, response PSDs, CPS and coherence functions. If multiple measurements were averaged, all spectral functions are the averaged quantities. The final time histories were also stored in these files.

Analyses of data recorded from this test structure are discussed in Sections 7., 7.8–7.11. Data from these tests can be downloaded from http://institutes.lanl.gov.ei/software-and-data/.

5.4 Simulated Building Structure

The three-storey building structure shown in Figure 5.14(a) was developed as a damage detection test bed. The structure consists of aluminium columns and plates assembled using bolted joints (Figueiredo

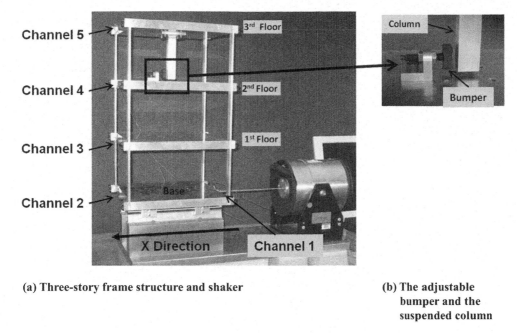

(a) Three-story frame structure and shaker

(b) The adjustable bumper and the suspended column

Figure 5.14   Three-storey test structure. (a) Three-storey frame structure and shaker; (b) the adjustable bumper and the suspended column.

et al., 2009). At each floor, four aluminium columns ($177 \times 25 \times 6$ mm^3, $7 \times 1 \times 0.25$ in^3) are connected to the top and bottom aluminium plates ($305 \times 305 \times 25$ mm^3, $12 \times 12 \times 1$ in^3), forming a structure that can accurately be represented as a 4-DOF system. Additionally, a centre column ($150 \times 25 \times 25$ mm^3, $6 \times 1 \times 1$ in^3) is suspended from the top floor, shown in Figure 5.14(b). Damage is simulated by the nonlinear phenomena introduced when this column impacts the bumper mechanism on the adjacent floor. This interference produces a repetitive impact-type nonlinearity similar to that produced by the previously described 8-DOF system. The position of the bumper can be adjusted to vary the extent of impacting that occurs during a test at a particular excitation level. This nonlinearity is intended to produce a small perturbation to an essentially stationary process, causing a nonlinear phenomenon called intermittency (Holger and Schreiber, 1997), where the system alternates between two conditions in an irregular way. This mechanism is intended to simulate a crack that opens and closes under dynamic loads ('breathing' crack) or loose connections that rattle. This design was motivated by several real-world examples that have reported bend and shear cracks that open and close under dynamic loads. For instance, in the prestressed bicellular box-girder bridge, the N. S. da Guia Bridge in Portugal, cracks were measured that open and close 0.12 mm (0.0047 in) under the loads produced by two vehicles passing side by side at 20 km/h (12 mph) (Pimentel, Santos and Figueiras, 2008). The structure slides on rails that allow movement in the *x* direction only, as shown in Figure 5.14(a). Figure 5.15 shows, schematically, the basic dimensions of the structure.

5.4.1 *Experimental Procedure and Data Acquisition*

A COTS data acquisition system was used to collect and process the data. The output channel of this system was input to a power amplifier to drive the electrodynamic shaker. The shaker provided a lateral excitation to the base floor along the centreline of the structure. The structure and shaker were mounted together on a rectangular aluminium base plate with dimensions $762 \times 305 \times 25$ mm^3 ($30 \times 12 \times 2$ in^3)

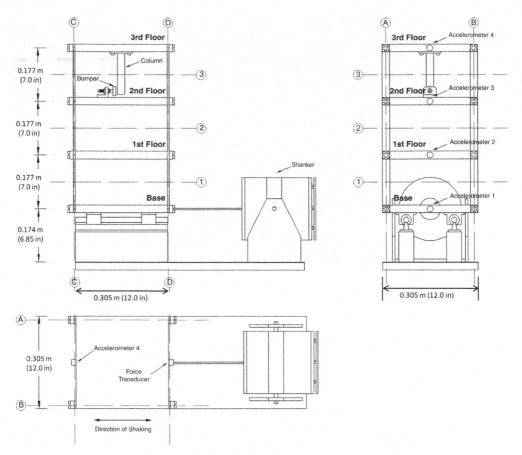

Figure 5.15   Dimensions of the three-storey frame structure.

and the entire system rested on rigid foam. The foam was intended to minimise extraneous sources of unmeasured excitation from being introduced through the base of the system. A force transducer with a nominal sensitivity of 2.2 mV/N (10 mV/lb) was attached at the end of a stinger to measure the input force from the shaker to the structure. As shown in Figure 5.14, four accelerometers with nominal sensitivities of 1 V/g were attached at the centreline of each floor on the opposite side from the excitation source to measure the system's response.

The analogue signals from the sensors were discretised with 8192 data points sampled at 3.1 ms intervals corresponding to a sample rate of 322.58 Hz. These sampling parameters yielded data windows that were 25.3 s in duration. When these data were transformed into the frequency domain, the spectra consisted of 3200 lines displaying the data up to a maximum frequency of 124.8 Hz at a resolution of 0.039 Hz. A band-limited random excitation in the range of 20–150 Hz was used to excite the system. This excitation signal was chosen in order to avoid the rigid-body modes of the structure that are present below 20 Hz.

5.4.2   Measured Data

Various time series samples with different structural conditions were collected. These state conditions are divided into four main groups. The first group is the baseline condition and corresponds to the structure in the condition shown in Figure 5.14 where the gap between the centre column and bumper mechanism was such that no impacting occurred.

The second group included a group of states where the structure had undergone changes in the mass and stiffness that simulate operational and environmental condition changes. Varying mass is intended to simulate operational changes such as different traffic loading on a bridge or changing fuel loads in an aircraft. Changing the stiffness is intended to simulate the effects of varying environmental conditions on the structure such as those produced by shifting temperature. Real-world structures have operational and environmental variability, which imposes difficulties in detecting and identifying structural damage (Alampalli, 2000; Peeters and De Roeck, 2001; Xia *et al.*, 2006). For instance, the influence of traffic loading on a 46-m-long simply-supported plate-girder bridge caused the natural frequencies to decrease by 5.4% (Kim *et al.* 2003). In a box-girder concrete bridge variations in the first natural frequency on the order of ± 10% were observed (Soyoz and Feng, 2009). Several investigations have suggested that the temperature plays the major role in the modal parameter variability. In the case of highway bridges, those variations can reach 5 to 10% (Ko and Ni, 2005). Therefore, in order to incorporate those variations into the frame structure, the operational and environmental effects are simulated through different mass and stiffness conditions, which will only cause the structural dynamic properties to change in a linear manner. The mass changes consisted of adding 1.2 kg (2.6 lb), approximately 19% of the total mass of each floor, to the first floor and to the base. The stiffness changes were introduced by reducing one or more of the column stiffnesses by 87.5%. This process was executed by replacing a column with one that had half the cross-sectional thickness in the direction of the applied force. These changes were designed to introduce variability in the fundamental natural frequency up to approximately 7% from the baseline condition and on the order of changes that have been observed in the field studies cited above. Impacting with the bumper mechanism did not occur in this group of tests.

The third group included the damaged state conditions with nonlinearities imposed by the bumper, as shown in Figure 5.16. The gap between the bumper and the suspended column was varied (0.05, 0.10, 0.13, 0.15 and 0.20 mm) in order to introduce different degrees of impacting for a given level of excitation.

The fourth and final group included the state conditions with nonlinearities mixed with the simulated operational and environmental condition changes. The bumper and the suspended column were included as members of the structure for both the damaged and undamaged conditions. For each structural state condition, ten acceleration time histories from each transducer were recorded.

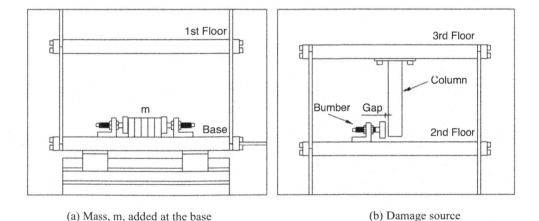

(a) Mass, m, added at the base   (b) Damage source

Figure 5.16   Structural details used to add varying mass and to simulate damage. (a) Mass, *m*, added at the base; (b) damage source.

Figure 5.17   The Alamosa Canyon Bridge near Truth or Consequences, New Mexico.

Analyses of data recorded from this test structures are discussed in Sections 7.2, 7.4, 7.12, 8.2, 10.5, 12.6, 12.7 and 13.10. Data from these tests can be downloaded from http://institutes.lanl.gov.ei/ software-and-data/.

5.5   The Alamosa Canyon Bridge

The Alamosa Canyon Bridge, which is located in southern New Mexico, has been tested numerous times in order to investigate the effects of environmental and experimental variability on the SHM process (Farrar *et al.*, 2002). The bridge has seven independent spans with a common pier between successive spans; it is shown in Figure 5.17. A drawing depicting an elevation view of the bridge's northern three spans is shown in Figure 5.18. Each span consists of a concrete deck supported by six W30 × 116 steel beams. The roadway in each span is approximately 7.3 m (24 ft) wide and 15.2 m (50 ft) long. A concrete curb and guardrail are integrally attached to the deck. Expansion joints are located at both ends of each span. The substructure of a span is shown in Figure 5.19. Between adjacent beams are four cross braces (C12 × 25 channel sections) equally spaced along the length of the span. Cross-sections of the span at locations showing the interior and end cross braces are shown in Figures 5.20 and 5.21.

5.5.1   Experimental Procedures and Data Acquisition

Experimental modal analyses were performed in order to investigate the temporal variability of modal parameters and to check the validity of the assumptions that the bridge would exhibit linearity and reciprocity. The data acquisition system consisted of a laptop computer, four eight-channel input modules that provided power to the accelerometers and performed A/D conversion of the accelerometer signals,

7 spans @ 15.2 m (50 ft)

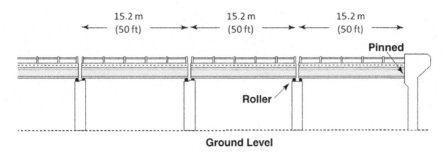

Figure 5.18   Elevation view of the Alamosa Canyon Bridge.

a signal processing module that performed the required FFTs and a commercial data acquisition/signal analysis software package. The system was powered by a 3500-watt AC generator.

The data acquisition system was set up to measure acceleration and force time histories and to calculate PSDs, CPS, FRFs and coherence functions. Thirty averages were typically used for the spectral estimates. Sampling parameters were specified that calculated the FRFs from a 16-s time window discretised with 2048 samples. The FRFs were calculated for a frequency range of 0 to 50 Hz at a frequency resolution of 0.0625 Hz. A force window was applied to the signal from the hammer's force transducer and exponential windows were applied to the signal from the accelerometers. AC coupling was specified to minimise DC

Figure 5.19   Alamosa Canyon Bridge substructure.

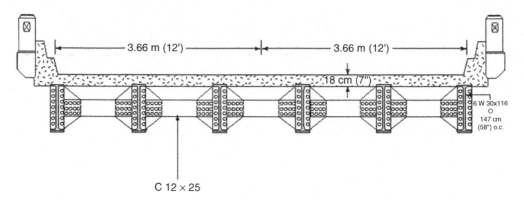

Figure 5.20   Cross-section of the Alamosa Canyon Bridge at the location of an interior cross brace.

offsets. With the sampling parameters listed above, data acquisition for a specific test usually occurred over a time period of approximately 30–45 minutes.

A COTS instrumented sledgehammer was used to provide the excitation source. The hammer weighed approximately 5.5 kg (12 lb) and had a 76-mm-diameter (3-in-diameter) steel head. The sensor in the hammer had a nominal sensitivity of 0.16 mV/N (0.73 mV/lb) and a peak amplitude range of 1120 N (5000 lb). A soft polymer hammer tip specially manufactured at Los Alamos National Laboratory was used to broaden the time duration of the impact and, hence, better excite the low-frequency response of the bridge.

A COTS piezoelectric accelerometer with a nominal sensitivity of 100 mV/g was used to make the driving point acceleration response measurement adjacent to the hammer impact point. This accelerometer had a specified frequency range of 5–15 000 Hz and a peak amplitude range of 50 g. All other measurements were made with COTS accelerometers having a nominal sensitivity of 1 V/g, a specified frequency range of 1–2000 Hz and an amplitude range of ±4 g's. All accelerometers were mounted to the bottom flange of the steel girders using magnetic mounts. A total of 31 acceleration measurements were made on the concrete deck and on the girders below the bridge, as shown in Figure 5.22. Two excitation points were located on top of the concrete deck. Point A was used as the primary excitation location. Point B was used to perform a reciprocity check.

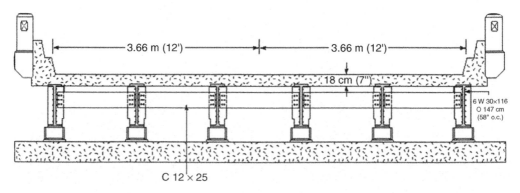

Figure 5.21   Cross-section of the Alamosa Canyon Bridge at the end of a span.

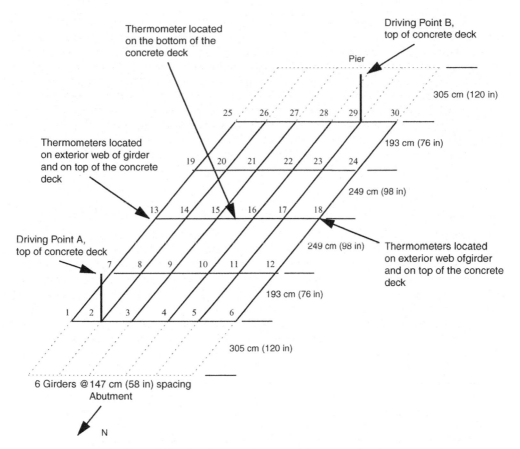

Figure 5.22   Accelerometer, impact and thermometer locations.

5.5.2   Environmental Measurements

Figure 5.23 shows the layout of the thermometers that were used to measure the influence of temperature at various locations on the bridge during the modal tests. Five indoor–outdoor digital-readout thermometers were located across the centre of the span. Two thermometers were positioned such that their outdoor

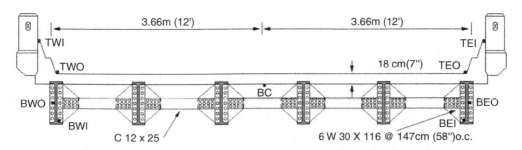

Figure 5.23   Detailed layout of thermometers.

sensor was taped to the outside web surface at mid-height of the exterior gird (BWO (T_8), BEO (T_7) in Figure 5.23). Note that the nomenclature (T_i) corresponds to that used in the discussion of the data normalisation procedure presented in Chapter 12. The indoor readings from these two thermometers were made on the inside bottom flanges of the exterior girders (BWI (T_6), BEI (T_5) in Figure 5.23). A third thermometer was taped to the underside of the concrete deck at the middle of the span (BC (T_9) in Figure 5.23). The outside sensor for this thermometer was located adjacent to the indoor sensor yielding almost identical temperature readings. The two remaining thermometers were located on the topside of the bridge. The outside sensors were taped to the bridge deck immediately adjacent to the concrete curbs (TWO (T_2), TEO (T_1) in Figure 5.23). The indoor sensor was located on top of the guardrail (TWI (T_4), TEI (T_3) in Figure 5.23). All sensors were shaded from direct sunlight and were read manually at the start and end of each modal test, with data being recorded in a test logbook.

5.5.3   Vibration Tests Performed to Study Variability of Modal Properties

In the preliminary tests on the Alamos Canyon Bridge it was observed that slightly different frequencies were obtained from the roving hammer impact modal tests and from the single excitation point impact tests when the tests were performed at different times of the day. For this reason, subsequent modal tests were designed specifically to examine the variability in modal parameters of the first span of the bridge caused by environmental effects. Hammer impact force data and acceleration response data were measured at two-hour increments over a 24-hour time period to investigate the change in the modal properties as a function of the time of day. The observed changes were assumed to be primarily caused by changes in the temperature of the structure. For this reason the temperature was measured at several locations on the structure throughout the test as described above. The Alamos Canyon Bridge is sensitive to temperature variations because of its north–south orientation where the sun heats one side of the bridge in the morning.

An additional set of measurements was made to study the variability in modal properties caused by vehicle weight. An impact modal test was performed on span 1 with four cars parked on the bridge and compared these results to impact tests without cars. For one span the concrete deck and reinforcing steel weighs approximately 5300 kg (118 000 lb) and the steel girders, cross bracing and gusset plates weigh 18 200 kg (40 000 lb), yielding a total span weight of 71 800 kg (158 000 lb). The four cars that were placed on the bridge weighed approximately 10 000 kg (22 000 lb). Assuming the parked cars have no other effects on the dynamics of the structure other than the addition of mass, they should lower the frequencies by a value proportional to the square root of the mass ratios, in this case approximately 6.4%.

Finally, subsequent tests were performed to compare the effects of using different excitation sources when identifying dynamic properties of the bridge. The excitation sources used were the hammer impact, ambient excitation from traffic on the Alamosa Canyon Bridge and from traffic on the nearby Interstate Highway Bridge, random excitation with an electrodynamics shaker and a swept sine lateral excitation provided by an eccentric mass shaker.

Analyses of data recorded from this test structures are discussed in Sections 7.12, 8.2, 12.3, 12.5 and 13.10. Data from these tests can be downloaded from http://institutes.lanl.gov.ei/software-and-data/.

5.6   The Gnat Aircraft

A Gnat trainer aircraft (shown in Figure 5.24) was made available by QinetiQ Ltd of Farnborough UK for a number of damage detection studies. The objectives of these studies were to detect and locate damage in the starboard wing. Unfortunately, at the time it was not considered possible to damage the aircraft. Because of this constraint, two alternate methods of simulating damage were investigated. These methods included adding an inspection panel with seeded damage for the detection study and removing various different panels from the wing for the damage location study.

Figure 5.24 Starboard wing of the Gnat aircraft.

5.6.1 Simulating Damage with a Modified Inspection Panel

The first method of simulating damage was adopted for the damage detection study and was accomplished by making ten copies of an inspection panel on the starboard wing of the aircraft. One of the replica panels was left intact and the remaining nine received controlled damage. The reason why one panel was reserved as a normal condition state was because it proved impossible to exactly match the aircraft aluminium of the true panel (although as close a match as possible was made). The geometry of the panel in question is shown in Figure 5.25.

The panel was fixed to the wing by 23 screws. There were originally 26 threaded holes, but a previous removal of the panel had involved drilling out some of the screws, resulting in damage to three of the threaded holes.

Problems were anticipated during the test as a result of the variability in the attachment of the panel. Because a different panel was used for each damage state, it had to be fixed to the wing prior to the test and then removed. This mounting scheme produced inevitable variations in the boundary conditions. An attempt was made to minimise the problem by using a constant-torque electric screwdriver. A more rigorous approach would have been to use a screwdriver head in a torque wrench and to make sure that the same torque was applied to each screw in each test. This was not carried out because: (a) it is simply

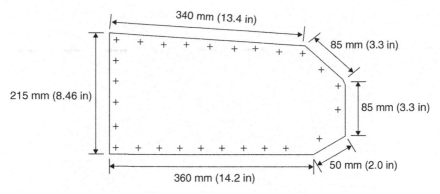

Figure 5.25   Inspection panel mounted on the starboard wing.

unrealistic to assume that that degree of care is used in removing and replacing inspection panels and (b) it would have been far too time-consuming.

Transmissibility data were used to monitor the region of interest locally. As discussed earlier in the context of the I-40 Bridge, the TF is a spectral quantity analogous to the FRF (see Appendix A). However, instead of normalising a response measurement by the input measurement, the response measurement is normalised by another reference response measurement. The sensors used were piezoelectric accelerometers. Four sensors were used in all: one pair to measure the responses (and subsequently estimate the TF) across the panel in the length direction (T1 and T2 in Figure 5.26) and one pair across the width (T3 and T4 in Figure 5.26). An additional pair was fixed across a panel diagonal, as shown in the photograph in Figure 5.27; however, the data capture proved too time-consuming to use this pair. The accelerometers were fixed to the wing with beeswax.

The wing was excited using an electrodynamic shaker attached directly below the inspection panel on the bottom surface of the wing. A Gaussian random excitation was generated within the data acquisition system and amplified using a power amplifier.

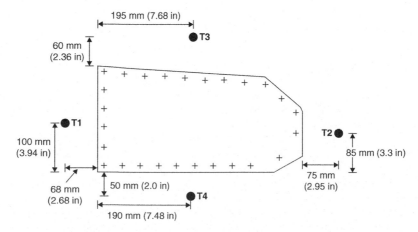

Figure 5.26   Location of piezoelectric sensors around the test panel.

Figure 5.27   Actual sensors mounted on the starboard wing.

The data used to estimate the TFs were measured using a 24-channel data acquisition system controlled by COTS data acquisition software running on a Unix workstation, as shown in Figure 5.28. Both real and imaginary parts of the functions were obtained. In order to resolve the defects in the inspection panel, it was assumed necessary to excite modes with appropriately short wavelengths and hence high frequencies. In order to select a suitable excitation band, TFs between transducers 1 and 2 were estimated from data acquired in the frequency range 0–2000 Hz with the undamaged inspection panel attached.

The panel was then completely removed (in order to give a worst-case damage state) and the measurement repeated. Comparison between the two TFs confirmed that the lower-frequency modes were insensitive to the damage, so the excitation band for the main body of tests was selected as 1000–2000 Hz. In all cases 2048 spectral lines were acquired.

The spectral estimation strategy for each TF was as follows. First, for the undamaged panel the function was obtained using 128 averages. This averaging was done to provide a clean reference signal to help with feature selection. Next, 110 measurements were taken sequentially using only a single average. Of these, 100 would be used to establish the statistics of the patterns for the damage detection (this is accomplished using outlier analysis – see Chapter 10) and 10 would be used for testing. The single-average data was very fast to acquire and considered to be the most likely candidates for an on-line system. The use of eight-average data was investigated, but was rejected on the grounds of acquisition time.

The second series of tests worked through the damaged panels in the order shown in Figure 5.29. Damage states (a), (b) and (c) were holes of diameter 20 mm, 38 mm and 58 mm, respectively. States (d), (e) and (f) were sawcuts across the width of the panel with (d) an edge cut of 50 mm and (e) and (f) central cuts of extent 50 mm and 100 mm, respectively. States (g), (h) and (i) were sawcuts along the long axis of the panel with (g) a 100-mm edge cut and (h) and (i) central cuts 100 mm and 200 mm

Figure 5.28   Test configuration and data acquisition system.

long, respectively. For each panel, the first function measured was based on 128 averages; again this was for reference and feature selection. The second set of tests on each panel recorded 10 single-average functions sequentially.

After all the damaged panels had been tested, another set of measurements was taken for the situation with the undamaged panel to check repeatability. These measurements were used to estimate the same 11 functions acquired for the damage states: one 128-average measurement and ten single-average measurements. The next sequence of tests took the same 11 measurements for the situation with the panel removed completely.

The penultimate series of tests also addressed the problem of repeatability. The undamaged panel was fixed and tested and then removed four times in order to investigate the effect of the fixing conditions on the measured features. Only 128-average patterns were estimated. Another important reason for this portion of the programme was to obtain a range of normal condition patterns that characterised the variation in the fixing conditions.

Finally, a 128-average TF was estimated with the panel completely removed again. The objective of this test was to investigate test variability that could not be associated with the fixing conditions, that is variability as a result of environmental changes and instrument drift. This variability is the least that could be expected.

5.6.2   Simulating Damage by Panel Removal

A second approach to simulating damage in the starboard wing was motivated by noting the presence of numerous inspection panels distributed over the wing and also the desire to extend the previously described study from one of identifying the presence of damage to one of locating the damage. It was

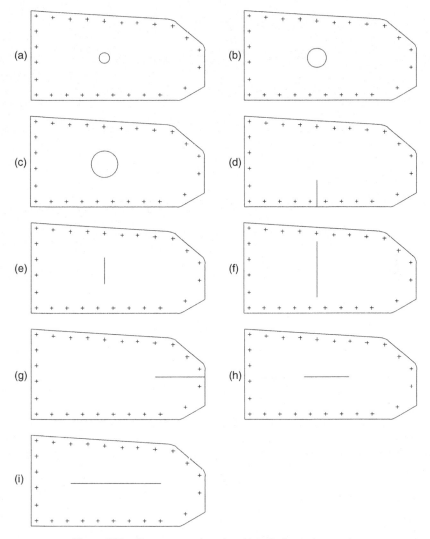

Figure 5.29	Damage cases introduced into the inspection panels.

therefore decided to simulate damage by sequentially removing these panels. As before, unlike real damage, this approach also had the distinct advantage that each damage scenario was reversible and it would be possible therefore to monitor the repeatability of the measurements.

Of the various panels available, nine were chosen, mainly for their ease of removal and also to cover a range of sizes. These panels were distributed as shown in Figure 5.30. The areas of the panels P1 to P9 are also given in Figure 5.30. Removal of any of these panels actually constitutes a rather large damage. Panels P3 and P6 are likely to give the most difficulty for a damage detection procedure because they are by far the smallest.

Each panel was fixed to the wing by a number of screws, the numbers varying between 8 and 26. On some of the panels, screws were missing as a result of damaged threads in the holes. In fact, during

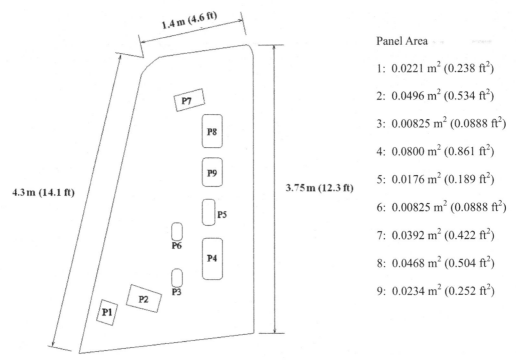

Figure 5.30   Panels that were removed and the respective sizes.

the repeated removal of the plates, further holes were damaged. This damage meant that there was some variation throughout the test, even for the normal condition (all panels attached). However, given experience gained during the previous experimental phase of the programme, it was assumed that this issue with the fasteners was unlikely to be a source of major variation, compared to the uncertainty in the fixing conditions of the remaining screws. The screws were secured and removed with an electric screwdriver with a controllable torque; the same torque setting was used throughout.

As with the procedures that introduced damage seeded in an inspection panel, it was decided to estimate TFs. The initial decision was to estimate a TF from data acquired across each plate. In order to use transducers effectively, the panels were split into three groups, A, B and C. Each group was allocated a centrally placed reference transducer, together with three other transducers, each associated with a specific plate. The transducer layout was as shown in Figure 5.31.

Although this network of sensors offered the possibility of forming many TFs, only those estimated from data acquired directly across the plates were used in the study. The TFs estimated from data acquired across plate Pn was denoted Tn. The reciprocals of these TFs were denoted Tn^*. Table 5.1 summarises the TFs that were estimated. The sensors used were standard COTS piezoelectric accelerometers.

The wing was excited using an electrodynamic shaker attached directly below the panel P4 on the bottom surface of the wing. The test procedure was similar to that described in the previous subsection for the damage detection problem based on the inspection panel. In all cases, 1024 spectral lines (both real and imaginary parts of the functions) were obtained. As the smaller panels were of the same order of size as the major defects in the previous inspection panel study, it was assumed necessary to excite modes with the same order of wavelengths and hence frequencies. The TFs were therefore estimated to be between 1024 and 2048 Hz.

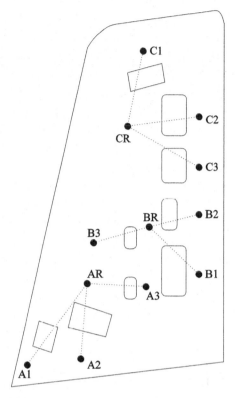

Figure 5.31   Transducer locations for the various panel removals.

In order to make use of the SIMO (single-input–multiple-output) mode of the data acquisition software, each configuration of the wing was tested three times, once each for transducer groups A, B and C. In all, 25 configurations were studied as follows:

1. Normal condition (all plates in place).
2. Plate P1 removed.

Table 5.1   Transducer pairs used to form the various transmissibility measurements

Plate	Associated transmissibility	Reference transducer	Response transducer
P1	T1	AR	A1
P2	T2	AR	A2
P3	T3	AR	A3
P4	T4	BR	B1
P5	T5	BR	B2
P6	T6	BR	B3
P7	T7	CR	C1
P8	T8	CR	C2
P9	T9	CR	C3

3. Plate P2 removed.
4. Plate P3 removed.
5. Normal condition.
6. Plate P1 removed.
7. Plate P2 removed.
8. Plate P3 removed.
9. Normal condition.
10. Plate P4 removed.
11. Plate P5 removed.
12. Plate P6 removed.
13. Normal condition.
14. Plate P4 removed.
15. Plate P5 removed.
16. Plate P6 removed.
17. Normal condition.
18. Plate P7 removed.
19. Plate P8 removed.
20. Plate P9 removed.
21. Normal condition.
22. Plate P7 removed.
23. Plate P8 removed.
24. Plate P9 removed.
25. Normal condition.

This programme meant that seven sets of measurements for the normal condition were made and two each for each damage state. This test sequence was done in order to investigate variability of the normal and damaged condition data between plate removals.

The measurement strategy for each TF was as follows. First, each function was obtained using 16 averages. This averaging was done to provide a clean reference signal that would help with feature selection. The previous inspection panel-based tests described earlier actually formed the TF from 128 averages; however, given the number of tests to be performed here this averaging was regarded as too time-consuming. A brief experiment showed that the 16-average TFs were adequately smooth. Next, 100 measurements were taken sequentially using only one average. Over the full sequence of 25 configurations, this gave 700 single-average measurements for the normal condition and 200 for each damage condition.

More details of the experimental programmes for this structure can be found in Manson, Worden and Allman (2003a, 2003b). Analyses of data recorded from this test structure are discussed in Sections 11.3 and 11.9.

References

Alampalli, S. (2000) Effects of testing, analysis, damage, and environment on modal parameters. *Mechanical Systems and Signal Processing*, **14**(1), 63–74.

Farrar, C.R., Baker, W.E., Bell, T.M. *et al.* (1994) Dynamic Characterization and Damage Detection in the I-40 Bridge over the Rio Grande, Los Alamos National Laboratory Report LA-12767-MS.

Farrar, C.R., Cornwell, P.J., Doebling, S.W. and Prime, M.B. (2002) Structural Health Monitoring Studies of the Alamosa Canyon and I-40 Bridges, Los Alamos National Laboratory Report LA-13635-MS.

Figueiredo, E., Park, G., Figueiras, J. *et al.* (2009) Structural Health Monitoring Algorithm Comparisons Using Standard Datasets, Los Alamos National Laboratory Report LA-14393.

Holger, K. and Schreiber, T. (1997) *Nonlinear Time Series*, Cambridge University Press, Cambridge, UK.

Kim, C.-Y., Jung, D.-S., Kim, N.-S. *et al.* (2003) Effect of vehicle weight on natural frequencies of bridges measured from traffic-induced vibration. *Earthquake Engineering and Engineering Vibration*, **2**(1), 109–115.

Ko, J. M. and Ni, Y.Q. (2005) Technology developments in structural health monitoring of large-scale bridges. *Engineering Structures*, **27**, 1715–1725.

Manson, G., Worden, K. and Allman, D.J. (2003a) Experimental validation of structural health monitoring methodology II: novelty detection on an aircraft wing. *Journal of Sound and Vibration*, **259**, 345–363.

Manson, G., Worden, K. and Allman, D.J. (2003b) Experimental validation of structural health monitoring methodology III: damage location on an aircraft wing. *Journal of Sound and Vibration*, **259**, 365–385.

Peeters, B. and De Roeck, G. (2001) One-year monitoring of the Z24-bridge: environmental effects versus damage events. *Earthquake Engineering and Structural Dynamics*, **30**, 149–171.

Pimentel, M., Santos, J. and Figueiras, J. (2008) Safety appraisal of an existing bridge via detailed modelling. International FIB Symposium 2008: Tailor Made Concrete Solutions, New Solutions for Our Society, Amsterdam, Netherlands.

Soyoz, S. and Feng, M.Q. (2009) Long-term monitoring and identification of bridge structural parameters. *Computer-Aided Civil and Infrastructure Engineering*, **24**, 82–92.

Xia, Y., Hao, H., Zanardo, G. and Deeks, A. (2006) Long term vibration monitoring of an RC slab: temperature and humidity effect. *Engineering Structures*, **28**, 441–452.

6

Introduction to Probability and Statistics

6.1 Introduction

The idea of this chapter is to begin to introduce the analytical machinery needed for the development of an SHM methodology based on the pattern recognition/machine learning paradigm. The analytical component of the SHM process begins with a signal recorded from a sensor of some form, as discussed in Chapter 4. The term *signal* will be understood here to mean a vector of numbers (usually ordered) which carries information. The classic example of a signal is of course a sampled time series. The objective of the SHM process will be to convert the signal data into diagnostic and prognostic information about the structure of interest. As discussed in Appendix A, one can broadly separate signals into two classes: *deterministic* and *probabilistic*. One can almost use the term deterministic as a synonym for the term *predictable*; basically a deterministic system is one such that, if the underlying physics is known and the initial conditions are known for the system, then in principle one can predict the behaviour of the system at any time following the initial time. For example, consider the linear SDOF mass–spring–damper system under single harmonic excitation:

$$m\ddot{y} + c\dot{y} + ky = F \sin(\omega t) \tag{6.1}$$

If the initial time is taken as $t = 0$, then knowing the initial conditions $y(0)$ and $\dot{y}(0)$ means that one can predict the value of $y(t)$ for any subsequent time. Such a system is a deterministic system and we can consider $y(t)$ to be a deterministic signal. Identifying determinism with predictability is not entirely straightforward. If Equation (6.1) were made nonlinear by the addition of a term $k_3 y^3$ to the left-hand side, a situation can arise where the response becomes extremely sensitive to the initial conditions – so sensitive that any error in the estimated initial conditions, however small, destroys predictability. This phenomenon is called *deterministic chaos*; it need not be of concern here, although the concept will later prove to be useful when one wishes to extract the signatures of damage from signals (Chapter 8).

The opposite of the deterministic system is a system where, even if one knows all the underlying physics of the system, it is not possible to predict the behaviour of the system as it evolves from given initial conditions. Such systems are commonly called *random* or *probabilistic*. As one might expect from the latter term, the only way to understand such systems properly is through the medium of probability theory. (The term *stochastic* is sometimes used interchangeably with the term random, although strictly speaking the term stochastic carries extra meaning.) Therefore the opposite of a deterministic signal is a random signal where one cannot predict, with certainty, the future values of the signal. The good news for such systems is that one can, in principle, select from a range of possible predictions with some degree of confidence. In other words, the most that can be learned is the *probability* of the signal taking on a certain value.

This concept is vital for an understanding of SHM because many (if not all) of the signals measured in an SHM context will be unpredictable or random for various reasons. One example of importance relates to measurements taken over different environmental conditions. If the result of a measurement depends on the ambient temperature, which is not measured but fluctuates between measurements, the sequence of measurements will display an unpredictable component that can only be accommodated in a probabilistic setting. Instrumentation noise will introduce an uncertain component to measurements even if the environment is constant; however, this is usually small if the instrumentation is selected carefully.

Before one can discuss these issues in detail, a review of the basic concepts of probability and statistics should be conducted. The theory of statistics is distinct from probability theory in that it is not fundamentally concerned with the likelihood or not of events; it is concerned with the extraction of summary information from bodies of data. As such, it is just as important for SHM purposes as probability theory. The very strong link between the two ideas is that the summary information from statistics can often be used to estimate probabilities. The term *statistic* as applied to a single number will be understood here to mean a summary figure that results from statistical analysis – the arithmetic mean of an array of numbers is a common example.

6.2   Probability: Basic Definitions

Defining the concept of probability is not straightforward. It is not trivial to explain the basis of the concept without using words that one later has to give a technical meaning to. In order to make some progress, let us argue that probability is ultimately concerned with *degrees of belief*. Consider a question of some importance in the context of SHM: what is the probability that bridge A will fall down tomorrow? Experts in the structural dynamics of bridges will formulate their answer on the basis of any evidence before them; therefore, for example, if a dynamic test carried out that morning showed that the dynamic properties of the bridge are the same as on the day the bridge opened, the expert may consider that this is strong evidence that no deterioration in the bridge structure has occurred and the bridge is very unlikely to fall down tomorrow. If small changes in the dynamic properties have occurred since commissioning, the expert may look into their body of experience or carry out some analysis and may be able to quantify their degree of belief in the safety of the structure; they may offer a judgement that they are '99% certain that the bridge is safe'. This process, although informed by evidence, is clearly subjective; however, it is perfectly possible to develop a probability theory on the basis of such concepts. Suppose now that the evidence in front of the expert takes the following form. The bridge in question is one of many that were built to the same design and the expert has the results of all structural tests on all these bridges available. On consulting the data, the expert finds that, of all the bridges that showed the same test results shown by bridge A on that day, 50% of them fell down the following day. On this evidence, the degree of belief in the safety of the bridge is likely to be equal to the degree of belief in disaster; with such a conclusion, the expert will very likely order the closure of the bridge. In this particular case, the evidence for a particular event is in the form of *frequencies of occurrence* and it can be argued that this is a more objective basis for the formulation of a prediction. Building a probability theory on the basis of such evidence is called the *frequentist approach*. One can argue that this is not the correct philosophical approach to applying probability in the context of SHM – many of the structures of interest are unique and it is not possible to look a frequencies of occurrence; however, in many ways the frequentist approach is the one that allows the simplest introduction to many of the concepts of probability theory that are common to all philosophies. For this reason, the following discussion is framed in terms of frequencies.

It is of interest to assess the outcomes after a certain situation has been observed. A possible outcome of a situation will be called an *event*. In this context 'bridge A experiences a fatigue crack 6 cm or larger in length' is a perfectly respectable event. Also acceptable as an event is 'bridge A generated the set of results B following a structural test'. This example is more useful because one can imagine carrying out the test many times and therefore being able to assess the relative frequency of the results B occurring. The observation of the results after a structural test is an example of what will be called an experimental trial. Probability can now be defined in terms of the frequentist approach.

The *probability* $P(E)$ of an event E occurring in a given situation or experimental trial, will be defined as

$$P(E) = \lim_{N(S) \to \infty} \frac{N(E)}{N(S)} \tag{6.2}$$

where $N(S)$ is the number of times the situation occurs or an experiment is conducted and $N(E)$ is the number of times the event E follows. The definition is formally given in terms of a limit; in practice *estimates* of the probabilities will be made by carrying out experiments a finite (but large) number of times.

With the definition given, it is clear that $1 \geq P(E) \geq 0$, with $P(E) = 1$ asserting the certainty of an event E and $P(E) = 0$ indicating its impossibility. Note that just because an event has not been observed does not necessarily imply that its probability is zero.

To give a concrete example of the assignment of probability, in a large number of throws of a true die, the result 6 would be expected $\frac{1}{6}$ of the time, so $P(6) = \frac{1}{6}$.

If two events E_1 and E_2 are *mutually exclusive* then the occurrence of one precludes the occurrence of the other. In this case,

$$P(E_1 \cup E_2) = P(E_1) + P(E_2) \tag{6.3}$$

where the set theory 'union' symbol $\cup$ represents the logical *or* operation, so $P(E_1 \cup E_2)$ is the probability that event E_1 *or* event E_2 occurs.

If E_1 and E_2 are not mutually exclusive, a simple argument leads to the relation

$$P(E_1 \cup E_2) = P(E_1) + P(E_2) - P(E_1 \cap E_2) \tag{6.4}$$

where the 'intersection' symbol $\cap$ represents the logical *and* operation.

If a set of mutually exclusive events $\{E_1, \ldots, E_N\}$ is exhaustive in the sense that one of the E_i *must* occur, it follows from the definitions above that

$$P(E_1 \cup E_2 \cup \cdots \cup E_N) = P(E_1) + P(E_2) + \cdots + P(E_N) = 1 \tag{6.5}$$

Therefore, in throwing a die,

$$P(1) + P(2) + P(3) + P(4) + P(5) + P(6) = 1 \tag{6.6}$$

(in an obvious notation). Also, if the die is true, all the events are equally likely:

$$P(1) = P(2) = P(3) = P(4) = P(5) = P(6) \tag{6.7}$$

and these two equations show that $P(6) = \frac{1}{6}$ as asserted earlier. The set of all possible outcomes or events from a trial is sometimes called the *sample space*.

Two events E_1 and E_2 are *statistically independent* (or often just *independent*), if the occurrence of one *in no way* influences the probability of the other. In this case,

$$P(E_1 \cap E_2) = P(E_1) \times P(E_2) \tag{6.8}$$

A simple example of this would be simultaneously tossing a coin and throwing a die.

This section has explained some of the common terms associated with events and their probabilities. In the context of SHM, as explained in the introduction, many of the events that are of concern are the occurrences of certain damage-sensitive features that are extracted from signals when measurements are taken. Such events are associated with continuous ranges of possible measurement values; in order to deal properly with this idea, one has to introduce the idea of a random variable.

6.3 Random Variables and Distributions

The outcome of an individual throw of a die is completely unpredictable. However, the value has a definite probability that can be determined. Variables of this type are referred to as *random variables* (RVs).

For the die, the RV can only take one of six values; it is *discrete*. Some of the random variables associated with SHM problems are also discrete; that is if one is only interested in characterising the state H of a structure as 'healthy' or 'damaged', then one has a discrete random variable. However, because the majority of signals measured for SHM purposes can take a continuous range of values, one will need to consider RVs that can similarly take a *continuous* range of values.

To give an example in the context of SHM, consider a composite plate that is impacted with a projectile as shown in Figure 6.1. The goal is to estimate the area of delamination that will occur when attempts are

Figure 6.1 Composite plate subjected to projectile impact.

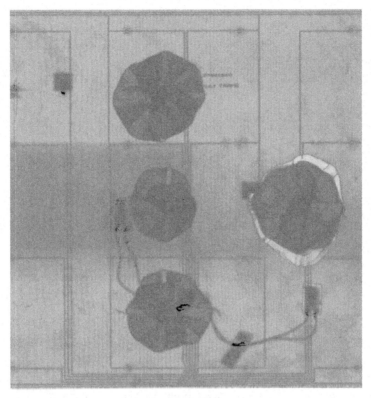

Figure 6.2   Ultrasonic scans show delamination areas after multiple projectile impacts. Note that the lines that can be seen in the scan are the wires connecting sensors to the data acquisition system.

made to impact nominally identical plates at a projectile velocity of 40 m/s. However, Figure 6.2 shows that variability in the layup and manufacturing of the test specimens along with variability in the control of the projectile velocity cause variability in the delamination areas that are detected by subsequent ultrasonic scans. Additional variability may enter into the measurement process when one attempts to quantify the delamination area. Assuming that there are no systematic changes to the manufacturing or experimental procedures, the resulting delamination area can be viewed as an RV. Therefore, for the engineers to make the best estimate of the delamination area resulting during the next test, they should be guided by probability.

Suppose that based on previous observations it can safely be assumed that $P(0.3 \text{ m}^2) = 0$ and $P(0.1 \text{ m}^2) = 0$. However, if it is assumed that all intermediate delamination values are possible, this gives an infinity of outcomes. A rough argument based on Equation (6.5) gives

$$P(x_1) + P(x_2) + \cdots + P(x_i) + \cdots = 1 \tag{6.9}$$

(where, for the moment, x is the delamination area) and so all the individual probabilities must be zero.

Right away there seems to be some difficulties; however, a moment's thought shows that this result agrees with common sense. If one estimates the next delamination area to be 0.18 m^2, there is no real chance of observing *exactly* this value (assuming there is an infinitely precise measurement system). However, if individual probabilities are all zero, how can probabilistic methods be applied?

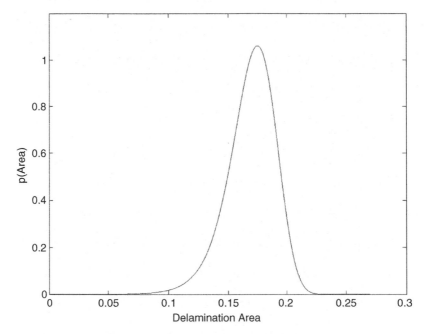

Figure 6.3   PDF for delamination area corresponding to 40 m/s projectile impact.

The solution to the apparent paradox is straightforward; in practice, one should specify a range of delamination areas centred on a particular value as the estimate. This approach points to the required mathematical structure; for an RV X, the *probability density function* (PDF) $p(x)$ is defined by

$$p(x)\mathrm{d}x \text{ is the probability that } X \text{ takes a value between } x \text{ and } x + \mathrm{d}x$$

(It is conventional to denote an RV by a capital letter, but any values it takes on by lower-case letters.) So what will $p(x)$ look like? Well it has already been observed that $P(0.3 \text{ m}^2) = P(0.1 \text{ m}^2) = 0$ and when these values are translated into a more precise framework, it can be stated that $p(x) = 0$ for all $x \geq 0.3 \text{ m}^2$ or $x \leq 0.1 \text{ m}^2$. Altogether the testing results on a large number of plates may yield a PDF like that in Figure 6.3. Intuitively, one might expect that the most probable delamination area for a subsequent test will correspond to the centroid of the PDF, which is around 0.18 m^2. The centroid of the PDF turns out to be one of the most important parameters that is used to describe an RV and will be discussed in more detail below. The important thing to take away from this discussion in the context of this book is that probability density functions arise very frequently and very naturally in the context of SHM. A few additional examples might be:

1. Suppose one were concerned with the remaining safe life of a concrete beam subjected to cyclic loading; the specific question being – if this beam has been subjected to 10 000 loading cycles, how many cycles C will it now endure before failure? Because any given beam will be unique due to the spatially random nature of the material, the best one could do is specify a PDF for the remaining life $p(C)$. (This example has an interesting characteristic. If one is counting cycles, then C is actually a discrete random variable; however, if a discrete random variable can take very many values – as in this case – it will often be approximated by a continuous one.)

2. Staying with the concrete beam, under the cyclic loading, the beam would be likely to generate very many localised cracks. These cracks would grow, multiply and coalesce as the loading continued and the beam would ultimately fail. As a means of assessing the condition before ultimate failure, one might try to establish the probability density function of the crack lengths.
3. Suppose one wished to measure the first natural frequency of an aircraft wing as a means of assessing condition. Suppose further that the natural frequency of the wing was sensitive to the ambient temperature. If the temperature during the test were not controllable (say the aircraft is in a large hangar with poor environmental control), the measured natural frequency would be a random variable. (This example is extremely important and a whole chapter will be devoted to related issues for SHM later.)

Returning to the composite plate, suppose one estimates that the delamination area corresponding to the next impact will be 0.175 ± 0.01 m^2. In this situation, with the PDF in hand, one can assign a true probability that the delamination falls in this *finite* range as follows. The analogue of the sum in Equation (6.5) is an integral, so

$$P(X = x; 0.165 \leq x \leq 0.185) = \int_{0.165}^{0.185} p(x)\mathrm{d}x \qquad (6.10)$$

In general the rule is

$$P(X = x; \ a \leq x \leq b) = \int_{a}^{b} p(x)\mathrm{d}x \qquad (6.11)$$

Geometrically, this probability is represented by the area under the PDF curve between a and b (as illustrated in Figure 6.2 for the delamination problem with the limits in Equation 6.10). This also illustrates why the probability of any given real number is zero – the area 'below' a single point on the curve is zero.

It is also clear that the total area under the curve, that is the probability of X taking *any* value, must be 1. In analytical terms,

$$\int_{-\infty}^{\infty} p(x)\mathrm{d}x = 1 \qquad (6.12)$$

Note that this condition requires that $p(x) \to 0$ as $x \to \pm\infty$.

In the context of the delamination problem, the engineers can potentially establish probabilities for their guesses. The question of how to optimise the guess is answered in the next section.

6.4 Expected Values

Suppose that the engineers have equipped themselves with a PDF for the delamination area. (Obtaining this PDF in itself is a difficult and fundamental question that will be addressed in more detail later.) For now, suppose that they have observed the delamination areas from lots of nominally identical tests and have counted how many values are in certain ranges – this is the histogram concept that may be familiar to readers from basic science and statistics courses. How can they use this information to compute a best guess or *expected value* for the RV?

To simplify matters by returning to a discrete RV, consider again the throw of a true die. In this case each outcome is equally likely and it is not clear what is even meant by expected value. However, consider a related question: if a die is cast N_c times, what is the expected value of the resulting sum? Well, this is clearly

$$N(1) \times 1 + N(2) \times 2 + N(3) \times 3 + N(4) \times 4 + N(5) \times 5 + N(6) \times 6 \qquad (6.13)$$

where $N(i)$ is the expected number of occurrences of the value i as an outcome. If N_c is small, say 12, statistical fluctuations will have a large effect. (Here the term *statistical fluctuations* means that a run of sixes may occur in a small set of throws and give an unduly high score. If very many throws are made, all possible occurrences are given a greater chance of happening with appropriate frequency.) However, for a true die, there is no better guess as to the numbers of each outcome than that given below. If N_c is large, then

$$N(i) \approx P(i) \times N_c \qquad (6.14)$$

and one would expect fluctuations to have a smaller effect. There will be a consequent increase in confidence in the expected value of the sum, which is clearly

$$E(\text{sum of } N_c \text{ die casts}) = \sum_{i=1}^{6} N_c P(i) i \qquad (6.15)$$

where (6.13) is rephrased and E is used to denote the expected value or *expectation*.

At a stretch of the imagination, one might consider that this expression contains a quantity independent of N_c that can be defined as the *expected value of a single cast*. If

$$E(\text{sum of } N_c \text{ die casts}) = N_c \times E(\text{single cast}) \qquad (6.16)$$

then it follows that

$$E(\text{single cast}) = \sum_{i=1}^{6} P(i) i \qquad (6.17)$$

and this value is simply a sum over the possible outcomes with each term weighted by its probability of occurrence. This definition quite happily deals with the case of a biased die $(P(i) \neq P(j), i \neq j)$ in the same way as for a true die.

In general, then, one might logically define

$$E(X) = \sum_{x_i} P(X = x_i) x_i \qquad (6.18)$$

where the RV X can take any of the discrete values x_i. For the throw of a true die, following Equation (6.17),

$$E(\text{single cast}) = \frac{1}{6} \times 1 + \frac{1}{6} \times 2 + \frac{1}{6} \times 3 + \frac{1}{6} \times 4 + \frac{1}{6} \times 5 + \frac{1}{6} \times 6 = 3.5 \qquad (6.19)$$

(This is odd semantically as the mathematically *expected* value is a value that cannot occur!) Writing the last expression as

$$E(\text{single cast}) = \frac{1 + 2 + 3 + 4 + 5 + 6}{6} \qquad (6.20)$$

the expected value in this case is the *arithmetic mean* taken over the possible values of the RV. It should be clear that this can *only* be used when all outcomes are equally likely; however, the expected value of an RV X will often be referred to as the *mean* and will be denoted $\overline{x}$.

The generalisation of Equation (6.18) to continuous RVs is conceptually straightforward. One defines

$$\overline{X} = E(X) = \int_{-\infty}^{\infty} xp(x)\mathrm{d}x \tag{6.21}$$

where $p(x)$ is the PDF for the random variable X. The integral need only be taken over the range of possible values for x; however, the limits are usually taken as $-\infty$ to ∞ as the PDF will be zero outside the allowed range. In signal processing in general and SHM in particular, it is often the case that the PDF corresponding to a particular signal is not known, in which case one would not be able to use Equation (6.21) in order to estimate the mean. A further restriction is that one normally has a finite number of samples from the distribution. In this case one assumes that the measurements one has are equiprobable and uses the discrete sum corresponding to Equation (6.20), that is the *arithmetic mean* defined by

$$\overline{X} \approx \frac{1}{n} \sum_{i=1}^{n} x_i \tag{6.22}$$

Note that the expected value of an RV need *not* be the same as the peak value of its PDF; this is clearly true for the distributions in Figures 6.3 and 6.4. The value of x for which the PDF is a maximum is termed the *mode*. Note also that a PDF can be any positive function that integrates to unity; it may have several maxima and therefore several modes. Such a PDF is termed *multimodal* in contrast to the

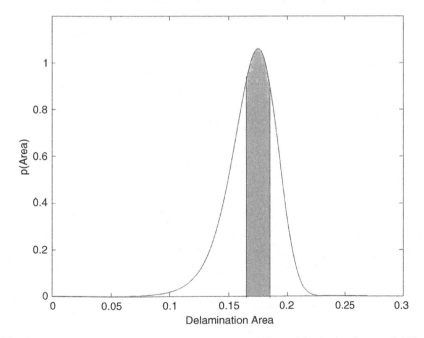

Figure 6.4   Area under the 40 m/s delamination area PDF corresponding to delaminations between 0.165 and 0.185 m².

single maximum – *unimodal* – case. The mean and the mode (in the unimodal case) are statistics that are associated with the centre of the PDF. Another central statistic that one can define is the *median*. This statistic is easiest to understand in the context of a discrete RV; to obtain the median, one simply produces an ordered list of all possible outcomes and takes the centre value. The median is important in certain respects. If one were to consider the ordered array of outcomes, one could imagine making the outer elements improbably large or small; while there would be no effect on the median, the mean of the array would be sensitive to such changes. Because of this property, the median is a more reliable estimate of the central tendency of the realisations of a RV if one allows for the fact that some of the outer elements may have been measured very wrongly. Such a statistic is called *robust*. The potentially erroneous outer elements of the array are called *outliers*.

Despite its sensitivity to outliers, the mean is arguably the most important statistic of a PDF. In the context of one of the example PDFs discussed earlier, the mean of the crack length distribution in the concrete beam would be likely to give a good summary of the health state of the beam; one can see that the mean crack length would increase as the lengths increased over the whole distribution. The other contender for the main summary statistic of a PDF is the *standard deviation*.

Consider the composite plate delamination experiments. The problem has been solved and the best estimate for the delamination area that results after the next impact is the expected value or mean. Now, when the new test is conducted and the delamination area is measured, there will certainly be some error in the estimate. So how good a guess is the mean?

Consider the two PDFs in Figure 6.5. For the first, the mean is always a good guess; for the second, the mean would often prove a bad estimate. In order to quantify this – the confidence in the guess – what is needed is the expected value $E(\varepsilon)$ of the error $\varepsilon = X - \overline{x}$:

$$E(\varepsilon) = E(X - \overline{x}) = E(X) - E(\overline{x}) \tag{6.23}$$

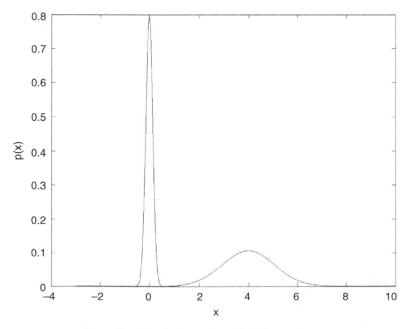

Figure 6.5 Distributions with small and large variance.

On ordinary numbers like $\bar{x}$ (as opposed to RVs), the expectation operator E has no effect as there is no uncertainty associated with a number; this means that

$$E(X) - E(\bar{x}) = \bar{x} - \bar{x} = 0 \tag{6.24}$$

Unfortunately, this result is not very useful; it arises because positive and negative errors are equally likely and average to zero.

The usual means of avoiding this problem is to consider the expected value of the error squared, that is $E(\varepsilon^2)$. This approach defines the statistic σ^2, known as the *variance*, as

$$\sigma^2 = E(\varepsilon^2) = E((X - \bar{x})^2) \tag{6.25}$$

$$E(X^2) - E(2\bar{x}X) + E(\bar{x}^2) = E(X^2) - 2\bar{x}E(X) + \bar{x}^2$$
$$E(X^2) - 2\bar{x}^2 + \bar{x}^2 = E(X^2) - \bar{x}^2 = E(X^2) - E(X)^2 \tag{6.26}$$

Equation (6.26) is often given as an alternative definition of the variance.

In the case of equally probable values for X, Equation (6.25) reduces to

$$\sigma^2 = \frac{1}{N} \sum_{i=1}^{N} (x_i - \bar{x})^2 \tag{6.27}$$

where the $x_i, i = 1, \ldots, N$ are the possible values taken by X. (This equation is the analogue of Equation (6.22) for the mean; actually a more careful analysis gives $N - 1$ in the denominator.)

In the case of X being a continuous RV, the appropriate form for the variance, corresponding to Equation (6.25) is

$$\sigma^2 = \int_{-\infty}^{\infty} (x - \bar{x})^2 p(x) \mathrm{d}x \tag{6.28}$$

The *standard deviation*, σ, of the distribution is stated more often and this parameter is simply the square root of the variance. It can be interpreted as the expected root-mean-square (RMS) error in using the mean as a guess for the value of an RV. It gives a measure of the width (or dispersion) of a distribution – the uncertainty associated with it – in the same way as the mean estimates where the centre is.

The mean and the variance are the simplest examples of *statistical moments* of the probability distribution. (The term *moments* comes from the similarity to area moments noting that the total area under the PDF is by definition equal to one.) Statistical moments come in two flavours, the basic moments,

$$m_i = E[X^i] \tag{6.29}$$

and the centred moments,

$$c_i = E[(X - \bar{x})^i] = E[(X - m_1)^i] \tag{6.30}$$

Therefore the mean of the distribution is the first moment and the variance is the second centred moment. In terms of continuous RVs,

$$m_i = \int_{-\infty}^{\infty} x^i p(x) \mathrm{d}x \tag{6.31}$$

and

$$c_i = \int_{-\infty}^{\infty} (x - \overline{x})^i p(x) \mathrm{d}x \tag{6.32}$$

As we have seen above, the lower-order moments of a given PDF give information about where the centre is and what the width of the distribution is. In fact, higher-order moments also offer information about the shape of the distribution. The two most important moments after the mean and variance are the third – the *skewness* – defined by

$$c_3 = \int_{-\infty}^{\infty} (x - \overline{x})^3 p(x) \mathrm{d}x \tag{6.33}$$

and the fourth – the *kurtosis* – defined by

$$c_4 = \int_{-\infty}^{\infty} (x - \overline{x})^4 p(x) \mathrm{d}x \tag{6.34}$$

The skewness is a measure of how asymmetric a PDF is – the distribution in Figure 6.3 is highly skewed. The kurtosis is a measure of how 'peaky' the PDF is. Both of these statistics are often used as diagnostic features in condition monitoring; however, they are usually used in a normalised form. Also in practice they are usually computed from sampled data and are therefore calculated using discrete sums. Defining the discrete or sampled version of the centred moment as

$$c_i = \frac{1}{N} \sum_{i=1}^{N} (x_i - \overline{x})^i \tag{6.35}$$

the normalised sample skewness is

$$\gamma = \frac{c_3}{c_2^{3/2}} \tag{6.36}$$

and the normalised sample kurtosis is

$$\kappa = \frac{c_4}{c_2^2} \tag{6.37}$$

It can even be shown that knowledge of all the moments of the PDF can fix the PDF form uniquely. One distribution is distinguished by the fact that its form is fixed uniquely by just its mean and variance. This distribution turns out to be distinguished for other reasons and will be singled out for discussion in the following section.

6.5   The Gaussian Distribution (and Others)

For reasons that will hopefully become clear later, the *Gaussian* or *normal* distribution is considered to be the most important of all PDFs. As discussed above, one of its many important properties is that it

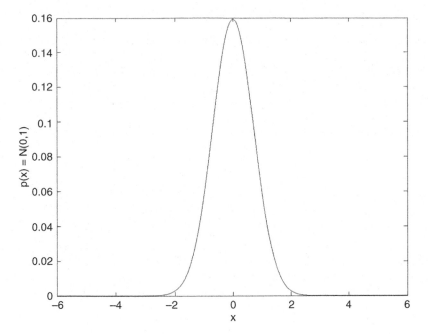

Figure 6.6   The Gaussian probability density function $N(0, 1)$.

is completely fixed by knowledge of its mean and variance. In fact the functional form of the Gaussian PDF is

$$p(x) = \frac{1}{\sqrt{2\pi\sigma^2}} \exp\left\{-\frac{1}{2}\left(\frac{x - \overline{x}}{\sigma}\right)^2\right\} \tag{6.38}$$

This equation is often denoted $N(\overline{x}, \sigma)$. As an example, $N(0, 1)$ is shown in Figure 6.6. Because of its shape, the Gaussian PDF is sometimes referred to as the *bell curve*.

One of the main reasons for the importance of the Gaussian distribution is provided by the *central limit theorem*, which states (roughly): if $X_i, i = 1, \ldots, N$ are N independent RVs, possibly with completely different distributions, then the RV X_Σ,

$$X_\Sigma = X_1 + X_2 + \cdots + X_N \tag{6.39}$$

approaches a Gaussian distribution as N tends to infinity. In practical terms one can say that adding random variables together tends to make their sum 'more Gaussian'. In signal processing, if there are many noises from different sources, it can often be assumed that the total noise is Gaussian.

Although the Gaussian distribution is the most important, one commonly encounters other distributions in engineering problems and in SHM in particular. Examples are:

1. *The Rayleigh distribution.* This distribution has a PDF given by

$$p(x) = \frac{x}{b^2} \exp\left[-\frac{1}{2}\left(\frac{x}{b}\right)^2\right] \tag{6.40}$$

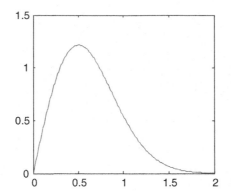

Figure 6.7   PDF for Rayleigh distribution ($b = 0.5$).

defined for $x > 0$. This distribution has a skewed form, as shown in Figure 6.7. One physical interpretation of this is that it is the distribution of distances travelled in unit time for particles whose velocities are drawn from a Gaussian distribution.

2. *The log-normal distribution.* As the name might suggest, this distribution describes the quantity $\log(X)$ when X itself has a Gaussian or normal distribution. The functional form is

$$p(x) = \frac{1}{\sqrt{2\pi}\,\sigma_{\log(x)}} \exp\left[-\frac{1}{2} \left[\frac{\log(x) - E[\log(x)]}{\sigma_{\log(x)}} \right]^2 \right] \qquad (6.41)$$

This distribution is again a skewed distribution defined for positive x and so resembles the Rayleigh distribution.

3. *The Weibull distribution.* This distribution has the functional form

$$p(x) = \frac{\beta}{\eta} \left(\frac{x}{\eta} \right)^{\beta-1} \exp\left[-\left(\frac{x}{\eta} \right)^{\beta} \right] \qquad (6.42)$$

It occurs frequently in modelling the failure of structures and components and will also appear in the discussion on extreme value statistics later in this chapter.

6.6   Multivariate Statistics

So far, the situation has been that the random variable is specified by a single number. In the case of the composite plate delamination, the engineers were trying to estimate the area of delamination that would result from the next experiment. Such a problem is called *univariate*. One might imagine an extension of such tests to impact damage location on the wing of an aircraft. Here the engineers might want to estimate the location and area of a delamination, so the problem becomes *multivariate*. In the context of SHM one is often interested in multivariate measurements. In one of the examples discussed earlier, the idea of diagnosing damage on the basis of a natural frequency measurement was raised. This example was not artificial as natural frequencies are known to be sensitive to certain types of damage; however, in practical applications, one usually measures a number of natural frequencies – say, the first five. If there is uncertainty in the measurements, this represents a multivariate random variable of five dimensions.

To simplify matters for now, consider the case of two measurement variables. One can either think of this in terms of two random variables or in terms of a random vector with two components:

$\{X\} = (X_1, X_2)$. It may be the case that the two components are independent; in the case of the aircraft wing, one might expect the contrary – there might be locations that are more susceptible to damage and in this case there may be a relation between the delamination area and location for a given projectile size and impact velocity. In this case the variables are referred to as *correlated*. The appropriate way to think about the (continuous RV) multivariate case is in terms of the joint probability distribution function $p_j(x_1, x_2)$, which is defined by saying that the probability of the variables simultaneously falling into the range $(x_1, x_1 + dx_1)$ and $(x_2, x_2 + dx_2)$ is $p_j(x_1, x_2)dx_1 dx_2$.

Where finite ranges are concerned, the appropriate result is a straightforward generalisation of Equation (6.11),

$$P(\{X\} = (x_1, x_2); \ a \leq x_1 \leq b, c \leq x_2 \leq d) = \int_a^b \int_c^d p_j(x_1, x_2)dx_1 dx_2 \tag{6.43}$$

with the result for more variables following straightforwardly. One can recover the individual or *marginal* distributions for the variables by integration as follows:

$$p(x_1) = \int_{-\infty}^{\infty} p_j(x_1, x_2)dx_2 \tag{6.44}$$

As one might imagine, if the variables x_1 and x_2 are statistically independent, then the expression in Equation (6.8) generalises to probability densities,

$$p_j(x_1, x_2) = p_1(x_1)p_2(x_2) \tag{6.45}$$

Some of the ideas of multivariate distributions are best explained in the context of a specific distribution and this is pursued in the next section.

6.7   The Multivariate Gaussian Distribution

The Gaussian is no less important in higher dimensions. However, the generalisation to random vectors $\{X\} = (X_1, \ldots, X_n)$ requires the introduction of a new statistic, the *covariance* $\sigma_{X_i X_j}$ defined by

$$\sigma_{X_i X_j} = E[(X_i - \overline{x}_i)(X_j - \overline{x}_j)] \tag{6.46}$$

which measures the degree of correlation between the random variables X_i and X_j.

Consider two independent random variables, X and Y. Then

$$\sigma_{XY} = E((X - \overline{x})(Y - \overline{y}))$$
$$= \int \int (x - \overline{x})(y - \overline{y})p_j(x, y)dxdy \tag{6.47}$$

Using the definition of independence in Equation (6.45), the joint PDF p_j factors so that

$$\sigma_{XY} = \int \int (x - \overline{x})(y - \overline{y})p_x(x)p_y(y)dxdy$$
$$= \left\{ \int (x - \overline{x})p_x(x)dx \right\} \left\{ \int (y - \overline{y})p_y(y)dy \right\} \tag{6.48}$$
$$= 0 \times 0$$

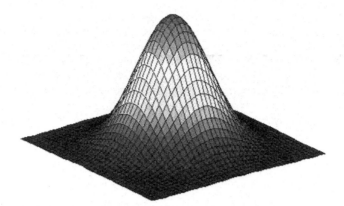

Figure 6.8   Bivariate Gaussian PDF.

So $\sigma_{XY} \neq 0$ indicates a degree of interdependence or correlation between X and Y. For a general random vector, the information is encoded in a matrix – the *covariance matrix* $[\Sigma]$ – where

$$\Sigma_{ij} = E[(X_i - \bar{x}_i)(X_j - \bar{x}_j)] \tag{6.49}$$

Note that the diagonals are the usual variances,

$$\Sigma_{ii} = \sigma_{X_i}^2 \tag{6.50}$$

As in the single-variable case, the vector of means $\{\bar{x}\}$ and the covariance matrix $[\Sigma]$ completely specify the multivariate Gaussian PDF. In fact,

$$p(\{x\}) = \frac{1}{(2\pi)^{N/2}\sqrt{|\Sigma|}} \exp\left\{-\frac{1}{2}(\{x\} - \{\bar{x}\})^{\mathrm{T}}[\Sigma]^{-1}(\{x\} - \{\bar{x}\})\right\} \tag{6.51}$$

for an N component random vector $\{X\}$. $|\Sigma|$ denotes the determinant of the matrix $[\Sigma]$.

Figure 6.8 illustrates the form of a bivariate Gaussian. In this case the covariance matrix is the unit matrix; thus contours of equal probability would be circles.

6.8   Conditional Probability and the Bayes Theorem

In general, random variables will not all be mutually independent; the probability of an event A may well depend on the previous or simultaneous occurrence of an event B. A is said to be *conditioned* on B. The need to incorporate this type of dependence into the theory results in the definition of the *conditional probability* $P(A|B)$, which is *the probability that A will occur given that B already has*. Note that the unconditional $P(A)$ can still be computed, but will have a different value.

The conditional probability is at the root of much of statistical reasoning used in SHM. Consider the situation discussed at the start of Section 6.2; the SHM expert is presented with the question: does bridge A have a 6 cm-long fatigue crack? Suppose further that the expert may or may not have the results B, of a dynamic test carried out that day. Now, we can define two events F = 'bridge A has a 6-cm fatigue crack' and T = 'the dynamic test on bridge A gave the results B today'. If the engineers are not given the results of the test, they are only in a position to estimate $P(F)$. One way to estimate this would be to recall how many bridges have been found to have 6-cm-long fatigue cracks based on previous inspections by the engineers and to calculate a frequency estimate. This approach provides an estimated probability without the benefit of all the available evidence. If, however, as in the scenario proposed earlier, the engineers are in possession of the test results and can make an estimate on the basis of how many bridges with similar test results have 6-cm-long fatigue cracks, then the quantity estimated is the conditional $P(F|T)$, that is 'the probability that bridge A has a 6-cm-long fatigue crack given that the test results B have been observed today'. This example is couched in somewhat simplistic terms; however, it is precisely this type of reasoning that is used in any data-based diagnostic system. In general, there will be a number of possible health states H_i that a given structure may be in (one of them will generally be the normal condition). The fundamental question of a statistical data-based SHM is: if a set of measurements $\{D\}$ have been observed, what are the probabilities $P(H_i|\{D\})$?

A mathematical definition of the conditional probability can be made via fairly intuitive (frequentist) reasoning. Suppose that N experiments are conducted; one can define the conditional probability $P(A|B)$ as

$$P(A|B) = \frac{N(A \cap B)}{N(B)} \tag{6.52}$$

where $N(A \cap B)$ is the number of times A occurs when B occurs and $N(B)$ is the total number of times B occurs. This equation can be rewritten as

$$P(A|B) = \frac{N(A \cap B)/N}{N(B)/N} = \frac{P(A \cap B)}{P(B)} \tag{6.53}$$

so that

$$P(A \cap B) = P(A|B)P(B) \tag{6.54}$$

where the right-hand side sort of decomposes the probability into two independent-looking events.

Interchanging variables in the defining equation gives

$$P(B|A) = \frac{P(A \cap B)}{P(A)} \tag{6.55}$$

and combining the two equations (6.53) and (6.56) produces the *Bayes theorem* (Lindley, 1980),

$$P(A|B)P(B) = P(B|A)P(A) \tag{6.56}$$

or the more usual form,

$$P(A|B) = \frac{P(B|A)P(A)}{P(B)} \tag{6.57}$$

The Bayes theorem will later prove critical in developing the machinery of statistical pattern recognition in Chapter 9.

Now suppose that A is actually conditioned on a random variable X; this will be the most frequently occurring situation in SHM. The question then concerns $P(A|X = x)$. Consider first the probability that the value taken by X lies in a small interval $[x, x + \Delta x]$. From the Bayes theorem one sees that

$$P(A|X \in [x, x + \Delta x])P(X \in [x, x + \Delta x]) = P(X \in [x, x + \Delta x]|A)P(A)$$

$$\Rightarrow P(A|X \in [x, x + \Delta x]) \int_x^{x+\Delta x} p(z)\mathrm{d}z \quad (6.58)$$

$$= \int_x^{x+\Delta x} p(z|A)\mathrm{d}z \, P(A)$$

where the conditional PDF is defined in the usual way. If Δx is small, the integrals in the last equation will be well approximated by assuming a rectangular area under the curves and

$$P(A|X \in [x, x + \Delta x])p(x)\Delta x \approx p(x|A)\Delta x \, P(A) \quad (6.59)$$

with the approximation getting better as Δx gets smaller. In fact, in the limit as $\Delta x \to 0$,

$$P(A|X = x)p(x) = p(x|A)P(A) \quad (6.60)$$

or

$$P(A|x) = \frac{p(x|A)P(A)}{p(x)} \quad (6.61)$$

in an obvious shorthand.

This equation is the basis of the Bayesian approach to pattern recognition described later in Chapter 9. $P(A)$ is called the *prior probability* of A, or the one before we have seen the data or evidence x. The term $p(x|A)$ is called the *likelihood* and $P(A|x)$ is called the *posterior probability*, or the one after we have seen the evidence. In the context of an earlier discussion $p(\{D\}|H_i)$ – the PDF value for observation of a measurement vector $\{D\}$ if a system is in a health state H_i – would be a likelihood; $P(H_i|\{D\})$ – the probability that a system is in a health state H_i given that measurements $\{D\}$ have been observed – is a posterior probability. In the latter case, the corresponding prior probability would just be $P(H_i)$.

If a set of events B_i are mutually exclusive and cover all eventualities, then clearly

$$P(A) = \sum_i P(A \cup B_i) \quad (6.62)$$

so that, using various results from above, one finds

$$P(A) = \sum_i P(A|B_i)P(B_i) \quad (6.63)$$

Finally, if A is conditioned on a continuous random variable X, the appropriate generalisation of the last equation is

$$P(A) = \int P(A|X = x)p(x)\mathrm{d}x \quad (6.64)$$

or if A is conditioned on a random vector $\{X\}$,

$$P(A) = \int \cdots \int P(A|\{X\} = \{x\})p(\{x\})\mathrm{d}\{x\} \quad (6.65)$$

where $\mathrm{d}\{x\}$ is an obvious shorthand for $\prod_{i=1}^{n} \mathrm{d}x_i$. The latter result is sometimes called the *total probability theorem*; a proof can be found in Papoulis and Pillai (2001).

Finally, it is important to mention that the Reverend Bayes is credited with initiating a new philosophy for probability (although the roots date back to Laplace). This philosophy is the one mentioned earlier in which one interprets probabilities as *degrees of belief*. As mentioned earlier, it is possible to build a logical and rigorous framework for probability based on subjective grounds and in this framework one assigns probabilities to events roughly by assessing what one would bet on the event occurring (Lindley, 1980). Reinforcing what was said earlier, one might argue that the Bayesian interpretation in terms of degrees of belief is precisely what is needed for SHM; if one were asked to estimate the probability that a given bridge has a certain level of damage present, it is clear that frequentist probability cannot give an answer as the bridge is unique. In the absence of any data, one would make an empirical estimate of the probability of damage $P(D)$ based perhaps on engineering experience; this is the *prior* probability. As measurements $\{x\}$ become available, the Bayes theorem allows one to update to more refined conditional probabilities $P(D|\{x\})$ that take the evidence into account.

6.9 Confidence Limits and Cumulative Distribution Functions

Anticipating some of the later developments in the book, let us consider the following situation. Suppose one has made many measurements of some quantity from a structure, say the first natural frequency f. Further suppose that one has been able to use this information to estimate the PDF $p(f)$ for this natural frequency given that the structure is undamaged. Now, during a monitoring phase for the structure, measurement of the natural frequency gives a value f'. Next, suppose that the engineer concerned has observed that f' is higher than the previously measured values and is worried that this indicates that the structure is damaged. How does one proceed? As mentioned earlier, one is well aware that the frequency will be subject to fluctuations due to variations in the measurement conditions and the environment of the structure, so one must allow that a given measured value may sometimes be higher than at other times. The question is: is the measured frequency *so high* that one believes the distribution $p(f)$ can no longer apply? In this case, one may infer that the structure is no longer in a normal condition, because it is no longer consistent with the statistical description of the normal condition. One might consider that f' is so far from the mean of the normal condition distribution that it is inconsistent or one might observe that $p(f')$ is so small that the value f' would never really arise in the context of an undamaged system. These questions are naturally answered in the framework of novelty detection and will be discussed in a great deal of detail in Chapter 10.

For now, it is useful to look at this problem in the context of a specific distribution. Suppose one has a specified Gaussian distribution for a variable with mean $\bar{x}$ and standard deviation σ and one is presented with a measurement x. What is the probability that the measurement belongs to the distribution or alternatively what is our *confidence* that it belongs?

The question becomes: what is the probability that $X \geq x$ within the given distribution? This amounts in this particular case to finding the area under the Gaussian PDF curve up to x, which can be evaluated as

$$P(X \geq x) = \frac{1}{\sqrt{2\pi}\sigma} \int_x^{\infty} \exp\left\{-\frac{1}{2}\left(\frac{x - \bar{x}}{\sigma}\right)^2\right\} \mathrm{d}x \tag{6.66}$$

or

$$P(X \geq x) = 1 - \frac{1}{\sqrt{2\pi}\sigma} \int_{-\infty}^{x} \exp\left\{-\frac{1}{2}\left(\frac{x - \bar{x}}{\sigma}\right)^2\right\} \mathrm{d}x \tag{6.67}$$

It is convenient at this point to change to the *standardised* variable z, where

$$z = \frac{x - \overline{x}}{\sigma} \tag{6.68}$$

so that z has the distribution $N(0, 1)$; under these circumstances, the integral in Equation (6.67) becomes

$$P(Z \geq z) = 1 - \frac{1}{\sqrt{2\pi}} \int_{-\infty}^{z} \exp\left(-\frac{1}{2}z^2\right) dz = 1 - \Phi(z) \tag{6.69}$$

where $\Phi(z)$ is called the *cumulative distribution function* (CDF) for $N(0, 1)$. The cumulative distribution has not been discussed until now because it has not been needed, but it has a simple definition in general. For any continuous random variable x, the cumulative distribution $F(x')$ is simply that probability that $X = x \leq x'$. It can be shown that the cumulative distribution is related to the PDF for $x - p(x)$ through the relationship

$$p(x) = \frac{dF(x)}{dx} \tag{6.70}$$

Figure 6.9 shows a bimodal PDF and its associated PDF; it is clear from this picture that the PDF is the derivative of the CDF. Finally on this matter, the CDF for the Gaussian distribution is so important that it is given its own specific symbol Φ.

Returning to Equation (6.68), one observes that the units of the standardised variable z are *standard deviations*; so that equation tells us *the probability that a measurement can be z standard deviations above the mean of the candidate distribution*. A more usual question is: *what is the probability that*

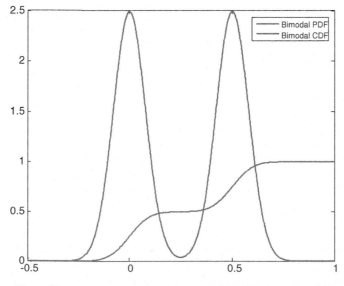

Figure 6.9   A general (in this case) bimodal PDF and its associated CDF.

a measurement can be z standard deviations away from the mean in either direction? In general, this question is answered by

$$P(Z \geq |z| \text{ or } Z \leq -|z|) = 1 - \frac{1}{\sqrt{2\pi}} \int_{-|z|}^{|z|} \exp\left(-\frac{1}{2}z^2\right) dz \tag{6.71}$$

For example if $z = 1.96$, then the area under the curve between the limits $-z$ and z is 0.95, so the probability that a measurement $z = 1.96$ can arise from $N(0, 1)$ is only 0.05 or 5%. Conversely, the 95% *confidence interval* for measurements is $[-1.96, 1.96]$ in the standardised variable z or $[\overline{x} - 1.96\sigma, \overline{x} + 1.96\sigma]$ in the original variable x.

The theory developed here allows one to give a reasoned answer to the question posed at the beginning of the section. Suppose that a value f' for the first natural frequency of a structure has been measured and it is known that the PDF for the first natural frequency under normal (undamaged) conditions is $p(f)$; furthermore, it has been previously established that this PDF is Gaussian with mean $\overline{f}$ and standard deviation σ_f. If it can be seen that if $f' > \overline{f} + 1.96\sigma_f$, this condition is known to only arise 5% of the time naturally (*by chance*) and one may wish to conclude that the measurement is no longer consistent with the distribution $p(f)$; that is the system is no longer in a normal condition. This assertion still leaves considerable room for doubt; however, a more stringent criterion can be defined where damage is inferred only if $f' > \overline{f} + 3\sigma_f$; then the properties of the Gaussian distribution dictate that this result could occur by chance only in 3 from 1000 measurements.

In the usual terminology of probability and statistics, one would say that a measurement that is not consistent with the usual distribution for that measurement is an *outlier*. Alternatively, one might say that the measurement is *discordant* with the known distribution or PDF. The identification of outliers is an important branch of probability and statistics in its own right; it is also very useful in the context of SHM and will be discussed in more detail in the next section.

Before ending the discussion of the CDF, one can make an interesting and important observation; in many ways the CDF is regarded in probability theory as a more fundamental quantity than the PDF. This observation is reflected in the fact that the CDF for a set of data is actually easier to visualise than the PDF. If one wishes to plot the PDF corresponding to a set of sampled data, one has to estimate the density first (various means of doing this are discussed in Section 6.11 later – a histogram is the most basic means). In contrast, an approximation to the CDF can be plotted directly from the sampled data; this is called the *empirical CDF*. Suppose that one has N points sampled from the distribution of interest $x_i; 1 = 1, \ldots, N$; the empirical CDF can be constructed and plotted as follows:

1. Order the x_i into ascending order to give an array $x_{oi}; 1 = 1, \ldots, N$.
2. Assign to each x_{oi}, a value $y_{oi} = (i - 1)/(N - 1)$.
3. Plot the graph generated by the pairs (x_{oi}, y_{oi}).

Figure 6.10 illustrates this process for data generated from the standard Gaussian distribution $N(0, 1)$. Figure 6.10(a) shows the empirical CDF constructed from a sample of 100 points, while Figure 6.10(b) shows the CDF generated by 10 000 points; in general, the more data points that are available, the smoother will be the empirical CDF.

In a way, the empirical CDF brings up another important subject – testing to see if data come from a Gaussian distribution. Because the Gaussian distribution is so fundamental in probability and statistics, if one believes that a given set of measurements is normally distributed, one immediately has the advantage of a large body of theory in support of further analysis. However, it is important to make sure that an assumption of Gaussianity is consistent with the evidence. There are many tests for Gaussianity, examples being the Wilks–Shapiro test, the Jarque–Bera test and the Kolmogorov–Smirnov test (Surhone, Timpleton and Marseken, 2010). An appealing (because simple) visual approach to testing for normality is based on plotting the empirical CDF on appropriately transformed axes. The reader will

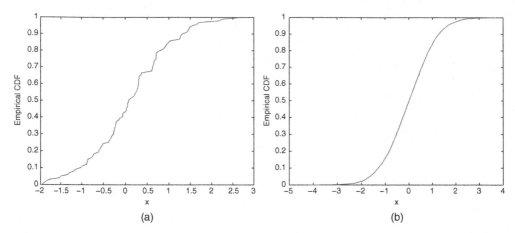

Figure 6.10 Empirical CDFs constructed from samples of normally distributed data: (a) 100 samples, (b) 10 000 samples.

be familiar with this idea in another context: when one wishes to fit an exponential curve $y = \exp(ax)$ to data, one simply plots the quantity $\log(y)$ against x. If the original relationship is truly exponential, the transformed result will be a straight-line graph. In fact, graph paper with logarithmic scales is commonly available to save the analyst the effort of computing the new abscissae. The situation is the same in the probabilistic context; Gaussian *plotting paper* is available so that the empirical CDF of the given data will appear as a straight line if the data are distributed normally. Rather than dwell on this here, the discussion of this topic will be postponed until Section 6.12, when the subject of plotting papers in the context of extreme value statistics is addressed.

6.10 Outlier Analysis

Outlier analysis is a thriving area of probability and statistics and there are many sophisticated means of trying to establish if a given observation is an outlier. Only the briefest description of the basic concepts will be given here; the reader can consult the classic monograph by Barnett and Lewis (1994) for more details.

6.10.1 Outliers in Univariate Data

As discussed earlier in the context of the mean and the median, a *discordant outlier* in a data set is an observation that is surprisingly different from the rest of the data and therefore is believed to be generated by an alternate mechanism to the other data. A discussion was provided in the last section describing why outlier detection is an important concept for SHM; the question posed in the previous section can be rephrased as: is the measured natural frequency f' a discordant outlier of the distribution $p(f)$? The *discordancy* of the candidate outlier is a measure that may be compared against some objective criterion allowing the outlier to be judged to be statistically likely or unlikely to have come from the assumed generating process.

The case of outlier detection in univariate data is relatively straightforward in that outliers must 'stick out' from one end or other of the data set. There are numerous discordancy tests but one of the most common, and the one whose extension to multivariate data will be employed later, is based on so-called *deviation statistics* and is given by

$$z_\zeta = \frac{|x_\zeta - \overline{x}|}{s} \tag{6.72}$$

where x_ζ is the potential outlier and $\overline{x}$ and s are the mean and standard deviation of the sample respectively. A point of notation arises here. Sometimes it is desirable to distinguish between the 'true' statistics of a variable and those estimated from actual observations: the latter are termed *sample statistics*. In the case of the standard deviation, one would use σ for the true statistic and s for the corresponding sample estimate. In the multivariate case, the corresponding symbols for the covariance matrix would be $[\Sigma]$ and $[S]$. In the case of Equation (6.72) the necessary statistics may be calculated with or without the potential outlier in the sample depending upon whether *inclusive* or *exclusive* measures are preferred. This discordancy value is then compared to some threshold value and the observation declared, or not, to be an outlier. Note that the principle introduced in the last section is being followed, that is asking how many standard deviations away from the mean is the measurement and then asking if that is likely.

6.10.2   Outliers in Multivariate Data

A multivariate data set consisting of n observations in p variables may be represented as n points in a p-dimensional object space. It becomes clear that detection of outliers in multivariate data is much more difficult than the univariate situation due to the potential outlier having more 'room to hide'.

The discordancy test which is the multivariate equivalent of Equation (6.72) is the *Mahalanobis squared-distance* measure given by

$$D_\zeta = (\{x\}_\zeta - \{\overline{x}\})^{\mathrm{T}}[S]^{-1}(\{x\}_\zeta - \{\overline{x}\}) \tag{6.73}$$

where $\{x\}_\zeta$ is the potential outlier, $\{\overline{x}\}$ is the mean of the sample observations (often called the training data in a pattern recognition context) and $[S]$ is the sample covariance matrix.

As with the univariate discordancy test, the mean and covariance may be inclusive or exclusive measures. In many practical situations the outlier is not known beforehand and so the test would necessarily be conducted inclusively. In condition monitoring or SHM, the potential outlier is always known beforehand and so it is more sensible to calculate a value for the Mahalanobis squared-distance without this observation 'contaminating' the statistics of the normal data. Whichever method is used, the Mahalanobis squared-distance of the potential outlier is checked against the threshold value, as in the univariate case, and its status determined.

6.10.3   Calculation of Critical Values of Discordancy or Thresholds

In order to label an observation as an outlier or an inlier there needs to be some threshold value against which the discordancy value can be compared. This value is dependent on both the number of observations and the number of dimensions of the problem being studied. In the univariate case, one might assume that the generating distribution for the normal condition is Gaussian and then compute the standard confidence intervals, where in only 5% of cases will a measurement be more than 1.96 standard deviations from the mean. In the multivariate case, the assumption of a Gaussian distribution again allows calculations of the thresholds in terms of a chi-squared-statistic (which is an important statistic in many contexts, but it is not used in this book; the curious reader can consult Grimmet and Stirzaker, 2001); however, it is sometimes useful to have a numerical method for generating the threshold.

A Monte Carlo method can be used to arrive at the threshold value. The procedure for this method is to construct an $n \times p$ (number of observations $\times$ number of dimensions) matrix with each element being a randomly generated number from a zero mean and unity standard deviation normal distribution. The Mahalanobis squared-distance is calculated for all the rows (which correspond to randomly generated feature vectors) where $\{\bar{x}\}$ and $[S]$ are inclusive measures and the largest value is stored. This process is repeated for at least 1000 trials whereupon the array containing all the largest Mahalanobis squared-distances is then ordered in terms of magnitude. The critical values for 5% and 1% tests of discordancy for a p-dimensional sample of n observations are then given by the Mahalanobis squared-distances in the array above which 5% and 1% of the trials occur.

The inclusive threshold and exclusive threshold are then related by a simple formula:

$$T_{\text{exc}} = \frac{(n-1)(n+1)^2 T_{\text{inc}}}{n(n^2 - (n+1)T_{\text{inc}})} \tag{6.74}$$

with an obvious notation. The use of outlier analysis for damage detection in the context of unsupervised learning will be covered in detail in Chapter 10.

6.11   Density Estimation

Again, consider the hypothesised SHM problem. At the beginning of Section 6.9, it was assumed that the SHM expert had been able to infer the density function $p(f)$ for the first natural frequency from a serious of previous measurements. In fact this density estimation is quite a difficult problem, one of the most difficult in the field of machine learning or learning from data. Density estimation problems naturally fall into two classes – *parametric* and *nonparametric*. Common to both classes is the need for a set of measured data for the RVs of interest; in the context of machine learning or pattern recognition, such data are referred to as *training data*. For simplicity, the assumption is made that there is a single variable of interest x so that the PDF is a function of this single variable $p(x)$. In this case, the training data are denoted as a set of N observations x_i; $i = 1, \ldots, N$. Despite the difficulty of the problem there are numerous established methods of estimating densities for univariate and multivariate data and many of these can be found in Silverman (1986) or Scott (1992).

The simplest approach to density estimation is the parametric approach; however, this approach presupposes that one knows which probability distribution the RV of interest belongs to. In that case, the only unknown quantities are the parameters of the distribution and these are estimated from the training data. Suppose that the data of interest are known to have a Gaussian distribution; in that case, the PDF has the functional form defined by Equation (6.38) and one needs only to know the appropriate mean and standard deviation in order to characterise the distribution fully. The required statistics can be estimated from the standard sample mean and variance for the training data:

$$\bar{x} = \frac{1}{N} \sum_{i=1}^{N} x_i \tag{6.75}$$

and

$$s = \frac{1}{N} \sum_{i=1}^{N} (x_i - \bar{x})^2 \tag{6.76}$$

In general, for distributions other than the Gaussian, estimation of the parameters from the training data will require distribution-specific formulae; these formulae can be derived by various principled methods.

One of the most popular approaches to generating parametric estimates is the *maximum likelihood* method, which will be discussed in more detail in Chapter 9.

Unfortunately, one is rarely in possession of knowledge as to the correct parametric form for the density of interest and one has to adopt nonparametric methods. As the name might suggest, these are general methods that make no assumptions as to the functional form of the distribution/density of interest.

The simplest nonparametric approach to extracting a PDF from data is one that most readers will be familiar with; it is simply to construct a *histogram*. As before, in order to illustrate this practice, it is useful to restrict the discussion to univariate data. The data used for the illustration here will be the Old Faithful geyser eruption data given in Silverman (1986); it is chosen as an illustration because it is bimodal and a naïve parametric approach based on a Gaussian assumption would fail miserably. The geyser data are composed of 107 observations of the lengths of eruptions.

Histograms are constructed by choosing an origin x_0 and a *bin width* h for the data. The *bins* are simply the intervals $[x_0 + mh, x_0 + (m + 1)h]$. The histogram is then defined by

$$\hat{p}(x) = \frac{1}{nh}(\text{number of points in same bin as } x) \tag{6.77}$$

where the caret denotes the estimate of the true density $p(x)$. Note that with all nonparametric approaches there are variables such as the bin width that one must choose and, as such, there is no unique density function that will be obtained from a given set of data. Such variables are often referred to as *hyperparameters* in the machine learning literature; some issues with hyperparameters will be discussed in Chapters 10 and 11.

A histogram constructed on this basis for the geyser data is given in Figure 6.11. For plotting purposes, each $\hat{p}(x)$ bar has been placed at the centre of the bin. The histogram immediately gives some very useful information. The distribution is clearly bimodal (i.e. has two peaks) and a coarse description of relative probabilities can be given. However, from the viewpoint of SHM or condition monitoring there is an

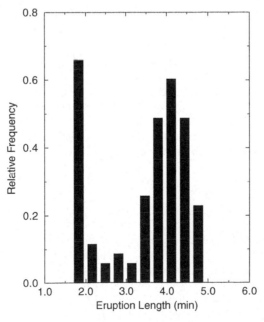

Figure 6.11   Histogram for Old Faithful geyser data.

immediate problem. If one is provided with a new x value, it is not possible to assign a probability to it unless it falls in one of the bins. This means that the only diagnostic that one can construct says that the data is 'normal' if it falls in a bin and 'novel' otherwise. This description of the situation is too coarse. Another problem with the histogram is that it is discontinuous at the bin boundaries.

A more principled approach to nonparametric density estimation is the so-called *kernel* method that overcomes both of these problems in a simple manner. The basic idea of the approach is that each point in the training data set contributes an 'atom' of probability density to the estimate. If the atoms are smooth functions, the overall density will be smooth; if the atoms are also defined away from the data points, the overall PDF will be defined away from the data also. It is clear that density will accumulate in regions where data are plentiful as required. In mathematical terms, the basic form of the estimate for general multivariate data is

$$\hat{p}(\{x\}) = \frac{1}{Nh} \sum_{i=1}^{N} K\left(\frac{\{x\} - \{x\}_i}{h}\right) \tag{6.78}$$

where $\{x\}_i$ is the ith data point, N is the number of points in the training set and h is the smoothing parameter that controls the width of the atoms; $\hat{p}(\{x\})$ is the estimate of the true density $p(\{x\})$. The *kernel function* $K(\{x\})$ can be any localised function satisfying the constraints $K(\{x\}) > 0$ and

$$\int_{-\infty}^{\infty} K(\{x\})\mathrm{d}\{x\} = 1 \tag{6.79}$$

Equation (6.78) ensures the necessary conditions on $\hat{p}(\{z\})$ that $\hat{p}(\{z\}) > 0$ and

$$\int_{-\infty}^{\infty} \hat{p}(\{z\})\mathrm{d}\{z\} = 1 \tag{6.80}$$

The most common choice of kernel function – and the one adopted for the illustrations here – is the multivariate Gaussian:

$$K(\{x\}) = \frac{1}{(2\pi)^{d/2}} \exp\left(-\frac{1}{2}||\{x\}||^2\right) \tag{6.81}$$

where d is the dimension of the data space. This function is analytic (essentially infinitely differentiable) and therefore defines PDFs that are analytic (actually, this analytic property is not necessary; the histogram can also be cast in a kernel form, but the 'atoms' of the decomposition are not differentiable; see Silverman, 1986). Various other kernel functions are given in Silverman (1986).

It will also prove useful to have a symmetric kernel, that is such that

$$K(\{x\}) = K(-\{x\}) \tag{6.82}$$

Once the estimate $\hat{p}(\{z\})$ is established, the PDF values at any new measurement points are trivially evaluated.

The kernel density estimation method is illustrated in Figure 6.12, which shows the density estimate for the Old Faithful data, obtained with a smoothing parameter of $h = 0.25$.

The quality of the estimate depends critically on two factors. The first factor is the size of the training set; discussions can be found in Silverman (1986) and Scott (1992), and the issue will be raised here a little later. The other factor of importance is the value of h. If h is too small, the estimated PDF will contain a lot of spurious local structure. If h is too large, the estimate will be oversmoothed and its decay rate with $\{x\}$ will be underestimated. The immediate effect of the latter issue in the context of SHM is

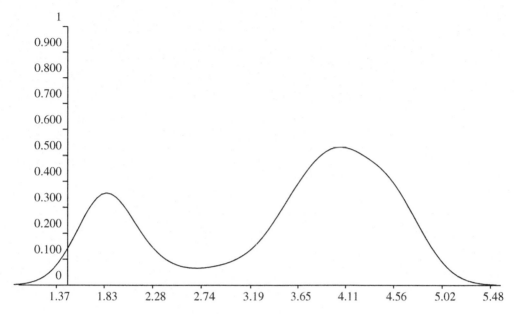

Figure 6.12   Kernel density estimate for Old Faithful geyser data.

that the normal condition set would appear to be larger than it is and the damage diagnostic may therefore suffer from false negatives. As one might expect, there is an optimum value for h and there are a number of ways of estimating it (Silverman, 1986).

The simplest method for optimising h and the one adopted for the illustrations here is *least-squares cross-validation* (Bowman, 1984). This method seeks to minimise the squared error between the density estimate and the true density:

$$J[\hat{p}] = \int [p(\{x\}) - \hat{p}(\{x\})]^2 d\{x\} \tag{6.83}$$

The true density $p(\{x\})$ is of course unknown a priori, so the equation above is somewhat limited in its usefulness. However, it can be shown under certain conditions (Silverman, 1986) that the value of h that minimises the quantity

$$M_1(h) - \frac{1}{n^2 h^d} \sum_i \sum_j K^*(\frac{\{Z\}_i - \{Z\}_j}{h}) + \frac{2}{n h^d} K(\{0\}) \tag{6.84}$$

also minimises the mean-squared error between $p(\{z\})$ and $\hat{p}(\{z\})$, where

$$K^*(\{x\}) = K^{(2)}(\{x\}) - 2K(\{0\}) \tag{6.85}$$

and

$$K^{(2)}(\{x\}) = \int K(\{z\} - \{x\}) K(\{z\}) d\{z\} \tag{6.86}$$

For the illustrations here, the minimisation was carried out using a simple quadratic-fit Newton-type method (Leuenberger, 1989). If the kernel is assumed symmetric as in Equation (6.81), one can halve the work by using the expression

$$M_1(h) = \frac{2}{n^2 h^d} \sum_i \sum_{j<i} K^*(\frac{\{Z\}_i - \{Z\}_j}{h}) + \frac{1}{nh^d} K^{(2)}(\{0\}) \qquad (6.87)$$

Any minimisation routine requires an initial estimate and the quality of the final result may depend critically on it; the analysis for the illustrations here followed the guidelines in Silverman (1986) and used the following procedure. An initial smoothing parameter is chosen, h^*; for a univariate distribution, this is

$$h^* = 0.9 A n^{-1/5} \qquad (6.88)$$

where

$$A = \min\left(\text{standard deviation}, \frac{\text{interquartile range}}{1.34}\right) \qquad (6.89)$$

where the interquartile range is a width parameter like the standard deviation. To compute it, one divides the PDF into four regions of equal area. The range encompassed by the two regions on either side of the mean is the interquartile range.

For a multivariate distribution, the appropriate quantity is

$$h^* = A n^{1/(d+4)} \qquad (6.90)$$

where

$$A = \begin{cases} 1, & d = 2 \\ \left(\frac{4}{d+2}\right)^{1/(d+4)}, & \text{otherwise} \end{cases} \qquad (6.91)$$

These estimates depend on the assumption that the true distribution is univariate or multivariate Gaussian and so the full cross-validation is needed for general distributions. Once h^* is established, the strategy for minimisation used here assumes a search interval $[h^*/4, 2h^*]$ and carries out a minimisation of $M_1(h)$ over this interval. This method proceeds by carrying out a coarse search over a mesh of 100 points in order to bracket the minimum, and then minimises $M_1(h)$ using the quadratic fit procedure as described above.

To illustrate the optimisation procedure, Figure 6.13 shows the kernel density estimate for a univariate standard Gaussian with the h value that minimised the cross-validation score in Equation (6.87) (the solid curve shows the reference Gaussian). In this procedure 1000 data points were used and the optimum value of the smoothing parameter was found to be $h = 0.314\,825$. For comparison, note that Equation (6.88) applies directly as the true distribution is Gaussian. This estimate of h^* is 0.220 575, and a comparison of the estimated distribution with the true distribution is given in Figure 6.14.

The formula estimate gives a mean-square error of 0.0319, while cross-validation gives 0.0324. The agreement is good.

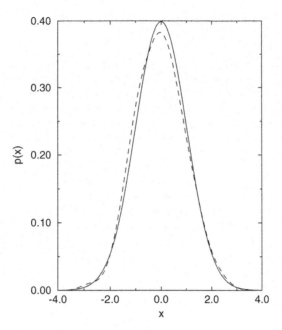

Figure 6.13 Density estimate for univariate Gaussian using least-squares cross-validation to estimate the smoothing parameter.

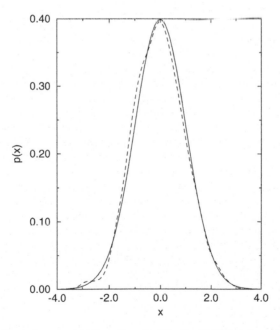

Figure 6.14 Density estimate for univariate Gaussian using Equation (6.88) to estimate the smoothing parameter.

6.12   Extreme Value Statistics

6.12.1   Introduction

Already in this chapter statistical methods have been discussed that are appropriate for SHM applications. Notably methods have been established for deciding if a measured data point is consistent with the normal condition of the structure. In general, one will infer damage if the new measurement is far from the mean of the undamaged system distribution or in a region of very low probability density. In both these cases, the implication is that the new measurement point will be somewhere in the tails of the distribution. This location means that our diagnostic, and in particular the confidence threshold associated with it, will be sensitive to how well we understand the distribution tails. The only means of determining the threshold that has been discussed so far is based on assuming that the normal condition distribution is Gaussian, which involves a gross assumption on the nature of the tails that may not be justified. There is a clear benefit here in exploring more refined methods of modelling the tails of distributions.

In fact, there is a large body of statistical theory that is explicitly concerned with modelling the tails of distributions and these statistical procedures can be applied in the context of SHM. The relevant field is referred to as *extreme value statistics* (EVS), a branch of *order statistics*. There are many excellent textbooks and monographs in this field. Some are considered classics (Gumbel, 1958; Galambos, 1978) and others are more recent (Embrechts, Kluppelberg and Mikosch, 1997; Kotz and Nadarajah, 2000; Reiss and Thomas, 2001). Castillo (1988) is notable in its concern with engineering problems in fields like meteorology, hydrology, ocean engineering, pollution studies, strength of materials and so on. Roberts (1998, 2000) introduced the ideas of EVS into novelty detection and applied them in the biosignal processing context.

6.12.2   Basic Theory

The Gaussian distribution occupies its central place in statistics for a number of reasons; not least is the central limit theorem that was previously discussed in Section 6.5. Although this theory is arguably the most important limiting theorem in statistics, it is not the only one. If the problem at hand is concerned with the tails of distributions, there is another theorem that is more appropriate.

Suppose that one is given a vector of samples $\{X_1, X_2, \ldots, X_n\}$ from an arbitrary *parent distribution*. The most relevant statistic for studying the tails of the parent distribution is the maximum operator, $\max(\{X_1, X_2, \ldots, X_n\})$, which selects the maximum value from the sample vector. Note that this statistic is relevant for the right tail of a univariate distribution only. For the left tail, the minimum should be used. The pivotal theorem of EVS states that in the limit as the number of vector samples tends to infinity, the induced distribution on the maxima of the samples can only take one of three forms: *Gumbel, Weibull* or *Frechet* (Fisher and Tippett, 1928). The rest of this section will be concerned with elaborating on this fact.

Before developing the EVS theory, it will be necessary to introduce a new probability distribution based on the idea of a *Bernoulli trial*. A Bernoulli trial or experiment is simply one that has two possible outcomes; for the sake of simplicity these outcomes are called *success* and *failure* for now. The probability of success is defined as p and because there are only two possible outcomes, this fixes the probability of failure as $q = 1 - p$. Suppose that the experiment is to toss a biased (i.e. $p \neq q$) coin N times and count the number of heads H (with tails denoted T). If the probability of getting a head on a single toss is p, then a given sequence of results $HTHHT \cdots TH$ will have probability

$$P(HTHHT \cdots TH) = p^{n_h}(1 - p)^{N - n_h}$$

where n_h is the number of heads and so on (This equation assumes that all tosses are statistically independent.) Allowing for permutations of the symbols H and T, the probability that a number of heads n_h will occur is

$$P(n_h) = \frac{N!}{n_h!(N - n_h)!} p^{n_h}(1 - p)^{N - n_h} \tag{6.92}$$

and this distribution is the Bernoulli distribution, a discrete probability distribution. If one sums over all possible n_h, one obtains

$$P(\text{`any result'}) = \sum_{n_h=1}^{N} \frac{N!}{n_h!(N-n_h)!} p^{n_h}(1-p)^{N-n_h} = (p+q)^N = 1$$

by using the binomial theorem that is familiar from elementary mathematics.

We return to the discussion of extreme values now. If the values of the general sequence of measurements $X_1, X_2, \ldots, X_n$ are arranged in ascending order $X_{1:n}, X_{2:n}, \ldots, X_{n:n}$, the rth element of this sequence $X_{r:n}$ is called the *r*th-*order statistic*. In order statistics it is customary to include the total sample size, n, in the notation. The basic question that now arises is what are the distributions of the order statistics, in particular the minimum, $X_{1:n}$, and the maximum, $X_{n:n}$.

Following Castillo (1988), let $m_n(x)$ be the number of samples for which $X_{r:n} \leq x$. Each time one chooses a value $X_{r:n}$ from the sample, one is essentially conducting a *Bernoulli trial*, with the two possible outcomes $X_{r:n} \leq x$ or $X_{r:n} > x$. If success is taken here to mean the first outcome, then the probability of success p will be given by $F(x)$, the CDF for $X_{r:n} \leq x$, and the failure probability q will be fixed by the complementary probability, $(1 - F(x))$, that $X_{r:n} > x$. The CDF of $m_n(x)$ is therefore a binomial distribution with $F^k(k)$ denoting the probability of k successes; this means that

$$F_{m_n(x)}(r) = \text{Prob}[m_n(x) \leq r] = \sum_{k=0}^{r} \binom{n}{k} F^k(x)[1 - F(x)]^{n-k} \tag{6.93}$$

Now, because the event $\{X_{r:n} \leq x\}$ is basically the same as the event $\{m_n(x) \geq r\}$ then $\text{Prob}[X_{r:n} \leq x] = \text{Prob}[m_n(x) \geq r] = 1 - \text{Prob}[m_n(x) < r]$ and $F_{X_{r:n}}(x) = 1 - F_{m_n(x)}(r - 1)$ or

$$F_{X_{r:n}}(x) = \text{Prob}[X_{r:n} \leq x] = \sum_{k=r}^{n} \binom{n}{k} F^k(x)[1 - F(x)]^{n-k} \tag{6.94}$$

If one is concerned with the maximum of the sample, the relevant order statistic is $X_{n:n}$ and the relevant distribution is

$$F_{X_{n:n}}(x) = F^n(x) \tag{6.95}$$

If one is concerned with the minimum of the sample, the relevant order statistic is $X_{1:n}$ and the appropriate distribution is

$$F_{X_{1:n}}(x) = 1 - [1 - F(x)]^n \tag{6.96}$$

Concentrating now on the maximum, consider what happens in the limit $n \to \infty$, where the limit distribution for the maximum will satisfy

$$\lim_{n \to \infty} F^n(x) = \begin{cases} 1, & \text{if } F(x) = 1 \\ 0, & \text{if } F(x) < 1 \end{cases} \tag{6.97}$$

Unfortunately, as it stands this distribution does not make sense because a CDF is developed on the assumption that it is continuous, but here the limit is discontinuous. The way around this discontinuity is to normalise the independent variable with a sequence of constants ($x \to a_n + b_n x$) in such a way that

$$\lim_{n \to \infty} F^n(a_n + b_n x) = H(x) \tag{6.98}$$

where $H(x)$ is a nondegenerate limit function. In fact, it is required that $H(x)$ be continuous. The situation for minima is similar. A sequence of normalisations is required such that

$$\lim_{n \to \infty} 1 - [1 - F(c_n + d_n x)]^n = L(x) \tag{6.99}$$

and $L(x)$ is a nondegenerate continuous limit function. Equations (6.97) and (6.98) are technical requirements needed for the theory to work; later it will be seen that in practical applications any complexities arising from them are not encountered.

The fundamental theorem of EVS (Fisher and Tippett, 1928) can now be stated.

Theorem 6.1. Feasible Limit Distributions for Maxima

The only three types of nondegenerate distributions $H(x)$ satisfying Equation (6.97) are

$$\text{FRECHET:} \quad H_{1,\beta}(x) = \begin{cases} \exp\left[-(\frac{\delta}{x - \lambda})^\beta\right], & \text{if } x \geq \lambda \\ 0, & \text{otherwise} \end{cases} \tag{6.100}$$

$$\text{WEIBULL:} \quad H_{2,\beta}(x) = \begin{cases} 1, & \text{if } x \geq \lambda \\ \exp\left[-(\frac{\lambda - x}{\delta})^\beta\right], & \text{otherwise} \end{cases} \tag{6.101}$$

$$\text{GUMBEL:} \quad H_{3,0}(x) = \exp\left[-\exp(-\frac{x - \lambda}{\delta})\right], \quad -\infty < x < \infty \text{ and } \delta > 0 \tag{6.102}$$

or in the appropriate form for minima.

Theorem 6.2. Feasible Limit Distributions for Minima

The only three types of nondegenerate distributions $L(x)$ satisfying Equation (6.99) are

$$\text{FRECHET:} \quad L_{1,\beta}(x) = \begin{cases} 1 - \exp\left[-(\frac{\delta}{\lambda - x})^\beta\right], & \text{if } x \leq \lambda \\ 1, & \text{otherwise} \end{cases} \tag{6.103}$$

$$\text{WEIBULL:} \quad L_{2,\beta}(x) = \begin{cases} 0, & x \leq \lambda \\ 1 - \exp\left[-(\frac{x - \lambda}{\delta})^\beta\right], & x > \lambda \end{cases} \tag{6.104}$$

$$\text{GUMBEL:} \quad L_{3,0}(x) = 1 - \exp\left[-\exp(\frac{x - \lambda}{\delta})\right], \quad -\infty < x < \infty \text{ and } \delta > 0 \tag{6.105}$$

where λ, δ and β are model parameters that are estimated from the data.

Now, given samples of maximum data from a number of n-point populations, it is possible to select an appropriate limit distribution and fit a parametric model to the data. It is also possible to fit models to portions of the parent distribution's tails as these models are equivalent in the tail to the appropriate extreme value distribution. Once the parametric model is obtained, it can be used to compute effective thresholds for outlier analysis based on the true statistics of the data as opposed to a blanket assumption of a Gaussian distribution.

6.12.3   Determination of Limit Distributions

In practical terms, the EVS theorems provide a very powerful tool. In general if one wants to know the density function for a distribution, a nonparametric estimate must be made or one must know which distribution is appropriate for a parametric approach. For EVS purposes one only needs to know which of three distributions apply in order to apply parametric methods. Previously, one problem was discussed in which one could test to see if data belongs to a given distribution, this being the important situation when one wishes to test for Gaussianity. It was indicated earlier that one can plot the data on Gaussian probability paper and then test to see if the result is a straight line. Happily, this approach works very well in the EVS context.

6.12.3.1   Probability Paper

Once a population of samples is obtained, it is a simple matter to plot the *empirical CDF*. The data are first placed in increasing order, with the data in this case being the order statistics $X_{r:n}$. One associates with each order statistic a *plotting position* or assignment of probability. A naïve approach assigns the value r/n to $X_{r:n}$. However, this assignment does not behave well under certain nonlinear transformations of the data that will be described later. A more robust choice is to assign the value $(r - 0.5)/n$ to $X_{r:n}$. There are numerous different formulae for plotting positions (Castillo, 1988).

One can usefully show the empirical CDF in a number of different coordinate systems, each appropriate for a given extreme value distribution. To illustrate, consider the Gumbel CDF for maxima in Equation (6.102). Let $y = H_{3,0}(x)$ be the formula for the CDF. If one makes the nonlinear transformation $g(x)$ and $h(y)$ of the x and y coordinates,

$$
\begin{aligned}
\xi &= g(x) = x \\
\eta &= h(y) = -\ln[-\ln(y)]
\end{aligned}
\tag{6.106}
$$

where $\ln(y)$ represents the natural logarithm. Then the new coordinates ξ and η satisfy

$$
\eta = \frac{\xi - \lambda}{\delta}
\tag{6.107}
$$

The Gumbel CDF will appear as a straight line in this coordinate system. Such a plot will be referred to as 'on Gumbel probability paper' or 'in Gumbel coordinates'. Figure 6.15 shows the empirical CDF for 1000 data points generated from the Gumbel maximum distribution with $\lambda = 50$ and $\delta = 10$. In Gumbel maximum coordinates one obtains a straight line as required. The final point in Figure 6.15 and some subsequent figures seem to stray greatly from the rest of the data. This aberration may be an artefact of slow convergence to the extreme value distribution for the most extreme points.

One can also superimpose a linear regression line together with the 95% confidence interval on the empirical CDF. If a straight line adequately models the data from an unknown distribution in Gumbel maximum coordinates, then this modelling provides support for the hypothesis that the unknown distribution is the Gumbel maximum. A similar transformation carries data from the Gumbel minimum distribution into a coordinate system where the empirical CDF is a straight line.

Figure 6.16 shows data sampled from a Weibull maximum distribution ($\lambda = 50$, $\delta = 100$, $\beta = 2$) plotted in Gumbel maximum coordinates. As one might expect, the result is not a straight line. In fact, there is definite curvature (concavity). Figure 6.17 shows data from a Frechet maximum distribution ($\lambda = 0$, $\delta = 30$, $\beta = 5$) plotted in Gumbel maximum coordinates. In this case, the curvature is not as marked as that in Figure 6.16, but it is clearly in the opposite sense (convexity). This curvature is one of the tests for the limiting distribution for maxima. First, the empirical CDF is plotted in Gumbel coordinates

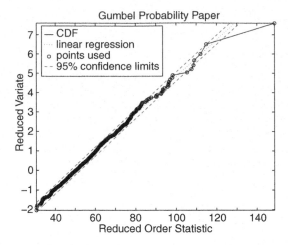

Figure 6.15 The empirical CDF for Gumbel maximum distributed data in Gumbel maximum coordinates.

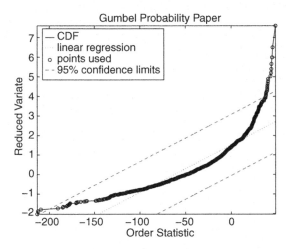

Figure 6.16 The empirical CDF for Weibull maximum distributed data in Gumbel maximum coordinates.

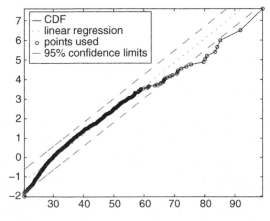

Figure 6.17 The empirical CDF for Frechet maximum distributed data in Gumbel maximum coordinates.

for maxima. The user then makes an assessment as to whether the curvature deviates significantly from zero and the limit distribution is assigned accordingly.

Although in principle one can infer the distribution type from a plot on Gumbel paper, one can also generate plotting papers for the other two limiting distributions. If the data are known to come from a Weibull distribution for maxima as in Equation (6.101) with the empirical CDF $y = H_{2,\beta}(x)$, then the transformation

$$\xi = g(x) = -\ln(\lambda - x)$$
$$\eta = h(y) = -\ln[-\ln(y)] \tag{6.108}$$

carries the empirical CDF into the straight line,

$$\eta = \beta(\xi - \ln \delta) \tag{6.109}$$

The difference in this situation is that the transformation requires an a priori estimate of the location parameter λ. As discussed above, in the Gumbel case, if the empirical CDF is plotted in the transformed coordinates, it will appear as a straight line.

If the data are known to come from a Frechet distribution for maxima as in Equation (6.100) with the empirical CDF $y = H_{1,\beta}(x)$, then the transformation

$$\xi = g(x) = \ln(x - \lambda)$$
$$\eta = h(y) = -\ln[-\ln(y)] \tag{6.110}$$

carries the empirical CDF into the straight line,

$$\eta = \beta(\xi - \ln \delta) \tag{6.111}$$

As before, an a priori estimate of λ is required.

6.12.3.2   Parameter Estimation

Having established the appropriate limit distribution, the next stage in the analysis is to estimate parameters of the chosen distribution. Algorithms are readily available that can fit the best parameters for both the least-squares and maximum likelihood cases (Castillo, 1988). In fact, one need only fit parameters to one canonical model form, the Gumbel distribution for minima; however, the following preprocessing is required before the actual curve fitting.

First, if the data are distributed as maxima, the transformations $x \to -x$ and $\lambda \to -\lambda$ carry each maximum CDF into the corresponding minimum CDF at least as far as optimisation is concerned. Next, suppose the data have the Weibull distribution for minima. Then, the transformation $Y = \ln(X - \lambda)$ carries the Weibull distribution X into the Gumbel distribution Y with the following relations between the parameters:

$$\lambda_G = \ln(\delta_W) \quad \text{and} \quad \delta_G = \frac{1}{\beta_W} \tag{6.112}$$

where the subscripts G and W denote Gumbel and Weibull distributions, respectively. As in the plotting problem, this transformation requires an a priori estimate of λ, but it can be obtained by optimising the linearity of the empirical CDF plot in Weibull coordinates.

If the data have the Frechet distribution for minima, the transformation $Y = -\ln(\lambda - X)$ carries the Frechet distribution X into the Gumbel distribution Y, with the following relations between the parameters:

$$\lambda_G = -\ln(\delta_F) \quad \text{and} \quad \delta_G = \frac{1}{\beta_F} \tag{6.113}$$

where the subscript F denotes a Frechet distribution. This relation means that the parameter estimation problem is reduced to fitting the data to the limit distribution of the form in Equation (6.102).

The optimisation estimates the parameters λ and δ by minimising some error criterion. The most straightforward error criterion is the weighted least-squares method, which seeks to minimise the following objective function G:

$$G = \sum_{i=1}^{n} w_i [p_i - L_{3,0}(x_i; \lambda, \delta)]^2 \tag{6.114}$$

where the training data are the points on the empirical CDF $\{(x_i, p_i), i = 1, \ldots, n\}$, the $p_i's$ are an appropriate choice of plotting positions and the w_i's are a set of weights. Although there are various possibilities, Castillo (1988) recommends

$$w_i = \frac{1}{p_i} \tag{6.115}$$

With these weights, the method is referred to as *least-squares probability relative error* (LSPRE). The other approach to optimisation is maximum likelihood (ML). The reader is referred to Castillo (1988) for details. As the objective function for optimisation is nonlinear in the parameters, the optimisation is nontrivial; again the reader can consult Castillo (1988) for guidance. The reference Worden *et al.* (2005) discusses how an evolutionary optimisation procedure based on the *differential evolution* algorithm can be very effective.

For the Gumbel distribution, a very simple but often inaccurate approach called the *method of moments* is available. It is possible to show that the mean, $\bar{x}$, and the variance, σ^2, of the Gumbel distribution maxima and minima are related to λ and δ by

$$\bar{x} = \lambda - \gamma\delta \quad \text{and} \quad \sigma^2 = \frac{\pi^2\delta^2}{6} \tag{6.116}$$

where γ is Euler's constant ($\approx 0.577\,72$). From the above mean and sample variance, the moment estimates of the parameters can be calculated as

$$\delta = \frac{\sigma\sqrt{6}}{\pi} \quad \text{and} \quad \lambda = \bar{x} + \gamma\delta \tag{6.117}$$

Figure 6.18 shows the LSPRE curve-fit to the Gumbel maximum data shown in Figure 6.15. The estimated parameters $\lambda = 50.12$ and $\delta = 10.16$ compare favourably with the exact values of $\lambda = 50$ and $\delta = 10$, respectively.

As discussed earlier, the importance of EV statistics is based on their use for estimating accurate thresholds for damage detection. As the application of the thresholds is central to the ideas of novelty detection, discussion of how EV statistics are used for their computation is postponed until Chapter 10.

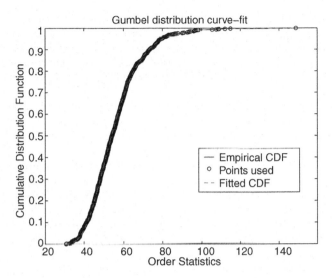

Figure 6.18   LSPRE curve-fit to Gumbel maxima data in Figure 6.15.

6.13   Dimension Reduction – Principal Component Analysis

One of the main problems associated with statistical and probabilistic techniques in general is their difficulty in dealing with data vectors of high dimensionality; this is the so-called *curse of dimensionality* (a phrase attributed to Richard Bellman). Many of the methods described above depend on the availability of *training data*, that is examples of the measurement vectors to be analysed or classified. The curse is simply that, in order to obtain accurate diagnostics, the amount of training data theoretically grows explosively with the dimension of the patterns. The most sobering discussion of this effect can be found in Silverman's text, which describes the use of density estimation techniques. Table 6.1 (reproduced from Silverman, 1986) gives the size of the training set needed to ensure a relative mean square error of less than 0.1 for a point on a kernel density estimate.

He also concludes, rather pessimistically, that if a global measure of fit were used, that is including the tails of the distribution, the sample sizes would probably go up. Unfortunately, as previously discussed, the tails are often the main areas of interest for fault detection. There is some evidence that Silverman

Table 6.1   Growth of training set size with dimension – kernel density estimation

Dimension	Training set size
1	4
2	19
3	67
4	223
5	768
6	2 790
7	10 700
8	43 700
9	187 000
10	842 000

is being unduly pessimistic; Scott (1992) observes that the global structure of the densities, that is multimodality, can be observed with much more modest sample sizes. A further ray of hope is provided by the fact that only the rank-ordered magnitudes of the densities need to be reproduced in order to obtain a sensible diagnostic. The case studies in Worden *et al.* (2000) illustrate successful applications of the density estimation method despite the fact that the training data are woefully inadequate under Silverman's criterion.

From a pragmatic point of view, there are two solutions to the problem. The first solution is to obtain adequate training sets. Unfortunately, this will not be possible in many engineering situations, due to limitations on the size and expense of testing programmes. The second approach is to reduce the dimension of the data to a point where the available data are sufficient. Generally speaking, dimension reduction can be achieved by either discarding uninteresting data in some (hopefully) principled manner or by making linear or nonlinear combinations of the data, again in some principled manner. The object of this section is to illustrate a couple of these techniques for reducing the dimension on a number of simulated and experimental data sets. Of course, it is not possible to reduce the dimension in general without discarding information, and where the methods illustrated here differ is in their criteria for deciding which information should be preserved. The methods discussed here will be simple projection and principal component analysis.

If it is possible to reduce the dimension to two or three, the methods will provide a means of visualising the data. Depending on which properties are preserved by the transformations, the important structure of the data may be accessible by visual inspection.

First, consider a couple of data sets that will be used in the subsequent discussion; the first data set is obtained simply by generating two fifteen-dimensional clusters of points with Gaussian distributions. Each component at each point is obtained by sampling the standard normal distribution $N(0, 1)$. Fifty of the points are centred at the origin and fifty are centred at the point $x_i = 10, i = 1, \ldots, 15$.

Throughout the rest of the section, vectors in the measured data space will be denoted by $\{x\}$ and the dimension of the space by p. Vectors in the reduced space will be denoted by $\{z\}$ and the dimension of that space will be denoted q.

6.13.1   Simple Projection

This approach is the simplest and most unprincipled means of data reduction. If a set of p-dimensional data vectors is given, $(x_1, x_2, \ldots, x_p)$, the q-dimensional projection is given by the set of $(x_{i_1}, x_{i_2}, \ldots, x_{i_q})$, where the indices i_j are chosen without repetition from the original p. For visualisation purposes q will be fixed at two here and the projections will be labelled P_{ij} to signify the selection of (x_i, x_j). This method is the most undiscriminating possible unless i and j are chosen wisely, as it simply discards all information from the last $p - 2$ channels of data ($p - q$ channels in general).

The first data set has actually been designed with ease of visualisation in mind, and in this case projection on to any two unequal components reveals the bimodal structure of the data. Figure 6.19 shows P_{12}, the projection onto the first two components. (As the figures from this point onwards are for visualisation of structure only, it is permissible to leave out axis labels.)

6.13.2   Principal Component Analysis (PCA)

This is a classical method of multivariate statistics and its theory and use are documented in any textbook from that field (e.g. Sharma, 1996). Only the briefest description will be given here. The principal components algorithm seeks to project, by a linear transformation, the data into a new p-dimensional set of Cartesian coordinates $(z_1, z_2, \ldots, z_p)$ called the principle component *scores*. The new coordinates

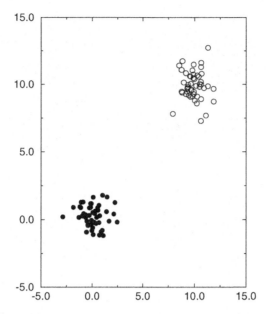

Figure 6.19 Visualisation by simple projection – synthetic data.

have the following property: z_1 is the linear combination of the original x_i with *maximal* variance, z_2 is the linear combination that explains most of the remaining variance and so on. It should be clear that if the original p coordinates are actually a linear combination of $q < p$ variables, the first q principal components will completely characterise the data and the remaining $p - q$ will be zero. In practice, due to measurement uncertainty, all the principal component scores (the coordinate values in the transformed dimensions) will be nonzero and the user should select the number of *significant* components for retention.

The calculation is carried out as follows: given data $\{x\}_i = (x_{1i}, x_{2i}, \dots, x_{pi})$, $i = 1, \dots, N$, form the covariance matrix $[\Sigma]$:

$$[\Sigma] = \sum_{i=1}^{n} (\{x\}_i - \{\overline{x}\})(\{x\}_i - \{\overline{x}\})^{\mathrm{T}} \qquad (6.118)$$

and decompose so that

$$[\Sigma] = [A][\Lambda][A]^{\mathrm{T}} \qquad (6.119)$$

where $[\Lambda]$ is diagonal. (The singular value decomposition can be used for this step.) The transformation to the principal component scores is then

$$\{z\}_i = [A]^{\mathrm{T}}(\{x\}_i - \{\overline{x}\}) \qquad (6.120)$$

where $\{\overline{x}\}$ is the vector of component-wise means of the x data. The $[\Lambda]$ matrix is interpreted as the relative contributions of the scores to the variance. One truncates the new vector at the q components that contribute the required proportion of the variance.

Considered as a means of dimension reduction then, PCA works by discarding those linear combinations of the data that contribute least to the overall variance or range of the data set. Note that this criterion is not necessarily the right one for amplifying the influence of the damage that one might wish to detect. This issue is a consideration with all methods of dimension reduction: one should bear in mind the criteria for discarding data.

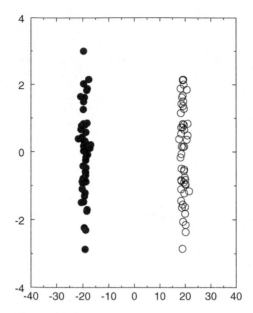

Figure 6.20   Visualisation by PCA – synthetic data.

Figure 6.20 shows a plot of the first two principal component scores for the synthetic data. As in the projections, the separation into two classes is evident. It is usual to standardise data before PCA visualisation to eliminate the possibility that the scores will be dominated by coordinates that simply have large amplitude. In the case of the data here, this results in elongation of the clusters in the vertical direction.

If one applies PCA to some four-dimensional acoustic emission data (for the moment this can be regarded as a four-dimensional set of data with three natural clusters; more details will be given in Chapter 9) and truncates at two dimensions, one can visualise the data and the structure becomes apparent. Figure 6.21 shows that there are effectively three clusters to the data.

Both approaches shown are successful to an extent. The simple projection method has the capability of displaying the structure of the data. The advantage of the method is its simplicity; the main disadvantage is that the analyst may generally be required to plot $d(d-1)/2$ projections for a d-dimensional data set. PCA is an attractive method as it is quick to implement and run and still yields good visualisation results. PCA is limited by the fact that it is fundamentally a linear transformation. Various nonlinear dimensional reduction techniques exist; however, they will not be discussed here.

6.14   Conclusions

The last section concludes this brief introductory chapter on probability and statistics. More material will be developed in later chapters in terms of the pattern recognition and machine learning to be discussed, but for now an adequate foundation has been provided. In many places in this chapter, considerable detail has been omitted, but all of this detail can be found in the references cited throughout the chapter. It will be clear from the discussion here that the acquisition of high-quality, low-dimensional data carrying information about damage is essential for the successful application of the various techniques. The types

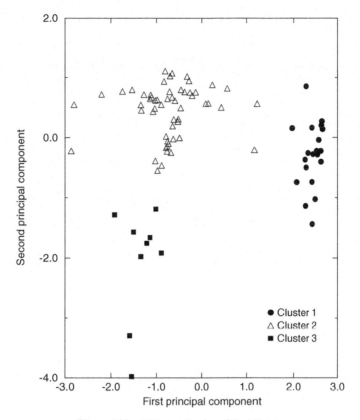

Figure 6.21 PCA visualisation of the AE data.

of data that carry appropriate information for SHM purposes will be the subject of a detailed discussion in the next two chapters.

References

Barnett, V. and Lewis, T. (1994) *Outliers in Statistical Data*, 3rd edn, Wiley-Blackwell.

Bowman, A.W. (1984) An alternative method of cross-validation for the smoothing of density estimates. *Biometrika*, **71**, 353–360.

Castillo, E. (1988) *Extreme Value Theory in Engineering*, Academic Press Series in Statistical Modeling and Decision Science, San Diego, CA.

Embrechts, P., Kluppelberg, C. and Mikosch, T. (1997) *Modeling Extremal Events*, Springer-Verlag, New York.

Fisher, R.A. and Tippett, L.H.C. (1928) Limiting forms of the frequency distributions of the largest or smallest members of a sample. *Proceedings of the Cambridge Philosophical Society*, **24**, 180–190.

Galambos, J. (1978) *The Asymptotic Theory of Extreme Order Statistics*, John Wiley & Sons, Inc., New York.

Grimmet, G. and Stirzaker, D. (2001) *Probability and Random Processes*, 3rd edn, Oxford University Press.

Gumbel, E.J. (1958) *Statistics of Extremes*, Columbia University Press, New York.

Kotz, S. and Nadarajah, S. (2000) *Extreme Value Distributions. Theory and Applications*, Imperial College Press, London.

Leuenberger, D.G. (1989) *Linear and Nonlinear Programming*, 2nd edn, Addison-Wesley.

Lindley, D.V. (1980) *Introduction to Probability and Statistics from a Bayesian Viewpoint, Part 1, Probability*, new edition, Cambridge University Press.

Papoulis, A. and Pillai, S.U. (2001) *Probability, Random Variables and Stochastic Processes*, 4th edn, McGraw-Hill.

Reiss, R.D. and Thomas, M. (2001) *Statistical Analysis of Extreme Values with Applications to Insurance, Finance, Hydrology and Other Fields*, Birkhäuser Verlag AG, Switzerland.

Roberts, S. (1998) Novelty detection using extreme value statistics. *IEE Proceedings in Vision, Image and Signal Processing*, **146**, 124–129.

Roberts, S. (2000) Extreme value statistics for novelty detection in biomedical signal processing. *IEE Proceedings in Science, Technology and Measurement*, **147**, 363–367.

Scott, D. (1992) *Multivariate Density Estimation: Theory, Practice and Visualization*, John Wiley & Sons, Inc., New York.

Sharma, S. (1996) *Applied Multivariate Techniques*, John Wiley & Sons.

Silverman, B.W. (1986) *Density Estimation for Statistics and Data Analysis*, Chapman and Hall, Monographs on Statistics and Applied Probability, p. 26.

Surhone, L.M., Timpleton, M.T. and Marseken, S.F. (2010) *Normality Test: Data Set, Normal Distribution, Random Variable, Model Selection, Interpretations of Probability*, Betascript Publishing.

Worden, K., Pierce, S.G., Manson, G. *et al.* (2000) Detection of defects in composite plates using Lamb waves and novelty detection. *International Journal of System Science*, **31**, 1397–1409.

Worden, K., Manson, G., Sohn, H. and Farrar, C.R. (2005) Extreme value statistics from differential evolution for damage detection. Proceedings of 23rd International Modal Analysis Conference, Orlando, Florida. On CD – Paper 327.

7

Damage-Sensitive Features

A damage-sensitive *feature* is some quantity extracted from the measured system response data that is used to indicate the presence of damage in a structure. Identifying features that can accurately distinguish a damaged structure from an undamaged one is the primary topic addressed in most of the SHM technical literature (Doebling *et al.*, 1996; Sohn *et al.*, 2004). These features are the focus of the discussions in this chapter and the next where they are presented independently from the rest of the pattern recognition process. Therefore, the reader should keep in mind that these features are the quantities that the pattern recognition and machine learning algorithms will subsequently analyse in an effort to identify and quantify the damage. If one makes a good choice in the features to use, the pattern recognition and machine learning process for feature classification can be made somewhat easy and often the presence of damage will be clear from simple visual inspection of the changes in the features. On the other hand, if features are used that are not correlated with the damage, even the most clever pattern recognition and machine learning algorithms will not significantly improve the damage detection process.

Feature extraction refers to the process of transforming the measured data into some alternative form where the correlation with the damage is more readily observed. Often in vibration-based SHM, the feature extraction process is based on fitting some model, either physics-based or data-based, to the measured system response data. The parameters of these models, quantities derived from the parameters or the predictive errors associated with these models then become the damage-sensitive features. An alternate approach is to identify features that directly compare the data waveforms, spectra of these waveforms or quantities derived from these waveforms or spectra. Many of the features identified for impedance-based and wave propagation-based SHM studies fall into this category (Park *et al.*, 2003; Cesnik, 2007)

Feature selection is the process of determining which feature to use in the damage detection process. Ideally one should select a feature that is sensitive to the presence of damage in the structure and insensitive to all forms of operational and environmental variability. However, in most real-world applications, features that are sensitive to damage are also sensitive to changes in the system response that are not related to damage (see Axiom IVB, Chapter 13). This issue will be addressed in the subsequent chapters on machine learning, pattern recognition and data normalisation. Finally, there is no one feature that will be applicable to all damage scenarios. If multiple types of damage are possible, it may require different features to be extracted from the data in an effort to identify these different types of damage.

One of the most common methods of feature selection is based on correlating observations of measured system response quantities with the first-hand observations of the degrading system made by the system operators or maintenance personnel. Another method of developing features for damage detection is to apply engineered flaws, similar to ones expected in actual operating conditions, to systems and develop an initial understanding of the parameters that are sensitive to the expected damage. The flawed system

Structural Health Monitoring: A Machine Learning Perspective, First Edition. Charles R. Farrar and Keith Worden.
© 2013 John Wiley & Sons, Ltd. Published 2013 by John Wiley & Sons, Ltd.

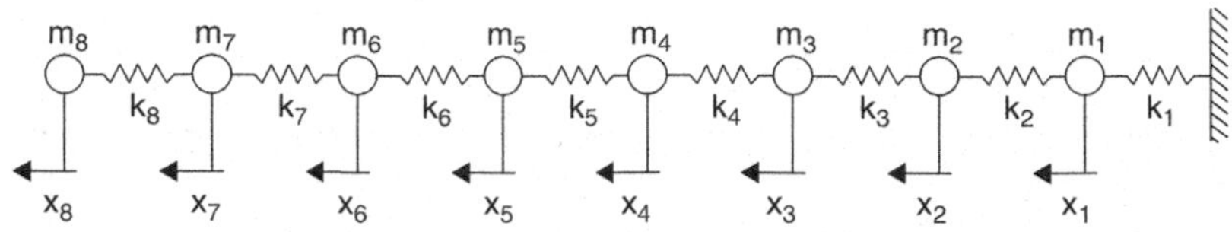

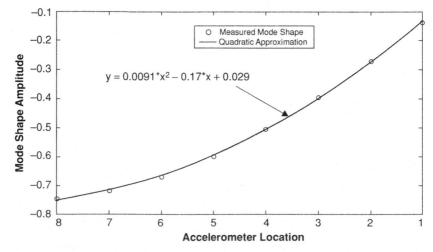

Figure 7.1 Eight degree-of-freedom spring mass system. All masses have a value of 0.4194 kg and all springs have a stiffness value of 56 700 N/m (see Section 5.3 for a detailed description).

can also be used to validate whether the diagnostic measurements are sensitive enough to distinguish between features identified from the undamaged and damaged systems. The use of analytical tools such as experimentally validated finite element models can be a great asset in this process. In many cases the analytical tools are used to perform numerical experiments where the flaws are introduced through computer simulation. Damage accumulation testing, during which significant structural components of the system under study are subjected to a realistic degradation, can also be used to identify appropriate features. This process may involve induced-damage testing, fatigue testing, corrosion growth or temperature cycling to accumulate certain types of damage in an accelerated fashion. Note that any such destructive testing approaches to feature identification can be costly and are typically prohibitively expensive for large capital expenditure equipment. Insight into the appropriate features can be gained from several sources and is usually the result of information obtained from some combination of these sources.

As will become more evident in the subsequent chapters on machine learning and statistical modelling for feature discrimination, one must be cognisant of the feature dimension. The feature dimension is the number of scalar quantities that is necessary to describe the feature. To illustrate the concept of feature dimension, consider the eight degree-of-freedom (DOF) spring–mass system as shown in Figure 7.1 with an accelerometer mounted on every mass (see Section 5.3 for a detailed description of the physical structure corresponding to this idealisation). This structure will be used throughout this chapter to illustrate various features and feature extraction issues. A mode shape might be used as a damage-sensitive feature and one can develop an estimate of the mode shapes based on the measured response at each DOF when this structure is subjected to a dynamic excitation. This estimate will produce an eight-dimensional feature vector, indicated by the circles shown in Figure 7.2. However, not all the

Figure 7.2 Measured mode shape and a quadratic approximation obtained by a least-squares fit to the measured data.

values in this feature vector are independent. Therefore, the *effective dimension* may be significantly less than the dimension of the mode shape vector. To illustrate this point, a quadratic function can be fitted to the discrete mode shape amplitudes and can accurately represent the mode shape with only the three constants, as shown in Figure 7.2. In this example it may be concluded that the effective dimension of this feature vector is three. The fact that the effective dimension is smaller than the dimension of the original feature vector implies that there is correlation between the different elements of the original feature vector.

The approaches to quantifying the effective dimension of a feature vector range from well-established procedures such as principal component analysis (Bishop, 1995) to more sophisticated procedures such as factor analysis (Fukunaga, 1990) and Sammon mapping (Ripley, 1996). In many cases the process of quantifying the actual dimension of a feature vector is still a research topic for people working in the information technology and machine learning fields. In general, because of statistical modelling issues, it is desirable to have feature vectors with as low a dimension as possible, as discussed in Chapter 6.

In an effort to obtain a low-dimensional feature vector, procedures are developed to fuse data from multiple sensors and compress these data. A common example of data fusion is the extraction of mode shapes from the relative amplitude and phase information contained in data from a sensor array. Similarly, the extraction of resonance frequencies from measured acceleration time histories can be thought of as a data compression process. As an example, a 1024-point acceleration time series may be collected from a sensor. Using a *fast Fourier transform* (FFT) algorithm the *power spectral density* (PSD) function (as described in Appendix A) can be calculated and estimates of the system's resonance frequencies can be identified from peaks in the PSD. If ten resonance frequency peaks are identified and these parameters are used as the damage-sensitive features, then the feature extraction process has condensed the 1024-point time series into a ten-dimensional feature vector.

Also, various forms of data normalisation are employed during the feature extraction process in an effort to separate changes in the measured response caused by varying operational and environmental conditions from changes caused by damage. The process of forming a frequency response function (FRF) (see Appendix A) whereby the measured responses are effectively normalised by the measured input can be viewed as one type of data normalisation process.

One must not confuse a *feature* with a *metric*. A metric is some quantity that defines the similarity or difference between two features. In general, metrics are some type of distance function, such as the Euclidean distance between two vectors, a statistical test or some form of correlation measure between two data sets. A feature can be extracted from a single data set while a metric quantifies the difference between features extracted from two different data sets.

At the risk of repetition it will be reiterated that the concept of *feature extraction* must be distinguished from *feature selection*. Feature extraction is the mathematical process of calculating a specific feature based on some set of data. Feature selection is the process of determining which subset of a given set of features is the best one to use for a specific damage detection problem. While the former is almost always done through some mathematically rigorous process, the latter can be done by a wide range of techniques ranging from equally mathematically rigorous processes to very heuristic or intuitive selection processes.

Damage-sensitive features and the feature extraction process is the primary focus of this chapter. A discussion of some approaches to feature selection is provided in Section 7.12 of this chapter and a specific example based on optimisation is discussed in detail at the end of Chapter 11. Finally, to begin the transition from feature extraction and selection to statistical classification of features, this chapter concludes with a summary of various metrics that are discussed throughout the chapter.

7.1 Common Waveforms and Spectral Functions Used in the Feature Extraction Process

This discussion and the subsequent feature extraction discussions will make use of data from the test structures described previously in Chapter 5, as well as many numerically generated data sets specifically developed to illustrate some of the feature extraction concepts presented herein. This discussion begins

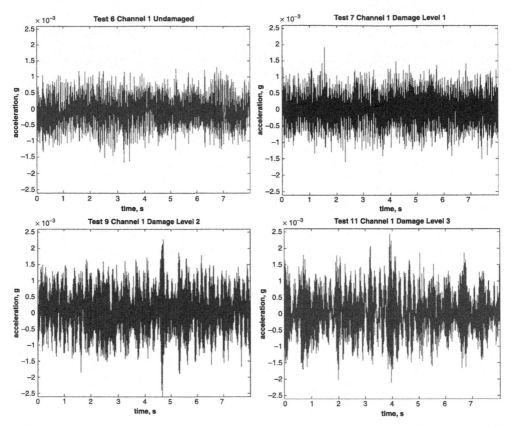

Figure 7.3 Acceleration–time histories measured on the concrete column in its undamaged state and at various damage levels.

by examining the acceleration time histories shown in Figure 7.3 that were recorded with sensor 1 on the concrete column (Section 5.2) that was subjected to dynamic excitations after various levels of damage had been introduced by quasi-static cyclic load tests. In the subsequent plots and discussions, damage level 0 corresponds to the undamaged column. Damage level 1 corresponds to a lateral displacement applied to the top of the column that caused incipient yield of the reinforcement and occurs after the concrete cracks. Damage level 2 corresponds to a lateral displacement at the top of the column that is 2.5 times the lateral displacement that caused incipient yield. Damage level 3 corresponds to a lateral displacement at the top of the column that is 7.0 times the lateral displacement that caused incipient yield.

7.1.1 Waveform Comparisons

The time histories that will be discussed are composed of 8192 discrete samples of acceleration measured in g's (i.e. the acceleration caused by gravity, 9.81 m/s^2) taken over an 8-second-long measurement window, which corresponds to a time interval between samples, $\Delta t = 0.000\,977$ s. The examination of the time series in Figure 7.3 represents one of the simplest forms of feature extraction and these features fall into the class referred to as *waveforms*. Although qualitatively some distinctions are evident from a visual inspection of these time histories, it is difficult to define quantitatively the features in a way that would allow one to classify subsequent data sets to either the undamaged or damaged condition simply through visual inspection. In fact, one might raise questions regarding the consistency of the input to the system that caused these responses or, more specifically, are the changes seen in these signals the result of varying inputs to the system? Also, these time histories represent high-dimensional feature vectors (an apparent dimension of 8192 in this case), which have already been identified as troublesome for statistical classification. Additionally, with visual inspection of a random signal it is difficult to discern if there is correlation in the data that would imply that the effective dimension might be smaller.

7.1.2   *Autocorrelation and Cross-Correlation Functions*

Some functions from signal processing discussed in Appendix A will now be introduced in an effort to reveal more structure in the time histories that might prove useful for identifying damage. It should be pointed out that signal processing does not create new information; it simply allows the time history to be viewed in a different manner whereby information contained in the signal is more easily visualised. Figure 7.4 shows the autocorrelation functions for the signals in Figure 7.3, respectively. Cross-correlation functions corresponding to the undamaged structure and three damage levels are shown in Figure 7.5. Here the cross-correlation is between channel 4 and channel 19. In both Figures 7.4 and 7.5, changes in

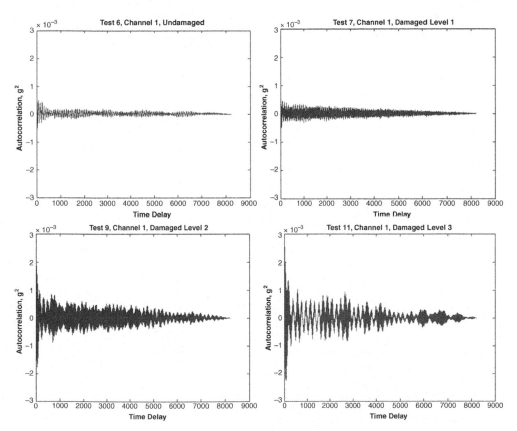

Figure 7.4   Autocorrelation functions from data measured on the concrete column in its undamaged state and at various damage levels.

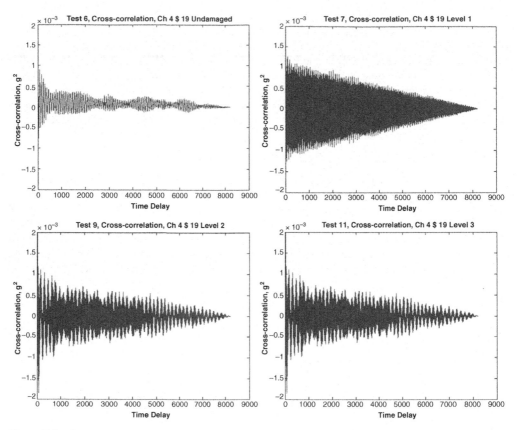

Figure 7.5 Cross-correlation function for the concrete column between sensors 4 and 19 for the undamaged state and each of the damaged cases.

the structure of these functions, which may be indicative of the changing structural condition, are evident from qualitative visual inspection. However, again without having any information about the input to the structure it is difficult to say that these changes are related to changing structural conditions or changes in the input to the system. Additionally, as with all the other waveform comparisons, there needs to be some way to quantify this change. In the autocorrelation plots it is seen that the correlation of the signal with itself diminishes at increasing time lags. This property will be used in subsequent time series modelling to identify the minimum number of time lags that should be used to develop an autoregressive model of a time series that will generalise to other measured time series acquired from the structure in the same condition. Finally, it should be noted that the cross-correlation process is also a form of data compression as this process has taken two 8192-point time series and formed a single 8192-point cross-correlation function. This function also provides relative information between sensor pairs.

7.1.3   The Power Spectral and Cross-Spectral Density Functions

A common function derived from the Fourier transform that is used to examine a signal in the frequency domain is the power spectral density (PSD) function introduced in Appendix A.

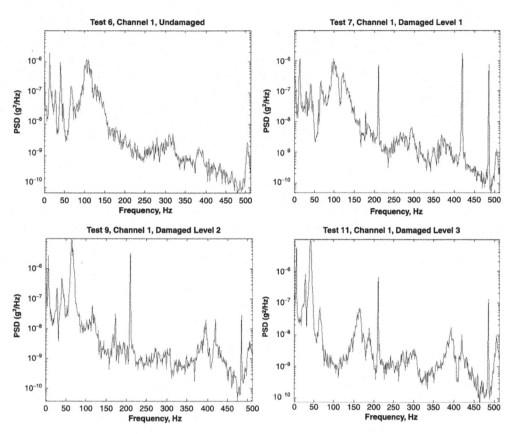

Figure 7.6 Power spectral density function of each signal shown in Figure 7.3 calculated from eight averages with a Hanning window (no overlap) applied to each average.

Figure 7.6 shows the PSD for the four sensor readings from Figure 7.3. Here the PSD was calculated from eight averages with no overlap and with a Hanning window applied to each average. Eight averages produce an estimate of the PSD at a frequency resolution of 1 Hz. Narrow spikes at just above 200 Hz can be seen in each damage case. Also, one can see distinct changes in the lower-frequency content of the response as the damage progresses. Note that such differences between the signals were not as obvious when the signals were examined in the time domain. However, for damage detection purposes there still needs to be a way to quantify these changes. The PSD can also be used to assess the consistency of the shaker input to the Concrete Column. Figure 7.7 shows the power spectral density of the shaker input signals that generated the responses shown in Figure 7.3. Although one would typically like to see a flat input PSD implying that all frequencies of the structure are being excited with equal energy, the plots in Figure 7.7 show that the input was consistent from one test to the other in terms of amplitude and frequency content. Therefore, the possibility that changes in the features are caused by inconsistent excitation of the structure can be ruled out.

Figure 7.8 shows the cross-spectral density function (CSD) of each response signal shown in Figure 7.5. Although the four plots in Figure 7.8 look somewhat similar, detailed examination of the peaks below 100 Hz shows that these peaks are shifting in frequency as the damage in the columns increases. However,

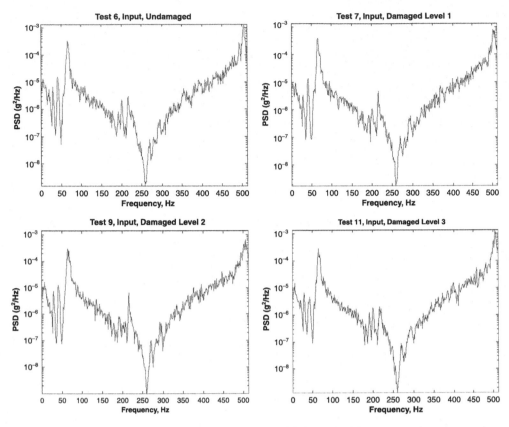

Figure 7.7   Power spectra of the input signal measured during each of the concrete column tests.

it is again emphasised that these spectra represent high-dimensional feature vectors and there is still a need to develop a quantifiable means of comparing these plots.

7.1.4   The Impulse Response Function and the Frequency Response Function

Figure 7.9 shows the FRF magnitudes corresponding to the response signals in Figure 7.3 and the corresponding inputs whose PSDs are shown in Figure 7.7. These FRFs were formed from eight averages with a Hanning window applied to each average. From this plot it can be seen that, once normalised for their respective inputs, there are clear regions of the spectra that show very distinct changes in response associated with the varying damage levels (from about 60 Hz to 130 Hz). Figure 7.10 shows the corresponding impulse response functions that were obtained by inverse Fourier transforming the respective FRFs in Figure 7.9. Again, there appears to be a clear distinction between the functions that were obtained with the structure in its different damage conditions.

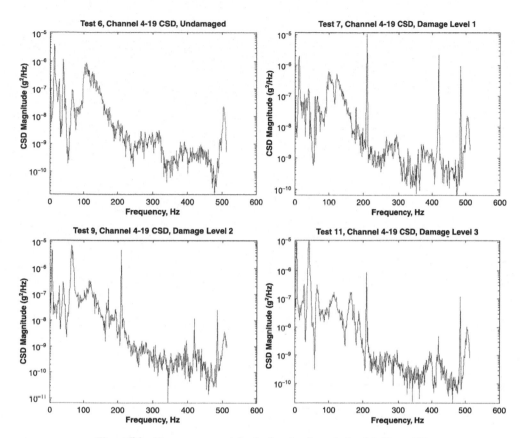

Figure 7.8   The cross-spectral density function for each signal in Figure 7.5.

7.1.5   The Coherence Function

As discussed in Appendix A, the coherence function provides a measure of how linearly related a system output is to the corresponding input. This function takes on a value between 0 and 1 with the value 1 indicating that the output can be completely attributed to the input through a linear process. At first this function seems ideally suited for damage detection, but one must realise that there are other factors that can influence the coherence function that are not related to damage. These factors include sources of unmeasured input that influence the output measurement, noise in the measurement system and certain aspects related to discrete signal processing of finite record time histories. As an example, when estimating the input–output relationship for lightly damped systems by averaging multiple data samples as discussed above, some of the response in the current sample may result from inputs associated with the previous sample that have not damped out. This issue is particularly acute at the system's resonance frequencies and can lead to drops in the coherence at these frequencies that are not associated with the loss of a linear input–output relation. The coherence function is shown in Figure 7.11 for each of the signals in Figure 7.3. These functions were again calculated from eight averages with no overlap and with a Hanning window applied to each average.

Ideally, from an experimental procedure perspective, one would like to see the coherence values close to unity across the entire frequency range of interest. Alternatively, one may wish to focus on portions

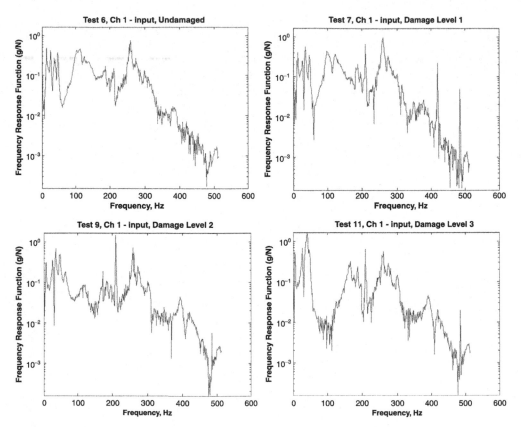

Figure 7.9   The frequency response function magnitudes for the four responses shown in Figure 7.3 and corresponding inputs shown in Figure 7.7.

of the spectra where there is good coherence in the undamaged case. The region from 60 Hz to 130 Hz shows good coherence for the undamaged case and considerable drops in coherence for the various damage cases. This region corresponds to the portion of the FRFs where one observed distinct changes associated with the different damage levels. These drops in coherence can themselves be indicators of damage.

7.1.6   *Some Remarks Regarding Waveforms and Spectra*

In summary all of the time-domain or frequency-domain waveforms associated with the signals presented above show changes with respect to the undamaged signal. However, it is difficult to quantify these changes because the feature vectors are of very high dimension. It should be noted that these features only give an indication that the system has changed. They do not indicate where the change has occurred, the type of change or the extent of that change. Also, with this limited amount of data it is difficult to determine how repeatable the waveforms and spectral quantities are should they be estimated from a subsequent set of measurements obtained from the structure in a similar condition. Finally, there is no information available from these data that would indicate how sensitive these feature vectors are to

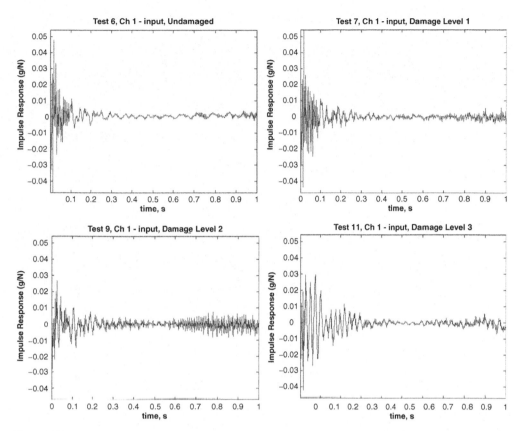

Figure 7.10 The impulse response functions obtained from the inverse Fourier transform on the functions shown in Figure 7.9.

changing operational and environmental conditions. Together these issues point to the need for lower-dimensional feature vectors whose statistical properties can be more readily assessed. Furthermore, there is a need for data normalisation procedures that can minimise the influence of operational and environmental variability of these features. In lieu of such data normalisation procedures, one might wish to model the influence of these sources of variability on the extracted features, as discussed in Chapter 12. In an effort to define alternate features that will start to alleviate some of the shortcomings associated with using high-dimensional waveforms as damage-sensitive features, some basic signal statistics will now be examined.

7.2   Basic Signal Statistics

Many basic signal statistics that were introduced in Chapter 6 can be used as damage-sensitive features. Examples of such statistics for an n-point discrete time series obtained from a sensor reading y_i, are summarised in Table 7.1. Note that all of the features in Table 7.1 reduce the n-dimensional time series into a feature of unit dimension. Additionally, more than one of these statistics can be used to form a multidimensional feature vector where different statistics are used to capture various aspects of

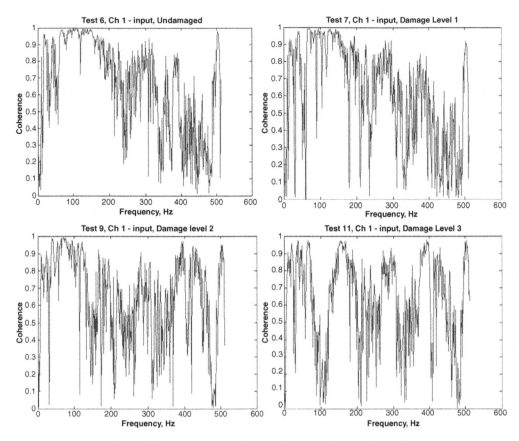

Figure 7.11   Coherence functions for the signals shown in Figure 7.3.

the original signal. Forming such a feature vector does not imply that these various features are not correlated – some will be.

For illustrative purposes these statistics have been applied to the acceleration time histories shown in Figure 7.3. Most of the statistics summarised in Table 7.1 are typically applied to random signals. However, two of these statistics, the crest factor and K-factor, are used to diagnose deviations from the sinusoidal response. Data acquired from the simulated building structure, discussed in Section 5.4, subjected to sinusoidal base excitations are used to illustrate these two features.

The first statistic discussed is the peak amplitude of the measured response. In cases where damage causes a reduction in stiffness and the random input to the system remains stationary, the peak amplitude of response will typically increase. This feature is plotted in Figure 7.12 for the time histories shown in Figure 7.3. The drop in the peak amplitude associated with damage level 3 relative to damage level 2 is not consistent with the expectation that increasing damage levels will result in a larger peak amplitude response when a comparable input is applied to the structure. Here one would want to verify that there were not differences in the input to the structure that caused this reduction in the peak amplitude. There are many other basic statistics that can also be used to identify when changes in the system response that are indicative of damage have occurred, as discussed below.

The mean (Equation (7.2) in Table 7.1) and root-mean-square (RMS) Equation (7.4) measure the central tendency and spread of the data, respectively. To illustrate the shift in the mean value caused

Table 7.1 Signal statistics used as damage-sensitive features

Peak amplitude (y_{peak})	$y_{peak} = \max \|y_i\|$	(7.1)
Mean $(\overline{y})$	$\overline{y} = \dfrac{1}{n} \sum_{i=1}^{n} y_i$	(7.2)
Mean square $(\overline{y}_{sq})$	$\overline{y}_{sq} = \dfrac{1}{n} \sum_{i=1}^{n} (y_i)^2$	(7.3)
Root-mean-square (rms)	$rms = \sqrt{\dfrac{1}{n} \sum_{i=1}^{n} y_i^2}$	(7.4)
Variance $(\sigma^2)^a$	$\sigma^2 = \dfrac{1}{n} \sum_{i=1}^{n} (y_i - \overline{y})^2$	(7.5)
Standard deviation $(\sigma)^a$	$\sigma = \sqrt{\dfrac{1}{n} \sum_{i=1}^{n} (y_i - \overline{y})^2}$	(7.6)
Skewness (dimensionless) $(\gamma)^a$	$\gamma = \dfrac{\dfrac{1}{n} \sum_{i=1}^{n} (y_i - \overline{y})^3}{\sigma^3}$	(7.7)
Kurtosis (dimensionless) $(\kappa)^a$	$\kappa = \dfrac{\dfrac{1}{n} \sum_{i=1}^{n} (y_i - \overline{y})^4}{\sigma^4}$	(7.8)
Crest factor (X_{cf})	$X_{CF} = y_{peak}/rms$	(7.9)
K-factor (X_k)	$X_{CF} = (y_{peak})(rms)$	(7.10)

[a]Note that these expressions produce biased estimates of the statistics. However, most time series studied in SHM applications have sufficient samples such that this bias is small.

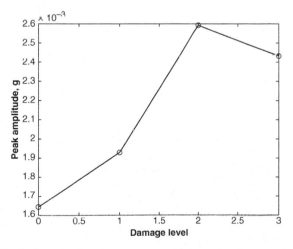

Figure 7.12 Peak amplitude of acceleration response measured at sensor 1 on the concrete column test plotted as a function of damage level.

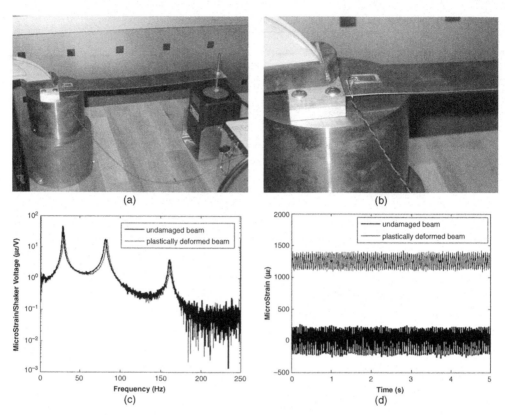

Figure 7.13 (a) Cantilever beam with shaker attached to the free end, (b) strain gauge used to measure the beam's dynamic response, (c) frequency response functions from the undamaged and plastically deformed beam and (d) strain time histories from the undamaged and plastically deformed beam showing the offset in the mean values caused by the plastic deformation.

by damage, consider the cantilever beam shown in Figure 7.13(a). A through hole is drilled in the free end of the beam, and that end is connected to an electrodynamic shaker with a threaded rod. A strain gauge was mounted 6 mm (1/4 in) from the aluminium clamp at the fixed end of the beam as shown in Figure 7.13(b). The beam was tested before and after it was plastically deformed. The plastic bending deformation was introduced near the strain gauge, which resulted in 1250 microstrain ($\mu\varepsilon$) of permanent static strain. Band-limited random noise with a frequency content between 0.2 Hz and 200 Hz was used as the dynamic excitation. Time- and frequency-domain data were captured for these random excitations. Figure 7.13(c) shows the FRF that relates the conditioned strain gauge signal to the excitation voltage supplied to the shaker's amplifier. Each FRF is averaged from 10 separate measurements using a Hanning window. Figure 7.13(c) illustrates that the deformation causes a shift in the resonance frequencies of the beam, increasing the first mode from 28 Hz to 30.5 Hz and the second mode from 81 Hz to 84 Hz. The strain time histories from the random excitations are shown in Figure 7.13(d) where the DC offset in the strain readings is very clear, as is evident in the distinct shift in the mean value of the signal, and this shift corresponds to the static plastic strain. This shift in the mean value will not be evident in an accelerometer reading because piezoelectric sensors do not measure DC response.

Note that the mean value is sensitive to outliers, so a few extreme data points can significantly influence this feature. In such cases it may be more effective to use the median (middle value of the feature samples

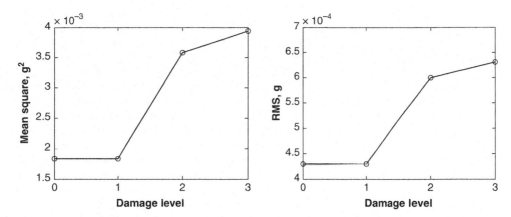

Figure 7.14   The mean squared values (left) and the root mean square values (right) for acceleration response measures at sensor 1 on the concrete column tests plotted as a function of damage level.

arranged in order of magnitude or, for an even number of feature samples, the mean of the two middle values, see Chapter 6) instead of the mean as this statistic is less sensitive to outliers. As illustrated in Figure 7.13, a shift in the mean value is a feature that can detect the permanent offset caused by yielding as long as the sensor can respond to static loads (e.g. an electric resistance strain gauge).

Returning to the concrete column data, the mean-squared values Equation (7.3) and root-mean-square values Equation (7.4) for the signals in Figure 7.3 are shown in Figure 7.14 for the various damage levels. Here one can see that these values increase with an increased damage level. There is a significant increase between the incipient damage case and the subsequent damage case.

The standard deviation Equation (7.6) measures the dispersion about the mean of the time-series amplitudes. For a fixed level of excitation, damage that reduces the stiffness of the system will, in general, cause an increase in the standard deviation of the measured dynamic response quantities such as acceleration or strain. Figure 7.15 shows the change in the standard deviation of the acceleration

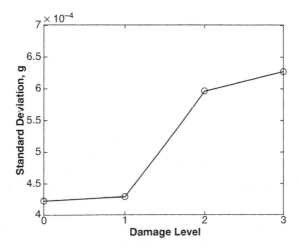

Figure 7.15   Standard deviation for acceleration response measured at sensor 1 on the concrete column tests plotted as a function of damage level.

response in Figure 7.3 as a function of increasing damage levels where the statistic is seen to increase with increasing damage levels. As with the other statistics, this series of tests shows that there is a significant increase in the standard deviation when the structure goes from the incipient damage level to the next higher damage level. Note that for a zero-mean random process the root-mean-square Equation (7.4) and the standard deviation Equation (7.6) will be equal, as can be seen in Figures 7.14 and 7.15.

The skewness Equation (7.7) is a measure of the symmetry in the distribution of a random variable. Any symmetric distribution such as the normal distribution will have a skewness value equal to zero. The skewness is sensitive to any asymmetry being introduced into an initially symmetric system such as the interference that produces the nonlinear response in the simulated building structure discussed in Section 5.4.

Figure 7.16 shows the changes in the probability density function (PDF) for the simulated building structure when an impact nonlinearity is designed such that impacts occur when relative deformations

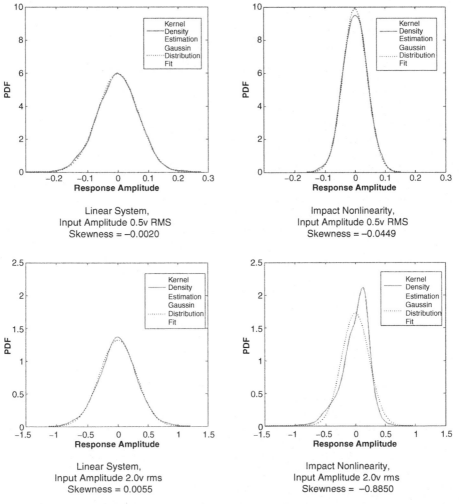

Figure 7.16 The change in skewness between the linear and symmetric simulated building structure and the same system when an asymmetric impact nonlinearity is present for two different levels of excitation.

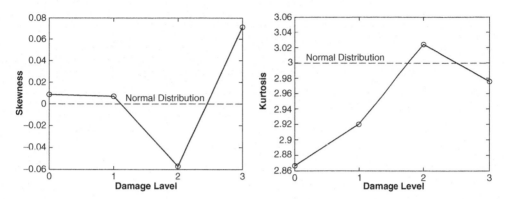

Figure 7.17   The skewness and kurtosis calculated for acceleration response measured at sensor 1 on the concrete column tests plotted as a function of damage level.

between the top two floors exceed a threshold value in one direction only. Tabulated below each PDF is the corresponding skewness value. Changes in these skewness values indicate that the nonlinearity introduced an asymmetry into the originally symmetric structure.

In Figure 7.16 the dashed lines correspond to a normal distribution fitted to the measured acceleration amplitudes of the third floor (mass 4) of the simulated building structure. The solid line corresponds to an approximation of the probability density function obtained with a kernel density estimator (as discussed in Chapter 6). When the system is linear, the kernel density estimate of the distribution function is nearly identical to the normal distribution that was fitted to the data, which illustrates the well-known property that a linear system subjected to a Gaussian random input will produce a Gaussian random output (Wirsching, Paez and Ortiz, 1995). In these cases the skewness value is close to zero, as would be expected for a symmetric distribution. At the low excitation level applied to the system with the nonlinearity, the estimated distribution function appears to overlay the normal distribution that was fitted to the data, but the skewness value has increased by more than an order of magnitude. When the higher-level excitation is applied to the nonlinear system, the estimated density function shows significant difference from the normal distribution and the skewness value increases by more than two orders of magnitude compared to the value obtained from the linear system.

Returning again to the concrete column data, Figure 7.17 shows the skewness values for the concrete column acceleration responses shown in Figure 7.3. Here it can be seen that there was some initial skewness present in the data from the undamaged column and this skewness did not change significantly after the incipient level of damage was introduced. The initial skewness is attributed to the shaker attachment, which resulted in a slightly asymmetric input to the system. At the higher levels of damage, significant increases in the skewness are observed as the concrete cracking and rebar yielding cause the column to exhibit a more significant asymmetric response. The severe nature of the last damage condition and its associated permanent deformation has resulted in the change in sign of the skewness values when damage levels 2 and 3 are compared. The kurtosis measures the peaked nature of the measured-response distribution. For a Gaussian distributed response the kurtosis will have a value of 3. The deviation from a value of 3 for the response data kurtosis for the undamaged column is again attributed to the shaker mounting scheme and associated feedback from the structure to the shaker. As with the skewness, the severe damage and permanent deformation associated with damage level 3 has contributed to the inconsistent trend in this feature with increasing damage.

The crest factor and K-factor are often used to assess the deviation from the sinusoidal response in rotating machinery. For linear systems responding to a harmonic input the crest factor has a theoretical value of $\sqrt{2}$. Table 7.2 summarises the crest and K-factors obtained from a force transducer at the

Table 7.2   Crest factor and K-factor for the simulated building structure subjected to harmonic base excitation

	Crest factor							
	53 Hz Excitation				70 Hz Excitation			
	No bumper		With bumper		No bumper		With bumper	
Sensor	Test 1	Test 2	Test 1	Test 2	Test 1	Test 2	Test 1	Test 2
Ch2	1.45	1.44	1.47	1.48	1.43	1.43	1.43	1.43
Ch3	1.42	1.42	1.42	1.42	1.44	1.44	1.41	1.41
Ch4	1.43	1.43	1.87	1.82	1.44	1.43	1.39	1.38
Ch5	1.42	1.42	1.03	1.02	1.43	1.42	1.32	1.32
Input (Ch 1)	1.43	1.44	1.43	1.43	1.43	1.44	1.43	1.42

	K-factor							
	53 Hz Excitation				70 Hz Excitation			
	No bumper		With bumper		No bumper		With bumper	
Sensor	Test 1	Test 2	Test 1	Test 2	Test 1	Test 2	Test 1	Test 2
Ch2	2.68	2.82	0.003 67	0.003 69	0.824	0.737	0.270	0.267
Ch3	2.32	2.43	0.352	0.349	3.80	3.31	0.206	0.206
Ch4	2.12	2.23	0.037 9	0.035 4	2.96	2.55	0.036 5	0.037 3
Ch5	2.28	2.38	0.062 7	0.060 3	0.518	0.446	0.007 29	0.007 14
Input (Ch 1)	821	792	2280	2270	1170	1210	1730	1720

Ch 2–5 are accelerometers mounted on each mass, Input (Ch 1) is the force transducer mounted between the bottom mass and shaker.
Test 1 corresponds to a 0.5 V rms input amplitude. Test 2 corresponds to a 2.0 V rms input amplitude.
Bumper is located between mass 4 and mass 3 (Ch 5 and Ch 4).

base (mass 1) and accelerometers at each floor (masses 1 to 4) of the simulated building structure summarised in Section 5.4 when the structure was subjected to 53 Hz and 70 Hz sinusoidal inputs at different amplitude levels. These input frequencies correspond to the first two resonance frequencies of the system. In Table 7.2 it can be seen that both the K-factor and the crest factor show a deviation from the expected values at the locations most influenced by the impacts. The crest factors obtained from sensor readings at the other locations show little change. The K-factor shows a significant change at all locations, but the changes are most pronounced at the locations closest to the impact location.

7.3   Transient Signals: Temporal Moments

Temporal moments provide an alternative to the basic statistics described above that have been used extensively to characterise transient dynamic signals (Smallwood, 1994). Temporal moments are analogous to statistical moments, but they are calculated for the signal amplitudes squared as opposed to the actual signal. Therefore, the kth-order temporal moments, M_k, about a reference time, t_s, can be defined as

$$M_k(t_s) = \int_{-\infty}^{+\infty} (t - t_s)^k y(t)^2 \mathrm{d}t, \ k = 0, 1, 2, 3, \ldots \tag{7.11}$$

Although the notation in this equation implies that the temporal moments are defined for a displacement quantity, they can be defined for any measurement quantity (e.g. strain or acceleration). With this definition the zero-order temporal moment for a history about the time, $t = 0$, is simply the area under the time history

$$M_0(0) = \int\limits_{-\infty}^{+\infty} y(t)^2 \mathrm{d}t \qquad (7.12)$$

By invoking Parseval's theorem discussed in Appendix A it can be shown that the zero-order temporal moment about time $t = 0$ is related to the area under the PSD by

$$M_0(0) = \int\limits_{-\infty}^{+\infty} y(t)^2 \mathrm{d}t = \frac{1}{2\pi} \int\limits_{-\infty}^{+\infty} |Y(f)|^2 \mathrm{d}f \qquad (7.13)$$

For a finite duration discrete time series this temporal moment is defined as

$$M_0 = \sum_{i=1}^{n-1} \frac{\Delta t}{2} \left[y_i^2 + y_{i+1}^2 \right] \qquad (7.14)$$

Note that the temporal moments are defined for the square of the time history. This definition yields moments that can be related to energy-like quantities should the time history have appropriate units. In fact, this first temporal moment is referred to as *energy* and is denoted E.

A normalised first-order temporal moment about time $t = 0$ is referred to as the *central time*, T, and is defined as

$$T = \frac{\int\limits_{-\infty}^{+\infty} t y(t)^2 \mathrm{d}t}{E} \approx \frac{\frac{1}{2} \sum\limits_{i=1}^{n-1} [\Delta t(i - 0.5)] \Delta t \left[y_i^2 + y_{i+1}^2 \right]}{E} \qquad (7.15)$$

This value can be thought of as the centroid of the area under the amplitude-squared time history and is directly analogous to the first statistical moment or mean value.

Next, a second-order temporal moment about the central time is referred to as the *mean-square duration*, D^2; it describes the dispersion of the signal's energy about the central time and is defined as

$$D^2 = \frac{\int\limits_{-\infty}^{\infty} (t - T)^2 y(t)^2 \mathrm{d}t}{E} \approx \frac{\frac{1}{2} \sum\limits_{i=1}^{n-1} [\Delta t(i - 0.5) - T]^2 \Delta t \left[y_i^2 + y_{i+1}^2 \right]}{E} \qquad (7.16)$$

This temporal moment is analogous to the second statistical moment or variance. The root-mean-square duration is the square root of this quantity and is designated D. It is analogous to the standard deviation. With these quantities, a group of other temporal moments can be defined as summarised in Table 7.3.

The temporal moments have been calculated from numerically generated accelerometer response measurements that were obtained from finite element analysis of the simulated building structure (see Section 5.4) when the structure was subjected to a transient excitation shown in Figure 7.18, which is the measured transient that was applied with a shaker in the experiment. Responses calculated with models that simulated the undamaged and damaged structures are shown in Figure 7.19. The corresponding temporal moments are summarised in Table 7.4. Note that to excite the nonlinearity that was used to simulate damage, a high level of input was necessary and the temporal moments are not normalised for

Table 7.3   Temporal moments

Symbol	Name	Units	Definition
E	Energy	$(EU)^{2*}$ seconds	$E = M_0$
A_e	Root energy amplitude	(EU)	$A_e^2 = E/D$
T	Central time (centroid)	Seconds	$T = M_1/E$
D	Root-mean-square duration	Seconds	$D = (M_2(T)/E)^{1/2}$
S_t	Central skewness	Seconds	$S_t^3 = M_3(T)/E$
S	Normalised skewness	Nondimensional	$S = S_t/D$
K_t	Central kurtosis	Seconds	$K_t^4 = M_4(T)/E$
K	Normalised kurtosis	Nondimensional	$K = K_t/D$

EU = Engineering Units associated with the particular time history (e.g. g's, m/s), assuming that time is expressed in seconds.

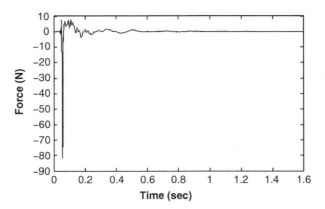

Figure 7.18   Transient input applied to the undamaged and damaged structure in the numerical simulation.

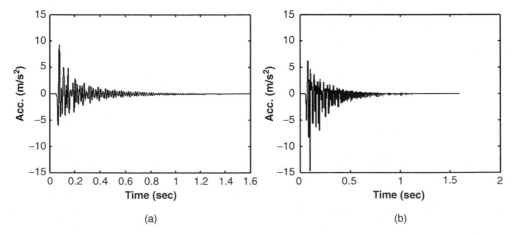

(a)   (b)

Figure 7.19   Response at channel 5 (see Figure 5.14) calculated for (a) the undamaged structure and (b) the damaged structure when they were subjected to the input shown in Figure 7.18.

Table 7.4 Temporal moments extracted from transient responses calculated for the simulated building structure

Undamaged			Damaged		
Energy	E	1.72	Energy	E	2.22
Central time	T	0.145	Central time	T	0.154, 0.199[a]
RMS duration	D	0.183	RMS duration	D	0.184
Central skewness	S	0.0230	Central skewness	S	0.020, 0.0260[a]
Central kurtosis	K	1.58	Central kurtosis	K	1.43, 1.86[a]
Root energy amplitude	A_e	3.06	Root energy amplitude	A_e	3.47

[a] Obtained using E and D from undamaged structure.

the input in any manner. For illustrative purposes the same input was applied to both the undamaged and damaged structures in this simulation. Table 7.4 shows that the energy, central time and root energy amplitude all increase significantly in the damaged system response. The RMS duration, central skewness and central kurtosis show much less change between the damaged and undamaged system responses. However, if the energy and RMS duration from the undamaged structure are used in the normalising process for the damaged response, these quantities also increase significantly and the central time shows a more pronounced increase.

7.4   Transient Signals: Decay Measures

Another feature that can be used to characterise changes in transient response caused by damage is the 10% duration time, which is defined as the time between the peak response and the time when the waveform has decayed to 10% of this peak value. Figure 7.20 illustrates this feature on the data acquired when a transient excitation was applied to the simulated building structure with a shaker. The time duration for this response decay is influenced by the structure's damping and stiffness characteristics; hence to be useful for damage detection, the assumption is that damage will change the structure's energy dissipation and/or stiffness characteristics, as is the case when a crack opens or closes, loose connections rattle or a structure yields as a result of the dynamic loading environment. Figure 7.21 shows the 10% duration features extracted from the transient response data corresponding to the undamaged and damaged structures. The data are from numerical simulations of the simulated building structure subjected to a

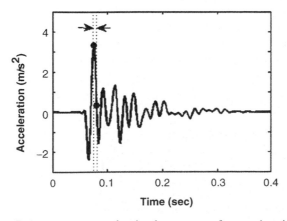

Figure 7.20   Ten percentage duration decay measure for a transient signal.

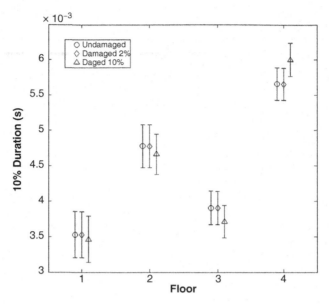

Figure 7.21 Mean and standard deviation for the 10% duration computed from ten numerical simulations of the simulated building structure.

transient excitation similar to the one shown in Figure 7.18. The undamaged case corresponds to the structure without the bumper. The first damage case is one with the bumper and with 2% damping added to the simulation. The second damage case increases the damping to 10% of critical. Note that damping was added because the physical energy dissipation mechanisms associated with hysteretic energy losses are not included in this model. In each case ten nominally similar inputs, but with small random perturbations, were applied to the structure. The results shown are the mean and standard deviation for the 10% duration for the response at each floor. For floors 1 and 2 there is minimal change in this feature even when 10% damping has been added. However, floors 3 and 4, which are on either side of the bumper, show a more pronounced change in the 10% duration for the 10% damping. It should be noted that this feature is based on a limited amount of the available data (only two points in the time history) and that it is not normalised in any manner for varying inputs.

Other features extracted from transient response data that are also related to the structure's energy dissipation characteristics are the parameters of an exponential decay function fitted to the transient response signal. Because the entire time history is used, this feature is expected to be a more global representation of the system's energy dissipation properties. The function that is fitted to the transient data is

$$y(t) = y_0 e^{at} \tag{7.17}$$

where the intercept y_0 and the decay constant a can be used as a two-dimensional feature vector. Similar to the 10% duration feature, these parameters are a function of the peak amplitude of the transient signal and the structure's global damping characteristic. Again, it must be noted that these features are not normalised in any manner to the input that was applied to produce the transient response. An estimated decay curve for one of the transient signals from floor 4 of the simulated building structure is shown in Figure 7.22. These figures correspond to the same numerical data described above for the 10% duration feature. In this study, the Hilbert transform (Lyons, 2011) was used to extract an envelope of fitting

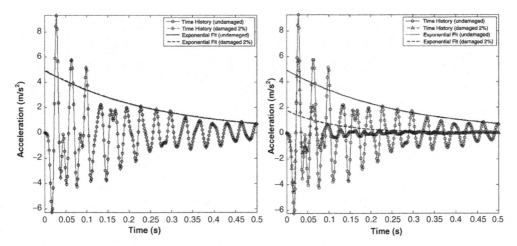

Figure 7.22   Exponential decay function fitted to undamaged and damaged data from numerical simulations of the simulated building structure: (a) damage is assumed to increase damping to 2% of critical and (b) damage is assumed to increase damping to 10% of critical.

points. The Hilbert transform (HT) is obtained by convolving the measured time history, $g(t)$, with the function $1/(\pi t)$ and is defined as

$$HT(g(t)) = \frac{1}{\pi} \int\limits_{-\infty}^{+\infty} \frac{g(\tau)}{(t - \tau)} d\tau \qquad (7.18)$$

Next, the envelope of the signal is designated $g(t)^e$ and can be obtained from the Hilbert transform as follows:

$$g(t)^e = |g(t) + iHT(g(t))| \qquad (7.19)$$

The exponentially decaying function was estimated by a least-squares fit of Equation (7.17) to this envelope. Figure 7.23 shows the coefficients of the decay functions obtained from the floor 4 data when the structure is in its undamaged and damaged conditions. In this figure the lines have been added and the values of the parameters have been slightly offset in the horizontal direction for visualisation purposes only. In general, the results of the decay function fitted to the data are similar to those obtained by the 10% duration feature. However, a closer examination of Figure 7.23, which again shows the mean and standard deviation from the ten simulations, reveals a significant change between the damaged and undamaged conditions, even for the 2% added damping damage case.

7.5   Acoustic Emission Features

Acoustic emission (AE) is the process of monitoring a structure for the release of transient elastic waves that can result from yielding, fracture, debonding or corrosion (Hellier, 2001). Because the AE signals are transient in nature, features similar to those described in the previous two sections are used to identify and quantify AEs that are associated with damage initiation and propagation. Common features that are used include the rise time, peak amplitude, duration, ring down count and the measured area under the rectified signal envelope (MARSE). These features are depicted in Figure 7.24. First, a detection threshold is set

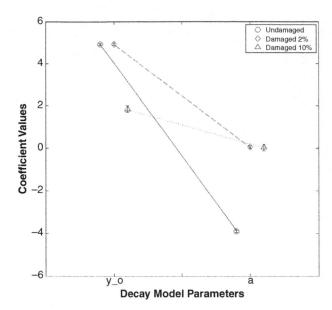

Figure 7.23 Estimated coefficients y_0 and a of the decay model fitted to numerical response data from the simulated building structure corresponding to the undamaged and damaged conditions.

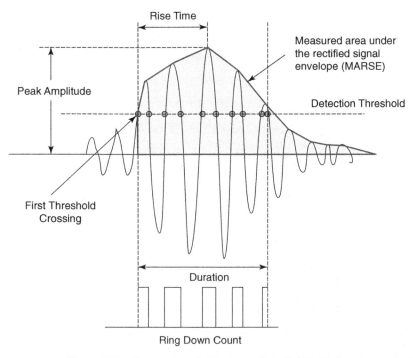

Figure 7.24 Features extracted from acoustic emission signals.

based on the statistics of the background acoustic environment (e.g. six standard deviations from the mean of the background noise). A meaningful AE is assumed to occur if the signal exceeds this threshold. The *peak amplitude* is the maximum value of the signal that occurs after the threshold has been crossed. Beginning at the time when the threshold is first crossed, the *rise time* is the time it takes to reach the peak amplitude. The total time that some portion of the signal remains above the threshold is designated as the *duration*. One can count the number of pulses where a portion of the signal remains above the threshold and this quantity is referred to as the *ring down count*. The MARSE provides a measure of the energy in the AE signal.

7.6   Features Used with Guided-Wave Approaches to SHM

Guided-wave approaches to SHM represent another class of damage detection methods that analyse transient signals. The literature on this subject is extensive and the use of guided waves for damage detection continues to be one of the most active areas of research in the SHM field. As such, this section is meant to briefly introduce the reader to some features and feature extraction procedures used in this branch of SHM. The discussion below is by no means an exhaustive survey of the subject. For a more complete treatment of this topic the reader is referred to various survey articles (Rose, 2002; Su, Ye and Lu, 2006; Cesnik, 2007) where the use of guided waves for damage detection can be traced back to studies reported in the 1950s (Worlton, 1957). These articles also cite the numerous books that have appeared on this subject beginning with Viktorov (1967). More recently, a study has been performed where many current approaches to guided-wave SHM are compared using a common data set (Flynn *et al.*, 2011).

The term *guided wave* refers to elastic waves that propagate along a path defined by the structure's boundaries (Worden, 2001). Most guided-wave approaches to SHM analyse Lamb waves, which are waves guided by the free surfaces of plates and have wavelengths of the same order of magnitude as the thickness of the plate. Lamb waves couple longitudinal and shear waves within a plate and propagate in a variety of either symmetric or antisymmetric modes. To use Lamb waves for SHM, it is useful to have a waveform that is easily recognisable both before and after propagation through the plate. For Lamb waves, the wave speed is frequency dependent, making them dispersive, whereby different frequency components travel at different velocities within the plate. To maximise the effectiveness of Lamb waves for SHM, it is useful to choose a driving frequency at which the various Lamb wave modes are initially well spaced and at which the modes of interest are relatively nondispersive (Kessler, Spearing and Soutis, 2002). Choosing a suitable driving frequency allows the receiving sensors to record the response to the input signal with a minimal amount of interference.

With the appropriate drive signal chosen, the approaches to SHM involve some form of either a pitch-catch actuation-sensing scheme or a pulse-echo scheme. In the pitch-catch scheme a Lamb wave is launched from an actuator and received by a sensor at another location. Change in the attenuation of these signals provides an indication of damage. An example of this attenuation can be seen in Figure 7.25 where corrosion has been introduced into the plate midway between sensor/actuators 2 and 3. This corrosion is causing significant attenuation of the signal produced by an actuator at location 2 and measured at location 3, as can be seen in the plot of the various measured response signals shown on the right of this figure.

With the pulse-echo approach the signal is generated at one location and the reflections of these waves off the free surfaces such as crack boundaries are measured at this same location. When the wave speed within the plate is known or measured, the arrival time of the reflected signals can be used to calculate the distance from the sensor to the damage. Almost all of the features used for guided-wave SHM are based in some manner on these attenuation and reflection characteristics.

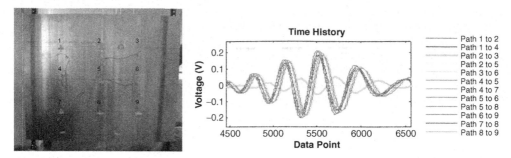

Figure 7.25   Test structure used to demonstrate attenuation-based features that can characterise changes in guided waves caused by damage.

7.6.1   Preprocessing

The feature extraction process for guided-wave SHM typically begins with signal preprocessing often involving filtering, enveloping, windowing, arrival-time estimation and/or temperature compensation. Note that these various signal preprocessing algorithms can be applied during other parts of the guided-wave feature extraction process as well. Filtering of the response signal is typically done with a band-pass filter centred on the excitation frequency and is used to reduce the influence of noise in the measured signals on the feature extraction process. Enveloping of the signal can be accomplished with a Hilbert transform as was previously discussed in Section 7.5 and has been used to minimise sensitivities to phase shifts caused by temperature variations (Croxford *et al.*, 2007). An exponential window has been used to reduce the effects of secondary reflections on the time-of-arrival estimate (Michaels and Michaels, 2007). The cross-correlation function between the input signal and response signal can be used to estimate the wave propagation time from the actuator to the sensor. The Hilbert transform has also been applied to this cross-correlation function to provide an alternate means of estimating this arrival time.

7.6.2   Baseline Comparisons

As Lamb waves propagate through a structure the mechanical energy is usually dissipated as a result of geometrical and internal damping in the structure, causing a decrease in the wave amplitude. The amount of attenuation between two points on a structure changes when damage is located in the sensor-actuator path. By comparing the amplitude of the wave packets in a baseline measurement to those in a test measurement, conclusions can be made about the existence of damage between the actuator and sensor. The attenuation feature can be used to identify the existence of damage; however, a single sensor–actuator pair can make it difficult to determine the exact location of the damage. Therefore, an array of transducers providing multiple sensor-actuator paths is usually employed to determine the damage location. Some form of comparison to the baseline waves measured on the undamaged structure is then made to establish the existence of damage. One of the most common approaches used to make this comparison is referred to as *baseline subtraction*. Here the baseline signal measured with a particular sensor–actuator pair on the undamaged plate is simply subtracted from the signal with the same transducers when the plate might potentially be a damaged system. If the two signals are from the undamaged plate in nominally similar environmental conditions this subtraction process should, in theory, yield a random noise-like signal. If damage is present, the baseline subtraction process should reveal the presence of the reflected wave or the increased attenuation associated with damage. In this case the time-domain, baseline-subtracted signal is the feature used to identify the presence of damage. Figure 7.26 shows baseline-subtracted signals for Lamb waves used to identify damage in a pipe (Thein, 2006). In the top figure the response measured in

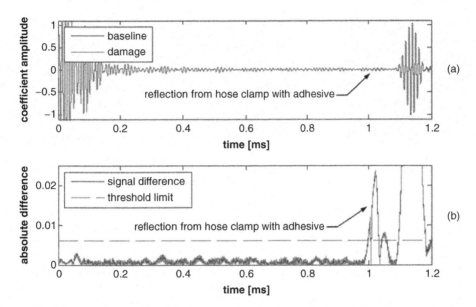

Figure 7.26   Measured response (a) and the baseline subtraction (b) used to identify the reflection of Lamb waves from simulated damage.

the undamaged and damaged conditions are overlaid where it is difficult to see the arrival of a reflected wave off the damage simulated by adding a hose clamp to the pipe. When the baseline subtraction is performed, as shown in the bottom plot, the arrival of the reflected wave is clearly identified. An alternative time-domain approach used to establish the presence of damage based on waves reflected from the damage boundary is to examine the cross-correlation function between the baseline and test measurement (Gao, Shi and Rose, 2005; Croxford *et al.*, 2007).

Alternatively, the existence of damage can be determined from frequency domain data. As an example the PSD function can be calculated for the response signal when the plate is in its undamaged and potentially damaged state. Here the assumption is that when a wave passes through damage, such as corrosion or a crack, the frequency content of the measured response signal is altered compared to a signal measured on the undamaged system. By looking at the degree of the frequency content change, one can infer that a path between a sensor and actuator contains damage. After measuring the propagated waves and band-pass filtering them about the excitation frequency, the PSD is calculated for both the data from the baseline condition and the test signals, which are designated S_{uu} and S_{tt}, respectively. A damage metric (DM) is then defined that is based on the cross-correlation coefficients of the two PSDs, which identifies the shape changes in the PSDs, implying that the frequency content has changed. This metric is defined as

$$DM_{psd} = 1 - \frac{\sigma_{S_{uu}S_{tt}}}{\sigma_{S_{uu}}\sigma_{S_{tt}}} \tag{7.20}$$

where the cross-correlation coefficient is defined as $\sigma_{S_{uu}S_{tt}} = E\left[S_{uu}S_{tt}\right] - \overline{S}_{uu}\overline{S}_{tt}$. $\overline{S}_{uu}$ and $\overline{S}_{tt}$ are the means of the baseline and test PSDs, respectively, and $\sigma_{S_{uu}}$ and $\sigma_{S_{tt}}$ are the standard deviations of the baseline and test PSDs, respectively.

The time-reversal property of Lamb waves has also been used to detect the presence of damage (Ing and Fink, 1998). Assume that an input has been applied at point A and the corresponding response measured at point B. The time-reversal property states that if the measured response is reversed in time

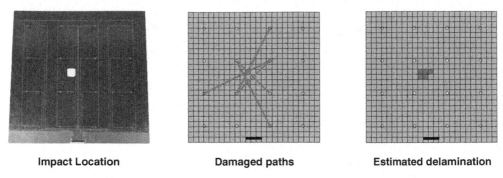

| Impact Location | Damaged paths | Estimated delamination |

Figure 7.27 A direct path approach to damage location.

and applied as an input at point B, then the response at point A should be the original input signal to within some scaling factor. If this property does not hold, then the assumption is that damage is present. Here the difference in the measured input and response at point A are the features that are being used for damage detection. There are numerous statements in the technical literature that the time-reversal approach can be used to develop a *baseline-free* damage detection method because one only needs to test for this time-reversal property with current data. However, it should be pointed out that one has to make an assumption that the plate will, indeed, exhibit this property and, as such, a baseline assumption about the system behaviour is being used to replace the baseline data.

7.6.3 Damage Localisation

In most guided-wave SHM applications, groups of sensor/actuators are placed on a plate to allow one to not only identify the presence of damage but to also locate the damage. To accomplish this goal, data from these sensor/actuators have to be combined through some form of data fusion.

A direct-path approach to damage localisation is based on analysing the Lamb wave attenuations along the various sensor/actuator paths. Once the damaged paths have been determined, by one of the previously discussed baseline comparison methods, they become the damage localisation feature and are plotted on a predefined grid (as shown in Figure 7.27 for a grid of 16 sensor/actuators from the study presented in Swartz *et al.*, 2006). The number of damaged paths intersecting at each grid element is divided by the number of undamaged paths intersecting at that point. The result is then normalised over the entire grid and the normalised values are plotted on the grid to indicate the most likely locations of damage, as shown in Figure 7.27.

Alternatively, a triangulation approach, or ellipse drawing, can be used to identify the location of damage based on the arrival times of waves reflected from the damage. One of the previously mentioned baseline comparison methods is used to identify the arrival time of the wave reflected from the damage. For a wave of known velocity, this arrival time can be used to calculate the distance from the sensor to the defect and this process can be repeated for all of the sensors. More recently, various imaging approaches to damage location have been proposed. These include the delay and sum methods (Michaels and Michaels, 2007), the hyperbola method (Croxford *et al.*, 2007) and a reconstruction algorithm for probabilistic inspection of damage (Gao, Shi and Rose, 2005), to name just a few.

7.7   Features Used with Impedance Measurements

Impedance-based SHM methods use swept-sine, high-frequency excitations, typically higher than 30 kHz, applied through surface-bonded piezoelectric patches to identify changes in the structure's mechanical impedance. As discussed in Chapter 4, the electrical impedance that is measured across a patch is a function of the mechanical impedances of the piezoelectric actuator and the host structure. Assuming that the mechanical impedance of the patch does not change over time, any changes in the electrical impedance measurement can be considered an indication of a change in the mechanical impedance of the host structure. These changes are attributed to damage such as the formation of a crack or the loosening of a bolted connection. The change in impedance caused by damage is exhibited in the real portion of the impedance signature (Park *et al.*, 2003, 2007; Bhalla and Soh, 2003; Giurgiutiu, Zagrai and Bao, 2004).

The impedance spectrum is a feature that is used to identify damage. Two basic statistical quantities have been adopted as metrics for a quantitative assessment of damage using the impedance method. The first is to find the correlation coefficient (ρ_{xy}) between the spectra of the signals corresponding to the undamaged and test conditions. This metric is defined as

$$\rho_{xy} = \frac{E\left[(Z_x - \overline{Z}_x)(Z_y - \overline{Z}_y)\right]}{\sigma_x \sigma_y} \tag{7.21}$$

The correlation coefficient is subtracted from unity yielding a damage index, $\overline{\rho}_{xy}$, that takes on values between 0 and 1, defined as

$$\overline{\rho}_{xy} = 1 - \rho_{xy} \tag{7.22}$$

A larger value of this damage index indicates a greater extent of damage.

The second method used to quantify the changes in impedance measurements is based on a frequency-by-frequency comparison and is referred to as the *root-mean-square deviation* (RMSD). This metric is defined as

$$DM_{RMSD} = \sum_{i=1}^{n} \sqrt{\frac{[\mathrm{Re}(z_{i_u}) - \mathrm{Re}(z_{i_t})]^2}{[\mathrm{Re}(z_{i_u})]^2}} \tag{7.23}$$

where DM_{RMSD} is the damage metric, Re designates the real part of a complex variable, z_{i_u} is the impedance of the baseline measurement at frequency line i, z_{i_t} is the impedance of the test measurement at frequency line i and n is the number of frequency lines over which the comparison is being made. Note that neither Equations (7.22) nor (7.23) are restricted to use with impedance measures and can be used to quantify the difference between other waveforms.

These metrics are demonstrated with impedance spectra obtained from a cantilevered beam described in Wait, Park and Farrar (2005). The aluminium beam, shown in Figure 7.28, is 101.5 cm × 10 cm × 3 mm thick. The beam is bolted, in a cantilevered condition, to two steel angles. These angles are fixed to unistrut columns, which are in turn bolted to an aluminium base plate.

The cantilevered aluminium plate is instrumented with four PZT patches and two macrofibre composite (MFC) patches. The MFC patches are a relatively new type of piezoelectric actuator that is more flexible than the conventional piezoceramic patches. The locations and the numbering schemes of these actuators/sensors are also shown in Figure 7.28. The impedance method was used to interrogate this structure using the combined network of PZT and MFC patches.

Damage was introduced by loosening one of the connection bolts in the clamped end of the test structure. For the first damage case, damage A, the bolt torque is reduced from 16.9 to 8.5 N m. Damage B refers to the condition where the same bolt has been further loosened to 1.1 N m. The real part of the impedance measurement corresponding to sensor P2, in the frequency range of 100–105 kHz, is shown in Figure 7.29 for the two different damage conditions. This portion of the spectra was chosen based on a trial-and-error procedure that showed it to be the portion where the most significant dynamic interaction between the PZT sensor and the host structure occurs. Qualitatively, this interaction is indicated by the number of peaks in this portion of the spectra, which is also indicative of local resonances.

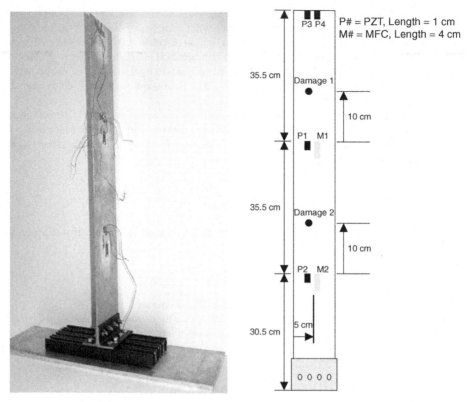

Figure 7.28   Cantilever beam instrumented with piezoelectric actuator/sensors.

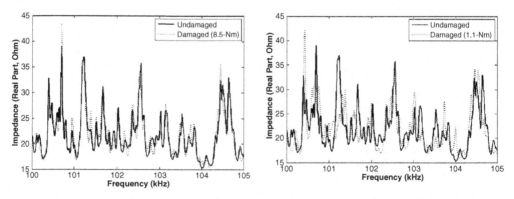

Figure 7.29   Impedance measurements (real part) measured on the test structure in Figure 7.28 corresponding to different bolt torque reductions.

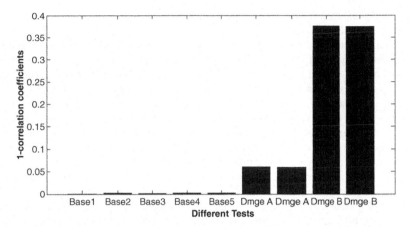

Figure 7.30   Correlation damage metric (Equation 7.22) extracted from impedance measurements made on the structure shown in Figure 7.28. Bases 1 to 5 are undamaged conditions. Damage A corresponds to a bolt torque of 8.5 N m. Damage B corresponds to a bolt torque of 1.1 N m.

It can be seen from the impedance signatures in Figure 7.29 that, with an increasing level of damage, the impedance signature shows a relatively large change in shape that clearly indicates a change in the system condition. For the first level of damage, only a small variation along the original signal (undamaged curve) is observed because there is still a clamping force being applied by the loosened bolt. When the bolt is further loosened to a torque of 1.1 N m, the impedance signature shows more pronounced variations as compared to previous readings where new peaks and valleys appear across the entire frequency range. This variation occurs because the bolt is sufficiently loosened so that it modifies the stiffness and damping properties of the joint.

A damage metric chart is illustrated in Figure 7.30 for both damage cases A and B and is based on the correlation coefficient defined by Equation (7.22). With the impedance spectrum feature, a damage metric value of zero, when compared to a baseline measurement, implies that there is complete correlation between the two impedance signatures. Such correlation between a given test measurement and a measurement from the undamaged structure, in turn, implies that there is no damage present when the test measurement was made. Increased values of this metric imply a certain degree of dissimilarity between the undamaged and test measurements. The first five measurements were made when no damage was present. Tests 6 and 7 correspond to damage A and tests 8 and 9 correspond to damage B. As can be seen in Figure 7.30, the baselines are repeatable, and when damage is introduced, there is an increase in the damage metric values. This chart provides a quantitative comparison between different data sets. Although the impedance method cannot precisely predict the exact nature and size of the damage, the method provides somewhat quantitative information about the condition of a structure by showing a damage metric that increases with increased damage severity.

7.8   Basic Modal Properties

The previous discussions of damage-sensitive features have focused primarily on some form of waveform comparison. This section introduces an alternative approach to feature extraction where physics-based models are fitted to the measured data and the parameters of these models become the damage-sensitive features. When properly applied, features based on changes in model parameters can potentially allow one to assess not only the existence and location of damage but also the extent of damage. Often such model parameters are used with an inverse modelling approach to damage detection and this approach will be discussed in this section.

Basic modal properties (resonance frequencies and mode shapes) can be readily extracted from measured dynamic input and response data by fitting parametric input–output models to these data. The well-documented field of experimental modal analysis has developed methods that fit parametric forms of the frequency response function or impulse response function to the measured input–output data (Juang and Pappa, 1985; Ewins, 1995; Maia, 1997). Also, methods for identifying basic modal properties solely from response measurements have been formalised (Andersen *et al.*, 1999; Peters and De Roeck, 2001; Zhang, Brincker and Andersen, 2005). However, when only response measurements are available, issues arise regarding methods to normalise the mode shapes and where the unmeasured inputs to the structure are only exciting a limited frequency band.

The underlying premise for using basic modal properties such as resonance frequencies and mode shapes as damage-sensitive features is that damage will alter the stiffness, mass or energy dissipation characteristics of a system. The measured dynamic response of the system and the associated modal properties that can be extracted from this response are a function of these characteristics. Although the basis for using modal properties as damage-sensitive features appears intuitive, its actual application poses many significant technical challenges. The most fundamental challenge is the fact that damage is typically a local phenomenon and may not significantly influence the lower-frequency global response of a structure that is normally measured during vibration tests.

Another challenge is that the process of extracting basic modal parameters from measured input and response data involves fitting a linear model to the data obtained before and after damage. Often there is no assessment of how accurate this fit is and if this fit will generalise to new data sets. Also, many types of damage introduce nonlinearities into the system and these nonlinearities may cause the system to violate the three basic assumptions for experimental modal analysis. These assumptions are:

1. The structure is linear.
2. The structure is time invariant.
3. The structure exhibits reciprocity.

Here reciprocity implies that if one excites a structure at one point, i, and the response is measured at another point, j, and then one subsequently reverses this process with the excitation now applied at the initial response measurement point and the subsequent response measurement made at the initial excitation point, the frequency response functions, $H(\omega)$, obtained from these two measurements processes will be identical, or

$$H_{ij}(\omega) = H_{ji}(\omega) \tag{7.24}$$

where the subscripts represent, in order, the input and response locations, respectively.

7.8.1   *Resonance Frequencies*

Structural systems will exhibit a maximal amplified response when excited by dynamic loads applied at certain frequencies. These frequencies are referred to as the *resonance frequencies* of the structure and are a function of the system's mass, stiffness and damping characteristics as well as the boundary conditions (see Appendix B). There is a large amount of literature that attempts to use shifts in resonance frequencies as a damage-sensitive feature with one of the earliest studies given in Cawley and Adams (1979). The basis for the various approaches that are summarised in the SHM literature reviews cited in the first paragraph of this chapter can be illustrated with the one- and two-degree-of-freedom (DOF) oscillators shown in Figure 7.31.

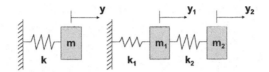

Figure 7.31   One- and two-degree-of-freedom oscillators.

For this discussion the damage is assumed only to influence the spring stiffness and the damping is neglected. The natural frequencies of the undamaged and damaged systems can be written as

$$\omega_n = \sqrt{\frac{k}{m}} \tag{7.25}$$

and

$$\omega_d = \sqrt{\frac{k_d}{m}} \tag{7.26}$$

where

ω_n = natural frequency of the undamaged system

ω_d = natural frequency of the damaged system

m = mass

k = undamaged spring stiffness and

k_d = damaged spring stiffness

It should be pointed out that throughout this chapter the subscript d will be used to denote a quantity associated with the damaged structure. It should not be confused with its use in Appendix B where this same subscript is used to denote the damped natural frequency. Also, the reader is reminded that because damping is being neglected, the system's natural frequencies and the resonance frequencies are identical.

Given an estimate of the natural frequencies corresponding to the undamaged and damaged systems and assuming that the stiffness of the damaged system will be less than the stiffness of the undamaged system, the stiffness value for the damaged system can be solved as

$$k_d = (\sqrt{k} - \Delta\omega\sqrt{m})^2 \tag{7.27}$$

where $\Delta\omega = \omega_n - \omega_d$ is the change in natural frequency derived from the measured system response data.

This derivation can be extended to the two degree-of-freedom (DOF) system shown in Figure 7.31 by following the multi-degree-of-freedom modal analysis procedure summarised in Appendix B. This analysis process yields the following homogeneous matrix equation of motion:

$$\left(-\omega^2 \begin{pmatrix} m_1 & 0 \\ 0 & m_2 \end{pmatrix} + \begin{pmatrix} k_1 + k_2 & -k_2 \\ -k_2 & k_2 \end{pmatrix} \right) \begin{pmatrix} y_1 \\ y_2 \end{pmatrix} = \begin{pmatrix} 0 \\ 0 \end{pmatrix} \tag{7.28}$$

For this system of equations to have a nonzero solution, the determinant of the coefficient matrix must be equal to zero. This determinant yields the following equation from which a quadratic in ω^2 emerges to solve for the system's two natural frequencies:

$$(-m_1\omega^2 + k_1 + k_2)(-m_2\omega^2 + k_2) - k_2^2 = 0 \tag{7.29}$$

If the system mass values are known and the two natural frequencies can be extracted from the measured system response data for the undamaged and damaged condition, then the following two sets of equations can be solved to obtain the undamaged and damaged stiffness values:

$$\begin{pmatrix} m_2\omega_1 & (m_1 + m_2)(\omega_1)^2 & -1 \\ m_2\omega_2 & (m_1 + m_2)(\omega_2)^2 & -1 \end{pmatrix} \begin{pmatrix} k_1 \\ k_2 \\ k_1 k_2 \end{pmatrix} = \begin{pmatrix} \omega_1^4 \\ \omega_2^4 \end{pmatrix}$$

$$\begin{pmatrix} m_2\omega_{1d} & (m_1 + m_2)(\omega_{1d})^2 & -1 \\ m_2\omega_{2d} & (m_1 + m_2)(\omega_{2d})^2 & -1 \end{pmatrix} \begin{pmatrix} k_{1d} \\ k_{2d} \\ k_{1d} k_{2d} \end{pmatrix} = \begin{pmatrix} \omega_{1d}^4 \\ \omega_{2d}^4 \end{pmatrix} \tag{7.30}$$

   With this information the damage can be identified, located and the changes in stiffness caused by the damage can be quantified. Note that identifying the change in stiffness is essentially providing an estimate of the severity of the damage and this level of damage detection can only be achieved because a model of the system has been employed and data from the damaged system are available (Axiom III). However, if an additional spring is added between the fixed base and m_2, then the two natural frequencies from before and after damage no longer provide sufficient information that will allow one to quantify all changes in spring stiffnesses resulting from damage.

   Although the equations become much more complicated, this approach can be extended to a general n-DOF system that includes known damping values with the number of spring elements equal to the number of DOFs. The reduced stiffness in each spring that is caused by the damage case can then be uniquely identified when there are as many measured resonance frequencies as springs (so that there are n equations and n unknowns). However, for large complex systems, there is usually more coupling between the spring elements associated with each DOF resulting in stiffness elements whose number is on the order of n^2 when, at best, only n resonance frequencies (i.e. damped natural frequencies) can be measured. This problem is confounded by the fact that usually only a limited number of the lower-frequency resonance frequencies are actually measured. Therefore, it is, in general, not possible to solve uniquely for the damaged stiffness values.

7.8.2   Inverse versus Forward Modelling Approaches to Feature Extraction

The method described above is an *inverse* approach to utilising changes in the measured resonance frequencies as a damage-sensitive feature where

$$\Delta k = f^{-1}(\Delta\omega) \tag{7.31}$$

Inverse problems refer to the general class of problems where the system parameters are estimated from measured quantities. In a *forward* formulation the system responses are normally predicted as some function of the system parameters. The entire field of system identification (e.g. Section A.11 of Appendix A) is defined by various forms of inverse problems. However, inverse problems can be ill-conditioned. Furthermore, these problems often cannot be solved directly as in the examples above because the number of unknown system parameters typically exceeds the number of available measured quantities. In these cases the inverse problem must be solved by some form of optimisation procedure.

7.8.3   Resonance Frequencies: The Forward Approach

Resonance frequencies can also be used as damage-sensitive features in a *forward* approach to damage detection where

$$\Delta \omega = f(\Delta k) \tag{7.32}$$

The forward approach consists of calculating frequency shifts from an expected type of damage. Typically, the damage type, location and extent are modelled mathematically, and then the measured changes in frequencies are compared to the predicted changes in frequencies resulting from damage. The assumption is that if the changes in measured frequencies correspond to those predicted by the simulated damage, then that damage is present in the structure. Such an assumption is predicated on damage being the only cause of changes in the measured resonance frequencies. Another underlying assumption behind using resonance frequencies as damage-sensitive features in this forward manner is that there is a one-to-one correspondence between measured frequency changes and candidate damage scenarios. Although in theory this approach can be used to locate and quantify the damage, it has typically been studied for cases where the investigator is only interested in establishing the presence of damage.

In an early example of the forward approach that makes use of resonance frequency changes as a damage-sensitive feature, the author examines changes in the frequencies associated with the first two bending modes and first torsion mode of an offshore light station tower (Vandiver, 1977). Structural members were systematically removed from a numerical model of the system and the corresponding reductions in resonance frequencies were calculated. The author went on to demonstrate how this procedure could be used to detect corrosion in the structural members. Clearly, there are significant practical limitations with such an approach. In addition to the assumptions discussed above, another limitation is the ability to model accurately the influence of the damage on a complex structure subjected to operational and environmental variability.

7.8.4   Resonance Frequencies: Sensitivity Issues

It should be noted that frequency shifts have significant practical limitations for applications to most *real-world* structures. In addition to modelling uncertainties discussed above, a primary issue is the low sensitivity of the lower-resonance frequencies, associated with global modes of the structure, to local damage. This insensitivity is illustrated in Figure 7.32 where a 10% reduction in stiffness has systematically been introduced at each spring in the 8-DOF system shown in Figure 7.1. The resonance frequencies for all eight modes have been plotted together in Figure 7.32. From this plot it is evident that there would have to be very accurate and consistent measures of these resonance frequencies to be able to definitively assess that damage was present in the system. In fact, for this noise-free simulation, the largest change in resonance frequency produced by the 10% stiffness change at any location is associated with mode 1 and corresponding to the 10% stiffness reduction at spring 1. Here one must be able to measure a 1.28% change in the resonance frequency to be able to identify that damage was present. For the frequency associated with this mode, the sampling parameters would have to be defined such that the data acquisition system could resolve a frequency change on the order of 0.13 Hz.

The somewhat low sensitivity of frequency shifts to damage requires either very precise measurements or large levels of damage for this feature to be an effective damage indicator. For example, in offshore platforms, damage-induced frequency shifts are difficult to distinguish from shifts resulting

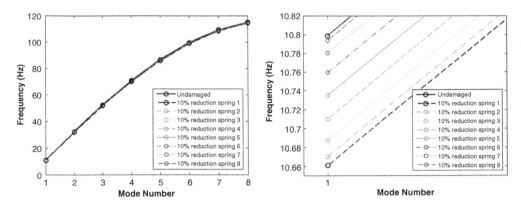

Figure 7.32 Resonance frequencies for eight modes of the system shown in Figure 7.1 and the corresponding changes in resonance frequencies caused by reductions in spring stiffness values at different locations. Plot on the right shows an expanded view of these changes associated with the lowest resonance frequency.

from increased mass caused by marine growth (Wojnarowski and Reddy, 1977). Results from experimental modal analyses carried out on the I-40 Bridge and concrete column data (Sections 5.1 and 5.2, respectively), which are summarised for mode 1 in Table 7.5, further illustrate this point. For the concrete column tests, which were conducted in a well-controlled laboratory setting, there is the expected consistent trend of decreasing first mode frequencies as the damage level increases. In contrast, the I-40 Bridge tests were conducted in the field under varying operational and environmental conditions. Table 7.5 shows much less consistent trends for the I-40 Bridge's first mode frequencies, where the removal of traffic and its associated mass produces almost as big a reduction in first mode frequency as the most severe damage case. Also, for damage levels 1 and 2 the resonance frequency is seen to increase, which is related to changing environmental conditions. Currently, using frequency shifts to detect damage appears to be more practical in applications where such shifts can be measured very precisely in a controlled environment, such as for quality control in manufacturing. As an example, a method known as *resonant ultrasound spectroscopy* has been used effectively to determine out-of-roundness of ball bearings (Migliori and Sarrao, 1997).

Also, because resonance frequencies are a global property of the structure, it is not clear that shifts in this parameter can be used to identify more than the mere existence of damage. In other words, the frequencies generally cannot provide spatial information about structural changes unless some form of model correlation is performed. An exception to this limitation occurs at higher modal frequencies

Table 7.5 Changes in resonance frequencies caused by damage

I-40 Bridge		Concrete column	
Damage case	First mode frequency	Damage case	First mode frequency
Undamaged (ambient traffic loading)	2.39 Hz	Undamaged (Test 6)	14.0 Hz
Undamaged (shaker)	2.48 Hz	First damage level (Test 7)	12.0 Hz
Damage level 1	2.52 Hz	Second damage level (Test 9)	7.0 Hz
Damage level 2	2.52 Hz	Third damage level (Test 11)	5.0 Hz
Damage level 3	2.46 Hz		
Damage level 4	2.30 Hz		

where the modes are associated with local responses. However, when one examines how the change in resonance frequency, $\delta\omega$, is related to a change in stiffness, δk, as given by the following relationship:

$$\delta\omega = \frac{\mathrm{d}\omega}{\mathrm{d}k}\delta k = \frac{1}{2\sqrt{km}}\delta k = \frac{1}{2m\omega}\delta k \tag{7.33}$$

it can be seen that proportionally larger changes in stiffness are needed at higher frequencies to produce a given level of frequency change. Also, the practical limitations involved with the excitation and identification of these local modes, caused in part by high modal density and the difficulties associated with providing enough energy to excite the higher-frequency response, can make these modal frequencies difficult to identify. As shown in the 2-DOF example in Section 7.8.1, multiple-frequency shifts can provide spatial information about structural damage because changes in the structure at different locations will cause different combinations of changes in the modal frequencies. However, there are often an insufficient number of frequencies experiencing significant changes to determine the location of the damage uniquely.

7.8.5   Mode Shapes

The SHM reviews cited at the beginning of this chapter report numerous studies that have used mode shapes as a damage-sensitive feature. Mode shapes provide spatially distributed information about the dynamic characteristics of the structure and therefore offer the ability to locate the damage as well as establish the existence of damage. When a linear structure that exhibits proportional damping (described in Appendix B) is excited by a harmonic function whose frequency corresponds to a resonance frequency of the structure, the structure will deform in a characteristic shape where the ratio of displacements for any two degrees of freedom is constant with time. These characteristic shapes are referred to as *modes of vibration*. In most cases one does not observe the mode shape directly. The operating deflection shapes are observed and these shapes are some linear combination of the modes of vibration.

For the subsequent discussion regarding the use of mode shapes as damage-sensitive features a couple of points must be emphasised. First, if a driving-point measurement is made, then mass-normalised mode shapes can be obtained. The term *unit-mass-normalised* implies that for a particular mode shape $\{\psi\}$ associated with a resonance frequency ω

$$\{\psi\}^{\mathrm{T}}[m]\{\psi\} = 1 \quad \text{and} \quad \{\psi\}^{\mathrm{T}}[k]\{\psi\} = \omega^2 \tag{7.34}$$

where $[m]$ and $[k]$ are the system mass and stiffness matrices, respectively. A driving-point measurement is one where a response measurement is made at the point of excitation during a vibration test. Clearly, one cannot obtain mass-normalised modes when extracting mode shape information solely from response measurements.

Second, proportional damping is not based on developing a realistic model of the physical energy loss mechanisms in a structure. However, it has been shown that for lightly damped systems this relationship gives an adequate approximation of the structure's observed dynamic behaviour. If the structure responds with higher levels of damping, then it will exhibit *complex modes* where the displacement ratio is no longer constant with time. A characteristic of the complex mode is that when the mode shape is animated it appears as if there is a travelling wave propagating through the structure.

Examples from the I-40 Bridge structure (Section 5.1) will be used to illustrate issues associated with using mode shapes as damage-sensitive features. Figure 7.33 shows the mode shapes obtained from experimental modal analyses performed on the I-40 Bridge in its undamaged state and after introducing the final level of damage. A qualitative inspection of these experimental mode shape changes reveals the damage location at the centre of the middle span. For this final level of damage the load path through

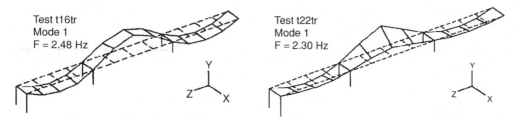

Figure 7.33 Mode 1 identified by experimental modal analysis. Plot on the left shows mode identified for the undamaged structure and the plot on the right shows the mode identified after the final level of damage.

the structure has been altered by the severance of the bottom flange, which is the key component for transferring the vertical loads through bending to the bridge piers. For the other damage cases associated with cuts in the girder web there is no discernible change in the mode shapes (Farrar *et al.*, 1994). This result points to a key observation regarding the effectiveness of mode shape changes to identify damage. *If the damage alters the load path through the structure, it will most likely produce measurable changes in the lower frequency global modes of the structure.*

It is interesting to examine the second bending mode obtained from the experimental modal analyses, as shown in Figure 7.34. This mode has a node point at the location where damage was introduced. A node point for a mode is a point on the structure that remains stationary when the structure responds in that particular mode. Even after the severe damage case, there is no discernible change in this mode when compared to the corresponding undamaged values. From these examples it becomes clear that the fidelity of a particular mode as a damage-sensitive feature is directly related to the location of damage relative to a node point for that mode.

Now the mode shapes associated with the 8-DOF system shown in Figure 7.1 will be analysed to investigate the use of mode shapes as damage-sensitive features in a more quantifiable manner. Figure 7.35 shows the changes in the first and eighth mode shapes, respectively, for the 8-DOF system when a 10% spring stiffness reduction was sequentially introduced at each spring. When plotted on a scale where the entire mode shape is visible, it appears that there is little change in these mode shapes when the 10% stiffness reduction is introduced at the various locations. However, percentage changes in mode shape amplitudes for mode 1 range from 0.015% to 8.8%, with the largest percentage change occurring at DOF 1 when the 10% reduction in stiffness has been introduced into spring 1. For mode 8 the largest changes were observed and varied between 0.33% and 30%, with the largest change occurring at DOF 7, and corresponded to the 10% reduction at spring 7. These examples indicate that in certain cases

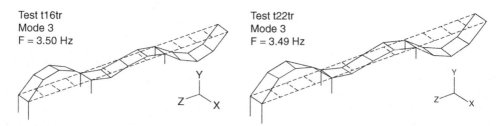

Figure 7.34 The second bending mode identified by experimental modal analysis. Plot on the left corresponds to the undamaged structure. Plot on the right was identified from data measured after the final level of damage had been introduced. Note that there is a node for this mode at the damage location.

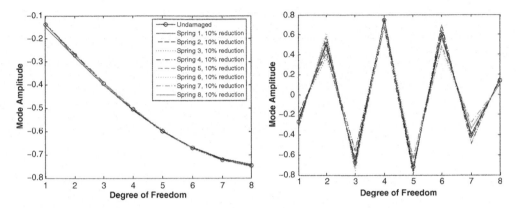

Figure 7.35 A comparison of mode shapes calculated for the 8-DOF system shown in Figure 7.1 after a 10% reduction in stiffness has been sequentially introduced to each spring. Plot on the left compares the first modes and the plot on the right compares the eighth modes.

mode shapes can be successfully used as a damage-sensitive feature, but there is a need to quantify these changes better.

In one of the earlier studies that systematically uses mode shape information for the location of structural damage, the modal assurance criteria (MAC) is used as a metric to determine the level of correlation between modes from the test of an undamaged space shuttle orbiter body flap and the modes from the test of the flap after it has been exposed to acoustic loading (West, 1984). The MAC that compares two complex modes, $\{\psi\}_x$ and $\{\psi\}_y$, is defined as

$$MAC(\{\psi\}_x, \{\psi\}_y) = \frac{\left|\{\psi\}_x^{\mathrm{T}} \{\psi^*\}_y\right|^2}{(\{\psi\}_x^{\mathrm{T}} \{\psi^*\}_x)(\{\psi\}_y^{\mathrm{T}} \{\psi^*\}_y)} \tag{7.35}$$

where * indicates the complex conjugate (Allemang and Brown, 1982). The MAC was developed to take advantage of the orthogonality properties of mode shapes and takes on a value between 0 and 1. A value of 1 indicates that the modes are identical up to some scalar normalisation factor and a value of 0 indicates that the two modal vectors being compared are orthogonal to each other. Essentially, the MAC is giving a measure of the cosine of the angle between the two vectors. It should be pointed out that the MAC represents a significant amount of data compression. It reduces the comparison of two mode shape vectors to a single scalar value. With this much compression there is often a loss of sensitivity to relatively small local changes in the mode shape vector. This metric is not limited to comparing mode shapes and can be applied to any vectors including the high-dimensional time series and spectral features introduced at the beginning of this chapter.

The matrix of MAC values in Table 7.6 compares all the modes for the 8-DOF system shown in Figure 7.1 corresponding to the undamaged system and the system that has the 10% spring stiffness reduction introduced into spring 1. The values along the diagonal of this matrix represent a comparison between a particular mode before and after damage. If the damage had produced no change in the mode shapes, the matrix of MAC values should be the identity matrix. Although examination of the diagonal terms in the matrix shows some difference between the undamaged and damaged mode shapes, the percentage changes in these values are much smaller than the individual mode shape amplitude changes. For this example the maximum percentage change in the MAC values is only 0.12% when the maximum amplitude change at a particular DOF for this damage scenario was previously found to be 8.8%. Note that the MAC value will not capture an overall scale change to the mode shape. As can be seen from

Table 7.6   Modal assurance criteria comparing the undamaged 8-DOF system modes to modes corresponding to a 10% spring stiffness reduction at spring 1

Mode	1	2	3	4	5	6	7	8
1	0.9999	0.0001	0.0000	0.0000	0.0000	0.0000	0.0000	0.0000
2	0.0001	0.9993	0.0005	0.0001	0.0000	0.0000	0.0000	0.0000
3	0.0000	0.0004	0.9984	0.0008	0.0002	0.0001	0.0000	0.0000
4	0.0000	0.0001	0.0008	0.9978	0.0010	0.0002	0.0001	0.0000
5	0.0000	0.0000	0.0002	0.0009	0.9978	0.0009	0.0001	0.0000
6	0.0000	0.0000	0.0001	0.0002	0.0008	0.9983	0.0005	0.0001
7	0.0000	0.0000	0.0000	0.0000	0.0001	0.0005	0.9991	0.0002
8	0.0000	0.0000	0.0000	0.0000	0.0000	0.0001	0.0002	0.9997

Figure 7.35, the 10% reductions in stiffness do not fundamentally change the *shape* of the mode and, hence, the orthogonality properties do not change significantly.

The MAC value can be illustrated with results from more complicated *real-world* structures such as the I-40 Bridge. An examination of the first mode shapes from this study shown in Figure 7.33 reveals that the final level of damage produces a significant change in the shape of the first mode. This change is indicated by the changes in MAC values given in Table 7.7, which compares modes obtained from experimental modal analyses performed on the structure when it was tested in the undamaged condition and similar analyses performed after the final level of damage had been introduced. For this example, there is an 18% change associated with the MAC value for mode 1. Note that for the second bending mode, mode 3 in this case, there is only a 0.3% change in the MAC value even after this severe level of damage has been introduced. This result again indicates that modes with a node at the damage location are relatively insensitive to even large amounts of damage. Table 7.7 also shows the MAC values corresponding to the third level of damage. Here it is seen that the biggest change in a MAC value is associated with mode 4

Table 7.7   Modal assurance criteria from modes obtained during undamaged and damaged forced vibration tests on the I-40 Bridge

Third level of damage (E-3)						
Mode	1	2	3	4	5	6
1	0.997	0.002	0.000	0.005	0.001	0.001
2	0.000	0.996	0.001	0.003	0.002	0.002
3	0.000	0.000	0.999	0.006	0.006	0.000
4	0.003	0.005	0.004	0.981	0.032	0.011
5	0.001	0.006	0.004	0.064	0.995	0.003
6	0.002	0.002	0.000	0.004	0.009	0.995
Final level of damage (E-4)						
Mode	1	2	3	4	5	6
1	0.821	0.168	0.002	0.001	0.000	0.001
2	0.083	0.884	0.001	0.004	0.001	0.002
3	0.000	0.000	0.997	0.005	0.007	0.001
4	0.011	0.022	0.006	0.917	0.010	0.048
5	0.001	0.006	0.003	0.046	0.988	0.002
6	0.005	0.005	0.000	0.004	0.009	0.965

and corresponds to only a 1.9% reduction from the undamaged case, which again provides evidence that until damage changes the load path through the structure the mode shapes are somewhat insensitive to local damage. Finally, it should be noted that although the MAC values quantify the change in mode shapes, because it is a scalar quantity it does not provide spatial information regarding the location of the damage.

Alternatively, to locate the damage better using mode shape data one can calculate a coordinate-by-coordinate MAC value, referred to as a coordinate modal assurance criteria (COMAC). For N modal vector pairs, $\{\psi\}_x$ and $\{\psi\}_y$, the COMAC for DOF p is defined as

$$COMAC(\{\psi\}_{x_p}, \{\psi\}_{y_p}) = \frac{\left| \sum_{r=1}^{N} \psi_{x_{pr}} \psi_{y_{pr}} \right|^2}{\sum_{r=1}^{N} \psi_{x_{pr}} \psi_{x_{pr}}^* \sum_{r=1}^{N} \psi_{y_{pr}} \psi_{y_{pr}}^*} \tag{7.36}$$

where the subscript r refers to the rth mode shape (Lieven and Ewins, 1988). Figure 7.36 shows contour plots generated from the 26 COMAC values calculated between the mode shape vectors obtained from the undamaged I-40 Bridge structure and the various sets of mode shapes obtained from the damaged bridge. These plots were generated with data from modes 1, 2, 3, 5 and 6. Here it is seen that at the highest three levels of damage the COMAC values corresponding to the locations closest to the damage show the largest discrepancy from the undamaged condition. For the first damage case there is no discernible change in the COMAC values except at the north girder end-point, which is attributed to the influence of noise in the measurements as this point is fixed in both the damaged and undamaged condition. There are many other studies that examine the MAC and COMAC values as damage-sensitive feature metrics (Kim, Jeon and Lee, 1992; Srinivasan and Kot, 1992; Ko, Wong and Lam, 1994; Salawu and Williams, 1994; Lam, Ko and Wong, 1995). A more recent detailed discussion of the MAC and COMAC as well as other related parameters can be found in Allemang (2003). Finally, as they have been presented above, the MAC and COMAC are being introduced as metrics for quantifying the similarity or difference between two features – the mode shape vectors in this case. However, one can also view changes in these metrics as damage-sensitive features. When the evolution of the COMAC values is viewed as shown in Figure 7.36, these changes are giving an indication of damage and its location.

The Yuen function provides another metric for quantifying the change in a mode shape (Yuen, 1985). This function is defined over the length of a beam-like structure and is constructed from the modal vectors and resonance frequency values for the damaged and undamaged systems. Suppose $\{\psi\}_{i_u}$ is the ith mode shape vector of the undamaged system and ω_{i_u} is the corresponding resonance frequency. Furthermore, suppose $\{\psi\}_{i_d}$ and ω_{i_d} are the corresponding quantities for the damaged system; then the ith Yuen function is a vector defined by

$$\{Y\}_i = \frac{\{\psi\}_{i_d}}{\omega_{i_d}} - \frac{\{\psi\}_{i_u}}{\omega_{i_u}} \tag{7.37}$$

This function is illustrated with numerical data from the system shown in Figure 7.1. The influence of 10% reductions in the spring stiffness at varying locations can be seen in Figure 7.35, where only a small distortion is evident. Again, it is noted that the distortion becomes more pronounced in the higher modes, although the low level of distortion suggests that it might be useful to consider further processing of the mode shapes in order to obtain features that give a clear indication of the damage location.

Figure 7.37 shows the function $\{Y\}_1$ (i.e. from the first mode) when the 10% stiffness reduction was introduced at each spring location. In this figure the damage location is signalled quite clearly as the

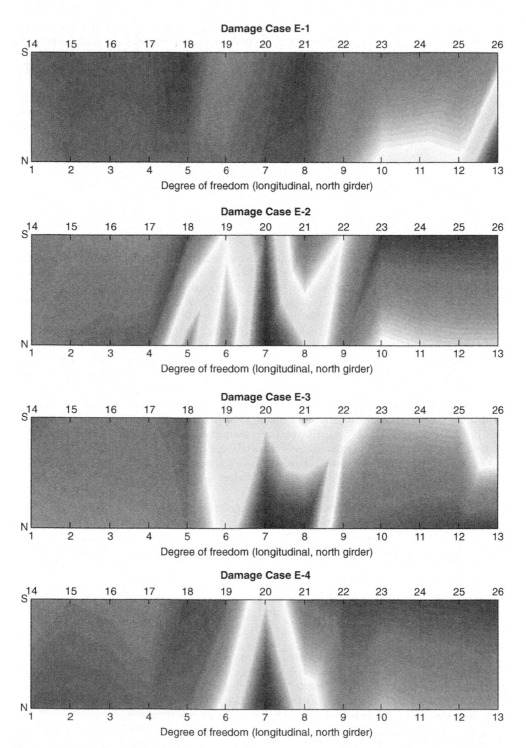

Figure 7.36 COMAC contour plots for the I-40 Bridge comparing the modes from the undamaged structure to modes corresponding to the structure in its different damage conditions.

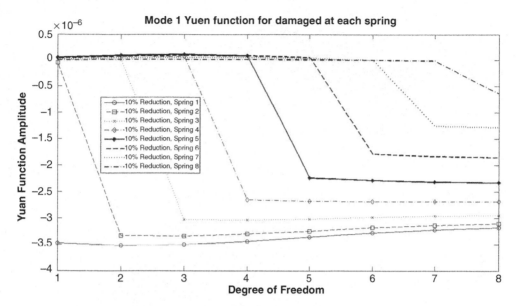

Figure 7.37   First mode Yuen function for the spring–mass system shown in Figure 7.1.

node at which the function first becomes nonzero. Note that the discontinuity in the function becomes less marked towards the free end of the system, which is to be expected as very little strain energy is concentrated in this region for vibrations at the first resonance frequency. As a consequence it is more difficult to distinguish between the damaged and undamaged mode shapes if the damage is in this region.

The process of subtracting the baseline mode from the mode corresponding to the damaged condition to accentuate the damaged area can be extended to other structural elements. Consider now the mode shapes of a two-dimensional aluminium cantilever plate structure as determined by a finite element simulation. Damage cases were simulated by lowering the modulus of elasticity to 1% of its baseline value at a specific location. Figure 7.38 shows the first mode shape for the undamaged system as a surface and in contour map form.

When damage was induced in one of the elements along the free edge and nearest to the clamped edge ($x = 200$ mm), there proves to be very little indication of damage when one solely examines the first mode shape from this damaged condition. However, if the mode shape from the undamaged plate is subtracted from the damaged condition mode shape, the damage location becomes much more pronounced, as can be seen in Figure 7.39. Note that this figure shows the normalised change in the mode shape, not the Yuen function.

Several other metrics for quantifying changes in the mode shape that have been proposed in the technical literature include a *node line* MAC based on measurement points close to a node point for a particular mode (Fox, 1992) and a structural translational and rotational error-checking procedure that calculates the ratios of relative changes in modal components (Mayes, 1992).

7.8.6   Load-Dependent Ritz Vectors

Mode shapes are a set of orthogonal vectors, defined by a modal model, that span the system's dynamic response space. These vectors are a function of the mass, stiffness and energy dissipation properties of the structure and are independent of the excitation. Load-dependent Ritz vectors represent an alternative set of orthogonal vectors that span the response space of dynamic systems and, as the name implies,

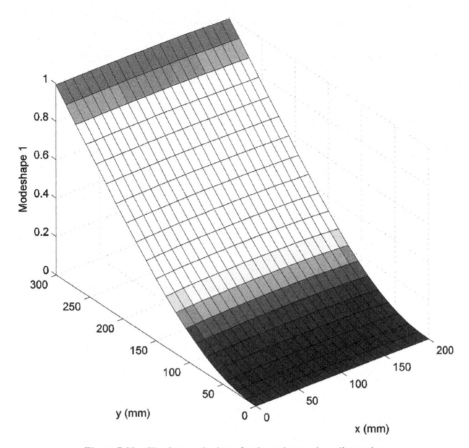

Figure 7.38 The first mode shape for the undamaged cantilever plate.

are a function of the applied loading to the system. Ritz vectors (or Lanczos vectors) have been used extensively for various types of dynamic analyses, but they had not until recently been applied to SHM problems because a procedure for experimentally identifying Ritz vectors from measured vibration testing data did not exist. In 1997 a system identification procedure for extracting Ritz vectors from measured vibration test data was developed. This procedure recursively identifies the Ritz vectors from the estimated state representation of the dynamic system determined from measured frequency response functions or impulse response functions (Cao and Zimmerman, 1997). Alternatively, the stiffness matrix can be estimated from the mass-normalised mode shapes by inverting the dynamic flexibility matrix (see Section 7.9.3 below) and the recursive process defined by Equations (7.39) to (7.43) can be employed to estimate the load-dependent Ritz vectors. Once this method of feature extraction was developed, the Ritz vectors were subsequently used in several SHM studies (Burton, Farrar and Doebling, 1998; Sohn and Law, 2001a, 2001b).

To calculate the Ritz vectors, recall the second-order differential equation of motion for a multi-degree-of-freedom (MDOF) lumped-parameter system subjected to a forcing function described by the loading vector $\{x(t)\}$

$$[m]\{\ddot{y}\} + [c]\{\dot{y}\} + [k]\{y\} = \{x(t)\} \tag{7.38}$$

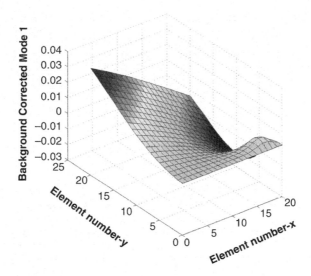

Figure 7.39   Baseline-normalised mode shape for the damaged cantilever plate.

An estimate of the first load-dependent Ritz vectors, $\{\overline{\lambda}\}_1$, is defined as the static displacement of the structure when subjected to forcing $\{x(t)\}$

$$\{\overline{\lambda}\}_1 = [k]^{-1} \{x(t)\} \tag{7.39}$$

This first estimate is then scaled to produce the first mass-normalised Ritz vector, $\{\lambda\}_1$, as follows:

$$\{\lambda\}_1 = \frac{\{\overline{\lambda}\}_1}{\left[\{\overline{\lambda}\}_1^T [m] \{\overline{\lambda}\}_1\right]^{1/2}} \tag{7.40}$$

The subsequent Ritz vector estimates are calculated using the recursive formula

$$\{\overline{\lambda}\}_i = [k]^{-1} [m] \{\lambda\}_{i-1} \tag{7.41}$$

During these calculations a Gram–Schmidt orthogonalisation process is applied to each newly derived Ritz vector to ensure that the vectors are linearly independent

$$\{\tilde{\lambda}\}_i = \{\overline{\lambda}\}_i - \sum_{j=1}^{i-1} \left(\{\lambda\}_j^T [m] \{\overline{\lambda}\}_i\right) \{\lambda\}_j \tag{7.42}$$

and each new orthogonalised vector, $\{\tilde{\lambda}\}_i$, is then normalised with respect to the mass matrix to produce the next Ritz vector

$$\{\lambda\}_i = \frac{\{\tilde{\lambda}\}_i}{\left[\{\tilde{\lambda}\}_i^T [m] \{\tilde{\lambda}\}_i\right]^{1/2}} \tag{7.43}$$

This process is repeated to estimate the subsequent Ritz vectors.

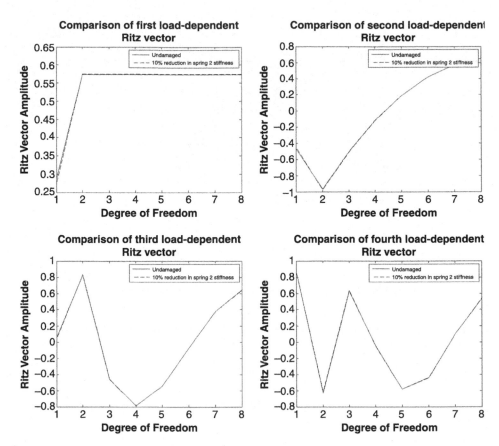

Figure 7.40   Load-dependent Ritz vectors for the 8-DOF system shown in Figure 7.1 with load applied at DOF 2.

To illustrate this process, the load-dependent Ritz vectors are calculated for the system shown in Figure 7.1 in its undamaged condition and with a 10% reduction in spring 2 stiffness. A unit load is assumed to be applied at node 2 for this calculation. These vectors are shown in Figure 7.40 where, as with the mode shape vectors, the larger differences are more apparent in the higher modes. The percentage change in the first Ritz vector amplitude for these cases ranges from −0.18% to 5.1% with the largest change occurring at DOF 1. The largest change not associated with a value close to zero is associated with the eighth Ritz vector at DOF 5 and is of the order of 50%. Note that the same metrics used to quantify changes in mode shape vectors (e.g. the MAC and COMAC) can also be used to quantify changes in the Ritz vectors.

7.9   Features Derived from Basic Modal Properties

In an effort to find features that have a higher sensitivity to damage, researchers working in the SHM field have identified alternate damage-sensitive features that are derived from the basic modal properties discussed above. In many cases these derived features were developed based on a physical interpretation of the relationship between the change in modal properties and the changes to the structural properties. The fact that all these features are derived from basic modal properties implies that they are based on fitting a linear modal model to the data obtained from the undamaged and damaged structure.

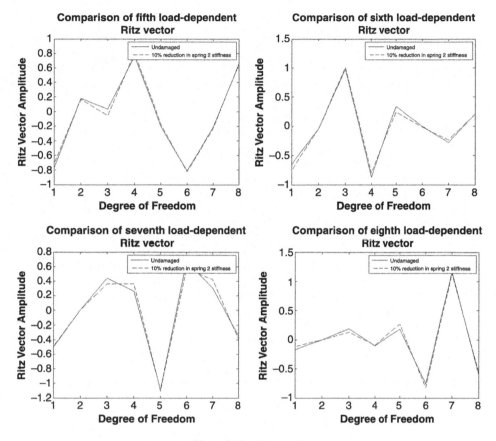

Figure 7.40 *(Continued)*

7.9.1 Mode Shape Curvature

For structures exhibiting bending behaviour, an alternative to using mode shapes to obtain spatial information about damage is to use mode shape derivatives, such as curvature (Pandey, Biswas and Samman, 1991). The derivative process has the effect of amplifying any discontinuities in the mode shape caused by localised damage. Given the mode shapes obtained from the undamaged and damaged structure, consider a beam cross-section at location x subjected to a bending moment $M(x)$. The curvature at location x, $v'(x)$, is approximated by

$$v''(x) \approx \frac{M(x)}{EI} \tag{7.44}$$

where E is the modulus of elasticity and I is the cross-sectional moment of inertia. From this equation, it is evident that the curvature is directly proportional to the inverse of the flexural stiffness, EI. Thus, for a given moment applied to the damaged and undamaged structure, a reduction of stiffness associated with damage, in turn, leads to an increase in curvature. Furthermore, for certain geometries an estimate of the extent of damage at a section can be obtained by measuring the amount of change in the mode shape curvatures. The larger the reduction in the flexural stiffness (i.e. higher level of damage), the larger the change in the mode shape curvatures from an undamaged to a damaged condition.

For a beam whose mode shapes are approximated by discrete measurement points equally spaced at a distance h along its length, the mode shape curvature can be computed using the central difference approximation to the second derivative at DOF i as

$$v''(\psi_i) \approx \frac{\psi_{i-1} - 2\psi_i + \psi_{i+1}}{h^2} \tag{7.45}$$

Backward and forward difference operators can be used to approximate the curvature at the ends of the beam.

This method can be used with mode shapes that have been normalised arbitrarily, but consistently, which implies that mode shapes obtained from ambient response data can be effectively analysed by this method. As in the case of using mode shapes themselves as a damage-sensitive feature, a method must be developed to quantify the change in mode shape curvature. For multiple modes, the absolute values of change in curvature associated with each mode can be summed to yield a damage metric for a particular location.

Figure 7.41 shows the first mode curvatures when a 10% stiffness reduction is sequentially enforced at springs 2–7 in the system shown in Figure 7.1. A clear maximum occurs near the location of the damage. This maximum is present for all damage locations. However, as with the Yuen functions, damage nearer the free end is not as well indicated.

The effect of the damage can be amplified by subtracting the mode shape curvature associated with the undamaged structure. Figure 7.42 shows the baseline-corrected mode shape curvature values for the spring–mass system in Figure 7.1 when damage was located at various locations along the length of the beam. The mode shape curvature estimate based on the central difference operator can be extended to plate-and-shell-type structures using a Laplacian operator (Ratcliffe, 1997).

As an alternative to applying the central difference operator in Equation (7.45) to the discrete mode shape values in an effort to estimate the curvature, one can fit a function to the mode shape and then analytically take the derivatives of this function as a means of defining a continuous curvature

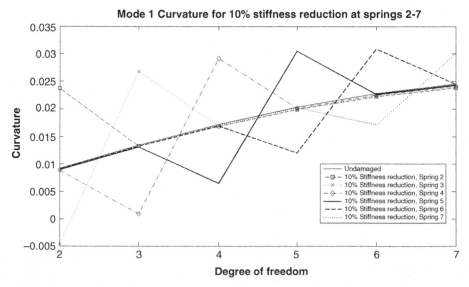

Figure 7.41 First mode shape curvature for the spring–mass system shown in Figure 7.1 corresponding to various damage locations.

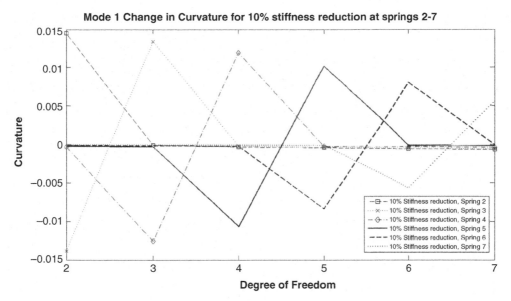

Figure 7.42   Baseline-corrected first mode shape curvature for the spring–mass systems shown in Figure 7.1 corresponding to various damage locations.

function along the length of the beam. This approach was taken when assessing changes in mode shape curvature for the I-40 Bridge. First, a cubic interpolation function was fitted through the measured modal amplitudes associated with the undamaged and damaged structure to obtain pre- and post-damage mode shape functions. Curvatures of these mode shapes, representative of the structure before and after damage, were then determined by differentiating the mode shape functions at 211 equally spaced locations along the length of the bridge girder. The absolute value of the difference between the respective mode shape curvatures was then calculated for each individual mode. Differences for each mode were added to form a final damage feature related to the changes in curvature at a particular location. The absolute differences between the mode shape curvatures for the undamaged and damaged bridge are plotted in Figure 7.43. Note that two sets of mode shape curvature data are presented in each plot. The first set is the sum of the absolute difference between curvatures corresponding to the undamaged and damaged conditions using only the first bending and first torsion modes. The second set represents an accumulation of the absolute difference in curvatures for the first six modes.

For all four damage cases, the change in curvature computed using only two modes reaches a peak at the correct damage location (node 106). As would be expected, the peak was most distinct for the final damage case. For the first three damage stages the maximum changes in curvatures at the true damaged location are accompanied by smaller peaks in adjacent areas. When all six modes were included, the maximum change in curvature for the first and third damage cases did not coincide with the actual location of damage. For the second and final damage cases (see plots (b) and (d) of Figure 7.43), the maximum change in curvature appears at the damaged region when all six modes are used in the calculation. Thus it was concluded that when damage is not severe, inaccurate predictions may be made if modes not significantly affected by damage are included in the calculation of the change in curvature.

An issue that must be considered when using the mode shape curvature as a damage-sensitive feature is that a polynomial fitted to the discrete mode shape data tends to smooth the data, which may mask a subtle change in curvature. As an example, when the change in mode shape curvatures for the I-40 Bridge is estimated using the central difference operator Equation (7.45) the changes at the measurement locations

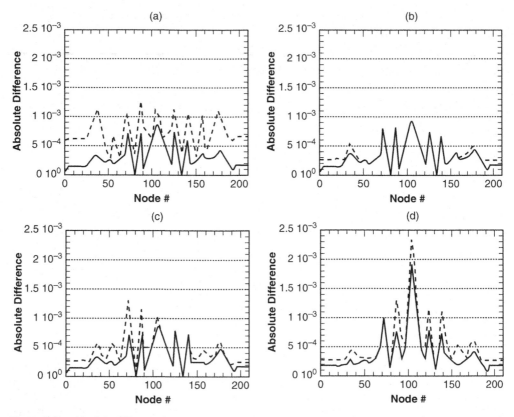

Figure 7.43 Absolute difference in mode shape curvatures between the undamaged and damaged modes corresponding to the I-40 Bridge's undamaged and damaged conditions. The solid line is based on the first two modes and the dashed line is based on all six identified modes.

are more pronounced than those obtained with the previously described polynomial fit. Additionally, it is well documented that the differentiation process amplifies high-frequency noise, so it can be expected that the variance associated with features obtained through a differentiation process might increase relative to the variance associated with the original data (Hamming, 1989). This potential for increased variance must be considered with the mode shape curvature feature and the strain energy features described in the next section.

7.9.2   Modal Strain Energy

The damage-sensitive feature described here is based on the concept of modal strain energy, which is the strain energy stored in a structure when it deforms in its mode shape pattern (Stubbs, Kim and Topole, 1992). The basic idea is that, when damage occurs, the distribution of strain energy originally stored in the structure will change in a more pronounced manner in the damaged areas. Once a structural member experiences a change in stiffness caused by damage, which is almost always a reduction, it can no longer absorb the same amount of energy as it did when undamaged. This effect results in a deviation from the original strain energy distribution when compared to the undamaged structure. Therefore, changes

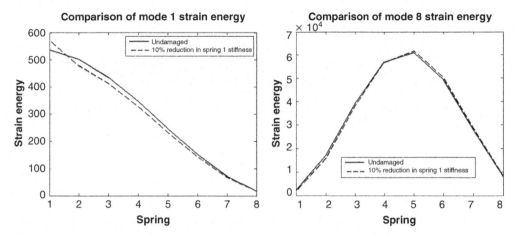

Figure 7.44   Change in modal strain energies for the system shown in Figure 7.1 subjected to a 10% reduction in stiffness at spring 1.

in the strain energy distributions of the undamaged and damaged structures can be used to detect and locate damage. Consider the structure shown in Figure 7.1. The strain energy stored in the spring when the structures deforms in one of its mode shapes is defined as

$$U = \frac{1}{2}k(\Delta x)^2 \tag{7.46}$$

where the values of Δx are the changes in length of the springs from its undeformed state (Duffey *et al.*, 2001). These values can be obtained from the mode shape vector by noting that the mode shape vector gives the change in the position of the DOF relative to its initial position.

Figure 7.44 shows the changes in the modal strain energies in each spring for mode shapes 1 and 8 corresponding to a 10% stiffness reduction in spring 1. Here the difference between mode shape amplitudes at adjacent degrees of freedom is being squared to arrive at the strain energy value. This simulated damage case produces observable changes in modal strain energy stored in the springs. When mode 1 is used, the largest change is a 6.5% increase in the modal strain energy associated with spring 1. The concept of modal strain energy is based on the idealisation that the structure is deforming purely in the pattern defined by a particular mode. It is not the true strain energy stored in the structure during a dynamic excitation, as the actual deformation is some scaled linear combination of the modal vectors and the true strain energy will be a time-varying quantity as the structure responds to the external dynamic load.

The modal strain energy feature can be extended to structures whose mode shapes resemble that of a beam characterised by a one-dimensional curvature (Stubbs, Kim and Farrar, 1995). This feature is closely related to the mode shape curvature discussed in Section 7.9.1. The strain energy, U, of a Bernoulli–Euler beam of length L is given by

$$U = \frac{1}{2} \int_0^\ell EI \left(\frac{\partial^2 w}{\partial x^2} \right)^2 dx \tag{7.47}$$

where EI is the flexural rigidity, w is the transverse displacement of the beam and x is the coordinate along the length of the beam.

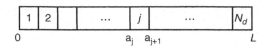

Figure 7.45   A schematic illustrating a beam's N_d subdivisions.

For a particular mode shape, $\{\psi\}_i$, the strain energy associated with the deformation in that mode shape pattern is

$$U_i = \frac{1}{2} \int_0^\ell EI \left(\frac{\partial^2 \{\psi\}_i}{\partial x^2} \right)^2 dx \qquad (7.48)$$

If the beam is subdivided into N_d divisions as shown in Figure 7.45, then the energy associated with each subregion j due to the ith mode is given by

$$U_{ij} = \frac{1}{2} \int_{a_j}^{a_{j+1}} (EI)_j \left(\frac{\partial^2 \{\psi\}_i}{\partial x^2} \right)^2 dx \qquad (7.49)$$

The fractional energy is therefore

$$F_{ij} = \frac{U_{ij}}{U_i} \qquad (7.50)$$

and

$$\sum_{j=1}^{N_d} F_{ij} = 1 \qquad (7.51)$$

Similar quantities can be defined for a damaged structure and are given by the following equations:

$$U_{i_d} = \frac{1}{2} \int_0^\ell EI_d \left(\frac{\partial^2 \{\psi\}_{i_d}}{\partial x^2} \right)^2 dx \qquad (7.52)$$

$$U_{ij_d} = \frac{1}{2} \int_{a_j}^{a_{j+1}} (EI_d)_j \left(\frac{\partial^2 \{\psi\}_{i_d}}{\partial x^2} \right)^2 dx \qquad (7.53)$$

$$F_{ij_d} = \frac{U_{ij_d}}{U_{i_d}} \qquad (7.54)$$

$$\sum_{j=1}^{N_d} F_{ij} = \sum_{j=1}^{N_d} F_{ij_d} = 1 \qquad (7.55)$$

By choosing the subregions to be relatively small, the flexural rigidity for the jth subregion, $(EI_d)_j$, is considered to be constant and F_{ij_d} becomes

$$F_{ij_d} = \frac{(EI_d)_i \displaystyle\int_{a_j}^{a_{j+1}} \left(\frac{\partial^2 \{\psi\}_{i_d}}{\partial x^2}\right)^2 dx}{U_{i_d}} \tag{7.56}$$

If it is assumed that the damage is primarily located at a single subregion then the fractional energy will remain relatively constant in the undamaged subregions and $F_{ij_d} = F_{ij}$ at these locations. For a single damaged location at subregion $j = k$ the following relationship can be established:

$$\frac{(EI)_k \displaystyle\int_{a_k}^{a_{k+1}} \left(\frac{\partial^2 \{\psi\}_i}{\partial x^2}\right)^2 dx}{U_i} = \frac{(EI_d)_k \displaystyle\int_{a_k}^{a_{k+1}} \left(\frac{\partial^2 \{\psi\}_{i_d}}{\partial x^2}\right)^2 dx}{U_{i_d}} \tag{7.57}$$

If it is assumed that EI is essentially constant over the length of the beam for both the undamaged and damaged conditions, Equation (7.57) can be rearranged to give an indication of the change in the flexural rigidity of the subregion as

$$\frac{(EI)_k}{(EI_d)_k} = \frac{\displaystyle\int_{a_k}^{a_{k+1}} \left(\frac{\partial^2 \{\psi\}_{i_d}}{\partial x^2}\right)^2 dx \Big/ \int_0^\ell \left(\frac{\partial^2 \{\psi\}_{i_d}}{\partial x^2}\right)^2 dx}{\displaystyle\int_{a_k}^{a_{k+1}} \left(\frac{\partial^2 \{\psi\}_i}{\partial x^2}\right)^2 dx \Big/ \int_0^\ell \left(\frac{\partial^2 \{\psi\}_i}{\partial x^2}\right)^2 dx} \equiv \frac{f_{ik_d}}{f_{ik}} \tag{7.58}$$

In order to use all m measured modes in the calculation, the damage index for subregion k is defined to be

$$\beta_k = \frac{\displaystyle\sum_{i=1}^m f_{ik_d}}{\displaystyle\sum_{i=1}^m f_{ik}} \tag{7.59}$$

One advantage of the formulations shown in Equations (7.58) and (7.59) is that the modes do not need be mass normalised although they do need to be normalised in a consistent manner. In its actual implementation, the cited authors then assume that the collection of the damage indices, β_k, represents a sample population of a normally distributed random variable, and a normalised damage index, Z_k, is defined as

$$Z_k = \frac{\beta_k - \overline{\beta}_k}{\sigma_k} \tag{7.60}$$

where $\overline{\beta}_k$ and σ_k represent the mean and standard deviation of the damage indices from all locations on the beam, respectively.

The modal strain energy feature for a beam-like structure is now illustrated with the I-40 Bridge data. An Euler–Bernoulli beam was selected to model the north damaged girder of the bridge for two reasons: (1) the fundamental behaviour of a bridge resembles that of a composite (concrete and steel) beam and (2) the mode shapes were estimated from vertical accelerations and these modes can be simulated with a one-dimensional beam approximation. In this feature extraction process the three-span girder was discretised into 210 elements defined by 211 equally spaced nodes. Discontinuities in the flange dimensions occur at elements 62, 80, 140 and 158. Splice plates, which also cause local stiffness discontinuities, are located at elements 47, 89, 122 and 163. The amplitudes of the mode shapes are known only at thirteen locations corresponding to the accelerometer locations. Therefore, in this study the magnitudes of the mode shapes at intermediate nodal locations were again estimated using a cubic polynomial interpolation scheme. As defined by Equation (7.60), a normal distribution was fitted to the damage localisation indicators and values more than two standard deviations from the mean were assumed to correspond to damage locations.

Figure 7.46 shows the values of Z_k for damage cases E-1 to E-4. Note that the elements span the entire length of the north girder and that element 106 corresponds to the location of damage. Two sets of Z_k values are plotted; one plot includes the influence of all six measured modes on the calculation of Z_k and

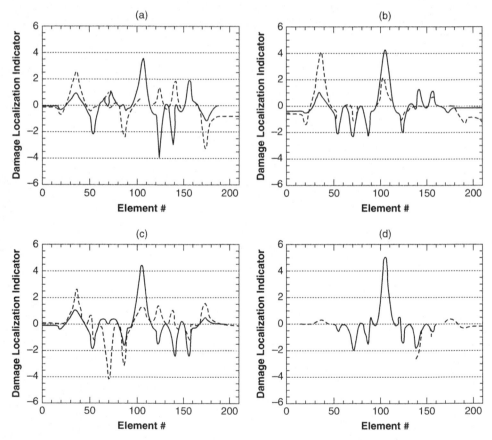

Figure 7.46   The damage index from changes in modal strain energy Equation (7.60) for the I-40 Bridge. The solid line is based on the first two modes and the dashed line is based on all six modes.

the other includes only the effect of the first bending and first torsion modes, which have been shown to be the modes most influenced by the damage (see Table 7.7). This comparison illustrates that modes not directly affected by damage may disrupt the performance of the damage localisation indicator. For damage cases E-1 to E-3, plots (a) to (c) show that a better indication of the actual location of damage was achieved with only two modes. By including the four higher frequency modes, erroneous indications of damage were produced. For damage case E-4, there is essentially no difference between the Z_k values calculated using six modes or two modes.

Extensions of the modal strain energy method have been made to structures exhibiting two-dimensional bending (Cornwell, Doebling and Farrar, 1999) and structures exhibiting axial and torsional responses (Duffey *et al.*, 2001).

One of the major difficulties associated with implementing the modal strain energy algorithms discussed here is the calculation of the derivatives and integrals when the mode shape is known at a relatively small number of discrete locations. With both the beam and plate algorithms additional intermediate points are obtained by curve-fitting the data. The derivatives and integrals required by the algorithms are then calculated analytically from this function. However, similar to the mode shape curvature feature, one must be aware that this curve-fitting process can smooth out the local changes in the mode shapes that are caused by damage.

Both the beam and the plate algorithms can be applied to detect damage in plate-like structures. The algorithm derived by assuming plate-like behaviour (two-dimensional curvature) can obviously be applied directly. To use the algorithm formulated by assuming one-dimensional curvature, the structure must be divided into slices and the algorithm is applied to each slice individually. The normalised damage index is then determined using the average and standard deviations of all the damage indices from all the slices. The advantage of this latter approach is that it is computationally more efficient than the two-dimensional algorithm. Regardless of the method chosen, several additional parameters must be chosen including the number of modes and the number of subdivisions to be used. An appraisal of the modal strain energy approach can be found in Worden, Manson and Allman (2001).

7.9.3   Modal Flexibility

The flexibility matrix, $[G]$, is defined as the inverse of the stiffness matrix, $[k]$:

$$\{f\} = [k]\{y\} \Rightarrow \{y\} = [k]^{-1}\{f\} = [G]\{f\} \tag{7.61}$$

where $\{f\}$ is the vector of static loads applied to the structure and $\{y\}$ is a vector corresponding to the deformation associated with this loading. The flexibility indices G_{ij} are defined as the displacement at DOF i caused by a unit load applied at DOF j. Therefore, each column of the flexibility matrix is the deformation pattern that the structure will assume when a unit load is applied at the DOF associated with that column. Figure 7.47 illustrates this definition of the flexibility matrix for a 2-DOF spring–mass system.

In this example the stiffness matrix and flexibility matrix are

$$[k] = \begin{pmatrix} k_1 + k_2 & -k_2 \\ -k_2 & k_2 \end{pmatrix} \quad \text{and} \quad [G] = \begin{pmatrix} \dfrac{1}{k_1} & \dfrac{1}{k_1} \\ \dfrac{1}{k_1} & \dfrac{k_1 + k_2}{k_1 k_2} \end{pmatrix} \tag{7.62}$$

For an undamaged structure with m mass-normalised modal vectors identified from experimental data obtained at n degrees of freedom, the $n \times n$ flexibility matrix, $[G]$, is derived from the modal data as follows (Pandey and Biswas, 1994).

$$[G] \approx [\Psi][\Lambda]^{-1}[\Psi]^T \approx \sum_{i=1}^{m} \frac{1}{\omega_i^2} \{\psi\}_i \{\psi\}_i^T \tag{7.63}$$

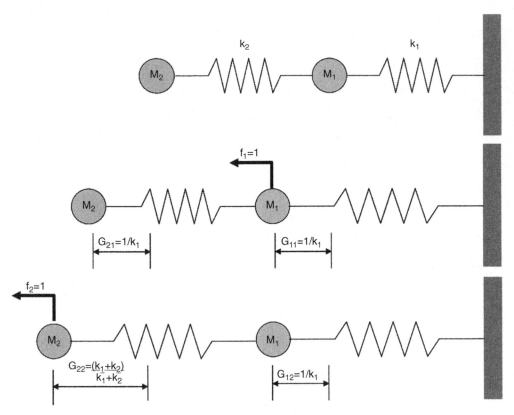

Figure 7.47 Development of the flexibility matrix for a 2-DOF system.

where

$\{\psi\}_i$ = the ith mass-normalised mode shape

$[\Psi]$ = the mode shape matrix = $[\psi_1, \psi_2, \ldots, \psi_m]$

ω_i = the ith modal frequency,

$[\Lambda]$ = the modal stiffness matrix = $\text{diag}(\omega_i^2)$

The approximations in Equation (7.63) come from the fact that typically the number of modes identified is less than the number of degrees of freedom that may be needed to accurately represent the motion of the structure (that is m is typically less than n). Additionally, Equation (7.63) shows that the outer vector product is scaled by the inverse of the resonance frequency squared associated with a particular mode. This scaling reduces the influence of higher-frequency modes on the estimate of the flexibility matrix.

From flexibility matrices corresponding to the undamaged and damaged structure, denoted by subscripts u and d, respectively, a metric for the flexibility change caused by the damage can be obtained from the difference of the respective matrices as

$$[\Delta G] = [G]_u - [G]_d \tag{7.64}$$

where ΔG represents the change in the flexibility matrix as determined from changes in the measured mode shapes and resonance frequencies; $\max_j$ is the maximum absolute value of the elements in the jth column. Hence,

$$\max_{j} = \max |g_{ij}|, \quad i = 1, \ldots, n \tag{7.65}$$

where the g_{ij} are elements of the matrix $[\Delta G]$ and are taken to be measures of the flexibility change at each measurement location. The column of the flexibility matrix corresponding to the largest $\max_j$ is indicative of the degree of freedom where damage is located. This method not only provides an estimate of the damage location but also provides a measure of the damage extent. It is re-emphasised that this method requires mass-normalised modes, which poses some challenges when using modal data obtained from ambient vibration tests. In such cases, the method can still be applied to locate damage if the modes are normalised in a consistent manner.

To illustrate this feature, the change in flexibility was calculated for the system shown in Figure 7.1. First, Equation (7.63) was used to calculate the flexibility matrix based on the analytically determined resonance frequencies and mode shapes and it was verified that when all eight modes were used in this calculation, the flexibility matrix obtained was equal to the inverse of the system stiffness matrix. Next, Equation (7.63) was used to calculate the flexibility matrix for the case where there is a 10% reduction in the stiffness associated with spring 1. Figure 7.48(a) shows an overlay of column 1 of the flexibility matrices corresponding to the undamaged and damaged structures as a function of the DOF (row) number. Remember that column 1 of the flexibility matrix corresponds to the deformed shape of the system when a unit load is applied at DOF 1. Figure 7.48(b) shows a similar plot comparing column 8 of the flexibility matrices, which corresponds to the deformed pattern when a unit load is applied at DOF 8. The uniform load deformed pattern corresponding to a unit load being applied at each DOF and obtained by summing the columns of the flexibility matrix is shown in Figure 7.48(c). Note that plots A to C in Figure 7.48 are based on a flexibility matrix formed from complete modal data, that is simulation of measurements at each DOF and using all eight mode shapes and resonance frequencies. Figure 7.48(d) shows the change in flexibility based on the first column of the flexibility matrix when only the first two modes are used to approximate this matrix. Finally, Figure 7.48(e) shows the change in flexibility based on the first column of the flexibility matrix when only the last two modes are used to approximate this matrix. This last plot shows the inverse weighting effects of the higher-mode resonance frequencies on the flexibility matrix estimate.

Flexibility changes evaluated with Equation (7.65) using one, two or six modes and corresponding to damage case E-4 for the I-40 Bridge are shown in Figure 7.49. Again, a polynomial has been fitted to the mode shape data to provide interpolated mode shape amplitudes at 211 equally spaced locations along the three spans. Damage cases E-1 and E-2 did not give any indications of the true damage location regardless of whether two or six modes were used to calculate the flexibility matrices. The change in flexibility derived from only two modes did reveal a peak in the area of the actual damage location for damage case E-3. However, for this damage case the change in flexibility obtained using all six modes did not give a clear indication of the damage location. The results from these damage cases are not shown, but can be found in Farrar and Jauregui (1996). For damage case E-4, clear indications of damage at the correct location are shown in Figure 7.49 for all cases (using one, two or six modes). For this damage case there is virtually no improvement in the change in flexibility when the final four modes are included in the calculation.

There are numerous studies that make use of changes in the flexibility matrix derived from measured mode shapes and resonance frequencies as a damage-sensitive feature (Aktan et al., 1994; Toksoy and Aktan, 1994; Mayes, 1995; Peterson, Doebling and Alvin, 1995), including approaches to account for the residual flexibility associated with out-of-band modes (Doebling, 1995). In one study the authors examine the curvature of a uniform load surface (Zhang and Aktan, 1995). This surface is constructed by

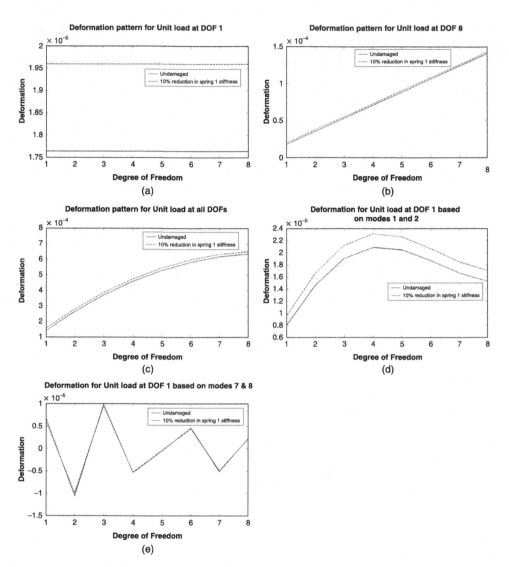

Figure 7.48   Changes in flexibility calculated for the system shown in Figure 7.1 with a 10% reduction in stiffness at spring 1.

summing the columns of the flexibility matrix. The resulting vector corresponds to the deformed shape of the structure when a unit load is applied simultaneously at each degree of freedom. The curvature of this surface is then calculated using a central difference operator as in Equation (7.45), and this change in curvature of the uniform load surface is used as a damage indicator.

7.10   Model Updating Approaches

Another class of inverse-modelling damage identification methods is based on the modification of structural mass, stiffness and damping matrices to reproduce, as closely as possible, the measured static

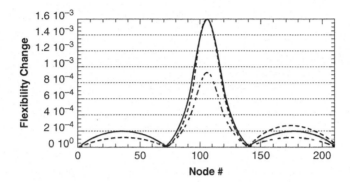

Figure 7.49   Change in flexibility identified from the I-40 Bridge mode shapes. The dash–dot line is based on only the first mode, the solid line is based on using the first two modes and the dashed line corresponds to using all six modes in the calculation.

or dynamic response data as summarised in Figure 7.50. These methods are also referred to as finite element model updating methods because finite element methods are often used to generate the matrices to be updated. Regardless of how the system matrices are generated, the approach is to solve for the updated matrices (or perturbations to the nominal model that produce the updated matrices) by forming a constrained optimisation problem based on the structural equations of motion, the nominal model and the measured data. Comparisons of the matrices updated with data acquired after a potentially damaging event to the original matrices correlated with data from the undamaged system provide an indication of damage and can be used to quantify the location and extent of damage. Therefore, the damage-sensitive features are the indices of the system matrices, and most often those of the stiffness matrix, or some

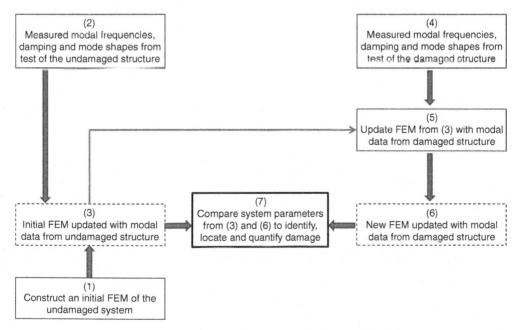

Figure 7.50   The model updating approach to damage detection.

residual error term as defined below. The methods use a common basic set of equations where the differences in the various algorithms can be classified as follows:

1. Objective function to be minimised.
2. Constraints placed on the problem.
3. Numerical scheme used to implement the optimisation.

7.10.1 Objective Functions and Constraints

There are several different physically-based equations that are used as either objective functions or constraints for the matrix update problem, depending upon the update algorithm. The structural equations of motion are the basis for the *modal force error equation*. The system's undamaged equation of motion as determined from an *n*-DOF finite element model (FEM) is defined by

$$[m]_u \{\ddot{y}\} + [c]_u \{\dot{y}\} + [k]_u \{y\} = \{x(t)\} \tag{7.66}$$

where, as before, the subscript *u* will designate matrices and vectors associated with the undamaged system. Following the discussion of modal analysis presented in Appendix B, the eigenvalue equation corresponding to Equation (7.66) is given as

$$\left(-\omega_{i_u}^2 [m]_u + i\omega_{i_u} [c]_u + [k]_u\right) \{\psi\}_{i_u} = \{0\} \tag{7.67}$$

where in the second term on the LHS of the equation: $i = \sqrt{-1}$, ω_{i_u} is the measured *i*th resonance frequency defined in units of radians per second (square root of the eigenvalue) and $\{\psi\}_{i_u}$ is the corresponding mode shape (eigenvector) of the undamaged structure. It is assumed that this equation is satisfied for all measured modes.

Now consider the resonance frequencies and mode shapes corresponding to the damaged state, ω_{i_d} and $\{\psi\}_{i_d}$. Substituting these quantities into Equation (7.67) yields

$$\left(-\omega_{i_d}^2 [m]_u + i\omega_{i_d} [c]_u + [k]_u\right) \{\psi\}_{i_d} = \{E\}_i \tag{7.68}$$

where $\{E\}_i$ is defined as the *modal force error*, or *residual force*, for the mode *i* of the damaged structure. This vector represents the harmonic force excitation that would have to be applied to the undamaged structure at the frequency of ω_{i_d} so that the structure would respond with mode shape $\{\psi\}_{i_d}$ (Ojalvo and Pilon, 1988).

There are several methods that have been used to compute the analytical model matrices of the damaged structure $[m]_d$, $[c]_d$, $[k]_d$ such that the resulting equation of motion (EOM) is balanced,

$$\left(-\omega_{i_d}^2 [m]_d + i\omega_{i_d} [c]_d + [k]_d\right) \{\psi\}_{i_d} = \{0\} \tag{7.69}$$

where the model matrices of the damaged system are defined as the model matrices of the undamaged structure minus a perturbation matrix:

$$[m]_d = [m]_u - [\Delta m]$$
$$[c]_d = [c]_u - [\Delta c] \tag{7.70}$$
$$[k]_d = [k]_u - [\Delta k]$$

Substituting Equation (7.70) into Equation (7.69) and moving the perturbation terms to the right side of the equation yields

$$\left(-\omega_{i_d}^2 [m]_u + i\omega_{i_d} [c]_u + [k]_u\right) \{\psi\}_{i_d} = \left(-\omega_{i_d}^2 [\Delta m] + i\omega_{i_d} [\Delta c] + [\Delta k]\right) \{\psi\}_{i_d} \qquad (7.71)$$

The left side of this equation consists of known quantities and has previously been defined as the modal force error, so the equation to be solved for the matrix perturbations can be written as

$$\left(-\omega_{i_d}^2 [\Delta m] + i\omega_{i_d} [\Delta c] + [\Delta k]\right) \{\psi\}_{i_d} = \{E\}_i \qquad (7.72)$$

The modal force error is used as both an objective function and a constraint in the various methods described below.

Preservation of the property matrix symmetry can be used as a constraint. This constraint can be written for each property matrix as

$$[\Delta m] = [\Delta m]^T$$

$$[\Delta c] = [\Delta c]^T \qquad (7.73)$$

$$[\Delta k] = [\Delta k]^T$$

Preservation of the property matrix sparsity (i.e. zero/nonzero pattern of the respective matrices) can also be used as a constraint. In the case of the stiffness matrix this pattern represents connectivity of the structural elements in the system. This constraint will be denoted as

$$sparse([m]_d) = sparse([m]_u)$$

$$sparse([c]_d) = sparse([c]_u) \qquad (7.74)$$

$$sparse([k]_d) = sparse([k]_u)$$

The preservation of sparsity is one way to maintain the allowable load paths through the structure as defined by the structural element connectivity in the updated model.

Preservation of the positive-definite property matrices can also be used as a constraint. This constraint can be written for each property matrix as

$$\{x\}^T [m]_d \{x\} \geq 0$$

$$\{x\}^T [c]_d \{x\} \geq 0 \qquad (7.75)$$

$$\{x\}^T [k]_d \{x\} \geq 0$$

where $\{x\}$ is any arbitrary vector. The discussion below will summarise how the objective functions and constraints are used to solve for the updated structural matrices.

7.10.2   Direct Solution for the Modal Force Error

One can directly solve for the modal force error vector defined in Equation (7.68) using the undamaged system matrices and the measured modal properties from the potentially damaged system. In this case the modal force error vector is the damage-sensitive feature and it is assumed that the undamaged system matrices accurately model the undamaged system. In fact, as shown in Figure 7.50, one should have obtained the undamaged system matrices through an initial model correlation step using modal properties measured on the undamaged system. This direct solution of the modal force error will be illustrated with the 8-DOF system shown in Figure 7.1. The vector $\{E\}_i$ can be tested to see if it departs significantly from zero and this provides a damage detection method. If the damage is local, one can further infer the location of the damage from the position of the nonzero entries within $\{E\}_i$.

In order to simulate damage in the system, consider the case where the stiffness of the spring joining masses 5 and 6 was reduced by 10%. When the residual force vector $\{E\}_i$ was computed for the first three modes using Equation (7.68), the results shown in Figure 7.51 were obtained.

The results show that the damage is located between the fifth and sixth masses. However, one should be aware that the data used in this example were generated from numerical simulation and, as such, have no variability that might be present in measured data. In order to investigate the effects of noise, the mode shapes from the damaged system were perturbed by Gaussian noise. The noise level was taken as 5% of the maximum mode shape amplitude, which is a realistic level. The resonance frequencies, which typically have less variance than the mode shape estimates, were left unperturbed. The residual vectors for the noisy modes are given in Figure 7.52.

With noise present the 10% reduction in spring 6 stiffness can still be identified based on modes 1 or 2, but mode 3 gives results that would be more difficult to properly interpret if one did not have previous knowledge of the damage location. The problem is that the update with the noisy mode shape vectors has smeared the stiffness change over the whole stiffness matrix.

7.10.3   Optimal Matrix Update Methods

Methods that use a closed-form, direct solution to compute the damaged model matrices or the perturbation matrices are commonly referred to as optimal matrix update methods (Smith and Beattie, 1991; Zimmerman and Smith, 1992; Hemez, 1993; Kaouk, 1993). The problem is generally formulated as a Lagrange multiplier or penalty-based optimisation, which can be written as

$$\min_{\Delta m, \, \Delta c, \, \Delta k} \{J(\Delta m, \, \Delta c, \, \Delta k) + \lambda R(\Delta m, \, \Delta c, \, \Delta k)\} \tag{7.76}$$

where J is the objective function, R is the constraint function and λ is the Lagrange multiplier or penalty constant. A common formulation of the optimal update problem minimises the Frobenius norm of global system matrix perturbations using a zero modal force error and places property matrix symmetry constraints on the updating process (Baruch, 1978).

In an effort to eliminate the problem of smearing the stiffness changes over the entire stiffness matrix, an approach to the optimal matrix update problem was developed that minimises the rank of the perturbation matrix, rather than the norm of the perturbation matrix, and is referred to as the minimum rank perturbation theory (MRPT) (Zimmerman and Kaouk, 1994). This approach is motivated by the assumption that damage will tend to be concentrated in a few structural members, rather than distributed throughout the structure. As such, the perturbation matrices will tend to be of small rank. The solution for the perturbation matrices is based on the theory that the unique minimum rank solution for the matrix of the underdetermined system defined by

$$[A]\{X\} = \{Y\}, \quad \text{with } [A]^{\mathrm{T}} = [A] \tag{7.77}$$

is given by

$$[A] = \{Y\}[B]\{Y\}^{\mathrm{T}}, \quad \text{where } [B] = (\{Y\}^{\mathrm{T}}\{X\})^{-1} \tag{7.78}$$

One implementation of this algorithm defines the modal force error from Equation (7.68) associated with the ith mode as the damage vector $\{d\}_i$, so that the perturbation error equation can be written as

$$\{d\}_i = [Z]_{i_d} \, \{\psi\}_{i_d} \tag{7.79}$$

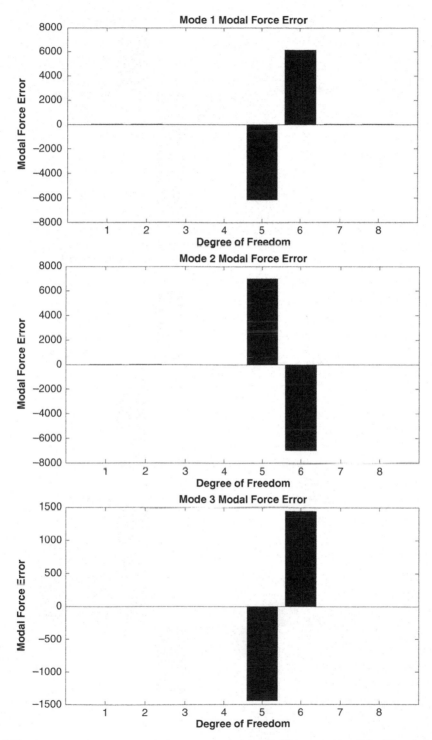

Figure 7.51   Residual force vectors for the first three modes of the 8-DOF system with a 10% stiffness reduction at spring 6.

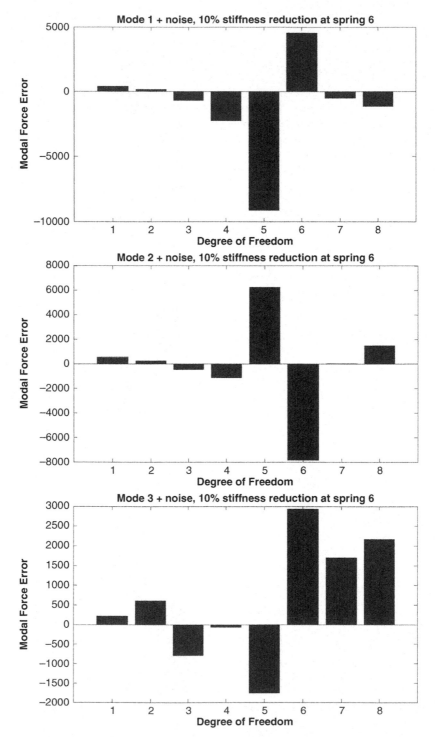

Figure 7.52 Residual vectors for the first three modes of the 8-DOF system corresponding to a 10% reduction in stiffness at spring 6 when random noise was added to the mode shapes.

where

$$[Z]_{i_d} = -\omega_{i_d}^2 [\Delta m] + i\omega_{i_d} [\Delta c] + [\Delta k] \tag{7.80}$$

or substituting the relationships in Equation (7.71)

$$[Z]_{i_d} = -\omega_{i_d}^2 [m]_u + i\omega_{i_d} [c]_u + [k]_u \tag{7.81}$$

By observing that the jth element of $\{d\}_i$ will be zero when the jth rows of the perturbation matrices are zero, a nonzero entry in $\{d\}_i$ is interpreted as an indication of the location of damage. However, changes in the perturbation matrices are not the only possible source of nonzero entries in $\{d\}_i$, as can be seen by rewriting the above equation at the qth structural DOF as

$$d_{iq} = z_{iq_d} \psi_{iq_d} = \|z_{iq_d}\| \; \|\psi_{iq_d}\| \; \cos(\theta_{iq}) \tag{7.82}$$

The deviation of the angle θ_{iq} from $90°$ is shown to be a better indicator of damage location than the nonzero entries of $\{d\}_i$, particularly when the row norms of $[Z]_{i_d}$ have different orders of magnitude.

In the case of a single nonzero perturbation matrix (for example $[\Delta k]$), the perturbation error equation can be solved using the MRPT equations as

$$[\Delta k] \{\psi\}_{i_d} = \{d\}_i , \quad \text{with} \; [\Delta k] = [\Delta k]^T \tag{7.83}$$

The solution to Equation (7.83) is

$$[\Delta k] = \{d\}_i [B] \{d\}_i^T , \quad \text{where} \; [B] = \left(\{d\}_i^T \{\psi\}_{i_d} \right)^{-1} \tag{7.84}$$

The resulting perturbation has the same rank as the number of modes used to compute the modal force error.

To illustrate this approach, data from the same simulated 8-DOF system with the 10% reduction in the spring 6 stiffness is again used. The perturbation matrix for the damaged system is first computed using all eight of the noise-free mode shapes from the damaged structure. The resulting perturbation is shown in Figure 7.53.

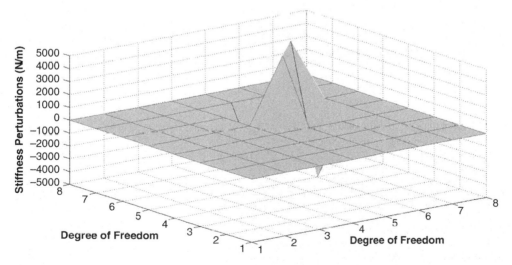

Figure 7.53 Stiffness perturbation matrix from the minimum rank update for the 8-DOF system with a 10% stiffness reduction at spring 6.

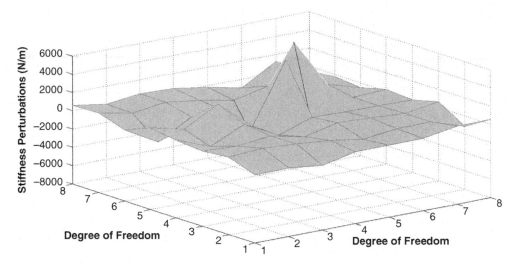

Figure 7.54 Stiffness perturbation matrix for the 8-DOF system with a 10% reduction in spring 6 stiffness and random noise added to the mode shapes.

The advantage of the minimum rank update approach is shown clearly when the method is then applied to the noisy mode shapes. The results are shown in Figure 7.54. The position and extent of the damage is shown almost as clearly in Figure 7.54 as in Figure 7.53. Kaouk and Zimmerman further extended the approach so that it could simultaneously estimate perturbations to the mass, damping and stiffness matrices (Kaouk and Zimmerman, 1994). This method also has been applied extensively to the I-40 Bridge data (Simmermacher *et al.*, 1995; Simmermacher, 1996).

7.10.4 *Sensitivity-Based Update Methods*

Another class of matrix update methods is based on the solution of a first-order Taylor series that minimises an error function of the matrix perturbations. Such techniques are known as sensitivity-based update methods. The basic approach begins with the determination of a modified parameter vector

$$\{p\}^{n+1} = \{p\}^n + \{\delta p\}^{n+1} \tag{7.85}$$

where the parameter perturbation vector $\{\delta p\}^{n+1}$ is computed from the Newton–Raphson iteration problem for minimising an error function

$$J(\{p\}^n + \{\delta p\}^{n+1}) \approx J(\{p\}^n) + \left[\frac{\partial J}{\partial p}(\{p\}^n)\right] \otimes \{\delta p\}^{n+1} = 0 \tag{7.86}$$

where $\otimes$ indicates the tensor or the Kronecker product and $J(\{p\})$ is the error function to be minimised. Typically the error function is selected to be the modal force error, as defined by Equation (7.68). Alternatively, other optimisation algorithms can be used.

A major difference between the various sensitivity-based update schemes is the method used to estimate the sensitivity matrix. Basically, either the experimental or the analytical quantities can be used in the differentiation. For experimental sensitivity, the orthogonality relations of the mass-normalised mode shape vectors (see Appendix B)

$$[\Phi]^T [m] [\Phi] = [I]$$
$$[\Phi]^T [k] [\Phi] = [\Lambda] \tag{7.87}$$

are used to compute the modal parameter derivatives (Norris and Meirovitch, 1989)

$$\left[\frac{\partial [\Lambda]}{\partial p} \right] \quad \text{and} \quad \left[\frac{\partial [\Phi]}{\partial p} \right] \tag{7.88}$$

Analytical sensitivity methods usually require the evaluation of the derivatives

$$\left[\frac{\partial [m]}{\partial p} \right] \quad \text{and} \quad \left[\frac{\partial [k]}{\partial p} \right] \tag{7.89}$$

which are less sensitive than experimental sensitivity matrices to noise in the data and to large perturbations of the parameters. Hemez presents a method for computing the global analytical sensitivity matrices based on assembly of the element-level analytical sensitivities (Hemez, 1993).

A sensitivity-based model updating approach to damage detection will now be demonstrated using an 8-DOF system similar to the one shown in Figure 7.1. Note that this example is slightly different from the previous 8-DOF examples discussed in this chapter. In this case the 8-DOF system does not have the spring connecting it to ground; that is it has free boundary conditions at each end and consequently will have one rigid-body mode. There are now only seven springs in the system with spring 1 now located between nodes 1 and 2. The process is started by defining the modal force error cost function that will subsequently be minimised in an effort to correlate the numerical model with measured modes and resonance frequencies. This modal force error cost function is similar to that given in Equation (7.68), but the damping term has been neglected. With this term removed, the modal force error for the ith mode of an undamped structure is

$$\{E\}_i = \left([K]_u - \omega_{i_d}^2 [M]_u \right) \{\psi\}_{i_d} \tag{7.90}$$

A column vector, $\{p\}$, corresponding to Equation (7.85) is defined containing seven entries that correspond to the percentage of the original spring stiffness values

$$\{p\} = [1, \ 1, \ 1, \ 1, \ 1, \ 1, \ 1]^T \tag{7.91}$$

where the unit entries imply that all stiffness values are 100% of their original value. This vector will be used to scale the individual spring stiffness values during the iterative updating process. The undamaged mass matrix is defined based on measured values from the system and the undamaged stiffness matrix is formed using the manufacturer's specified nominal stiffness values for the springs. In this example mass 1 has a considerably different value from the other masses because of the additional hardware needed to attach the shaker to this first mass when experimental modal analyses were performed. The slight variations in mass at the other locations are based on measurements made on the actual physical system as described in Chapter 5.

The undamaged mass and stiffness matrices are entered into Equation (7.90) and the eigenvalues and eigenvectors corresponding to these matrices are calculated. The rigid-body mode and its associated resonance frequency are eliminated from further consideration in this process and this yields a 7×7 diagonal matrix for $[\Lambda]_u$ and an 8×7 matrix for $[\Psi]_u$. This process is repeated for the system with a 14% reduction in stiffness assigned to spring 5 to simulate a set of modal data corresponding to the damaged condition. All mode shapes are mass-normalised according to Equation (7.87).

The modal force error matrix $[E]$ is calculated using the first three non-rigid-body modes and their associated resonance frequencies corresponding to the damaged structure as

$$[E]_{8 \times 3} = [k]_{u_{8 \times 3}} [\Psi]_{u_{8 \times 3}} - \left([m]_{u_{8 \times 3}} [\Psi]_{d_{8 \times 3}} \right) [\Lambda]_{d_{3 \times 3}} \tag{7.92}$$

Next, a cost parameter, C, is defined as the Frobenius norm of the modal force error matrix

$$C = \sqrt{\sum_{i,j} (E_{i,j})^2} \tag{7.93}$$

Now a nonlinear optimisation algorithm (Nelder and Mead, 1965; Avrie, 2003) is used to solve Equation (7.86). This solution procedure, which is implemented in a standard commercial mathematics software package, is used to minimise the cost parameter C by modifying the initial vector p that scales the stiffness values and then recalculates the modal force error based on modal parameters obtained with the updated stiffness values. For this problem the initial estimate, p_0, corresponds to all springs with 100% of their initial values. Note that this method of scaling the individual stiffness values in terms of this parameter vector (as opposed to scaling stiffness indices in the assembled stiffness matrix) preserves symmetry, sparseness and the positive-definite properties of the initial stiffness matrix Equations (7.73 to 7.75), but requires a new stiffness matrix to be assembled each time a new value of p is calculated.

The results of this model updating approach to damage detection are shown in Table 7.8, where the initial and final values of the vector p are summarised. The updating procedure yielded a 14.1% reduction in stiffness for spring 5 as compared to the actual damage value of a 14% reduction. Note there are also some other small changes to other stiffness values (less than or equal to 0.33%) that do not correspond to the actual damaged condition.

For the analytically simulated damage case, even when using only the first three modes and 70 iterations of the updating process, this sensitivity-based model updating process identified the correct spring as damaged and estimated the amount of damage to be within 1% of the actual damage. There are slight variations in the updated stiffness values at other locations. All these results indicate that when the damaged structure can be accurately modelled as a linear system, both before and after damage, and there is a one-to-one correspondence between model DOFs and measurement DOFs, then this model updating approach works very well. One primary advantage of such methods is that they identify the existence, location and extent of the damage.

Next, the sensitivity-based updating approach was applied to the same 8-DOF system, but this time actual experimental modal analysis results were used in the updating process. For this case damage

Table 7.8   A comparison of the p vector and the associated stiffness changes produced by the updating procedure

Spring number	Initial value (p vector)	Optimised value (p vector)	Percentage adjustment
1	1.000 00	1.000 01	0.00%
2	1.000 00	1.002 01	0.20%
3	1.000 00	1.000 03	0.00%
4	1.000 00	1.002 52	0.25%
5	1.000 00	0.858 60	−14.1%
6	1.000 00	1.000 26	0.03%
7	1.000 00	1.003 34	0.33%

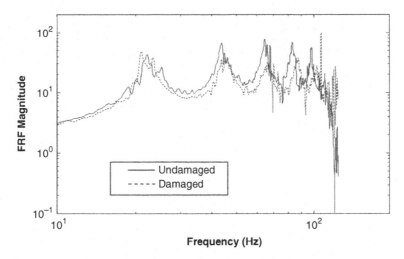

Figure 7.55   Typical FRFs used to identify modal properties of the undamaged and damaged 8-DOF system.

was introduced experimentally by substituting spring 5 in the baseline system with a new spring that had a manufacturer-specified 14% stiffness reduction. The model updating process follows the previous example with the notable exception that there is a two-step updating process. First, the baseline model of the system is updated with measured modal data from the undamaged condition. This updated baseline model is then updated again with the modal data measured on the damaged system.

It should be noted that the measured FRFs from which the modal parameters were extracted did not show the distinct peaks that are typically associated with a linear structure's resonances. Figure 7.55 shows a typical FRF used in this example where even for the undamaged system one does not see well-defined resonance peaks. These distortions are caused by the masses not only translating along the centre rod but also rotating slightly out of plane. As a result difficulties arose when trying to accurately identify the modal properties needed for both updating steps.

When the initial analytic model is updated with the undamaged data, there are increases in stiffness values at all locations, with a large increase for spring 1. The increase in the value at spring 1 is related to additional stiffness provided by the shaker attachment at this location. Without measuring the individual spring stiffness values, which was not done in the experimental study, one cannot tell if the updated spring stiffness values are, in fact, a more accurate representation of the actual physical system.

With the analytical model now updated based on the undamaged measured modal properties, the experimentally identified modal parameters from the damaged system were next used in the second updating step, where the process starts with stiffness values corresponding to the updated undamaged model. The stiffness values that result from this second updating step are summarised in Table 7.9.

When the model that has been updated based on modal data from the undamaged system was further updated in a second step with data from the damaged system using the first three modes, there was a clear indication of damage at the correct location, but the updating process overestimated the stiffness reduction at this location by more than 40%. There were also further small increases and some small decreases in stiffness values at other locations. The increases in stiffness are not physically realistic as there is no physical mechanism for increasing any of the spring stiffness values. One way to improve this updating procedure would be to perform a constrained optimisation problem that prevents any of the stiffness values from increasing beyond their initial value. This example is meant to illustrate the difficulties that can be encountered with model updating methods when there are difficulties accurately identifying the modal parameters used in the updating process.

Table 7.9 A comparison of the p vector and the associated stiffness changes produced by the updating procedure using the experimental modal parameters obtained from the damaged structure

Spring number	Initial value (p vector)	Optimised value (p vector)	Percentage adjustment
1	1.171 38	1.141 30	−2.57%
2	1.063 48	1.059 26	−0.40%
3	1.097 02	1.124 14	2.47%
4	1.056 07	1.069 51	1.27%
5	1.024 85	0.822 79	−19.7%
6	1.047 21	1.059 30	1.15%
7	1.078 38	1.085 06	0.62%

7.10.5 Eigenstructure Assignment Method

Another matrix update method, known as an *eigenstructure assignment*, is based on the design of a fictitious controller that minimises the modal force error (Lim, 1995). The controller gains are then interpreted as parameter matrix perturbations to the undamaged structural model. Consider the basic dynamic structural system equation of motion with a controller

$$[m]_u \{\ddot{y}\} + [c]_u \{\dot{y}\} + [k]_u \{y\} = [F] \{u\} \tag{7.94}$$

where

$$\{u\} = -[G][F]^{\mathrm{T}} \{x\} \tag{7.95}$$

Suppose that the control gains are selected such that the modal force error between the nominal structural model and the measured modal parameters from the damaged structure is zero

$$\left(-\omega_{i_d}^2 [m]_u + i\omega_{i_d} [c]_u + \left[[k]_u + [F][G][F]^{\mathrm{T}}\right]\right) \{\psi\}_{i_d} = \{0\} \tag{7.96}$$

with the definition

$$[L]_{ij} = \left(-\omega_{i_d}^2 [m]_u + i\omega_{i_d} [c]_u + [k]\right)_u^{-1} [F]_j [F]_j^{\mathrm{T}} \tag{7.97}$$

Then the *best achievable eigenvectors* $\{\psi\}_{i_{d(a)}}$ (which lie in the subspace spanned by the columns of $[L]_{ij}$) can be written in terms of the measured eigenvectors as

$$\{\psi\}_{i_{d(a)}} = [L]_{ij} [L]_{ij}^{\mathrm{T}} \{\psi\}_{i_d} \tag{7.98}$$

The relationship between the best achievable eigenvectors and the measured eigenvectors is then used as a measure of damage location. Specifically, if damage is in member j, then the measured and best achievable eigenvectors are identical. Thus, the angle between the vectors gives an indication of how much a particular member contributes to the change in a particular mode. This information can be used to hypothesise the location of the structural damage. The magnitude of the damage is then computed using the eigenstructure assignment technique such that the best achievable eigenvectors, model matrices for the undamaged structure and controller satisfy the modal force error equation. This approach has been applied to damage detection in truss and beam structures (Zimmerman and Kaouk, 1992; Lindner and Goff, 1993; Lim and Kashangaki, 1994).

7.10.6   Hybrid Matrix Update Methods

There are additional studies reported in the literature that make use of a hybrid matrix updating approach to the damage detection problem. As an example, a two-step damage detection procedure for large structures with limited instrumentation has been developed where the first step uses an optimal matrix update to identify the region of the structure where damage has occurred (Kim and Bartkowicz, 1994). The second step is a sensitivity-based method, which locates the specific structural element where damage has occurred. The first advantage of this approach lies in the computational efficiency of the optimal update method in locating which structural parameters have changed. The second advantage lies in the small number of parameters updated by the sensitivity-based technique.

7.10.7   Concluding Comment on Model Updating Approaches

The various model updating approaches have all been introduced based on a residual vector defined by the modal force error term given in Equation (7.68). However, it should be noted that alternative residual vectors can be defined using frequency responses and static displacements. For a system where damping is neglected, the residual vector associated with frequency response measures is

$$(-(\omega)^2 [m]_u + [k]_u) \{Y(\omega)\}_d - \{X(\omega)\}_d = \{E\}_i \tag{7.99}$$

where ω is a variable as opposed to designating a resonance frequency.

When a force, $\{F\}$, is applied to the structure and the resulting deflections, $\{\delta\}$, are measured, a residual vector can be defined as

$$[k]_u \{\delta\}_{i_d} - \{F\}_{i_d} = \{E\}_i \tag{7.100}$$

Note that this residual only provides information regarding the stiffness matrix. Additionally, as was discussed in Section 7.9.3, the static deformation can be obtained from the mass-normalised mode shapes.

Because all the model updating approaches discussed use measured modal properties in the updating process, they are inherently based on the three main assumptions associated with experimental modal analysis: (1) the structure responds in a linear manner, (2) the structure exhibits reciprocity and (3) the structural properties are time-invariant during a specific data acquisition process. As such, these methods have not generally been applied to structures that exhibit nonlinear or nonstationary response characteristics either in their undamaged or damaged condition.

For the examples shown above and most of the studies cited, the modelling updating approaches have been applied to fairly well-defined problems that match the model updating assumptions very well and, most importantly, there is often a one-to-one correspondence between the measured and analytical DOFs. In such cases these methods were shown to perform well. However, when one examines variability, either simulated or associated with *real-world* data, as in the example summarised in Section 7.10.4, the performance of the methods appears to degrade, even in relatively well-controlled laboratory experiments.

If one extends these model updating methods to larger-scale structures with lots of joints and interfaces, it may require a large number of elements to model the structure in sufficient detail to identify incipient damage conditions. In such cases there will typically be a significant mismatch between the number of analytical DOFs and the experimentally measured DOFs. This situation requires either a condensation of the analytical DOFs or some interpolation scheme to expand the experimental DOFs. Both approaches can cause further difficulties with the updating procedures. The process of condensing the analytical DOFs will smear the modelling errors through the system, making it difficult to establish the DOFs associated with the damage. Therefore, a modal expansion technique is usually preferred where the mass, damping and stiffness matrices are partitioned into subsets corresponding to the measured and unmeasured DOFs:

$$[m_p] = \begin{pmatrix} m_{11} & m_{12} \\ m_{21} & m_{22} \end{pmatrix}, \quad [c_p] = \begin{pmatrix} c_{11} & c_{12} \\ c_{21} & c_{22} \end{pmatrix} \text{ and } [k_p] = \begin{pmatrix} k_{11} & k_{12} \\ k_{21} & k_{22} \end{pmatrix} \tag{7.101}$$

where p designates partitioned and the 1 and 2 designate the measured and unmeasured DOFs, respectively. The modal force error defined in Equation (7.68) can now be written as

$$\left(-\omega_{i_{d_j}}^2 [M_p] + i\omega_{i_{d_j}} [C_p]_u + [K_p]_u \right) \begin{pmatrix} \psi_{i_d} \\ \psi_2 \end{pmatrix}_j = \{E_p\}_j \tag{7.102}$$

where the subscript i_d refers to the identified mode shape and resonance frequency. The vector $\{\psi_2\}_j$ is now considered an additional optimisation variable in the solution procedure. Finally, there has been no discussion of the numerical implementation issues related to the various updating procedures, which can be significant when attempting to apply them to large numerical models. The reader is referred to the cited references for a more detailed discussion of these issues.

7.11 Time Series Models

Another class of models that can be employed for damage detection is linear time series models. Three different time series models were introduced in Appendix A. The first is the autoregressive model, usually designated AR(p), where p is the order of the model and defined as

$$y_i = \sum_{j=1}^{p} a_j y_{i-j} \tag{7.103}$$

Here the current value of the response, y_i, is defined as a linear combination of the p previous response values and this linear combination is defined by the coefficients a_j.

Next is the moving-average model of order q, MA(q), which is defined as

$$y_i = \sum_{j=1}^{q} b_j x_{i-j} \tag{7.104}$$

This model is very similar in form to the AR model, but now the response is predicted from q previous inputs, x_i, and the model is defined by the coefficients b_j. If the current input is available, this sum can begin at $j = 0$.

Finally, these two models can be combined to form an autoregressive moving-average model (ARMA(p,q)), designated

$$y_i = \sum_{j=1}^{p} a_j y_{i-j} + \sum_{j=1}^{q} b_j x_{i-j} \tag{7.105}$$

In Appendix B it was shown that applying a finite difference approximation to the single DOF equation of motion for a spring–mass system leads to an ARMA(2,1) model of that system.

These models will not be exact and Equations (7.103) to (7.105) are usually written with an error term on the right-hand side, which is assumed to be zero-mean white noise. Also, similar to the previously discussed modal models, the assumption is made that the system is stationary during the time that the data used in the parameter estimation process are acquired. A key to employing this method is to determine the appropriate model order. A higher-order model may end up fitting the noise in the data and, hence, will not generalise to other data sets and therefore will have a tendency to give false positive indications of damage. On the other hand, if one selects a low-order model it will not necessarily capture the underlying physical system response. There are a variety of techniques for choosing the model order, such as Akaike's information criterion (Bishop, 1995) or a partial autocorrelation function (Box, Jenkins and Reinsel, 1994), that are often employed to help decide the appropriate model order. A comparative summary of various techniques for model order selection is given in (Figueiredo *et al.*, 2011).

It should be noted that the time series models given in Equations (7.103) to (7.105) are sometimes defined in a slightly more general form with a constant term added to the right-hand side, but that term has been omitted in this discussion. Also note that the concepts of these time series models can be extended to predicting the response from linear combinations of multiple inputs and responses.

In principle, the use of these time series models for damage detection is no different than the approach taken with the model updating described in the previous sections. A time series model is fitted to measured response and/or input data from the system being monitored when the system is known to be in its undamaged condition and the coefficients of the model become the damage-sensitive features. The process is repeated with data measured at a subsequent time after a potentially damaging event. Changes in the model parameters are then attributed to the presence of damage. As with all other features one must ensure that these changes are the result of damage and not the result of changing operational and environmental conditions.

This feature extraction process will now be illustrated with data from the concrete column. A fifth-order AR model (AR(5)) is applied to the 8192-point acceleration time series data from sensor 4. This model has the form:

$$y_i = a_1 y_{i-1} + a_2 y_{i-2} + a_3 y_{i-3} + a_4 y_{i-4} + a_5 y_{i-5} + \varepsilon_i \qquad (7.106)$$

Typical time series lead to an overdetermined set of equations that must be solved to obtain estimates of the AR coefficients. There are a variety of methods that can be used to solve the coefficients including the Yule-Walker approach (Brockwell and Davis, 2006) or a least-squares method that minimises the error term in Equation (7.106). For the 8192 point time series in this example the coefficients a_i are solved by applying the pseudo-inverse technique to the following equation to obtain a least-squares solution for the AR coefficients:

$$\begin{pmatrix} y_1 & y_2 & y_3 & y_4 & y_5 \\ y_2 & y_3 & y_4 & y_5 & y_6 \\ \vdots & \vdots & \vdots & \vdots & \vdots \\ y_{8187} & y_{8188} & y_{8189} & y_{8190} & y_{8191} \end{pmatrix} \begin{pmatrix} a_5 \\ a_4 \\ \vdots \\ a_1 \end{pmatrix} = \begin{pmatrix} y_6 \\ y_7 \\ \vdots \\ y_{8192} \end{pmatrix} \qquad (7.107)$$

This procedure yields the model parameters shown in Figure 7.56 for the undamaged data (Test 6) and the data corresponding to incipient damage (Test 7). Additionally, Figure 7.56 shows the changes in the AR model parameters for an order 15 model. Here it can be seen that the two sets of parameters representing the feature vectors extracted from data corresponding to the undamaged and damaged systems, respectively, are quite distinct for both the order 5 and order 15 models. However, unlike the new parameters of the stiffness matrix that were obtained through the finite element updating procedures, it is more difficult to assign a physical meaning to the changes in a time series model.

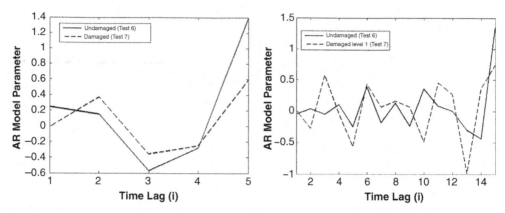

Figure 7.56   Autoregressive model parameters fitted to data from the concrete column (sensor 4).

7.12   Feature Selection

Given the wide variety of features that one can extract from any given data set, the question arises as how to select the best features to use for a given damage detection problem. There are a variety of approaches that can be used to address this question and a few will be outlined in this section. The methods that will be discussed are sensitivity analyses, feature information content and assessment of robustness. Reference will be made to a fourth method based on machine learning that is discussed in more detail in Chapter 11. Note that feature selection is challenging when one does not have data available from a damaged structure or when there is not an a priori definition of the damage that is to be detected.

7.12.1   Sensitivity Analysis

Sensitivity methods were previously introduced with the finite element updating approach to feature extract. They are approaches that are based on estimating how a given output variable changes with respect to changes in the input variables. This concept can be extended to feature selection as one would be able to discard those features with low sensitivity to damage and choose the features that show the greatest sensitivity to damage. One simple approach to sensitivity analysis is to estimate the derivative of an output variable with respect to the input variables. If an analytical expression relating inputs and outputs is available, one can find mathematical expressions for the sensitivities, or alternatively numerical estimates of the sensitivities can be computed. Note that this estimate is typically made with some form of numerical simulation because, in general, when selecting features, data will not be available from the damaged structure. Sensitivity approaches can be used to determine if a single feature is adequate for the damage detection process in an absolute sense or it can be used to compare the relative performance of multiple features.

A sensitivity analysis is demonstrated with data from the Alamosa Canyon Bridge discussed in Section 5.5. Dynamic response measurements from the Alamosa Canyon Bridge were analysed to determine the 95% statistical uncertainty bounds on three features: the modal frequencies, mode shapes and mode shape curvatures. These uncertainty bounds were based on the propagation of random errors on the frequency response function (FRF) estimates through the modal identification process to the modal parameters using a Monte Carlo simulation procedure described in Doebling and Farrar (2001) and Farrar *et al.* (2000). Changes in the modal frequencies, mode shapes and mode shape curvatures that are expected as a result of damage were computed using a correlated finite element model. These predicted

Table 7.10 Uncertainty bounds on the modal parameters

Mode number	Error on modal frequency	Average error on mode shape	Average error on mode shape curvature
1	0.06%	1.7%	560%
2	0.73%	45%	5200%
3	0.06%	1.7%	6.8%
4	0.24%	24%	13%
5	0.50%	160%	640%
6	0.06%	5.6%	37%
7	0.09%	3.6%	34%
8	0.11%	5.5%	9.5%
9	0.19%	160%	37%

changes were compared to the 95% confidence bounds on these same quantities computed from the experimental data, to determine which feature changes could be classified as statistically significant. The 95% uncertainty bounds on the modal frequencies, mode shapes and mode shape curvatures resulting from random disturbances and noise associated with the measurement process, as computed by the Monte Carlo analysis, are presented in Table 7.10, where the value indicates the 95% confidence limit in terms of a percentage of the mean value. As an example, for the first mode frequency the 95% confidence limits are ±0.06% of the mean value for this frequency. For the mode shape and mode shape curvatures these confidence limits were calculated for the 30 measurement DOFs and the values summarised in Table 7.10 are the average over all 30 DOFs. It is observed from these results that the uncertainty bounds on the modal frequencies are much smaller than on the mode shapes and the mode shape curvatures have the largest uncertainties.

The damage case that was simulated for the Alamosa Canyon Bridge was the complete failure of the bolted connection of two cross members at an interior girder. The damage was simulated by a 99% reduction in the modulus of elasticity of the cross members on either side of the connection. Thus, their ability to carry loads was lost, but their mass was still contained in the model, as would be the case in an actual connection failure. The changes in the modal frequencies, mode shapes and mode shape curvatures as a result of damage are presented in Table 7.11. It is observed in this table that the relative change of the mode shapes is larger than that of frequencies and the relative change of mode shape curvatures is typically the largest.

Table 7.11 Changes in FEM modal parameters resulting from the simulated damage

Mode number	Change in modal frequency	Average change in mode shape	Average change in mode shape curvature
1	0.00%	0.03%	4.6%
2	0.02%	0.16%	2.4%
3	0.27%	0.87%	5.8%
4	1.11%	3.9%	3.5%
5	0.00%	0.07%	8.1%
6	0.03%	0.25%	1.8%
7	0.24%	1.5%	24%
8	0.76%	5.4%	6.5%
9	1.2%	21%	4.8%

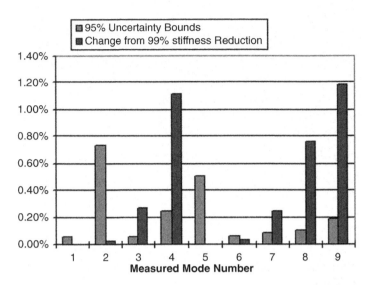

Figure 7.57 Comparison of modal frequency 95% confidence bounds to changes predicted as a result of damage.

A comparison of the estimated 95% confidence bounds determined by the Monte Carlo method and the predicted changes as a result of damage for the modal frequencies are shown in Figure 7.57. The modal frequencies of modes 3, 4, 7, 8 and 9 undergo a change that is significantly larger than the corresponding 95% confidence bounds. The relative magnitudes of the changes indicate that the frequency changes of these modes could be used with confidence in a damage identification analysis. It should be noted from the *y*-axis scale of Figure 7.57 that the overall changes in frequency as a result of damage are quite small (<1.2%), but as a consequence of the extremely low uncertainty bounds on the modal frequencies (many less than 0.2%), these small changes can be considered to be statistically significant.

A comparison of the average 95% confidence bounds and the predicted mode shape changes as a result of damage are shown (on a semilog scale) in Figure 7.58. Although many of the mode shapes undergo a significant (>5%) average change, none of the mode shapes undergo an average change over all degrees of freedom that is larger than the 95% confidence bounds caused by random variations in the measurements. This figure only indicates the average change over all of the mode shape components. A more complete study would need to examine if there are local changes that are statistically significant.

The statistical significance of changes to the mode shape curvature can be evaluated in a manner analogous to the analysis of the mode shapes. A comparison of the average 95% confidence bounds and the predicted changes as a result of damage for the mode shape curvature components for each mode are shown in Figure 7.59. Although many of the mode shape curvatures undergo a significant (>5%) average change, none of the mode shape curvatures undergoes an average change over all degrees of freedom that is larger than the 95% confidence bounds caused by random variations in the measurements.

In summary, this sensitivity analysis shows that changes in the resonance frequencies are the best indicators of the particular damage scenario that was investigated. It also shows that the sensitivities by themselves are not adequate to effectively select the features. Instead, these sensitivities must be compared to the relative uncertainty with which one can estimate a particular feature. In this example, although the changes in the frequencies are proportionally smaller than changes in the mode shapes or mode shape curvature, there is less uncertainty in the estimate of these features. As a result, of these three candidate features, the resonance frequencies associated with modes 3, 4, 7, 8 and 9 are the ones that should be selected to identify this particular type of damage.

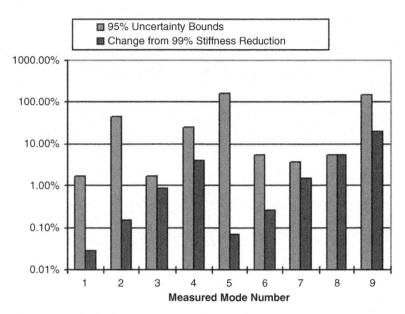

Figure 7.58 Comparison of average mode shape component 95% confidence bounds to changes predicted as a result of damage.

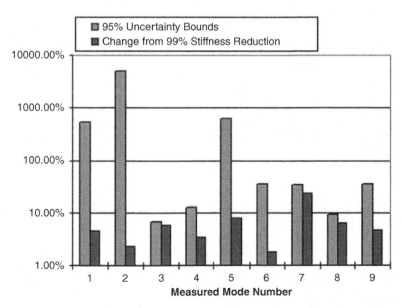

Figure 7.59 Comparison of average mode shape curvature component 95% confidence bounds to changes predicted as a result of damage.

7.12.2   Information Content

A number of measures have been developed to assess the informativeness of features, and these can be used as a criterion for feature selection. Feature selection methods based on information measures aim to reduce the dimensionality of the feature space by eliminating features with low information content or with high redundancy with respect to other features.

The *mutual information* measure derived from information theory assesses the information content of random variables and can be used as a criterion for feature selection (Shannon and Weaver, 1959; Battiti, 1994). In the context of damage detection, the mutual information, $I(C,F)$, between a set of damage conditions C and a set of features F can be defined as

$$I(C, F) = H(C) - H(C \mid F) \tag{7.108}$$

where the entropy $H(C)$ and conditional entropy $H(C \mid F)$ are given by

$$H(C) = - \sum_C p(C) \log p(C) \tag{7.109}$$

and

$$H(C \mid F) = - \sum_F p(F) \left(\sum_C p(C \mid F) \log p(C \mid F) \right) \tag{7.110}$$

respectively. Here $p(C)$ is the probability of the class C and $p(C|F)$ is the conditional probability of the class C given that one has observed the feature input vector F. The entropies given in the above equations essentially measure the uncertainty of the output classes C.

To illustrate the use of mutual information in the feature selection process, consider the simulated building structure summarised in Chapter 5. First, the bivariate mutual information was calculated between all the sensor pairs where it was seen that the least similar information was associated with accelerometers 2 and 3, as summarised in Table 7.12. Next, a feature vector was defined that included the standard deviation, variance, skewness and kurtosis of the relevant time series. The bivariate mutual information was computed for each feature pair using data corresponding to the damaged system. One would expect the mutual information assessment of the standard deviation relative to the variance to yield a larger mutual information value because the standard deviation is simply the square root of the variance, and that is indeed the case as shown Table 7.13. Furthermore, the skewness shows the least similarity and this result is also to be expected because the damage introduces an asymmetry into the response.

This mutual information estimates arbitrary dependences between random variables and is capable of evaluating even nonlinear relationships between features and different damage classes. The method is also independent of the learning process. It has been successfully applied to feature selection for a neural

Table 7.12   Mutual information measured between sensors

	Force transducer	Accel 1	Accel 2	Accel 3	Accel 4
Force transducer	3.4	**0.25**	**0.04**	**0.03**	**0.03**
Accel 1	0.25	3.4	**0.25**	**0.11**	**0.09**
Accel 2	0.04	0.25	3.6	**0.04**	**0.13**
Accel 3	0.03	0.11	0.04	3.4	**0.70**
Accel 4	0.03	0.09	0.13	0.70	3.3

Table 7.13 Mutual information measure between features extracted from accelerometers 2 and 3

	A2 variance	A2 standard deviation	A2 skewness	A2 kurtosis
A3 variance	1.1	0.91	0.19	0.17
A3 standard.deviation	1.0	0.94	0.15	0.19
A3 skewness	0.24	0.19	0.080	0.18
A3 kurtosis	0.22	0.21	0.13	0.21

network-based classification problem in Battiti (1994) and to sensor optimisation for an impact location problem in Wong and Staszewski (1998).

7.12.3 Assessment of Robustness

Another method that has been proposed for feature selection is based on assessing the robustness of the selected features to uncertainty (Stull, Hemez and Farrar, 2012). Anchored in information-gap decision theory (IGDT) (Ben-Haim, 2006), this approach assumes a nonprobabilistic description of the variability in the measured dynamic response data from which the features will be extracted. Using this as a point of departure, the proposed framework aims to assess the robustness of the selected features to sources of uncertainty that cannot necessarily be described in a probabilistic manner. In the study described in Stull, Hemez and Farrar (2012) this approach to feature selection was applied to detecting damage in the drive mechanism of a telescope (see Section 10.7.2), where it was used to determine the appropriate autoregressive model order to use for damage detection.

7.12.4 Optimisation Procedures

A variety of optimisation procedure can be employed to help select the best feature from among a candidate group of possible features. Common to all such approaches to feature selection will be the definition of some observability criteria that will form the basis for the objective function that is to be minimised. As an example, the feature selection can be based on the Fisher information matrix (Middleton, 1960). This matrix is the inverse of the covariance matrix associated with the analysed features. Maximising the trace or determinant and minimising the condition number of the Fisher information matrix is a useful optimisation-based approach to feature selection. A detailed discussion of another optimisation procedure for feature selection based on genetic algorithms will be presented in Section 11.9.

7.13 Metrics

The concept of a feature metric that was introduced at the beginning of the chapter is briefly reviewed here. Again, a metric is some parameter that can be used as a measure of the similarity or difference between two feature vectors. Some of the simplest metrics are simply the percentage change between two scalar features (e.g. the signal's mean amplitude) or the Euclidean distance between two feature vectors. Throughout this chapter additional metrics have been introduced including the PSD correlation coefficient (guided waves), root-mean-square deviation (impedance measures), the modal assurance criteria (modal vectors), the coordinate modal assurance criteria (modal vectors), the Yuen function (beam modal vectors) and the modal strain energy ratios for beams and plates. Note that several of these metrics (e.g. the modal assurance criteria) can be extended to other types of high-dimensional features. These metrics will be a key component of the classification processes that will be discussed in subsequent chapters. Finally, recall that changes in these metrics can also be used as damage-sensitive features.

7.14   Concluding Comments

This chapter introduced many types of damage-sensitive features that have been used for SHM. First, the concepts of feature extraction, feature dimension, feature selection and metrics for feature comparison were discussed in general terms. The rest of this chapter then described features that can be generally classified as either waveforms or model parameters. It is re-emphasised that this chapter and the next by no means presents an exhaustive summary of all features that can be used for SHM as new features are continually being reported in the technical literature.

Waveforms were shown to be high-dimensional features, which pose difficulties when trying to quantify their changes. Therefore, basic statistics and temporal moments were introduced as low-dimensional features that can be derived from the waveforms. Also, several metrics that are used to quantify changes in the high-dimensional waveforms were introduced. In general, extraction and comparison of features based on waveforms makes no assumptions regarding the linear or nonlinear nature of the system generating the waveform data. Typically, waveform comparisons are limited to only identifying and possibly locating the damage.

Next, damage-sensitive features related to various model parameters derived from linear structural dynamics were discussed. Modal models and discrete-parameter spatial models (e.g. finite element models) were the two primary types of models. In general, these features were estimated by some form of inverse-modelling procedure. In all cases, a linear model is fitted to data before and after damage has occurred. When there is a one-to-one correspondence between analytical DOFs and experimental measurements and the damage produces changes in the structure that can be accurately simulated with these linear models, the methods can identify the existence, location and extent of the damage. The ability to model large-scale structures with adequate fidelity at the joints and interfaces and the practical limitations of obtaining extensive response measurements pose difficulties for these feature extraction approaches. Next, Chapter 8 will extend both the waveform and model parameter features to systems that can be accurately simulated with linear structural dynamics models in their undamaged condition, but then exhibit nonlinear response in their damaged condition.

References

Aktan, A., Lee, K., Chuntavan, C. and Aksel, T. (1994) Modal testing for structural identification and condition assessment of constructed facilities. Proceedings of the 12th International Modal Analysis Conference, Honolulu, HI, Society for Experiment, pp. 462–468.

Allemang, R. (2003) The modal assurance criteria: twenty years of use and abuse. *Journal of Sound and Vibration*, **37**(8), 14–21.

Allemang, R.J and Brown, D.L (1982) A correlation coefficient for modal vector analysis. 1st International Modal Analysis Conference, Society for Experimental Mechanics, Orlando, FL, pp. 110–116.

Andersen, P., Brincker, R., Peeters, B. *et al.* (1999) Comparison of system identification methods using ambient bridge test data. 17th International Modal Analysis Conference, Society for Experimental Mechanics, Kissimmee, FL, pp. 1035–1041.

Avrie, M. (2003) *Nonlinear Programming: Analysis and Methods*, Dover Publishing.

Baruch, M. (1978) Optimization procedure to correct stiffness and flexibility matrices using vibration tests. *AIAA Journal*, **16**(11), 1208–1210.

Battiti, R. (1994) Using mutual information for selecting features in supervised neural network learning. *IEEE Transactions on Neural Network*, **5**(4), 537–550.

Ben-Haim, Y. (2006) *Info-Gap Decision Theory: Decisions under Severe Uncertainty*, Academic Press, Oxford.

Bhalla, S. and Soh, C. (2003) Structural impedance based damage diagnosis by piezo-electric transducers. *Earthquake Engineering and Structural Dynamics*, **32**, 1897–1916.

Bishop, C. (1995) *Neural Networks for Pattern Recognition*, Oxford University Press, Oxford, UK.

Box, G., Jenkins, G. and Reinsel, G. (1994) *Time Series Analysis, Forecasting and Control*, 3rd edn, Prentice Hall, Englewood Cliffs, NJ.

Brockwell, P. and Davis, R. (2006) *Timer Series: Theory and Methods*, Springer Science + Business Media, LLC, New York, NY.

Burton, T., Farrar, C. and Doebling, S. (1998) Two methods for model updating using damage Ritz vectors. 16th International Modal Analysis Conference, Society for Experimental Mechanics, Santa Barbara, CA, pp. 973–979.

Cao, T. and Zimmerman, D. (1997) A procedure to extract Ritz vectors from dynamic testing data. 15th International Modal Analysis Conference, Society for Experimental Mechanics, Orlando, FL, pp. 1036–1042.

Cawley, P. and Adams, R. (1979) The locations of defects in structures from measurements of natural frequencies. *Journal of Strain Analysis*, **14**(2), 49–57.

Cesnik, C. (2007) Review of guided-wave structural health monitoring. *The Shock and Vibration Digest*, **39**(2), 91–114.

Cornwell, P., Doebling, S. and Farrar, C. (1999) A strain-energy-based damage detection method for plate-like structures. *Journal of Sound and Vibration*, **224**(2), 359–374.

Croxford, A., Wilcox, P., Drinkwater, B. and Konstantinidis, G. (2007) Strategies for guided-wave structural health monitoring. *Philosophical Transactions of the Royal Society A: Mathematical, Physical and Engineering Sciences*, **463**(2087), 2961–2981.

Doebling, S. (1995) Measurement of structural flexibility matrices for experiments with incomplete reciprocity. Doctoral Dissertation, Department of Aerospace Engineering, University of Colorado, Boulder, CO.

Doebling, S. and Farrar, C. (2001) Estimation of statistical distributions for modal parameters identified from averaged frequency response function data. *Journal of Vibration and Control*, **7**(4), 603–624.

Doebling, S., Farrar, C.R., Prime, M. and Shevitz, D. (1996) Damage Identification and Health Monitoring of Structural and Mechanical Systems from Changes in Their Vibration Characteristics: A Literature Review, Los Alamos National Laboratory, Los Alamos, NM.

Duffey, T., Doebling, S., Farrar, C. *et al.* (2001) Vibration-based damage identification in structures exhibiting axial and torsional response. *Journal of Vibration and Acoustics*, **123**(1), 84–91.

Ewins, D. (1995) *Modal Testing: Theory and Practice*, John Wiley & Sons, Inc., New York, NY.

Farrar, C. and Jauregui, D. (1996) Damage Detection Algorithms Applied to Experimental and Numerical Modal Data From the I-40 Bridge, Los Alamos National Laboratory Report LA-13074-MS, Los Alamos, NM.

Farrar, C.R., Baker, W.E, Bell, T.M *et al.* (1994) Dynamic Characterisation and Damage Detection in the I-40 Bridge over the Rio Grande, Los Alamos National Laboratory Report LA-1276, Los Alamos, NM.

Farrar, C., Cornwell, P., Doebling, S. and Prime, M. (2000) Structural Health Monitoring Studies of the Alamosa Canyon and I-40 Bridges, Los Alamos National Laboratory Report LA-13635-MS, Los Alamos, NM.

Figueiredo, E., Figueiras, J., Park, G. *et al.* (2011) Influence of autoregressive model order on damage detection. *International Journal of Computer-Aided Civil and Infrastructure Engineering*, **26**(3), 225–238.

Flynn, E., Todd, M., Wilcox, P. *et al.* (2011) Maximum-likelihood estimation of damage location in guided-wave structural health monitoring. *Proceedings of the Royal Society A: Mathematical, Physical and Engineering Science*, **467**(2133), 2575–2596.

Fox, C.H (1992) The location of defects in structures: a comparison of the use of natural frequency and mode shape data. 10th International Modal Analysis Conference, Society for Experimental Mechanics, San Diego, CA, pp. 522–528.

Fukunaga, K. (1990) *Introduction to Statistical Pattern Recognition*, 2nd edn, Academic Press, San Diego, CA.

Gao, H., Shi, Y. and Rose, J. (2005) Guided wave tomography on an aircraft wing with leave in place sensors, in *Review of Progress in Quantitative Nondestructive Evaluation* (eds D.O. Thompson and D.E. Chimenti), American Institute of Physics, Goldon, CO, pp. 1788–1794.

Giurgiutiu, V., Zagrai, A. and Bao, J. (2004) Damage Identification in Aging Aircraft Structures with Piezoelectric Wafer Active Sensors. *Journal of Intelligent Material Systems and Structures*, **15**, 673–688.

Hamming, R. (1989) *Digital Filters*, 3rd edn, Prentice Hall, Englewood Cliffs, NJ.

Hellier, C.J (2001) *Handbook of Nondestructive Evaluation*, McGraw-Hill, New York.

Hemez, F. (1993) Theoretical and experimental correlation between finite elementmodels and modal tests in the context of large flexible space structures. PhD Dissertation, Department of Aerospace Engineering Sciences, University of Colorado, Boulder, CO.

Ing, R. and Fink, M. (1998) Time-reversed Lamb waves. *IEEE Transactions on Ultrasonics, Ferroelectrics and Frequency Control*, **45**(4), 1032–1043.

Juang, J. and Pappa, R. (1985) An eigensystem realization algorithm for modal parameter identification and model reduction. *Journal of Guidance, Control and Dynamics*, **8**(5), 620–627.

Kaouk, M. (1993) Finite element model adjustment and damage detection using measured test data. PhD Dissertation, Department of Aerospace Engineering and Mechanics and Engineering Science, University of Florida, Gainesville, FL.

Kaouk, M. and Zimmerman, D. (1994) Assessment of damage affecting all structural properties. Proceedings of the 9th VPI & SU Symposium on Dynamics and Control of Large Structures, Blacksburg, VA, pp. 445–455.

Kessler, S., Spearing, S. and Soutis, C. (2002) Damage detection in composite materials using Lamb wave methods. *Smart Materials and Structures*, **11**, 269–278.

Kim, H. and Bartkowicz, T. (1994) A two-step structural damage detection using a hexagonal truss structure. Proceedings of 35th AIAA/ASME/ASCE/AHS/ASC Structures, Structural Dynamics and Materials Conference, pp. 318–324.

Kim, J.-H., Jeon, H.-S. and Lee, C.-L. (1992) Application of the modal assurance criteria for detecting and locating structural faults. 10th International Modal Analysis Conference, Society for Experimental Mechanics, San Diego, CA, pp. 536–540.

Ko, J., Wong, C. and Lam, H. (1994) Damage detection in steel framed structures by vibration measurement approach. 12th International Modal Analysis Conference, Society for Experimental Mechanics, Honolulu, HI, pp. 280–286.

Lam, H., Ko, J. and Wong, C. (1995) Detection of damage location based on sensitivity analysis. 13th International Modal Analysis Conference, Society for Experimental Mechanics, Nashville, TN, pp. 1499–1505.

Lieven, N. and Ewins, D. (1988) Spatial correlation of mode shapes: the coordinate modal assurance criterion (COMAC). 6th International Modal Analysis Conference, Orlando, FL, Society for Experimental Mechanics, pp. 690–695.

Lim, T. (1995) Structural damage detection using constrained eigenstructure assignment. *Journal of Guidance, Control, and Dynamics*, **18**(3), 411–418.

Lim, T. and Kashangaki, T.-L. (1994) Structural damage detection of space truss structure using best achievable eigenvectors. *AIAA Journal*, **32**(5), 1049–1057.

Lindner, D. and Goff, R. (1993) Damage detection. Location and estimation for space trusses. Proceedings of Smart Structures and Materials 1993: Smart Structures and Intelligent Systems, Albuquerque, NM, vol. 1917, SPIE, pp. 1028–1039.

Lyons, R. (2011) *Understanding Digital Signal Processing*, 3rd edn, Pearson Education, Inc., Boston, MA.

Maia, N. (ed.) (1997) *Theoretical and Experimental Modal Analysis*, Research Studies Press Ltd, Taunton, Somerset, England.

Mayes, R. (1992) Error localization using mode shapes – an application to a two link robot arm. 10th International Modal Analysis Conference, Society for Experimental Mechanics, San Diego, CA, pp. 886–891.

Mayes, R. (1995) An experimental algorithm for detecting damage applied to the I-40 Bridge over the Rio Grande. Proceedings of the 13th International Modal Analysis Conference, Society for Experimental Mechanics, pp. 219–225.

Michaels, J. and Michaels, T. (2007) Guided wave signal processing and image fusion for *in situ* damage localization in plates. *Wave Motion*, **44**, 482–492.

Middleton, D. (1960) *An Introduction to Statistical Communication Theory*, McGraw-Hill.

Migliori, A. and Sarrao, J. (1997) *Resonant Ultrasound Spectroscopy: Applications to Physics, Materials Measuremnets and Nondestructive Evaluation*, John Wiley & Sons, Inc., New York, NY.

Nelder, J. and Mead, R. (1965) A simplex method for function minimization. *Computer Journal*, **7**(4), 308–313.

Norris, M. and Meirovitch, L. (1989) On the problem of modeling for parameter identification in distributed structures. *International Journal for Numerical Methods in Engineering*, **28**, 2451–2463.

Ojalvo, I. and Pilon, D. (1988) Diagnostics for geometrically locating structural math model errors from modal test data. Proceedings of 29th AIAA/ASME/ASCE/AHS/ASC Structures, Structural Dynamics, and Materials Conference, Williamsburg, VA, pp. 1174–1186.

Pandey, A.K, Biswas, M. and Samman, M.M (1991) Damage detection from changes in curvature mode shapes. *Journal of Sound and Vibration*, **145**(2), 321–332.

Pandey, A. and Biswas, M. (1994) Damage detection in structures using changes in flexibility. *Journal of Sound and Vibration*, **169**(1), 3–17.

Park, G., Sohn, H., Farrar, C. and Inman, D. (2003) Overview of piezoelectric impedance-based health monitoring and path forward. *Shock and Vibration Digest*, **35**(5), 451–463.

Park, S., Lee, J., Yun, C. and Inman, D. (2007) A built-in active sensing system-based structural health monitoring technique using statistical pattern recognition. *Journal of Mechanical Science and Technology*, **21**, 896–902.

Peters, B. and De Roeck, G. (2001) Stochastic system identification for operational modal analysis: a review. *Journal of Dynamic Systems, Measurement and Control*, **123**(4), 659–667.

Peterson, L., Doebling, S. and Alvin, K. (1995) Experimental determination of local structural stiffness by disassembly of measured flexibility matrices. Proceedings of 36th AIAA/ASME/ASCE/AHS/ASC Structures, Structural Dynamics and Materials Conference.

Ratcliffe, C.P (1997) Damage detection using a modified Laplacian operator on mode shape data. *Journal of Sound and Vibration*, **204**(3), 505–517.

Ripley, B. (1996) *Pattern Recognition and Neural Networks*, Cambridge University Press, Cambridge, UK.

Rose, J. (2002) A baseline and vision of ultrasonic guided wave inspection potential. *Journal of Pressure Vessel Technology*, **124**, 273–282.

Salawu, O. and Williams, C. (1994) Damage location using vibration mode shapes. 12th International Modal Analysis Conference, Society for Experimental Mechanics, Honolulu, HI, pp. 933–939.

Shannon, C. and Weaver, W. (1959) *The Mathematical Theory of Communications*, The University of Illinois Press, Champaign, IL.

Simmermacher, T. (1996) Damage detection and model refinement of coupled structural systems. Doctoral Dissertation, Department of Mechanical Engineering, University of Houston, Houston, TX.

Simmermacher, T., Zimmerman, D., Mayes, R. *et al.* (1995) The effects of finite element grid density on model correlation and damage detection of a bridge. Proceedings of the 1995 AIAA Adaptive Structures Forum, New Orleans, LA, pp. 2249–2258.

Smallwood, D. (1994) Characterization and simulation of transient vibrations using band limited moments. *Journal of Shock and Vibration*, **1**(6), 507–527.

Smith, S. and Beattie, C. (1991) Secant-method adjustment for structural models. *AIAA Journal*, **29**(1), 119–126.

Sohn, H. and Law, K. (2001a) Damage diagnosis using experimental Ritz vectors. *ASCE Journal of Engineering Mechanics*, **127**(11), 1184–1193.

Sohn, H. and Law, K. (2001b) Extraction of Ritz vectors from vibration test data. *Mechanical Systems and Signal Processing*, **15**(1), 213–226.

Sohn, H., Farrar, C., Hemez, F. *et al.* (2004) *A Review of Structural Health Monitoring Literature from 1996–2001*, Los Alamos National Laboratory Report LA-13976-MS, Los Alamos, NM, available online from Los Alamos Laboratory Library, http://library.lanl.gov/.

Srinivasan, M. and Kot, C. (1992) Effects of damage on the modal parameters of a cylindrical shell. 10th International Modal Analysis Conference, Society for Experimental Mechanics, San Diego, CA, pp. 529–535.

Stubbs, N., Kim, J.-T. and Farrar, C. (1995) Field verification of a nondestructive damage localization and severity estimation algorithm. 13th International Modal Analysis Conference, Society for Experiment, Nashville, TN, pp. 210–218.

Stubbs, N., Kim, J.-T. and Topole, K. (1992) An efficient and robust algorithm for damage localization in offshore structures. 10th ASCE Structures Conference, American Society of Civil Engineers, New York, NY, pp. 543–546.

Stull, C., Hemez, F. and Farrar, C. (2012) On assessing the robustness of structural health monitoring technologies. *International Journal of Structural Health Monitoring* (accepted).

Su, Z., Ye, L. and Lu, Y. (2006) Guided Lamb waves for identification of damage in composite structures: a review. *Journal of Sound and Vibration*, **295**(3–5), 753–780.

Swartz, R., Flynn, E., Backman, D. *et al.* (2006) Active piezoelectric sensing for damage identification in honeycomb aluminum panels. 24th International Modal Analysis Conference, Society for Experimental Mechanics, St Louis, MO.

Thein, A. (2006) *Pipeline Structural Health Monitoring Using Micro-fiber Composite Active Sensors*, Los Alamos National Laboratory, Los Alamos, NM.

Toksoy, T. and Aktan, A. (1994) Bridge-condition assessment by modal flexibility. *Experimental Mechanics*, **34**(3), 271–278.

Vandiver, J. (1977) Detection of structural failure on fixed platforms by measurement of dynamic response. *Journal of Petroleum Technology*, **29**, 305–310.

Viktorov, I. (1967) *Rayleigh and Lamb Wave – Physical Theory and Applications*, Plenum Press, New York.

Wait, J., Park, G. and Farrar, C. (2005) Integrated structural health assessment using piezoelectric active sensors. *Shock and Vibration*, **12**(6), 389-405.

West, W. (1984) Fault Detection in Orbiter OV-101 Structure and Related Structural Test Specimens, Loads and Structural Dynamics Branch Report, NASA-JSC, Houston, TX.

Wirsching, P.H, Paez, T. and Ortiz, K. (1995) *Random Vibrations Theory and Practice*, John Wiley & Sons, Inc, New York, NY.

Wojnarowski, M.E., Stiansen, S.G. and Reddy, N.E. (1977) Structural integrity evaluation of a fixed platform using vibration criteria. Proceedings of the Offshore Technology Conference, Houston, TX, pp. 247–256.

Wong, C. and Staszewski, W. (1998) Mutual information approach to sensor location for impact detection of composite structures. The International Conference on Noise and Vibration Engineering, ISMA 23, Leuven, Belgium, pp. 1417–1422.

Worden, K. (2001) Rayleigh and Lamb waves – basic principles. *Strain*, **34**(4), 167–172.

Worden, K., Manson, G. and Allman, D. (2001) An experimental appraisal of the strain energy damage location method, in *Damage Assessment of Structures* (eds K.M. Holford, J.A. Brandon, J.M. Dulieu-Barton, M.D. Gilchrist and K. Worden), Transtech Publications, Cardiff, Wales, UK, pp. 35–46.

Worlton, D. (1957) Ultrasonic testing with Lamb waves. *Nondestructive Testing*, **15**, 218–222.

Yuen, M.M. (1985) A numerical study of the eigenparameters of a damaged cantilever. *Journal of Sound and Vibration*, **103**, 301–310.

Zhang, Z. and Aktan, A. (1995) The damage indices for the constructed facilities. Proceedings of the 13th International Modal Analysis Conference, Society for Experimental Mechanics, pp. 1520–1529.

Zhang, L., Brincker, R. and Andersen, P. (2005) An overview of operational modal analysis: major development and issues. Proceedings of the 1st International Operational Modal Analysis Conference, Department of Building Technology and Structural Engineering, Aalborg University, Copenhagen.

Zimmerman, D. and Kaouk, M. (1992) Eigenstructure assignment approach for structural damage detection. *AIAA Journal*, **30**(7), 1848–1855.

Zimmerman, D. and Kaouk, M. (1994) Structural damage detection using a minimum rank update theory. *Journal of Vibration and Acoustics*, **116**(2), 222–230.

Zimmerman, D. and Smith, S. (1992) Model refinement and damage location for intelligent structures, in *Intelligent Structural Systems* (ed. H.T.G.L. Anderson), Kluwer Academic Publishers, pp. 403–452.

8

Features Based on Deviations from Linear Response

The features described in Chapter 7 have several issues associated with them that must be addressed when they are used in 'real-world' applications. To recap some points brought out in the previous chapter, most of the features that are based on modal properties involve fitting a linear physics-based model to the measured data from both the healthy and potentially damaged structure. Often these models do not have the fidelity to accurately predict the local response at boundaries and connections, which are prime locations for damage accumulation. Also, this process does not take advantage of changes in the system response that are caused by nonlinear effects. As a result, nonlinear effects tend to be 'smeared' throughout the linear model fitting process. From a more practical perspective, real-world structures' modal properties have been shown to be sensitive to changing environmental and operational conditions (Farrar *et al.*, 2002) and such sensitivity can lead to false indications of damage, as will be discussed in Chapter 12. Furthermore, modal properties associated with lower-frequency global modes are somewhat insensitive to local damage. In contrast, at higher frequencies the modal properties are associated with local response. However, the practical limitations with the excitation and identification of the modal properties associated with these local modes, caused in part by high modal density and low participation factors, can make them difficult to identify. Finally, the inverse approaches to damage identification do not necessarily produce unique solutions and the degree-of-freedom (DOF) mismatch between the numerical model and the measurement locations can severely limit the ability to accurately perform the required matrix updates. Based on these limitations and the observation that many damage scenarios cause a previously linear structure to exhibit nonlinear behaviour, researchers have developed damage-sensitive features that take advantage of the nonlinear response exhibited by a damaged structure.

8.1 Types of Damage that Can Produce a Nonlinear System Response

There are many types of damage that can cause an initially linear structural system to respond to its operational and environmental loads in a nonlinear manner. One of the most common types of damage is cracks that subsequently open and close under operational loading. This type of damage may include fatigue cracks that form around rivets in airframes, cracks that occur in brittle materials such as concrete and cracks that result from excessive deformation such as those found in some moment-resisting steel frame connections after the Northridge earthquake. It is reiterated that a nonlinear response to operational and environmental loading will only be observed in these cracked structures if the loading causes the

Structural Health Monitoring: A Machine Learning Perspective, First Edition. Charles R. Farrar and Keith Worden.
© 2013 John Wiley & Sons, Ltd. Published 2013 by John Wiley & Sons, Ltd.

cracks to open and close while data are being acquired. Otherwise the crack simply results in a change in geometry of the structure and the structure will continue to respond as a linear system, but in a different configuration. In some cases it is difficult to produce cracks in a controlled manner such that they exhibit the same dynamic characteristics as the cracks that occur in the *in situ* structure. Figure 5.4 shows a 'crack' being introduced into the I-40 Bridge structure to simulate fatigue cracks that occur at the welds of cross-beam seats to the web of the main load-carrying plate girder. However, this crack, which was introduced with a torch, results in a cut through the plate girder web that was approximately 10 mm wide. This crack did not open and close under any subsequent loading and, therefore, did not exhibit the same dynamic characteristics as the type of crack it was intended to simulate.

Many of the common damage types observed in rotating machinery produce nonlinear responses to the harmonic excitation associated with the machine's operational frequency. There are numerous detailed charts of characteristic faults for a variety of machines and machine elements (e.g. see the charts in Crawford, 1992, and Mitchell, 1993). Damage to bearing races, loose bearings and chipped gear teeth are examples of damage that can produce a nonlinear response in rotating machinery. Commercially available software specifically designed for the isolation of faults in rotating machinery based on vibration signatures is readily available. For example, an automated expert diagnostic system is evaluated in Watts and Van Dyke (1993).

Another common type of damage that results in nonlinear system response is that of loose connections. This damage can include debond of glued connects or other types of chemical debond such as that between concrete and reinforcing steel, loose bolts and interference fits that loosen because of material deformation. This type of damage overlaps with cracking as cracks in welded connections can result in a loose connection. The associated rattling or impacting that results when these loose connections are subject to sufficient loading causes the structure to exhibit nonlinear dynamic response characteristics. In some cases, such as the insertion of the femoral component of an artificial hip (shown in Figure 8.1), engineers are attempting to utilise the transition from a nonlinear response to a linear response as an indicator of a successful implant in a cementless procedure that utilises an interference fit (Crisman *et al.*, 2007).

Another type of damage that can result in a nonlinear dynamic response is delamination in bonded, layered materials such as fibre-reinforced composite plates and shells when such structures are subjected to dynamic loading. Often such delaminations are introduced by impact loading. Damage from this type of loading is difficult to detect because the delaminations often occur underneath the surface of the

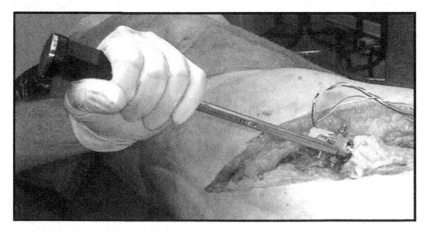

Figure 8.1   The femoral component of an artificial hip being inserted into the femur of a cadaver with accelerometers monitoring the insertion process.

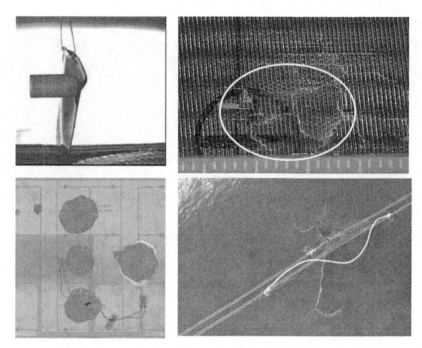

Figure 8.2 Upper left: high-speed projectile impacting a plate. Upper right: photo showing minimal surface damage (within the ellipse) where a projectile impacted the plate. Lower left: ultrasonic scan showing more extensive delamination areas after four different impacts. Lower right: more significant damage on the side opposite the impact.

plate on the side opposite from the impact location and these locations may not be readily accessible for visual inspection. Delamination can be accompanied by matrix cracking and fibre breakage, which add to the nonlinear dynamic response characteristics exhibited by the plate or shell if it is subsequently loaded to levels that cause the delamination to open and close or if there is relative motion across the ply faces. Figure 8.2 shows a composite plate that was subject to a projectile impact. Although some surface damage is visible on the side of the plate opposite from the impact, an ultrasonic scan reveals more extensive delaminations that cannot be seen by surface inspection.

If one extends the concepts of SHM to manufacturing processes, then machine tool chatter is a phenomenon that can produce a nonlinear dynamic response of the cutting tool. Chatter is an unstable vibration that results during turning and facing operations. If not controlled, chatter can lead to poor surface finish and parts that will not meet their design tolerances. Poor surface finish, in turn, can lead to the premature formation of fatigue cracks. The onset of chatter can potentially be detected by the nonlinear dynamic response characteristics measured on the cutting tool. Figure 8.3 shows a piezoelectric sensor that has been mounted on a cutting tool for the purpose of monitoring chatter during facing and turning operations.

Finally, material nonlinearities associated with excessive deformation such as yielding of steel can cause a structure to respond to dynamic loading in a nonlinear manner. This type of damage is particularly difficult to detect because in most cases yielding does not alter the stiffness or mass distribution of a structure once the loading has been removed. Yielding is accompanied by permanent deformation and in some cases this permanent deformation can lead to a nonlinear system response if it results in subsequent interference with neighbouring components when the structure is dynamically loaded.

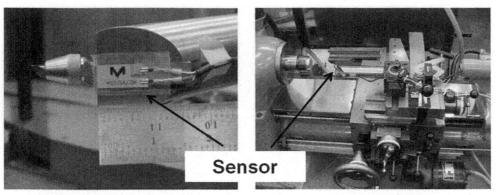

(a) Close-up of Cutting Tool and Sensor   (b) Experimental Setup

Figure 8.3   Sensor mounted on a cutting tool to monitor chatter during facing and turning operations.

8.2   Motivation for Exploring Nonlinear System Identification Methods for SHM

Research aimed at defining damage-sensitive features based on the concepts of nonlinear system identification is motivated by an examination of the two simple single-degree-of-freedom (SDOF) systems shown in Figure 8.4. Also shown in this figure are the force–displacement properties of their stiffness element. The response of the system on the left when subjected to a harmonic forcing function of amplitude X_0 at a frequency of ω can be described by the solution to the second-order linear differential equation presented in Appendix B,

$$m\ddot{y} + c\dot{y} + ky = X_0 \cos \omega t \tag{8.1}$$

where m is the system mass, c is the viscous damping coefficient and k is the spring stiffness. Note that for this case m, c and k do not vary with either time or position of the mass.

The solution to this system is readily obtained in closed form for the given forcing function and, in principle, can be obtained for an arbitrary forcing function with a convolution integral described

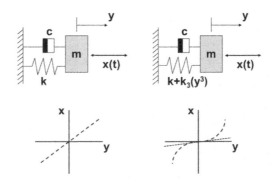

Figure 8.4   Linear and nonlinear spring–mass systems and their associated force–displacement characteristics.

in Appendix B. If the excitation is a Gaussian random input, the responses will also have a Gaussian distribution. For systems made up of multiple interconnected SDOF systems, normal modes are defined and through proper coordinate transformation the response of the multi-degree-of-freedom (MDOF) system can be defined as the superposition of responses from a set of SDOF systems. The MDOF system will also have a unique frequency-domain system input–output relationship defined by the frequency response function (FRF) matrix.

Furthermore, the system will exhibit reciprocity, meaning that if one excites at one location and measures the response at another location, that frequency response function will be the same as the one obtained by reversing that process. A structure exhibiting reciprocity is characterised by symmetric stiffness and FRF matrices.

The system shown in the right in Figure 8.4 corresponds to a Duffing oscillator. With this system the stiffness is not only a linear function of the mass displacement, but also has a stiffness term that is proportional to the cube of this displacement and the equation of motion becomes

$$m\ddot{y} + c\dot{y} + ky + k_3 y^3 = X_0 \cos \omega t \tag{8.2}$$

This seemingly small perturbation to the system's stiffness element brings about some significant changes to the system dynamics. First, a closed-form solution is no longer available even for a harmonic input. Superposition no longer applies and the FRF no longer defines a general unique input/output relation in the frequency domain. Instead, the FRF is now a function of the system's input. If a Gaussian random input is applied, the responses will not have a Gaussian distribution. For MDOF nonlinear systems, normal modes are not defined, the system response cannot be determined through modal superposition and reciprocity no longer applies.

From this example, one can infer that when damage causes an initially linear system to respond in a nonlinear manner, the new dynamic response characteristics exhibited by the now nonlinear system can be used as indicators of damage. However, there are confounding factors that make identifying such changes a challenge. First, many systems exhibit nonlinear response characteristics in their undamaged state. In this case it is even more imperative that these characteristics are accurately quantified if changes in these nonlinear properties are to be used as damage indicators. Second, as shown in the force–displacement curve corresponding to the Duffing oscillator on the right side of Figure 8.4, if the excitation is small, the system will exhibit linear characteristics as indicated by the dashed straight line whose slope would define this low-amplitude stiffness value. Therefore, it may be difficult to identify the onset of nonlinear response characteristics without proper excitation. A similar situation arises with metallic structures, as shown in Figure 8.5. If damage is defined as yielding of the material, this damage can only be detected if measurements are made during the yielding process or if the measurements are sensitive to the permanent deformation in the material that results from the yielding process. The strain hardening characteristics of such materials make it difficult to detect damage with low level vibration-based techniques because such damage may not significantly alter the mass, stiffness or energy dissipation properties when load has been removed from the material.

A final challenge is the somewhat obvious statement that there is no single general method to model and identify all types of nonlinearities. Different nonlinearities have different characteristics, as indicated by the two types of nonlinearities shown on the right of Figure 8.4 and in Figure 8.5, one of which has continuous load–displacement characteristics and the other that has a discontinuous load–deformation relationship. The implication is that one needs some knowledge of the nonlinearity type that the damage will produce in order to effectively identify the onset of that nonlinearity and the associated damage. With this motivation various methods of nonlinear system identification and modelling that have been applied to SHM are summarised next.

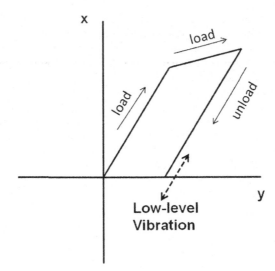

Figure 8.5   Load–deformation for an idealised metallic structure.

8.2.1   Coherence Function

The coherence function was introduced as a possible damage-sensitive feature in Chapter 7 and will be revisited here. It is a spectral function that quantifies the extent to which the response power is *linearly* correlated with the input; it can therefore also detect nonlinearity. It can provide a quick visual inspection of the quality of an FRF and, in many cases, is a rapid indicator of the presence of nonlinearity in specific frequency bands or resonance regions. It is arguably the most often used test of nonlinearity by virtue of the fact that almost all commercial spectrum analysers come with software that calculates this quantity. The coherence function is derived and explained in detail in Appendix A, where its application to nonlinearity detection is demonstrated with a numerical example of an SDOF system with a cubic stiffness term. As the excitation is increased in that example, the nonlinear terms begin to play a part and the coherence is shown to drop. This type of situation will occur for all polynomial nonlinearities and systems exhibiting piecewise-linear stiffness, like those induced by a breathing crack (Friswell and Penny, 2002). However, if one considers Coulomb friction, the opposite occurs. At high excitation, the friction breaks out (i.e. stick response gives way to slip response) and a nominally linear response will be obtained and hence coherence values closer to unity will result.

Changes in coherence measured on the four-storey simulated building structure introduced in Chapter 5 are shown in Figure 8.6. The structure was excited with a random base input. Coherence functions are shown for the undamaged (left) and damaged (right) structures subjected to two different levels of excitation. The coherence for the undamaged case shows drops at 60 Hz, which is typically associated with electric noise in the United States. There is also a drop slightly above 60 Hz, which corresponds to an antiresonance of the structure where the signal-to-noise ratio is poorer than at other frequencies. For the case where damage is present, similar results as those produced numerically in Appendix A are seen. At the low excitation levels, the impacts are not occurring very often and the coherence does not deviate significantly from the undamaged case although there is some degradation just above 100 Hz. When the amplitude of excitation is increased and the impacts occur more frequently, the coherence deteriorates significantly.

It is important to stress again that in order to use the coherence function for detecting nonlinearity and hence damage, it is necessary to realise that a reduction in the level of coherence can be caused by a range of problems, such as noise on the output and/or input signals that may in turn be caused by

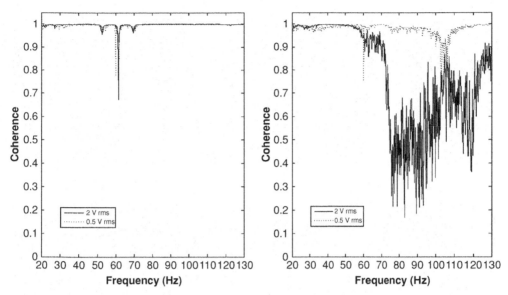

Figure 8.6 Coherence function measured on the simulated building structure when it was subject to different random excitation levels. These plots correspond to responses measured with the accelerometer on the first floor (Channel 3 in Figure 5.14). No damage was present for the plot on the left. For the plot on the right damage is simulated with the bumper located between the second and third floors.

incorrect gain settings on amplifiers, leakage and extraneous sources of unmeasured input to the system. Such obvious causes should be checked before structural nonlinearity and, hence, damage is suspected as the cause of loss of coherence. As an example, changes in coherence for the sensor located closest to the damage in the I-40 Bridge tests are shown in Figure 8.7. In this figure the coherence function for each damage level is compared to the coherence function obtained from the test corresponding to the undamaged structure. At first one might conclude that although the coherence is in general poor, it degrades further with increasing damage levels, as indicated by the shaded area between the coherence curves. However, when one also views the input power spectra for the respective tests (Figure 8.8), it is clear that the degradation in coherence is directly correlated with the level of excitation, which influences the signal-to-noise ratio in the measurements. For lower levels of excitation the traffic on the adjacent bridge was producing an unmeasured source of input and this input proportionally contributed more to the measured response at these excitation levels. The response to these unmeasured inputs caused poorer coherence measurements for those tests. In fact, if the excitation remained consistent between different tests, one would not expect any significant change in coherence associated with the damage introduced during the I-40 Bridge tests because the cuts in the girder only change the geometry of the structure. This damage did not cause the structure to respond in a nonlinear manner for the levels of excitation applied during this study.

8.2.2 Linearity and Reciprocity Checks

One method to demonstrate that a structure is responding in a nonlinear manner is to excite the structure at two different levels and show that the response does not scale linearly by the ratio of the excitation levels. A common approach to visualise this property is to overlay the FRFs from the excitations at the

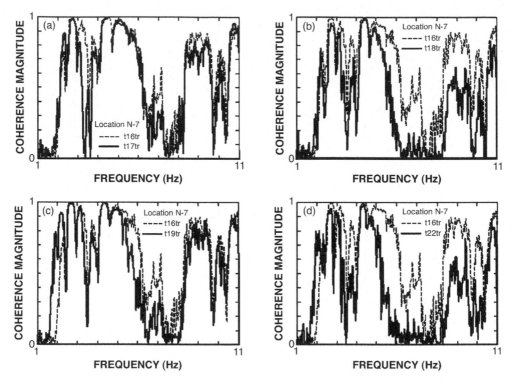

Figure 8.7 Coherence functions obtained with sensor N-7 (see Figure 5.5) corresponding to various levels of damage in the I-40 Bridge. The labelling scheme is t16tr–undamaged, t17tr–damage level E-1, t18tr–damage level E-2, t19tr–damage level E-3, t22tr–damage level E-4.

two different levels. For linear systems the two FRFs obtained with the different excitation levels should overlay because the response spectrum is normalised by the input spectrum.

Figure 8.9 shows an overlay of the input impulse power spectral density functions from a linearity check performed on the Alamos Canyon Bridge summarised in Chapter 5 and the corresponding driving point FRFs. In this case the two impact tests where the power spectral density amplitudes were different by a factor of approximately five were performed within 2 hours of each other and no changes were introduced to the structure's condition within that time period. Ideally, a linearity check should be performed over the expected range of operating loads, but this linearity test was limited to what could be obtained using the instrumented hammer. These plots highlight a difficulty with interpreting data from a linearity check. Between 7 and 25 Hz the structure exhibits linear characteristics when excited at the levels shown in this figure. However, above 25 Hz there is a noticeable difference in the two measurements, suggesting the possibility that nonlinearities were excited in this frequency range or that signal-to-noise ratios were poor, thus providing the appearance of a nonlinear response. This frequency range also corresponds to lower coherence in the lower-excitation level measurement, as shown in Figure 8.10. The coherence was observed to improve for the higher-magnitude input, suggesting that possibly the lower signal-to-noise ratio in the low-amplitude measurements might be causing the linearity test discrepancies shown in Figure 8.9. Therefore, in addition to the FRFs overlaying, one would also like to see the corresponding coherence functions overlay as well to have confidence that the structure's response scales linearly with increasing amplitude.

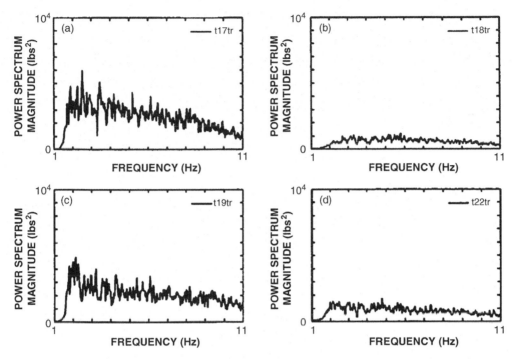

Figure 8.8 Input power spectra measured with the accelerometer mounted on the reaction mass of the shaker used in the I-40 Bridge tests and corresponding to the excitations applied after the different levels of damage had been introduced.

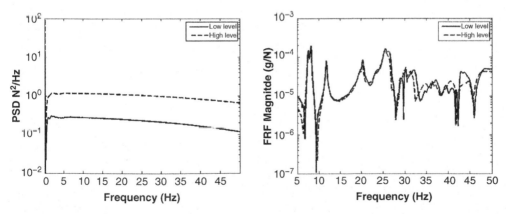

Figure 8.9 Results of a linearity check performed on the Alamosa Canyon Bridge structure shows that different portions of the FRF are more sensitive to the nonlinear system response.

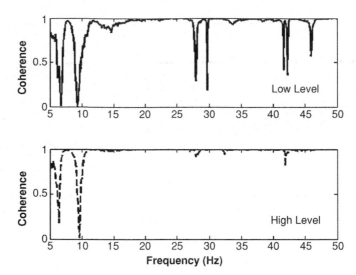

Figure 8.10 The coherence functions corresponding to the linearity tests shown in Figure 8.9.

Figure 8.11 shows similar results for measurements made at the top mass in the simulated building structure when excited with different amplitude band-limited random base inputs. This figure shows the cases when the system is undamaged and when an impact nonlinearity is present between the top two floors. The figure also shows the corresponding coherence functions. For the linear case the FRF magnitudes overlay across the entire spectrum. With an exception for the drop in coherence associated

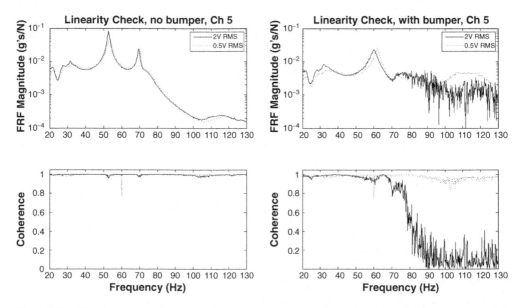

Figure 8.11 The linearity test FRF magnitudes and corresponding coherence functions for a measurement made at the third floor of the simulated building structure in a linear condition (left) and when an impact nonlinearity is present (right) between floors 2 and 3.

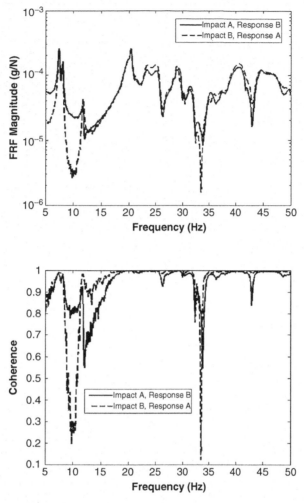

Figure 8.12 The FRFs and corresponding coherence functions measured during a reciprocity test on the Alamosa Canyon Bridge. Measurements were conducted using the procedure depicted in Figure 8.13.

with 60 Hz electric noise in the low-amplitude tests, the coherence functions for the two linear systems also overlay.

When the impact nonlinearity used to simulate damage is present, there are significant distortions in the FRFs when compared to the corresponding linear cases. Also, there is considerable change in these FRFs and the corresponding coherence functions associated with the impacting system as a function of excitation level, particularly in the higher-frequency portions of the spectra.

Reciprocity checks are also performed to assess the linearity of a system's response. Figure 8.12 shows the FRF magnitudes for an impact applied at one point on the Alamosa Canyon Bridge deck and a response measured at one of the most distant measurement points on that same deck. Also shown in this figure is the FRF magnitude for an impact applied at point B and a response measured at point A. Coherence plots corresponding to the reciprocity results are also shown in Figure 8.12. Because the data acquisition system channels were moved from one point to another (i.e. the input measurement channel

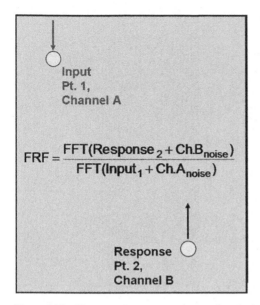
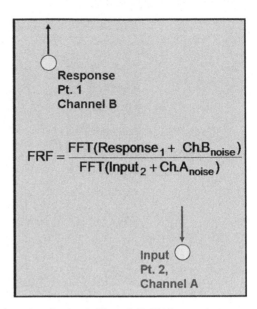

Figure 8.13   The measurement procedure used to obtain the reciprocity data in Figure 8.12. The input and response sensors remain connected to their original data acquisition channel. This approach measures the reciprocity of the structure and the data acquisition system.

is the same for inputs at both locations, as shown in Figure 8.13), these figures show the reciprocity of the entire measurement system, including the structure and the data acquisition system.

Alternatively, one can keep the data acquisition channel associated with a particular location as depicted in Figure 8.14. A plot of the FRF magnitudes when the accelerometers at points A and B have been switched in this manner is shown in Figure 8.15 along with the corresponding coherence functions. By switching the accelerometers the reciprocity being measured is that of the structure alone. From Figures 8.12 and 8.15 it is evident that the structure itself is exhibiting reciprocity in all portions of the spectrum where there is good coherence. Also, when these two figures are compared, it is evident that the noise in the measurement electronics of the data acquisition system is contributing to the loss of reciprocity, particularly in the frequency range between the second and third natural frequencies (8.5–11.5 Hz).

Figure 8.16 shows a reciprocity check from the simulated building structure when an impact excitation was applied at the first floor and an acceleration response was measured at the third floor. The testing procedure used corresponds to that shown in Figure 8.14. In this figure the FRF magnitudes and corresponding coherence functions are shown for the cases when the structure is in a linear configuration and for the case where the excitation excites the impact nonlinearity used to simulate damage. The undamaged structure exhibits reciprocity over all portions of the spectrum where high coherence between the two measurements was obtained. When the damage is present, there are considerable distortions of the FRFs as compared to those obtained from the linear system and the structure no longer exhibits reciprocity, as indicated by FRFs that do not overlay.

8.2.3   Harmonic Distortion

Harmonic or waveform distortion is one of the clearest indicators of nonlinearity. Such distortions are the primary damage-sensitive features used to diagnose the condition of rotating machinery (Rao, 2000), which as previously mentioned is the SHM application that has made the most significant transition

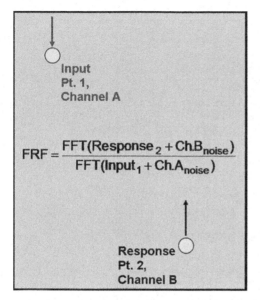 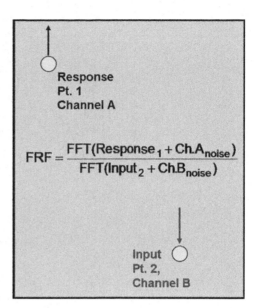

Figure 8.14 Reciprocity measurement procedure where the input and response measurement points are reversed. The input and response sensors are connected to data acquisition channels associated with the measurement point. This measurement procedure gives a measure of the reciprocity of only the structure.

from research to practice. If the excitation to a linear system is a mono-harmonic signal, that is a sine or cosine wave of frequency ω, the response will be mono-harmonic at the same frequency (after any transients have died out), as discussed in Appendix B. In addition, the response of a linear system to some initial disturbance (displacement or velocity impulse) can be represented by a sum of decaying sinusoids after transients have died out, as described by the well-known superposition approach referred to as the convolution integral (Thomson and Dahleh, 1998).

It is not always true to say that a sine wave input to a damaged system exhibiting nonlinear characteristics will not produce a sine wave output. However, this is usually the case and is the basis of a simple and powerful test for nonlinearity as sine wave inputs are simple signals to generate in practice. The form of the distortion is caused by the appearance of higher harmonics in the response. Similar higher harmonics are observed in the response of nonlinear systems excited by an initial displacement or by an initial velocity impulse.

Distortion can be easily detected by observing the input and output time response signals. Figure 8.17 shows an example of harmonic distortion where a sinusoidal acceleration response signal is altered by, in this case, a cubic nonlinearity, as described by Equation (8.2).

In Figure 8.17 the output response from a nonlinear system is shown in terms of the displacement, velocity and acceleration. The reason that the acceleration is more distorted compared with the corresponding velocity or displacement is easily explained. Let $x(t) = \sin(\omega t)$ be the input to the nonlinear system. The output, $y(t)$, will generally (at least for weak nonlinear systems) be represented as a Fourier series composed of harmonics, written as

$$y(t) = A_1 \sin(\omega t + \phi_1) + A_2 \sin(2\omega t + \phi_2) + A_3 \sin(3\omega t + \phi_3) \ldots \tag{8.3}$$

and the corresponding acceleration response, $\ddot{y}(t)$, is

$$\ddot{y}(t) = -\omega^2 A_1 \sin(\omega t + \phi_1) - 4\omega^2 A_2 \sin(2\omega t + \phi_2) - 9\omega^2 A_3 \sin(3\omega t + \phi_3) \ldots \tag{8.4}$$

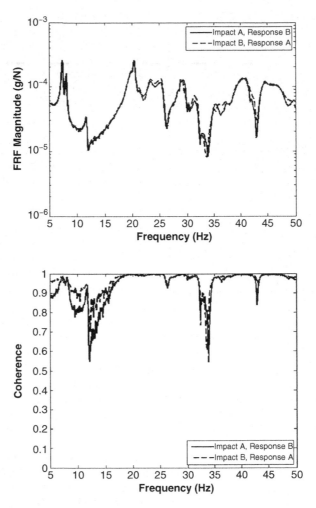

Figure 8.15 The FRFs and corresponding coherence functions measured during a reciprocity test on the Alamosa Canyon Bridge. Measurements were conducted using the procedure depicted in Figure 8.14.

Thus the nth output acceleration term is weighted by the factor n^2 compared to the fundamental.

Harmonic generation can also be seen in systems subjected to nonsinusoidal inputs. Figure 8.18 shows the Wigner–Ville time-frequency transform (Mallat, 1999) applied to free-vibration response data measured at the free end of both uncracked and cracked cantilever beams subject to an initial displacement (Prime and Shevitz, 1996). The crack is located at the mid-span and penetrates half the thickness of the beam. The time-frequency plots in Figure 8.18 show the generation of natural frequency harmonics in the freely vibrating, cracked cantilever beam as well as the change in stiffness state as the crack opens and closes. A comparison of the two plots in Figure 8.18 clearly shows that the presence of nonlinearity adds considerable complexity to the frequency response characteristics of the system (Axiom VIII) and measures of this complexity relative to the undamaged system response can be used to infer the presence of damage.

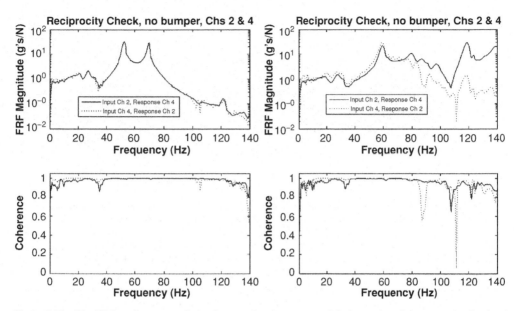

Figure 8.16 The FRFs and corresponding coherence functions measured during reciprocity tests on the simulated building structure without damage (left) and with damage simulated by an impact nonlinearity (right). Measurements were conducted using the procedure depicted in Figure 8.14. Measurement locations were on the first and third floors.

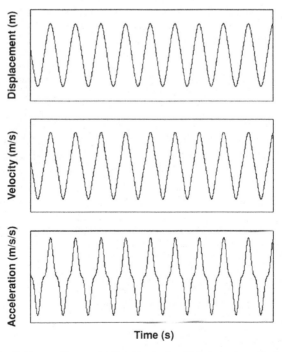

Figure 8.17 Evidence of harmonic distortion in time histories as a result of nonlinearity.

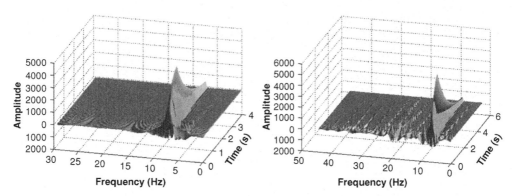

Figure 8.18 Wigner–Ville transforms of the free-vibration acceleration–time histories measured on an uncracked cantilever beam (left) and a cracked cantilever beam (right) subjected to an initial displacement.

Another illustration of harmonic distortion can be seen in data obtained from the simulated building structure when the system is subjected to a 53-Hz harmonic base excitation. In one case the system is in an undamaged configuration and in the other case the excitation excited the impact nonlinearity that is used to simulate damage in this structure. Figure 8.19 shows a portion of input time histories and corresponding power spectra for the inputs from the two tests. This figure shows that the input remained constant between the two tests. Figure 8.20 shows a portion of the response time histories and corresponding power spectra (normalised to a peak value of 1) from the top floor for both the undamaged and simulated damage cases. This response measurement shows distortion from a harmonic response. Furthermore, harmonics of the driving frequency are visible in the power spectra when the simulated damage is present. These harmonics are accompanied by a significant drop in the amplitude of response even though Figure 8.19 shows that the base excitation remained consistent in each test.

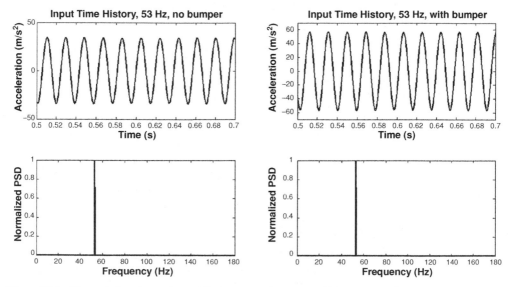

Figure 8.19 Harmonic base input signals (53 Hz) used to drive the simulated building structure in its undamaged (left) and damaged configuration (right).

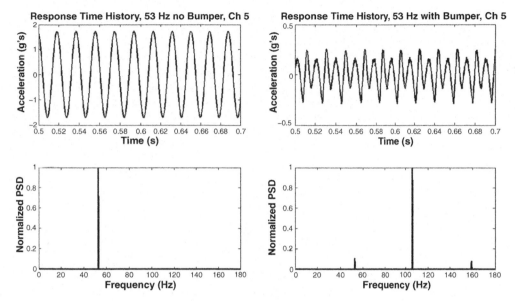

Figure 8.20   Top floor response to a 53 Hz harmonic base input signal used to drive the simulated building structure. Left: undamaged configuration; right: damaged configuration simulated by a bumper between floors 2 and 3. Peak amplitudes in the spectra have been scaled to 1 to show the harmonics that were generated.

In general, if the system inputs are nonsinusoidal waveforms, such as band-limited random signals, waveform distortion associated with the nonlinear response resulting from damage is difficult to detect and alternate features are required such as those that can be obtained from the coherence function or the probability density function.

8.2.4   *Frequency Response Function Distortions*

As previously discussed, damage can be inferred from the loss of amplitude invariance of any number of measured quantities if it is assumed that the system responds in a predominantly linear manner when in its undamaged state. One of the most fundamental constructs for modal analysis is the FRF; this takes a specific form for linear MDOF systems and departures from this form as the excitation level changes can yield information about the type of nonlinearity present in the system. The study of *FRF distortion* was the subject of some of the earliest attempts to reconcile nonlinearity with modal analysis (Ewins, 1995).

Before discussing FRF distortion, recall the principal definitions of the FRF; first for a stepped-sine test. If a signal $x(t) = X \sin(\omega t)$ is input to a linear system, it will result in a response $y(t) = Y(\omega) \sin(\omega t + \phi)$. For $X > 0$ the FRF, which is a function of ω, is defined as

$$H(\omega) = \frac{|Y(\omega)|}{X} e^{i\phi(\omega)} \tag{8.5}$$

This quantity is very straightforward to obtain experimentally. Over a range of excitation frequencies, sinusoids $X \sin(\omega t)$ are applied sequentially to the system. At each frequency, the time histories of the input and response signals are recorded after transients have died out and transformed in the frequency domain using the Fourier transform. The ratio of the response spectrum to the input spectrum yields the FRF value at the frequency of interest. In the case of a linear system, the FRF in Equation (8.5) summarises the input/output process in its entirety and does not depend on the amplitude of excitation X. In such a situation, the FRF is referred to as *pure*.

In the case of a nonlinear system, it is well known that sinusoidal forcing generates response components at frequencies other than the excitation frequency. In particular, the distribution of energy among these frequencies depends on the level of excitation X, so the measurement process described above will also lead to a quantity that depends on X. However, because the process is simple, it is often carried out experimentally in an unadulterated fashion for nonlinear systems. The FRF resulting from such a test is referred to as *composite* and denoted by $\Lambda_s(\omega)$. Strictly speaking $\Lambda_s(\omega, X)$ is more appropriate, but the amplitude argument will always be clear in this context. In the control literature, $\Lambda_s(\omega)$ is often called a *describing function*.

The form of the composite FRF also depends on the type of excitation used. If white noise of constant power spectral density amplitude P is used and the FRF is obtained by taking the ratio of the cross- and auto-spectral densities (S_{yx}, S_{xx}) for a linear system,

$$H(\omega) = \frac{S_{yx}(\omega)}{S_{xx}(\omega)} \tag{8.6}$$

it can be shown that this quantity is identical to the FRF obtained from stepped-sine testing. If however, the same approach is used for a nonlinear system, another composite FRF is obtained,

$$\Lambda_r(\omega, P) = \frac{S_{yx}(\omega)}{S_{xx}(\omega)} \tag{8.7}$$

and the function $\Lambda_r(\omega, P)$ is distinct from the $\Lambda_s(\omega, X)$ obtained from a stepped-sine test. The object of this section is to outline some ways of computing FRFs, in the hope that the analysis can be used to identify nonlinearities that are indicative of damage.

The first calculation will illustrate a well-known means of approximating $\Lambda_s(\omega)$, which is the analytical analogue of the stepped-sine test and is referred to as the method of *harmonic balance* (Nayfeh and Mook, 1979). An SDOF nonlinear system with a cubic stiffness term will again be analysed; this system is defined by Duffing's equation,

$$m\ddot{y} + c\dot{y} + ky + k_3 y^3 = X(t) \tag{8.8}$$

Harmonic balance simply assumes that the response to a sinusoidal excitation is a sinusoid at the same frequency. The trial solution $Y \sin(\omega t)$ is substituted in the equation of motion and to simplify matters the phase is transferred on to the input to allow Y to be taken as real. This substitution yields

$$-m\omega^2 Y \sin(\omega t) + c\omega Y \cos(\omega t) + kY \sin(\omega t) + k_3 Y^3 \sin^3(\omega t) = X \sin(\omega t - \phi) \tag{8.9}$$

or, after a little trigonometry,

$$-m\omega^2 Y \sin(\omega t) + c\omega Y \cos(\omega t) + kY \sin(\omega t) + k_3 Y^3 \left\{ \frac{3}{4} \sin(\omega t) - \frac{1}{4} \sin(\omega t) \right\}$$
$$= X \sin(\omega t) \cos\phi - X \cos(\omega t) \sin\phi \tag{8.10}$$

Equating the coefficients of $\sin(\omega t)$ and $\cos(\omega t)$ (the *fundamental* components) yields the equations

$$-m\omega^2 Y + kY + \frac{3}{4} k_3 Y^3 = X \cos\phi \tag{8.11}$$

and

$$c\omega Y = -X \sin \phi \tag{8.12}$$

The required gain and phase follow routinely as

$$\left| \frac{Y}{X} \right| = \frac{1}{\left[\left\{ -m\omega^2 + k + \frac{3}{4}k_3 Y^2 \right\}^2 + c^2\omega^2 \right]^{1/2}} \tag{8.13}$$

$$\phi = \tan^{-1} \frac{-c\omega}{-m\omega^2 + k + \frac{3}{4}k_3 Y^2} \tag{8.14}$$

which can be combined into the complex FRF

$$\Lambda_s(\omega) = \frac{1}{k + \frac{3}{4}k_3 Y^2 - m\omega^2 + ic\omega} \tag{8.15}$$

At a fixed level of excitation, the FRF has natural frequency

$$\omega_n = \sqrt{\frac{k + \frac{3}{4}k_3 Y^2}{m}} \tag{8.16}$$

which depends on Y and, hence, indirectly on X. If $k_3 > 0$, the natural frequency increases with X and such a system is referred to as *hardening*. If $k_3 < 0$ the system is *softening* and the natural frequency decreases with increasing X.

This approximation reproduces experimental results showing that the natural frequency changes with amplitude. Furthermore, the actual form of the FRF can depart substantially from the linear form. For given X and ω, Y is obtained by solving the (essentially) cubic equation (8.11). As complex roots occur in conjugate pairs, Equation (8.11) will either have one or three real solutions. The effect of this observation is well known, but remains striking.

At low levels of excitation, the FRF is a barely distorted version of that for the underlying linear system as the k term dominates for $Y \ll 1$. A unique response amplitude (a single real root of Equation (8.11)) is obtained for all ω. As X increases, the FRF becomes more distorted and departs from the linear form, but a unique response is still obtained for all ω. This result continues until X reaches a critical value X_{crit}, where the FRF has a vertical tangent. Beyond this point a range of ω values [$\omega_{\text{low}}, \omega_{\text{high}}$] is obtained over which there are three real solutions for the response. This is a *bifurcation point* of the parameter X. As the test or simulation steps past the point ω_{low}, two new responses become possible and persist until ω_{high} is reached and two solutions disappear. The graph of the response is shown in Figure 8.21.

The relevant experimental circumstances for this analysis occur during a stepped-sine or a sine-dwell test. Consider an upward sweep. A unique response exists up to ω_{low}. However, beyond this point, the response stays on branch $y^{(1)}$ essentially by continuity. This condition persists until, at frequency ω_{high}, $y^{(1)}$ ceases to exist and the only solution is $y^{(3)}$; a jump to this solution occurs giving a discontinuity in the FRF. Beyond ω_{high} the solution stays on the continuation of $y^{(3)}$, which is the unique solution in this range. The amplitude follows path ABD in Figure 8.21. The type of FRF obtained from such a test is shown in Figure 8.22. The downward sweep is similarly described and follows path DCA.

If $k_3 > 0$, the resonance peak moves to higher frequencies and the jumps occur on the right-hand side of the peak as described above. If $k_3 < 0$, the jumps occur on the left of the peak and the resonance

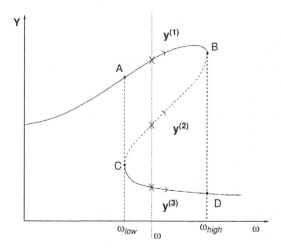

Figure 8.21   Possible response amplitudes for the Duffing oscillator under high-level excitation.

shifts downward in frequency. These discontinuities are frequently observed in experimental FRFs and can be attributed to certain types of nonlinearities brought on by damage.

8.2.5   *Probability Density Function*

In Chapter 7 it was noted that when a linear system is subject to a random input whose amplitudes are described by a Gaussian distribution, the response amplitudes will also have a Gaussian distribution. Also, it is known that a Gaussian distribution has well-defined properties such as skewness equal to zero and the normalised fourth moment (kurtosis) equal to three. When the input is known to be Gaussian in nature, deviations of the acceleration response statistics can result when nonlinearities caused by damage are present. An example of this change in the probability density function (PDF) was previously shown with data from the simulated building structure in Figure 7.16. Similar estimates of the acceleration response PDFs obtained with a kernel density estimator from the concrete column are shown in Figure 8.23. In this figure one can see that the acceleration PDFs show significant changes for the measurement made at

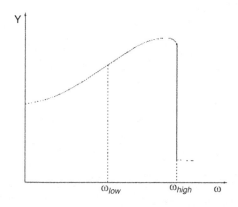

Figure 8.22   Composite FRF for Duffing oscillator at high excitation (upward sweep).

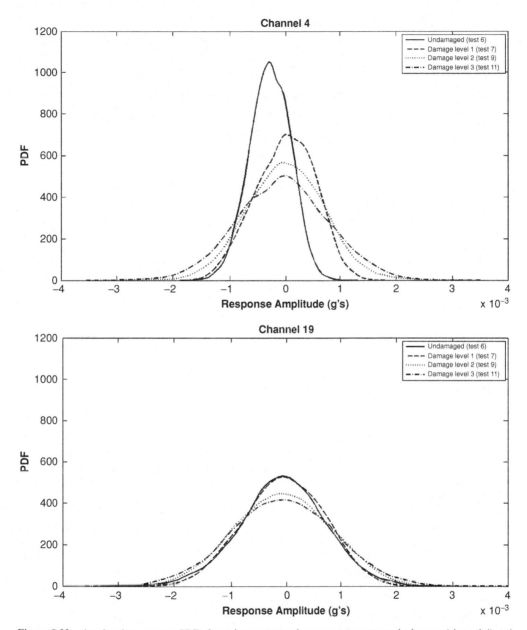

Figure 8.23 Acceleration response PDFs from the concrete column measurements at the bottom (channel 4) and top (channel 19) of the column.

the bottom of the column close to the damage location. The PDF corresponding to data from the top of the column shows relatively little change as damage accumulates at the bottom of the column. If baseline statistics are available from the undamaged system, examination of the PDF and the associated statistics can be an indicator of nonlinearities associated with damage. These statistics are used extensively as low-dimension features to diagnose damage to roller bearings (Martin, 1989).

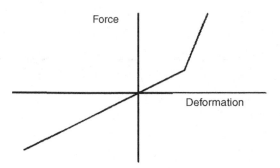

Figure 8.24   Offset bilinear stiffness system.

8.2.6   Correlation Tests

There are two simple correlation tests that can signal nonlinearity by analysing measured time history data. If records of both input x and response y are available, a correlation function, $\varphi_{x^2 y'}(j)$, can be defined as

$$\varphi_{x^2 y'}(j) = E[x(i)^2 y'(i+j)] \tag{8.17}$$

where E designates the expectation operator. The function $\varphi_{x^2 y'}(j)$ vanishes for all discrete time delays, j, if and only if the system is linear (Billings and Tsang, 1990). The apostrophe signifies that the mean has been removed from the response signal.

If only sampled response data are available, it can be shown that under certain conditions, the correlation function defined as

$$\varphi_{y' y'^2}(j) = E[y'(i+j)y'(i)^2] \tag{8.18}$$

is zero for all j if and only if the system is linear (Billings and Fadzil, 1985). In practice, these functions will never be identically zero; however, confidence intervals for a near-zero result can be calculated. As an example, acceleration data were obtained from an SDOF system with an offset bilinear stiffness whose force–deformation plot is shown in Figure 8.24. The system was subjected to a random excitation. The correlation functions for the response at both low and high excitations are shown in Figure 8.25. The dashed lines are the 95% confidence limits for a zero result. The function shown in the bottom of Figure 8.25 indicates that the data from the high-excitation test arises from a nonlinear system. The low-excitation test did not excite the nonlinearity and the corresponding function (top of Figure 8.25) shows that only approximately 5% of the points are outliers, as would be expected.

Note that Equation (8.18) as it stands only detects even nonlinearities such as quadratic stiffness. In practice, to identify odd nonlinearity, the input signal should contain a DC offset. This type of nonlinearity offsets the output signal and adds an even component to the nonlinear terms. A further restriction on Equation (8.18) is that it cannot detect odd damping nonlinearity as it is not possible to generate a DC offset in the velocity to add an even component to the nonlinearity. There are implications for damage detection here as the occurrence of damage in a structure may result in the appearance of a friction nonlinearity when the damage causes two surfaces to slide over each other.

8.2.7   The Holder Exponent

The Holder exponent is a damage-sensitive feature that identifies nonlinearities associated with discontinuities introduced into dynamic response data resulting from certain types of damage such as breathing

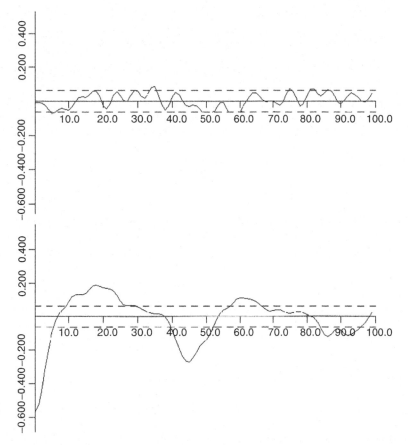

Figure 8.25 Nonlinear correlation functions for a linear system (top) and nonlinear system (bottom).

cracks (Robertson, Farrar and Sohn, 2003). This feature, also known as the Lipschitz exponent, provides information about the regularity of a signal (Mallat and Hwang, 1992). In essence, the regularity identifies to what order the signal is differentiable. For instance, if a signal is differentiable at $t = 0$, it has a Holder exponent of 1 or greater. If the signal is discontinuous, but bounded in the neighbourhood of $t = 0$, such as a step function, then the Holder exponent is 0. The Dirac delta function then has a Holder exponent of -1 because it is unbounded at $t = 0$. From these examples, one can see that there is a relationship between the Holder exponent of a function and its derivatives and primitives. Taking the derivative of a function decreases its regularity by 1 and integrating increases it by 1. For SHM applications the assumption is that a linear system will have a continuous response to random, transient or steady state excitation. When nonlinearity is introduced, for example by a crack opening and closing, the response signals will have discrete portions that are discontinuous and the Holder exponent will identify such discontinuities.

To define the Holder regularity, assume that a signal $y(t)$ can be approximated locally at t_0 by a polynomial of the form (Struzik, 2001)

$$
\begin{aligned}
y(t) &= c_0 + c_1(t - t_0) + \cdots + c_n(t - t_0)^n + C\,|t - t_0|^\alpha \\
&= P_n(t - t_0) + C\,|t - t_0|^\alpha
\end{aligned}
\tag{8.19}
$$

where P_n is a polynomial of order n and C is a coefficient. The term associated with the exponent α can be thought of as the residual that remains after fitting a polynomial of order n to the signal. The local regularity of a function at t_0 can then be characterised by this 'Holder' exponent:

$$|y(t) - P_n(t - t_0)| \le C |t - t_0|^\alpha \tag{8.20}$$

A higher value of α indicates a better regularity or a smoother function. In order to detect singularities, a transform is needed that ignores the polynomial part of the signal. A wavelet transform that has n vanishing moments is able to ignore polynomials up to order n:

$$\int\limits_{-\infty}^{\infty} t^n \psi(t)\mathrm{d}t = 0 \tag{8.21}$$

Transformation of Equation (8.20) using a wavelet with at least n vanishing moments then provides a method for extracting the values of the Holder exponent in time:

$$|W (y(a, b))| \le C a^\alpha \tag{8.22}$$

The wavelet transform of the polynomial is zero and so what remains is a relationship between the wavelet transform of $y(t)$ and the error between the polynomial and $y(t)$, which relates to the regularity of the function. When a complex wavelet such as the Morlet wavelet is used, the resulting coefficients are also complex. Therefore, the magnitude of the modulus of the wavelet transform must be used to find the Holder exponent. As shown below, the exponent α can be calculated at a specific time point by finding the slope of the log of the modulus at that time versus the log of the scale vector a.

For SHM applications, singularities are defined as points in the response time history that are discontinuous. As discussed above, bounded discontinuities have a Holder exponent of 0. Therefore, measuring the regularity of the signal in time can be used to detect these singularities. The Holder exponent can pertain to the global regularity of a function or it can be found locally. A common method for finding its value is through the use of the Fourier transform. The asymptotic decay of a signal's frequency spectrum relates directly to the uniform Holder regularity. However, the Fourier transform approach only provides a measure of the minimum global regularity of the function and cannot be used to find the regularity at a particular point in time. Wavelets, on the other hand, are well localised in time and can therefore provide an estimate of the Holder regularity both over time intervals and at specific time points. The wavelet method for estimating the Holder exponent is similar to that of the Fourier transform. The wavelet provides a time–frequency map called a scalogram. By examining the decay of this map at specific points in time across all scales (frequencies), the pointwise Holder regularity of the signal can be determined.

Any time–frequency transform can be used for Holder exponent extraction, but certain characteristics of the wavelet transform make it particularly well adapted for this application. Specifically, these characteristics are: the decay of the wavelet basis functions in the frequency domain, which is associated with the number of vanishing moments, and the variability of the bandwidth of the wavelet transform in time and frequency. The order of the wavelet limits the degree of regularity that can be measured in a function. Therefore, wavelets can be tuned to the signals that are being analysed. Also, the variability of the time and frequency bandwidths provides a finer time resolution at the higher frequencies, which can be helpful in detecting the point in time when sudden changes occur in a signal.

The easiest way to identify a discontinuity in a signal is by looking for a distinct downward jump in the Holder exponent versus time plot. As previously mentioned, a discontinuous point should have a Holder exponent value of zero, but resolution limitations of the wavelet transform will result in slightly different values. Therefore, identifying time points where the Holder exponent dips from positive values towards zero, or below, will identify when the discontinuities in the signal occur. The steps for calculating the

Holder exponent in time are as follows. First, take the wavelet transform of the given signal and take the absolute value of the resulting coefficients to obtain the wavelet transform modulus:

$$|W(y(a, b))| = \left| \frac{1}{\sqrt{s}} \int\limits_{-\infty}^{\infty} y(t) \, \psi^* \left(\frac{t-b}{a} \right) dt \right| \tag{8.23}$$

Arrange the coefficients in a two-dimensional time–scale matrix. One dimension of the time–scale matrix (b) represents a different time point in the signal and the other dimension denotes a different frequency scale (a). Take the first column, which represents the frequency spectrum of the signal at the first time point, and plot the log of it versus the log of the scales, a, at which the wavelet transform was calculated. This calculation can be shown mathematically by taking the log of each side of Equation (8.22):

$$\log |W(y(a, b))| = \log(C) + \alpha \log(a) \tag{8.24}$$

Ignoring the offset due to the coefficient C, the slope m of a straight-line fit to the spectrum is then the decay of the wavelet modulus across its scales. Negating the slope will give the decay versus the frequencies of the transform rather than the scales, because of the inverse relationship between scale and frequency. The Holder exponent α is then simply the slope m defined as

$$m = \frac{\log |W(y(a, b))|}{\log(a)} = \alpha \tag{8.25}$$

This process provides the Holder exponent for the first time point in the signal. To find the Holder exponent at all time points, repeat this process for each time point of the wavelet modulus matrix.

Now the Holder exponent will be applied to two different experimental data sets. First, Figure 8.26 shows the Holder exponent for data from the simulated building structure where impacts caused by the bumper produce discontinuities in the measured response signal. Four distinct impact events can be seen in this figure. Here the initial gap was set relatively wide (0.20 mm), which resulted in few impacts. In Figueiredo *et al.* (2009) these results showed exact agreement to other time-domain approaches for

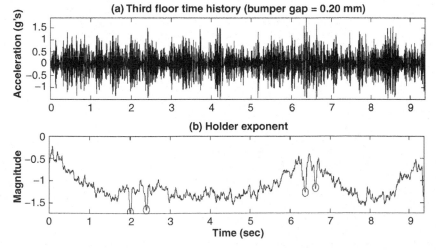

Figure 8.26   Holder exponent calculated from the simulated building structure data.

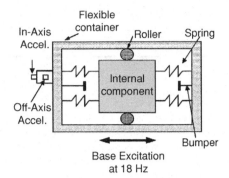

Figure 8.27 Schematic diagram of the test structure with a loose internal part.

identifying the specific impact events. This study also showed that frequency-domain approaches were not able to detect these discontinuities because such approaches provide no localised time information and this limited number of impacts had a negligible impact on the spectra when compared with undamaged data.

The next example shows the application of the Holder exponent to data from a mechanical system with a loose part that was subjected to a harmonic input on a shake table. A schematic of the test structure is shown in Figure 8.27. The nonsymmetric bumpers cause the structure to exhibit a rattle during one portion of the excitation. Figure 8.28 shows the response of the structure as measured by accelerometers mounted on the outer structure in the in-axis and off-axis directions. The rattle produced by these impacts is evident in the sensor measurements that are off-axis from the excitation. The short oscillations of increased magnitude in these measurements are indicative of the rattle. These same oscillations are not readily apparent in the in-axis data, particularly if one does not have the off-axis measurements for reference.

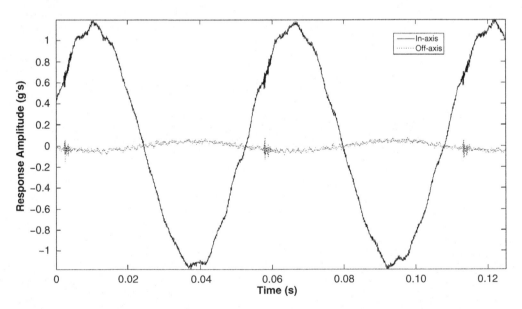

Figure 8.28 Acceleration response of the test structure as measured in the in-axis and off-axis directions by accelerometers mounted on the outer structure.

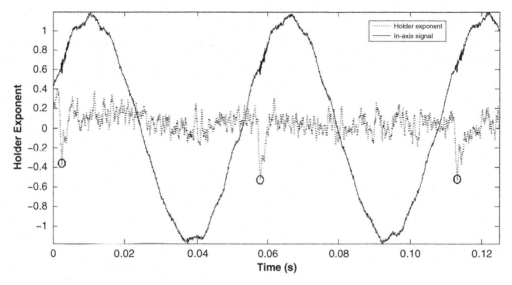

Figure 8.29   The Holder exponent extracted from the wavelet modulus for the in-axis acceleration data.

Extraction of the Holder exponent was then performed using the wavelet transform for the in-axis data, as shown in Figure 8.29. Values of the Holder exponent associated with the discontinuities are highlighted with circles. When compared to the acceleration data overlaid on the Holder exponent plot, the singularities associated with the rattle are clearly visible in this plot at each time they occur during the oscillatory cycles.

8.2.8   Linear Time Series Prediction Errors

A general approach to feature extraction for systems that exhibit the transition to nonlinear response characteristics as a result of damage is to first fit a linear model to the data obtained from the undamaged structure that is assumed to be responding in a linear manner. This model is then used to predict the measured system response and a residual error is calculated at each time step, which is the difference between the measured and predicted response. This same model is then used to predict the response from data obtained in the potentially damaged state. The assumption is that this linear model will no longer accurately predict the response of the damaged system that is exhibiting a nonlinear response and there will be a significant increase in the residual errors associated with this latter prediction. This approach can also signal damage even if the damage does not cause a nonlinear response. In this case the procedure does not necessarily detect nonlinearity, but rather a deviation from the baseline linear model.

In theory, this approach can be taken with either physics-based models such as finite element models or with data-driven models such as time series models, both of which were introduced in Chapter 7. The time series models are relatively simple to fit to measured response data and the application of one such model to damage detection using the concrete column data is illustrated below.

Recall from Chapter 7 that an pth order autoregressive (AR) time series model is defined as

$$y_i = \sum_{j=1}^{p} a_j(y_{i-j}) + \varepsilon_i \qquad (8.26)$$

where y_i is the ith time series value, p is the model order, $x_{i-j} = j$ previous measured time series values, a_j are the AR coefficients and ε_i is assumed to be an unobservable random error with zero mean and constant variance. If $\hat{y}_i$ denotes the estimated acceleration measurement obtained from the fitted AR model, then the residual at time i is

$$\varepsilon_i = y_i - \hat{y}_i \tag{8.27}$$

If the estimated model accurately represents the measured signal, the residuals should be nearly uncorrelated. Note that for a fitted pth-order model residuals cannot be computed for $i < p$. From a pattern recognition perspective, the residuals can be thought of as features derived from the measured time histories. This feature is of dimension 1 and many estimates of this feature are obtained from a typical time series. In the example below, a fifth-order AR model (AR(5)) is applied to the 8192-point acceleration time series data from sensor 4 on the concrete column tests described in Chapter 5. This model has the form

$$y_i = a_1 y_{i-1} + a_2 y_{i-2} + \cdots + a_5 y_{i-5} + \varepsilon_i \tag{8.28}$$

For an 8192-point time series the coefficients a_i are estimated by applying a pseudo-inverse technique to obtain a least-squares solution for the AR coefficients. The AR model is used to estimate the measured time series and based on Equation (8.27) the residuals are simply calculated as

$$\varepsilon_i = y_i - (a_1 y_{i-1} + a_2 y_{i-2} + \cdots + a_5 y_{i-5}) \tag{8.29}$$

The PDFs of the residuals obtained with a kernel density estimator are plotted in Figure 8.30. This figure compares the residual errors obtained when the model constructed from data measured on the undamaged structure was used to predict subsequent data from the damaged structures. Here it can be

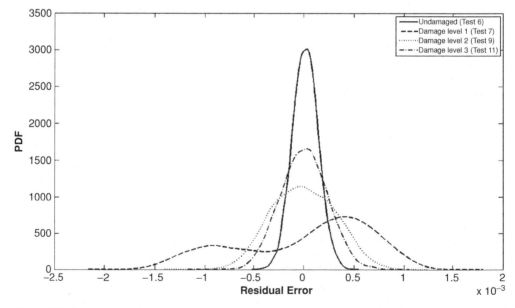

Figure 8.30 Probability density functions of residual errors from time series models based on the undamaged response data from the concrete column (channel 4).

seen that the residual errors from the undamaged time series model predictions are near zero mean and appear to have a Gaussian distribution. At the first damage level the residual errors now appear to have bimodal distribution with considerably higher values for the residuals. At the higher damage levels the distributions return to near zero mean and are Gaussian-like in shape, but with generally higher values, as indicated by the wider distributions. A more detailed statistical analysis of these results using the control charts described in Chapter 10 can be found in Fugate, Sohn and Farrar (2001).

An explanation for the trends shown in Figure 8.30 is related to yielding of the rebar. The first damage level corresponds to incipient yielding of the rebar. Because of strain compatibility, this incipient yielding is accompanied by cracking of the concrete. The shaker, which was driven with the same input voltage level during each test, is most likely to cause a nonlinear response associated with the cracks opening and closing at this first damage level. At the higher damage levels, the rebar has yielded significantly. When load was removed from the column, this yielding tended to hold the cracks in an open configuration during the subsequent dynamic tests. The excitation provided by the shaker, which was maintained at consistent amplitude, could not excite the nonlinearities associated with cracks opening and closing to the same degree when the rebar had yielded significantly. Therefore, the column responded in a more linear manner, but with sufficiently different properties to produce generally higher amplitude residual errors.

8.2.9   Nonlinear Time Series Models

In the previous section a linear time series model was fitted to measured system data and then that model was used to predict system response when damage was present. The underlying assumption was that if the damage caused the structure to respond in a nonlinear manner, the model would not predict data from the damaged system well, which would manifest itself as a relative increase in the residual errors between the measured and predicted responses. In this section a method is described to predict the time series with a nonlinear model. This approach has the advantage over the previous linear approach in that it can be better applied to a system that responds in a nonlinear manner in its initial undamaged state. The parameters of the model as well as the residual error can be used as damage-sensitive features. In application of this approach to damage detection, one must ensure that the model accurately represents the response of the system over the entire range of input. If this process is not done, damage could be inferred when in reality the model does not accurately predict the response to some new loading scenario.

Suppose one is interested in the SDOF linear system

$$m\ddot{y} + c\dot{y} + ky = x(t) \tag{8.30}$$

As discussed in Section A.11, this equation of motion can be converted by a process of discrete approximation to the discrete-time form

$$y_i = a_1 y_{i-1} + a_2 y_{i-2} + b_1 x_{i-1} \tag{8.31}$$

where a_1, a_2 and b_1 are constants and functions of the original constants m, c, k and the sampling interval $\Delta t = t_{i+1} - t_i$, where the t_i are the sampling instants. In a more general form,

$$y_i = F(y_{i-1}, y_{i-2}, x_{i-1}) \tag{8.32}$$

Equation (8.32) represents an autoregressive with exogenous (ARX) inputs model. The advantage of adopting this model form is that only the two states y and x need to be measured in order to estimate all the model parameters a_1, a_2 and b_1 in Equation (8.31) and thus identify the system. It is a simple matter to show that a general MDOF linear system has a discrete-time representation

$$y_i = \sum_{j=1}^{n_y} a_j y_{i-j} + \sum_{j=1}^{n_x} b_j x_{i-j} \tag{8.33}$$

or

$$y_i = F(y_{i-1}, \ldots, y_{i-n_y}; x_{i-1}, \ldots, x_{i-n_x}) \tag{8.34}$$

As before, all parameters $a_1, \ldots, a_{n_y}; b_1, \ldots, b_{n_x}$ can be estimated using measurements of the y and x data only.

The extension to nonlinear systems is straightforward. Considering the Duffing oscillator represented by Equation (8.8), one can pass to the discrete-time representation

$$y_i = a_1 y_{i-1} + a_2 y_{i-2} + b_1 x_{i-1} + c y_{i-1}^3 \tag{8.35}$$

The model defined by Equation (8.35) is now termed a nonlinear ARX (NARX) model. The regression function $y_i = F(y_{i-1}, y_{i-2}; x_{i-1})$ is now nonlinear; it contains a cubic term. If *all* terms of order three or less were included in the model structure, such as $(y_{i-2})^2 x_{i-1}$, a much more general model would be obtained:

$$y_i = F^{(3)}(y_{i-1}, y_{i-2}; x_{i-1}) \tag{8.36}$$

where the superscript above F denotes the highest-order product terms that would be sufficiently general to represent the behaviour of any dynamical systems with nonlinearities up to third order (i.e. containing terms of the form $\dot{y}^3$, $\dot{y}^2 x$ and others).

The most general polynomial NARX model (including products of order $\leq n_p$) is denoted by

$$y_i = F^{(n_p)}(y_{i-1}, \ldots, y_{i-n_y}; x_{i-1}, \ldots, x_{i-n_x}) \tag{8.37}$$

It has been proved that, under very mild assumptions, any input/output process has a representation by a model of this form (Leontaritis and Billings, 1985a, 1985b). If the system nonlinearities are polynomial in nature, this model will represent the system well for all levels of excitation. If the system nonlinearities are not polynomial, they can be approximated arbitrarily with accuracy by polynomials over a given range of their arguments (Weierstrass approximation theorem; Simmons, 1963). This means that the system can be accurately modelled by taking the order n_p high enough. However, the model would be input-sensitive as the polynomial approximation required would depend on the data. This problem can be removed by including nonpolynomial terms in the NARX model (Billings and Chen, 1989). The NARX model can even be cast as a neural network (Billings, Jamaluddin and Chen, 1991).

The preceding analysis unrealistically assumes that the measured data is free of noise. As shown below, if the system is nonlinear the noise process can be very complex; multiplicative noise terms with the input and output are not uncommon, but can be easily accommodated in the discrete-time models (Leontaritis and Billings, 1985a, 1985b; Korenburg, Billings and Liu, 1988; Chen, Billings and Liu, 1987).

Suppose the measured output has the form

$$y(t) = y_c(t) + \zeta(t) \tag{8.38}$$

where $y_c(t)$ is the 'clean' output from the system. If the underlying system is the Duffing oscillator $m\ddot{y} + c\dot{y} + ky + k_3 y^3 = X(t)$ Equation (8.8), the equation satisfied by the measured data is now

$$m\ddot{y} + c\dot{y} + ky + k_3 y^3 - m\ddot{\zeta} - c\dot{\zeta} - k\zeta - k_3\zeta^3 - 3x^2\zeta + 3x\zeta^2 = x(t) \tag{8.39}$$

and the corresponding discrete-time equation will contain terms of the form ζ_{i-1}, ζ_{i-2}, $\zeta_{i-1}y_{i-1}^2$. Note that even simple additive noise on the output introduces cross-product terms if the system is nonlinear. Although these terms all correspond to unmeasurable states, they must be included in the model. If they are ignored the parameter estimates will generally be biased. The system model described by Equation (8.37) is therefore extended again by the addition of a *noise model* and takes the form

$$y_i = F^{(3)}(y_{i-1}, y_{i-2}; x_{i-1}; \zeta_{i-1}, \zeta_{i-2}) + \zeta_i \tag{8.40}$$

This type of model is referred to as nonlinear autoregressive moving average with exogenous (NARMAX) inputs (Leontaritis and Billings, 1985a, 1985b).

Finally, the term 'moving average' requires some explanation. Generally, for a linear system a moving-average model for the noise process takes the form

$$\zeta_i = e_i + c_1 e_{i-1} + c_2 e_{i-2} + \cdots \tag{8.41}$$

implying that the system noise is assumed to be the result of passing a zero-mean white noise sequence $\{e_i\}$ through a digital filter with coefficients $c_1, c_2, \ldots$. The terminology comes from the literature of time series analysis. Equation (8.28) can now be extended to the nonlinear case. This generalisation is incorporated in the NARMAX model that takes the final form

$$y_i = F^{(n_p)}(y_{i-1}, \ldots, y_{i-n_y}; x_{i-1}, \ldots, x_{i-n_x}; e_{i-1}, \ldots, e_{i-n_e}) + e_i \tag{8.42}$$

In this form the noise sequence or *residual sequence* e_i is now zero-mean white noise. This formulation allows the model to accommodate a wide class of possible nonlinear noise terms.

The input and output variables x_i and y_i are usually physical quantities like force input and displacement response, respectively. An interesting alternative approach to this was followed by Thouverez and Jezequel (1996) who fitted a NARMAX model using modal coordinates.

Having obtained a NARMAX model for a system, the next stage in the identification procedure is to determine if the structure is correct and the parameter estimates are unbiased. It is important to know if the model has successfully captured the system dynamics so that it will provide good predictions of the system output for different input excitations or if it has simply fitted the model to a specific data set. In the latter case the model will be of little use because it will only be applicable to one data set. Three basic tests of the validity of a model have been established and they are described below in increasing order of stringency (Billings and Voon, 1983).

1. One-step-ahead prediction. For the NARMAX representation of a system given by Equation (8.42), the one-step-ahead prediction of y_i is made using measured values for all past inputs *and* outputs. Estimates of the residuals are obtained from the expression $\hat{e}_i = y_i - \hat{y}_i$, that is

$$\hat{y}_i = F^{(n_p)}(y_{i-1}, \ldots, y_{i-n_y}; x_{i-1}, \ldots, x_{i-n_x}; \hat{e}_{i-1}, \ldots, \hat{e}_{i-n_e}) \tag{8.43}$$

The one-step-ahead series can then be compared to the measured outputs. Good agreement is clearly a necessary condition for model validity.

2. Model outputs. If the inputs, x_i, are the only measured quantities used to generate the model output, then Equation (8.43) takes on the form

$$\hat{y}_i = F^{(n_p)}(\hat{y}_{i-1}, \ldots, \hat{y}_{i-n_y}; x_{i-1}, \ldots, x_{i-n_x}; 0, \ldots, 0) \tag{8.44}$$

The zeros are present because the prediction errors will not generally be available when one is using the model to predict output. In order to avoid a misleading transient at the start of the record for $\hat{y}$, the first n_x values of the measured output are used to start the recursion. As above, the estimated outputs must be compared with the measured outputs, with good agreement a necessary condition for accepting the model. It is clear that this test is stronger than the previous one; in fact the one-step-ahead predictions can be excellent in some cases when the model predicted output shows complete disagreement with the measured data.

3. Correlation. These assessments represent the most stringent of the validity checks (Billings, Chen and Backhouse, 1989). The correlation function $\phi_{uv}(k)$ for two sequences of data u_i and v_i with N data points is defined as

$$\phi_{uv}(k) = E(u_i v_{i+k}) \approx \frac{1}{N-k} \sum_{i=1}^{N-k} u_i v_{i+k} \tag{8.45}$$

For a linear system the necessary conditions for model validity are

$$\phi_{ee}(k) = \delta_{0k} \tag{8.46}$$

and

$$\phi_{xe}(k) = 0 \; \forall k \tag{8.47}$$

The first of these conditions is true only if the residual sequence e_i is a white noise sequence. It is essentially a test of the adequacy of the noise model whose job it is to reduce the residuals to white noise. If the noise model is correct, the system parameters should be free from bias. The second of the conditions above states that the residual signal is uncorrelated with the input sequence x_i; that is the model has completely captured the component of the measured output that is correlated with the input. Another way of stating this requirement is that the residuals should be unpredictable from the input.

In the case of a nonlinear system it is sometimes possible to satisfy the requirements above even if the model is invalid. An exhaustive test of the fitness of a nonlinear model requires the evaluation of three additional correlation functions (Billings, Chen and Backhouse, 1989). The extra conditions are,

$$\phi_{e(ex)}(k) = 0, \; \forall k \geq 0 \tag{8.48}$$

$$\phi_{x^{2'}e}(k) = 0, \; \forall k \tag{8.49}$$

and

$$\phi_{x^{2'}e^2}(k) = 0, \; \forall k \tag{8.50}$$

Again, the dash that accompanies x^2 above indicates that the mean has been removed. Normalised estimates of all the correlation functions above are usually obtained so that confidence limits for a null result can be added.

8.2.10   Hilbert Transform

The Hilbert transform (HT) introduced in Section 7.4 can be used to diagnose nonlinearity on the basis of measured FRF data. If the structure is assumed to respond in a linear manner in its undamaged state, then this diagnosis of nonlinearity can be used as a damage indicator. To begin, the map on an FRF $H(\omega)$ is

$$HT\,[H(\omega)] = \hat{H}(\omega) = \frac{1}{i\pi} \int_{-\infty}^{\infty} \frac{H(\Omega)}{\Omega - \omega} d\Omega \tag{8.51}$$

This mapping reduces to the identity on the FRFs of linear systems. Suppose $H(\omega)$ is decomposed so that

$$H(\omega) = H^+(\omega) + H^-(\omega) \tag{8.52}$$

where $H^+(\omega)$ (respectively $H^-(\omega)$) has poles only in the upper (respectively lower) half of the complex ω-plane. Then one can show (Worden and Tomlinson, 2001) that

$$HT[H^\pm(\omega)] = \pm H^\pm(\omega) \tag{8.53}$$

The HT distortion is

$$\Delta H(\omega) = HT[H(\omega)] - H(\omega) = -2H^-(\omega) \tag{8.54}$$

This means that FRF distortion only appears after the Hilbert transform if the original FRF had poles in the lower half of the complex plane. It is argued in Worden and Tomlinson (2001) that linear system FRFs will always have poles in the upper half-plane and will therefore be invariant under the Hilbert transform. On the contrary, if the naïve approach to estimating FRFs for nonlinear systems in Section 8.2.4 is followed, the resulting functions are not constrained in the same way and will in the general case have poles in the lower half-plane; such functions will then suffer Hilbert transform distortions.

The Hilbert transform can be calculated in a number of ways, from direct numerical approximation of the integral in Equation (8.51) to more sophisticated frequency-domain methods (Worden and Tomlinson, 2001); however, whatever method is used, a major problem in computing the HT occurs when band-limited data are employed. As this situation is often encountered experimentally, a more recent development in computation is given here.

The HT can be recast in a slightly different form to that described above:

$$\Re \tilde{H}(\omega) = -\frac{2}{\pi} \int_{0}^{\infty} \frac{\Im H(\Omega)\Omega}{\Omega^2 - \omega^2} d\Omega \quad \text{and} \quad \Im \tilde{H}(\omega) = \frac{2\omega}{\pi} \int_{0}^{\infty} \frac{\Re H(\Omega)}{\Omega^2 - \omega^2} d\Omega \tag{8.55}$$

If zoomed data from $(\omega_{min}, \omega_{max})$ is measured, data are missing from the intervals $(0, \omega_{min})$ and (ω_{max}, ∞). The problem is usually overcome by adding correction terms to the HT (Simon and Tomlinson, 1984).

The alternative approach to the HT exploits the pole-zero form of the FRF. A general FRF may be expanded into a rational polynomial representation,

$$H(\omega) = \frac{Q(\omega)}{P(\omega)} \tag{8.56}$$

Once the rational polynomial model, H_{RP}, is established, it can be converted into a pole-zero form,

$$H_{RP}(\omega) = \frac{\displaystyle\prod_{i=1}^{n_Q} (\omega - q_i)}{\displaystyle\prod_{i=1}^{n_P} (\omega - p_i)} \tag{8.57}$$

Long division and partial-fraction analysis produce the decomposition (King and Worden, 1997)

$$H_{RP}^{+}(\omega) = \sum_{i=1}^{N_+} \frac{C_i^{+}}{\omega - p_i^{+}}, \quad H_{RP}^{-}(\omega) = \sum_{i=1}^{N_-} \frac{C_i^{-}}{\omega - p_i^{-}} \tag{8.58}$$

Once this decomposition is established, the HT follows.

Considering again the Duffing oscillator, but now with slightly different coefficients,

$$\ddot{y} + 20\dot{y} + 10000y + 5 \times 10^9 y^3 = X \sin(\omega t) \tag{8.59}$$

Data were generated from 0 to 38.4 Hz. The data were then truncated by removing data above and below the range 9.25–32.95 Hz.

Figure 8.31 shows the Hilbert transforms of the FRF calculated by the rational polynomial method on the truncated data and by a standard numerical integration method that used the full range of the data. As one can see, the two methods agree well; however, the important thing to note in Figure 8.31 for the purposes of this chapter is the departure from the normal circular appearance of the appropriately scaled Nyquist plot – this is the Hilbert transform distortion that allows a potential diagnosis of damage. The pole-zero decomposition method can also be used to compute analytical expressions for the Hilbert transform (King and Worden, 1997).

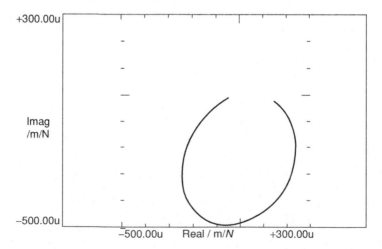

Figure 8.31 Comparison of Hilbert transforms from a rational polynomial approach and standard integral.

8.2.11 Nonlinear Acoustics Methods

Other approaches that capture the nonlinearity caused by structural damage are nonlinear acoustic methods. Damaged materials exhibit a nonlinear response because of wave distortion caused by the irregularities of the damage. This nonlinear response can manifest itself as the generation of harmonics or sum and difference (modulated) frequencies. The premise of the method is the same as those described in the previous sections, but the frequency ranges used in these nonlinear acoustic methods span up to tens of MHz.

Typical signal processing methods employed by these techniques are the 'harmonic distortion' and 'frequency modulation' methods. First, the phenomena of generating higher harmonics caused by nonlinear damage will be briefly summarised (Krohn, Stoessel and Busse, 2002). The tangential movement of a crack causes hysteretic energy dissipation, resulting in distortion of sinusoidal waves in a symmetrical manner. This dissipation will produce strong odd harmonics. On the other hand, upon the application of a perpendicular load, the measured frequency will have components of a sinusoidal part (excitation) and a square signal with a time-dependent load duty cycle (side clapping). This clapping nonlinearity, resulting in a (sin ω)/ω decay, generates even harmonics. These types of nonlinearity are highly localised in the vicinity of the defects, so that imaging methods utilising a scanning laser vibrometer can identify the location of damage (Krohn, Stoessel and Busse, 2002).

Another interesting approach based on nonlinear acoustics is the 'frequency modulation' method described in Donskoy and Sutin (1999), Vladimir *et al.* (2006) and Ballad *et al.* (2004). The idea is to excite a structure with two sinusoidal excitation sources of the frequencies f_h and f_l. When the response time history is converted to a spectrum, if the system is linear, the spectrum contains a single line at f_h, as shown in Figure 8.32(a). If the system is nonlinear, the nonlinearity will result in the appearance of sidebands at the frequencies $f_h \pm f_l$, as shown in Figure 8.32(b). The appearance of the sidebands are the indicator of nonlinearity and, hence, damage. An index of damage extent can then be formed by recording the height of the sidebands (Donskoy *et al.*, 2006) or the spread of the sidebands (Ryles *et al.*, 2006). In order to excite the nonlinearity efficiently, the low-frequency excitation, f_l, can come from ambient excitation, operational frequencies such as pumps or an impact excitation. Some researchers utilise the two different types of actuators, piezoelectric stack and patch actuators, for low- and high-frequency excitations, respectively (Ryles *et al.*, 2006). The frequency modulation method has its application in fatigue damage accumulation monitoring (Campos-Pozuelo, Vanhille and Gallego-Juarez, 2006; Cantrell and Yost, 2001; Zagrai *et al.*, 2006) and an excellent review of this method can be found in Donskoy (2009).

Although the nonlinear acoustic methods are much more sensitive to small defects than linear methods, there are some limitations. For instance, the level of the 'background' nonlinearities, such as those caused by material nonlinearity, joints or structural contacts, must be well understood and quantified in order to distinguish them from the nonlinearities caused by structural damage. The calibration of modulated

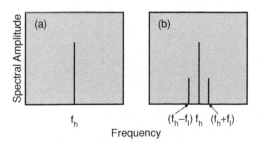

Figure 8.32 (a) Response to excitation at high frequency f_h. (b) Response to simultaneous excitation at high (f_h) and low frequencies (f_l).

frequencies is also important to achieve repeatable measurements as it is obviously dependent on the magnitude of the low-frequency excitation. Furthermore, most modern instrumentation generates harmonics; therefore, the method requires higher-cost data acquisition equipment that suppresses these sources of noise.

8.3   Applications of Nonlinear Dynamical Systems Theory

The main concept of this chapter is that, if a given type of damage converts a linear system into a nonlinear system, then any observed manifestations of nonlinearity serve to indicate that damage is present. This section discusses one of the most dramatic manifestations of nonlinearity, namely chaos. It is well-known that, under certain conditions of forcing, many nonlinear systems can be driven into a chaotic regime. Here a chaotic regime is defined as one where there is sensitive dependence on initial conditions. Over the years, the dynamical systems community has derived many indicators of chaotic behaviour, such as fractal attractor dimensions and Lyapunov exponents (Thompson and Stewart, 2001). As only nonlinear systems can behave chaotically, any feature that detects chaos necessarily detects nonlinearity and might potentially be put to service in detecting damage. The aim of this section is to illustrate the use of such features. It will be shown that these features are suboptimal for damage detection as they do not produce an indicator that monotonically increases with the damage extent. The latter part of the section illustrates a feature motivated by chaotic dynamics that does have this desired property.

Consider a simply supported beam. In its undamaged state an assumption that the beam can be modelled as a linear system is quite adequate, but consider what happens when a crack is introduced halfway along its length, as shown in Figure 8.33.

When the beam sags, the effects of the crack are negligible because the two faces of the crack come together and the beam behaves as though the crack was not there. When the beam hogs, however, the presence of the damage must affect the beam because the crack opens and the effective cross-sectional area of the beam is reduced. Under these circumstances, an appropriate model of the beam would perhaps be that shown in Figure 8.34, which has the general equation of motion

$$m\ddot{y} + c\dot{y} + ky = x(t) \tag{8.60}$$

where

$$k = \begin{cases} k, & y < 0 \\ \alpha k, & y \geq 0 \end{cases} \tag{8.61}$$

When the displacement y of the mass m is positive, the stiffness of the system k is reduced by a factor α. The two stiffnesses produce an overall restoring force F_k that is bilinear. This type of model can

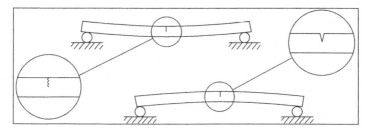

Figure 8.33   A cracked beam under positive and negative deflections.

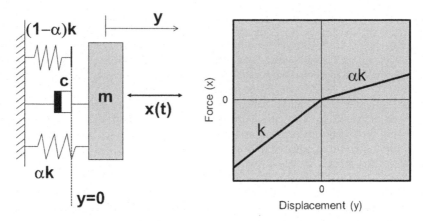

Figure 8.34   Single-degree-of-freedom bilinear system.

be applied to a number of mechanical systems in which moving parts make contact with each other at intermittent points in time, such as the one shown in Figure 8.27.

Various theoretical and experimental studies have shown that bilinear oscillators are capable of a variety of responses to a sinusoidal force (Moon and Shaw, 1983; Thompson, Bokaian and Ghaffari, 1983; Shaw and Holmes, 1983; Shaw, 1985; Natsiavas, 1990; Todd and Virgin, 1997; Thompson and Stewart, 2001). As well as regular harmonic motions, the system can display subharmonic motions and in certain circumstances the system can also exhibit chaotic behaviour. This response has even been evident in experimental studies (e.g. Todd and Virgin, 1997).

The question of importance for damage detection purposes is whether the induced bilinearity from a crack can be used to signify the presence of damage. More specifically, can one or more features be extracted from the recorded dynamics that detect the crack with a degree of statistical confidence? Further, can a feature be obtained that shows the *severity* of the damage? One might argue that a good feature should at least grow monotonically with the damage extent. Examples of features proposed in the past have been the correlation dimension of the attractor (from embedding) and the highest Lyapunov exponent. The use of such features is discussed below.

8.3.1   Modelling a Cracked Beam as a Bilinear System

The single-degree-of-freedom system shown in Figure 8.34 was used to model a simply-supported aluminium beam, 0.8 m long, 50 mm wide and 10 mm thick. Different levels of damage were given to the system by varying the value of the stiffness ratio coefficient α. Putting the bilinear system into the context of a cracked beam made it possible to explore the types of behaviour that could be expected from such a beam as the damage level was increased.

A density of $\rho = 2700$ kg/m^3 and a Young's modulus of $E = 70$ GPa were assumed as material properties. Using an elementary theory of elasticity the undamaged stiffness of the beam was found to be $k = 27\ 346$ N/m. With a mass of $m = 0.6$ kg, the undamaged beam had a natural frequency of $\omega_n = 214$ rad/s, which corresponds to 34 Hz. A damping constant of $c = 2.6$ N s/m was used, which corresponds to 1% of critical damping. A sinusoidal force with an amplitude of 10 N was applied to the system so that $x(t) = 10 \sin(\omega t)$. As the system is homogeneous the amplitude of the force merely scales the output of the system.

A numerical algorithm based on a fourth-order Runge–Kutta integration scheme was used to integrate the equation of motion and calculate the dynamic response of the beam. The key to the success of the algorithm was its ability to detect each point in time when the displacement of the system crossed $x = 0$. This capability means that the algorithm knows exactly when to change the value of k.

Examples of the phase portraits obtained from the algorithm, after transient motions have decayed, are given in Figure 8.35. They show the displacement–velocity trajectories of the system in response to three different forcing frequencies, but for the same value of the stiffness reduction factor, $\alpha = 0.6$. In each case the system has settled on to a periodic motion or attractor.

Forcing the system at the undamaged natural frequency, $\omega = 214$ rad/s, resulted in a harmonic motion, but for $\omega = 384$ rad/s and $\omega = 561$ rad/s subharmonic at one-half and one-third of the forcing frequency are found. This subharmonic generation manifests itself in the time series as a distortion from the sinusoidal form that would be expected from a linear system.

As the forcing frequency is increased from $\omega = 214$ rad/s to $\omega = 384$ rad/s, at some point there must be a qualitative change in the topology of the attractor for the period of the motion to change. The point at which such a change occurs is a *bifurcation*. Bifurcations have already been discussed in this chapter in the context of FRF discontinuities or 'jumps' and, as observed there, bifurcations can only occur in nonlinear systems.

If the response of the system is sampled at the forcing frequency and the resulting points are plotted in the phase plane, one obtains the *Poincare map* of the response. A moment's thought shows that the map associated with a period 1 motion will be a single point, a period 2 motion will produce two points and so on. If a system is chaotic, the Poincare map is usually much more complicated than this, and considered as a point set in the phase plane may actually have a fractal (noninteger) dimension. This point set can loosely be called the *chaotic attractor* of the system. A *bifurcation diagram* is obtained by varying a system parameter and plotting against this a one-dimensional projection of the Poincare map associated with each value of the parameter. The bifurcation diagram in Figure 8.36 is obtained by varying the forcing frequency and plotting the displacement values from the associated Poincare maps.

The diagram in Figure 8.36 is easy to read. If there is only one point on a given vertical line (value of frequency) the system exhibits a period 1 motion at that frequency. If there are n points at a given frequency this is characteristic of a period n motion.

Figure 8.36 shows the bifurcation diagram for $\alpha = 0.2$. There are many regions of subharmonic motion. From $\omega \approx 79$ rad/s to $\omega \approx 83$ rad/s there is a region where the period appears to be very high. In fact, it is infinite and this is characteristic of chaotic motions that do not repeat even if the excitation is periodic. All the period n motions are indicators of nonlinearity. However, only in the limited chaotic regions will features like the chaotic attractor dimension show nontrivial behaviour (i.e. fractional dimension).

The simulations show the limitations of parameters like the attractor dimension for the detection and quantification of damage. Certainly if the undamaged system is linear, the presence of a fractional attractor dimension would indicate nonlinearity and thus damage. However, such parameters will not be reliable. The simulations show that chaotic response is not present at all frequencies. Similar runs have shown that fixing the frequency and varying the stiffness ratio produces the same pattern of movement between subharmonic responses with isolated regions of chaos. This result means that dimension parameters will not vary monotonically with the damage extent. The same will be true of any detectors of chaos such as a Lyapunov exponent. The next section investigates a recently proposed feature motivated by chaos theory that does appear to have the required monotonicity property.

8.3.2   Chaotic Interrogation of a Damaged Beam

The use of a chaotic signal to excite a dynamical system presents some interesting features. The structure can be thought of as a filter acting on the signal. Any changes to the system resulting from damage alter the way in which the signal is filtered. The benefits of using a chaotic excitation of the system relate to the way attractor stability (in the Lyapunov sense) and geometry (dimension) interact, which is discussed in detail in Todd *et al.* (2001).

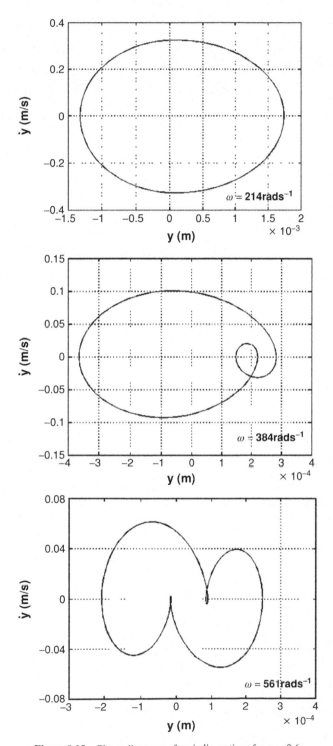

Figure 8.35 Phase diagrams of periodic motions for $\alpha = 0.6$.

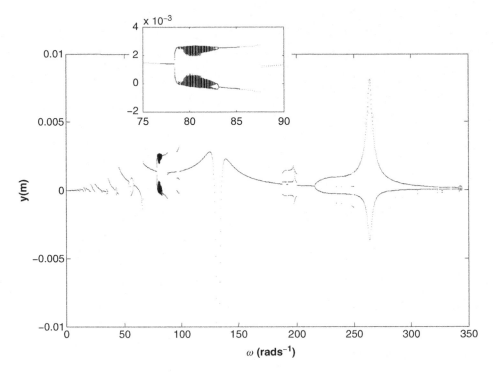

Figure 8.36   Bifurcation diagram for $\alpha = 0.2$.

8.3.3   Local Attractor Variance

The effects of the damage are evident when the attractors of the input and the output are compared. The average local attractor variance ratio (ALAVR) is a simple means of achieving this comparison. It is designed to analyse the geometry changes at a local level. In many real situations it is only (practically) possible to obtain a single phase space variable from a system. This limitation is not a problem because this variable will still be affected by all of the relevant dynamical variables and will therefore contain a relatively complete historical record of the system's dynamics. The ALAVR can be calculated as long as a discrete time series of the input signal $\{x\}$ and output variable $\{y\}$ can be obtained. To achieve a correspondence between the input and output attractors, the two time series must be obtained over the same period of time and using the same sampling frequency, such that

$$\{x\} = \{x_1, x_2, x_3, \ldots, x_N\} \tag{8.62}$$

and

$$\{y\} = \{y_1, y_2, y_3, \ldots, y_N\} \tag{8.63}$$

From these time series input and output attractors can be reconstructed by projecting the time series into a pseudo-phase space. In each dimension of this space the time series are delayed by some number of time steps τ and a geometric structure is created that recaptures the dynamics of the system. The reconstructed attractor is usually topologically equivalent to the actual attractor of the system. The dimensionality of the pseudo-phase space required to capture the dynamics fully is not known in advance, but as the

ALAVR is designed to merely detect simple geometric changes, a two-dimensional space is adequate. The coordinates of the two reconstructed attractors, $\{x\}$ and $\{\underline{y}\}$, are then

$$\underline{x}_n = (x_n, x_{n+\tau}) \tag{8.64}$$

$$\underline{y}_n = (y_n, y_{n+\tau}) \tag{8.65}$$

for $n = 1, 2, 3, \ldots, N - \tau$.

The local geometry of the attractors is analysed by first selecting, at random, a point on the reconstructed forcing attractor $\underline{x}_i$. The Euclidean distance from this point to the other points on the attractor is calculated and the coordinates of the points nearest to $\underline{x}_i$ are extracted from the attractor to form a subset of points $\{\underline{x}_i\}_j$ (where $j = 1, 2, 3, \ldots, N_j$) that express the local geometry of the forcing attractor around the chosen point. There are some subtleties to this process that are explained in Todd *et al.* (2001).

The points with the same indices i and j are then extracted from the response attractor to give another subset of points $\{\underline{y}_i\}_j$ that contains the local geometry of the corresponding region on the response attractor.

The variance of the coordinates in each set of points $\sigma^2_{\{\underline{x}_i\}_j}$ and $\sigma^2_{\{\underline{y}_i\}_j}$ is used to describe the geometry of each set of points, which means the ratio of the two variances $R(i)$ can be used to describe the degree to which the local geometry of the input attractor has been altered by the system

$$R(i) = \frac{\sigma^2_{\{\underline{x}_i\}_j}}{\sigma^2_{\{\underline{y}_i\}_j}} \tag{8.66}$$

This process is then repeated for a total of N_i random points and the ALAVR, $\overline{R}$, is calculated as

$$\overline{R} = \frac{1}{N_i} \sum_{i=1}^{N_i} R(i) \tag{8.67}$$

The ALAVR metric discussed above is one of several nonlinear features that have been presented recently as damage indicators. Attractor-based auto- and cross-prediction error metrics have been constructed to detect loss of preload in bolted joints (Nichols *et al.*, 2003; Nichols, Todd and Wait, 2003; Nichols *et al.*, 2004; Todd *et al.*, 2004; Overbey, Olson and Todd, 2007). The concept here is that a reconstructed attractor may be used to auto-predict itself (or cross-predict another simultaneously sampled attractor) in a baseline condition, and subsequently this prediction will fail as the system changes due to damage.

The general concept of correlation and predictability has been used to build a continuity statistic to detect changes in an electrical circuit that was used to simulate a lumped-mass structural system (Moniz *et al.*, 2004).

Other researchers have focused on attractor-based measures derived straight from the local probability structure of the data. Estimates of the local probability density were computed and the resulting distribution on the attractor was then used as the damage-indicating feature for a nonlinear aeroelastic plate subject to stiffness degradation (Epureanu and Yin, 2004). In another study the local probability structure of data in detecting static load shifts in a reinforced-concrete plate was examined (Trendafilova, 2006). Phase space measures of probability have also been used to distinguish between healthy and damaged motor-driven systems (Hively and Protopopescu, 2004).

Finally, some attention has been given to using Lyapunov exponents as a damage-sensitive feature. A system's Lyapunov exponents capture the rate at which trajectories in phase space diverge or converge in each of the phase space directions. Global Lyapunov exponents have been used to detect backlash in robot

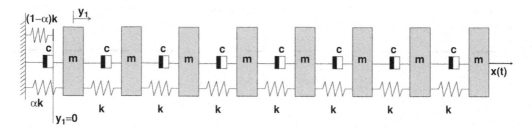

Figure 8.37   Eight-DOF system with bilinear stiffness to simulate damage.

joints (Hively and Protopopescu, 2004) and in gears (Golnaraghi, Lin and Fromme, 1995), respectively. A more complete discussion on the use of Lyapunov exponents in health monitoring, including a discussion of local Lyapunov exponent measures, can be found in Overbey and Todd (2007).

8.3.4   Detection of Damage Using the Local Attractor Variance

An attractor-based approach for damage detection is now applied to a simulated cantilever beam. An eight-degree-of-freedom system similar to that shown in Figure 7.1 was used and a crack at the root of the beam was simulated by placing a bilinear stiffness between the wall and the mass element adjacent to it, as shown in Figure 8.37. The equation of motion of this system is

$$[m]\{\ddot{y}\} + [c]\{\dot{y}\} + [k]\{y\} = \{B\}x(t) \qquad (8.68)$$

where $[m]$, $[c]$ and $[k]$ are the respective mass, damping and stiffness matrices formed from the elements in the model, with

$$k_{11} = \begin{cases} k, & y_1 < 0 \\ \alpha k, & y_1 \geq 0 \end{cases} \qquad (8.69)$$

The vector $\{B\}$ simply selects where the force $x(t)$ is applied to the system, which in this case is the end mass.

A chaotic signal was created from the displacement response of a Duffing oscillator, which is known to be chaotic (Todd *et al.*, 2001). Details of this process can be found in Moon (1987). The attractor associated with the force is shown in Figure 8.38. A time step of 0.0785 s was used in the integration. The system parameters of the structure were chosen as $m = 0.01$, $c = 0.15$ and $k = 2$. Again, a fourth-order Runge–Kutta integration scheme was employed for the simulation. The displacement y_1 of the first mass element was chosen as the variable to use in the calculation of the ALAVR because it was the most sensitive to the bilinear stiffness.

The ALAVR was calculated at each damage level over $N_i = 5000$ randomly selected points and local subsets were created using the $N_i = 40$ nearest points. The values of ALAVR obtained were normalised with respect to the ALAVR of the undamaged system so that

$$\Delta \overline{R} = \frac{|\overline{R} - \overline{R^*}|}{\overline{R}} \qquad (8.70)$$

where $\overline{R^*}$ refers to the ALAVR of the undamaged system. The results are shown in Figure 8.39.

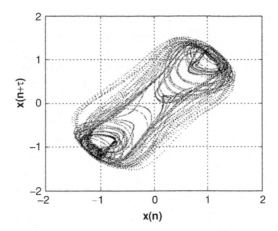

Figure 8.38   Reconstructed forcing attractor.

There is a change in the geometry of the attractors and this is reflected in the values of ALAVR. As the level of damage increased, the value of the ALAVR increased, indicating that the local variance of the attractors was also increasing. It was shown in Todd *et al.* (2001) that the ALAVR can be used in a statistical procedure to flag damage with a given statistical confidence. The same procedure is expected to work here. Moreover, the same input/output attractor variance 'mapping' implied by this procedure may be applied to output-only data. In other words, two measured responses can be compared in a similar way, forming their own cross-variance ratio, which detects *relative* geometric changes between responses. Such a formulation is useful when either one cannot measure the input or it may reasonably be hypothesised that the damage mechanism is closely related to synchronisation or coupling between two output measurements, for example a connectivity loss. A related attractor-based correlation metric was also discussed in Nichols, Todd and Wait (2003).

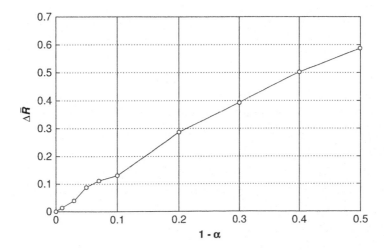

Figure 8.39   Normalised values of ALAVR at different damage levels.

8.4   Nonlinear System Identification Approaches

Nonlinear system identification is the process that identifies the parameters of nonlinear system models based on the measured inputs and/or responses of the system. Structural health monitoring is achieved by performing this parameter identification using data from the undamaged system to obtain a baseline set of parameters and then the process is repeated with a subsequent data set obtained from the structure in a possibly damaged state. A comparison of this subsequent set of parameters and the baseline set provides the damage assessment. Features based on nonlinear system identification not only can identify the presence of damage but they also have the potential to locate and quantify damage. This process is directly analogous to linear model updating approaches presented in Chapter 7, particularly if physics-based model structures for the nonlinear systems are adopted. One should note that the success of this method is predicated on assuming the correct nonlinear model form to represent the system response in both an undamaged and damaged condition. The discussion herein provides a brief summary of one simple nonlinear system identification method as an illustration of what is achievable; readers are referred to an extensive review of this topic for a much more complete discussion of nonlinear system identification (Kerschen *et al.*, 2006).

8.4.1   Restoring Force Surface Model

In order to illustrate the nonlinear system identification methodology, the restoring force surface approach to identification will be discussed (Masri and Caughey, 1979; Crawley and Aubert, 1986). The equation of motion of a standard SDOF mass–spring–damper can be written as

$$m\ddot{y} + g(y, \dot{y}) = x(t) \tag{8.71}$$

where m is the mass of the system and $g(y, \dot{y})$ is the internal restoring force that acts to return the oscillator to equilibrium when disturbed. The function g can be a quite general function of position $y(t)$ and velocity $\dot{y}(t)$. Note that, in the sense that the model here is based on a lumped-mass approximation, the identification is directly analogous to the model updating approaches discussed in Chapter 7 once the analysis is extended to MDOF systems. In the special case when the system of interest is linear,

$$g(y, \dot{y}) = c\dot{y} + ky \tag{8.72}$$

where c and k are the damping constant and stiffness, respectively. Because g is assumed to be dependent only on y and $\dot{y}$, it can be represented by a surface over the phase plane, that is the $(y, \dot{y})$ plane. A trivial rearrangement of Equation (8.71) gives

$$g(y, \dot{y}) = x(t) - m\ddot{y} \tag{8.73}$$

If the mass m is known and the excitation $x(t)$ and acceleration $\ddot{y}(t)$ are measured, all the quantities on the right-hand side of this equation are known and, hence, so is g. If $t_i = (i - 1)/\delta t$ denotes the ith sampling instant, then at t_i Equation (8.73) gives

$$g_i = g(y_i, \overline{y}_i) = x_i - m\ddot{y}_i \tag{8.74}$$

where $x_i = x(t_i)$, $\ddot{y}_i = \ddot{y}(t_i)$ and, hence, g_i are known at each sampling instant. If the velocities $\dot{y}_i$ and displacements y_i are also known (i.e. from direct measurement or from numerical integration of the sampled acceleration data), at each instant $i = 1, \ldots, N$ a triplet $(y_i, \dot{y}_i, g_i)$ is specified. The first two values indicate a point in the phase plane and the third gives the height of the restoring force surface above that point.

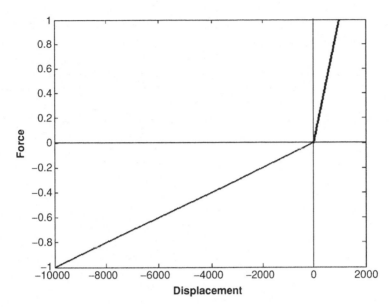

Figure 8.40   Stiffness section of an SDOF system with a bilinear stiffness representing a breathing crack.

Given this scattering of force values above the phase plane, there are a number of methods for interpolating a continuous surface on a regular grid. The various interpolation and plotting procedures are discussed in some detail in Worden and Tomlinson (2001). Two simple approaches are shown here. In the first case, a force surface is obtained by dividing the phase plane into a regular grid and averaging the force values above the resultant cells. These average values are plotted above the centres of the cells. The main disadvantage of this approach is that there may be many phase space cells that are empty of data points and, therefore, do not have force values. Such cells will produce 'holes' in the surface, although there are some simple heuristics that can be used to fill in some of the holes (Worden and Tomlinson, 2001). The main advantage of this approach is that a regular plotting grid is obtained that is amenable to standard graphics software. The second plotting procedure is to show a slice of the force data on a plane of constant velocity or displacement. Such plots give the damping or stiffness *sections* respectively.

In order to illustrate the approach in the context of damage identification, consider a bilinear system of the form shown in Figure 8.40 that simulates a breathing crack. The k value used was 10^4 and the stiffness ratio α was chosen as 10. As in some of the previous examples, the time data were simulated using a fourth-order Runge–Kutta integration scheme. In this case, because random excitation is preferred for system identification, the forcing $x(t)$ was taken as Gaussian white noise. For the purposes of this simulation, the velocity and acceleration data were taken directly from the numerical solution of Equation (8.60), as described above. In practice one would probably measure acceleration data and numerically integrate these data to obtain the corresponding velocities and displacements. The simplest representation of the force surface is the stiffness section, which is shown in Figure 8.40.

A more comprehensive representation of the nonlinearity is given by the full restoring force surface shown in Figure 8.41. Both representations show the bilinear nature of the internal force and thus detect the presence of the 'crack'. Once the restoring force surfaces or stiffness sections are computed, they can be expressed as polynomial expansions in displacement and velocity and coefficients can be estimated by least-squares methods or otherwise. The coefficients or parameters can then be used as features for SHM; if a subsequent system identification produces (statistically) significantly different coefficients for the restoring force surface then one might infer that damage has occurred. At the risk of labouring

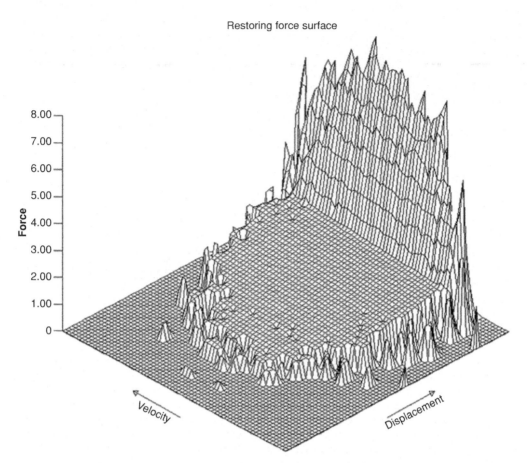

Figure 8.41   Restoring force surface for an SDOF bilinear stiffness system.

the point, this is the equivalent procedure to the linear model updating approaches described in the last chapter, particularly when the nonlinear restoring force functions are treated as extensions to a linear MDOF physics-based model. The SDOF approach described above would ordinarily prove too uninformative for any levels of damage identification beyond detection; for example, in order to extract location information for the damage it is necessary to move to a more sophisticated MDOF model of the system like the link models described in Chapter 7 of Worden and Tomlinson (2001).

However, even the SDOF restoring force method can sometimes be put to good use and an example can be given here that comes from an experimental study. A fatigue crack was grown at the centre point of a 3-m-long concrete beam. A random excitation was applied using an electrodynamic shaker immediately below mid-span and the response was measured at various points on the top surface. In order to simplify matters, an SDOF restoring force surface analysis was carried out using the measured force and the response from the nearest accelerometer to the crack. The resulting stiffness section is shown in Figure 8.42 with a ninth-order least-squares polynomial fit superimposed. Although the section is rather noisy (as a result of the neglected degrees of freedom), the smoothed curve fit shows a quite clear bilinear nature, as one might expect from a breathing crack.

The NARMAX model discussed earlier in Section 8.2.9 is one of the most general and versatile nonlinear system identification models and can be applied in the MDOF case as simply as in the SDOF

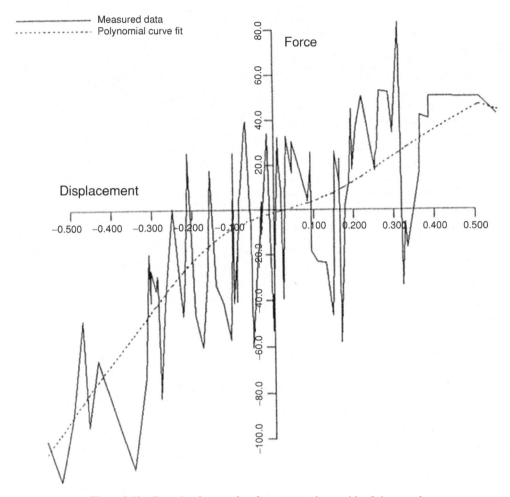

Figure 8.42 Restoring force surface for a concrete beam with a fatigue crack.

case at the expense of adding more lagged (usually linear) terms. Although the model is not physics-based insofar as it is a discrete-time rather than continuous-time structure, the coefficients or parameters of the models are just as acceptable as features for damage detection, as is the residual model error.

8.5 Concluding Comments Regarding Feature Extraction Based on Nonlinear System Response

This chapter discussed many approaches where nonlinearity is being exploited in an effort to identify damage-sensitive features. In Sections 8.2 and 8.3 the assumption is that damage will cause a system that can accurately be modelled as a linear system in its undamaged condition and to exhibit nonlinear behaviour when damaged. Here many of the methods proposed simply try to identify the deviation from a linear response or the existence of such nonlinearities. As such, these features are usually only successful in identifying the presence of damage (level 1 damage detection). In Section 8.4 the nonlinearity is being explicitly modelled, which allows for the possibility of locating and quantifying the damage (levels 2 and 4 damage detection), but this approach requires a priori knowledge of the functional form of the nonlinearity, which is analogous to having level 3 damage detection before the process has been implemented.

There are still many other outstanding issues for the application of nonlinear dynamic system concepts to SHM problems. In most cases reported herein, almost all studies have been applied to laboratory structures of relatively simple geometry or to numerical simulations of similarly simple structures. If these approaches to SHM feature extraction are to make the transition from a research demonstration to practical application, studies must be conducted on more complicated systems.

Furthermore, the discussion in this chapter has in most cases been limited to applications where the structure of interest can accurately be modelled as a linear system when in its undamaged state. Many real-world structures, particularly those with numerous complicated joints and interfaces, will exhibit a nonlinear response in their undamaged state. Therefore, approaches are needed that can distinguish changes in nonlinear response associated with varying damage levels from the inherent nonlinearities associated with the system in its undamaged condition. The nonlinear system identification approaches can potentially be used to address this problem and additional examples of such methods are presented in Chapter 11.

The studies reported herein have not addressed many other issues associated with using nonlinear system characteristics for damage detection. These issues include the amplitude dependence of many nonlinearities and the ability of the excitation sources (ambient or prescribed) to excite the nonlinearity. In most studies quantifying the location of the nonlinearity, identifying the type of nonlinearity and identifying the extent of damage associated with the nonlinearity and the time-rate-of-change of the damage based on changes of the measured system response have not been addressed. Furthermore, most studies of systems with nonlinearities caused by damage tend to focus on discrete-source nonlinearities such as a local crack or a single nonlinear stiffness element. Few studies have focused on distributed sources of nonlinearities such as a distributed crack field that may occur in a concrete structure.

Almost all SHM studies that make use of nonlinear system characteristics do so in a passive sensing mode. By passive sensing, it is implied that a prescribed input with a known waveform that is designed to enhance the damage observability has not been used. The active sensing approaches associated with the nonlinear acoustic methods and the local attractor variance methods are examples where researchers are starting to couple active sensing with concepts from nonlinear dynamics to enhance the damage detection process.

Finally, most of the experimental studies reported herein have been conducted in well-controlled laboratory settings. The ability of these methods to distinguish changes in system response associated with damage from changes associated with varying operational and environmental conditions must be established for all applications that are beyond a proof-of-concept research study.

References

Ballad, E., Vezirov, S., Pfleiderer, K. *et al.* (2004) Nonlinear modulation technique for NDE with air-coupled ultrasound. *Ultrasonics*, **42**, 1031–1036.

Billings, S. and Chen, S. (1989) Extended model set, global data and threshold model identification of severely non-linear systems. *International Journal of Control*, **50**(5), 1897–1923.

Billings, S. and Fadzil, M. (1985) The practical identification of systems with nonlinearities. Proceedings of IFAC Symposium on System Identification and Parameter Estimation, York.

Billings, S. and Tsang, K. (1990) Spectral analysis of block-structured nonlinear systems. *Mechanical Systems and Signal Processing*, **4**, 117–130.

Billings, S. and Voon, W. (1983) Structure detection and model validity tests in the identification of nonlinear systems. *Control Theory and Applications, IEEE Proceedings D*, **130**(4), 193–199.

Billings, S., Chen, S. and Backhouse, R. (1989) Identification of linear and nonlinear models of a turbocharged automotive diesel engine. *Mechanical Systems and Signal Processing*, **3**(2), 123–142.

Billings, S., Jamaluddin, H. and Chen, S. (1991) A comparison of the back propagation and recursive prediction error algorithms for training neural networks. *Mechanical Systems and Signal Processing*, **5**(3), 233–255.

Campos-Pozuelo, C., Vanhille, C. and Gallego-Juarez, J. (2006) Comparative study of the nonlinear behavior of fatigued and intact samples of metallic. *IEEE Transaction on Ultrasonics, Ferroelectrics, and Frequency Control*, **53**, 175–184.

Cantrell, J. and Yost, W. (2001) Nonlinear ultrasonic characterization of fatigue microstructures. *International Journal of Fatigue*, **23**, 487–490.

Chen, S., Billings, S. and Liu, Y. (1987) Orthogonal least-squares methods and their application to nonlinear system identification. *International Journal of Control*, **50**(5), 1873–1896.

Crawford, A.R. (1992) *The Simplified Handbook of Vibration Analysis*, Computational Systems, Inc., Knoxville, TN.

Crawley, E. and Aubert, A. (1986) Identification of nonlinear structural elements by force-state mapping. *AIAA Journal*, **24**, 155–162.

Crisman, A., McCuskey, M., Yoder, N. *et al.* (2007) Femoral component insertion monitoring using human cadaveric specimens. Proceedings of the 25th IMAC Conference on Structural Dynamics, Society for Experimental Mechanics, Orlando, FL.

Donskoy, D. (2009) Nonlinear acoustics methods, in *Encyclopedia of Structural Health Monitoring* (eds C. Boller, F.-K. Chang and Y. Zujino), vol. **1**, John Wiley & Sons, Ltd., pp. 321–332.

Donskoy, D. and Sutin, A. (1999) Vibro-acoustic modulation nondestructive evaluation technique. *Journal of Intelligent Material Systems and Structures*, **9**, 765–777.

Donskoy, D., Zagrai, A., Chudnovsky, E. *et al.* (2006) Nonlinear vibro-acoustic modulation technique for life prediction of aging aircraft components. Proceedings of the Third European Workshop on Structural Health Monitoring, Granada, Spain.

Epureanu, B. and Yin, S. (2004) Identification of damage in an aeroelastic system based on attractor deformations. *Computers and Structures*, **82**(31–32), 2743–2751.

Ewins, D. (1995) *Modal Testing; Theory and Practice*, John Wiley & Sons, Inc., New York, NY.

Farrar, C., Cornwell, P., Doebling, S. and Prime, M. (2002) *Structural Health Monitoring Studies of the Alamosa Canyon and I-40 Bridges*, Los Alamos National Laboratory Report LA-13635-MS, Los Alamos, NM.

Figueiredo, E., Park, G., Figueiras, J. *et al.* (2009) *Structural Health Monitoring Algorithm Comparisons Using Standard Datasets*, Los Alamos National Laboratory Report LA-14393, Los Alamos, NM.

Friswell, M. and Penny, J. (2002) Crack modelling for structural health monitoring. *International Journal of Structural Health Monitoring*, **1**(2), 139–148.

Fugate, M., Sohn, H. and Farrar, C. (2001) Vibration-based damage detection using statistical process control. *Mechanical Systems and Signal Processing*, **15**(4), 707–721.

Golnaraghi, M., Lin, D. and Fromme, P. (1995) Gear damage detection using chaotic dynamics techniques: a preliminary study. Proceedings of the 1995 ASME Design Technology Conference, Symposium on Time-varying Systems and Structures, Boston, MA, pp. 121–127.

Hively, L. and Protopopescu, V. (2004) Machine failure forewarning via phase-space dissimilarity measures. *Chaos*, **14**(2), 408–419.

Kerschen, G., Worden, K., Vakakis, A. and Golinval, J.-C. (2006) Past, present and future of nonlinear system identification for structural dynamics. *Mechannical Systems and Signal Processing*, **20**(3), 505–592.

King, N. and Worden, K. (1997) A rational polynomial technique for calculating Hilbert transforms, in *Structural Dynamics: Recent Advances*, Institute of Sound and Vibration Research, Southampton, U.K., pp. 1669–1683.

Korenburg, M., Billings, S. and Liu, Y. (1988) An orthogonal parameter estimation algorithm for nonlinear stochastic systems. *International Journal of Control*, **48**(1), 193–210.

Krohn, N., Stoessel, R. and Busse, G. (2002) Acoustic nonlinearity for defect selective imaging. *Ultrasonics*, **40**, 633–637.

Leontaritis, I. and Billings, S. (1985a) Input–output parametric models for nonlinear systems. Part I: deterministic nonlinear systems. *International Journal of Control*, **41**(2), 303–328.

Leontaritis, I. and Billings, S. (1985b) Input–output parametric models for nonlinear systems. Part II: stochastic nonlinear systems. *International Journal of Control*, **41**(2), 329–344.

Mallat, S. (1999) *A Wavelet Tour of Signal Processing*, 2nd edn, Academic Press, San Diego, CA.

Mallat, S. and Hwang, W. (1992) Singularity detection and processing with wavelets. *IEEE Transactions on Information Theory*, **38**, 617–643.

Martin, H. (1989) Statistical moment analysis as a means of surface damage detection. Proceedings of the International Modal Analysis Conference, Society for Experimental Mechanics, Las Vegas, NV, pp. 1016–1021.

Masri, S. and Caughey, T. (1979) A nonparametric identification technique for nonlinear dynamic problems. *Journal of Applied Mechanics*, **46**, 433–447.

Mitchell, J. S. (1993) *Introduction to Machinery Analysis and Monitoring*, PenWel Books, Tulsa, OK.

Moniz, L., Pecora, L., Nichols, J. *et al.* (2004) Dynamical assessment of structural damage using the continuity statistic. *Structural Health Monitoring: An International Journal*, **3**(3), 199–212.

Moon, F. (1987) *Chaotic Vibrations*, John Wiley & Sons, Inc.

Moon, F. and Shaw, S. (1983) Chaotic vibrations of a beam with non-linear boundary conditions. *International Journal of Non-linear Mechanics*, **18**, 465–477.

Natsiavas, S. (1990) On the dynamics of oscillators with bi-linear damping and stiffness. *International Journal of Non-linear Mechanics*, **25**, 535–554.

Nayfeh, A. and Mook, D. (1979) *Nonlinear Oscillations*, Wiley Interscience, New York.

Nichols, J., Nichols, C., Todd, M. *et al.* (2004) Use of data driven phase space models in assessing the strength of a bolted connection in a composite beam. *Smart Materials and Structures*, **13**(2), 241–250.

Nichols, J., Todd, M. and Wait, J. (2003) Using state space predictive modeling with chaotic interrogation in detecting joint preload loss in a frame structure experiment. *Smart Materials and Structures*, **12**(3), 580–601.

Nichols, J., Todd, M., Seaver, M. and Virgin, L. (2003) The use of chaotic excitation and attractor property analysis in structural health monitoring. *Physical Review E*, **67**(1), 016209-1–016209-8.

Overbey, L. and Todd, M. (2007) Analysis of local state space models for feature extraction in structural health monitoring. *Structural Health Monitoring: An International Journal*, **6**(2), 145–172.

Overbey, L., Olson, C. and Todd, M. (2007) Parametric investigation of stochastically-driven, state space-based prediction methods for structural health monitoring. *Smart Materials and Structures*, **16**(5).

Prime, M. and Shevitz, D. (1996) Linear and nonlinear methods for detecting cracks in beams. Proceedings of the 14th International Modal Analysis Conference, Society of Experimental Mechanics, Dearborn, MI, pp. 1437–1443.

Rao, J. (2000) *Vibratory Condition Monitoring of Machines*, CRC Press, Boca Raton, FL.

Robertson, A., Farrar, C. and Sohn, H. (2003) Singularity detection for structural health monitoring using Holder exponents. *Mechanical Systems and Signal Processing*, **17**(6), 1163–1184.

Ryles, M., McDonald, I., Ngau, F. and Staszewski, W. (2006) Ultrasonic wave modulations for damage detection in metallic structures. Proceedings of the Third European Workshop on Structural Health Monitoring, Granada, pp. 275–282.

Shaw, S. (1985) Forced vibration of a beam with one-sided amplitude constraint: theory and experiment. *Journal of Sound and Vibration*, **99**, 199–212.

Shaw, S. and Holmes, P. (1983) A periodically forced piecewise linear oscillator. *Journal of Sound and Vibration*, **90**, 129–155.

Simmons, G. (1963) *Introduction to Topology and Modern Analysis*, McGraw-Hill, New York, NY.

Simon, M. and Tomlinson, G. (1984) Use of the Hilbert transform in modal analysis of linear and nonlinear structures. *Journal of Sound and Vibration*, 421–436.

Struzik, A. (2001) Wavelet methods in (financial) time-series processing. *Physica A*, **296**, 307–319.

Thompson, J. and Stewart, H. (2001) *Nonlinear Dynamics and Chaos*, John Wiley & Sons, Ltd, Chichester.

Thompson, J., Bokaian, A. and Ghaffari, R. (1983) Subharmonic resonances and chaotic motions of a bilinear oscillator. *Journal of Applied Mechanics*, **31**, 207–234.

Thomson, W. and Dahleh, M. (1998) *Theory of Vibrations with Applications*, Prentice-Hall, Englewood Cliffs, NJ.

Thouverez, F. and Jezequel, L. (1996) Identification of NARMAX models on a modal base. *Journal of Sound and Vibration*, **189**(2), 193–213.

Todd, M. and Virgin, L. (1997) An experimental impact oscillator. *Chaos, Solitons and Fractals*, **8**, 699–714.

Todd, M., Nichols, J., Pecora, L. and Virgin, L. (2001) Vibration-based damage assessment utilizing state space geometry changes: local attractor variance ratio. *Smart Materials and Structures*, **10**, 1000–1008.

Todd, M., Erickson, K., Chang, L. *et al.* (2004) Using chaotic interrogation and attractor nonlinear cross-prediction error to detect fastener preload loss in an aluminum frame. *Chaos: An Interdisciplinary Journal of Nonlinear Science*, **14**(2), 387–399.

Trendafilova, I. (2006) Vibration-based damage detection in structures using time series analysis. *Proceedings of the Institution of Mechanical Engineers Part C: Journal of Mechanical Engineering Science*, **220**(C3), 261–272.

Vladimir, Z., Nazarov, V., Gusev, V. and Castagnede, B. (2006) Novel nonlinear-modulation acoustic technique for crack detection. *NDT&E Journal*, **39**, 184–194.

Watts, B. and Van Dyke, S.J. (1993) An automated vibration-based expert diagnostic system. *Journal of Sound and Vibration*, **27**(9), 14–20.

Worden, K. and Tomlinson, G. (2001) *Nonlinearity in Structural Dynamics: Detection, Identification and Modelling*, Institute of Physics Publishing, Bristol.

Zagrai, A., Donskoy, D., Chudnovsky, A. *et al.* (2006) Micro/meso scale fatigue damage accumulation monitoring using nonlinear acoustic vibro-modulation measurements, in *Testing, Reliability, and Application of Micro- and Nano-Material Systems IV, Proceedings of SPIE* (eds R.E. Geer, N. Mevendorf, G.Y. Baaklini and D.W. Vogel), vol. 6175, SPIE.

9

Machine Learning and Statistical Pattern Recognition

9.1 Introduction

The preceding chapters have focused on the first three steps of the four-step SHM process introduced in Chapter 1: *operational evaluation*, *data acquisition* and *feature extraction*. At this point one is now faced with the challenge of making an accurate assessment of the damage condition of a given structure based on any extracted features. The objective of this chapter is to argue that there is one particular approach that is particularly well suited to this decision-making process. This approach is based on the discipline of *machine learning* or often, more specifically, the *pattern recognition* aspects of machine learning. The idea is that one can *learn* relationships from data. In the context of SHM, this means that one can learn to assign a damage state or *class* to a given measurement vector from the structure or system of interest. The measurement vectors must be formed from measurements that are sensitive to the damage; in the normal terminology of pattern recognition, they are referred to as *features*, as discussed extensively in the last two chapters. To recap, an example of a feature vector for SHM might simply be the first five natural frequencies of a structure or it might be more sophisticated, for example a set of wavelet coefficients. Once features have been established, the map between the features and the diagnosis can be constructed and a principled means of doing this is the subject of this and the next two chapters. The use of pattern recognition offers the possibility of automating the SHM process, that is, removing the need for the intervention of human experts as far as possible. So that one does not lose this benefit, the rest of the SHM process should be automated as far as possible, which leads one to the idea of *intelligent damage detection*.

9.2 Intelligent Damage Detection

In order to facilitate the discussion, the reader is referred back to Section 1.3 where the uses of the terms *defect*, *damage* and *failure* were defined. It will be assumed that the inevitable presence of defects is adequately accommodated by current best practice in engineering design and this leaves only damage and failure as a concern for SHM.

The first observation one might make is that failure detection is, in a sense, trivial, as failure was previously defined to be a change in the condition of the structure that produces an unacceptable

---

Structural Health Monitoring: A Machine Learning Perspective, First Edition. Charles R. Farrar and Keith Worden.
© 2013 John Wiley & Sons, Ltd. Published 2013 by John Wiley & Sons, Ltd.

reduction in quality. By implication, such a change will be evident. Thus, *intelligent* damage detection actually entails detecting the damage that will, if not corrected, lead to a failure.

Damage detection is a facet of the broader problem of damage awareness or damage identification. The objective of an SHM system must be to accumulate sufficient information about the damage for appropriate remedial action to be taken to restore the structure or system to high-quality operation or at least to ensure safety. Also, efficiency demands that only the necessary information should be returned by the monitor. With this in mind, it is helpful to think of the identification problem as the hierarchical structure that was defined in Section 1.4.8. At the risk of repetition, the hierarchy is:

- Level 1: detection. The method gives a qualitative indication that damage might be present in the structure.
- Level 2: localisation. The method gives information about the probable position of the damage.
- Level 3: classification. The method gives information about the type of damage.
- Level 4: assessment. The method gives an estimate of the extent of the damage.
- Level 5: prediction. The method offers information about the safety of the structure, for example estimates a residual life.

The vertical structure of that hierarchy is clear; each level (usually) requires that all lower-level information is available. In the hierarchy it should be noted that the classification associated with level 3 is important, if not vital, for effective identification at level 5 and possibly at level 4. Level 5 is distinguished from the others in that it cannot be accomplished without an understanding of the physics of the damage propagation. Level 1 and possibly level 2 are also distinguished in the sense that they can be accomplished with no prior knowledge of how the system will behave when damaged. In order to explain this, a digression on the terminology of pattern recognition or machine learning is needed.

The classification approach to SHM that will be discussed here is based on the idea of pattern recognition (PR). In the broadest sense, a PR algorithm is simply one that assigns to a sample of measured data a class label, usually from a finite set. In the case of damage identification, the measured data or derived features could be vibration mode shapes, full-field thermoelastic data or scattered wave profiles, and so on. The appropriate class labels would encode damage information (e.g. type, location). In order to carry out the higher levels of identification using PR, it is almost certainly necessary to obtain examples of data or features corresponding to each class; that is, in order to establish that a given set of measurements from a composite panel shows the presence of a delamination, the algorithm must have prior knowledge of what data from a delaminated panel looks like as opposed to one with, say, a resin-rich area. Each possible fault class should usually have a training set of measurement vectors or features that are associated uniquely with it. Many PR algorithms work by *training* a diagnostic. For example, a neural network (see Chapters 10 and 11) can learn by example – it is shown the measurement data and asked to produce the correct class label; if the result differs from the desired label, the network is corrected. Typically many presentations of data are required. This type of learning algorithm in which the diagnostic is trained by showing it the true label for each data set is called *supervised learning*.

If supervised learning is required, there will be serious demands associated with it; data from every conceivable damage situation should be available. The two possible sources of such data are physics-based modelling (i.e. from finite element analysis) and experiment. Modelling presents problems if the structure or system of interest is geometrically or materially complex; for example, finite element analysis of structures requiring a fine mesh can be extremely time consuming even if the material is well understood. Structures with composite or viscoelastic elements may not even have accurate constitutive models. The damage itself may be difficult to model; it may also make the structure dynamically nonlinear, like an opening–closing fatigue crack, and this also presents a formidable problem (even if the nonlinearity means that useful features are available, as discussed in the last chapter). Finally, it might be difficult to anticipate the future loading for the system or structure. Unfortunately, the situation is no better for experiment. In order to accumulate enough training data, it would be necessary to make copies of the system of interest and damage it in all the ways that might occur naturally; for high-value structures like aircraft, this is simply not possible.

Fortunately there is an alternative to supervised learning – *unsupervised learning*. However, this mode of learning almost exclusively only applies to level 1 and in some cases level 2 diagnostics; that is, it can only be used for detection and possibly locating the damage. The techniques are often referred to as *novelty detection* or *anomaly detection* methods (Bishop, 1994; Worden, 1997; Markou and Singh, 2003a, 2003b). The idea of novelty detection is that only training data from the normal operating condition of the structure or system is used to establish the diagnostic. A model of the normal condition is created. Later, during monitoring, newly acquired data are compared with the model. If there are any significant deviations the algorithm indicates novelty. The implication is that the system has departed from the normal condition, that is, acquired damage. The advantage of such an approach is clear; if the training data is generated from a model, only the unfaulted condition is required and this will simplify matters considerably. From an experimental point of view, there is no need to damage the structure of interest. Although novelty detection is only a level 1 (or sometimes level 2) approach, there are many situations where this suffices, for example safety-critical systems where any fault on the system would require it to be taken out of service. Novelty detection will be discussed in much more detail in the next chapter of this book.

It is an important qualifier that the novelty detectors should flag only significant deviations from the normal operating condition. All real systems are subject to measurement noise and usually operate in a changing environment; the monitor must be able to distinguish between a statistical fluctuation in the data and a real deviation from normality. This means that of the various flavours of pattern recognition[1] existing (Schalkoff, 1991), the most appropriate one is *statistical pattern recognition* (SPR). Another important observation is that there may be variations in the normal condition that are not statistical; that is, the characteristics of the structure may vary with changing environmental conditions, and this must be addressed. This issue is discussed in much more detail in Chapters 11 and 12 of this book. In general, it is important that the algorithms used for damage identification should account properly for sources of uncertainty and variation in the data. The algorithms should also, as far as possible, return a confidence interval with their diagnosis.

The term *normal operating condition* requires some discussion. As implied earlier, the nature of engineering materials means that there will always be defects and sometimes even damage present in a structure to some extent. The normal operating condition therefore means a state of the system when there is some assurance, statistical or otherwise, that the system is fit-for-purpose. In some cases there may be macroscopic damage, like a fatigue crack; however, if it is known that the crack will not grow under the standard loadings on the system, this state qualifies as a normal operating condition. Novelty detection will then look for new cracks or unexpected growth of the old crack.

The discussion above is intended to show that there is often a trade-off between the level of a diagnostic system and the expense of training it adequately. Given this fact, the main requirement of an intelligent damage detection system is that it should return information at the apposite level for the context. It should measure the appropriate data and process these data with the appropriate algorithm. It should take proper account of uncertainty in the data and return a confidence level in its diagnosis.

---

[1] There are (at least) three entirely separate theoretical frameworks for developing pattern classifiers. The most recent, based on the use of neural networks, will be described in detail in Chapter 11. The other two, the *statistical* and *syntactic* approaches, are arguably more well established. In some respects, the three approaches complement each other, so in attacking a particular problem the ideal situation would be to know something of all three. This is the approach taken in the book by Schalkoff (1991), a book that has had a major influence on the presentation of the subject here. However, as the syntactic approach is rather abstract and calls for a rather different armoury of mathematical tools than one would normally acquire in a career in engineering, the discussion of this chapter is restricted to the statistical approach with neural methods discussed later.

9.3 Data Processing and Fusion for Damage Identification

As discussed in Chapter 1, once an operational evaluation stage has passed and a sensor network has been designed (and these are by no means trivial processes), the SHM system can begin to deliver data. The choice and implementation of algorithms to process the data and carry out the identification is arguably the most crucial ingredient of an intelligent damage detection strategy. Before even choosing the algorithm, it is necessary to choose between two complementary approaches to the problem:

- Damage identification is an *inverse problem*.
- Damage identification is a *pattern recognition problem*.

The first approach usually adopts a physics-based model of the structure and tries to relate changes in measured data from the structure to changes in the model; sometimes locally linearised models are used to simplify the analysis. The algorithms used are mainly based on linear algebra or optimisation theory. Some examples like finite element updating and minimum rank perturbation theory have been discussed in the context of the linear features presented in Chapter 7; a good further reference is Friswell (2008).

The second approach is the main subject of this book, whereby measured data from the system of interest are assigned a damage class by a pattern recognition algorithm. This is the approach that is chosen here for detailed discussion. There is no implied criticism of the inverse problem approach; the authors here are simply concentrating on an alternative and self-consistent framework. For a critical appraisal of inverse problem approaches to damage identification, the reader can consult Friswell and Penny (1997).

Returning to ideas of data fusion introduced earlier in Chapter 4, the data processing element of a monitoring system comprises all actions on the data upstream from the point of acquisition by the sensors. The ultimate product of the analysis is a decision as to the health of the system. The analysis has been neatly summed up by Lowe (2000) as the D2D (data to decision) process; the basic steps are summarised in Figure 4.16(A). The main steps are: *sensing, preprocessing, feature extraction, postprocessing, pattern recognition* and *decision making*.

Beyond the sensing level, which generates the raw data, the first stage is *preprocessing*. The purpose is to prepare the data for feature extraction, but more of that later. The preprocessing stage can encompass two tasks. The first of these is data cleansing. Examples of cleansing processes are: filtering to remove noise, spike removal by median filtering, removal of outliers (care is needed here as the presence of outliers is one indication that the data are not from a normal condition) and treatment of missing data values. The second (optional) preprocessing stage is a preliminary attempt to reduce the dimensions of the data vectors and further de-noise the signal. For example, given a random time series with many points, it is often useful to convert the data to a spectrum by Fourier transformation. The number of points in the spectrum can be much lower than in the original time history and noise can be averaged away. Another advantage of treating the time signal in this way is that the data vector obtained should be independent of time. If the original time series is random, it makes little sense to compare measurements at different starting times. The preprocessing is usually carried out on the basis of engineering judgement and experience. At this stage, the aim would be to reduce the dimension of the data set from possibly many thousands to perhaps a hundred.

The second stage is *feature extraction*. At the risk of repeating here material from Chapter 7, the term *feature* comes from the pattern recognition literature and is short for 'distinguishing feature'. The fundamental problem of pattern recognition is to assign a class label to a vector of measurements; this task is made simple if the data contain dominant features that distinguish them from data from other classes. In general, the components of the signal that distinguish the various damage classes are difficult to distinguish from the normal operating condition data, particularly when the damage is not yet severe. The task of feature extraction is to magnify the characteristics of the various damage classes and suppress the normal background. Suppose the raw data from the sensors is a time series of accelerations from the outside of a gearbox casing. Further suppose that the time data has been preprocessed and converted into an averaged spectrum. Feature extraction in this situation could be extracting only the spectral lines at the meshing frequency and its harmonics as these lines are known to be sensitive to damage. Thus feature extraction can be carried out on the basis of engineering judgement also. Alternatively, statistical algorithms like principal component analysis (PCA), as discussed in Chapter 6, can be used to reduce the dimension of the data (an example of the use of PCA will be given later in this chapter). The resulting low-dimensional data set is the feature vector or pattern vector that the pattern recognition algorithm will use to assign a class. The aim of this stage would be to generate a feature vector of dimension less than ten. A low-dimensional feature vector is a critical element in any pattern recognition problem as the number of data examples needed for training grows explosively with the dimension of the problem. This explosion is a manifestation of the *curse of dimensionality* first mentioned in Section 6.13. Care must be taken at this stage that the information discarded in the dimension reduction is not relevant for diagnosing the damage. Feature extraction should only discard components of the data that are irrelevant for the purposes of identifying damage.

There may need to be a subsequent *postprocessing* step after feature extraction has been performed and before the pattern recognition step. This step will typically involve further data cleansing, compression, fusion or normalisation applied to the extracted features. As an example, once the modal parameter features were extracted from the I-40 Bridge data discussed in Chapter 5, one might have elected to discard all but the resonance frequencies and mode shapes associated with the first two modes because the other modes had a node point at the suspected damage location. This selection process can be thought of as a subsequent data cleansing and compression step implemented after the feature extraction process.

The next stage is *pattern recognition*. This step is the application of an algorithm that can decide the damage state on the basis of the given feature vector. An example would be a neural network that has been trained to return the damage type and severity when presented with, say, condensed spectral information from a gearbox. In the data-based approach to damage detection that is the main theme of this book, the algorithms are largely taken from the discipline of machine learning or pattern recognition. Three types of algorithm can be distinguished depending on the desired diagnosis:

- Novelty detection. In this case, the algorithm is required to simply indicate if the data come from a normal operating condition or not. This is a two-class problem, which has the advantage that unsupervised learning can be used. Novelty detection will be discussed in considerable detail in Chapter 10.
- Classification. In this case, the output of the algorithm is a discrete class label. In its most basic form, this algorithm might simply assign a 'damage' or 'not damage' label to the features. In order to apply such an algorithm more generally, the damage states must be quantised; that is, for location, the structure should be divided into labelled substructures. In this case, the algorithm could only locate to within a substructure, so resolution of what is essentially a continuous parameter may not be good unless many labels are used. However, this type of algorithm is useful in the sense that the algorithms can be trained to give the probability of class membership; this gives an inbuilt confidence factor in the diagnosis. In the case where the desired diagnosis is from a discrete set, for example for diagnosing damage type, this class of algorithms is singled out. Classification is usually expressed as a pattern recognition problem and will be discussed further in the latter part of this chapter; it will also form one of the main subjects for discussion in Chapter 11, which concentrates on SHM as a supervised learning problem.
- Regression. In this case the outputs of the algorithm are one or more continuous variables. For location purposes, the diagnosis might be the Cartesian coordinates of the fault, while for severity assessment it could be the length of a fatigue crack. The regression problem is often nonlinear and is particularly suited to neural networks or other machine learning algorithms. Because regression is a supervised learning problem it will be discussed in Chapter 11.

In all cases, pattern processing is subject to an important limitation. There is a trade-off between the resolution of the diagnosis and the noise-rejection capabilities of the algorithm. Put simply, if the data is always noise-free, there will be very little fluctuation in the measurement from a normal operating condition; in this case, small damages will cause detectable deviations. If there is much noise on the training data, it will be difficult to isolate deviations due to damage unless the damage is severe. This problem is encoded in SHM Axiom VI, as discussed in Chapter 13. One of the tasks of feature extraction is to eliminate, as far as possible, fluctuations on the normal condition data. This optimisation for performance is a requisite feature of intelligent damage detection or *intelligent SHM* generally.

The final stage in the D2D chain is the *decision*. This stage is a matter of considering the outputs of the pattern recognition algorithm and deciding whether action needs to be taken and what that action should be.

Having stressed the importance of adopting an intelligent approach to SHM, the next sections will outline in greater detail how machine learning, and pattern recognition (PR) in particular, facilitate this process. The focus is on PR as the discussion will be limited (for now) to classification problems to allow a number of important general concepts to be introduced and discussed in the most direct manner. These important ideas are most easily explained from the perspective of statistical pattern recognition (SPR) so the remainder of the current chapter will provide a background in the technology of SPR appropriate for use in the SHM context. To recap, the basic philosophy of PR is to associate classes with features extracted from measured data. The classes of interest for SHM are those that distinguish between the normal condition of a structure and any of its possible damage states. Given that the diagnosis (class assignment) will be made on the basis of measured data, it is important that the PR algorithm is able to accommodate a degree of imprecision commensurate with expected levels of measurement error and noise. If one works under the assumption that measured quantities are drawn from probability distributions of possible measurements, it is clear that some probabilistic approach to PR is desirable.

9.4   Statistical Pattern Recognition: Hypothesis Testing

In brief, the problem of pattern recognition is to associate classes C_i, $i = 1, \ldots, N_c$ with measured data usually expressed in terms of feature vectors $\{x\}$. In the simplest possible situation there will be just two classes, but in a sense this is the most important problem in SHM as one will fundamentally wish to distinguish between the classes 'undamaged' U and 'damaged' D. It will be assumed here that data are available for the damaged situation and thus the problem is one of supervised learning. In a probabilistic context, one is interested in the probability that the system is damaged or undamaged given that one has observed the feature vector $\{x\}$, and this is expressed in terms of the conditional probabilities $P(C_i | \{x\})$, as discussed in Chapter 6. If these probabilities are computable, it is clearly natural to diagnose the class on the basis of which class has the highest probability. This problem is a natural example of the statistical discipline of *hypothesis testing* (Kay, 1998).

Hypothesis testing arose as a means of trying to determine if a given effect can be inferred from measured data. This is clearly a two-class problem; the effect is either present or not. In the terminology of the subject, one has two hypotheses: the *null hypothesis* H_0 is that the effect of interest is not present; the *alternate hypothesis* H_1 asserts the contrary. In the context of SHM it is natural to assume as the null hypothesis that damage is not present, so in the following discussion H_0 will be denoted by U and H_1 by D. Given the measured data $\{x\}$, the problem is to establish which is the greater, $P(U | \{x\})$ or $P(D | \{x\})$. For the two-class problem under discussion this is clearly encoded in the ratio

$$\gamma = \frac{P(U | \{x\})}{P(D | \{x\})} \tag{9.1}$$

Unfortunately, the two probabilities in question here are not usually directly available. If probabilities are available from the accumulation of previous data – *training data* – they are usually based on observing data when the structure is known to be in a given condition; that is, one has estimated the *densities* $p(\{x\}|U)$ and $p(\{x\}|D)$. Accurate estimation of these densities is, in itself, a nontrivial problem, but it is assumed here that one can accomplish this. (In the simplest cases of density estimation one could form a parametric estimate or histogram; alternatively more sophisticated methods like kernel density estimation can be applied and these matters are discussed in Chapter 6). Now, two applications of Bayes theorem Equation (6.57) gives

$$P(U|\{x\}) = \frac{p(\{x\}|U\}P(U)}{P(\{x\})} \tag{9.2}$$

and

$$P(D|\{x\}) = \frac{p(\{x\}|D)\,P(D)}{p(\{x\})} \tag{9.3}$$

As the *evidence term* $p(\{x\})$ is common to both expressions, the ratio in Equation (9.1) can be expressed as

$$\gamma = \frac{p(\{x\}|U)P(U)}{p(\{x\}|D)P(D)} \tag{9.4}$$

This latter expression involves the observable densities and the prior probabilities for the two classes (see Chapter 6). In the context of SHM, one will not know the prior probabilities $P(U)$ and $P(D)$, although one would hope that $P(U) \gg P(D)$. In a general hypothesis test, one sometimes assumes $P(U) = P(D)$; that is, in the absence of any observed data each of the classes is equally likely. This latter assertion is sometimes called the *principle of indifference*. If the principle of indifference is thought to apply, the ratio in Equation (9.4) is simplified to

$$\gamma = \frac{p(\{x\}|U)}{p(\{x\}|D)} \tag{9.5}$$

and, as the terms remaining in the RHS are now just the likelihoods of observing the data under the two hypothesis, this is called the *likelihood ratio* (Kay, 1998). Assignment of a diagnosis is now straightforward; one assigns the class U if $\gamma > 1$ and class D otherwise.

Assuming that the likelihoods for the two classes are available in a simple damage detection problem, this leads to a diagnostic assignment via the likelihood ratio. Unfortunately, as the arguments are probabilistic there is room for error in the diagnosis and it is very important to discuss such errors. A moment's thought shows that two types of error can occur:

1. Type I error. This is an error whereby the test rejects the null hypothesis even though it is true; it is also called a *false-positive* (FP). In the case of SHM, a type I error could result in a system being withdrawn from service even though it is actually healthy and would therefore have potentially severe consequences for the cost of ownership of the structure of interest as well as for confidence in the SHM system. Apart from revenue lost from needless downtime, it is sometimes the case that the stripping down of machines or structures for detailed physical inspection following an alarm actually results in the introduction of damage.
2. Type II error. This is an error whereby the test accepts the null hypothesis even though it is false; it is also called a *false-negative* (FN). The consequences of such an error are potentially even more severe than a type I error in the context of SHM. If a damaged system is allowed to continue to operate in a misdiagnosed damaged state, there will be a threat to safety and there may even be complete loss of the structure and consequent loss of life.

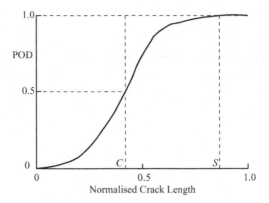

Figure 9.1 Probability of detection (POD) curve.

Because of the possibility of error and because of the (possibly) different consequences of the different types of error, it is important to assess how effective a given classifier is. The coarsest measure one could use is to estimate the probability of misclassification; however, this would not distinguish between the types of error. As an alternative, one could also estimate the probabilities of false-positives and false-negatives separately. One should then apply a weighting to the severity of consequences of different errors in order to form a final judgement on the classifier. One can also use graphical representations of classifier performance; one simple graphical method is to show the *probability of detection* (POD) curve. This simply shows the probability of correctly identifying a class as a function of some scale parameter. Figure 9.1 shows an illustration where the classifier is being used to detect a fatigue crack. In this case, the scale parameter is the crack length. There are two important points on the abscissa: the first is the critical scale parameter C; when the crack has greater length than C, the probability of detection is better than 0.5. The second important point S is the threshold crack length for sure detection, that is, the POD is 1.0.

Another important graphical performance measure is the *receiver operating characteristics* (ROC) curve. This graphs the POD against the probability of false alarm (PFA) for a particular detection threshold. A better performing classifier is indicated by a curve that approaches the upper left corner indicating a high POD and an associated small PFA. An example is presented in Figure 9.2.

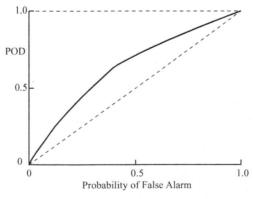

Figure 9.2 Receiver operating characteristics (ROC) curve.

The framework of the hypothesis test is not sufficiently general for multiclass classification problems; however, there is a natural generalisation that will be discussed in the next section.

9.5 Statistical Pattern Recognition: General Frameworks

As discussed earlier, the most general problem of pattern recognition is to associate multiple (more than two) classes C_i, $i = 1, \ldots, N_c$ with the measured data. In the context of SHM these classes could correspond to different substructures in which there may be damage or to different damage mechanisms. Statistical methods allow two distinct approaches:

- SPR1. Given a feature vector $\{x\}$, calculate the probability that this vector is associated with a given class C_i; this is the conditional probability $P(C_i|\{x\})$. Repeat for all possible classes and choose that which gives the highest probability.
- SPR2. Form a measure of the error associated with choosing a particular class and then pick the class that minimises it.

The discussion below concentrates on SPR1 and, in fact, this is the approach that makes most direct contact with the neural approaches to PR, which will be discussed in Chapter 11 and later. As discussed above, statistical methods are necessary because, in general, many different measurement vectors will correspond to noisy or distorted versions of the same basic pattern or *template* and thus require assignment to the same class.

It is assumed as usual that *training data* are available; that is, a sequence of measurement vectors $\{x^{(k)}\}$, $k = 1, \ldots, N_t$ are known, together with the correct class for each vector $C_i^{(k)}$. As discussed earlier, this allows the construction of the a priori conditional probability density functions $p(\{x\}|C_i)$ or *likelihoods*, which specify the probability that a measurement vector $\{x\}$ can arise from a class C_i. Applications of Bayes' theorem again yields the required $P(C_i|\{x\})$ via

$$P(C_i|\{x\}) = \frac{p(\{x\}|C_i)P(C_i)}{p(\{x\})} \tag{9.6}$$

where $p(\{x\})$ is the unconditional density function, which can also be computed from the training set if necessary by concatenating all the feature vectors over all classes. $P(C_i)$ is the *prior* probability of finding an example from class C_i without considering any measurement information. As always, the requirement of a training set makes this approach one of *supervised learning*.

To illustrate how this leads to a decision rule, at the risk of repetition one can return to a two-class example (following Schalkoff, 1991). (The repetition is justified here by the opportunity to frame the problem in the general notation and to introduce one or more new quantities of interest along the way.) Suppose the two classes C_1 and C_2 have equal prior probabilities; that is, $P(C_1) = P(C_2)$. Further suppose that there is a single distinguishing feature x (like the first natural frequency of a structure) and that the densities $p(x|C_1)$ and $p(x|C_2)$ have been established as Gaussians with the same variance σ^2 but different means μ_1 and μ_2. So

$$p(x|C_i) = \frac{1}{\sqrt{2\pi\sigma^2}} \exp\left\{-\frac{1}{2}\left(\frac{x - \mu_i}{\sigma}\right)^2\right\}, \quad i = 1, 2 \tag{9.7}$$

The basic decision rule is

Choose C_1 if $P(C_1|x) > P(C_2|x)$, otherwise choose C_2

Bayes' theorem shows that $P(C_1|x) > P(C_2|x)$ implies

$$\frac{p(x|C_1)P(C_1)}{p(x)} > \frac{p(x|C_2)P(C_2)}{p(x)} \tag{9.8}$$

so using the fact that $P(C_1) = P(C_2)$ yields

$$P(C_1|x) > P(C_2|x) \Rightarrow p(x|C_1) > p(x|C_2) \tag{9.9}$$

allows a decision rule

Choose C_1 if $p(x|C_1) > p(x|C_2)$ otherwise choose C_2

As before, this rule is based only on likelihoods calculated from the training set. In this case, because Gaussian densities have been assumed, fixing the likelihoods is a parametric problem as determining the probability density functions simply requires an estimation of μ_1, μ_2 and σ. Now, one can define a *decision boundary* (in this case a single number) α, fixed by the condition

$$p(\alpha|C_1) = p(\alpha|C_2) \tag{9.10}$$

which gives, on using the functional form for the Gaussian distribution,

$$\alpha = \frac{\mu_1 + \mu_2}{2} \tag{9.11}$$

Therefore α is mid-way between μ_1 and μ_2. This fact agrees entirely with intuition as the distributions have the same width (same σ). The situation is illustrated in Figure 9.3. The *decision regions* R_1 and R_2 corresponding to when one accepts the relevant classes are simply the half-lines meeting at α.

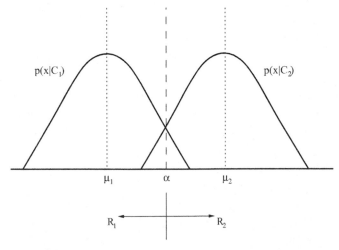

Figure 9.3   Decision boundaries for a simple classification problem. (This figure is a schematic only; the tails of the densities are not those of true Gaussian distributions.)

As the probability density functions for the two classes overlap, there will always be the possibility of choosing incorrectly on the basis of a given measurement x. It is possible to compute the probability of making such a wrong decision. By the total probability theorem Equation (6.65) it follows that

$$
\begin{aligned}
P(\text{error}) &= P(\text{error}|C_1)P(C_1) + P(\text{error}|C_2)P(C_2) \\
&= P(x \in R_2|C_1)P(C_1) + P(x \in R_1|C_2)P(C_2) \\
&= P(x > \alpha|C_1)P(C_1) + P(x < \alpha|C_2)P(C_2)
\end{aligned}
\tag{9.12}
$$

so that, finally,

$$
P(\text{error}) = P(C_1) \int_{\alpha}^{\infty} p(z|C_1)\mathrm{d}z + P(C_2) \int_{-\infty}^{\alpha} p(z|C_2)\mathrm{d}z
\tag{9.13}
$$

This probability can be estimated from a priori information, that is the training set.

As an illustration of SPR2, suppose now that the decision boundary is *not known*. Equation (9.13) can still be obtained, except that α is now a variable. $P(\text{error})$ can then be minimised with respect to α, the relevant condition being

$$
\frac{\partial P(\text{error})}{\partial \alpha} = 0 = P(C_1)\{p(\infty|C_1) - p(\alpha|C_1)\} + P(C_2)\{p(\alpha|C_2) - p(-\infty|C_2)\}
\tag{9.14}
$$

It is known that $p(\pm\infty|C_i)$ must vanish, so

$$
- P(C_1)p(\alpha|C_1) + P(C_2)p(\alpha|C_2) = 0
\tag{9.15}
$$

Finally, using $P(C_1) = P(C_2)$, an implicit equation for α,

$$
p(\alpha|C_1) = p(\alpha|C_2)
\tag{9.16}
$$

is recovered.

In the general problem where multivariate measurements/features are used to distinguish between many classes, the approach SPR1 can be summarised as follows:

1. Establish a training set $\{\{x^{(j)}\}, C_i^{(j)}\}$, $j = 1, \ldots, N$, for each class C_i.
2. Compute the a priori information $p(\{x\}|C_i)$, $P(C_i)$ and $p(\{x\})$.
3. Given a new unclassified measurement $\{y\}$, use Bayes' theorem to obtain the measurement conditioned probability or *posterior probability* $P(C_i|\{y\}) = p(\{y\}|C_i)P(C_i)/p(\{y\})$ for each class C_i.
4. Choose C_i such that $P(C_i|\{y\}) > P(C_j|\{y\})$ for all $i \neq j$.

As observed earlier in the discussion of the hypothesis test, in step 1 of the scheme, the denominator of the expression, $p(\{x\})$, is common to all of the measurement-conditioned probabilities, so step 4 above can be modified to

$$
\text{Choose } C_i \text{ if } p(\{y\}|C_i)P(C_i) > p(\{y\}|C_j)P(C_j) \quad \text{for all } i \neq j
\tag{9.17}
$$

and the overall feature density need not be computed at step 2.

As noted earlier, step 2 may be far from trivial. There are two main possibilities here: a parametric form can be assumed for the probability density with the training set used to determine the values of the parameters or a nonparametric form can be obtained. In the latter case, the probability density can be constructed from a frequency histogram or kernel density estimate (see Section 6.11) over the training data and stored as an array of values. It is arguably most usual to use the former approach and in the vast majority of cases, the probability density is assumed to be Gaussian, that is,

$$p(\{x\}|C_i) = \frac{1}{(2\pi)^{\frac{N}{2}} \sqrt{|\Sigma_i|}} \exp\left\{-\frac{1}{2}(\{x\} - \{\mu\}_i)^T [\Sigma]_i^{-1}(\{x\} - \{\mu\}_i)\right\} \tag{9.18}$$

for an N-component measurement/feature vector. $[\Sigma]_i$ and $\{\mu\}_i$ are respectively the covariance matrix and mean vector for the measurement vectors associated with class C_i. $|\Sigma_i|$ is the determinant of $[\Sigma]_i$. In this case, the problem of parameter estimation is reduced to obtaining $[\Sigma]_i$ and $\{\mu\}_i$, and one method for achieving this will be discussed later.

9.6   Discriminant Functions and Decision Boundaries

The discussion has now reached a point where it is possible to determine the decision boundaries for the Bayesian classifiers defined above in a number of cases. This is accomplished by defining appropriate *discriminant functions*.

Implementation of the SPR1 strategy depends on maximising the discriminant function

$$g_i(\{x\}) = P(C_i|\{x\}) \tag{9.19}$$

over the set of classes. Suppose for now that the classes are equally likely to occur, that is, $P(C_i) = P(C_j)$ for all i, j. In this case, Equation (9.17) shows that it is equally valid to maximise the function

$$g_i(\{x\}) = p(\{x\}|C_i) \tag{9.20}$$

in order to make a decision. Finally, note that maximising *any* monotonically increasing function of g_i leads to the same decision rule. In the case of Gaussian distributed measurement vectors, as in Equation (9.18), a particularly simple rule is obtained by taking

$$g_i(\{x\}) = \log p(\{x\}|C_i) \tag{9.21}$$

in which case,

$$g_i(\{x\}) = -\frac{1}{2}(\{x\} - \{\mu\}_i)^T [\Sigma]_i^{-1}(\{x\} - \{\mu\}_i) - \frac{N}{2}\log(2\pi) - \frac{1}{2}\log|\Sigma_i| \tag{9.22}$$

The second term on the right-hand side is class-independent and can be dropped, which gives

$$g_i(\{x\}) = -\frac{1}{2}(\{x\} - \{\mu\}_i)^T [\Sigma]_i^{-1}(\{x\} - \{\mu\}_i) - \frac{1}{2}\log|\Sigma_i| \tag{9.23}$$

Note that maximising g_i corresponds to minimising $-g_i$. Ignoring the $-\frac{1}{2}\log|\Sigma_i|$ term for the moment, this leads to a decision rule based on minimising

$$||\{x\} - \{\mu\}_i||_{\Sigma^{-1}}^2 = \frac{1}{2}(\{x\} - \{\mu\}_i)^T [\Sigma]_i^{-1}(\{x\} - \{\mu\}_i) \tag{9.24}$$

where, as the notation suggests, this has the form of a distance squared, based on a norm weighted by $[\Sigma]^{-1}$. The distance $|| \ ||_{\Sigma^{-1}}$ is basically the *Mahalanobis squared-distance* that was introduced in Chapter 6 in the context of outlier analysis. The geometrical content of the argument here is that, under restricted circumstances, a classification can be obtained by minimising the distance between a measurement/feature vector $\{x\}$ and the class mean-vector $\{\mu\}_i$ found from the Mahalanobis squared-distance. In this case, the $\{\mu\}_i$ are clearly to be regarded as *templates* for the class patterns.

To further the geometrical interpretation of Equation (9.24), a simplifying assumption can be made. Suppose that it is assumed that the covariance matrices are independent of class, so that $[\Sigma]_i = [\Sigma]$ for all i; in that case the discriminant function in Equation (9.24) reduces to

$$
\begin{aligned}
g_i(\{x\}) &= -\frac{1}{2}(\{x\} - \{\mu\}_i)^{\mathrm{T}}[\Sigma]^{-1}(\{x\} - \{\mu\}_i) \\
&= -\frac{1}{2}\{x\}^{\mathrm{T}}[\Sigma]^{-1}\{x\} + 2\{\mu\}_i^{\mathrm{T}}[\Sigma]^{-1}\{x\} - \frac{1}{2}\{\mu\}_i^{\mathrm{T}}[\Sigma]^{-1}\{\mu\}_i
\end{aligned}
\tag{9.25}
$$

In this case the first term is class-independent and so can safely be ignored. The result is a *linear discriminant function*,

$$
g_i(\{x\}) = \{\alpha\}_i^{\mathrm{T}}\{x\} + \beta_i
\tag{9.26}
$$

where

$$
\{\alpha\}_i = [\Sigma]^{-1}\{\mu\}_i
\tag{9.27}
$$

and

$$
\beta_i = \{\mu\}_i^{\mathrm{T}}[\Sigma]^{-1}\{\mu\}_i
\tag{9.28}
$$

The decision boundaries (unlike the simple univariate case portrayed in Figure 9.3, these will be multi-dimensional) for a classification based on Equation (9.26) are specified by the conditions

$$
g_i(\{x\}) = g_j(\{x\}), \quad i \neq j
\tag{9.29}
$$

or

$$
(\{\alpha\}_i - \{\alpha\}_j)^{\mathrm{T}}\{x\} + (\beta_i - \beta_j) = 0
\tag{9.30}
$$

Therefore the boundaries are simply hyperplanes,

$$
\{\alpha\}_{ij}^{\mathrm{T}}\{x\} + \beta_{ij} = 0
\tag{9.31}
$$

Although certain simplifying assumptions were made, one conclusion remains valid in the general case; if a linear discriminant can be found, the decision boundaries are hyperplanes. The effect of relaxing the restrictions are summarised below:

- $[\Sigma]_i$ class-specific. In this case, the first term in Equation (9.25), $-\frac{1}{2}\{x\}^{\mathrm{T}}[\Sigma]_i^{-1}\{x\}$, would be class-specific and could not be dropped from the discriminant. This would result in decision boundaries that are general quadratic functions. Such discriminants will be discussed later in the context of a case study.
- Unequal a priori probabilities. In this case, where $P(C_i) \neq P(C_j)$, the basic discriminant function is modified to $g_i(\{x\}) = \log\{p(\{x\}|C_i)P(C_i)\}$. In the case of Gaussian statistics, this leads to a discriminant function

$$
g_i(\{\underline{x}\}) = -||\{x\} - \{\mu\}_i||_{\Sigma^{-1}}^2 - \frac{1}{2}\log|\Sigma_i| + \log P(C_i)
\tag{9.32}
$$

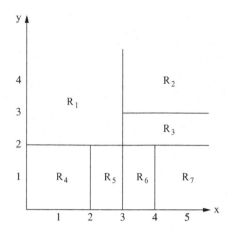

Figure 9.4 Partition of measurement space into simple decision regions.

9.7 Decision Trees

If the classification problem has been reduced to the specification of decision boundaries, classification can be reduced to a sequence of logical operations on the measurement/feature vector. Consider the decision regions shown in Figure 9.4. (Following Schalkoff, 1991, again, it is assumed that the decision boundaries are parallel to the measurement axes. Extension to the case with general plane boundaries is straightforward if not completely trivial.)

The decision problem is built into the pseudo-code shown below. Algorithms for extracting the decision boundaries and reducing the classification problem to a decision tree are reasonably well understood (Breiman *et al.*, 1984).

```
if(y > 2)
        {
        if(x > 3)
                {
                if(y > 3)
                        {
                        Class = 2;
                        }
                else
                        {
                        Class = 3;
                        }
                else
                {
                Class = 1;
                }
        }
else . . .
```

9.8 Training – Maximum Likelihood

It is perhaps opportune to make a short digression now on the more general problem of *machine learning*. This deals in general with a class of problems where one wishes to discover the underlying relationships between certain sets of data, using only samples of data from the sets of interest. Cherkassky and Mulier (1998) describe the three main problems of machine learning as:

- Classification. One wishes to associate a class label from a discrete set, to a given sample of data.
- Regression. One wishes to associate a value of a continuous variable with a given sample of data.
- Density estimation. One wishes to discover the underlying probability density function from which a sequence of data samples have been drawn.

(It is clear that these problems map naturally on to the three main issues in SHM with novelty detection in its most direct form accomplished by density estimation.) It is generally agreed that these problems are given here in order of difficulty. Now a general principle of scientific research should be that one should never attempt to convert a given problem into a more difficult one. However, in earlier discussions this principle appears to have been violated by the wish to solve classification problems by estimating probability density functions. In fact, this is not necessarily the case. As observed above, in order to compute the probability density function of a given set of data, one can pursue either a *parametric* route or the *nonparametric* route.

Nonparametric density estimation methods like kernel density estimation (see Chapter 6) make very few a priori assumptions about the nature of the desired density. As a result of this, they are extremely general. They also suffer considerably from the *curse of dimensionality* and the number of training samples needed in order to establish accurate density estimates grows explosively with dimension. It is these types of method that underlie the statement that density estimation is the hardest machine learning problem. Alternatively, parametric approaches make the assumption that the functional form of the density is known, up to a few parameters, which are then estimated from the data. In general, the number of parameters characterising a probability density function is small (it is only two for the univariate Gaussian distribution) and the parameters can be accurately estimated from comparatively small sets of training data. The parametric approach is therefore the one advocated for classification problems if there is evidence as to the nature of the functional form of the density (except perhaps when the dimension of the data set is low).

Having decided on a course of parametric density estimation, one is faced with two problems: to decide on the appropriate functional form and then to estimate the relevant parameters. Contextual information may provide a clue to the functional form, but in the absence of such a priori knowledge a Gaussian function is often adopted. The central limit theorem lends support for this choice and it has the further advantage that only the mean and covariance parameters require estimation. The remainder of this section comprises a discussion of how to obtain the required parameters $\{\mu\}_i$ and $[\Sigma]_i$ for the Gaussian density functions, which are assumed to represent the class likelihoods. The exercise is not carried out with the intention of deriving the results for a Gaussian as they will turn out to be very well known; the exercise is rather to expose a general method by which parameters can be estimated for any functional form of density.

Learning of the parameters is accomplished using training data sets that will be denoted by D. Again, the main reference for this section is Schalkoff (1991). It is assumed that the statistics for each class C_i can be obtained separately using only training data from the subset D_i of D related to that class.

For notational convenience the required parameters $\{\mu\}$ and $[\Sigma]$ are assumed to be ordered in a single vector $\{w\} = (\mu_1, \ldots, \mu_N, \Sigma_{11}, \ldots, \Sigma_{NN})$. (From this point onwards in this section, the class label will be omitted except on the training data. The reader should remember that the estimation procedure should be carried out for each class using the training data appropriate to that class.) Throughout the remainder of this book, parameters learned from data will be denoted by a vector $\{w\}$. This draws a general analogy with the neural network where the tunable parameters are traditionally referred to as *weights* (see Chapter 11).

The general parameter estimation principle adopted here is the *maximum likelihood* (ML) approach; for a description of an alternative Bayesian approach see Schalkoff (1991). Bayesian approaches offer a number of advantages over ML; however, they are generally a little more difficult to explain.

Given a set of training data $\{\{x^{(i)}\}, i = 1, \ldots, N_t\}$, the likelihood, can be defined by

$$p(D_i|\{w\}) = \prod_{j=1}^{N_t} p(\{x^{(j)}\}|\{w\}) \tag{9.33}$$

This determines the likelihood that the given training set actually arose from the distribution with parameters $\{w\}$. It is intuitively obvious that a meaningful definition of the *best* set of distribution parameters is that which causes the likelihood in Equation (9.33) above to be a maximum, and hence the name *maximum likelihood* for this estimator.

The most direct approach to finding the required maximum is to use the condition

$$\nabla_{\{w\}} p(D_i|\{w\}) = 0 \tag{9.34}$$

where $\nabla_{\{w\}}$ is the gradient operator in the space of parameters. However, note that

$$\nabla_{\{w\}} f(p(D_i|\{w\})) = 0 \tag{9.35}$$

will also yield the correct maximum as long as the function f is monotonically increasing. As in the case of discriminant functions described above, when dealing with the Gaussian distribution, it turns out to be most useful to work with the *log likelihood* function

$$l(\{w\}) = \log[p(D_i|\{w\})] \tag{9.36}$$

so that, applying Equation (9.33), one finds

$$l(\{w\}) = \sum_{j=1}^{N_t} \log[p(\{x^{(j)}\}|\{w\})] \tag{9.37}$$

and in terms of this quantity, the condition for a maximum becomes

$$\nabla_{\{w\}} l(\{w\}) = 0 = \sum_{j=1}^{N_t} \nabla_{\{w\}} \log[p(\{x^{(j)}\}|\{w\})] \tag{9.38}$$

The following example will show how the condition (9.38) leads to an estimator for the mean $\{\mu\}$ in the specific case of Gaussian statistics where

$$p(\{x\}|\{\mu\}) = \frac{1}{(2\pi)^{N/2}\sqrt{|\Sigma|}} \exp\left\{-\frac{1}{2}(\{x\} - \{\mu\})^{\mathrm{T}}[\Sigma]^{-1}(\{x\} - \{\mu\})\right\} \tag{9.39}$$

and the covariance matrix is assumed known. For the particular sample from the training data $\underline{x}^{(k)}$, it follows from Equation (9.39) that

$$\log\{p(\{x^{(k)}\}|\{\mu\})\} = -\frac{1}{2}(\{x^{(k)}\} - \{\mu\})^{T}[\Sigma]^{-1}(\{x^{(k)}\} - \{\mu\}) - \frac{1}{2}\log\{(2\pi)^{N}|\Sigma|\} \tag{9.40}$$

Substituting this last expression into Equation (9.38) gives

$$\sum_{k=1}^{N_t} \nabla_{\{\mu\}} \left\{ (\{x^{(k)}\} - \{\mu\})^T [\Sigma]^{-1} (\{x^{(k)}\} - \{\mu\}) \right\} = 0 \qquad (9.41)$$

or

$$\sum_{k=1}^{N_t} \nabla_{\{\mu\}} \left\{ \{x^{(k)}\}^T [\Sigma]^{-1} \{x^{(k)}\} - \{x^{(k)}\}^T [\Sigma]^{-1} \{\mu\} - \{\mu\}^T [\Sigma]^{-1} \{x^{(k)}\} + \{\mu\}^T [\Sigma]^{-1} \{\mu\} \right\} = 0 \quad (9.42)$$

Differentiating a function with respect to a vector is similar to differentiating with respect to a complex variable z. In the latter case z and its complex conjugate $\bar{z}$ must be treated as independent quantities so that $\partial \bar{z} / \partial z = 0$. Similarly, $\{\mu\}$ and $\{\mu\}^T$ can be treated as independent so that $\nabla_{\{\mu\}} \{\mu\}^T = 0$. (The reader may encounter this idea again in the section of Appendix A dealing with least-squares parameter estimation.) Applying this principle to Equation (9.42) leads to

$$\sum_{k=1}^{N_t} \left\{ -\{x^{(k)}\}^T [\Sigma]^{-1} + \{\mu\}^T [\Sigma]^{-1} \right\} = 0 \qquad (9.43)$$

Transposing ($[\Sigma]$ is symmetric, so this follows from its definition) gives

$$\sum_{k=1}^{N_t} \left\{ -[\Sigma]^{-1} \{x^{(k)}\} + [\Sigma]^{-1} \{\mu\} \right\} = 0 \qquad (9.44)$$

Now, premultiplying by $[\Sigma]$ yields

$$\sum_{k=1}^{N_t} \left\{ -\{x^{(k)}\} + \{\mu\} \right\} = 0 \qquad (9.45)$$

or

$$\sum_{k=1}^{N_t} \{x^{(k)}\} = N_t \{\mu\} \qquad (9.46)$$

As one might expect, the maximum likelihood estimate of the mean is simply the sample arithmetic mean taken over all the training vectors,

$$\{\mu\} = \frac{1}{N_t} \sum_{k=1}^{N_t} \{x^{(k)}\} \qquad (9.47)$$

Fortunately in this case, the covariance matrix cancelled out as this needs to be estimated also. A more complex calculation yields the ML estimate of the covariance estimate. It is found to be the sample covariance

$$[\Sigma] = \frac{1}{N_t} \sum_{k=1}^{N_t} (\{x^{(k)}\} - \{\mu\})(\{x^{(k)}\} - \{\mu\})^T \qquad (9.48)$$

as expected. (In fact this estimate is known to be biased and the correct prefactor should be $1/(N_t - 1)$; however, the bias only makes itself felt for very small samples of data in any case.

9.9   Nearest Neighbour Classification

Finally in the discussion of theory, it should be mentioned that it is possible to devise classifiers that use the training data in a direct manner without computing distributions for the classes. Arguably the most direct methods are those based on *nearest neighbours*. The simplest of these, which shall be referred to as $1 - NN$, operates as follows: given a new measurement $\{y\}$, this is assigned the same class as the *nearest* point in the training set. (This approach assumes that the training vectors are naturally clustered according to their classes. The procedure is known to be suboptimal; in fact, it can be shown that (Schalkoff, 1991)

$$P(\text{error})_{Bayes} < P(\text{error})_{1-NN} < 2P(\text{error})_{Bayes} \tag{9.49}$$

in the limit as N_t, the number of training vectors, tends to infinity. However, if the Bayes error is small in any case, nearest neighbour methods can be remarkably effective. One disadvantage of the method, which should be weighed against the fact that there is no learning involved, is that it may be time-consuming to search for the nearest neighbours if the training set is large.

In order to reduce the possibility of misclassifications due to outlying measurements, the approach can be generalised to consider the k nearest neighbours ($k - NN$); the classification is obtained by giving each neighbour a vote for its class. To avoid 'ties' in the voting scheme, k is usually taken to be an odd number.

The number k here is one of the first examples of a *hyperparameter* for a classifier that has been encountered in this chapter. As discussed in Chapter 6, it is necessary to specify hyperparameters for an algorithm before the training phase can be accomplished. In fact, it is possible to learn hyperparameters from the data in much the same way that one learns parameters. This is discussed in more detail in later sections and chapters.

9.10   Case Study: An Acoustic Emission Experiment

This study concerns the classification of experimental acoustic emission data obtained from a box girder of a bridge. The data here were obtained from a test at Cardiff University in the United Kingdom concerned with the classification of acoustic emission (AE) data from a box girder of a bridge. Five sensors (WDI-PCA Ltd) of resonance frequency 100–1000 kHz were attached around the diaphragm using magnetic clamps, as shown in Figure 9.5. The surface of the box girder was smoothed by light sanding to remove irregularities, prior to installation of the sensors. Grease was used as an acoustic couplant and the integrity of the sensor was checked using a Hsu–Nielson (H-N) source (a controlled pencil lead break). Digitised AE signal waveforms were recorded using a 12-channel data acquisition system under cyclic loading of 0.1–85 kN.

Figure 9.6 shows one of the AE samples from the box girder. Each sample comprised 2048 points sampled at a frequency of 4 MHz.

The object of the exercise was to distinguish the various AE sources – in particular, to distinguish the AE from crack growth from any others (i.e. AE generated by crushing at the clamps). This would be a prerequisite for carrying out any damage prognosis on the basis of AE activity.

The AE bursts were separated from the background by establishing a detection threshold on the initial data at mean plus six standard deviations and classifying any activity above this threshold as burst activity. Each signal contained one AE burst and as each burst was identified, the basic features were recorded. Figure 7.22 illustrates these traditional AE features selected, namely *rise time*, *peak amplitude*, *duration* and *ring down count*.

A total of 91 AE bursts were available for classification based on tabulated values of the four aforementioned parameters. The tabulated values for each burst are assembled as the signal's *feature vector*.

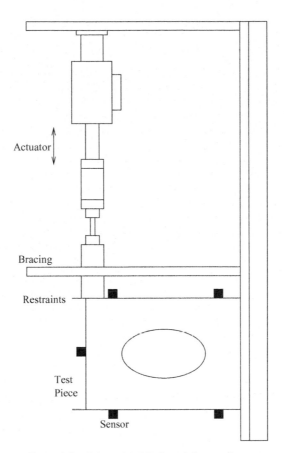

Figure 9.5 Schematic of the box girder experiment.

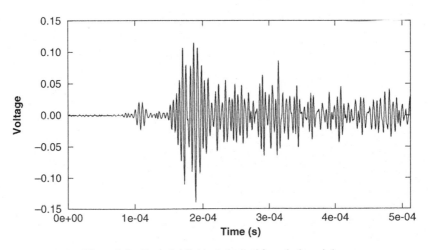

Figure 9.6 Typical AE signal obtained from the box girder.

9.10.1   Analysis and Classification of the AE Data

9.10.1.1   Principal Component Analysis

The first analysis applied to the AE data was PCA (see Chapter 6) for dimensionality reduction and visualisation. To recap, as a means of dimension reduction, PCA works by linearly transforming the data to retain the maximum variance in the principal components while discarding those combinations of the data that contribute least to the overall variance of the data set. There is a caveat associated with this transformation. The components of least variance may in practice be most relevant for damage identification and may be erroneously discarded. In the more general case of dimension reduction or feature extraction, more advanced techniques may be applied in order to preserve the features of interest.

PCA was used here to reduce the dimension of the feature vectors to a level where visualisation of the data was possible. Before transformation to principal components, the data were standardised by dividing the mean-corrected data for each parameter by its respective standard deviation. This procedure ensures that all parameters receive equal weighting. The technique was used to reduce the dimension of the four AE parameters to two corresponding to the first two principal components. The first two principal component scores (i.e. coordinates with respect to the PCA basis) of the standardised AE feature vectors are shown plotted in Figure 9.7.

A complete analysis of this can be followed in Manson *et al.* (2002). The key point to note from this figure is the three sizeable clusters of reduced feature vectors labelled in the figure. The work in Manson *et al.* (2002) concluded that the three clusters blindly identified using PCA each related to a different AE activity known from the experimentation procedure. Those observations in cluster 1 arose from crack-related events, all those in cluster 2 from frictional processes away from the crack and all those in cluster 3 from crack-related events detected at a distance away from the sensor. Figure 9.8 shows an example of each of the three distinct AE signals.

The three clusters of AE data identified formed the basis for the later analysis: an attempt to classify an AE feature vector automatically according to the generation mechanism of the AE activity. Note the simplicity of the problem. It is encouraging that a situation of real engineering interest breaks down into a

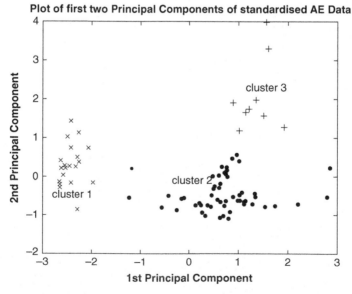

Figure 9.7   PCA reduction of standardised AE feature vectors from the box girder (three clusters are identified).

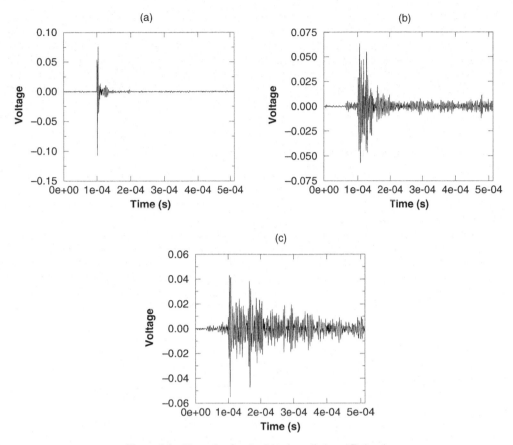

Figure 9.8   Example of each of the three distinct AE signals.

separable classification problem. This will allow the use – in subsequent sections – of simple processing procedures. However, this is usually specific only to laboratory conditions where background noise, that is the number of competing sources, is controllable. Because simple classifiers are likely to succeed with such data, it provides an ideal opportunity to illustrate some of the simpler discriminant-based classifiers described earlier in this chapter.

9.10.1.2   Training and Validation Data

In each of the pattern recognition techniques, it is advantageous to employ the maximum amount of available information in designing the pattern recognition system. The generation mechanism of each AE burst is known a priori so the clusters are assigned as in Figure 9.7. The complete data set is then split into a training set and a validation set with equal numbers of feature vectors belonging to each class assigned to each. The training data set is split further into three separate clusters, each containing AE feature vectors representing a different generation mechanism. Availability of the class assignments for the training data means that supervised learning is possible.

The classifier learnt from the training data can then be used to make a decision when testing data are presented. As described in Bishop (1998), a rigorous procedure requires that three data sets are distinguished, namely *training*, *validation* and *testing*. The idea of the training set is to fix the parameters of the learning algorithm, for example, the weights of a neural network. The validation set is used to set hyperparameters, for example the number of hidden units in a neural network or the optimal number of neighbours in nearest neighbour analysis. For true rigour, a third testing set is required in order to judge generalisation performance (i.e. how well the classifier can deal with previously unseen data). However, given the sparsity of the data set used here, only a training and validation set were selected. The visualisation results show later that this course of action is not too destructive.

9.10.1.3 Discriminant Analysis and Decision Boundaries

The approach to pattern recognition applied here uses the parametric methodology based on the assumption that the data takes a multivariate Gaussian distribution for each class as described by Equation (9.18). The learning from data in that case is therefore the estimation of the relevant class-conditioned mean vectors and covariance matrices.

As discussed earlier in this chapter, the multivariate Gaussian distribution assumption is equivalent to using a *discriminant function*, that is a function that signals class membership. In the first case discussed here, the full four-dimensional feature vectors were used. For each of the three training data sets, the class-dependent mean vector $\{\mu\}_i$ and covariance matrix $[\Sigma]_i$ were calculated using the usual formulae, as established by the principle of maximum likelihood. Also, as discussed earlier, if different covariance matrices are allowed for each class, the resulting discriminant function will be quadratic. Estimation of the mean and covariance parameters also allowed the computation of probability density functions for each class, that is each AE waveform type.

The Gaussian distributions for each of the classes, $p_1(\{x\})$, $p_2(\{x\})$ and $p_3(\{x\})$, could then be evaluated for each feature vector, $(x_1, x_2, x_3, x_4)^T = \{x\}$, in the test data set. Based on these probabilities, each test data point $\{x\}$ was assigned to the cluster i that gave the largest value of p_i. The results are presented in Table 9.1 in the form of a confusion matrix, displaying assigned class against true class (known from the experimental technique).

This shows that only one of the test data points was incorrectly assigned using the Gaussian discriminant function. Considering that the objective of engineering interest here is to separate class 1, that is crack-related events, the single error is no real cause for concern.

The data point that was incorrectly assigned is identified in Figure 9.9. This shows that the incorrectly assigned point is that of cluster 2 that is closest to cluster 3 where one may expect some confusion. The numerical results show that the technique did not confidently assign the point to either cluster as the probability densities given for assignment to clusters 2 and 3 were extremely close, being 0.0213 and 0.0220, respectively.

In order to illustrate the determination of decision boundaries, the quadratic discriminant analysis was also applied to the two-dimensional data derived from PCA. This is essentially another visualisation technique in that it displays the data in relation to the class-labelled decision regions. The border of each region is a decision boundary and the data are classified by assigning each point according to the decision

Table 9.1 Confusion matrix of results from quadratic discriminant analysis

		Assigned class (cluster)		
		1	**2**	**3**
True	**1**	11	0	0
class	**2**	0	28	1
(cluster)	**3**	0	0	5

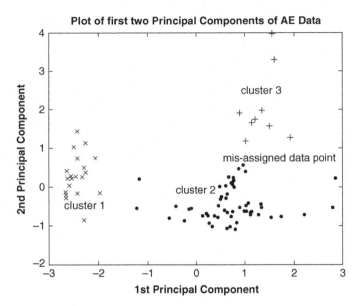

Figure 9.9 PCA reduction showing a misassigned data point by quadratic discriminant analysis.

region in which it falls. If two clusters are labelled i and j, the corresponding decision boundary between them, $D(i, j)$, is specified by the condition

$$g_i(\{x\}) = g_j(\{x\}), \qquad i \neq j \qquad (9.50)$$

where the quadratic discriminant function g_i in each case is defined by Equation (9.25) using the appropriate class-conditioned mean vector and covariance matrix. Three equations were defined and used to plot the decision boundaries between the three clusters. The decision boundaries can be seen in Figure 9.10 in relation to the reduced-dimension AE test data set.

The plot shows that the decision boundaries have accurately separated the AE data into its true classes. As with the four-dimensional analysis, slight confusion results between clusters 2 and 3 whereby one data point is visibly positioned on the decision boundary. This point cannot therefore be blindly assigned with any degree of confidence to a specific cluster. The single misclassification is there purely because of the simplicity of the classification procedure, that is because each cluster distribution was assumed Gaussian.

9.10.1.4 Kernel Discriminant Analysis

As the dimension of the data is very low in this case, it is feasible to try a nonparametric means of density estimation. The algorithm forms a kernel density estimate (as described in Section 6.11), estimates the probability density function at each point in the training set and reports the range of values encountered. The density can then be estimated on a test data set that is not used in the construction of the density. Kernel discriminant analysis (KDA) is accomplished by calculating the density value for the test point on the basis of each class and then assigning to the class that gives the highest density value.

The analysis was conducted here using the first two principal component scores of the feature vectors, again for visualisation purposes. The density estimates were made using the software KDE (Worden,

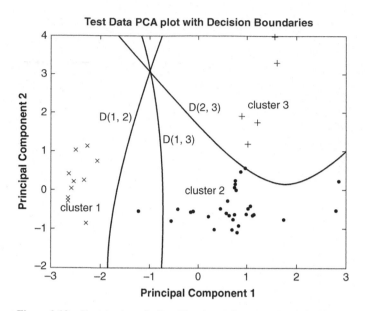

Figure 9.10   Decision boundaries with reduced dimension AE training data.

1998). The analysis was first conducted using estimates of the smoothing parameter given by least-squares cross-validation (as described in Section 6.11). The results are summarised as a confusion matrix in Table 9.2. The analysis was repeated but this time the kernel density estimates were made using the values of smoothing parameters given by applying the Gaussian assumption Equation (9.18). The results produced were the same as those in Table 9.2. The values of the smoothing parameter used were in the range of 0.5 to 1. Only one misclassified test data point is present in both cases.

Figure 9.11 shows a density plot of the test data generated by KDE applying a value of the smoothing parameter of 0.5 (top plot). Inspection of this plot suggested that the values of the smoothing parameter being generated were too large, resulting in oversmoothing of the data. The analysis was repeated by determining a value of smoothing parameter using cross-validation over the complete training data set, rather than over the individual clusters. The results can be seen in Table 9.3. This shows completely correct classification of the AE test data set. The suggested value of the smoothing parameter is therefore 0.2, as opposed to approximately 0.5 estimated over the individual training data sets. The difference in density plots can be seen by comparing the two respective plots in Figure 9.11. These demonstrate the criticality in setting the value of smoothing parameter for the kernel density estimation.

Table 9.2   Confusion matrix for classification of AE test data using KDA with the h value given by cross-validation

		Assigned class (cluster)		
		1	**2**	**3**
True	**1**	11	0	0
class	**2**	0	28	1
(cluster)	**3**	0	0	5

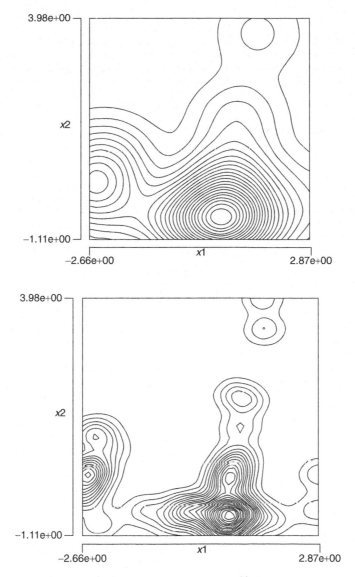

Figure 9.11   KDE density estimations of AE test data applying smoothing parameter values of 0.5 (top plot) and 0.2 (bottom plot).

Table 9.3   Confusion matrix for classification of AE test data using KDE taking $h = 0.2$

		Assigned class (cluster)		
		1	**2**	**3**
True	**1**	11	0	0
class	**2**	0	29	1
(cluster)	**3**	0	0	5

9.11 Summary

The purpose of this chapter has been to introduce the main ideas of statistical pattern recognition (SPR) as it has been argued that this is the appropriate framework for data-based SHM. The following chapters will provide detailed explanations of how SPR and machine learning algorithms can be applied in the context of specific SHM problems from detection to prognosis.

References

Bishop, C.M. (1994) Novelty detection and neural network validation. *IEEE Proceedings – Vision and Image Signal Processing*, **141**, 217–222.

Bishop, C.M. (1998) *Neural Networks for Pattern Recognition*, Cambridge University Press.

Breiman, L., Friedman, J., Stone, C.J. and Olsen, R.A. (1984) *Classification and Regression Trees*, Chapman and Hall/CRC.

Cherkassky, V. and Mulier, F. (1998) *Learning from Data, Concepts, Theory and Methods*, Wiley Interscience.

Friswell, M.I. (2008) Damage identification using inverse methods, in *Dynamic Methods for Damage Detection in Structures* (eds A. Morassi and F. Vestroni), Springer Wien, New York.

Friswell, M.I. and Penny, J.E.T. (1997) Is damage detection using vibration measurements practical? Proceedings of the 2nd International Conference on Structural Damage Assessment using Advanced Signal Processing Procedures (DAMAS 97), Sheffield, UK, pp. 351–362.

Kay, S.M. (1998) *Fundamentals of Statistical Signal Processing Detection Theory*, Prentice Hall.

Lowe, D. (2000) Feature extraction, data visualisation, classification and fusion for damage assessment. Oral Presentation at EPSRC SIDANet Meeting, Derby, UK.

Manson, G., Worden, K., Holford, K.M. *et al.* (2002) Visualisation and dimension reduction of acoustic emission data for damage detection. *Journal of Intelligent Material Systems and Structures*, **12**, 529–536.

Markou, M. and Singh, S. (2003a) Novelty detection – a review. Part I: statistical approaches. *Signal Processing*, **83**, 2481–2497.

Markou, M. and Singh, S. (2003b) Novelty detection – a review. Part II: neural network based approaches. *Signal Processing*, **83**, 2499–2521.

Schalkoff, R.J. (1991) *Pattern Recognition: Statistical, Structural and Neural Approaches*, John Wiley & Sons, Ltd.

Worden, K. (1997) Structural fault detection using a novelty measure. *Journal of Sound and Vibration*, **201**, 85–101.

Worden, K. (1998) *KDE – Kernel Density Estimator Version 1.1 – A User's Manual* (obtainable via k.worden@sheffield.ac.uk).

10

Unsupervised Learning – Novelty Detection

10.1 Introduction

As discussed in previous chapters, the property that sets apart damage *detection* from the higher levels of damage identification is that it can often be approached using *unsupervised learning*. To recap, this means that the class labels of the acquired data are not needed for the machine learning operation. Despite this, one is still able to establish a two-class classifier, which can distinguish data from a normal condition from data corresponding to a damage state. The body of techniques concerned are called *novelty detection* or *anomaly detection* methods within the machine learning community; within the statistics community they are often referred to as outlier detection methods. Within this book, these terms will often be used interchangeably. The philosophy of novelty detection is very simply stated; if one is in possession of data guaranteed to be from the normal condition of a system or structure, one can construct a statistical (or other) model of that data. Any subsequent data from the system can be tested to see if it conforms in some strict sense with the model of normality; noncomformity can then be said to infer damage.

The main advantage of novelty detection is substantial. At the risk of repetition, the major problem in machine learning approaches to SHM is the question of where the data corresponding to damage are obtained from. If the data are to be obtained from modelling, it is clear that for, say, wave scattering from damage in composite structures, one will have a formidable modelling task that may be very difficult and may be expensive (in terms of development cost or computer resource) to run and validate. Alternatively, unless the structures of interest are extremely inexpensive, it will not be possible to obtain the data from an experimental programme; one clearly could not contemplate making a number of copies of an aircraft wing, for example, and then damaging them all in different ways. Under some conditions, data from damage conditions may be available; for example, in the domain of condition monitoring, it may be that historical data are available from the failure of nominally identical machines or components. The problem here is with the word 'nominally'; no two machines or components will be truly identical and statistical variations or uncertainties would have to be taken very carefully into account. One may also have historical or legacy data from the failure of a structure of interest; again one must take into account variability between structures and one should also bear in mind that if the structure has been repaired, it is very likely a different structure anyway. In any case, the nature of the problem means that damage data will be very sparse, if available at all.

The nature of novelty detection means that one only ever needs to model or take measurements from the undamaged structure; this is an enormous advantage over supervised learning in terms of SHM. One of

Structural Health Monitoring: A Machine Learning Perspective, First Edition. Charles R. Farrar and Keith Worden.
© 2013 John Wiley & Sons, Ltd. Published 2013 by John Wiley & Sons, Ltd.

the main caveats with novelty detection is that it is sometimes accomplished by inferring the probability density function of the normal condition data in some way or another. If this exercise is attempted in its full generality, it is essentially the hardest of all machine learning problems. Alternatively, if a parametric form for the density is assumed, the problem can be made more tractable at the expense of generality.

The objective of this chapter is to illustrate the technique of novelty detection in a number of different contexts. First, it is shown how one of the simplest possible techniques – one based on outlier analysis – is applied to a simulated system. More general methodologies based on auto-associative neural networks and density estimation are then discussed and illustrated using case studies: in the former case, a simple lumped-mass simulation and in the latter a more complicated experiment. Following these, a framework generated within the process control community is discussed and illustrated in the context of one of the test structures from Chapter 5. Finally, the question of a threshold or alarm level setting for novelty detectors is addressed and a principled framework based on extreme value statistics is discussed.

This chapter is by no means intended as a comprehensive review of novelty detection methods; such an exercise has already been conducted with success by Markou and Singh (2003a, 2003b). The reader is referred to these papers for a survey of the many alternative techniques available.

10.2   A Gaussian-Distributed Normal Condition – Outlier Analysis

As discussed above, one can make progress in novelty detection by assuming a particular form for the probability distribution of the features that define the normal condition. The method discussed in this section is tailored to the Gaussian distribution and therefore depends on the implicit assumption that the data can be characterised by its first two statistical moments – the mean and variance (or covariance for multidimensional feature vectors). This limitation reduces the problem to one where the probability density function of the normal data is fixed by the estimation of a small number of parameters; the main drawback is that the assumption of Gaussianity is by no means always merited. However, one can, of course, test the hypothesis that the data are Gaussian by means like those discussed in Chapter 6.

In the literature of statistics, the problem of novelty detection has long been considered in the context of *outlier analysis* as discussed in detail in Section 6.10. The basic idea is to compute discordancy measures for data and then compare the discordancy with a threshold; if the measure exceeds the threshold the data are flagged as discordant or novel. The discordancy measures are based on statistics extracted from normal condition data as discussed in Section 6.10; the measure appropriate to univariate data is the deviation statistic,

$$z = \frac{|x_\zeta - \overline{x}|}{\sigma_x} \tag{10.1}$$

where x_ζ is the candidate outlier and $\overline{x}$ and σ_x are the mean and standard deviation of the data sample respectively. The latter two values may be calculated with or without the potential outlier in the sample depending upon whether inclusive or exclusive measures are preferred (as discussed in Chapter 6). The discordancy measure in Equation (10.1) is not in itself restricted to a Gaussian normal condition; it is simply a scaled distance from the mean and will work to some extent for data from any unimodal distribution. However, if one is confident of Gaussianity, then a rigorous definition of the confidence interval or threshold is available; for example, the 95% confidence level for an outlier is given by the limits ± 1.96 for z.

The discordancy test, which is the multivariate equivalent of Equation (10.1), is the Mahalanobis squared-distance measure given by

$$D_\zeta^2 = (\{x\}_\zeta - \{\overline{x}\})^{\mathrm{T}} [\Sigma]^{-1} (\{\underline{x}\}_\zeta - \{\overline{x}\}) \tag{10.2}$$

where $\{x\}_\zeta$ is the potential outlier, $\{\overline{x}\}$ is the mean of the normal condition features and $[\Sigma]$ is the normal condition feature covariance matrix. (Strictly speaking one should specify the sample covariance matrix $[S]$; however, the notation here is unlikely to cause confusion.) As with the univariate discordancy test, the mean and covariance may be inclusive or exclusive measures. As observed above, for SHM purposes the potential outlier is always known beforehand and so it is more sensible to use the exclusive measures. Whichever method is used, the Mahalanobis squared-distance of the potential outlier is checked against the threshold value, as in the univariate case, and its status as an outlier determined.

As in the univariate case, in order to label an observation as an outlier or an inlier there needs to be some threshold value against which the discordancy value can be compared. A Monte Carlo method can be used to compute the threshold as discussed in Section 6.10.

As an illustration of the approach in the multivariate case, a data set from a computer simulation described in Worden, Manson and Fieller (2000) is considered. It is formed from the responses of the three-degree-of-freedom (3-DOF) lumped-parameter system shown in Figure 10.1. The equations of

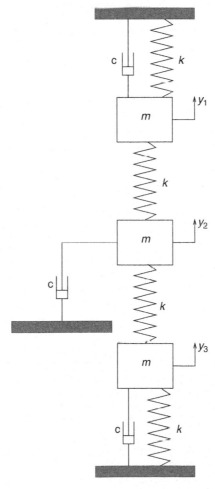

Figure 10.1 The simulated 3-DOF simulated system.

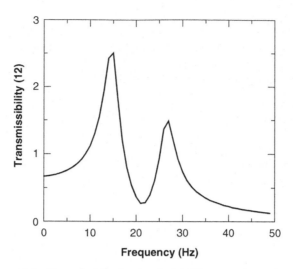

Figure 10.2   Transmissibilty function for 3-DOF system without damage.

motion of this system are

$$m\ddot{y}_1 + c\dot{y}_1 + k(2y_1 - y_2) = x_1(t)$$
$$m\ddot{y}_2 + c\dot{y}_2 + k(2y_2 - y_1 - y_3) = x_2(t) \qquad (10.3)$$
$$m\ddot{y}_3 + c\dot{y}_3 + k(2y_3 - y_2) = x_3(t)$$

The values $m = 1$, $c = 20$ and $k = 10^4$ were used for the undamaged condition.

The feature that was used for the detection process was the transmissibility function (the ratio of acceleration or displacement response spectra) between the two top masses. It was computed by simulating the responses to a harmonic excitation on mass 3 for a frequency range between 0 Hz and 50 Hz. The relative gain and phase between the responses of masses 1 and 2 were extracted in each case. However, only the magnitude was used for the process. The transmissibility function was sampled at 50 regularly spaced points over the frequency range of interest to give the pattern or feature vector to be used as the undamaged condition in the analysis. This feature vector is shown in Figure 10.2.

The damage in the lumped-mass system was then simulated by reducing the stiffness between the top two masses in Figure 10.1 by 1, 10 and 50% of the original value and the three feature vectors from the damaged system were calculated in the same manner as above, with the stiffness altered in the equations of motion.

In order to construct a suitable mean vector and covariance matrix for the normal condition, the undamaged feature was copied 1000 times and each copy was subsequently corrupted with different Gaussian noise vectors of RMS 0.05. This procedure was repeated for the three testing features. These three data sets were then concatenated on to the normal data to give a 4000 observation testing data set.

The exclusive Mahalanobis squared-distances for each of these 4000 observations were then calculated using Equation (10.2) and the results plotted as shown in Figure 10.3. The 99% threshold value for a 1000-observation, 50-dimensional problem was found to be 101 after 1000 trials for the Monte Carlo threshold-setting approach. The plot shows that the undamaged data set (first 1000 observations) were all correctly labelled as inliers, as expected, and all the observations corresponding to the 10 and 50% stiffness reductions (third and fourth sets of 1000 observations respectively) were correctly diagnosed as outliers. Unfortunately, the method is unable to classify virtually any of the 1% reduction observations

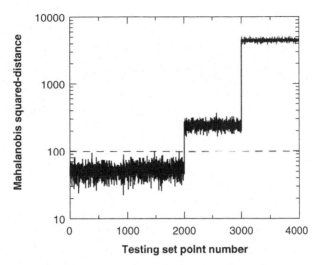

Figure 10.3   The outlier statistic (discordancy) for the undamaged and damaged simulated 3-DOF system.

(second set of 1000 observations) as outliers. This result illustrates a general property of novelty detectors, as mentioned previously in terms of the 'axioms' of SHM, and in more detail later in Chapter 13; there is a trade-off between the noise rejection capabilities and the sensitivity to damage.

10.3   A Non-Gaussian Normal Condition – A Neural Network Approach

The outlier detection approach described in the last section made an implicit assumption of the Gaussian nature of the normal condition data through the use of the Mahalanobis squared-distance and also through the construction of the detection threshold. In many cases, this assumption will be unwarranted and more general techniques will be needed. As described above, the Mahalanobis squared-distance may not be a bad approach as long as the distribution of the normal condition data looks 'elliptical' in the feature space and its probability density function is unimodal (only has one peak). If the data departs radically from these assumptions, for example, if the distribution is not convex or, in the worst case, actually separated into distinct unconnected regions, the outlier approach may fail badly. An example of this situation is illustrated in Figure 10.4, where in both cases depicted, the point shown by a cross would be classified as normal, despite the fact that it falls well outside the normal condition set.

If it should seem implausible that the normal condition set should take such shapes, one can consider Figure 10.5. The figure shows the normal condition set for features extracted from propagated waves used to diagnose damage in composite plates (Worden, 2001). The nonconvex distribution is due to the fact that the normal condition in this case encompasses a range of temperatures for the plate specimen. The data set in this case has been projected down from 50 dimensions to two using principal component analysis (PCA). Because PCA is a linear projection, the nonconvex nature of the original data is preserved. The example considered in the remainder of this section will show how a disconnected normal condition set can arise.

One approach that is able to cope with non-Gaussian normal conditions is that based on the idea of an auto-associative neural network (AANN); the application of this network was first proposed in Pomerleau (1994) within a slightly different context. A more detailed discussion of neural networks will be presented in Chapter 11. All that is required to know at this point is that a neural network maps an

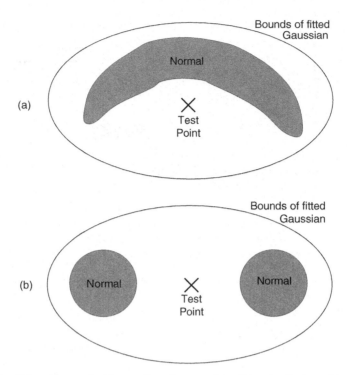

Figure 10.4 Potential problems raised by non-Gaussian normal condition distributions: (a) nonconvex normal condition, (b) disconnected normal condition.

input (typically the damage-sensitive feature vector) on to some output quantity (e.g. damage class). This mapping is achieved with a layered set of connected processing units (neurons). Information flows through the network from an input layer to an output layer. Each neuron carries out a local nonlinear calculation involving its particular inputs from the previous layer and passes the result to the neurons in the next layer. The various parameters of the network are estimated by an iterative process that minimises the error between network predictions and desired outputs using known input–output pairs. The parameter estimation stage is called *learning* or *training*.

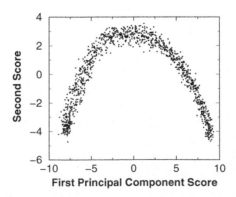

Figure 10.5 Nonconvex normal condition set extracted from wave features.

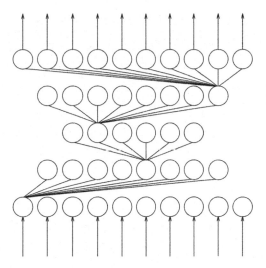

Figure 10.6   Auto-associative neural network.

The approach taken for novelty detection is simply to train an AANN on the normal condition features. As will be discussed further in Chapter 11, this training is accomplished by means of a feedforward multilayer perceptron (MLP) network, which is asked to reproduce at the output layer those patterns that are presented at the input. This process would be a trivial exercise except that the network structure has a 'bottleneck'; that is, the features are passed through hidden layers that have fewer nodes than the input layer (Figure 10.6). This architecture forces the network to learn the significant characteristics of the features; the output values or *activations* of the smallest, central layer correspond to a compressed representation of the input feature vectors. Training proceeds by presenting the network with many versions of the features corresponding to the normal condition corrupted by noise or other sources of variability with the same features prescribed as the output.

The novelty index $v(\{x\})$ corresponding to a feature vector $\{x\}$ is then defined as the Euclidean distance between the feature vector and the result of presenting it to the trained network $\{\hat{x}\}$:

$$v(\{x\}) = ||\{x\} - \{\hat{x}\}|| \tag{10.4}$$

If learning has been successful, then for all data in the training set one has $v(\{x\}) \approx 0$ if the feature vector $\{x\}$ represents the normal condition. If $\{x\}$ corresponds to damage then $v(\{x\}) \neq 0$. Note that there is no guarantee that $v(\{x\})$ will increase monotonically with the level of damage and this is why this form of novelty detection only gives a level 1 diagnostic in Rytter's terms. Note also that the universal approximation property of the neural network (see Chapter 11 or Bishop, 1998), means that the novelty detector can learn the properties of any normal condition distribution; it does not have to be Gaussian or even unimodal. In fact, the case study following will consider the extreme case of a normal condition set defined on two disconnected components.

As an illustration, the three-degree-of-freedom system with concentrated masses described in Equation (10.3) and shown in Figure 10.1 was simulated again but with two normal conditions. In the first normal condition (NCI) the following values were used: $m = 1$, $c = 20$ and $k = 10^4$. In order to simulate a second distinct normal condition (NCII), the same parameters were used with the exception that the bottom mass was reduced by 50%. This problem with two normal conditions with different masses is intended to mimic the situation where, for example, an aeroplane drops a store. One would not wish to diagnose damage on the basis of an acceptable, in fact anticipated, change of structural properties. The

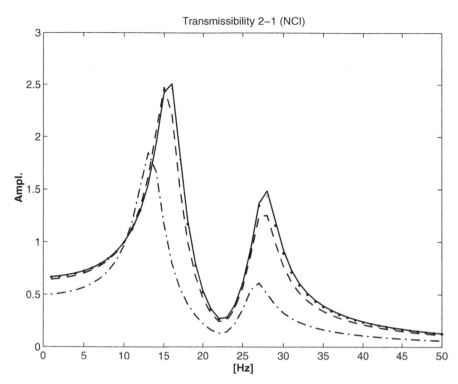

Figure 10.7 Transmissibilities for normal condition 1 (solid line) and corresponding damages: 2% (dotted line), 10% (dashed), 50% (dot-dashed).

damage in the system was simulated by decreasing the stiffness between the top two masses in Figure 10.1 by different degrees.

In both conditions, the response transmissibility function between the top two masses was calculated (as before) in terms of the transverse displacement for a harmonic excitation applied at the top mass. In the simulations performed, the frequency of the harmonic exciting force was varied in the range from 0 to 50 Hz (again, as before), in order to encompass the three natural frequencies of the system.

Figure 10.7 illustrates the transmissibility function corresponding to the structure in NCI together with the functions relative to three damage conditions, denoted DCI, with 2, 10 and 50% reductions in the stiffness of the spring of interest, respectively. Note that the 2% damage case (dotted line) is indistinguishable from the undamaged case (solid line). Figure 10.8 shows equivalent functions corresponding to NCII together with damaged cases with the same reductions in stiffness, denoted DCII.

Note that the normal conditions NCI and NCII are significantly different from each other compared to the differences between a normal condition and its corresponding damage states. This result means that an outlier analysis based on the Mahalanobis squared-distance conducted on a given normal condition would strongly indicate damage for data from the other normal condition.

The training data for the neural network diagnostic system were obtained by making 500 copies of a 50-spectral line range from the transmissibility functions corresponding to the undamaged structure for each of the two normal conditions, and then polluting each of these independently by adding Gaussian noise with an RMS value equal to 1% of the peak value of the transmissibility function. The testing sets were constructed by adding 500 noise-corrupted copies of the transmissibility functions relative to the different damage scenarios considered for both operating conditions.

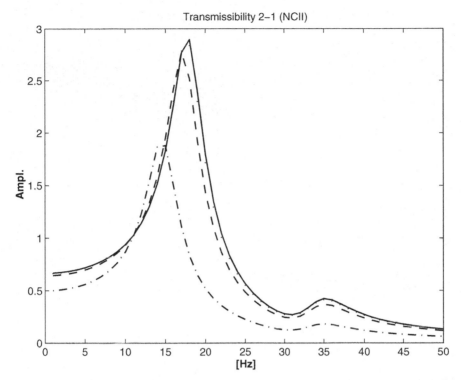

Figure 10.8 Transmissibilities for normal condition 2 (solid line) and corresponding damages: 2% (dotted linc), 10% (dashed), 50% (dot-dashed).

For the task of pattern recognition, an auto-associative neural network with 5 layers and node structure 50:40:30:40:50 was selected and trained for 100 000 cycles, presenting the patterns in random order. A slight modification of the novelty index in Equation (10.4) was used in which the two normal condition sets were normalised to the same peak amplitude; details can be found in Surace and Worden (1997).

The results shown in Figure 10.9 correspond to a 10% stiffness reduction. As one can see, the novelty index permits unambiguous identification of the presence of the fault and gives the required near-zero response on both normal conditions. As in the case of outlier analysis, a confidence threshold can be computed (Worden, 1997), but it would be superfluous here.

As the feature vectors are passed through hidden layers, which have fewer nodes than the input layer, the network is forced to learn just the significant prevalent features of the patterns. This process means that the output of the training data is the transmissibility function of the undamaged structure without any noise contamination.

This example illustrates a case where it is necessary to design a novelty detector that does not fire when anticipated and significant operational variations occur. In the field of SHM it is often even more important to be able to account for environmental changes such as changes due to variations in temperature, and this issue will be discussed in considerable detail in Chapter 12.

10.4   Nonparametric Density Estimation – A Case Study

As discussed above, a more direct approach to novelty detection is to estimate the probability density function (PDF) for the feature vectors over the normal condition set. Once the PDF is known, new data

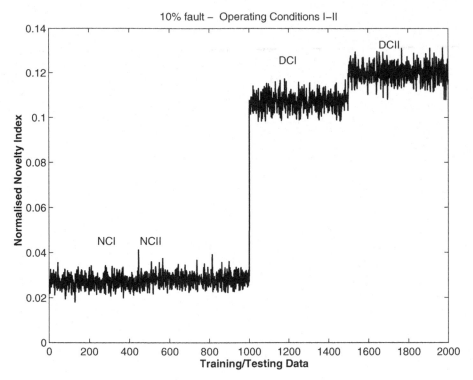

Figure 10.9 Novelty detection of the two 10% damage conditions.

can be accepted or rejected as normal on the basis of the PDF magnitude for the feature. Features with very low values are considered unlikely to have come from the undamaged distribution and are diagnosed as damaged. The outlier analysis approach discussed above performs a parametric density estimation based on the somewhat significant assumption that the normal condition data has a multivariate Gaussian distribution. There are numerous *nonparametric* methods of estimating densities for multivariate data (Silverman, 1986); however, one should bear in mind the caveat that density estimation is arguably the hardest of the machine learning problems. The approach discussed here is the standard kernel density estimation (KDE) method as discussed previously in Chapter 6. To recap, the basic form of the estimate is

$$p(\{x\}) = \frac{1}{Nh} \sum_{i=1}^{N} K\left(\frac{\{x\} - \{x\}_i}{h}\right) \tag{10.5}$$

where $\{x\}_i$ is the ith training data vector or point (i.e. point in a multidimensional space), N is the number of points in the training set and h is the smoothing parameter that controls the width of the individual kernels. As discussed in Chapter 6, the most common choice of kernel function, and the one that will be adopted for the analysis here, is the multivariate Gaussian,

$$K(\{x\}) = \frac{1}{(2\pi)^{d/2}} \exp\left(-\frac{1}{2}||\{x\}||^2\right) \tag{10.6}$$

where d is the dimension of the data space.

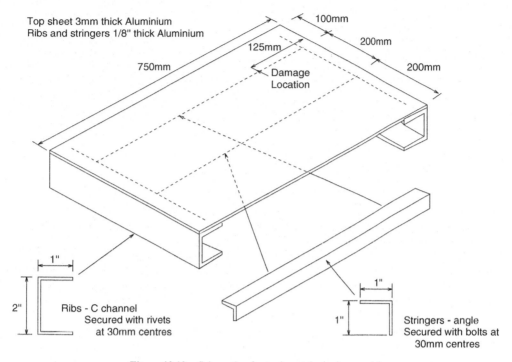

Figure 10.10   Schematic of experimental wingbox model.

Once the estimate is established, the PDF values at any new measurement points are trivially evaluated. Again as discussed in Chapter 6, the quality of the estimate depends critically on two factors: the size of the training data set and the value of h. The method used to establish the 'correct' h in the following is least-squares cross-validation using the error measure in Equation (6.87).

10.4.1   The Experimental Structure and Data Capture

The data for this case study were acquired from an experimental structure that was designed and constructed in order to have realistic structural dimensions and complexity. It is a physical model of an aircraft wingbox. The upper surface is $750 \times 500 \times 3 \text{ mm}^3$ aluminium sheet (Figure 10.10, which is a schematic figure and is not to scale; it is simply intended to show the construction of the wingbox). This plate is stiffened by the addition of two ribs composed of lengths of C-channel riveted to the short edges. Two stiffening stringers composed of angle sections run along the length of the sheet. The tests were all conducted with free–free boundary conditions for the panel, which was suspended from a substantial frame using springs and nylon line.

Damage was simulated by the introduction of a saw cut in the outside stringer 125 mm from the edge of the panel (Figure 10.10). Nine levels of damage were investigated from 10% height to 90%. As the stringer is 2.54 cm (1 inch) in height, each level corresponds to 2.5 mm of damage.

The analysis was based on transmissibility features, again as discussed earlier in this chapter. At each stage of damage, transmissibilities were calculated from data measured on the panel. Three transmissibility paths were identified: AB, AC and DC, as seen in Figure 10.11. AB is along the line of the damaged stringer, while DC is offset by 100 mm. (Again note that Figure 10.10 is a schematic only.) Only the data

Figure 10.11 Schematic showing transmissibility 'paths' on the wingbox top surface.

for path AB will be discussed here; more details of the experiment and further analysis can be found in Worden, Manson and Allman (2003).

The system was excited using a Gearing and Watson electrodynamic shaker driven by broadband white noise amplified by a Gearing and Watson power amplifier. The responses were measured using PCB piezoelectric accelerometers and sampled using a DIFA/Scadas acquisition system running LMS software under the control of an HP computer. The DIFA system was also used to form the random excitation.

The frequency range of the transmissibilities was 0 to 250 Hz and in all cases 2048 spectral lines were obtained. In order to have clean data to identify which modes were sensitive to the damage, an averaged transmissibility was calculated for each path. 128 samples were used for the average in each case. In order to accumulate a reasonable size of normal condition set, 128 transmissibilities (not averaged) were then calculated for each path. This procedure was followed in order to validate the damage detection methods, which require reasonably sized data sets without resorting to a priori assumptions regarding the extent and colour (i.e. spectral characteristics) of any measurement noise. Each damage case was characterised by one transmissibility averaged over 128 samples as well as an additional 10 unaveraged transmissibilities. Figures 10.12 and 10.13 show the test facility. (Note the extra accelerometers, which were used to perform a top surface modal analysis for other purposes.)

10.4.2   Preprocessing of Data and Features

Preprocessing of the patterns was kept to a minimum and it was decided to use the raw transmissibility functions for the basic features. Figure 10.14 shows the magnitude of the normal condition transmissibility for the path AB under investigation. As one might expect, the functions only proved sensitive to the damage in the stringer in the immediate vicinity of the peaks and there are four dominant peaks.

The next step in the preprocessing was to reduce the dimension of the features as far as possible as 2048 spectral lines is excessive. It was decided to single out the four dominant peaks and examine which ones showed highest sensitivity to the damage. Peak 4, around line 1900 in Figure 10.14, showed marked *and* systematic variation in its form as a function of the damage (Figure 10.15), so it was selected as the basic multivariate feature. Spectral lines 1886 to 1935 were selected to form a 50-dimensional feature vector.

The averaged transmissibilities took over 20 minutes to acquire. In order to allow a fast diagnostic with potential for on-line use, it was decided to use unaveraged data to train and test the novelty detectors. Figure 10.16 shows three examples of unaveraged features from peak 4 of the transmissibility; the level of noise is clearly substantial in the region of the maximum. In order to train the various novelty detectors, 118 of these features were used to form the training set. A further 10 normal condition feature vectors were held back for a test set with the remaining 90 features in the test set acquired from the damaged conditions, with each subset consisting of 10 feature samples from the nine damage levels.

Figure 10.12   Layout of wingbox experiment showing sensor placement.

10.4.3   Novelty Detection

As the KDE approach is known to be more sensitive to small data sets, a pseudo-synthetic training set of 1000 points was obtained by computing the mean and covariance matrix of the training set and then generating 1000 feature vectors around the mean with the corresponding Gaussian distribution. A first attempt using the smoothing parameter from least-squares cross-validation gave results that were clearly undersmoothed. This result is understandable, as the cross-validation calculation itself may suffer if the data set is sparse, that is, not large enough to adequately characterise the underlying PDF. The value obtained of $h = 0.8$ was increased to 2.0 in order to give more smoothing. The results on the testing set are shown in Figure 10.17 in the form of a *novelty score*, which is $-\log p(\{x\})$. The damage is only detected unambiguously beyond the 40% stage. Note that the novelty score appears to saturate at an upper bound. At these points, the density was returned as zero (to machine precision) and an arbitrary value of 10^{-100} was assigned.

The thresholds shown in Figures 10.17 correspond to 99 and 99.99% confidence. They were computed as follows. Given a 50-dimensional Gaussian density, the radius was calculated, which would bound 99 or 99.99% of the probability mass. The values of the density at these radii were found and used to scale the estimated density, that is, to find the value of density corresponding to 99 or 99.99% confidence. The results from KDE give some cause for concern, which is understandable given that, of the novelty detection methods discussed so far, it is arguably the most sensitive to using a sparse training set. Applying the guidelines from Silverman (1986) reproduced in Table 10.1 here, 1000 training points would appear to be a ridiculously small number for a 50-dimensional data set. In fact Silverman's guidelines are rather conservative and sparse data sets can still be acceptable if the intrinsic dimensionality of the data is much less than 50. When the outlier analysis and AANN approaches discussed in earlier sections of this chapter were applied to this problem, the results were as shown in Figures 10.18 and 10.19, respectively.

In Worden, Manson and Allman (2003) all three novelty detection approaches discussed so far were applied to the wingbox data. All three approaches appeared to work on the data. The fact that outlier analysis seemed to work would indicate that the normal condition data for this example do not deviate

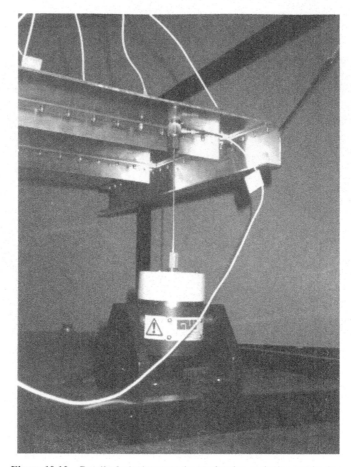

Figure 10.13 Detail of wingbox experiment showing excitation mechanism.

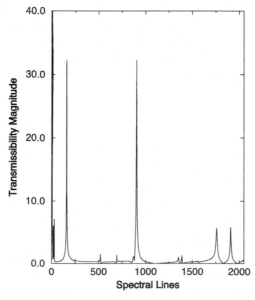

Figure 10.14 Overall transmissibility for 'path' AB: undamaged system.

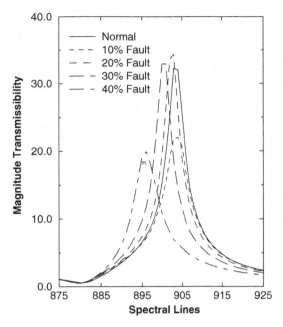

Figure 10.15   Variation in transmissibility peak 4 with increasing damage.

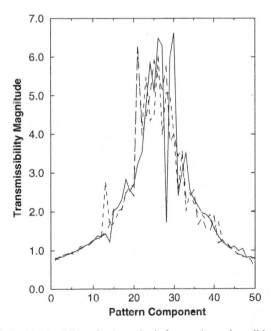

Figure 10.16   Effect of noise on basic features (normal condition).

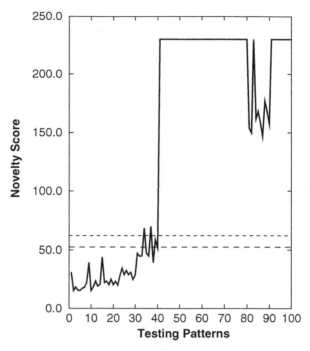

Figure 10.17   Novelty score from KDE for undamaged and damaged wingbox.

Table 10.1   Data labels of the structural state conditions for the simulated building structure

Label	State condition	Description
State 1	Undamaged	Baseline condition
State 2	Undamaged	Mass = 1.2 kg at the base
State 3	Undamaged	Mass = 1.2 kg on the first floor
State 4	Undamaged	87.5% stiffness reduction in column 1BD
State 5	Undamaged	87.5% stiffness reduction in column 1AD and 1BD
State 6	Undamaged	87.5% stiffness reduction in column 2BD
State 7	Undamaged	87.5% stiffness reduction in column 2AD and 2BD
State 8	Undamaged	87.5% stiffness reduction in column 3BD
State 9	Undamaged	87.5% stiffness reduction in column 3AD and 3BD
State 10	Damaged	Gap = 0.20 mm
State 11	Damaged	Gap = 0.15 mm
State 12	Damaged	Gap = 0.13 mm
State 13	Damaged	Gap = 0.10 mm
State 14	Damaged	Gap = 0.05 mm
State 15	Damaged	Gap = 0.20 mm and mass = 1.2 kg at the base
State 16	Damaged	Gap = 0.20 mm and mass = 1.2 kg on the first floor
State 17	Damaged	Gap = 0.10 mm and mass = 1.2 kg on the first floor

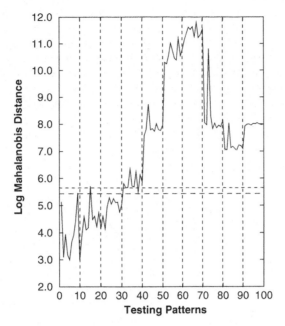

Figure 10.18 Novelty index from outlier analysis for undamaged and damaged wingbox. The vertical lines delimit the testing set components; the first 10 points are from the normal condition and then follow 9 subsets corresponding to 10% damage through to 90% damage.

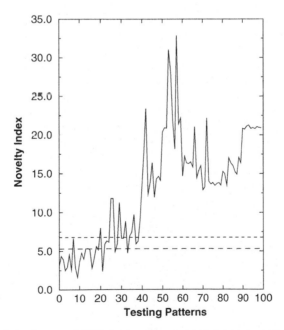

Figure 10.19 Novelty index from an AANN for an undamaged and damaged wingbox. The 90% confidence threshold is given as a dashed line; the 99% threshold is shown dotted.

too much from the assumed Gaussian-distributed form. The most sensitive of the approaches appeared to be the AANN, which started to detect damage in the range of 10–20% cut depths (the results are shown in Figure 10.19). Given the reservations about the nonparametric density estimation approach, it is encouraging that the other methods work well. Of course, as the problems encountered are probably associated with the high dimension of the feature vectors, one could use a dimension reduction approach, like principal component analysis, to bring the dimension down to a reasonable value for the amount of training data acquired. Density estimation has proved its worth on lower-dimensional features in other diagnostic contexts than SHM (Tarassenko *et al.*, 1995).

10.5   Statistical Process Control

So far in this chapter, novelty detection has been accomplished by using various feature vectors extracted from measured data that will, in some sense, encapsulate the dynamical system condition under consideration. All of these features have required a degree of preprocessing – sometimes a considerable amount. One might argue that a more direct approach, if possible, would be to monitor directly one of the response variables, for example a measured acceleration. In such a case, sudden changes in the response mean or variance could be used to infer damage. In fact, such an approach to novelty detection has long been adopted in the process engineering community where time series are monitored directly using statistical novelty criteria; the relevant discipline is that of *statistical process control* (SPC), or *statistical quality control*. The fundamental problem of detecting a sudden change in the statistical behaviour actually dates back to the 1930s where it was concerned with monitoring the quality of manufacturing processes (Wetherill and Brown, 1991). The acknowledged standard reference in the field of SPC is Montgomery (1996), although Oakland (1999) provides a more gentle introduction to the subject. The question of detecting changes in measured time series has been the subject of rigorous statistical analysis for many years; an excellent reference is Basseville and Nikiforov (1993). A shorter but more up-to-date overview of statistical methods can be found in Poor and Hadjiliadis (2009). Among the topics discussed in the latter reference is the use of the sequential probability ratio test (SPRT), which is used for SHM to good effect in Sohn *et al.* (2003).

In this section, SPC and how it can be applied to vibration-based damage detection is discussed. Thinking in terms of direct responses, acceleration measurements, $x(t)$, taken when the structure is in good condition, will have some distribution with mean μ and variance σ^2. Recall that if the response is Gaussian, these two moments are sufficient to fully characterise the distribution. If the structure is damaged, the mean, the variance, or both might change. SPC provides a framework for monitoring future acceleration measurements and for identifying new data that are inconsistent with past data. In particular, *quality control charts* can be proposed to monitor the mean, the variance or some other function of the acceleration measurements.

If the mean and standard deviation for normal condition data are known, one type of control chart can be constructed by drawing a horizontal line at μ and two more horizontal lines representing the upper and lower *control limits* (the UCL and LCL respectively). The upper limit is drawn at $\mu + k\sigma$ and the lower limit at $\mu - k\sigma$. The number k is chosen so that when the structure is in a good condition a large percentage of the observations will fall between the control limits; in later examples k is chosen so that at least 99% of the charted values fall between the control limits. If the response is known to be Gaussian, then 95% of the data will fall between the limits defined by $k = 1.96$; if $k = 3$, then the limits should include 99.7% of the data. These values are standard confidence limits for the Gaussian distribution (see Chapter 6).

As each new measurement is made, it can be plotted versus time or observation number. If the condition of the structure has not changed, almost all of these measurements should fall between the upper and lower control limits, the exact percentage being determined by the choice of k. In addition, there should be no obvious pattern in the charted data; for example, there should not be a repeated pattern of five observations above the mean followed by five observations below the mean. If the structure is damaged there might be a shift in the mean acceleration, which could be indicated by an unusual number of charted values beyond the control limits. Plotting the individual measurements on a control chart is referred to in the SPC literature as an *X chart* (Montgomery, 1996). Note that observing an unusual number of observations outside the control limits does not imply that the structure *is* damaged but only that something has happened to cause the distribution of the current acceleration measurements to change. However, if data outside the control limits cannot be accounted for by operational or environmental factors, the structure should probably be inspected for damage.

To detect a change in the mean of the acceleration measurements, an intuitively appealing idea is to form rational subgroups of size n, compute the sample mean within each subgroup and chart the sample means. The centreline for this control chart will still be μ but the standard deviation of the charted values would be $\sigma/\sqrt{n}$. Therefore, the upper and lower control limits would be placed at $\mu \pm k\sigma/\sqrt{n}$. This type of control chart is referred to as an *X-bar chart* or *Shewhart chart* (Montgomery, 1996). The subgroup size n is chosen so that observations within each group are, in some sense, more similar than observations between groups. If n is chosen too large, a drift that may be present in the mean can possibly be obscured or averaged-out. An additional motivation for charting sample means, as opposed to individual observations, is that the distribution of the sample means can, by an application of the central limit theorem, be approximated with greater confidence by a normal distribution.

When μ and σ are unknown, they can be estimated from observed data taken when the structure is in good condition. An obvious estimate of μ is the sample mean of the acceleration measurements. Several different methods have been proposed for estimating σ. For example, the sample standard deviation of the data or some function of the range of the data could be used. Alternatively, if rational subgroups are constructed, the sample standard deviation, or a function of the range within each subgroup, could be computed and the estimates from each subgroup pooled. Montgomery (1996) has a more complete discussion of estimating a standard deviation for use in control charts. The discussion of control charts presented herein has assumed that the acceleration measurements are uncorrelated. In practice the observed acceleration measurements are likely to be autocorrelated. When the data are correlated, the control limits as just described are inappropriate because the estimate of σ is inappropriate. In the following section a technique for constructing control charts when data are autocorrelated is discussed.

10.5.1   Feature Extraction Based on Autoregressive Modelling

If the acceleration measurements are autocorrelated, constructing a control chart that ignores the correlation can lead to charts that give many false alarms and charts that fail to signal when the process being monitored has changed significantly (see Montgomery, 1996, and the references therein). Because the acceleration measurements will be autocorrelated, it is suggested that an X chart or an X-bar chart is constructed by using the residuals obtained from fitting an autoregressive (AR) model to the observed data. If the fitted AR model is approximately correct, the residuals from the fit should be (nearly) uncorrelated with no systematic pattern.

As introduced in Chapter 7, an AR model with p autoregressive terms, an AR(p) model, can be written in the form

$$x(t) = a + \sum_{j=1}^{p} a_j x(t - j) + \varepsilon(t) \tag{10.7}$$

The variables $x(t)$ are acceleration measurements observed at time t and $\varepsilon(t)$ is an unobservable random error with zero mean and constant variance. The mean of $x(t)$ is μ for all t and $a \equiv (1 - \sum a_j)\mu$. The a_j''s and μ are unknown parameters that must be estimated. In this analysis the AR coefficients are estimated by the Yule–Walker method (Box, Jenkins and Reinsel, 1994).

To fit an AR model to observed data, the order of the model, p, needs to be chosen and the parameters need to be estimated. There are a variety of techniques for choosing the model order, such as Akaike's information criterion (AIC), and for estimating the parameters. These techniques can be found in most textbooks on time series analysis; for example, Box, Jenkins and Reinsel (1994) and Figueiredo *et al.* (2011) specifically discuss the issue of model-order selection in the context of an SHM example.

If $\hat{x}(t)$ denotes the estimated acceleration measurement from the fitted AR model, then the residual at time t is $\varepsilon(t) = x(t) - \hat{x}(t)$. An X or an X-bar chart can be constructed as previously described with the residuals as data. If the fitted model is approximately correct, the residuals should be nearly uncorrelated. Note that for a fitted AR(p) model, residuals cannot be computed for $t \leq p$. From a pattern recognition perspective, the residuals can be thought of as another type of feature derived from the measured time histories. When new data become available, the current acceleration measurement is predicted using these data and the AR model whose estimated parameters were based on the initial or undamaged data. A new vector of residual errors is then determined. These new residual errors or the average residual error from rational subgroups are then charted. If there is damage, the AR model should not fit the new acceleration measurements very well and a statistically significant number of residuals or average residuals should chart beyond the control limits.

As mentioned previously, it is possible for environmental factors such as wind, rain and temperature to affect the acceleration measurements on *in situ* structures. Observed environmental and operational factors can be incorporated directly into the AR model by simply adding a parameter for each factor of interest along with observed values of the factor. Alternatively, the observed acceleration measurements can be regressed on the observed environmental and operational factors and the residuals computed. An AR model could then be fitted to the residuals.

10.5.2   The X-Bar Control Chart: An Experimental Case Study

In this section, the construction of an X-bar control chart for data from the simulated building structure (Section 5.4) is discussed, using the time-history acceleration measurements taken from channel 5 before the structure was damaged, denoted as state 1. Both AR(5) and AR(30) models were fitted to these data in order to investigate the effect of AR model order on the process. X-bar control charts were constructed based on the residuals from the fitted models using subgroups of size 4, based on discussions in Montgomery (1996). Before dividing the residuals into subgroups, they were normalised by subtracting the mean and dividing by the standard deviation of the residuals from the baseline condition. The purpose of the normalisation was to minimise the effects of environmental variations over the period of data acquisition. This issue has been discussed previously and will be considered in more detail later in Chapter 12. The upper and lower control chart limits were calculated as described earlier, based on the sample mean and standard deviation of the baseline condition. Note that in the X-bar control chart presented later, each subgroup is represented by one data point and the centreline of each chart is zero because the sample mean of the associated normalised residual errors is zero, as described above. The upper and lower control chart limits correspond to 99.73% (3σ) confidence intervals, implying that approximately 0.27% of the data points from normal condition states can be expected to fall outside the control limits. For 2046 data points, these limits imply that about six points should fall outside those limits.

As mentioned above, AR models of order 5 and 30 were fitted; however, only the control charts for the $p = 30$ case will be given in detail with only summary information for the $p = 5$ case given. It was shown in Figueiredo *et al.* (2011) that the order 5 model did not include enough lags to accurately reproduce the measured time histories and resulted in residual errors that were autocorrelated. Figures 10.20 and 10.21 show the X-bar control charts using the grouped AR(30) residual errors for the undamaged (with simulated variability) and damaged cases respectively. Points falling outside the control limits and therefore indicating the possibility of damage are marked by crosses in the relevant figures. In addition, Figure 10.22 summarises the number of outliers falling beyond the control limits for both model orders considered. The threshold horizontal dashed line in Figure 10.22 is determined based on the maximum

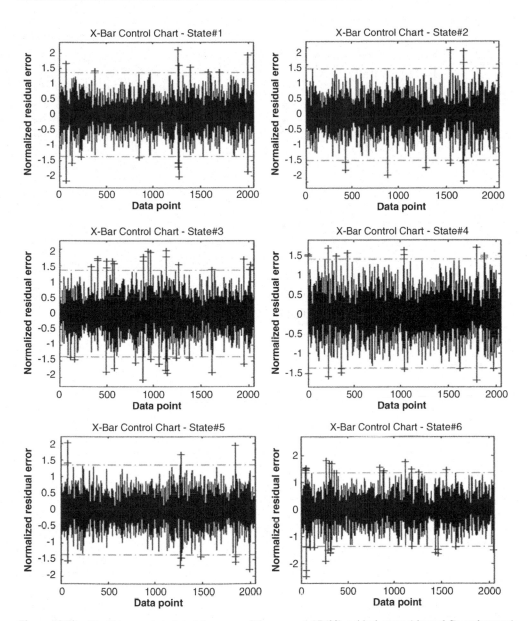

Figure 10.20   The X-bar control plot of the mean of the grouped AR(30) residual errors (channel 5), undamaged states 1 to 9.

number of outliers present in all the undamaged states (states 1 to 9). These states are considered to represent the normal condition, but with simulated environmental or operational variations. The relevant state descriptions are given in Table 10.1. For the AR(5) and AR(30) residuals, this threshold is defined by state 9.

As shown in Figure 10.22, for the residual error data obtained with either the AR(5) or AR(30) models, the number of outliers beyond the control limits seems to increase for the damaged state conditions. These results indicate that some unusual source of variability is present in the damaged states. The threshold

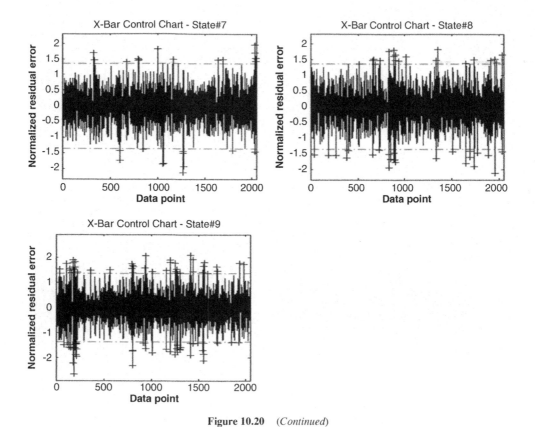

Figure 10.20   (*Continued*)

horizontal dashed line in this figure can be used to examine the relative performance of the AR(5) and AR(30) models. In the case of the AR(30) model, the number of outliers associated with all damaged states is greater than that the ones associated with undamaged states (state 16 is on the transition border). In the case of the AR(5) model, some damaged states associated with the lower damage levels have a number of outliers that are within those found for the undamaged states, such as states 10, 11 and 16.

In summary, the results from this analysis have shown that, in general, a statistically significant number of outliers result when the control limits were determined from the baseline structural condition and then sources of simulated environmental and operation variability were added as indicated in Figure 10.20. If not properly accounted for in the SHM system training phase, these sources of variability could lead to false indications of damage. Setting the outlier threshold as shown in Figure 10.22, which was based on all the data obtained from the undamaged structure, is one approach to accounting for the environmental and operational variability in the damage detection process. In general, the number of outliers beyond the control limits was seen to increase for the damage states relative to the undamaged states even when they are affected by the simulated operational and environmental variations. In terms of the influence of AR model order on the approach, more consistent results were obtained using the grouped AR(30) residual errors. The AR(5) model produced residual errors that were autocorrelated, which violates the basis for formulating the X-bar control chart, and this model was known to be insufficient to accurately model the measured response. These results were shown simply to illustrate the importance of selecting the appropriate model order when performing this type of outlier analysis.

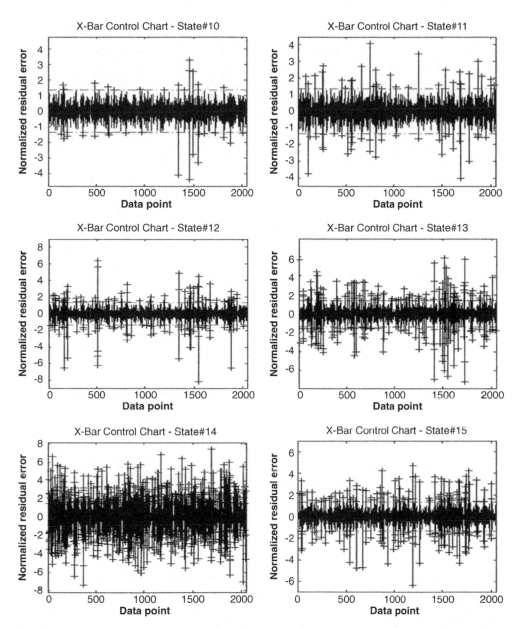

Figure 10.21   The X-bar control plot of the mean of the grouped AR(30) residual errors (channel 5), damaged states 10 to 17.

10.6   Other Control Charts and Multivariate SPC

The control chart described in the previous section is one of the most commonly used options. In the univariate case, where only one time series is monitored, three other control charts are also used, which have proved particularly effective for detecting small changes (Montgomery, 1996). In addition,

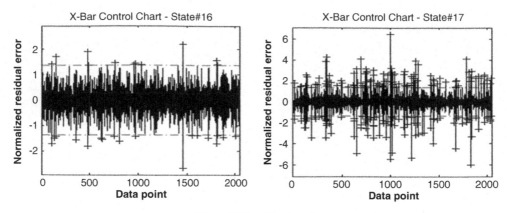

Figure 10.21 (*Continued*)

multivariate control charts have proved very powerful and some of the commoner variants are also summarised here. In many respects, this section follows the informative article of Kullaa (2009).

10.6.1 The S Control Chart

To monitor variability within subgroups and to potentially detect changes in variance, an S control chart can be used. For each subgroup the sample standard deviation of the (normalised) residuals is computed. These sample standard deviations from each subgroup, s_j for $j = 1, \ldots, 2046$, become the charted values. The upper and lower control limits (UCL and LCL) are obtained from (Montgomery, 1996)

$$\text{UCL} = \overline{S}\sqrt{\frac{\chi_{1-\alpha/2,n-1}}{n-1}} \quad \text{and} \quad \text{LCL} = \overline{S}\sqrt{\frac{\chi_{\alpha/2,n-1}}{n-1}} \tag{10.8}$$

where $\chi^2_{p,n}$ denotes the pth quantile for a chi-square random variable with n degrees of freedom and $\overline{S}$ is the average of s_j for $j = 1, \ldots, 2046$. Fugate, Sohn and Farrar (2001) demonstrate the application of an S control chart in the context of SHM where it is applied to data from the concrete column described in Chapter 5.

10.6.2 The CUSUM Chart

Unlike the X and X-bar charts discussed previously, which only use information from the current sample of the time series (be it a measured variable or model residual), the CUSUM, or *cumulative sum (cusum)*, chart accumulates information for the whole series of monitored values. The use of such charts dates back to Page (1954). It is assumed that, when the process is under control, the monitored variable has a Gaussian distribution with mean $\overline{x}$ and standard deviation σ. In the process control community, $\overline{x}$ is sometimes referred to as the *target value*, as the objective will often be to keep the monitored variable as close to some desired value as possible. The *tabular cusum* counts deviations from the mean that are above some control limit in the statistic C^+ and counts excursions below a lower control limit in the statistic C^-. These statistics are charted and incremented using the rules

$$C_i^+ = \max\{0, x_i - (\overline{x} + k\sigma) + C_{i-1}^+\}$$
$$C_i^- = \max\{0, (\overline{x} + k\sigma) - x_i + C_{i-1}^-\} \tag{10.9}$$

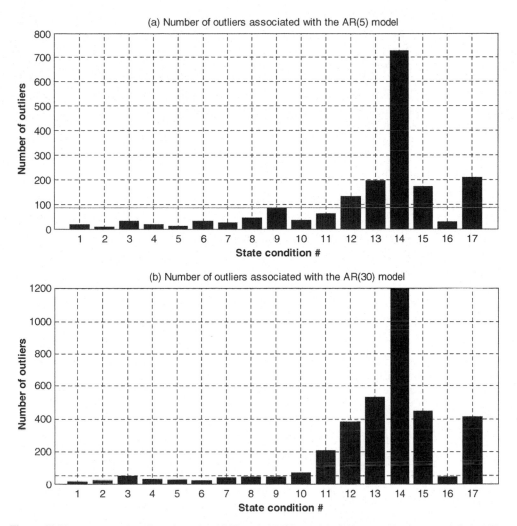

Figure 10.22 Number of outliers from the AR(5) and AR(30) models falling outside the control limits. The horizontal dashed line corresponds to the maximum number of outliers found in the undamaged states (channel 5).

and $C_0^+ = C_0^- = 0$. These statistics are called upper and lower one-sided cusums respectively. The constant k is referred to as the *reference value* and fixes the sensitivity of the diagnostic; it is often chosen to give a control limit halfway between the target value and the out-of-control value that it is desired to flag. If either cusum becomes negative, it is reset to zero. The idea of the cusum is fairly clear; if the process is in control (i.e. Gaussian), it is unlikely to spend extended periods on one particular side of the mean. The process is considered to be out of control, or to show novelty, when one of the cusums exceeds a decision value $H = h\sigma$.

10.6.3   The EWMA Chart

This control chart is the *exponentially weighted moving-average* chart and is essentially an adaptive mean value for the monitored variable, which 'forgets' previous data points at an exponential rate. The chart was introduced in Roberts (1959). If z denotes the EWMA and x is the monitored variable, one has

$$z_i = \lambda x_i + (1 - \lambda)z_{i-1} \tag{10.10}$$

where λ is a 'forgetting factor' chosen between zero and unity; choosing $\lambda = 1$ results in an on-line estimate of the usual mean value. It can be shown that if the monitored values x_i are independent variables with mean $\bar{x}$ and standard deviation σ, then the variance of z_i is

$$\sigma_{z_i}^2 = \sigma^2 \left(\frac{\lambda}{2 - \lambda} \right) \left[1 - (1 - \lambda)^{2i} \right] \tag{10.11}$$

This result leads to definitions for the control limits

$$UCL, LCL = \bar{x} \pm L\sigma \sqrt{\frac{\lambda}{2 - \lambda}[1 - (1 - \lambda)^{2i}]} \tag{10.12}$$

where the design parameter L controls the sensitivity of the chart.

Both of the previous charts have proved popular in the field of *syndromic surveillance* (Lombardo and Buckeridge, 2007). This field is a subfield of general health informatics, which seeks to determine, as quickly as possible, if a threat to public health is emerging, such as a disease epidemic. As one might imagine, the field shares a number of concerns with SHM and the connections are explored in Deering *et al.* (2008).

All the control charts so far are univariate; that is, a single scalar is monitored. As one might imagine, improved sensitivity to damage can be obtained if multiple features are observed at the same time; this leads to *multivariate SPC (MSPC)* control charts.

10.6.4   The Hotelling or Shewhart T^2 Chart

This *Hotelling or Shewhart T^2* chart is the multivariate generalisation of the *X*-bar chart and the *X* chart when the subgroup size $n = 1$. In the latter case, it is essentially the Mahalanobis squared-distance discussed earlier,

$$T^2 = (\{x\}_i - \{\bar{x}\})^{\mathrm{T}}[\Sigma]^{-1}(\{x\}_i - \{\bar{x}\}) \tag{10.13}$$

where $\{x\}_i$ is the current observation, $\{\bar{x}\}$ is the mean of an in-control set of observations and $[\Sigma]$ is the corresponding sample covariance matrix. In the general case when the subgroup size is greater than unity, one obtains

$$T^2 = n(\{\bar{\bar{x}}\}_i - \{\bar{x}\})^{\mathrm{T}}[\Sigma]^{-1}(\{\bar{x}\}_i - \{\bar{\bar{x}}\}) \tag{10.14}$$

where the subgroup average $\{\bar{x}\}_i$ is now the monitored variable and the corresponding mean and co-variances are computed as before from an in-control sample. This statistic is positive semi-definite and therefore one only needs an upper control limit. If the samples are assumed to be drawn from a distribution with a fixed covariance matrix, it is possible to derived an analytical expression for the control limit

$$UCL = \frac{p(m + 1)(n - 1)}{mn - m - p + 1} F_{\alpha, p, mn-m-p+1} \tag{10.15}$$

where p is the dimension of the feature vector or the number of monitored variables, n is the subgroup size, m is the number of subgroups in the in-control sample used to estimate the mean and covariance statistics and $F_{\alpha,p,mn-m-p+1}$ is the α percentage point of the F distribution with p and $mn - m - p + 1$ numbers of degrees of freedom. In the case of the X chart or Mahalanobis squared-distance, the equivalent expression to Equation (10.15) is

$$UCL = \frac{p(m^2 - 1)}{m^2 - mp} F_{\alpha,p,m-p} \tag{10.16}$$

10.6.5   The Multivariate CUSUM Chart

The multivariate analogue of the CUSUM chart was proposed in Crosier (1988); only the basic definitions will be given here. A scalar Y_i is charted, which is defined by

$$Y_i^2 = \{s\}_i^T [\Sigma]^{-1} \{s\}_i \tag{10.17}$$

where

$$\underline{s}_i = \begin{cases} 0, & \text{if } C_i \leq k \\ (\{s\}_{i-1} + \{x\}_i - \{\overline{x}\})(1 - k/C_i), & \text{if } C_i > k \end{cases} \tag{10.18}$$

$$C_i^2 = (\{s\}_{i-1} + \{x\}_i - \{\overline{x}\})^T [\Sigma]^{-1} (\{s\}_{i-1} + \{x\}_i - \{\overline{x}\}) \tag{10.19}$$

and k (greater than zero) is defined in a similar manner as for the CUSUM chart. The iteration is begun with $\{s\}_0 = 0$. An alarm is raised when the charted value $Y_i > h$, where the threshold h depends on the dimension of the feature vector.

10.6.6   The Multivariate EWMA Chart

The multivariate analogue of the EWMA chart was introduced in Lowry et al. (1992). The update rule for the weighted average is directly generalised from the univariate case,

$$\{z\}_i = \lambda\{x\}_i + (1 - \lambda)\{z\}_{i-1} \tag{10.20}$$

The charted quantity is

$$T_i^2 = \{z\}_i^T [\Sigma]^{-1}_{\{z\}_i} \{z\}_i \tag{10.21}$$

where the relevant covariance matrix is obtained from

$$[\Sigma]_{\{z\}_i} = \frac{\lambda}{2 - \lambda}[1 - (1 - \lambda)^{2i}][\Sigma] \tag{10.22}$$

The MEWMA chart raises an alarm when $T_i^2 > h$, where the threshold h depends again on the dimension of the feature vector.

This takes the discussion as far as it needs to go for the immediate purposes of the book; readers wishing to know more about MSPC will find Montgomery (1996) useful.

10.7   Thresholds for Novelty Detection

10.7.1   Extreme Value Statistics

As discussed in the previous sections, one of the critical issues in designing or building a novelty detector is the determination of a detection threshold or alarm level. In the sections on outlier analysis and SPC, it was observed that appropriate thresholds could be established if the feature vector for the normal condition was assumed to have a Gaussian distribution. Unfortunately, as discussed in Section 10.3, the Gaussian assumption may not be justified, particularly when environmental or operational variations in the normal condition come into play. The objective of this section is to show how appropriate thresholds can be obtained for non-Gaussian feature vectors using the extreme value statistics (EVS) discussed in Chapter 6. The discussion will be based on an experimental case study, the structure in question being an ancestor of the simulated building structure described in Chapter 5. The structure was a three-storey one composed of unistrut columns and aluminium floor plates joined by bolted connections, as illustrated in Figure 10.23. The structure was excited horizontally at the base and accelerometers were placed in the vicinity of all the bolted joints in order to record responses. Damage was simulated by loosening bolted connections at various points. More details of the structure can be found in Worden *et al.* (2002).

The signal chosen for analysis was the raw acceleration time series from channel 21, which was associated with an accelerometer on the third floor of the structure (see Figure 10.23); damage in this case was simulated by loosening the bolts in the joint nearest the accelerometer. Data from the highest level of excitation were selected. If the data are plotted in series, data corresponding to the damaged state are observed to have a lower variance and this feature would be picked up immediately by an appropriate hypothesis test (see Chapter 9). Figure 10.24 shows the data from the undamaged condition followed by

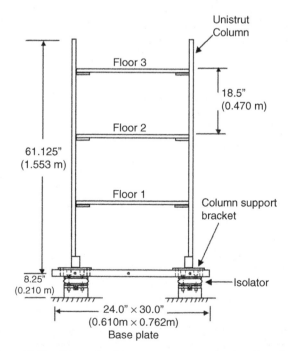

Figure 10.23   Schematic of three-storey structure showing dimensions.

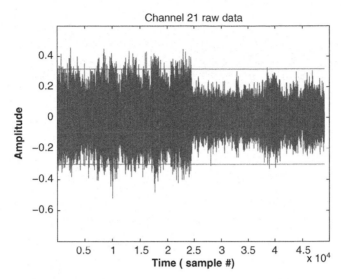

Figure 10.24 Raw time series data from channel 21; undamaged system data immediately followed by damaged system data.

the data from the damaged condition; the two horizontal lines in the figure denote the 99% confidence interval computed from the undamaged data on the basis of a Gaussian assumption. (Note that if one is interested in a simple threshold-crossing criterion, one could not deduce damage from the data as the low variance data clearly falls within the 99% confidence interval of the undamaged data.) Note also that there are many more excursions outside the thresholds than one would normally associate with 99% confidence; this is probably due to the fact that the variables are not, in reality, Gaussian.

The fact that a threshold is not of use here is not directly of concern as the raw time series data would not usually be used in this manner, for one would probably adopt an S-bar control chart. There is also no reason to suppose that, in an SHM context, general damage would produce an increase in the amplitude of the signal, although an increase is often observed in the case of vibration-based condition monitoring. The real object of this exercise is to show that one can construct an effective threshold-crossing diagnostic by using extreme value statistics.

The samples of data for the undamaged conditions were each composed of 4096 points, so a moving window of width 64 samples was stepped through each data set to generate 128 maxima for each condition. When the empirical cumulative distribution function (CDF) was plotted in coordinates appropriate for a Gumbel maximum probability paper (see Section 6.12), the results in Figure 10.25 were obtained. A straight-line fit to the 64 highest-order statistics (see Section 6.12) gave a satisfactory degree of agreement with the data. This fit was interpreted as evidence that the maxima were Gumbel distributed. The apparent bilinear nature of the plot in Figure 10.25 might be regarded as some cause for concern. However, an analysis of the parent distribution indicated that the raw accelerations were very close to Gaussian and it is known that the maxima from Gaussian distributions slowly converge to a Gumbel distribution (Castillo, 1988).

The next stage in the analysis was to estimate parameters for a Gumbel maximum distribution fit to the empirical CDF. The maximum likelihood method was used (again, the reader should refer to Section 6.12) and the parameters obtained were $\lambda = 0.2637$ and $\delta = 0.0373$. The results of this process are shown in Figure 10.26 and an excellent fit to the CDF in the right tail of the distribution is obtained as required.

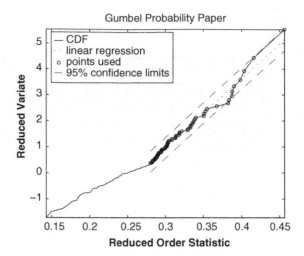

Figure 10.25   Plot of windowed acceleration maxima in Gumbel coordinates.

Having obtained estimates for the parameters, it is a trivial matter to use the CDF to generate values for the 100α percentiles of the distribution to give confidence limits for the data. Using the CDF from Section 6.12, the value of the 100α percentile is given by

$$x = \lambda - \delta \ln(-\ln(\alpha)) \tag{10.23}$$

In this case, the upper and lower thresholds were taken by setting $\alpha = 0.995$ and $\alpha = 0.005$ so that they should enclose 99.5% of the data. Plotting the maxima with these thresholds gave the results in Figure 10.27. The first set of points in the figure comes from the undamaged condition, while the second half comes from the damaged condition.

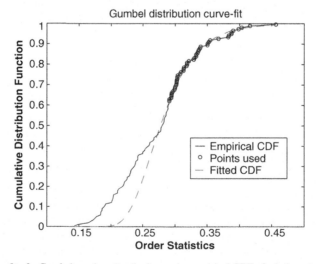

Figure 10.26   Curve-fit of a Gumbel maxima distribution to the empirical CDF of windowed maxima observations.

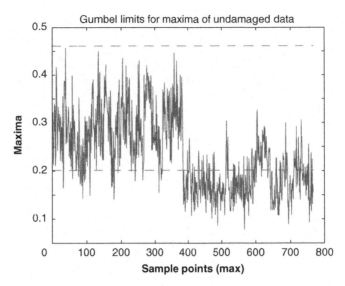

Figure 10.27 Windowed maxima from channel 21 accelerations. Upper and lower thresholds are 99.5 and 0.5 percentiles from the Gumbel distribution for maxima. Note that only the upper threshold is appropriate for novelty detection.

The results in Figure 10.27 show that a good upper threshold appears to have been obtained for the higher values. In the figure there is also a lower threshold that serves to partially separate the undamaged and damaged conditions. It is very important to note that this threshold is in fact *incorrect*, which is indicated clearly from the number of threshold crossings on the normal condition data set. The problem is of course that the lower threshold has also been taken here from the Gumbel distribution for maxima (here the concern is with *maxima of maxima*); the lower threshold should actually be computed from an EV distribution for minima (here the *minima of the subgroup maxima*).

In order to find the appropriate distribution for a lower threshold, the empirical CDF of the data was first plotted in coordinates appropriate for a Gumbel minimum probability paper resulting in a figure that had marked curvature in both tails and thus showed that the Gumbel distribution was inappropriate. An important point here is that a set of data need not have the same limit distribution for both maxima and minima. The next attempt used Weibull minimum coordinates. The location parameter λ was adjusted by trial and error until using a value of $\lambda = 0.12$ gave the plot shown in Figure 10.28. This plot shows that, oddly, the *whole population* of the channel 21 windowed maxima was well described by a Weibull distribution for minima. However, this fact should be regarded as irrelevant and one should only use the threshold derived for the left tail.

That an appropriate distribution had been found was confirmed by the maximum likelihood curve fit shown in Figure 10.29. As one might expect from Figure 10.28, although only points in the left tail were used to fit parameters, the curve describes the empirical CDF over the entire range. The remaining parameters for the distribution were estimated as $\delta = 0.1696$ and $\beta = 2.6976$. Using the whole data for curve fitting gave parameter estimates of $\delta = 0.1720$ and $\beta = 2.7056$. Once the parameters for the distribution were obtained, the thresholds could once again be computed using the inverse CDF, in this case (see Section 6.12)

$$x = \lambda + \delta[-\ln(1 - \alpha)]^{1/\beta} \tag{10.24}$$

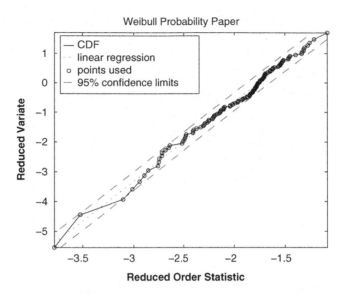

Figure 10.28 Plot of channel 21 windowed maxima in coordinates appropriate for a Weibull minimum distribution.

This parameter estimation gave upper and lower threshold values of 0.4347 and 0.1438, respectively, as shown in Figure 10.30.

As one would expect, the lower threshold appears to be much more consistent with the undamaged data than the previous attempt based on the (spurious) Gumbel maximum distribution shown in Figure 10.27. The important, if not critical, point here is that the lower threshold allows detection of the damage, as many points in the damaged data set cross the threshold. The possibility of improved threshold estimation appears to be the major factor in favour of using EVS for damage identification. Extreme

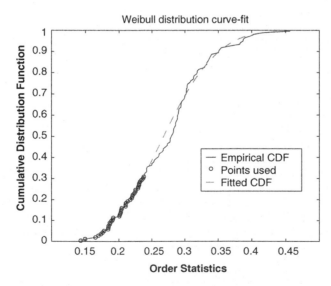

Figure 10.29 Weibull minimum distribution fit to channel 21 windowed maxima.

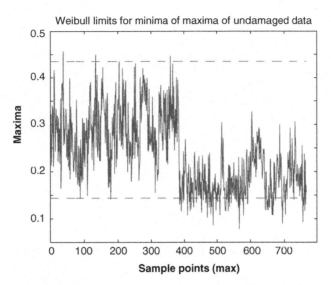

Figure 10.30 Windowed maxima from channel 21 accelerations. Upper and lower thresholds are 99.5 and 0.5 percentiles from the Weibull distribution for minima. Note that only the lower threshold is appropriate for novelty detection.

value statistics allow for much better control of the thresholds that are used to signify novelty. In order to drive the point home, Figure 10.31 shows the windowed acceleration feature with thresholds computed on the assumption that the data are Gaussian. With the Gaussian assumption, the 0.5 and 99.5 percentiles are clearly very conservative; many more of the damaged condition maxima would be judged normal. The Gaussian assumption here makes the analysis equivalent to a univariate outlier analysis, as discussed earlier.

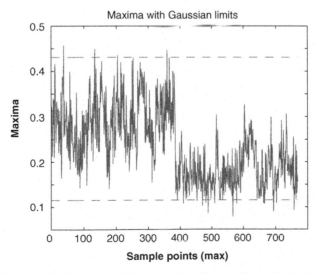

Figure 10.31 Windowed maxima from channel 21 accelerations. Upper and lower thresholds are 99.5 and 0.5 percentiles from the Gaussian distribution.

10.7.2   Type I and Type II Errors: The ROC Curve

The last section showed how one can compute thresholds based on percentiles of the extreme value distributions for feature data. While these percentiles will be much more accurate than those based on assuming a Gaussian distribution for the features, they may still not be completely appropriate for designing an SHM system. The reason for this observation is the fact that type I errors (false positives) and type II errors (false negatives) may well have different costs, as discussed in Chapter 9. The consequences of a false positive in an SHM system may be economically serious in the sense that a perfectly healthy system or structure may be taken out of service and thus cause loss of revenue. However, the consequences of a false negative could include loss of life. Unless the measured features for a given problem completely separate the undamaged and damaged condition, there will be type I and type II errors and the placing of the threshold will mediate their relative frequency. The experimental example of the last section provides a neat illustration of the situation under discussion here. The relevant threshold for discussion, that which separates the undamaged and damaged data, is of course the lower one. Consider the three thresholds:

1. The threshold in Figure 10.27. Setting aside the fact that this threshold was computed spuriously from the Gumbel maximum distribution, it provides an example of a threshold that understates the (lower) limit of the normal condition. The effect of the placing of this threshold is that a great many of the damage states are detected at the cost of a high number of false positives.
2. The threshold in Figure 10.31. This threshold was also computed from an incorrect distribution; however, in this case it is a distribution – the Gaussian – which would often be assumed in the absence of other prior knowledge. The threshold here overstates the extent of the normal condition and thus generates no false positives but very many false negatives.
3. The threshold in Figure 10.30. This is the 'correct' threshold as computed using the appropriate EV distribution. This threshold provides a compromise between the other two, yielding minimal false positives and a high enough number of true positives to confirm the detection of damage.

In the absence of any reason to assign different costs to type I and type II errors, it is clear that the principled approach based on the percentiles of the correct EV distribution is the most effective. However, if the error types have radically different costs, it is likely that one would wish to change the threshold to accommodate this fact. One way of choosing an appropriate threshold in such a case can be based on the ROC curve discussed in Section 9.4.

To illustrate the use of the ROC curves, consider an SHM system that was deployed to detect damage in the drive mechanism of an automated telescope (Stull *et al.*, 2012). The *rapid telescopes for optical response* (RAPTOR) observatory network consists of several ground-based, autonomous, robotic astronomical observatories primarily designed to search for gamma-ray bursts. Currently, the telescopes are maintained in an ad hoc manner, often in a run-to-failure mode. The required maintenance logistics are further complicated by the fact that many of the observatories are situated in remote locations.

An individual RAPTOR 'observatory' consists of an automated enclosure and the telescope system itself. In the context of this study, damage is defined as being any changes in the mechanical characteristics of the telescope mount that reduce, limit or prevent its capacity for capturing gamma-ray burst events. Experience indicates that damage to the telescope mount manifests itself primarily in the drive mechanism that positions the telescope optics. At the heart of this drive mechanism, for both the right ascension (RA) and declination (DEC) axes (see Figure 10.32 (top)), are components referred to as 'capstans', which provide the friction interfaces between the motors that drive the mounts and the drive wheels. The capstan itself is an 8.58-cm-long stainless steel rod, 6.35 mm in diameter, with a urethane coating, which will wear with use, resulting in irregular travel of the drive wheel and the eventual inability to accurately position the wheel. Figure 10.32 (bottom) illustrates a drive motor with a capstan in place, accompanied by three capstans at various levels of wear.

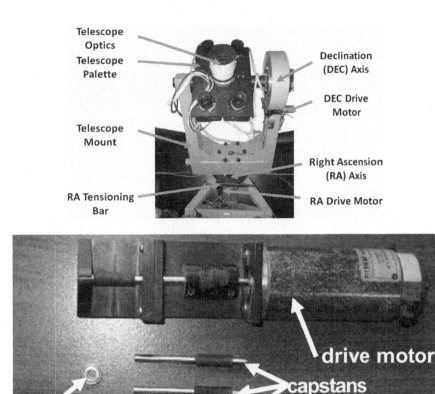

Figure 10.32 Telescope system with mount, drive mechanisms (top); drive motor unit, capstans with various amounts of wear (bottom).

The capstans, which are almost invariably the first component to fail, are often replaced only after an inability to control the telescope optics is observed. Left unchecked, the inevitable wearing away of the urethane coating can cause damage to other, considerably more expensive components (e.g. the telescope optics). Given the variability in both the geospatial locations and the duty cycles associated with these telescope systems, the rates at which the capstans will experience wear is extremely difficult to predict, precluding the a priori development of cost-effective time-based maintenance schedules for individual telescope systems or for the network as a whole. Conversely, replacing the capstans at a conservatively chosen interval without regard to their condition is hampered by the man-hours involved in accessing the oftentimes remotely located systems. While past performance indicates that capstan wear presents an issue for both the RA and DEC axes, the higher inertial loads imparted to the RA axis typically result in higher incidences of damage to this capstan and, therefore, this discussion focuses on results related to the RA axis only.

A single accelerometer (PCB Piezotronics model 352A24) was attached to the tensioning bar (see Figure 10.32), a mechanism that allows for adjustments to be made in the contact force between the RA drive wheel and the capstan. Experimental tests were conducted by recording accelerations during clockwise and counterclockwise rotations of the telescope mount. This procedure was executed a total

Table 10.2   Expert assessments of capstan conditions

Training capstans		Validation capstans	
Capstan number	Assessment	Capstan number	Assessment
1	Undamaged	1	Borderline
2	Undamaged	2	Borderline
3	Undamaged	3	Damaged
4	Damaged	4	Damaged
5	Damaged	5	Undamaged
6	Damaged	6	Undamaged

of 10 times for each of 12 capstans and each time a 4096-point acceleration–time history was recorded at a sampling rate of 640 Hz. The first six capstans, in which the condition is known, were employed in the training of the damage classifier (denoted 'training capstans') and the latter six capstans were used to validate the classifier (denoted 'validation capstans'). An expert's visual assessments of each capstan's condition are summarized in Table 10.2. Note that two of the validation capstans were, in the expert's assessment, borderline between a damaged and undamaged state.

The damage-sensitive features used in this study were the parameters of an order-18 autoregressive (AR(18)) model as described by Equation (10.7) and an outlier analysis was performed using the Mahalanobis squared-distance metric Equation (10.2). For the approach adopted herein, the statistics $\{\bar{x}\}$ and $[\Sigma]$ in Equation (10.2) are estimated from features associated with the training capstans that are known to be undamaged (training capstans 1 to 3). To accomplish the outlier detection process, the acceleration time histories are first divided into equal-size records, each having a length of 256 time points. Subsequently, the AR model parameters are computed for each of these smaller records. In this way, variability exhibited between individual tests of a capstan, and from one capstan to the next, can be assessed. This process yields 160 sets of AR parameters for each capstan.

Figure 10.33 shows the Mahalanobis squared-distances for all the capstans used in this study. In the figure, the log of the squared-distances has been plotted to visualise the results better. The dashed lines shown in this figure are the mean values of the 160 Mahalanobis squared-distances for each capstan. This figure illustrates again that, in general, damaged versus undamaged conditions can be distinguished using the Mahalanobis squared-distance metric. It can be seen that the fifth training capstan exhibits less distinction from the undamaged cases. It does appear, however, that this capstan falls into more of a borderline condition, particularly when compared to the validation data. Another ambiguous result is that related to the fifth validation capstan. In addition to being assessed as undamaged during visual inspection, this particular capstan was brand new when installed into the drive mechanism, yet its outlier statistics indicate a more borderline condition. It is speculated that general classifier definitions, while able to detect whether capstans are damaged or undamaged, may be less able to discern incremental changes in the capstan condition. Thus, it may be that instead of adopting a general classifier definition for the system-wide implementation, individual classifiers that are trained as new capstans are installed and will form the primary mechanism by which *in situ* condition assessments are made.

Figure 10.34 illustrates the PDFs of the Mahalanobis squared-distances associated with each capstan condition, undamaged, damaged and borderline, where the vertical dashed lines correspond to the mean of these three distributions. These density functions were estimated with a Gaussian kernel density estimator of the type discussed in Chapter 6. There is a clear separation between the mean scores for the undamaged and damaged conditions, but there is also considerable overlap between the distributions. Therefore, any choice of a decision threshold will lead to the possibility of false positive and false negative indications of damage. In fact, when the distributions overlap there is no way to simultaneously

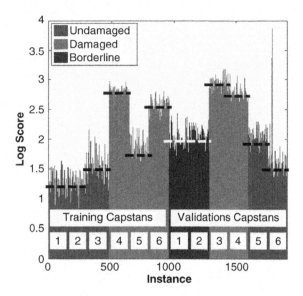

Figure 10.33   The Mahalanobis squared-distance metric for the various training and validation capstans.

minimise both false positive and false negative indications of damage. As discussed in Section 9.4, ROC curves provide a means of assessing the trade-off between false positive and false negative classifications for various decision thresholds.

The ROC curve for the AR(18) model parameter feature vector is shown in Figure 10.35. As summarised in Section 9.4, each point on an ROC curve is defined by the ordered pair, $[(t), \mathrm{TPR}(t)]$, where t is a fixed decision threshold corresponding to the scores associated with the classification method. TRP

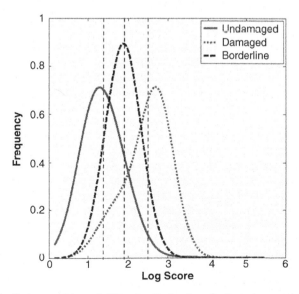

Figure 10.34   Estimates of the probability density functions for the various capstan conditions.

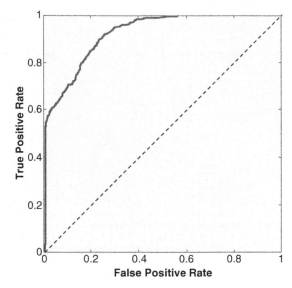

Figure 10.35 The receiver–operator characteristic curve for the AR(18) model parameter features used to identify damage to the capstans.

stands for the 'true positive rate' or 'probability of detection' and is computed by summing the number of damage feature vectors that are classified as damaged and dividing that sum by the total number of damage feature vectors. The FPR, standing for the 'false-positive rate', is then computed by summing the number of undamaged instances classified as damaged and dividing that sum by the total number of undamaged instances. This process is repeated for a range of decision thresholds, resulting in the curve shown in Figure 10.35. As a reference, the dashed line orientated at 45° is representative of a purely random (i.e. coin-flip) decision process. An ROC curve located above and to the left of the 45° line is indicative of good classifier performance with better performance indicated by curves that approach the upper left corner.

Graphically this process can be viewed as starting by placing a vertical line (or decision threshold) at the right tail of the distribution corresponding to the normal condition shown in Figure 10.34. This location corresponds to an FPR value of zero. The corresponding TPR is the area under the distribution corresponding to damage that is to the right of the threshold. The threshold is incrementally shifted to the left and a new FPR value is calculated as the area under the normal condition distribution to the right of the vertical line and the new TPR value is again calculated as the area under the damaged system distribution curve to the right of this new decision threshold. The process is repeated until the entire damaged system distribution lies to the right of the decision boundary.

While improvements could possibly be made through the selection of alternative features, the ROC curves illustrated in Figure 10.35 give evidence that the Mahalanobis squared-distance metric used in conjunction with the AR(18) model parameters offer acceptable performance. The ROC curve provides a quantifiable means of assessing the performance of a classifier for a particular decision threshold. Often the approach taken to defining the appropriate threshold is to define an acceptable rate of false positives and then maximise the true positives or probability of detection. This process is one of the fundamental problems discussed in detection theory (Kay, 1998) and is addressed through the likelihood ratio test defined by the Neyman–Pearson theorem. It is important to note that the ROC curve can only be constructed if examples of features from both the undamaged and damaged systems are available, implying a supervised learning approach. This approach could be taken because previously damaged capstans were available and because the telescope owners were amenable to running the telescopes for short times with these damaged capstans in place.

10.8 Summary

The methods of analysis discussed in this chapter – unsupervised learning methods – are germane to the case where one wishes to carry out damage detection but only possesses training data from the undamaged system or structure of interest. As discussed earlier, this will often be the situation for engineering problems. The disadvantage of such methods is that they do not allow much diagnostic capability beyond simple detection. The next chapter will discuss the situation where data from damaged systems is available and it will be shown that the relevant algorithms in that case yield much more diagnostic information.

References

Basseville, M. and Nikiforov, I.V. (1993) *Detection of Abrupt Changes – Theory and Application*, Prentice-Hall Ltd.

Bishop, C.M. (1998) *Neural Networks for Pattern Recognition*, Oxford University Press.

Box, G.E., Jenkins, G.M. and Reinsel, G.C. (1994) *Time Series Analysis: Forecasting and Control*, Prentice-Hall, Inc.

Crosier, R.B. (1988) Multivariate generalisations of cumulative sum quality control schemes. *Technometrics*, **30**, 291–303.

Deering, S., Manson, G., Worden, K. *et al.* (2008) Syndromic surveillance as a paradigm for SHM data fusion. Proceedings of the 4th European Workshop on Structural Health Monitoring, Krakow.

Figuciredo, E., Figueiras, J., Park, G. and Farrar, C.R. (2011) Influence of model order selection on damage detection. *International Journal of Computer-Aided Civil and Infrastructure Engineering*, **26**, 225–238.

Fugate, M.L., Sohn, H. and Farrar, C.R. (2001) Vibration-based damage detection using statistical process control. *Mechanical Systems and Signal Processing*, **15**, 707–721.

Kay, S.M. (1998). *Fundamentals of Statistical Signal Processing, Volume II: Detection Theory*, Prentice-Hall, Upper Saddle River, New Jersey.

Kullaa, J. (2009) Vibration-based structural health monitoring under variable environmental or operational conditions, in *New Trends in Vibration-Based Structural Health Monitoring* (eds A. Deraemaeker and K. Worden), Springer-Verlag.

Lombardo, J.S. and Buckeridge, D.L. (2007) *Disease Surveillance: A Public Health Informatics Approach*, Wiley Interscience.

Lowry, C.A., Woodall, W.H., Champ, C.W. and Rigdon, S.E. (1992) A multivariate exponentially weighted moving average control chart. *Technometrics*, **34**, 46–53.

Markou, M. and Singh, S. (2003a) Novelty detection: a review. Part 1: statistical approaches. *Signal Processing*, **83**, 2481–2497.

Markou, M. and Singh, S. (2003b) Novelty detection: a review. Part 2: neural network based approaches. *Signal Processing*, **83**, 2499–2521.

Montgomery, D.C. (1996) *Introduction to Statistical Quality Control*, John Wiley & Sons, Inc., New York.

Oakland, J.S. (1999) *Statistical Process Control*, 4th edn, Butterworth-Heineman.

Page, E.S. (1954) Continuous inspection schemes. *Biometrics*, **41**, 100–115.

Pomerleau, D.A. (1994) Input reconstruction reliability information, in *Neural Information Processing Systems 5* (eds S.J. Hanson, J.D. Cowan and C.L. Giles), Morgan Kaufman Publishers.

Poor, H.V. and Hadjiliadis, O. (2009) *Quickest Detection*, Cambridge University Press.

Roberts, S.W. (1959) Control chart tests based on geometric moving averages. *Technometrics*, **42**, 97–102.

Silverman, B.W. (1986) Density Estimation for Statistics and Data Analysis. Chapman and Hall Monographs on Statistics and Applied Probability, 26.

Sohn, H., Allen, D.W., Worden, K. and Farrar, C.R. (2003) Statistical damage classification using sequential probability ratio tests. *Structural Health Monitoring*, **2**, 57–74.

Stull, C.J., Taylor, S.G., Wren, J. *et al.* (2012) Real-time condition assessment of the RAPTOR telescope system. *ASCE Journal of Structural Engineering* (in press), DOI: http://dx.doi.org/10.1061/(ASCE)ST.1943-541X.0000567.

Surace, C. and Worden, K. (1997) Some aspects of novelty detection methods. Proceedings of the 3rd International Conference on Modern Practice in Stress and Vibration Analysis, Dublin, pp. 89–94.

Tarassenko, L., Hayton, P., Cerneaz, N. and Brady, M. (1995) Novelty detection for the identification of masses in mammograms. Proceedings of the 4th IEE International Conference on Artificial Neural Networks, Cambridge, UK, IEE Conference Publication 409, pp. 442–447.

Wetherhill, G.B. and Brown, D.W. (1991) *Statistical Process Control*, Chapman and Hall, London.

Worden, K. (1997) Structural fault detection using a novelty measure. *Journal of Sound and Vibration*, **201**, 85–101.

Worden, K. (2001) Inferential parametrisation using principal curves. Proceedings of the 2nd International Conference on Identification in Engineering Systems, Swansea, UK.

Worden, K., Manson, G. and Allman, D.J. (2003) Experimental validation of structural health monitoring methodology I: novelty detection on a laboratory structure. *Journal of Sound and Vibration*, **259**, 323–343.

Worden, K., Manson, G. and Fieller, N.R.J. (2000) Damage detection using outlier analysis. *Journal of Sound and Vibration*, **229**, 647–667.

Worden, K., Allen, D.W., Sohn, H. and Farrar, C.R. (2002) Extreme Value Statistics for Damage Detection in Mechanical Structures. Los Alamos National Laboratory Report LA-13903-MS.

11

Supervised Learning – Classification and Regression

11.1 Introduction

The last chapter discussed a framework for detecting damage when only data from the normal condition of a structure or system are available for training an algorithm. The framework – novelty detection based on unsupervised learning – is undeniably powerful, but fails to provide any diagnostic capability beyond detection (and occasionally location). These are the lowest levels of capability in Rytter's hierarchy (Rytter, 1993) and in general the SHM practitioner would always wish to have further information. One would always like to be able to locate and size any damage; this information is also critical in progressing to prognostics. As observed in Chapter 9, it is usually rather difficult to obtain data from damaged systems and structures because of cost and practicality constraints. However, when such data are available, a whole new range of algorithms based on supervised learning become possible. The objective of this chapter is to introduce and develop the theory of supervised learning algorithms. It will begin with a discussion of one of the oldest and most reliable learning algorithms, one based on the idea of the *artificial neural network*.

11.2 Artificial Neural Networks

11.2.1 Biological Motivation

One could argue that the discipline of machine learning really began with the development of artificial neural networks (ANNs). The original motivations for the research were mainly: (a) a wish to shed light on the actual learning and computing processes in the human brain and (b) a desire to find and exploit an alternative computing paradigm to the serial (von Neumann) computer. It is arguably the latter motivation that has driven much of recent research and this has been of considerable benefit to the engineering community. The subject is extremely mature now and substantial textbook and monograph accounts exist; excellent examples are Bishop (1998) and Haykin (1994). The current chapter can make no serious attempt at a comprehensive survey of the field of ANNs; however, the discussion is intended to provide the necessary background for applications in SHM.

The foundations for the study of ANNs begin with the pioneering work on neurons as structural constituents of the brain in the 1910s (Ramón y Cajál, 1911). The early work established that the basic processing unit of the brain is the nerve cell or *neuron*, and the structure and operation of such neurons

Structural Health Monitoring: A Machine Learning Perspective, First Edition. Charles R. Farrar and Keith Worden.
© 2013 John Wiley & Sons, Ltd. Published 2013 by John Wiley & Sons, Ltd.

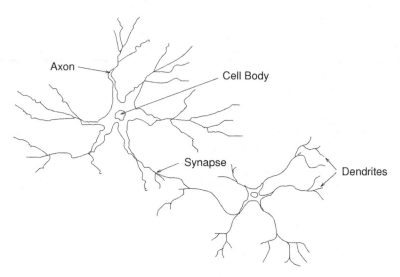

Figure 11.1   The biological neuron.

will be the subject of this section. In brief, the neuron acts by summing stimuli from connected neurons. If the total stimulus or *activation* exceeds a certain threshold, the neuron 'fires'; that is, it generates a stimulus that is passed on into the network. The essential components of the neuron are shown in the schematic of Figure 11.1.

The cell body, which contains the cell nucleus, carries out those biochemical reactions that are necessary for sustained functioning of the neuron. Two main types of neuron are found in the cortex (the part of the brain associated with the higher reasoning capabilities); they are distinguished by the shape of the cell body. The predominant species have a pyramid-shaped body and are usually referred to as *pyramidal* neurons. The majority of the remaining nerve cells have star-shaped bodies and are referred to as *stellate* neurons. The cell bodies are typically a few micrometres (microns) in diameter. Fine tendrils called *dendrites* surround the cell body; they typically branch profusely in the neighbourhood of the cell and extend for a few hundred microns. The *nerve fibre* or *axon* is usually much longer than the dendrites, sometimes extending for up to a metre. These connect the neurons with distant parts of the nervous system via the spinal cord; they are not connections within the brain. The axon only branches at its extremity where it makes connections with other cells.

The dendrites and axon serve to conduct signals to and from the cell body. In general, input signals to the cell are conducted along the dendrites, while the cell output is directed along the axon. Signals propagate along the fibres as electrical impulses. Connections between neurons, called *synapses*, are usually made between axons and dendrites although they can occur between dendrites, between axons and between an axon and a cell body.

Synapses operate as follows: the arrival of an electrical nerve impulse at the end of an axon, say, causes the release of a chemical – a *neurotransmitter* into the synaptic gap (the region of the synapse, typically 0.01 microns). The neurotransmitter then binds itself to specific sites – *neuroreceptors* – usually in the dendrites of the target neuron. There are distinct types of neurotransmitters: *excitatory* transmitters, which trigger the generation of a new electrical impulse at the receptor site, and *inhibitory* transmitters, which act to prevent the generation of new impulses.

Table 11.1 (constructed from data in Abeles, 1991) gives the statistics for and typical properties of neurons within the cerebral cortex (the term *remote sources* in the table refers to sources outside the cortex). One of the first things one sees from the table is that the network is far from fully connected.

Table 11.1 Properties of the cortical neural network

Variable	Value
Neuronal density	40 000/mm^3
Neuronal composition	Pyramidal 75%/Stellate 25%
Axonal length density	3200 m/mm^3
Dendritic length density	400 m/mm^3
Synapses per neuron	20 000
Inhibitory synapses per neuron	2000
Excitatory synapses from remote sources per neuron	9000
Excitatory synapses from local sources per neuron	9000
Dendritic length per neuron	10 mm

The operation of the neuron in reality is not simple as the dynamics are those of a complex electrochemical dynamical system. However, in broad terms, the cell body carries out a sort of *summation* of all the incoming electrical impulses directed inwards along the dendrites. The effectiveness of the connection between two neurons is determined by the chemical environment in the synapse, so the elements of the summation over neuronal connections are individually weighted by the strength of the connection or synapse. If sufficient energy is directed into a neuron within a certain interval of time from its neighbours, it will itself discharge an electrical pulse along the axon. This process can be put into terms that will make the design of the artificial neuron seem clear. If the value of the summation over incoming signals – the *activation* of the neuron – exceeds a certain threshold, the neuron fires and directs an electrical impulse outwards via its axon. From synapses with the axon, the signal is communicated to other neurons. If the activation is less than the threshold, the neuron remains dormant.

A mathematical model of the neuron, exhibiting the essential features of this restricted view of the biological neuron was developed as early as 1943 by McCulloch and Pitts (1943). This model forms the subject of a later section; the remainder of this section is concerned with those properties of the brain that emerge as a result of its massively parallel nature.

11.2.1.1 Memory

Information is actually stored in the brain in the network connectivity and the strengths of the connections or synapses between neurons. In this case, knowledge is stored as a distributed quantity throughout the entire network. The act of retrieving information from such a memory is rather different from that for an electronic computer. In order to access data on a PC, say, the processor is informed of the relevant address in memory, and it retrieves data from that location. In a neural network, a stimulus is presented (i.e. a number of selected neurons receive an external input) and the required data are encoded in the subsequent pattern of neuronal activations. Potentially, recovery of the pattern is dependent on the entire distribution of connection weights or synaptic strengths.

One advantage of this type of memory retrieval system is that it has a much greater resistance to damage. If the surface of a PC hard disk is damaged, all data at the affected locations may be irreversibly corrupted. In a neural network, because the knowledge is encoded in a distributed fashion, local damage to a portion of the network may have little effect on the retrieval of a pattern when a stimulus is applied.

11.2.1.2 Learning

According to the argument in the previous section, knowledge is encoded in the connection strengths between the neurons in the brain. The question arises of how a given distributed representation of data is obtained. One way is that the initial state of the brain at birth is gradually modified as a result of its interaction with the environment. This development is thought to occur as an evolution in the connection strengths between neurons as different patterns of stimuli and appropriate responses are activated in the brain as a result of signals from the sense organs.

The first explanation of such learning in terms of the evolution of synaptic connections – *Hebbian learning* – was given by Hebb (1949):

> When a cell *A* excites cell *B* by its axon and when in a repetitive and persistent manner it participates in the firing of *B*, a process of growth or of changing metabolism takes place in one or both cells such that the effectiveness of *A* in stimulating and impulsing cell *B* is increased with respect to all other cells which can have this effect.

It was considered that, if some similar mechanism could be established for computational models of neural networks, there would be the attractive possibility of 'programming' these systems simply by presenting them with a sequence of stimulus–response pairs so that the network could learn the appropriate relationship by reinforcing some of its internal connections.

11.2.2   The Parallel Processing Paradigm

As observed earlier, it was the massively parallel nature of the brain as a computing facility that led researchers to believe that it could motivate a new and powerful paradigm for artificial computing. It was assumed that the most important elements of this paradigm would include (Haykin, 1994):

- Nonlinearity. Neurons are nonlinear devices and therefore neural networks are also nonlinear. Moreover, the nonlinearity can be distributed throughout the network according to its structure. ANNs are usually designed to be nonlinear.
- Adaptivity and learning. ANNs can modify their behaviour in response to the environment. Once a set of inputs with desired outputs is presented to the network, the connections between the neurons self-adjust to produce consistent responses. This training is repeated for many examples until the network reaches a stable state.
- Evidential response. A neural network can be designed to provide information not only about the response but also about the confidence in the response; examples include pattern recognition analysis where patterns can be classified with a confidence level.
- Fault tolerance. If a neuron or its connections are damaged, processing of information is impaired. However, a neural network can exhibit a graceful degradation in performance rather than total failure.
- Very large scale integration (VLSI) implementation. The parallel nature of neural networks make them ideally suited for implementation using microchip technology.

The power of ANNs in terms of SHM research is in their suitability for sensor data processing problems that require parallelism and optimisation due to high dimensionality of the problem space and complex interactions between the analysed variables. Broadly speaking, ANNs can offer solutions to four different problems:

1. Autoassociation. A signal is reconstructed from noisy or incomplete data.
2. Regression/heteroassociation. Input–output mapping – that is, for a given input data produce a required output characteristic.
3. Classification. Assign input data to given classes.
4. Novelty detection. Detect statistical abnormalities in the input data.

The first three tasks are often associated with modelling applications using neural networks. The application of the second and third tasks will be discussed extensively in the remainder of this chapter. The final category includes the problem of *novelty detection*, which was discussed extensively in the last chapter. There also exists a number of general data fusion and signal processing applications within each group of these tasks. In general, ANNs can be used for (Luo and Unbehauen, 1998): filtering, detection, reconstruction, array signal processing, system identification, signal compression and adaptive feature extraction. Case studies will be used later in this section to illustrate some of the main uses of ANNs as they apply to SHM. Before discussing some of the most commonly used variants of ANNs, it is useful to spend a little time discussing how a single neuron might be modelled in a mathematical way.

11.2.3   The Artificial Neuron

Despite the diversity of artificial neural network paradigms, nearly all consist of very similar building blocks – the artificial neurons. The structure of these neurons has changed very little since the first study by McCulloch and Pitts (1943). The neurons receive a set of inputs or stimuli (regarded as emerging from neighbouring neurons) and produce a single output or response. In the McCulloch–Pitts (MCP) model, both the inputs and outputs are considered to be binary (reflecting the fact that biological neurons either fire or do not fire). The MCP neuron structure is considered to consist of two blocks concerned with summation and activation, as shown in Figure 11.2.

The input values $x_i \in 0, 1$ are weighted by a factor w_i before they are passed to the body of the neuron. The weighted inputs are then summed to produce an *activation* signal z,

$$z = \sum_{i-1}^{n} w_i x_i \tag{11.1}$$

This signal is then passed through a nonlinear *activation* or *transfer function*. Any one of a number of functions could be used for processing. For example, if the output signal y is described as

$$y = kz \tag{11.2}$$

where k is a constant coefficient, the neuron is linear and consequently networks made up from these types of neurons would be linear networks. (Such neurons have considerable limitations; however, in the literature, classes of linear neurons and networks have found some use and are called *Adeline* and *Madeline* (Widrow and Hoff, 1960), respectively.)

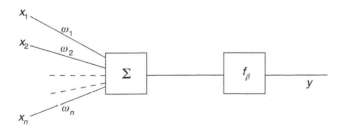

Figure 11.2   The McCulloch–Pitts neuron. In this diagram, signals flow from left to right.

The *nonlinear* activation function used in the MCP model is a hard threshold function. The MCP neuron fires (i.e. produces a nonzero output) if the weighted sum z exceeds some predefined threshold β, that is if

$$z > \beta \tag{11.3}$$

and does not fire if

$$z \leq \beta \tag{11.4}$$

Initial studies of the MCP neuron indicated that its computational capabilities were extremely limited if one were used alone. This result was no surprise; a brain with a single neuron would not be expected to work in any way. The solution was to move to networks of neurons and the first serious learning machine designed on this basis made its appearance in 1950 in the form of the *perceptron*.

11.2.4   The Perceptron

The first serious study of artificial neural *networks* (as opposed to single neurons) was carried out by Rosenblatt (1962), who proposed a three-layered network structure – the *perceptron* (Figure 11.3). This network was not homogeneous; the first layer was an input layer that simply distributed signals to the processing layers. The first processing layer was the *associative* layer (also referred to as the *hidden* layer because it had no external connections); the second, which output signals to the outside world, was termed the *decision* layer. In the classic form of the perceptron, only the connections between the decision and associative nodes were adjustable in strength; those between the input and associative nodes had to be preset before training took place. One of the main motivations of the perceptron was that it might take inputs, black and white, from an image in order to recognise patterns within that image; for this reason, it was sometimes considered to be an artificial retina. It was possible to prove a number of nice theorems for the perceptron, including a proof that if a given pattern were learnable, a learning rule existed that would converge in finite time. The problems proved to be related to discovering which patterns were learnable.

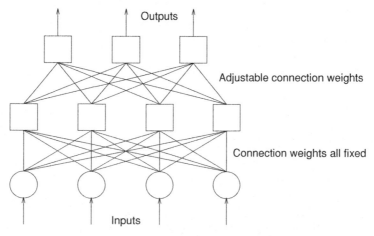

Figure 11.3   Structure of the Rosenblatt perceptron.

Although perceptrons were initially received with enthusiasm, they were soon associated with many problems; a completely rigorous investigation into the capability of perceptrons was given in Minsky and Papert (1969: later edition 1988). In representing a function with N arguments, the generic perceptron was shown to need 2^N elements in the associative layer; that is, the networks appeared to grow exponentially in complexity with the dimension of the problem. It was initially hoped that real problems could be solved that did not require the maximal structure, but Minsky and Papert (1988) showed that many fundamental problems required full complexity. For example, it was shown that a perceptron could not determine if a given pattern was connected (not falling into disjoint components) without the full number of neurons. One way out of the dilemma appeared to be the possibility of adding and training other layers of weights in the perceptron; however, it was not possible to establish an algorithm that would allow this. Only the layer of weights between the outermost hidden layer and the output layer could be trained using the perceptron learning rule (which was based on Hebbian learning as described above). With the perceptron structure as it stood, further progress proved impossible. The effect of Minsky and Papert's book was to discourage research in ANNs and the field lay dormant for many years.

The period of inactivity ended with the work of Hopfield (1984). He considered the network from the point of view of dynamical systems theory; the outputs of the constituent neurons in Hopfield's networks were regarded as dynamical states that could evolve in time. The Hopfield network proved capable of solving a number of practical problems (many of them optimisation problems) and reinvigorated ANN research. An immediate result of the resurgence in activity was the solution by various groups of the problems associated with Rosenblatt-type perceptrons. The problem of finding a learning rule for the multilayer structures turned out to be the result of using the hard threshold as an activation function in the individual neurons. The solution proved to hinge on a matter as simple as replacing the hard threshold with a continuous function such as the sigmoidal function

$$y = \frac{1}{1 + e^{-z}} \tag{11.5}$$

or hyperbolic tangent function

$$y = \tanh(z) \tag{11.6}$$

Once the activation function became continuous, the solution to the whole problem turned out to centre on the chain rule of partial differentiation. The *backpropagation* learning rule was discovered simultaneously by a number of research groups and was reported in Rumelhart, Hinton and Williams (1986) and LeCun (1986). In fact, the learning rule had been discovered as early as 1974 in the PhD work of Paul Werbos (1974); however, this work lay undiscovered by the machine learning community, partly because of the lack of activity in the field as a result of the disappointment in perceptrons. The backpropagation rule is essentially the gradient descent algorithm for optimisation and in this sense appears to have had antecedents in the control engineering literature as early as the 1960s (Bryson, Denham and Dreyfuss, 1963). The existence of the backpropagation algorithm allowed the development of the multilayer perceptron (MLP) algorithm, which has proved to be one of the most commonly used and influential machine learning paradigms so far discovered. Interest in ANNs flowered and a number of large programmes of research were initiated.

11.2.5   *The Multilayer Perceptron*

The multilayer perceptron (MLP) network is a natural generalisation of the perceptrons described in the last section. The main references for this material are Bishop (1998) and Haykin (1994); only a brief discussion is given here for the sake of completeness.

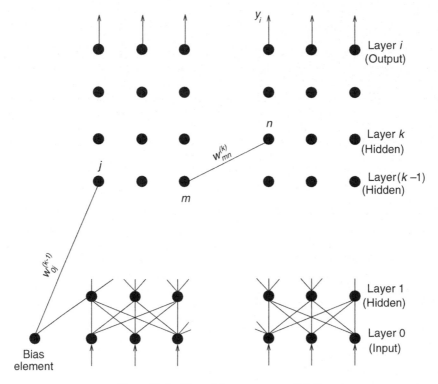

Figure 11.4   The multilayer perceptron (MLP).

The MLP is a feedforward network with the neurons arranged in layers (Figure 11.4). Signal values pass into the *input layer* nodes, progress forward through the network *hidden layers*, and the result finally emerges from the *output layer*. Each node i is connected to each node j in the preceding and following layers through a connection of weight w_{ij}. Signals pass through the nodes as follows: in layer k a weighted sum is performed at each node i of all the signals $x_j^{(k-1)}$ from the preceding layer $k-1$, giving the excitation $z_i^{(k)}$ of the node; this is then passed through a nonlinear *activation function* f to emerge as the output of the node $x_i^{(k)}$ to the next layer, that is

$$x_i^{(k)} = f\left(z_i^{(k)}\right) = f\left(\sum_j w_{ij}^{(k)} x_j^{(k-1)}\right) \tag{11.7}$$

As discussed earlier, various choices for the function f are possible (as long as they are continuous and satisfy some other mild conditions); the hyperbolic tangent function $f(x) = \tanh(x)$ is a good choice. A novel feature of this network is that the neuron outputs can take any value in the interval $[-1, 1]$. There are also no explicit threshold values associated with the neurons. One node of the network, the *bias* node, is special in that it is connected to all other nodes in the hidden and output layers; the output of the bias node is held fixed throughout in order to allow constant offsets in the excitations z_i of each node.

The first stage of using a network is to establish the appropriate values for the connection weights w_{ij}, that is the *training* phase. The type of training usually used is a form of *supervised* learning and makes use of a set of network inputs for which the desired network outputs are known. At each training step, a set of inputs is passed forward through the network yielding trial outputs that can be compared with the desired outputs. If the comparison error is considered small enough, the weights are not adjusted. If, however, a significant error is obtained, the error is passed *backwards* through the net and the training algorithm uses the error to adjust the connection weights so that the error is reduced. The learning algorithm used is usually referred to as the *backpropagation* algorithm and can be summarised as follows. For each presentation of a training set, a measure J of the network error is evaluated where

$$J(t) = \frac{1}{2} \sum_{i=1}^{n^{(l)}} (y_i(t) - \hat{y}_i(t))^2 \qquad (11.8)$$

and $n^{(l)}$ is the number of output layer nodes. J is implicitly a function of the network parameters $J = J(\{w\})$ where the parameters $\{w\}$ are *all* the connection weights, ordered into a vector in some appropriate manner. The integer t labels the presentation order of the training sets. After presentation of a training set, the standard steepest-descent algorithm requires an adjustment of the parameters according to

$$\Delta w_i = -\eta \frac{\partial J}{\partial w_i} = -\eta \nabla_i J \qquad (11.9)$$

Where ∇_i is the gradient operator in the parameter space. The parameter η determines how large a step is made in the direction of steepest descent and therefore how quickly the optimum parameters are obtained. For this reason η is called the *learning coefficient*. Detailed analysis (Haykin, 1994) gives the update rule after the presentation of a training set,

$$w_{ij}^{(m)}(t) = w_{ij}^{(m)}(t-1) + \eta \delta_i^{(m)}(t) x_j^{(m-1)}(t) \qquad (11.10)$$

where $\delta_i^{(m)}$ is the error in the output of the ith node in layer m. This error is not known a priori but must be constructed from the known errors $\delta_i^{(l)} = y_i - \hat{y}_i$ at the output layer l. This process is the source of the name *backpropagation*; the weights must be adjusted layer by layer, moving backwards from the output layer.

There is little guidance in the literature as to what the learning coefficient η should be. If it is taken too small, convergence to the correct parameters may take an extremely long time. However, if η is made large, learning is much more rapid but the parameters may diverge or oscillate. One way around this problem is to introduce a *momentum* term into the update rule so that previous updates persist for a while, that is,

$$\Delta w_{ij}^{(m)}(t) = \eta \delta_i^{(m)}(t) x_j^{(m-1)}(t) + \alpha \Delta w_{ij}^{(m)}(t-1) \qquad (11.11)$$

Where α is termed the *momentum coefficient*. The effect of this additional term is to damp out high-frequency variations in the backpropagated error signal. This learning algorithm is termed *first order* because the first derivative of the objective function is used. The current state-of-the-art for training MLPs uses second-order algorithms that require evaluation of the second derivatives of the objective function. These algorithms are computationally more expensive on an iteration by iteration basis, but converge (much) faster and have better properties with respect to trapping in local minima. The two most often-used variants of the second-order approach are the *scaled conjugate gradient* approach (Moller, 1993) and the Levenburg–Marquardt approach (Press *et al.*, 1992). The training of neural networks raises a number of issues that should be discussed.

11.2.5.1 Existence of Solutions

Before advocating the use of neural networks in representing functions and processes, it is important to establish what they are capable of. As described above, artificial neural networks were all but abandoned as a subject of study following Minsky and Papert's book (1988), which showed that perceptrons were incapable of modelling very simple logical functions. In fact, recent years have seen a number of rigorous results (Cybenko, 1989, is a good example), which show that an MLP network is capable of approximating any given function with arbitrary accuracy, even if possessed of only a single hidden layer. Unfortunately, the proofs are not constructive and offer no guidelines as to the complexity of the network required for a given function. A single hidden layer may be sufficient but might require many more neurons than if two hidden layers were used.

11.2.5.2 Uniqueness of Solutions

This is the problem of local minima. The error function for an MLP network is an extremely complex object. Given a converged MLP network, there is no way of establishing if it has arrived at the global minimum. Some attempts to avoid the problem are centred around the association of a temperature with the learning schedule. Roughly speaking, at each training cycle the network may randomly be given enough 'energy' to escape from a local minimum. The probable energy is calculated from a network temperature function that decreases with time. (Recall that molecules of a solid at high temperature escape the energy minimum that specifies their position in the lattice.) An alternative approach is to seek network paradigms with less severe problems, for example radial-basis function networks (Broomhead and Lowe, 1988).

11.2.5.3 Generalisation and Regularisation

One of the main problems encountered in practice is that of *generalisation* or avoidance of *overfitting*. This problem is essentially one of rote learning the training data rather than learning the underlying function of interest. It occurs when there are too many parameters in the model compared to the number of training points or patterns. Consider a simple one-dimensional regression problem with 10 points of training data. Suppose that the true relationship between x and y is linear, but the presence of noise means that the plot of y against x is far from a straight line. If one were to fit a nine-dimensional least-squares polynomial to the data, there would be 10 tunable parameters that could be set so that the function passes through the data with no error. The problem here is that the polynomial fit would very probably deviate badly from the true linear form away from the training points. If one now applied the model to different data, generated by the same physical process as the training data, the predictions could be very bad indeed and the model would fail to generalise. At the heart of this problem is the availability of too many parameters. In the context of a neural network, one is likely to *overfit* the data if there are too many weights compared to training data points. The simplest solution to the problem is to always have enough data; the rule-of-thumb espoused in Tarassenko (1998) is that one should have 10 training patterns for each network weight (this rule is not arbitrary and some theoretical motivation is given in Bishop, 1998). In SHM problems, data are often the result of expensive experimentation and will be in short supply; in this case, the only way to ensure generalisation is to restrict the number of weights in the network. In any case, if one has fitted a neural network to training data, one should always evaluate performance on an independent test set in order to assess generalisation. It should also be said that there are situations where data are available in large quantities for SHM analysis, for example in the continuous monitoring of some bridges.

One way of controlling the number of adjustable weights in the network is to control the number of hidden units; this is accomplished by the use of *cross-validation* on an independent data set. One divides the available data into three sets for *training*, *validation* and *testing*. For all numbers of hidden units between 1 and some maximum, one trains a network on the training data and then also computes the error on the validation data. When the number of hidden units has reached the point where overtraining is beginning, the error on the validation set will begin to rise even though the error on the training set continues to decrease. One then fixes the number of hidden units at the point where the minimum error on the validation set occurred. Now, the network has been tuned to *both* the training and validation sets and the independent testing set is brought in to assess generalisation.

Another way of thinking about overfitting is in terms of the magnitude of the weights. In situations where there are as many weights as data points, one often finds that the weights have very high values and the accurate predictions on the training set are the result of cancellations between large positive and large negative weights. This suggests that better generalisation can be achieved by controlling the size of the weights. Alternatively, one can regard the issue as being one of *smoothness* of the fitted function. A high-order polynomial model as discussed before can dance rapidly between noisy data points, where the true underlying linear system response is much smoother. It can be shown that smaller weights generate a smoother response. The science of controlling the smoothness of the network is generally called *regularisation*. Having established that smaller weights are desirable, one simple way to achieve this is to add a term to the neural network objective/error function that penalises large weights, the simplest such term being

$$\sum_{i=1}^{W} w_i^2 = ||\{w\}||^2 \tag{11.12}$$

This prescription is commonly called *weight decay* regularisation. Two other methods of regularisation are *early stopping*, where one stops training before the algorithm has tuned the weights to the point of overfitting, and adding noise to the data during training in order to stop the algorithm learning exact training data points. It has been shown that these three methods of regularisation are theoretically equivalent (Bishop, 1994). One of the most advanced theoretical frameworks for assessing generalisation capacity of models is that of *statistical learning theory* (Vapnik, 1995); aspects of the theory will be discussed later in Section 11.5.

11.2.5.4   Choice of Output Activations Functions

This is a straightforward but important matter concerned with the nonlinear activation functions of the neurons in the MLP. In order to have a universal approximator, it is necessary that the hidden units of the MLP adopt a nonlinear activation function. As discussed above, the hyperbolic tangent and sigmoid functions are those most often used as activation functions. One also needs to specify a form for the activation functions of the output layer and it turns out that best practice demands that this is problem specific (Bishop, 1998). For regression problems, one should use linear activation functions; for classification problems, one should use a nonlinear activation. Furthermore, if the classifier is trained using what is called the *1 of M* rule (which will be discussed in the next section), a *softmax* activation should be used. In this case, there will be M output neurons, each with an associated activation $z_i, i = 1, \ldots, M$. The output of the ith neuron is defined as

$$y_i = \frac{\exp(z_i)}{\sum_{j=1}^{M} \exp(z_i)} \tag{11.13}$$

This rule forces all the network outputs to sum to unity, a necessary condition for interpreting the outputs as probabilities of class membership.

11.3   A Neural Network Case Study: A Classification Problem

As discussed in the introduction to this chapter, two of the main problems in machine learning are classification and regression. This section will illustrate how the MLP neural network provides a powerful means of building a classifier in the context of an SHM problem.

The problem that forms the basis of this case study is that of locating damage in the aircraft wing discussed as one of the main benchmark structures of this book in Section 5.6. In some ways, this should be regarded as a regression problem where one would like to map the measured features into the actual spatial coordinates of damage. However, in the study here, it was not allowed to truly damage the aircraft, so *damage* was simulated by removing inspection panels one at a time. As there were only nine panels considered, the problem is reduced to a discrete classification problem. As in all machine learning problems, choice of features proved critical. In previous work on the Gnat structure, which was restricted to damage detection only, it proved very useful to use features based on measured *transmissibilities* from the structure (Manson, Worden and Allman, 2003a, 2003b). As discussed in the last chapter, these quantities are simply the ratios of acceleration spectra from the responses at two points on the structure. This frequency domain information was further refined by reducing each transmissibility feature to a single scalar *novelty index*, as described below. The details of the construction of the novelty indices used ideas from the statistical discipline of *outlier analysis*, as described in Chapter 10. The location problem required a number of such novelty indices covering the wing in such a way that the patterns of novelty values could be used to indicate the location of damage when they were used to train a classifier.

Figure 5.30 shows a schematic of the wing and the panels that were removed in order to simulate damage. As discussed there, the area of the panels varied from about 0.008 m^2 to 0.08 m^2 with panels P3 and P6 the smallest. Transmissibilities were again used and were recorded in three groups, A, B and C, as shown in Figure 5.31. Each group consisted of four sensors (a centrally placed reference transducer and three others). Only the transmissibilities directly across the plates were measured in this study. The training data for the classifier were obtained from a programme of tests in which each panel was removed twice (in order to check repeatability).

The full details of feature selection were rather involved; the details can be found in Manson, Worden and Allman (2003b). The main steps were as follows:

1. The transmissibilities were examined visually in order to find frequency ranges over which the functions separated the undamaged and damaged conditions. In order to simplify matters, only the group A transmissilibilities were considered to construct features for detecting the removal of the group A panels, and similarly for groups B and C. At this stage, the features were classified as *weak*, *fair* or *strong* using visual criteria. In total, 44 feature vectors were judged to be either fair or strong. To recap, these features are vectors of transmissibility line magnitudes over small regions of frequency; the number of lines in each case varied from 10 to 50. These features were denoted the *primary* features for the problem.
2. The fair and strong features were then evaluated using outlier analysis. The best features were chosen according to their ability to identify correctly the 200 (per panel) damage condition features as outliers while correctly classifying those features corresponding to the undamaged condition as inliers.

At this stage the actual novelty indices associated with each of the primary features were saved to form the set of scalar *secondary* features. Note that these secondary features already served a useful purpose in that they were solutions to the damage *detection* problem.

3. Using the results of the outlier analysis, nine final secondary features were chosen from the 44 candidates. An attempt was made to select one feature per panel that was sensitive to the removal of that panel and no others. Finding such a feature was not possible, the main problems being associated with panels 3 and 6, the smallest of the set. However, the inability to find such a feature was not considered as a cause for concern as the objective of the exercise was to interpret the secondary features using the neural network.

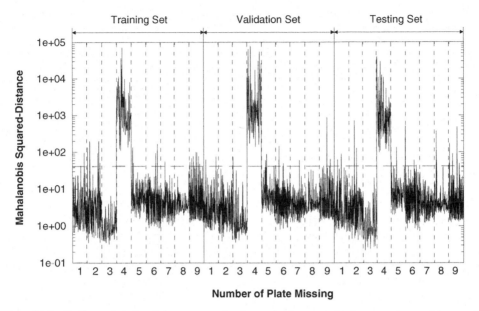

Figure 11.5   Outlier statistic for all damage states for the novelty detector trained to recognise panel 4 removal.

Next, the data were divided into training, validation and testing sets in anticipation of presentation to the neural network classifier. As there were 200 patterns for each damage class, the total number of patterns was 1800. These were divided evenly between the training, validation and testing sets, so (with a little wastage) each set received 594 patterns, comprising 66 representatives of each damage class. The plot in Figure 11.5 shows one of the more successful secondary features that was chosen to represent the removal of panel 4. The plot shows the discordancy values returned by the novelty detector over the whole set of damage states separated into training, validation and testing sets. The horizontal dashed lines in the figures are the thresholds for 99% confidence in identifying an outlier, they were calculated according to the Monte Carlo scheme described in Section 10.2. The novelty detector substantially fires only for the removal of panel 4, for which it has been trained.

The final stage of the analysis was to produce the damage location classifier. The algorithm chosen was a standard multilayer perceptron (MLP) neural network as described in the last section. The neural network was presented with nine novelty indices at the input layer and was required to predict the damage class at the output layer. The procedure for training the neural network followed the guidelines in Tarassenko (1998), in that the data were divided into a training set, a validation set and a testing set. The training set was used to establish weights, while the network structure and training time and so on were optimised using the validation set. The testing set was then presented to this optimised network to arrive at a final classification error. For the network structure, the input layer necessarily had nine neurons, one for each novelty index, and the output layer had nine nodes, one for each class. A single hidden layer was assumed, as it is known that such networks are universal approximators (Cybenko, 1989).

In terms of a *pseudo-code*, the training strategy was:

for number of hidden layer neurons = 1 to 50

for different random initial conditions = 1 to 10

train network on training data

evaluate on validation data

terminate training at minimum in validation set error

Table 11.2   Confusion matrix from the best neural network on the Gnat aircraft location problem: testing set

Predicted class	1	2	3	4	5	6	7	8	9
True class 1	62	1	0	0	2	0	0	1	0
True class 2	0	61	0	0	5	0	0	0	0
True class 3	0	1	52	0	7	4	0	2	0
True class 4	1	0	3	60	0	1	0	1	0
True class 5	2	1	0	0	60	3	0	0	0
True class 6	2	0	6	0	8	52	0	0	0
True class 7	1	0	4	0	1	1	58	1	0
True class 8	0	0	0	0	1	1	0	62	2
True class 9	2	1	1	0	0	0	0	15	47

The training phase used Tarassenko's implementation of the *1 of M strategy*. This strategy is very important in the training of neural network classifiers as it allows contact with statistical pattern recognition (SPR) on an important point (see Chapter 9). In SPR, the best strategy for a classifier is to establish the posterior probability of the damage class C_i, conditioned on the measured features $\{x\}$. The posterior probability for each class is denoted $P(C_i|\{x\})$. It can be shown that an optimal strategy for class assignment is to choose the class with the highest posterior probability given the features (see Chapter 9). The important point here is that MLP neural classifiers can estimate the posterior probabilities if trained appropriately. The means of accomplishing this training is called the *1 of M strategy* and is quite straightforward in principle. When designing the MLP classifier, one assigns one network output per class. During training, if a feature vector corresponding to class i is presented, the desired outputs are set to zero except for the output i associated with the true class, which is set to unity. It can be shown that, after training when the network is presented with new data, it will not actually predict ones and zeros, but actual posterior probabilities at each output. This result is remarkable; the proof can be found in Bishop (1998).

The best network for the problem discussed here had 10 hidden units and resulted in a testing classification error of 0.135; that is, 86.5% of the patterns were classified correctly. The *confusion matrix* is given in Table 11.2. The confusion matrix is an excellent way of summarising classifier performance. The matrix is populated as follows: it is initially set to zero throughout; for each data point of class i that is labelled as class j, the (i, j)th entry of the matrix is incremented. It is clear now that a confusion matrix for a perfect classifier will be zero everywhere except on the diagonals. In the current case, the main errors were associated with the two small panels P3 and P6 and the panels P8 and P9 whose novelty detectors sometimes fired when either of the two panels was removed.

In a later section, it will be shown how a more principled approach to feature selection leads to much better classification results on this problem.

11.4   Other Neural Network Structures

Although the MLP is possibly the most commonly used ANN, there are many other paradigms and structures. In general terms, ANNs can be classified into the following categories.

11.4.1   Feedforward Networks

Signals are passed through the network in only one direction; there are always connections between the neurons in adjacent layers and there may or may not be more far-reaching connections. Feedforward networks have no memory; their output is solely determined by the current inputs and the values of the weights. Apart from the MLP, another very popular feedforward network is the *radial basis function* network (Broomhead and Lowe, 1988).

11.4.2   *Recurrent Networks*

Feedback connections are possible between network elements. Some recurrent networks recirculate outputs back to the inputs. Since their output is determined both by their current input and their previous outputs, recurrent networks have internal dynamics and can exhibit properties similar to short-term memory in the human brain. Training of recurrent networks is usually more complicated than for 'static' networks (no feedback) and stability issues can arise (Haykin, 1994). Recurrent variations on the MLP have proved popular, as have the so-called *Elman networks* (Elman, 1990).

11.4.3   *Cellular Networks*

A basic cellular network has a two-dimensional structure. The processing units – called cells – are connected only to their neighbouring cells. Cells contain linear and nonlinear processing elements. Although only adjacent cells can interact directly with each other, the cells not directly connected can still affect each other due to a propagation effect. Cellular networks resemble structures found in cellular automata. The most well-known example of such a network is the *Kohonen* network or *self-organising map* (SOM) (Kohonen, 2000).

11.5   **Statistical Learning Theory and Kernel Methods**

11.5.1   *Structural Risk Minimisation*

The material of this section and the next has been discussed in considerable detail in textbooks – notably Cherkassky and Mulier (1998) and Cristianini and Shawe-Taylor (2000) are followed here; the review paper by Burges (1998) is also a valuable source of information.

The basic principle of *statistical learning theory* (SLT) essentially coincides with the aims of machine learning in general in that an algorithm is defined that *learns* a relationship between two sets of data. The first is defined on a d-dimensional *input space* and is denoted $\{x\}$ and the second is taken (without much loss of generality) as one-dimensional and denoted y. As always, the relationship is induced by or constructed on the basis of training data $(\{x\}_k, y_k)$; $k = 1, \ldots, N$. If y is a continuous variable, the problem is one of regression and if y is a class label, the problem is one of classification. The mapping between the two will be denoted f.

The theoretical basis of SLT stems from a very careful consideration of the learning problem in terms of the minimisation of some objective (error) function. Suppose a relationship $y = f(\{x\}, \{w\})$ is learned from the data where $\{w\}$ (analogous to the neural network weights) is a vector of free parameters that are adjusted to give a best-fit of the model. The expected error in predicting the function is denoted by

$$R(\{w\}) = \int |y - f(\{x\}, \{w\})| p(\{x\}, y) \mathrm{d}\{x\} \mathrm{d}y \tag{11.14}$$

and is termed the *actual risk*. This function measures the acceptability of the model; unfortunately, it is impossible to compute. First, only a finite set of data will usually be available for training or testing.

Second, the joint density of the inputs and outputs, $p(\{x\}, y)$, is not normally known. In the absence of the actual risk, the *empirical risk*,

$$R_{\mathrm{emp}}(\{w\}) = \frac{1}{N} \sum_{i=1}^{N} |y_i - f(\{x\}_i, \{w\})| \qquad (11.15)$$

is usually computed. Unfortunately it is impossible to tell on the basis of this quantity whether the map f will generalise well, that is will give low errors on different data sets generated with the joint density $p(\{x\}, y)$. The basis of SLT, largely developed by Vapnik and co-workers (Vapnik, 1995; Cherkassky and Mulier, 1998), is to construct *bounds* on the actual risk of the form

$$R(\{w\}) \leq R_{\mathrm{emp}}(\{w\}) + \Phi(h) \qquad (11.16)$$

where the *confidence interval* Φ is a function of the parameter h, which is the *Vapnik–Chervonenkis* (*VC*) *dimension*. This parameter scales with the complexity of the model basis used for fitting f (and is usually rather difficult to compute). The important point here is that, if the model basis is very versatile and can represent huge classes of functions, it is likely to overfit the data and generalise badly. Thus, the confidence interval will be increasing with h. An important point about Equation (11.16) is that it is independent of $p(\{x\}, y)$ – it depends only on the training data and h. (This is rather a simplification. The bound in Equation (11.16) holds only with probability $1 - \eta$, where η measures the deficit, and the confidence Φ also depends on η.) Bounds of this type form the basis for the idea of *structural risk minimisation* (SRM).

The idea of the SRM *inductive principle* is to form a nested group of model bases as illustrated in Figure 11.6. Each model has the VC dimension h_i with $h_i \leq h_{i+1}$. For each group, one computes h_i (or estimates an upper bound) and thus determines the confidence interval. The model is selected that minimises the sum of the empirical risk and the confidence interval. The principle is that better generalisation will result if the complexity of the model is controlled carefully. Consider fitting a function with a neural network; if enough weights are included, the network will fit any relationship without error on the training data (minimise the empirical risk), but will be unable to generalise. The complexity of the neural network basis can be controlled in two ways:

1. one can restrict the number of weights or hidden units or
2. one can regularise as discussed in Section 11.2.5 and control the smoothness of the mapping.

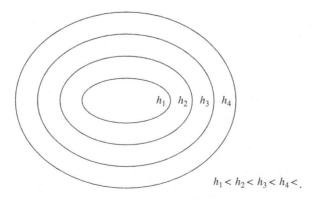

$$h_1 < h_2 < h_3 < h_4 < .$$

Figure 11.6 A hierarchy of models parameterised by the VC dimension.

In fact, Vapnik (1995) indicates two main ways of minimising the structural risk:

1. Fix the confidence interval and minimise the empirical risk; in a sense this is the 'classical' strategy used in neural network learning.
2. Fix the empirical risk and minimise the confidence interval. As will become clear, this is the basis of the *support vector machine* (SVM) concept.

11.5.2   Support Vector Machines

It is simplest to begin the discussion of support vector machines (SVMs) with a description of linear discriminant analysis. This is a statistical technique that seeks to separate two classes of data using a hyperplane (i.e. a straight line in two dimensions). Suppose that the two classes of data are indeed linearly separable as shown in Figure 11.7(a). In general there will be many separating hyperplanes. The problem with many of them is that they will not generalise well, the reason being that they pass too close to the data. The idea of an SVM for linear discrimination is that it will select the hyperplane that generalises best, and in this case this means the one that is furthest from the data in some sense.

Suppose the hyperplane has the equation

$$D(\{x\}) = < \{w\}, \{x\} > = 0 \qquad (11.17)$$

(For convenience the input vector has been extended to $d + 1$ dimensions by adding a 0th component that is held constant at unity. This means that the offset from the origin of the hyperplane can be added as the 0th component of $\{w\}$. Note that $<$, $>$ is the standard Euclidean scalar (dot) product. The constant term actually requires a little more subtlety than is implied in the following discussion; the reader should consult Vapnik, 1995, or Cherkassky and Mulier, 1998.) The separation condition is given by

$$\{x\}_k \in C_1 \Rightarrow D(\{x\}_k) = < \{w\}, \{x\}_k > \geq 1$$

$$\{x\}_k \in C_2 \Rightarrow D(\{x\}_k) = < \{w\}, \{x\}_k > < -1 \qquad (11.18)$$

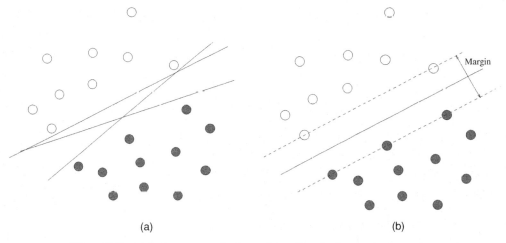

(a)   (b)

Figure 11.7   (a) Arbitrary separating hyperplanes, (b) optimal separating hyperplane.

where C_1 and C_2 are the two classes, or, more concisely,

$$D(\{x\}_k) = y_k < \{w\}, \{x\}_k >\geq 1 \tag{11.19}$$

where y_k in this case is a class label $y_k = 1$ (respectively $y_k = -1$) if $\{x\}_k \in C_1$ (respectively $\{x\}_k \in C_2$).

It is a simple matter to show that the distance of each point in the training set $\{x\}_k$ from the separating hyperplane is $|D(\{x\}_k)|/\|\{w\}\|$. Now, τ is a *margin* (an interval containing the hyperplane but excluding all data – Figure 11.7b) if

$$y_k \frac{D(\{x\}_k)}{\|\{w\}\|} > \tau, \quad \forall k \tag{11.20}$$

Now, note that the parametrisation of the hyperplane is currently arbitrary. This can be fixed by specifying

$$\|\{w\}\|\tau = 1 \tag{11.21}$$

and this converts Equation (11.20) into the separation condition of Equation (11.19). It is clear now that maximising the margin will place the hyperplane at the furthest point from the data and this can be accomplished – in the light of Equation (11.21) – by minimizing $\|\{w\}\|$ subject to the constraints of Equation (11.19). An appropriate objective function is

$$Q(\underline{w}) = \frac{1}{2}\|\{w\}\|^2 - \sum_{i=1}^{N} \alpha_i y_i(< \{w\}, \{x\}_i > -1) \tag{11.22}$$

where the parameters α_i are Lagrange multipliers. Now this is a quadratic programming problem and is *convex* (Cristianini and Shawe-Taylor, 2000). The parameters $\{w\}$ can be expressed in terms of the α_i by using the Kuhn–Tucker (KT) conditions (again, see Cristianini and Shawe-Taylor, 2000), which leads to the *dual* formulation of the problem, that is

$$\frac{\partial Q}{\partial \{w\}} = 0 \Rightarrow \{w\} = \sum_{i=1}^{N} \alpha_i y_i \{x\}_i \tag{11.23}$$

at the optimum. The KT conditions also assert that an α_i can only be nonzero if the corresponding constraint holds with equality, that is

$$\alpha_i y_i(< \{w\}, \{x\}_i > -1) = 0 \tag{11.24}$$

and the corresponding data points $\{x\}_i$ are seen to be on the limits of the margin – they are termed *support vectors*. It may well be the case that only a small number of support vectors are needed for a given problem.

The dual formulation of the problem is found by substituting Equation (11.23) into the objective function (11.22) and it is required to maximise the following function with respect to the α_i:

$$Q(\{w\}) = \sum_{i=1}^{N} \alpha_i - \frac{1}{2}\sum_{i=1}^{N}\sum_{j=1}^{N} \alpha_i\alpha_j y_i y_j < \{x\}_i, \{x\}_j > \tag{11.25}$$

subject to the constraints

$$\sum_{i=1}^{N} y_i \alpha_i = 0, \quad \alpha_i \geq 0, \ i = 1, \ldots, N \tag{11.26}$$

Once the optimum is found, the optimal separating hyperplane is given by Equation (11.23) as

$$D(\{x\}) = \sum_{i=1}^{N} \alpha_i y_i < \{x\}_i, \{x\} > \tag{11.27}$$

(up to a term involving the constant offset, see Vapnik, 1995, or Cherkassky and Mulier, 1998). This shows how a classifier with maximum margin is constructed, *provided* the data are linearly separable.

If the data are indeed linearly separable and there are potentially many appropriate decision boundaries, all of them will have zero empirical risk and it is clear here that maximising the margin is actually satisfying some other principle; in fact, it is minimising the confidence interval and thus implementing SRM. It can be shown that maximising the margin can be regarded as a proxy for controlling the VC dimension of the SVM (Vapnik, 1995). In terms of the SRM inductive principle, one has the nested sets S_k, where S_k is characterised by $||\{w\}_k|| \leq c_k$ and $\ldots c_k \leq c_{k+1} \ldots$ (recall that a lower weight norm implies a larger margin). The complexity of the SVM is also reflected in the number of support vectors needed. Vapnik showed that on a sample of test data (not used in training), one has the profound result (Cherkassky and Mulier, 1998)

$$E_n[P(\text{error})] = \frac{1}{N-1} E_n[\text{number of support vectors}] \tag{11.28}$$

where E_n is an expectation over all possible training sets with N points. This equation shows that generalisation (performance on unseen test data) does not depend on the dimension of the problem, but only on whether one has a small number of support vectors compared to the number of training points.

Returning now to the classification problem, in the case when the two data classes of interest are *not* separable, the procedure of training the SVM requires a little modification. The basic approach is to add *slack variables* ξ_i for the nonseparable data such that

$$D(\{x\}_k) = y_k < \{w\}, \{x\}_k > \geq 1 - \xi_k \tag{11.29}$$

As the ξ_i quantify the nonseparability, a penalty term of the form $\sum_{i=1}^{N} \xi_i^p$ can be added to the objective function. If $p = 1$ or 2, the optimisation problem remains one of quadratic programming. If $p = 1$, there is a further advantage, which is that the ξ_i do not appear in the dual problem. The basic objective function is

$$Q(\{w\}) = \frac{1}{2}||\{w\}||^2 + C\sum_{i=1}^{N} \xi_i - \sum_{i=1}^{N} \alpha_i y_i(< \{w\}, \{x\}_i > -1 + \xi_i) - \sum_{i=1}^{N} \mu_i \xi_i \tag{11.30}$$

where the parameters μ_i are Lagrange multipliers that force the ξ_i to be positive. C is a variable parameter that encodes the cost of nonseparation. The corresponding dual problem is found to be a maximisation of

$$Q(\{w\}) = \sum_{i=1}^{N} \alpha_i - \frac{1}{2}\sum_{i=1}^{N}\sum_{j=1}^{N} \alpha_i \alpha_j y_i y_j < \{x\}_i, \{x\}_j > \tag{11.31}$$

subject to the constraints

$$\sum_{i=1}^{N} y_i \alpha_i = 0, \quad C \geq \alpha_i \geq 0, \ i = 1, \ldots, N \tag{11.32}$$

The corresponding expression for the hyperplane is exactly as before. Such a classifier is said to have a *soft margin*. In a sense, the value of C controls the complexity of the model, so maximising the performance of the classifier over a range of C values is consistent with the philosophy of structural risk minimisation.

The analysis so far has assumed that the classifier is linear. The technique can be made nonlinear by the use of a clever trick. This essentially exploits *Cover's theorem* (Haykin, 1994), which basically asserts that a classification problem embedded nonlinearly in a high-dimensional space is more likely to be separable than in a low-dimensional space. The input data are mapped into a higher-dimensional *feature space* using a nonlinear mapping φ, that is $\{z\}_k = \varphi(\{x\}_k)$. In the high-dimensional space, the optimisation problem has the objective function

$$Q(\{w\}) = \sum_{i=1}^{N} \alpha_i - \frac{1}{2} \sum_{i=1}^{N} \sum_{j=1}^{N} \alpha_i \alpha_j y_i y_j < \varphi(\{x\}_i), \varphi(\{x\}_j) > \tag{11.33}$$

(for the separable case) where $< , >$ denotes the inner product on the feature space. (The same notation for an inner product will be used throughout; this is unlikely to cause confusion as the particular space of interest will be clear from the contents of the brackets.) Because the feature space is high – possibly infinite – dimensional, the inner product evaluations are costly and would render the approach impractical. However, the aforementioned trick comes to the rescue. For most classes of nonlinear functions φ of interest – polynomials, sigmoidal neural networks and radial basis functions – there exists a *kernel function* $k(\{x\}_i, \{x\}_j)$ *evaluated in the input space* such that

$$k(\{x\}_i, \{x\}_j) = < \varphi(\{x\}_i), \varphi(\{x\}_j) > \tag{11.34}$$

and this makes the optimisation tractable. The resulting *nonlinear* discriminant function is given by

$$D(\{x\}) = \sum_{i=1}^{N} \alpha_i y_i k(\{x\}_i, \{x\}) \tag{11.35}$$

(again, beware of the constant terms). As always, $\{x\}$ is assigned to class 1 if $D(\{x\}) > 0$ and to class 2 if $D(\{x\}) < 0$.

The discussion has so far ignored the practical issue of solving the optimisation problem presented by Equation (11.33), and in some respects this is justified by the fact that specific software for the solution is freely available. However, it is useful to make some remarks. The simplest, naive solution would be to apply gradient descent (Frieβ, Cristianini and Campbell, 1998) as in so many other machine learning approaches; however, this fails to exploit the fact that the optimisation problem here is a particular quadratic one and that powerful techniques are available for that class of problem. Even so, the SVM problem presents difficulties because of the number of variables concerned; the training problem requires that the kernel matrix be held in memory and this means that the cost of optimisation scales quite badly with the number of points in the training set. The SVM community devised a number of algorithms based on working, at any given time, with subsets of the training data, with these subsets chosen because they dominate the problem in some sense. The simplest heuristic of this type is known as *chunking* (Cristianini and Shawe-Taylor, 2000). If one takes this idea to its extreme, one arrives at the *sequential minimal optimisation* (SMO) algorithm, which works with only two data points at each iteration. The SMO algorithm was devised by Platt (1998); it is discussed in some detail in Cristianini and Shawe-Taylor (2000) who also include a pseudo-code for the algorithm.

The SVM can also be used for unsupervised learning; in this form, the algorithm is referred to as a *one-class* SVM (Schőlkopf *et al.*, 2000). The idea of the algorithm is to separate the training data from the origin using a hyperplane with the largest margin possible. The algorithm is a natural means of carrying out novelty detection and has been effectively applied to a jet engine condition monitoring problem by Hayton *et al.* (2007) and to an interesting SHM problem relating to bridge monitoring by Oh and Sohn (2009).

11.5.3 *Kernels*

A critical ingredient in the construction of the classifier for nonlinearly separable data in the last section was the map into a high-dimensional space $\varphi(\{x\})$ and its associated kernel $k(\{x\}, \{x'\})$. It is worth spending a little time discussing this map; a much more detailed discussion of the issues raised here can be found in Cristianini and Shawe-Taylor (2000). For simplicity, a two-dimensional input space will be assumed for the moment so that $\{x\} = (x_1, x_2)$. Suppose that the problem of interest is regression; with the vector $\{x\}$ alone, one can form the regressor

$$y = \{w\}^{\mathrm{T}}\{x\} + w_0 = w_1 x_1 + w_2 x_2 + w_0 \tag{11.36}$$

which will clearly be of little use in learning a real function if that function is highly nonlinear. However, suppose that one maps $\{x\}$ into the higher-dimensional space $\varphi(\{x\})$ where

$$\varphi(\{x\}) = \left(1, x_1, x_2, x_1^2, x_1 x_2, x_2^2\right) \tag{11.37}$$

then one can form a regressor of the form

$$y = \{w\}^{\mathrm{T}}\varphi(\{x\}) = w_{11} x_1^2 + w_{12} x_1 x_2 + w_{22} x_2^2 + w_1 x_1 + w_2 x_2 + w_0 \tag{11.38}$$

which is at least beginning to accommodate possible nonlinearity in the problem of interest. This simple step has increased the dimension of the input space from 2 to a dimension of 6 for the regression problem. While this may seem reasonable for the gain of the nonlinear representation, it is not difficult to convince oneself that the dimension of the higher-dimensional space D increases very quickly with the dimension of the input space d and the polynomial order of the desired regression p. If one increases the dimension of the input space to 3, then D becomes 10; if one instead increases the polynomial order to 3, then D becomes 10 also. It is not difficult to show that the general case gives

$$D = d^p \tag{11.39}$$

This means that a regression up to fifth power on an input space of 10 dimensions – not unreasonable requirements – would require $D = 100\,000$. The SVM methodology requires the evaluations of scalar products in the high-dimensional space and would therefore require 100 000 multiplications in the example just discussed; this motivates the use of kernels. Following Cristianini and Shawe-Taylor (2000), consider the kernel

$$k(\{x\}, \{x'\}) = (<\{x\}, \{x'\}> +1)^2 \tag{11.40}$$

Expanding the right-hand side of this equation gives

$$\left(\sum_{i=1}^{d} x_i x_i' + 1 \right) \left(\sum_{j=1}^{d} x_j x_j' + 1 \right) = \sum_{i=1}^{d} \sum_{j=1}^{d} (x_i x_j)(x_i' x_j') + \sum_{i=1}^{d} (\sqrt{2} x_i)(\sqrt{2} x_i') + 1 \quad (11.41)$$

and one can see that the expression is the scalar product of higher-dimensional vectors,

$$\varphi(\{x\}) = \left(1, \sqrt{2} x_1, \ldots, \sqrt{2} x_d, x_1^2, \frac{1}{2} x_1 x_2, \ldots, x_d^2 \right) \quad (11.42)$$

Now, these vectors include all monomial terms in the d input variables up to second order and form a perfectly acceptable basis for a quadratic regression, as discussed earlier; the fact that the terms have picked up some constant multipliers (to account for multiple counting in Equation (11.41)) is irrelevant in the regression context. Importantly, the kernel in Equation (11.40) has allowed the scalar product in D dimensions to be calculated from the scalar product in d dimensions.

In the general case, the use of the *polynomial kernel*

$$k(\{x\}, \{x'\}) = (< \{x\}, \{x'\} > +1)^p \quad (11.43)$$

allows for maps $\varphi(\{x\})$ with all terms up to polynomial order p. The SVM approach is of course not limited to polynomial kernels. The kernel

$$k(\{x\}, \{x'\}) = \exp \left(-\frac{1}{r^2} < \{x\} - \{x'\}, \{x\} - \{x'\} > \right) \quad (11.44)$$

allows the fitting of radial basis functions and the kernel,

$$k(\{x\}, \{x'\}) = \tanh(a < \{x\}, \{x'\} > +b) \quad (11.45)$$

allows the fitting of an MLP structure. (It is important to note that the latter two kernels contain hyperparameters, which would have to be established in some manner.)

One cannot choose the kernels arbitrarily. They should satisfy a technical condition derived by Mercer (1909), which is

$$\iint k(\{x\}, \{x'\}) \phi(\{x\}) \phi(\{x'\}) d\{x\} d\{x'\} > 0, \quad \forall \phi \neq 0, \quad \int \phi(\{x\})^2 d\{x\} < \infty \quad (11.46)$$

The Mercer condition is not trivial; for example, the hyperbolic tangent kernel in Equation (11.45) only satisfies Mercer's condition for certain values of a and b (working values are $a = 2$, $b = 1$ (Cherkassky and Mulier, 1998).

The discussion above has been framed in terms of a regression problem and should make sense even though SVM regression has not been discussed yet. The ideas discussed above apply quite happily to the classification SVMs of the last section.

This concludes the basic theory of support vector machines and the discussion can now move on to a case study.

11.6   Case Study II: Support Vector Classification

The data under consideration are exactly the same as those discussed in Section 11.3; the problem is that of damage location in the Gnat aircraft wing (see Section 5.6 also). To recap, there are nine damage locations each corresponding to the removal of an inspection panel from the wing. As before, prior to training the SVM classifier, the data were divided into training, validation and testing sets. Each set had 594 training vectors, 66 for each location class as described in Section 11.3.

The first issue one faces here is that the SVMs described in the last section are designed to separate two classes, that is to form a *dichotomy*. However, the problem of interest here is to assign data to nine classes. The usual method of using the SVM for multiclass problems is to train one SVM for each class so that each separates the class in question from all the others. This means that nine quadratic programming problems have to be solved for the current problem. There are strategies for multiclass problems that overcome the need for several SVMs; some are summarised in Bennett (1999). The software used in this study was SVM^{light} (Joachims, 1999).

The SVM expresses the discriminant function in terms of a series of basis functions; in this study it was decided to adopt a radial basis kernel. The basis functions are

$$\varphi_i(\{x\}) = \exp(-\gamma \|\{x\} - \{x\}_i\|^2) \tag{11.47}$$

and the discriminant function is thus equivalent to a radial-basis function network (Bishop, 1998) with the centres at each of the training data points and the radii of each of the basis functions equal. The parameter γ is to be specified for each of the classifiers, along with the strength of the margin C. (As discussed above this is the parameter – introduced in Equation (11.30) – that decides how severely to punish misclassifications in the event that the data are not separable.) The 'best' values for these constants were determined by optimising the performance of the classifiers on the validation sets. The final performance of the classifiers was determined by their effectiveness on the testing sets. A fairly coarse search was carried out with candidate values of γ taken from a range between 0.01 and 10.0 and with C taken from the range 1.0 to 1024.0 in powers of two.

The results from the nine classifiers on the testing set of are given in Table 11.3. Note that this is different from the usual confusion matrices in one important respect. When a neural network is used to create a classifier with the *1 of M* scheme, each data point is assigned to one and only one class of the *M* possibilities. In the case of the support vector machine, each data point is presented to all of the *M* SVMs and could in principle be assigned to all or none of the classes. This process means that the numbers in the rows of the class matrices do not add up to the number of points in the class.

Overall, on the testing set, the multiple SVM scheme gives 89.2% correct classifications (including multiple classifications where the correct class is identified) and 7.2% misclassifications. As discussed above, the classifier does not respond to 5.6% of the data and 2.7% points give multiple class labels. Thus, the SVM appears to outperform the MLP neural network (see Table 11.2). (If one only counts unambiguous classifications, the SVM achieves 86.5% – the same as the neural network.) Also, the SVM indicates which data points are the source of confusion. One of the other strengths of the SVM approach discussed here is that it is *universal* and the same algorithm embedded in the same software could just as

Table 11.3   Confusion matrix for testing data – SVM on the Gnat damage location problem

Predicted class	1	2	3	4	5	6	7	8	9
True class 1	55	1	0	0	1	0	0	1	0
True class 2	0	60	2	0	2	0	0	0	2
True class 3	0	1	58	0	0	5	0	2	0
True class 4	1	0	0	60	0	1	0	1	0
True class 5	0	2	2	0	61	0	0	0	0
True class 6	2	0	9	0	0	53	0	0	0
True class 7	2	0	1	0	0	0	63	0	1
True class 8	0	0	0	0	0	2	0	59	1
True class 9	1	0	0	1	0	0	0	2	61

easily have fitted polynomial or neural network discriminants. This offers the possibility of optimising the performance over the model basis.

One of the often-cited strengths of the SVM approach is that the quadratic programming problem is convex and thus has a single global minimum. However, in the case described above, this is only true once appropriate values for γ and C have been fixed. If these values had been left free for optimisation, the problem would no longer be quadratic.

11.7   Support Vector Regression

The formulation of the regression problem is almost exactly as expected, given the existence of a training set comprising N corresponding inputs $\{x\}_i$ and outputs y_i from some function $y = f(\{x\})$ (possibly/probably corrupted by noise). One assumes that the function can be approximated by a linear superposition from an appropriate set of basis functions $\{g_i(\{x\}); i = 1, \ldots, M\}$,

$$f(\{x\}, \{w\}) = \sum_{i=1}^{M} w_i g_i(\{x\})$$
(11.48)

As in the classification case, this can be achieved by minimising an appropriate risk functional. The standard risk in regression problems is the least-squares empirical risk given by

$$R_{\text{emp}}^{LS}(\{w\}) = \frac{1}{N} \sum_{i=1}^{N} (y_i - f(\{x\}_i, \{w\}))^2$$
(11.49)

In fact, SVM regression uses a different risk function, motivated by the idea of *robust regression* (Huber, 1964), which is designed to be less sensitive to outliers. The LS formulation of the problem is known to be optimal only when the errors have a Gaussian distribution; in circumstances when Gaussianity fails (but not too badly) it can be better to use the risk/error function

$$R_{\text{emp}}^{RR}(\{w\}) = \frac{1}{N} \sum_{i=1}^{N} |(y_i - f(\{x\}_i, \{w\}))|$$
(11.50)

SVM regression goes one step further than this in assuming an ε-*insensitive* risk function of the form

$$R_{\text{emp}}^{\varepsilon}(\{w\}) = \frac{1}{N} \sum_{i=1}^{N} |y_i - f(\{x\}_i, \{w\})|_\varepsilon$$
(11.51)

where

$$|y_i - f(\{x\}_i, \{w\})|_\varepsilon = \begin{cases} |y_i - f(\{x\}_i, \{w\})|, & \text{if } |y_i - f(\{x\}_i, \{w\})| > \varepsilon \\ \varepsilon, & \text{else} \end{cases}$$
(11.52)

Starting with linear regression (as the discussion of SVM classification began with a linear discriminant), Vapnik (1995) defines SVM regression as having three essential components:

1. Representation as a linear regression, $y = \sum_{i=0}^{d} w_i x_i = \{w\}^T\{x\}$.
2. Minimisation with respect to the ε-insensitive loss function.

3. Use of the SRM principle. In this case, it is implemented by defining S_k to be a family of regressions where $||\{w\}_k||^2 \leq c_k$ and $\ldots c_k \leq c_{k+1} \ldots$. Complexity of the model (and therefore capacity for generalisation) is controlled by keeping the weights as small as possible.

Now, it can be shown that minimisation of Equation (11.51) with respect to $\{w\}$ for a given training set is equivalent to the problem of minimising

$$F(\{\xi^*\}, \{\xi\}) = \sum_{i=1}^{N} (\xi_i^* + \xi_i) \tag{11.53}$$

(the $\{\xi\}$ and $\{\xi^*\}$ are slack variables again), subject to the constraints

$$\xi_i^* \geq 0 \tag{11.54}$$

$$\xi_i \geq 0 \tag{11.55}$$

$$y_i - \{w\}^T\{x\}_i \leq \varepsilon + \xi_i^* \tag{11.56}$$

$$\{w\}^T\{x\}_i - y_i \leq \varepsilon + \xi_i \tag{11.57}$$

$$||\{w\}||^2 \leq c_n \tag{11.58}$$

(for all $i = 1, \ldots, N$). Introducing appropriate Lagrange multipliers leads to an objective function

$$L(\{w\}, \{\xi^*\}, \{\xi\}, \{\alpha^*\}, \{\alpha\}, C^*, \{\gamma\}, \{\gamma^*\}) = \sum_{i=1}^{N} (\xi_i^* + \xi_i) - \sum_{i=1}^{N} \alpha_i^*[\{w\}^T\{x\}_i - y_i + \varepsilon + \xi_i^*]$$

$$- \sum_{i=1}^{N} \alpha_i[y_i - \{w\}^T\{x\}_i + \varepsilon + \xi_i]$$

$$- \sum_{i=1}^{N} (\gamma_i^*\xi_i^* + \gamma_i\xi_i) - \frac{1}{2}C^*(c_n - ||\{w\}||^2) \tag{11.59}$$

Minimising the latter function with respect to $\{w\}$, $\{\xi^*\}$ and $\{\xi\}$ and substituting back the resulting equations generates the objective function

$$W(\{\alpha^*\}, \{\alpha\}, C^*) = -\varepsilon \sum_{i=1}^{N} (\alpha_i^* + \alpha_i) + \sum_{i=1}^{N} y_i(\alpha_i^* - \alpha_i)$$

$$- \frac{1}{2C^*} \sum_{i=1}^{N} \sum_{j=1}^{N} (\alpha_i^* - \alpha_i)(\alpha_j^* - \alpha_j)\{x\}_i^T\{x\}_j - \frac{1}{2}c_nC^* \tag{11.60}$$

via a condition

$$\{w\} = \sum_{i=1}^{N} \left(\frac{\alpha_i^* - \alpha_i}{C^*}\right)\{x\}_i := \sum_{i=1}^{N} \beta_i\{x\}_i \tag{11.61}$$

(which serves to define the β_i). It transpires that (as in the classification problem) only a small subset of the β_i will be nonzero in many cases; these define the support vectors in the regression context.

The optimisation problem defined above can be shown to be convex. Vapnik (1995) discusses how a simple modification to the objective function can render the optimisation problem quadratic.

As in the classification case, one can generalise the problem to the case of a nonlinear regression by mapping the data into a higher-dimensional feature space via $\varphi(\{x\})$, such that

$$f(\{x\}, \{w\}) = \sum_{i=1}^{P} w_i K(\{x\}, \{x\}_i) = \sum_{i=1}^{P} w_i < \varphi(\{x\}), \varphi(\{x\}_i) > \tag{11.62}$$

Again, the 'kernel trick' saves any heavy computation in calculating the scalar products in the high-dimensional feature space. In terms of this new problem, the convex form of the objective function becomes simply

$$W(\{\alpha^*\}, \{\alpha\}, C^*) = -\varepsilon \sum_{i=1}^{N} (\alpha_i^* + \alpha_i) + \sum_{i=1}^{N} y_i(\alpha_i^* - \alpha_i)$$

$$- \frac{1}{2C^*} \sum_{i=1}^{N} \sum_{j=1}^{N} (\alpha_i^* - \alpha_i)(\alpha_j^* - \alpha_j) K(\{x\}_i, \{x\}_j) - \frac{1}{2} c_n C^* \tag{11.63}$$

and $w_i = (\alpha_i^* - \alpha_i)/C^*$. As in the linear case, the optimisation problem can be made quadratic with a small change in the objective. Choosing the polynomial kernel in Equation (11.43) will give a polynomial regression, while the kernel in Equation (11.44) will give an RBF expansion.

11.8   Case Study III: Support Vector Regression

As discussed in Chapter 9, regression is the problem of constructing a map between inputs and a continuous output variable. As such, it is one of the less common techniques used in SHM studies; classification is much more common. However, there are some problems of SHM where regression is vital. One example is in *prognosis*, the highest level of damage identification. In the prognosis problem, one wishes to predict the remaining safe life of the structure of interest from measured data and this leads naturally to a regression problem in many cases. Prognosis will be discussed in much more detail in Chapter 14 of this book. Another situation where regression is important is in the building of data-based models of structural behaviour as a necessary part of damage detection or location. A good example of this is given in Chapter 10 in the section on statistical process control, where an AR model is fitted to data from a structure and the residuals of the model are then monitored in order to assess whether the structure remains in or leaves the normal condition.

The AR residuals approach works because damage either results in the structure behaving as a different linear system or in it becoming a nonlinear system. In both cases, the initial AR model is no longer able to describe the behaviour of the structure and the model errors or residuals will increase so that damage can be inferred. One restriction of the AR approach is that it is based on a linear model and this would not be appropriate if the structure of interest were *actually nonlinear, even before the occurrence of damage*. Of course, one could proceed with the approach anyway and fit a best linear model; however, it is likely that the initial model would then have large residuals and this would reduce the sensitivity to subsequent damage. In addition, one also runs the risk that this linear model would be specific to the input associated with the training data and would not generalise well to other data, which will lead potentially to false indications of damage. A better approach is to use a class of models that is able to accommodate the possibility of an initially nonlinear structure. As support vector machines are capable in principle of modelling any relationship, linear or nonlinear, they provide an appropriate model class.

The natural approach to the problem of an initially nonlinear structure is simply to build a nonlinear AR (NAR) model of the form

$$y_i = f(y_{i-1}, \ldots, y_{i-p}) + \varepsilon_i \tag{11.64}$$

for the response variable of interest and then monitor the residual ε_i using a control chart or some other means. A model of the form given in Equation (11.64) will be referred to as a NAR(p) model. Such nonlinear models were introduced in Section 8.2.9. Now it will be shown how SVM regression can be used to establish such nonlinear time series models. An example of simulated damage will be used here as an illustration.

Following the discussion by Bornn *et al.* (2009), the performance of an SVM-based damage detection method will be compared to one based on a traditional AR model with coefficients estimated by the Yule–Walker method (Brockwell and Davis, 1991). The data are generated as follows for discrete time points $t_i = 1, \ldots, 1200$:

$$y_i = \sin^2\left(\frac{\pi t_i}{3}\right) + \sin^2\left(\frac{\pi t_i}{3}\right) + \sin\left(\frac{\pi t_i}{6}\right) + \sin\left(\frac{\pi t_i}{6}\right) + \Psi_i + \varepsilon_i \tag{11.65}$$

where the sequence ε_i is Gaussian random noise with mean 0 and standard deviation 0.1 and Ψ is a 'damage' term. Three different damage cases are added to this time series at various times, as defined by

$$\Psi_i = \begin{cases} \varepsilon_{1i}, & t_i = 600, \ldots, 650 \\ \dfrac{1}{2}\sin(\pi t_i/6), & t_i = 800, \ldots, 850 \\ \varepsilon_{2i}, & t_i = 1000, \ldots, 1050 \\ 0, & \text{otherwise} \end{cases} \tag{11.66}$$

where the sequences ε_{1i} and ε_{2i} are Gaussian random noises with mean 0 and 1 and standard deviation 0.5 and 0.2, respectively. Through the use of Ψ it is attempted to simulate several different types of damage to compare the AR and NAR models' performance handling each. This raw signal is plotted in Figure 11.8, where it can be seen that the changes caused by the 'damage' are somewhat subtle.

The order p for both the AR and NAR models was set at five, as determined from the autocorrelation plot in Figure 11.9. This plot is the measure of correlation between successive time points for a given time lag. One can see from the plot that after a lag of 5, the correlation is quite small and hence little information is gained by including a longer past history p. This is a standard method for determining model order for traditional AR models, and as such should maximise the performance of the method, ensuring that the SVM-based model is not afforded an unfair advantage.

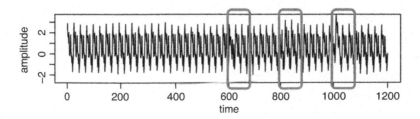

Figure 11.8   Raw simulated data with highlighted artificial damage.

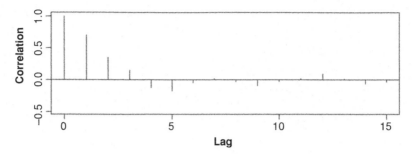

Figure 11.9   Autocorrelation plot of simulated data.

The results of applying both the SVM model and a traditional AR model to the undamaged portion of the signal between time points 400 and 600 are shown in Figure 11.10, where the signals predicted by these models are overlaid on the actual signal. A qualitative visual assessment of Figure 11.10 shows that the SVM more accurately predicts this signal. A quantitative assessment is made by examining the distribution of the residual errors obtained with each model. The standard deviation of the residual errors from the SVM model is 0.26 while for the traditional AR it is 0.71, again indicating that the SVM is more accurate at predicting the undamaged portion of this time series.

In order for a model to excel at detecting damage, it must fit the undamaged data well (i.e. small and randomly distributed residual errors) while fitting the damaged data poorly, as identified by increased residual errors with possibly nonrandom distributions. In other words, the model must be sensitive to distributional changes in the data that result from damage. To quantify such changes a control chart is developed based on the undamaged portion of the time series to establish statistically based thresholds for the damage detection process (see Chapter 10). This control chart is calculated based on the fit to the undamaged data, specifically 99% confidence lines are drawn based on the undamaged residual error data and carried forward for comparison on the potentially damaged data. It is in this part of the process that the ability of the SVM to represent the data more accurately enhances the damage detection process. The 99% confidence lines for the SVM are much closer to the mean value of the residual errors and, hence, will more readily identify small perturbations to the underlying system that produce changes in the residual error distribution. In addition, the traditional AR model shows a trend in the residuals, indicating a lack of model fit, even in the undamaged case. One can see that during the times of damage,

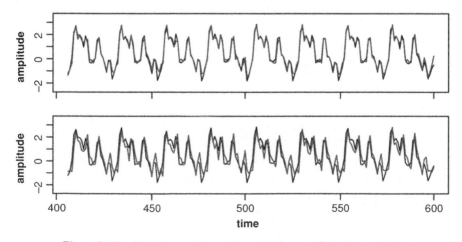

Figure 11.10   SVM (top) and linear AR models (bottom) fit to subset of data.

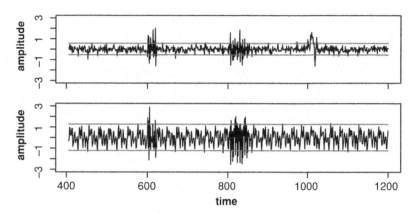

Figure 11.11 Residuals from SVM (top) and linear AR models (bottom) applied to simulated data. The 99% control lines based on the residuals from the undamaged portion of the signal are shown in red.

the residuals for the SVM-based model exceed the control limits more than occurs with the residuals from the traditional AR model. In fact, the latter method would be likely to miss the damage between time points 1000 and 1050, where only one point exceeds the threshold versus over 10 for the SVM-based model. This result can be seen in Figure 11.11.

To conclude this discussion of SVM regression and its comparison with linear AR models, it should be noted that each method performs differently for different sources of damage. Therefore, it is of interest to determine when each method will be successful in indicating damage. The traditional AR model fits a single model to the entire data and as such the model fit will be very poor if the data is nonstationary (for instance if the excitation is in the form of hammer impacts). Additionally, since the traditional AR model as presented above does not contain a moving average term, it will continue to fit when damage is in the form of a shift up or down in the raw time series (as demonstrated by the third damage scenario above). Conversely, the SVM-based method works by comparing each length of p data to all corresponding sets in the training set. Thus, if a similar sequence exists in the training set, one can expect the fit to be quite good. One can imagine two scenarios in which the SVM-based method will perform poorly. First, if there is some damage in the 'undamaged' scenario, and similar damage occurs in the testing set, the model will be likely to fit this portion quite well. Second, if damage manifests itself in such a way that the time series data is extremely similar to that of the undamaged time series, the SVM methodology will be unable to detect it. However, one should emphasise that other methods, including the AR model, will suffer in such scenarios as well. As an attempted solution when the sensitivity of the method to a given type of damage is unknown and simulation tests are impossible, various damage detection methods could potentially be combined to boost detection power.

11.9   Feature Selection for Classification Using Genetic Algorithms

As discussed in the case studies shown earlier in this chapter, the process of feature selection is not always carried out in an optimal manner; in fact, it is often done in a somewhat ad hoc manner, as discussed in Chapter 7. In the case of the damage location problem for the Gnat aircraft wing (Section 11.3), the extraction/selection process was carried out in stages, with some stages more objective than others. The feature selection process can thus only be termed *semi-objective* in that case. The aim of the current section is to show how a principled optimisation procedure can be used to select features. In fact, the exercise will begin with the set of 44 candidate features (each a scalar novelty index) that had been classified as fair or strong according to engineering judgement. This classification does not invalidate in any way the analysis to follow as a method of optimal subset selection.

11.9.1   Feature Selection Using Engineering Judgement

First of all, the 44 strong/fair features were given an integer identifier in the range 1 to 44. As the baseline, the semi-objective feature selection procedure gave the following feature set for neural network training: 2:3:5:7:14:X:28:33:43. For example, feature 2 is from the transmissibility reciprocal T_2^* (see Section 5.6 for the nomenclature), lines 667 to 676 (10-dimensional), while feature 7 is from T_4^*, lines 249 to 258 (again 10-dimensional). The feature denoted X is a weak feature that was originally selected on a visual basis as the best vector for distinguishing the panel 6 removal. There were no fair or strong features deemed suitable for this purpose.

When the networks were trained using the semi-objective feature data, the one that gave the lowest validation error had 10 hidden units and gave a misclassification error of 0.155 (15.5% misclassification rate), corresponding to a training error of 0.158 (15.8% misclassification). The best results were obtained after 150 000 presentations of training data. The network weights were updated after each presentation; that is, the training epoch was 1. When the best network was tested, it gave a generalisation error of 0.135; that is, 86.5% of the patterns were classified correctly. The confusion matrix was given earlier as Table 11.2 and showed that the main errors were associated with the removal of panels 3 and 6, the smallest ones. There was also some confusion in diagnosing the removal of panels 8 and 9. Although the results were excellent, there was certainly room for improvement and it was postulated that an advance could be made by using a more principled optimisation scheme for the feature selection process.

11.9.2   Genetic Feature Selection

For a more detailed analysis of Genetic Algorithms (GAs), the reader is referred to Goldberg (1989). Very briefly, GAs are search procedures based on the mechanism of Darwinian natural selection. A *population* of solutions is iterated, with the fittest solutions propagating their genetic material into the next generation by combination with other solutions. In the simplest form of GA, each possible solution is coded into a binary bit-string, which constitutes the individual. Mating is implemented by the operation of *crossover*, which means exchanging corresponding sections of pairs of individuals. *Mutation* is also simulated by the occasional random switching of a bit. In order to make sure the fittest solutions in a population are not accidentally destroyed by the selection and mating processes, an *elite* can be used whereby a certain proportion of the fittest individuals are automatically copied into the next generation. Finally, in order to avoid stagnation of the population a certain number of new randomly chosen individuals – the *new blood* – can be inserted in each new generation. The new blood is not allowed to overwrite the elite.

It was shown by Worden and Burrows (2001) that the binary representation is unsuitable for the feature selection problem. As a result a modified GA is used, where the gene is a vector of integers, each specifying a feature; that is, the individual 2:14:17 represents a three-feature distribution, with each integer a feature index (in the range 1 to 44 here). The operations of reproduction, crossover and mutation for such a GA are straightforward modifications of those for a binary GA (Goldberg, 1989). The initial population for the GA was generated randomly as standard.

The individuals in the population – in this case feature distributions – are propagated according to their fitness and this was evaluated as follows. For each feature distribution, a neural network diagnostic was trained and the probability of error was evaluated on the validation set. As GAs are designed to maximise fitness, the reciprocal of the error probability was used as the individual fitness. The training data for each network were generated by restricting the full 44-feature candidate patterns to the subset specified by the individual. In the GA runs, distributions with repeated features were penalised; that is, the fitness was set to a very low value so that the solutions had very low probability of propagating.

The parameters used for the GA runs were as follows: population size of 50, number of generations 100, probability of crossover 0.75 and probability of mutation 0.05. A single-member elite was used and five new blood individuals were added at each iteration. Single-point crossover was used. For more detailed explanations of all these terms, the reader can refer to Goldberg (1989). Because the algorithm is stochastic by nature, five runs of the GA were made and the best fitness taken to indicate the solution. It was decided to optimise the feature selection and the network structure separately, so the number of hidden units in the neural network was set at 10 and the network training was conducted for a fixed number of 100 000 iterations (with an epoch of 1).

After five runs of the GA, the best fitness obtained on the validation set was 21.2143, corresponding to a probability of error of 0.047. The classification rate was therefore 95.29% and this rate represents a marked improvement on the semi-objective feature selection used in Section 11.3. However, note that this rate was obtained on the validation set. One might argue that this classification rate is a valid measure of network generalisation as the network structure and training time were fixed. However, this is not the case as the network has been tuned to the given feature set and should still be assessed on an independent data set. The best nine-feature distribution was 2:3:4:7:22:23:24:34:42; that is, the optimisation procedure chose only three of the features (2, 3 and 7) chosen by the semi-objective method.

The network structure (starting conditions, number of hidden units and stopping time) were then optimised according to Tarassenko's scheme, discussed earlier (Tarassenko, 1998), using the best nine-feature set above. The best network had 24 hidden units and was trained for 120 000 cycles. The training error was 0.044, the validation error was 0.040 and the final testing error was 0.066. This result corresponds to a classification rate of 93.4%, an improvement of 6.9% on the semi-objective method. The confusion matrix on the testing set is given in Table 11.4.

The confusion matrix in Table 11.4 shows that not only has the overall performance of the locator been improved but the panel 3 removals no longer prove problematical and the ambiguity between the removals of panels 8 and 9 has been eliminated. The performance relating to the panel 6 removals could still be improved.

Before testing was carried out, it was observed that the novelty indices had very large ranges (note that Figure 11.5 is plotted on a log scale) and that the input scaling of the neural network on to the interval $[-1, 1]$ might well be weighting out lower values of the indices with potentially useful information content. It was therefore decided to transform the initial features by taking their natural logarithm. Clearly, there was no need to change the indexing. Five runs of the GA were made with the same GA parameters as before and the same neural network parameters. This time, the GA found the same maximum fitness of 99.0001 in two out of the five runs. This corresponds to a probability of error of 0.0101 on the validation set or a classification rate of 98.99%. The two fittest feature distributions were 1:2:3:7:9:28:31:33:44 and 1:2:3:6:15:25:32:33:34. The first agrees with the semi-objective method (Section 11.3) on five features and the second on three features. For the purposes of brevity, only the first of these distributions will be discussed in any detail.

The network structure (starting conditions, number of hidden units and stopping time) were then optimised according to Tarassenko's scheme, discussed earlier, using the first of the two nine-feature sets above. The best network had 10 hidden units and was trained for 80 000 cycles. The training error was 0.0034, the validation error was 0.0084 and the final testing error was 0.0185. This result corresponds to

Table 11.4 Confusion matrix for the best neural network using linear features – testing set

Predicted class	1	2	3	4	5	6	7	8	9
True class 1	62	1	1	0	1	0	0	0	1
True class 2	0	62	4	0	0	0	0	0	0
True class 3	0	0	64	0	0	1	0	0	1
True class 4	1	0	0	62	0	3	0	0	0
True class 5	1	1	1	0	59	1	3	0	0
True class 6	2	3	0	0	0	58	2	1	0
True class 7	1	0	0	0	0	0	64	0	1
True class 8	0	0	0	0	3	0	0	62	1
True class 9	1	1	1	0	0	1	0	2	62

Table 11.5 Confusion matrix for the best neural network using log features – testing set

Predicted class	1	2	3	4	5	6	7	8	9
True class 1	65	0	0	0	0	0	0	0	1
True class 2	0	65	0	1	0	0	0	0	0
True class 3	1	0	62	0	0	1	0	1	1
True class 4	0	0	0	66	0	0	0	0	0
True class 5	0	0	0	0	66	0	0	0	0
True class 6	0	3	0	0	0	62	0	1	0
True class 7	0	0	0	0	0	0	66	0	0
True class 8	1	0	0	0	0	0	0	65	0
True class 9	0	0	0	0	0	0	0	0	66

a classification rate of 98.1%, an improvement of 11.6% on the semi-objective method. As an alternate measure of performance, the network made only 11 misclassifications on the testing set of 594 patterns. The confusion matrix on the testing set is listed in Table 11.5.

Figure 11.12 shows the evolution of the fitness throughout the optimisation for the first log feature distribution using nine features given above. The GA converged on a good solution in 22 generations, corresponding to 1100 function evaluations. This is an excellent performance given that the set of possible distributions has in excess of 10^8 elements. As each neural network takes approximately 4 seconds to train, the optimum was arrived at in about 73 minutes (it may be noted that five runs were carried out, so the overall time to arrive at the solution would be five times this duration).

The first optimal nine-feature distribution used transmissibilities measured across panels 1, 2 (twice), 4, 5, 7 (twice), 8 and 9. This result is very interesting and supports the idea used in Manson, Worden and Allman (2003b) that one should maximise geometrical coverage of the wing. The only transmissibilities omitted are those measured over the small panels 3 and 6. This omission is a little surprising as one might expect them to yield vital information. However, it appears that there is enough information in the other features to diagnose removal of the small panels adequately.

The improvement in classifier performance with a more principled feature selection strategy allows for two possibilities. First, one can maintain the use of nine features for the classifier, in which case one obtains a marked improvement in performance. A second possibility is to accept a lower level of

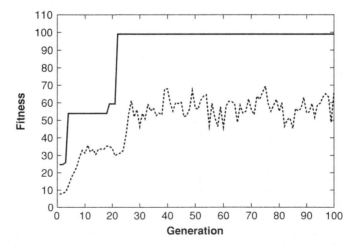

Figure 11.12 Evolution of fitness in GA run: average fitness (dotted) and maximum fitness (solid).

Table 11.6   Best sensor distributions using linear features

Number of features	Feature distribution
1	34
2	4:34
3	4:34:35
4	3:4:34:40
5	2:3:4:24:34
6	2:3:4:27:33:34
	2:3:4:7:24:34
7	2:3:4:7:23:28:34
8	2:3:4:6:27:34:42:44
9	2:3:4:7:22:23:24:34:42
10	2:3:4:6:9:23:24:34:42:44
	2:3:4:6:12:24:25:34:35:42

performance with a reduced feature set. Although the second option would not be appropriate for a safety-critical damage identification problem, it is interesting to see how smaller feature sets, chosen on the basis of optimisation, can perform. In order to investigate this performance, the GA was allowed to optimise feature sets containing between one and ten features. In each case, five runs of the GA were made with different initial populations. The resulting optimal distributions are listed in Tables 11.6 and 11.7 for the linear and log features, respectively.

The selection of features indicated in Tables 11.6 and 11.7 is summarised in Figures 11.13 and 11.14, which show frequency histograms for selection of the linear and log features, respectively. Two points are apparent. The first is that the two histograms agree on the most important features (2, 3 and 34) as one would hope. The second point is that there is a wider selection of log features chosen than the linear ones. Again, this result is to be expected as the operation of taking logs was intended to reveal information at low levels of novelty that were not visible in the linear features.

Once the optimal distribution was found for each feature number, the neural network structure was optimised for the number of hidden units, initial conditions and stopping time following best practice as before. The resulting error rates for each number of features are listed in Table 11.8 together with the number of hidden units selected with the consequent number of network weights.

Table 11.7   Best sensor distributions using log features

Number of features	Feature distribution
1	43
2	23:44
3	3:15:35
4	3:4:23:39
5	2:4:7:34:35
6	2:3:6:15:28:35
7	2:3:7:15:28:34:39
8	1:2:3:7:9:25:34:43
	2:3:4:7:15:26:27:35
	2:3:4:6:15:33:34:35
9	1:2:3:6:15:28:32:33:34
	1:2:3:7:9:28:31:33:44
10	1:2:3:6:9:23:25:26:34:39
	1:2:3:4:7:9:28:33:34:39
	2:3:4:7:15:18:28:29:34:39

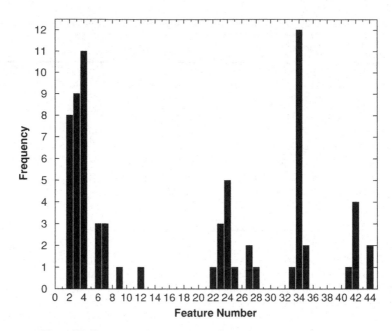

Figure 11.13   Frequency histogram for feature selection: linear features.

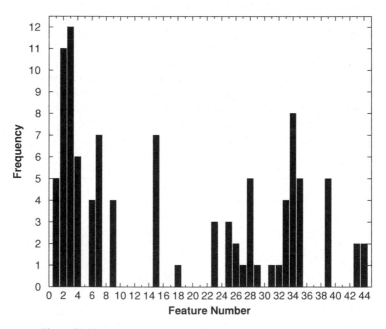

Figure 11.14   Frequency histogram for feature selection: log features.

Table 11.8   Best neural networks for each feature number: linear and log features

Number of features	Linear features			Log features		
	Error rate	Number of units	Number of weights	Error rate	Number of units	Number of weights
1	0.700	2	31	0.549	27	306
2	0.525	4	57	0.357	6	81
3	0.337	30	399	0.189	26	347
4	0.283	20	289	0.118	19	275
5	0.163	30	459	0.066	28	429
6	0.146	24	393	0.054	16	265
7	0.091	29	502	0.022	17	298
8	0.074	25	459	0.020	14	261
9	0.066	24	465	0.019	10	199
10	0.060	14	289	0.020	10	209

Noting that the error rate corresponding to the semi-objective method of choosing nine features was 0.135 (on the testing set), one can see that a reduction to seven linear features or four log features is possible with no reduction in performance when the features are optimised in a principled manner.

11.9.3   Issues of Network Generalisation

The results listed in Table 11.8 indicate that neural networks with quite a large number of weights were necessary to give the required performance. This result would normally raise questions about the generalisation capacity of the resulting networks, given that there is a comparatively small amount of training data – 594 patterns to be specific. As mentioned earlier, the often-quoted rule-of-thumb (Tarassenko, 1998) is that, to avoid overtraining of neural networks, one should have ten training patterns for each weight in the network. This rule-of-thumb is clearly not the case here. A more analytical guide, based on Vapnik's statistical learning theory (Vapnik, 1995), is given in Bishop (1998). It is stated that, to classify a fraction $1 - \varepsilon$ of new patterns correctly, the training set should contain a number of patterns n given by at least

$$n = \frac{W}{\varepsilon} \tag{11.67}$$

where W is the number of weights in the network. For an error rate of $\varepsilon = 0.1$ or a 90% success rate, this analytical guide agrees with the ten-pattern rule discussed above. In order to achieve the success rate indicated by the semi-objective approach, one would require a minimum number of patterns given by

$$n = 7.41W \tag{11.68}$$

whereas the ratio actually found for the optimum nine-feature log network is $n = 2.98W$. In other cases for different numbers of features the ratio is worse than this, particularly for the linear feature sets. There are two comments that can be made in defence of the analysis here and one to the contrary.

In defence of the analysis, one might first say that the theory that leads to Equation (11.67) actually guarantees only *worst-case bounds* on generalisation performance and it would be hoped in practice that one could achieve good generalisation with fewer training patterns than suggested by Equation (11.67). The second comment in support of the results here is that the neural network training has actually been carried out in a principled manner that incorporates the practice of 'early stopping'; that is, networks are chosen that have been trained to the point where the validation error reaches its minimum and begins to rise again. As mentioned earlier, such an approach has been shown to be equivalent to weight-decay regularisation and is an aid to generalisation.

The problem with the analysis here relates to the second observation above. Early stopping relies on the existence of a validation set that is appropriately 'independent' of the training set. If the training and validation sets are too highly correlated, then the validation error will decrease as long as the training error decreases and this decrease will invalidate the practice of early stopping. In fact, the independence of the training and validation sets used here could be called into question. The reason for this concern is that the training, validation and testing sets were obtained by taking all the measured patterns for the fault cases and then cycling through this set, taking every third pattern into the training data and so on. The three data sets are thus interleaved. One might argue that this process would maximise the correlations between the training and validation sets. This strategy was used here for a good reason. When the damage cases were examined, 100 patterns were recorded for each panel removed and then later another 100 patterns were recorded for a completely independent removal of the same panel. One might argue that 66 of the first 100 points should have been used for training while 66 of the second 100 points should be reserved for validation. In fact, this approach would have been doomed to failure because of the large operational variations in the data as a result of fixing the panels down each time with different boundary conditions. The work by Worden, Manson and Allman (2003) showed clearly that the effects of the boundary conditions are able to conceal the effects of the faults. As there was not enough time in the experimental programme to consider many panel removals and thus sample the whole range of operational conditions, it was decided to interleave the training, validation and testing sets as described above, in order to make the most of the available data.

Because there is evidence both for and against overtraining here, it was decided to investigate the effect on network performance of reducing the allowed number of hidden neurons. In the exercises described earlier, the neural networks were allowed to have 10 different initial conditions, stopping times up to 200 000 presentations of data and up to 30 hidden units. The exercise was repeated, using the linear features as they generally produced the networks with the most weights, but only allowing up to 10 hidden neurons. The results obtained are listed in Table 11.9.

Note that the results in Table 11.9 are the *testing* errors, so it is not impossible for lower errors to be obtained than in the earlier simulations. As one can see from Table 11.9, there is no marked deterioration in the results when the number of hidden units are restricted. In fact, the overall best error is obtained for a ten-feature network and is lower than that previously obtained with more hidden units (Table 11.8).

Table 11.9 Best neural networks for each feature number: linear features, maximum of ten hidden units

Linear features			
Number of features	Error rate	Number of units	Number of weights
1	0.700	2	31
2	0.525	4	57
3	0.340	10	139
4	0.279	7	107
5	0.180	8	129
6	0.124	10	169
7	0.091	7	128
8	0.074	9	171
9	0.072	9	180
10	0.057	6	129

Table 11.10 Best neural networks for each feature number:
linear features, maximum of five hidden units

	Linear features		
Number of features	Error rate	Number of units	Number of weights
1	0.700	2	31
2	0.525	4	57
3	0.337	4	61
4	0.278	4	65
5	0.187	5	84
6	0.170	4	73
7	0.099	5	94
8	0.104	5	99
9	0.104	5	104
10	0.079	5	109

The best results are obtained with a network with six hidden units, for which $n = 4.6W$. In the case of the networks trained with log features, the best performance was actually obtained with a network with ten hidden units, so a similar exercise would be unnecessary.

A further exercise was carried out with a restriction to a maximum of five hidden units and the results are listed in Table 11.10. As one might expect, this restriction does lead to a marked deterioration in performance, although the best error rate obtained (0.079 at ten features) is still much better than that obtained with the semi-objective approach to feature selection. The log features fare better here and the nine-feature network with five hidden nodes achieves an error rate of 0.04 corresponding to a 96% success rate, which is radically better than that achieved using the semi-objective approach. The nine-feature log network has $n = 5.7W$.

In summary, even taking into account concerns about the generalisation capacity of some of the networks in Table 11.8, it is clear that superior results are obtained using the principled evolutionary optimisation procedure, even with marked restrictions on the number of hidden units. In the case of the nine-feature log networks, the performance only falls from 98.1% to 96% when the number of hidden neurons is reduced from ten to five. Note that low model complexity is not a guarantee in itself of generalisation capability; however, it is at least a necessary condition.

11.9.4 Discussion and Conclusions

Since its introduction, the GA has proved itself to be extremely powerful at solving combinatorial optimisation problems in general and sensor/feature optimisation problems in particular (a recent example from the condition monitoring field is that of Jack and Nandi, 2000. It is no surprise that it proves advantageous in the SHM context described here. Together with a simple feature transformation (taking logarithms), the GA feature selection algorithm provides a nine-feature distribution with a classification rate of 98.1% (on testing data) compared with a rate of 86.5% based on a semi-objective selection strategy and linear features. The optimisation procedure is reasonably inexpensive (computationally) and there is every reason to recommend its use for general feature selection problems. There is still work to be done in removing the element of subjectivity present in the current approach to feature selection. This reduction in subjectivity can be addressed by folding the selection of the transmissibility frequency ranges into the overall optimisation procedure. This modification to the optimisation process is a more difficult task and work is in progress (Manson, Papatheou and Worden, 2008). A further step in the direction of complete automation of the feature selection/diagnosis process would be to also include the optimisation of the sensor positions. This sensor optimisation process is even more difficult and an effective strategy will require considerable further research.

The second main conclusion is not directly concerned with the optimisation, but has been forced by the re-evaluation of the data and is more a matter of generalisation of data-driven models like neural networks. It is an undisputed fact that the study here has a small amount of data available for training. This limitation is ultimately due to the fact that the aircraft concerned was only available in a one-week window. This time constraint meant that despite the fact that the boundary conditions of the panel were known to be a matter of some concern, only two repetitions of each panel removal were possible. The way that the training, validation and testing sets were constructed here means that the neural networks were able to distinguish different fault cases, despite the fact that each case spanned two boundary conditions. However, experience with the problem suggests that two samples is not enough to span the range of possible operational conditions in this case and that the neural networks produced may not generalise to unseen data from different fixing conditions. One is forced to the conclusion that if one is to pursue data-driven approaches to damage identification, one must take the time to acquire appropriate data spanning all possibilities. This observation is even more important in the context of novelty detection, where one must ensure that the diagnostic system does not signal an alarm because a previously unobserved form of the normal condition occurs. The question of accounting for all possible normal conditions given natural environmental and operational variations will be discussed in much more detail in Chapter 12.

11.10   Discussion and Conclusions

There are many lessons to be learned from the examples in this chapter. Perhaps the overwhelming message is that an SHM system based on machine learning can only perform as well as the data that has been used to train the diagnostic; the adage *garbage in–garbage out* is particularly apt.

If one wishes to conduct supervised learning, one is immediately faced with the most important question for SHM based on machine learning; how does one acquire data corresponding to any damage states? However, one should bear in mind that this question is already raised at the lowest level of detection. The reason that the question is already pertinent at the detection level is that it is necessary to decide features that distinguish between the normal condition of the system or structure and the damaged conditions, and this is not possible without examples of data from the damage conditions. In the examples in Sections 11.3 and 11.9 above, the features were based on selected regions of certain transmissibility functions. In the absence of examples from the damage cases it would not be possible to assess if a given transmissibility peak was sensitive to a given type of damage, or in fact any type of damage. The GA feature selection procedure discussed in Section 11.9 is only possible because one can form an objective function based on the error of the classifier. In the case of damage *detection*, this problem can potentially be overcome by training novelty detectors for each candidate feature and then monitoring all of them for threshold crossings on new data. This approach would be tedious, but effective. In the case of a damage location system, data for each class of damage becomes essential. As discussed previously at various points in this book, such data can only be acquired in two ways, by modelling or from experiment. Both approaches have potential problems. If one considers modelling, one must hope that a low-cost model should suffice; otherwise one simply invests all the effort that a model-driven approach like finite element model updating would require anyway. If one considers experiment, it will not generally be possible to accumulate data by damaging the structure in the most likely ways unless the structure is extremely cheap and mass-produced. Unless safety-critical, such structures are typically not good candidates for SHM systems. It is obvious that a testing programme based on imposed damage could not be used on, say, an aircraft wing. If one cannot impose real damage, one might be able to simulate the effects of damage experimentally. This latter suggestion shows some promise; it has been shown by Papatheou *et al.* (2010) that a procedure as simple as adding masses nondestructively to the structure can give guidance on the effectiveness of features for novelty detection.

The second major problem in damage identification was also raised in Section 11.9. Without careful feature selection, the variations in the measured data caused by boundary condition changes in the structure swamped the changes caused by damage. (This is also a problem for model-based approaches.) This observation is particularly pertinent for civil engineering infrastructure. If one wishes to carry out a program of automatic monitoring for an aircraft, it is conceivable that one might do it off-line in the reasonably well-controlled environment of a hangar. Such a controlled testing environment is not possible for a bridge that is at the mercy of the elements. As will be discussed in Chapter 12, it is known that changes in the natural frequencies of a bridge as a result of daily temperature variation are likely to be larger than the changes in these features caused by damage. Bridges will also have a varying mass as a result of taking up moisture from rain and from varying traffic loading. There are two possible solutions to this problem. The first is to accumulate normal condition data spanning all the possible environmental conditions. Such data acquisition is time-consuming and will generate such a large normal condition set that it is likely to be insensitive to certain types of damage. The second solution is to determine features that are insensitive to environmental changes but sensitive to damage. Both of these strategies will be discussed in detail in the next chapter.

A third problem relating to data-driven approaches is that the collection or generation of data for training the diagnostic is likely to be expensive; this means that the data sets acquired are likely to be sparse. This situation puts pressure on the feature selection activities as sparse data will usually require low-dimensional features if the diagnostic is ever going to generalise beyond the training set. There are possible solutions to this situation; for example, regularisation can be used in the training of neural networks in order to aid generalisation and this process can be as simple as adding noise to the training data. Other possibilities are to use learning methods like the support vector machines discussed in Section 11.5, which are implicitly regularised and therefore better able to generalise on the basis of sparse data.

One issue that applies equally to data-driven and model-driven approaches is that they are more or less limited to levels 1 to 4 of the damage classification heirarchy presented in Chapter 1 (Rytter, 1993). If one is to pursue damage prognosis, it is necessary to extrapolate rather than interpolate and this is difficult for machine learning solutions. If prognosis is going to be possible, it is likely to be very context specific and to rely critically on understanding the physics of the damage progression. In certain simple cases, it is already possible to make such calculations. For example, for a crack in a metallic specimen with a simple enough geometry to allow the theoretical specification of a stress intensity, one can use the Paris–Erdogan law (for details, the reader can consult any basic textbook on fatigue and fracture) to predict the development of the crack given some estimate of the future environmental and operational loading history. Even here there are problems. First of all, the loading future is uncertain and it may only be possible to specify bounds. Second, the constants of the Paris–Erdogan equation are strongly dependent on material microstructure and would probably have to be treated as random variables in a given prediction. These observations are intended to show that prognosis is only likely to be possible in the framework of a statistical theory where the uncertainty in the calculation is monitored at all stages. Another major stumbling block in the application of prognosis is that most realistic situations will not be backed up by applicable theory; that is, the laws of damage progression are not known for 'simple' materials with complicated geometry or for 'complicated' materials like composite laminates, even with simple geometries. These matters will be discussed in much more detail in Chapter 14 of this book.

The overall conclusion for this chapter is that if the conditions are favourable, machine learning algorithms can be applied to great effect on damage identification problems. In the light of the comments above, 'favourable conditions' largely means that data are available in order to adequately train the machine learning diagnostics and the optimal conditions would mean that data are available from damage states so that supervised learning could be used. Even if the conditions seem to exclude such a solution, one should bear in mind that even a model-driven approach will need appropriate data for model validation will be subjected to many of the same sources of uncertainty persistent in the machine learning methods discussed in this chapter.

References

Abeles, M. (1991) *Corticonics – Neural Circuits of the Cerebral Cortex*, Cambridge University Press.

Bennett, K.P. (1999) Combining support vector and mathematical programming methods for classification, in *Advances in Kernel Methods – Support Vector Learning* (eds B. Schőlkopf, C.J.C. Burgess and A.J. Smola), MIT Press.

Bishop, C.M. (1994) Training with noise is equivalent to Tikhonov regularization. *Neural Computation*, **7**, 108–116.

Bishop, C.M. (1998) *Neural Networks for Pattern Recognition*, Oxford University Press.

Bornn, L., Farrar, C.R., Park, G. and Farinholt, K.M. (2009) Structural health monitoring using autoregressive support vector machines. *ASME Journal of Vibration and Acoustics*, **131**(2), 021004-1–9.

Brockwell, P. and Davis, R. (1991) *Time Series Analysis: Forecasting and Control*, Prentice-Hall.

Broomhead, D.S. and Lowe, D. (1988) Multivariable functional interpolation and adaptive networks. *Complex Systems*, **2**, 321–355.

Bryson, A., Denham, W. and Dreyfuss, S. (1963) Optimal programming problem with inequality constraints. I: necessary conditions for extremal solutions. *AIAA Journal*, **1**, 25–44.

Burges, C.J.C. (1998) A tutorial on support vector machines for pattern recognition. *Data Mining and Knowledge Discovery*, **2**, 1–47.

Cherkassky, V. and Mulier, F.M. (1998) *Learning from Data: Concepts, Theory and Methods*, John Wiley & Sons, Inc.

Cristianini, N. and Shawe-Taylor, J. (2000) *An Introduction to Support Vector Machines and Other Kernel-Based Learning Methods*, Cambridge University Press.

Cybenko, G. (1989) Approximation by superpositions of sigmoidal functions. *Mathematics of Control, Signals and Systems*, **2**, 303–314.

Elman, J.L. (1990) Finding structure in time. *Cognitive Science*, **14**, 179–211.

Frie β, T.-T., Cristianini, N. and Campbell, C. (1998) The kernel adatron: a fast simple learning procedure for support vector machines. Machine Learning: Proceedings of the Fifth International Conference (ed. J. Shavlik), Morgan Kaufman.

Goldberg, D.E. (1989) *Genetic Algorithms in Search, Optimization, and Machine Learning*, Addison-Wesley Publishing Co., Inc., Reading, MA.

Haykin, S. (1994) *Neural Networks. A Comprehensive Foundation*, Macmillan College Publishing Company.

Hayton, P., Utete, S., King, D. *et al.* (2007) Static and dynamic novelty detection methods for jet engine health monitoring. *Philosophical Transactions of the Royal Society, Series A*, **365**, 493–514.

Hebb, D.O. (1949) *The Organisation of Behaviour*, John Wiley & Sons, Inc., New York.

Hopfield, J.J. (1984) Neural networks and physical systems emergent collective computational abilities. *Proceedings of the National Academy of Sciences, USA*, **52**, 2554–2558.

Huber, P. (1964) Robust estimation of location parameters. *Annals of Mathematical Statistics*, **35**, 73–101.

Jack, L.B. and Nandi, A. (2000) Genetic algorithms for feature selection in machine condition monitoring with vibration signals. *IEE Proceedings – Vision, Image and Signal Processing*, **147**, 205–212.

Joachims, T. (1999) Making large-scale supportvector machine learning practical, in *Advances in Kernel Methods - Support Vector Learning* (eds B. Schőlkopf, C.J.C. Burgess and A.J. Smola), MIT Press.

Kohonen, T. (2000) *Self-Organising Maps*, 3rd edn. Springer.

LeCun, Y. (1986) Learning processes in an asymmetric threshold network. *Disordered Systems and Biological Organisations*, Les Houches, France, Springer, pp. 233–240.

Luo, F.L. and Unbehauen, R. (1998) *Applied Neural Networks for Signal Processing*, CambridgeUniversity Press.

Manson, G., Papatheou, E. and Worden, K. (2008) Genetic optimisation of a neural network damage diagnostic. *The Aeronautical Journal*, **112**, 267–274.

Manson, G., Worden, K. and Allman, D.J. (2003a) Experimental validation of a structural health monitoring methodology. Part II: novelty detection on an aircraft wing. *Journal of Sound and Vibration*, **259**, 345–363.

Manson, G., Worden, K. and Allman, D.J. (2003b) Experimental validation of a structural health monitoring methodology. Part III: damage location on an aircraft wing. *Journal of Sound and Vibration*, **259**, 365–385.

McCulloch, W.S. and Pitts, W. (1943) A logical calculus of the ideas immanent in nervous activity. *Bulletin of Mathematical Biophysics*, **5**, 115–133.

Mercer, J. (1909) Functions of positive and negative type and their connection with the theory of integral equations. *Philosophical Transactions of the Royal Society, Series A*, **209**, 415–446.

Minsky, M.L. and Papert, S.A. (1988) *Perceptrons*, MIT Press.

Moller, M.F. (1993) A scaled conjugate gradient algorithm for fast supervised learning. *Neural Networks*, **6**, 525–533.

Oh, C.K. and Sohn, H. (2009) Damage diagnosis under environmental and operational variations using unsupervised support vector machine. *Journal of Sound and Vibration*, **325**, 224–239.

Papatheou, E., Manson, G., Barthorpe, R.J. and Worden, K. (2010) The use of pseudo-faults for novelty detection in SHM. *Journal of Sound and Vibration*, **329**, 2349–2366.

Platt, J.C. (1998) Sequential Minimal Optimisation: A Fast Algorithm for Training Support Vector Machines. Technical Report MSR-TR-98-14, Microsoft Research.

Press, W.H., Teukolsky, S.A., Vetterling, W.T. and Flannery, B.P. (1992) *Numerical Recipes in C*, Cambridge University Press.

Ramón y Cajál, S. (1911) *Histologie du Systémenerveux de l'Hommeet des Vertébrés*, Maloine, Paris.

Rosenblatt, F. (1962) *Principles of Neurodynamics*, Spartan.

Rumelhart, D.E., Hinton, G.E. and Williams, R.J. (1986) Learning representations by back propagating errors. *Nature*, **323**, 533–536.

Rytter, A. (1993) Vibration-based inspection of civil engineering structures. PhD Thesis, Department of Building Technology and Structural Engineering, University of Aalborg, Denmark.

Schőlkopf, B., Williamson, R.C., Smola, A.J. *et al.* (2000) Support vector method for novelty detection, in *Advances in Neural Information Processing Systems*, vol. 12 (eds S.A. Solla, T.K. Leen and K. Műller), MIT Press, pp. 582–588.

Tarassenko, L. (1998) *A Guide to Neural Computing Applications*, Arnold.

Vapnik, V. (1995) *The Nature of Statistical Learning Theory*, Springer-Verlag.

Werbos, P.J. (1974) Beyond regression: new tools for prediction and analysis in the behavioural sciences. PhD Thesis, Applied Mathematics, Harvard University.

Widrow, B. and Hoff, M.E. (1960) Adaptive Switching Circuits. IRE Wescon Convention Records, Part 4, pp. 96–104.

Worden, K. and Burrows, A.P. (2001) Optimal sensor placement for fault diagnosis. *Engineering Structures*, **23**, 885–901.

Worden, K., Manson, G. and Allman, D.J. (2003) Experimental validation of a structural health monitoring methodology. Part I: novelty detection on a laboratory structure. *Journal of Sound and Vibration*, **259**, 323–343.

12

Data Normalisation

12.1 Introduction

The first eleven chapters of this book have focused on approaches that can be used to determine if damage is present in a structure based on changes in some measured system dynamic response characteristics. In addition to the ones presented in the previous chapters, there are many other damage detection techniques that have been proposed in the technical literature. However, as originally reported, many of these existing methods neglect the important effects of *environmental and/or operational variations* (EOVs) on the underlying structure, even though studies have existed for quite some time showing that such EOVs can significantly alter the dynamic response characteristics of a structure (Loland and Dodds, 1976; Askegaard and Mossing, 1988). This situation arises from the fact that in many of these studies the damage detection procedures are validated on numerical simulations of the structure's response to dynamic loading or with experiments conducted in a well-controlled laboratory environment. Several of the test structures described in Chapter 5 were tested in this manner. These idealised situations are a necessary first step in the development of an SHM procedure, but for *in situ* structures the changes in dynamic response that can result from time-varying EOVs must be considered in the SHM process. As such, this chapter will focus on *data normalisation* procedures. As it applies to SHM, data normalisation is defined as the process of separating changes in the features derived from sensor readings that are caused by damage from those changes caused by EOVs. The authors believe that the ability to perform robust data normalisation is one of the biggest challenges facing SHM when attempting to transition this technology from research to field deployment and practice on *in situ* structures.

To begin the discussion of data normalisation consider Figure 12.1, which shows a hypothetical case where the damage-sensitive feature distribution (represented by ellipses) is influenced by some EOV parameter denoted T_i, where the index i labels one of four environmental or operational conditions. T_i causes the mean and variances of the two-dimensional feature distribution to shift, as indicated by the change in the location of the ellipses' centroids and their size and orientation. If EOVs produce changes in the damage-sensitive features that are similar to the changes produced by damage, as depicted in Figure 12.1, additional measurements of this EOV parameter may be required to provide the information necessary to remove this effect from the damage detection process.

The alternative (shown in Figure 12.2) is that the damage may produce changes in the measured system response data and corresponding damage-sensitive features that are in some way orthogonal to the changes produced by EOVs. In this case there is the possibility that data normalisation can be achieved without direct measurement of the EOV parameters that are causing the changes. However, it is often difficult to discern a priori how the changing EOVs will influence the features used in the SHM process relative to how damage will influence these features, particularly when one initially only

Structural Health Monitoring: A Machine Learning Perspective, First Edition. Charles R. Farrar and Keith Worden.
© 2013 John Wiley & Sons, Ltd. Published 2013 by John Wiley & Sons, Ltd.

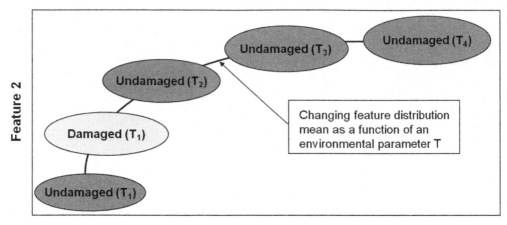

Feature 1

Figure 12.1 A hypothetical case where changes in the feature distribution caused by damage are similar to those caused by a varying environmental condition.

has data available from the undamaged condition, as will be the case for many SHM applications. This situation corresponds to the unsupervised learning mode of SHM discussed in Chapter 10. In actual practice the SHM system may need to be deployed for a long period of time before enough data are collected such that robust data normalisation procedures can be developed. Even for the case depicted in Figure 12.2, it is always preferable to have actual measures of the EOV parameters that are influencing the features.

Regardless of whether the situation shown in Figure 12.1 or 12.2 prevails, data normalisation has to be accomplished with data analysis procedures or through the use of appropriated measurement hardware and data acquisition procedures. In most cases, some combination of hardware and software will be developed to address this issue.

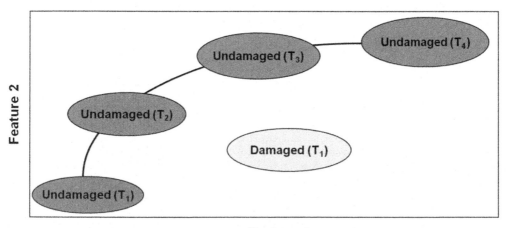

Feature 1

Figure 12.2 Hypothetical case where the change to the feature distribution caused by damage is in some way orthogonal to changes caused by varying operational and environmental conditions.

This chapter first provides motivation for considering the effects of EOVs on a measured system response. Then, six approaches to data normalisation are presented:

1. Experimental approaches.
2. Regression modelling.
3. Look-up tables.
4. Machine learning approaches.
5. Intelligent feature selection.
6. Cointegration.

For approaches 2 to 6, their effectiveness will be directly dependent on the training data used to develop these various data normalisation methods.

12.2   An Example Where Data Normalisation Was Neglected

Although many damage detection techniques have successfully been applied to numerical simulations of structural response, scale models or specimens tests in controlled laboratory environments, the performance of these techniques in the field requires additional validation. As an example, one can revisit the I-40 Bridge study introduced in Chapter 5 where modal parameters were used to identify structural damage. Four different levels of damage were introduced to the bridge by gradually cutting one of the girders. The change of the bridge's fundamental frequency is plotted for the undamaged case (labelled 0) and with respect to the four damage levels (labelled 1 to 4 in order of increasing severity), as shown in Figure 12.3. Because the magnitude of the bridge's natural frequency is proportional to its stiffness, a decrease in frequency is expected as the damage progresses. However, the results in Figure 12.3 contradict this expectation. In fact, the frequency value increased for the first two damage levels and then eventually decreased for the remaining two damage cases. Subsequent studies identified the ambient temperature variations as a possible source of differences in the bridge's measured dynamic response. This temperature effect was not considered at the time of the tests and no effort was made to either measure the temperature conditions or to conduct the tests at the same time of day in order to minimise this source of variability. Subsequent studies of other bridges have noted similar modal property sensitivities

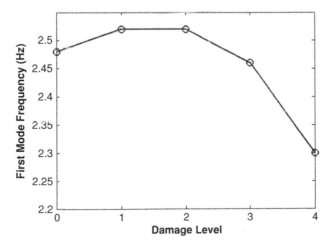

Figure 12.3   First-mode frequency change for the I-40 Bridge as a function of the four damage levels shown in Figure 5.4.

to EOVs (Roberts and Pearson, 1998; Helmicki *et al.*, 1999; Williams and Messina, 1999; Rohrmann *et al.*, 1999; Alampalli, 2000; Farrar *et al.*, 2000; Peeters and De Roeck, 2001; Xia *et al.*, 2006). As a further example, the influence of traffic loading on a 46-m-long simply supported plate-girder bridge caused the natural frequencies to decrease by up to 5.4% (Kim *et al.*, 2003). In another study of highway bridges, variations in modal properties on the order of 5% to 10% were observed and again attributed to changing thermal conditions (Ko and Ni, 2005). In a more recent study of a box-girder concrete bridge, variations in the first natural frequency on the order of ±10% were observed (Soyoz and Feng, 2009).

12.3   Sources of Environmental and Operational Variability

Common environmental conditions that can cause variability in the measured system response include aerodynamic (turbulence experienced by aircraft) and hydrodynamic (sea states experienced by ships and offshore oil platforms) loading, radiation environments, varying temperature, varying wind conditions and changing moisture and humidity levels. Changing operational conditions depend on such things as varying live-load conditions, power fluctuations, start-up and shut-down transients, changing operational speed, vehicle manoeuvres, roadway conditions and varying mass loading including changing fuel levels and changing payloads. Almost always, a structure will be simultaneously subjected to multiple EOVs and the influence of all sources of variability must be assessed.

Temperature variations influence almost all material properties (e.g. elastic modulus, yield stress and mass density) and typically result in geometric distortion of the structure from some nominal condition. As an example, modal testing of the Alamosa Canyon Bridge (summarised in Chapter 5) revealed that the first-mode frequency of the bridge varied by approximately 5% during the 24-hour cycle, as shown in Figure 12.4. This variation in frequency was not correlated with the absolute air temperature, but rather with the temperature differential across the bridge deck. It was assumed that the north–south orientation of this bridge and the fact that expansion joints had not been maintained as shown in Figure 12.5 contributed to the sensitivity of the modal frequencies to varying thermal conditions. Similar results were observed when these tests were repeated a year later (Farrar *et al.*, 2000). The importance of accounting for such temperature variability in the damage detection process is recognised when one considers that the final damage condition for the I-40 Bridge (Figure 5.4), which most people would consider somewhat severe,

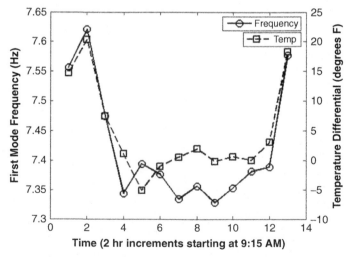

Figure 12.4   Correlation between temperature differential across the Alamos Canyon Bridge and changes in the first-mode resonance frequency.

Figure 12.5   Debris in the Alamosa Canyon Bridge expansion joint.

only produced a 7% reduction in the first-mode resonance frequency. Furthermore, tests performed with and without traffic on the I-40 Bridge showed a 4% change in the first-mode frequency. Clearly the changes in frequency produced by severe damage are of the same order of magnitude as changes that can be expected by varying environmental conditions (thermal) or operational conditions (traffic loading).

A subsequent study on the Alamosa Canyon Bridge was carried out in 2008 and it was found that the bridge frequencies again changed over a 24-hour cycle in a manner similar to that observed 10 years earlier. This more recent study also identified an asymmetrical variation in the first-mode shape that changed throughout the day, as had been identified 10 years previously (Figueiredo *et al.*, 2011). The asymmetry along the longitudinal axis shown in Figure 12.6 was correlated with the time of day and the associated thermal environments. If not properly accounted for, such changes in the dynamic response characteristics can potentially result in false indications of damage. If the mode in Figure 12.6(a) is considered to be the baseline undamaged condition, a novelty detection algorithm as described in Chapter 10 would identify the mode in Figure 12.6(b) as anomalous. This novelty could inappropriately be associated with a damaged structure if the environmental variability influencing this feature was not taken into account in the novelty detection process.

To highlight the difficulties posed by varying environmental conditions further, consider the ship structure shown in Figure 12.7, which was instrumented with a fibre optic strain measurement system to monitor the response of the composite hull during sea trials. Figure 12.8 shows responses of the structure during various sea trials. The top two figures correspond to the structure in the 'undamaged' condition, but measured during vastly different sea states. The bottom figure corresponds to the structure

(a)   (b)

Figure 12.6   First-mode shape of one span of the Alamosa Canyon Bridge during two distinct times of the day: (a) in the morning (7.75 Hz); (b) in the afternoon (7.42 Hz).

Figure 12.7 Composite hull ship monitored with a fibre optic strain sensing system (courtesy of the US Navy).

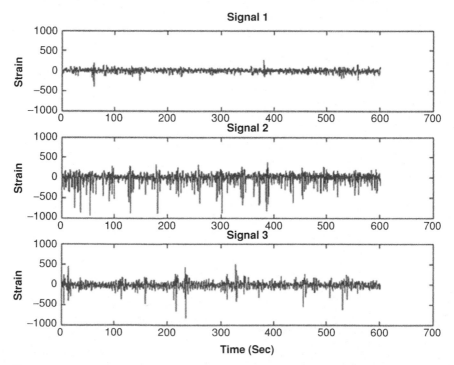

Figure 12.8 Strain time histories measured on the ship shown in Figure 12.7 when it was in different conditions and operating in varying sea states.

Figure 12.9   Ice build-up on a ship that alters its mass and stiffness properties (courtesy of the US National Oceanic and Atmospheric Administration).

in a different condition and measured during a third sea state. Clearly, it would be quite challenging to separate the effects of the varying sea states from the changes in the dynamic response measurements that are associated with the different system conditions.

As a final example, consider the ice build-up shown on the ship in Figure 12.9. The ice can alter the structural response of the ship in several manners. First, the ice can produce substantial mass changes up to the point where the ship can become unstable. The ice can produce local gradients in material properties where material near the ice is maintained at one temperature and exposed material will experience ambient air temperatures. Additionally, the ice can stiffen the structure locally, further changing the dynamic response characteristics of the system.

In summary, there is a wide array of EOVs that can influence the dynamic response characteristics of a particular structure and often it is not clear how all these sources of variability interact to produce the observed response. Nonetheless, it is imperative that these sources of variability are addressed in the damage detection process in an effort to avoid incorrect diagnoses. The subsequent sections of this chapter will introduce various procedures that can be used to minimise the influence of EOVs on the SHM process.

12.4   Sensor System Design

As discussed in Chapter 3, an important part of the operational evaluation step in the SHM process is to define the constraints under which the SHM system needs to perform. Included in these constraints must be the EOVs. Once these constraints have been defined, then it is always advantageous to make direct measurements of any EOV parameters that might influence the dynamic responses that are being monitored as part of the SHM process. As an example, the ship shown in Figure 12.10 was monitored during sea trials. In addition to the strain and acceleration measurements that were used to assess the structural condition, many operational and environmental parameters were also measured, including: speed, propeller shaft rotational speed, heading relative to the wind, wave height and fuel level (Brady, 2004). Clearly, these EOV parameters will be useful when attempting to discern if any changes in the

Figure 12.10   US Navy ship being monitored during sea trials (courtesy of the US Navy).

measured structural response are the result of damage. Therefore, the SHM sensor system should be designed to accommodate the necessary measurements needed to characterise the EOVs accurately. However, it is not always straightforward to determine the necessary measurements a priori. As in the Alamosa Canyon Bridge example above, it was first assumed that the absolute air temperature was the most important environmental parameter that would capture the influence of varying thermal conditions on the bridge's dynamic response. As previously mentioned, after subsequent studies the results shown in Figure 12.4 revealed that, in fact, the temperature differential across the bridge was the important environmental parameter to monitor. This parameter required additional data acquisition channels to those needed for the single air temperature measurement.

Several sensing approaches offer some advantages when significant changes to the structure's ambient vibration conditions arise. As an example, recall the impedance method that was described in Chapter 4 as an approach to local damage detection. With this active sensing approach, the same piezoelectric patches are used to both excite the structure with a swept-sine signal, typically in the 30 to 200 kHz frequency range, and to measure the response to this excitation. These excitation frequencies are sufficiently high that they are not influenced by most mechanical vibrations arising from typical EOVs. As an example, the test structure shown in Figure 12.11(a) was subjected to a variety of base inputs including random, transient and sinusoidal excitations with frequency contents ranging from 20 Hz to 10 kHz. The baseline impedance measurements between 40 kHz and 50 kHz (see Figure 12.11b) that were measured when these various inputs were applied are shown to essentially overlay, implying that the lower-frequency vibrations from the shaker are having no influence on the higher-frequency impedance measurements. When damage was simulated by loosening a bolt at one of the joints, there was a clear change in the impedance signature that is distinct from any of the measurements made with the tight bolt that were accompanied by the shaker excitations shown in Figure 12.11(c). It should be noted that although the impedance measurements were shown to be insensitive to lower-frequency vibrations, they have been shown to be sensitive to varying thermal environments and several studies have focused on compensating for the thermal effect on impedance measurements (Pardo de Vera and Guemes, 1997; Park, Cudney and Inman, 1999). In many instances, guided wave and acoustic emission approaches to SHM have also been shown to have similar insensitivity to lower-frequency ambient vibration.

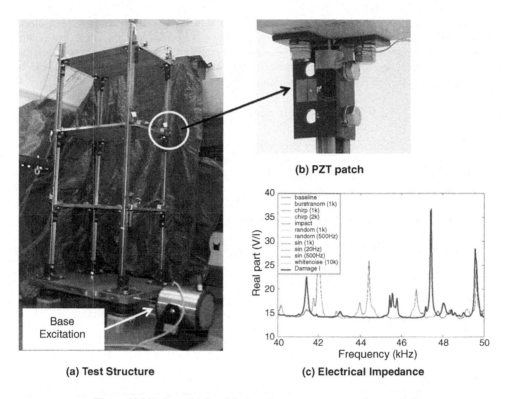

Figure 12.11 Insensitivity of the impedance measurement to base excitations.

In summary, the data normalisation process begins with the operational evaluation portion of the SHM process where one attempts to identify any EOVs that might influence the process. With this information, one can proceed to design the data acquisition system so that appropriate environmental and operational parameters are monitored. Additionally, for certain types of EOV, the structural monitoring system can be designed to minimise the influence of the EOVs on the structural response measurements. The next section will discuss how measured environmental and operational parameters can be used in the data normalisation process.

12.5 Modelling Operational and Environmental Variability

If measurements of the environmental and operational parameters are available, various types of models can be developed to predict the influence of these parameters on the measured response and, hence, damage-sensitive features. Almost all modelling that attempts to capture the influence of measured environmental and operational parameters on the system response and associated damage-sensitive features is based on a variety of regression techniques. To begin, consider one of the simplest types of models – those obtained by linear regression – that can be used to predict the relationship between a measured environmental and operational parameter, t_i (e.g. temperature or vehicle speed) on a vector of damage-sensitive features, $\{f\}_i$. This model is described by a linear function of the form

$$\{f\}_i = \{a\} + \{b\}t_i \tag{12.1}$$

Here the subscript i designates the discrete point of time at which the measurements were obtained. After acquiring various feature vectors and the corresponding EOV parameters, the coefficient vectors $\{a\}$ and $\{b\}$ can be estimated through a linear least-squares process. Clearly, there is no restriction that the feature vector must be a simple linear function of the EOV parameter and Equation (12.1) can be extended to the more general polynomial relationship defined by

$$\{f\}_i = \{a\} + \{b\}t_i + \{c\}t_i^2 + \cdots \tag{12.2}$$

These models assume there is no history dependence between the feature vector and the EOV parameter; that is, in temporal terms they are *static* rather than *dynamic*. However, history dependence can be important when inertial effects influence the relationship between the EOV parameters and the damage-sensitive features. History dependence can be incorporated by considering previous values of the measured EOV parameter resulting in the following model form when the feature vector is considered to be a history-dependent quadratic function of an EOV parameter:

$$\{f\}_i = \{a\} + \{b\}t_i + \{c\}t_i^2 + \{d\}t_{i-1} + \{e\}t_{i-1}^2 + \{f\}t_{i-2} + \{g\}t_{i-2}^2 + \cdots \tag{12.3}$$

In general, more time delays as well as higher-order terms can be incorporated into the model.

These models can be further extended to the case where multiple EOV parameters have been measured to produce an EOV vector $\{t_i\}$, which for a history-dependent quadratic function becomes

$$\{f\}_i = [a] + [b]\{t\}_i + [c]\{t^2\}_i + [d]\{t\}_{i-1} + [e]\{t^2\}_{i-1} + [f]\{t\}_{i-2} + [g]\{t^2\}_{i-2} + \cdots \tag{12.4}$$

Note that in its most general form Equation (12.4) would include cross-terms such as a term that is a function of $\{t\}_i$ and $\{t\}_{i-1}$. As the model increases in complexity, more coefficients are needed and therefore more data must be collected in order to estimate these coefficients accurately. However, with sufficient feature vector–EOV parameter pairs available, the coefficients of the models defined by Equations (12.1) to (12.4) can still be estimated by linear least-squares techniques. Several investigators have applied regression techniques to model the influence of EOVs on damage-sensitive features including data from *in situ* structures with particular emphasis on bridge structures (Rohrmann *et al.*, 1999; Peeters and De Roeck, 2001; Worden, Sohn and Farrar, 2002; Fritzen, Mengelkamp and Guemes, 2003; Ni *et al.*, 2005; Ni, Zhou and Ko, 2009; Cross *et al.*, 2010).

As an example consider the Alamosa Canyon Bridge described in Chapter 5. In 1996, experimental modal analyses were performed every two hours for a 24-hour period. In addition, temperature measurements were made at various locations when each modal analysis was performed. Equation (12.4) can be trained with these data to develop a model that can predict the modal frequencies based on the temperature reading. In this example a model was developed that predicts the first two modal frequencies as a function of temperature measurements t_1 and t_2, which are located on the east and west sides of the bridge deck. This choice of temperature readings is based on the results shown in Figure 12.4 where it was seen that the modal frequency is related directly to the temperature differential across the deck. Additionally, the model included the previous temperature readings in an effort to account for the thermal inertia of the system. This model then has the form

$$\begin{Bmatrix} f_1 \\ f_2 \end{Bmatrix}_i = \begin{Bmatrix} a_1 \\ a_2 \end{Bmatrix} + \begin{bmatrix} b_{11} & b_{12} \\ b_{21} & b_{22} \end{bmatrix} \begin{Bmatrix} t_1 \\ t_2 \end{Bmatrix}_i + \begin{bmatrix} c_{11} & c_{12} \\ c_{21} & c_{22} \end{bmatrix} \begin{Bmatrix} t_1 \\ t_2 \end{Bmatrix}_{i-1} \tag{12.5}$$

There are 10 unknown coefficients that must be estimated from the 13 sets of frequency–temperature data. Note that because temperature data from a previous time are being used, only 12 training equations

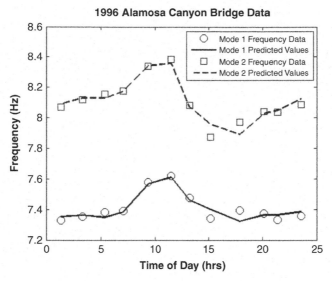

Figure 12.12   Modal frequencies predicted from temperature readings using a linear model with time delay.

can be generated. A linear least-squares solution for the coefficients yields a model that can predict the resonance frequencies reasonably well, as shown in Figure 12.12.

The model developed from the 1996 tests on the Alamosa Canyon Bridge was used to predict the frequencies measured during another series of modal tests performed over a 24-hour period in 1997. The results shown in Figure 12.13 reveal that the model again predicts the frequencies well.

This example illustrates the concept of applying regression analysis to develop a model that predicts how the measured EOVs influence the feature vector. Clearly, a much larger set of measurements that

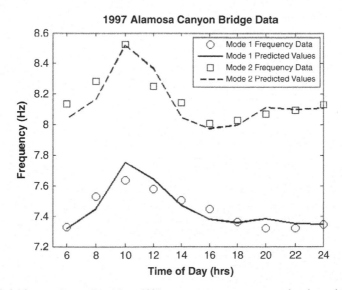

Figure 12.13   Modal frequencies predicted from 1997 temperature measurements using the model trained with 1996 data.

encompasses seasonal variability would be necessary to obtain a more general predictive capability. Also, there are more temperature measurements available and these additional data may produce a more accurate predictive model. A detailed statistical analysis of the full set of temperature measurements has been performed in an effort to identify the optimal set of measurements to use in the modelling process (Sohn *et al.*, 1998). This study yielded a model similar in form to the one described above, but that used a different set of measurements and this model provided improved predictive accuracy over the results shown in Figure 12.12.

After developing these models with training data that are assumed to span the possible EOVs that can be expected in a given SHM application, the damage detection would proceed in much the same manner as previously described for novelty detection based on features derived from model parameters. To begin the process, new dynamic system response data are acquired from the structure in its unknown condition along with the measurements of the associated EOV. Next, a check is made to make sure that these EOV conditions are within the range of those used to develop the functional relationship between features and EOV parameters; that is, the data have not been obtained when the structure experiences new EOVs not used to train the model. Then the model is used to predict the feature vector for the particular EOV and a residual error between the measured and predicted feature vector is obtained. A Mahalanobis squared-distance (MSD) metric can then be used to determine if this residual vector is an outlier to those obtained with the training data. Because EOVs have been accounted for in this process, any outlier is assumed to be caused by damage.

If Equation (12.1) is now modified such that the feature vector is modelled as a different nonlinear function of t_i, in this case an exponential function where the decay coefficient, c, is unknown, the following equation results:

$$\{f\}_i = \{a\} + \{b\}e^{ct_i} \tag{12.6}$$

This functional relationship poses some challenges in terms of estimating the coefficients because it now must be solved as a nonlinear least-squares problem using a more general optimisation process. Many of the machine learning approaches described in Chapters 10 and 11 can be used to develop general nonlinear regression models for predicting the influence of EOV parameters on the damage-sensitive features.

12.6   Look-Up Tables

The methods described above that explicitly model the influence of EOVs on the damage-sensitive features require a direct measurement of the EOV parameters influencing the features. When measures of these parameters are not available, one of the conceptually simplest approaches to data normalisation is to monitor the structure under EOVs while in its undamaged state and create a table of feature vectors that were acquired under these varying conditions. This training phase is complete once one is confident that the features in the look-up table are representative of the range of EOVs the structure is likely to experience. When new data become available from a potentially damaged condition, the feature vector extracted from these new data is compared to the one in the look-up table that is closest to it in terms of a Euclidean distance metric. This process is depicted in Figure 12.14, where the feature vector extracted from the new data (shown as 'damaged?') are shown to be closest in the Euclidean distance sense to the feature from the undamaged condition collected under EOV parameter T_3. Note that one can also consider using the MSD when such comparisons are made with multidimensional feature vectors. However, the Euclidean distance was chosen in an effort to compare the features from as similar environmental and operational conditions as possible rather than to include the influence of those conditions on the distribution of the feature vectors. Novelty detection is then performed using these two sets of features. Note that this approach is giving the new feature vector the 'best chance' of

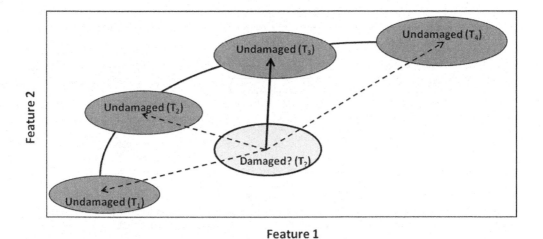

Feature 1

Figure 12.14 Look-up table approach to data normalisation where the test data feature vector acquired under unknown EOV is found to be closest to the undamaged data acquired under EOV T_3.

not being classified as novel and this will minimise false-positive indications of damage. Alternatively, if one wished to minimise false-negative indications of damage, then the comparison can be made to the features that are furthest apart in the Euclidean distance sense. As with all data normalisation procedures, the effectiveness of this method will depend on how representative the features in the look-up table are of the EOVs experienced by the structure. There is no guarantee that features extracted from data acquired under new EOVs not represented in the look-up table will not be classified as an outlier. In fact, the authors would recommend that any outliers identified after such a data normalisation procedure should first be checked to confirm that they do not correspond to some new EOVs not previously observed. This step will be difficult if measures of the EOV parameters are not available.

The look-up table procedure is now illustrated with data from the simulated building structure summarised in Chapter 5. Recall that this structure was designed such that changes in modal frequencies resulting from adding masses to the floor and changing the size of the columns are on the order of those that have been observed during modal tests of civil infrastructure subjected to *in situ* EOV variability. The data set for this example corresponds to the structure being subjected to a random excitation applied with the shaker attached to the base floor. Acceleration measurements that are analysed come from the accelerometer on the third (top) floor. Fifty 8192-sample time histories were acquired at 17 different structural states for a total of 850 time histories. Table 12.1 describes each state and Figure 12.15 shows some example time histories from various damaged and undamaged conditions.

First, each measured time history is standardised (as described in Chapter 6) by subtracting its mean and dividing by its standard deviation such that

$$\hat{x} = \frac{x - \overline{x}}{\sigma_x} \tag{12.7}$$

where $\hat{x}$ is the normalised signal and $\overline{x}$ and σ_x are the sample mean and sample standard deviation of the training values of x, respectively. Standardisation is a process of normalisation in itself. Typically, the mean is subtracted to remove DC offsets in the data. With the acceleration data it is not anticipated that the damage being considered in this example will cause a DC offset in the measured response. However, if one was making strain measurements on a structure where damage resulted in permanent deformation or yielding, this step should not be used because the damage will cause meaningful DC

Table 12.1   Data sets used in the data normalisation studies

Label	State condition	Description[a,b]	Testing data time history number
State 1	Undamaged	Baseline condition	1–50
State 2	Undamaged + variability	Mass = 1.2 kg added at the base	51–100
State 3	Undamaged + variability	Mass = 1.2 kg added on the first floor	101–150
State 4	Undamaged + variability	87.5% stiffness reduction in column 1BD	151–200
State 5	Undamaged + variability	87.5% stiffness reduction in column 1AD and 1BD	201–250
State 6	Undamaged + variability	87.5% stiffness reduction in column 2BD	251–300
State 7	Undamaged + variability	87.5% stiffness reduction in column 2AD and 2BD	301–350
State 8	Undamaged + variability	87.5% stiffness reduction in column 3BD	351–400
State 9	Undamaged + variability	87.5% stiffness reduction in column 3AD and 3BD	401–450
State 10	Damaged	Gap = 0.20 mm	451–500
State 11	Damaged	Gap = 0.15 mm	501–550
State 12	Damaged	Gap = 0.13 mm	551–600
State 13	Damaged	Gap = 0.10 mm	601–650
State 14	Damaged	Gap = 0.05 mm	651–700
State 15	Damaged + variability	Gap = 0.20 mm and mass = 1.2 kg added at the base	701–750
State 16	Damaged + variability	Gap = 0.20 mm and mass = 1.2 kg added on the first floor	751–800
State 17	Damaged + variability	Gap = 0.10 mm and mass = 1.2 kg added on the first floor	801–850

[a] See Figure 5.15 for column locations.
[b] Bumper simulating damage is located between floors 2 and 3, as shown in Figure 5.14.

offsets in the strain measurements. Division by the standard deviation is used to minimise the influence of input-level variations on the measured response. However, one should be aware that if the anticipated damage reduces the stiffness of the structure and this in turn causes higher levels of response when the input does not vary, then this division may be detrimental to the damage detection process. These normalised signals are used in all of the subsequent steps listed below.

Next, a time series modelling approach similar to the one described in Chapter 7 is used to extract damage-sensitive features. Autoregressive models of order ten (AR(10)) are fitted to all time histories in the training data, which consists of the odd-numbered measurements from undamaged states 1 to 9 (i.e. measurements $1, 3, 5, \ldots, 449$). These AR coefficients are the feature vectors that form the look-up table. Similar features are extracted from the time histories $y(t)$ in the testing data, which consists of all 850 measurements. This process produces 225 sets of AR(10) coefficients from the training data and 850 sets of AR(10) coefficients from the testing data. Note that the 225 training feature vectors are also contained in the testing feature vector set simply to verify that the look-up table process picks the identical feature vector in these cases. Figure 12.16 shows a typical comparison between the measured signal and the associated AR model prediction for a measurement from the undamaged system (state 9, measurement 401) and the damaged system (state 14, measurement 651). In the subsequent discussion these two measurements will be referred to as states U and D, respectively. Also shown in this figure are plots of the corresponding residual error histograms. It can be seen that the residuals from the undamaged case are well represented by a zero-mean normal distribution while the residuals from the damaged case

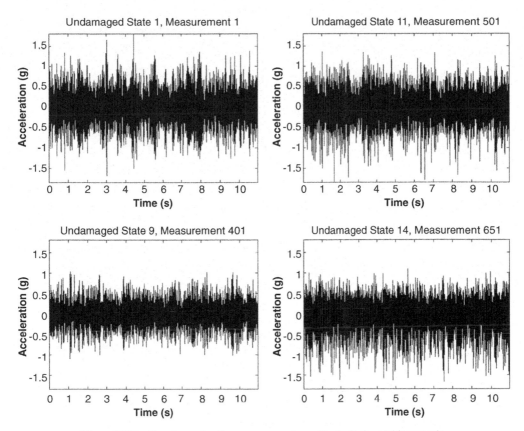

Figure 12.15   Sample acceleration measurements used in the look-up table example.

are showing increased skewness as a result of the impact (this is manifested by the heavier lower tail on the density function estimate).

For the AR parameters forming a feature vector from a given testing data time history $y(t)$, the training data feature vector in the look-up table is found such that the Euclidean distance between the two feature vectors is minimised. This process is depicted in Figure 12.17 where the following steps are used to select a training feature set for the novelty detection process:

1. Fit order p AR models to all m training data sets and the testing data set

$$x_i^k = \sum_{j=1}^{p} a_j^k x_{i-j}^k, \ k = 1, \ldots, m, \quad \text{and} \quad y_i = \sum_{j=1}^{p} a_j^y y_{i-j} \qquad (12.8)$$

where a^k are the AR coefficients for the training data set $\{x\}_k$.

2. Find the training data set, $\{x\}_k, \ k = c$, whose AR coefficients are closest to the AR coefficients from the testing data as determined by a Euclidean distance measure

$$\{x\}_c = \{x\}_k : \min_k \sum_{j=1}^{p} \left(a_j^k - a_j^y\right)^2, \ k = 1, \ldots, m \qquad (12.9)$$

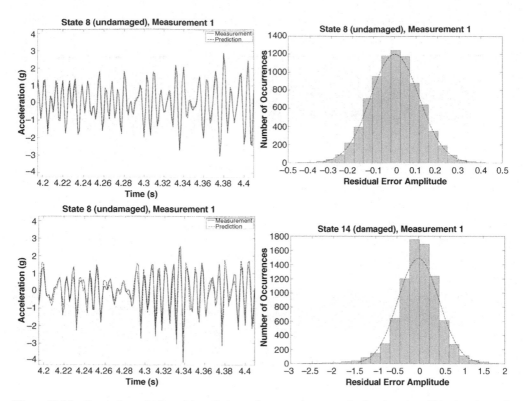

Figure 12.16 Comparison of AR model prediction and measured response for data from state U (top) and state D (bottom), along with the corresponding residual error histograms.

3. For $\{x\}_c$ and $\{y\}$ containing n data samples calculate the residual errors for the respective AR model approximations to these measured time histories using the AR coefficient corresponding to $\{x\}_c$,

$$e_i^c = x_i^c - \sum_{j=1}^{p} a_j^c x_{i-j}^c; \quad e_i^y = y_i - \sum_{j=1}^{p} a_j^c y_{i-j}, \quad i = p+1, \ldots, n \qquad (12.10)$$

4. Define a damage-sensitive feature, f_y, that is the ratio of the standard deviations of the residual errors, and use this feature in the outlier detection process

$$f_y = \frac{\sigma(e_i^y)}{\sigma(e_i^c)} \qquad (12.11)$$

Figure 12.18 shows a plot for the Euclidean distances when the features from state U are compared to all features in the look-up table. Note that the errors are smallest for the measurements taken from the same state (training measurements 201 to 225). Also, the AR parameters from this measurement are part of the look-up table and, hence, the Euclidean distance for AR parameter set 201 in the training data is zero. A similar plot of the Euclidean distances for state D is also shown in this figure where much larger distances are evident, as would be expected because of the different system state resulting from the simulated damage. For this case the AR parameters from state 5, measurement 213

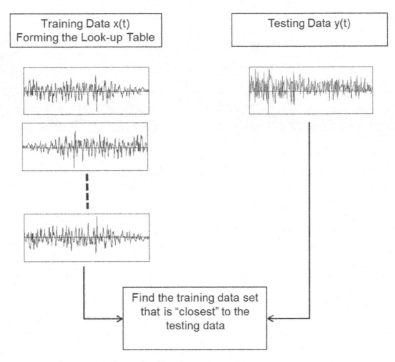

Figure 12.17   The look-up table concept.

(measurement 107 in the training data) are closest to the testing data and will be used in the subsequent novelty detection.

Figure 12.19 shows the standard deviation ratio features for the damaged and undamaged testing cases where it can be clearly seen that the damaged conditions are well separated from the undamaged conditions even when simulated sources of EOVs are present for the two most benign damage conditions (0.20 mm bumper gap, measurements 701 to 800). Note that for the undamaged data that were used

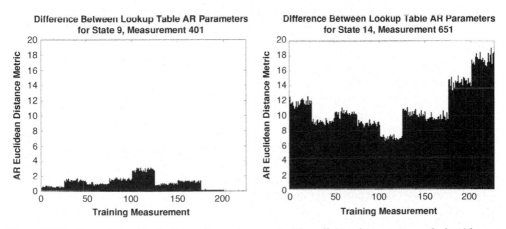

Figure 12.18   Comparison of the Euclidean distance between the AR coefficients feature vectors calculated for state U (left) and state D (right) to all AR coefficient feature vectors in the training data.

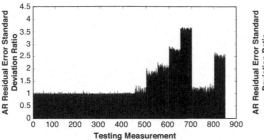 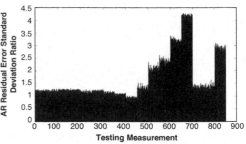

Figure 12.19 Standard deviation ratios of residual errors for undamaged (measurements 1 to 450) and damaged (measurements 451 to 850) data. Left: with look-up table data normalisation. Right: using state 9, measurement 401 for a baseline.

to form the look-up table (the odd numbered measurements between 1 and 450 in Figure 12.19), the value of the standard deviation ratios should be equal to unity as these same data are contained in the testing cases. Also, shown in Figure 12.19 is a similar plot where the look-up table approach is not taken and state 9, measurement 401 is used for a baseline. Both plots show that the damage cases can be distinguished from the undamaged cases. However, when only state 9, measurement 401 is used as the baseline, it can be seen that most of the other undamaged cases associated with different simulated EOVs would be incorrectly classified as outliers. Also, it should be pointed out that for a rigorous classification process, thresholds would need to be established as discussed in Section 6.10. However, it was felt that with these data the look-up table process could be adequately illustrated without this step.

A similar time series modelling approach was applied to the data shown in Figure 12.8 to separate the effects of varying sea states from the effects of changing system conditions. Here the unmeasured input to the ship associated with ship speed, heading relative to the wind and wave height is a major source of variability that was not measured. Therefore, as an additional data normalisation step, a second time series modelling process was employed after the training data set closest to the test data had been selected based on Equation (12.9). The residual errors from the AR models (Equation 12.10) were assumed to be correlated with the unmeasured wave impact inputs to structure. These residuals were then used as inputs to an autoregressive with exogenous (ARX) input model, described in Section 7.13, yielding a new estimate of measured time history given by

$$x_i^c = \sum_{j=1}^{p} \alpha_j^c x_{i-j}^c + \sum_{j=1}^{q} \beta_j^c e_{i-j}^c \tag{12.12}$$

where the residuals $\{e\}_c$ are obtained from Equation (12.10). The α^c and β^c values are the AR and exogenous input parameters for the 'closest' training data, respectively.

A new residual error, ε_i^c, for the 'closest' training data can then be defined as

$$\varepsilon_i^c = x_i^c - \sum_{j=1}^{p} \alpha_j^c x_{i-j}^c - \sum_{j=1}^{q} \beta_j^c e_{i-j}^c \tag{12.13}$$

Now using the ARX input parameters from the training data 'closest' to the testing data, calculate a new residual error, ε_i^y, for this approximation to the testing data as

$$\varepsilon_i^y = y_i - \sum_{j=1}^{p} \alpha_j^c y_{i-j} - \sum_{j=1}^{q} \beta_j^c e_{i-j}^y \tag{12.14}$$

and form a new damage-sensitive feature from the ratio of the standard deviations of these residual errors as

$$f_y = \frac{\sigma(\varepsilon_i^y)}{\sigma(\varepsilon_i^c)} \qquad (12.15)$$

These features were subsequently used for the novelty detection.

The only data available were the time series shown in Figure 12.8. Therefore, these three time histories were first divided into two halves. In the first analysis the first halves of signals 1 and 2 were considered as the training data and the second halves of these signals plus the entire third signal were considered as the testing data. Within these halves 40 series of 2048 data samples beginning at random times were selected for analysis. To verify that this approach was not sensitive to using the first halves of these signals for training, this process was repeated by considering the second halves of signals 1 and 2 as the training data and the first halves of these signals as well as the entire signal 3 as the testing data. This use of training and testing data was predicated on the a priori knowledge that signals 1 and 2 corresponded to the undamaged system and signal 3 was obtained from the ship in a different condition.

Results showing the standard deviation ratios of the ARX model residual errors are shown in Figure 12.20, where it can be seen that for the selected threshold there is a 1/40 (2.5%) probability of false-positive and a 3/40 (7.5%) probability of false-negative damage classifications. Considering the difficulty of the data normalisation problem associated with the unmeasured varying sea states, these results were considered quite good. A more detailed summary of this example can be found in Sohn, Farrar and Worden (2001).

12.7 Machine Learning Approaches to Data Normalisation

Machine learning algorithms can also be used for data normalisation and these methods will be illustrated with the same simulated building structure data used in the look-up table example as described in

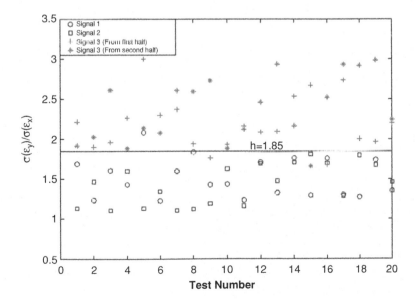

Figure 12.20 Residual error standard deviation ratios for the patrol boat data.

Figueiredo (2010) and Figueiredo *et al.* (2011). The algorithms are implemented in a manner that allows for direct comparison of their relative data normalisation performance. First, each algorithm is trained using the same feature vectors extracted from time series data acquired from the undamaged structure. However, these data were collected under different simulated EOVs. Similarly to the look-up table approach, these algorithms do not need direct measures of the EOVs. Second, in the test phase all the machine learning algorithms will transform each input feature vector into a scalar feature referred to as a damage index (DI) and then perform the damage classification using a novelty detection approach applied to this DI. If adequate data normalisation has been achieved, the DIs should be nearly invariant when calculated from feature vectors corresponding to the undamaged condition when EOVs are present. Additionally, robust data normalisation will allow the DIs to be classified as outliers when the features correspond to the damaged condition even with EOV present.

The four different machine learning algorithms used to illustrate this approach are first briefly summarised. All the algorithms assume that training data from the undamaged structure consisting of n-dimensional feature vectors from m different operational and environmental conditions are available. Subsequently, an $n \times k$ test data matrix consisting of k samples of feature vectors obtained from a potentially damaged system under varying EOV are analysed after data normalisation has been performed. The parameters of an order 10 AR model fit to the measured response time histories are again the feature vectors that will be used in this example.

12.7.1 Auto-Associative Neural Networks

As described in Chapter 10, an auto-associative neural network (AANN) is a feedforward, multilayer perceptron neural network that maps the input on to itself through a hidden bottleneck layer that has fewer nodes than the input layer (Kramer, 1991; Sohn, Worden and Farrar, 2003). In this example the AANN is trained to characterise the underlying dependency of the features on the unobserved EOVs by treating that dependency as the hidden variables in the network architecture. If the number of nodes in the bottleneck layer is chosen such that it represents the number of independent EOVs that cause variability in the feature vectors, then the training process will allow the AANN to learn this functional dependency, even without a measure of the EOV. The network architecture contains three hidden layers: the mapping layer, the bottleneck layer and the demapping layer. The mapping and demapping layers consist of neurons with hyperbolic tangent sigmoid transfer functions while the bottleneck and output layers are formed using linear neurons. Once the network has been trained using features from the undamaged condition, the assumption is that the prediction error of the AANN will grow when the feature vectors that come from a damaged condition are fed to the network. In both the training and testing portions of the process, feature vectors will have been obtained from the structure when it is subjected to EOVs. For the test matrix $[Y]$ (a set of AR parameters possibly corresponding to a damaged condition and also obtained from data measured in the presence of EOV), the residual error matrix $[E]$ is given by

$$[E] = [Y] - [\hat{Y}] \tag{12.16}$$

where $[\hat{Y}]$ corresponds to the estimated feature vectors that are the output of the AANN. Following the discussion in Section 10.3, a quantitative measure of damage for each of the k testing vectors, DI_i, is defined as the Euclidean distance between the respective input and output feature vectors

$$DI_i = \|\{Y\}_i - \{\hat{Y}\}_i\| \tag{12.17}$$

If the feature vector i is related to an undamaged condition, $Y_i \approx \hat{Y}_i$ and then $DI_i \approx 0$. Alternatively, if the features are extracted from data corresponding to the damaged condition, the neural network should not be able to map the inputs on to themselves, which is the same for the undamaged case. The residual errors will then increase, resulting in DI deviations from zero, which are used to identify an abnormal condition in the structure that cannot be attributed to the EOVs represented in the training data.

12.7.2   Factor Analysis

Following an earlier summary (Kullaa, 2003), factor analysis (FA) is a technique used to describe the correlation among observed dependent variables (feature vectors of dimension m) in terms of a small number of independent variables of dimension f called *factors* ($f < m$). The factors are not directly observed and are assumed to be independent with zero mean and unit variance. In the current context, the factors are intended to correspond to latent variables driving the EOVs (e.g. temperature, traffic loading). The linear factor model can be written as

$$\{x\} = [\Lambda]\{\xi\} + \{\varepsilon\} \tag{12.18}$$

where $\{x\}$ is an $n \times 1$ feature vector obtained from the measured data (in this case the AR model parameters obtained from data measured on the structure in its undamaged state under EOVs), $\{\xi\}$ is an $f \times 1$ vector of *factor scores* (in this case the environmental and operational variables such as temperature that influence the observed feature vectors), $[\Lambda]$ is an $n \times f$ matrix of *factor loadings* that model the influence of the unobserved environmental and operational parameters on the observed features and $\{\varepsilon\}$ is an $n \times 1$ vector of *unique factors* (or error terms). In the current context, the unique factors are the residual errors that result when one attempts to reconstruct the data from the lower-dimensional representation provided by the factors.

First, the undamaged condition is defined by the $n \times n$ covariance matrix, $[\Sigma]$ (see Section 6.7), formed from the feature vectors obtained from data measured on the undamaged structure with EOVs present. The covariance matrix of the features is related to the factor loading by the following relation:

$$[\Sigma] = [\Lambda][\Lambda]^{\mathrm{T}} + [\Psi] \tag{12.19}$$

The unique factors are assumed to be normally distributed with zero mean and a diagonal covariance, $[\Psi]$.

A maximum likelihood procedure is then used to estimate $[\Lambda]$ and $[\Psi]$ (Joreskog, 1967). Once the factor loading and the unique factor covariance matrices have been obtained based on data from the undamaged condition, the factor scores for the testing data, $\{\hat{\xi}\}$, are estimated using a linear regression technique (Sharma, 1996) as

$$\{\hat{\xi}\} = [\Lambda]^{\mathrm{T}}([\Psi] + [\Lambda][\Lambda]^{\mathrm{T}})^{-1}\{\hat{x}\} \tag{12.20}$$

Finally, the unique factors in Equation (12.18) are computed from,

$$\{\hat{\varepsilon}\} = \{\hat{x}\} - [\Lambda]\{\hat{\xi}\} \tag{12.21}$$

Note that the factor loading matrix in Equation (12.18) has been defined based on the features associated with the undamaged structure. This factor loading matrix is subsequently used with the test data, as indicated by the hats in Equations (12.20) and (12.21). The assumption is that when features from a damaged condition feed the model trained with features obtained from the data acquired from the undamaged structure, the unique factors will increase in magnitude. Although this development has been shown for a single feature vector, the process can be directly extended to a matrix of feature vectors, which was the approach used in the example summarised below. Finally, the DIs are estimated from the matrix of unique factors using Equation (12.17).

12.7.3   Mahalanobis Squared-Distance (MSD)

Another method for performing data normalisation when direct measures of the environmental and operational parameters that are causing variability in the damage-sensitive features are not available is based on the Mahalanobis squared-distance (MSD), previously discussed in Chapters 6, 7 and 10. The training data matrix $[X]$ (in this case the AR model parameters), where each column contains a feature vector extracted from data obtained when the undamaged structure is under EOVs, is used to calculate a mean feature vector $\{\overline{x}\}$ and covariance matrix $[\Sigma]$. These two quantities are used to quantify the influence of the EOVs on the feature vectors. The MSD is now used as in standard outlier analysis (see Section 10.2) to provide a damage index, DI_i, and this index is calculated for all feature vectors in the training and testing sets using the equation

$$DI_i = (\{x\}_i - \{\overline{x}\})^{\mathrm{T}}[\Sigma]^{-1}(\{x\}_i - \{\overline{x}\}) \tag{12.22}$$

where $\{x\}_i$ is the particular feature vector being tested. The assumption is that if a feature vector is obtained from data collected on the damaged system that includes similar sources of operational and environmental variability as that represented by $\{\overline{x}\}$ and $[\Sigma]$, this vector will be further from the mean feature vectors corresponding to the undamaged condition as quantified by the MSD. The effectiveness of this approach depends on capturing all possible EOVs in the training set.

12.7.4   Singular Value Decomposition

The use of the *singular value decomposition* (SVD) technique relies on the determination of the rank of a matrix containing feature vectors measured under EOVs (Ruotolo and Surace, 1997). Introduced in Section 6.13, the SVD is a factorisation of a rectangular real or complex matrix $[M]$ defined by

$$[M] = [U][\Lambda][V]^{\mathrm{H}} \tag{12.23}$$

where $[U]$ and $[V]$ are two orthogonal matrices and $[\Lambda]$ contains the singular values of the matrix $[M]$ along its diagonal, sorted in descending order (if $[M]$ is square, the singular values are simply its eigenvalues). (The superscript H denotes the Hermitean conjugate (complex conjugate transpose); if $[M]$ is real, this simply becomes the superscript T.) The rank of the matrix $[M]$ can be estimated based on the number of nonzero singular values obtained from the SVD.

In order to use SVD to remove EOVs, first $[M]$ is formed from the set of n-dimensional feature vectors corresponding to data acquired from the undamaged structure under the varying EOV conditions, $[X]$, and the rank of this matrix is estimated and designated R. Next, a new matrix, $[M']$, is formed by augmenting $[M]$ with a feature vector, $\{y\}_i$, corresponding to a possible damage condition, placed in the last column:

$$[M'] = [[X], \{y\}_i] \tag{12.24}$$

If the potential outlier comes from an undamaged condition, it is assumed that it can be represented as some linear combination of the feature vectors forming $[X]$ and therefore the ranks of $[M]$ and $[M']$ will be identical. Alternatively, if the new feature vector $\{y\}_i$ comes from a damaged condition and it is independent from the feature vectors corresponding to the undamaged condition, the rank of $[M']$ will increase to $R + 1$.

Theoretically, if $[M]$ is rank-deficient (i.e. some feature vectors are linear combinations of the others), some of the singular values should be zero. However, real-world data have variability that can affect the estimate of the rank by resulting in residual, rather than zero, singular values.

In order to use the SVD as a data normalisation technique for damage detection, a three-step process is followed. Note that it is assumed that the number of feature vectors (m) in $[X]$ is greater than or equal to the dimension of the feature vector (n). If this were not the case the method could then fail as the rank of $[X]$ could then be no greater than m. If the true rank of the undamaged data was actually $R > m$, adding a new test feature vector to form $[M']$ might then increase the rank, even if the test data came from the undamaged condition. The steps taken are:

1. The singular values of the training matrix $[M]$, containing feature vectors (the AR model parameters in this case) obtained from the undamaged structure acquired under different EOVs, are calculated and stored in a vector $\{s\}_M$. It is desirable to have many estimates of the vectors from the undamaged condition corresponding to different EOVs in an effort to minimise the influence of measurement noise on the estimate of the singular values. The vector $\{s\}_M$ defines the normal condition under different EOVs.
2. For each potential outlier $\{y\}_i$ the singular values are calculated for $[M']$ and stored in a new vector $\{s\}_{M'}$. This step identifies whether a feature vector from the potential damaged condition is some linear combination of the previous undamaged ones. If the new feature vector $\{y\}_i$ is related to damage in the structure, and therefore not represented by the feature vectors in the training matrix $[M]$, a plot of the singular values contained in $\{s\}_{M'}$ should not overlap with those contained in $\{s\}_M$. The assumption here is that damage produces a new feature vector, $\{y\}_i$, that is linearly independent of the feature vectors from the undamaged condition that were measured under varying EOV.
3. The vectors of singular values become the features and the DI is calculated based on the difference between $\{s\}_M$ and $\{s\}_{M'}$:

$$DI_i = \| \{s\}_m - \{s\}_{m'} \| \tag{12.25}$$

12.7.5 Application to the Simulated Building Structure Data

For each test, four individual AR(10) models are used to fit the corresponding time series from the accelerometers on each of the four floors and these parameters are used as damage-sensitive features in a concatenated format, yielding 40-dimensional feature vectors. Note that AR parameters should be constant when estimated based on time series data obtained from time-invariant systems. However, in the presence of the simulated EOV as well as damage, the parameters are expected to change. Figure 12.21(a) shows the change in the feature vector extracted from one test corresponding to each state condition listed in Table 12.1. Here the feature vectors are divided according to structural conditions into the undamaged and damaged conditions. This figure shows more significant changes in the AR model parameter values at channels 4 and 5 when the features are from the damaged conditions. Note that these channels are closest to the source of damage. Figure 12.21(b) highlights the changes in the AR parameters of channel 4. Although it cannot be discerned from Figure 12.21, the results showed that the higher the level of damage (i.e. the smaller the initial bumper gap), the lower the amplitude of the AR parameters.

The next step is to carry out the feature classification in conjunction with the previously described data normalisation procedures. First, the AANN, FA, MSD and SVD algorithms outlined above are implemented with features from all the undamaged conditions (states 1 to 9 in Table 12.1). The undamaged data sets are split into two groups: the training and the test data. The matrix of training data is again composed of AR parameters obtained from the odd numbered data sets corresponding to each undamaged state condition (state 1 to 9) and so it has a dimension of 40×225. The test data (dimension 40×625) are composed of AR parameters from the remaining 25 tests of each undamaged state condition together with AR parameters from all the 50 tests of each damaged state condition (states 10 to 17). With this procedure the performance of the machine learning algorithms is assessed in an exclusive manner because features extracted from the time series used for testing are not included in the training data. During the test phase,

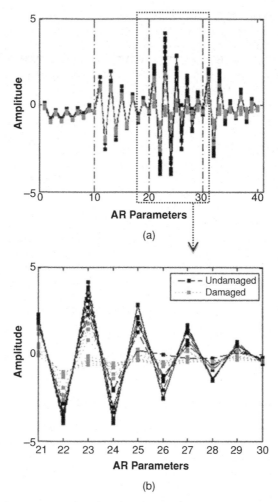

Figure 12.21 Features vectors from one test of each state condition (states 1 to 17): (a) AR parameters from channels 2 to 5 in concatenated format; (b) only the AR(10) parameters from channel 4.

the algorithms are expected to detect deviations from the normal condition when features come from damaged conditions even with the presence of EOVs.

The AANN-based algorithm is constructed with a feedforward neural network to perform the mapping and demapping. Based on recommendations by other investigators (Kramer, 1991), the network contained ten nodes in each mapping and demapping layer and two nodes (factors) in the bottleneck layer. The nodes in the bottleneck layer were intended to represent the two underlying unobserved variables (changing mass and stiffness) causing the variability in the features. A Levenberg–Marquardt backpropagation algorithm was used to train the network. The FA-based algorithm assumes two factors to represent the number of underlying unobserved variables affecting the features. Because the undamaged and damaged conditions are known a priori, the MSD algorithm was developed in an exclusive manner by using only data from the undamaged conditions when forming the estimates of the mean and covariance. Therefore, the normal condition of the structure is defined by the mean vector (40×1) and covariance matrix (40×40) obtained from the training data. In the case of the SVD-based algorithm, the state matrix $[M']$

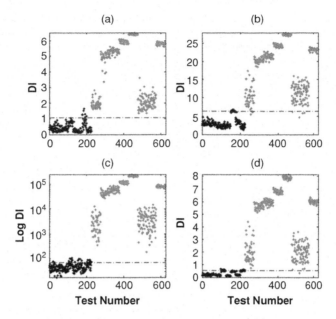

Figure 12.22 DIs calculated based on feature vectors from the undamaged (black) and damaged (grey) conditions using the four machine learning algorithms along with thresholds defined by the 95% confidence limit for the training data: (a) AANN-based algorithm; (b) FA-based algorithm; (c) MSD algorithm; (d) SVD-based algorithm.

for each feature vector potentially corresponding to the damaged condition has a dimension of 40×226. As a consequence, the number of singular values computed for each matrix is equal to 40.

Figure 12.22 plots the DI for each feature vector of the entire test data along with thresholds defined based on the 95% threshold value over the training data. Equation (12.17) is used to compute the DIs from the residuals of the AANN-, FA- and SVD-based algorithms. All the algorithms show a monotonic relationship between the level of damage and the amplitude of the DI, as shown in Figure 12.22, even when EOVs are present. However, one common way of reporting the performance of a binary classification (undamaged and damaged structural conditions) is using the type I (false-positive indication of damage) and type II (false-negative indication of damage) errors. This technique recognises that a false-positive classification may have different consequences than false-negatives. In order to evaluate the performance of the classifiers, Table 12.2 summarises the number of type I and type II errors for each algorithm. The table shows that the AANN-based algorithm has the best performance in terms of a reduced number of total misclassifications (only 1.1%). Additionally, the AANN- and MSD-based algorithms avoid any

Table 12.2 Number and percentage of type I and type II errors for each algorithm

Algorithm	Error		
	Type I	Type II	Total
AANN	7 (3.1%)	0	7 (1.1%)
FA	13 (5.8%)	8	21 (3.4%)
MSD	24 (11%)	0	24 (3.8%)
SVD	14 (6.2%)	1	15 (2.4%)

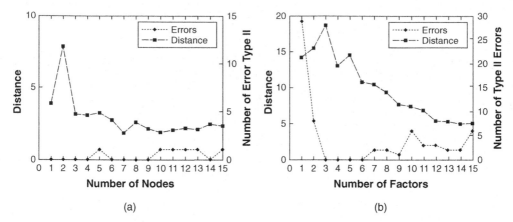

Figure 12.23 Distance between the mean of the DI distributions on the undamaged and damaged conditions along with the number of type II errors as a function of (a) number of bottleneck node for the AANN; (b) number of factors for the FA model.

false-negative indications of damage. However, the percentage of false-positive indications of damage for the MSD algorithm deviates significantly from the 5% tolerance assumed for the threshold. Although not rigorously verified, this result is most likely due to the fact that systematic variability has been introduced in the training data to simulate EOV and, therefore, the normal condition was probably not Gaussian. Nevertheless, all the algorithms perform well when applied to these same data sets, with the percentage of total misclassifications (both type I and type II errors) ranging between 1.1% and 3.8%.

The classification performance of the AANN- and FA-based algorithms could possibly be improved by adjusting the number of unobserved variables (i.e. the number of factors or bottleneck nodes) in the formulation (in this case two unobserved variables were used). To highlight the influence of the assumed number of factors on the performance of both algorithms, Figure 12.23 shows the distance between the mean of the undamaged and damaged DI distributions along with the number of type II errors as a function of the number of factors. Based on a large distance between means and a corresponding small number of errors, this figure indicates that the appropriate number of factors is 2 and 3 for the AANN- and FA-based algorithms, respectively. This result shows that the number of factors assumed in the previous analyses is nearly optimal. The difference in the number of factors can be attributed to the fact that the AANN-based algorithm is able to learn nonlinear relationships between the features while the factor analysis is fundamentally linear. Furthermore, the number of errors also indicates that assuming a higher than suitable number of factors increases the model complexity and, consequently, reduces their ability to generalise to other data sets. Note that in the case of the AANN-based algorithm, the performance can also be improved by increasing the number of nodes in the mapping and demapping layers.

All four algorithms offer some advantages over other parametric data normalisation techniques because EOVs (e.g. traffic loading and temperature) do not need to be measured to reveal their influence on the structure's response. However, these algorithms have potential problems if the training data are only characteristic of a limited range of EOVs. Hence, all sources of variability must be well characterised by the training data in order for the algorithms to accurately learn their influence on the system's response and to separate changes in the measured response caused by damage from changes caused by the EOVs. Thus, one should note that with these algorithms there is no guarantee that they will work effectively when they are used to analyse new data corresponding to EOVs not used in the training phase. Also, if the damage produces changes in the system's dynamic response characteristics that are similar to those produced by the sources of variability, it is not at all guaranteed that these algorithms will be able to separate changes in the features caused by damage from changes caused by the EOVs.

12.8   Intelligent Feature Selection: A Projection Method

As an alternative to the data normalisation procedures described above one can instead attempt to intelligently select the damage-sensitive features such that they are insensitive to the EOVs while retaining their sensitivity to damage. This process is directly related to Axiom IVb presented in Chapter 1. One example of this approach uses principal component analysis (PCA) to identify such EOV-insensitive features (Manson, 2002b). As discussed in Chapter 6, PCA is a projection technique used to reduce the dimension of features such that the most variance present in the original data is retained. This dimension reduction is achieved by projecting the original features into a space spanned by the eigenvectors associated with the largest eigenvalues of the feature covariance matrix. Alternatively, Manson proposed to project the features into a space spanned by the eigenvectors associated with the smallest eigenvalues, or minor components. In this case the covariance matrix is formed from data acquired from the undamaged structure under EOVs. The idea is that most of the variance in the features results from the EOV and this process projects the data into a space that minimises this variance.

To illustrate this process a wave-propagation-based SHM example will be given in contrast to the vibration-based examples considered so far in this chapter. The example is based on an experimental study where a 300-mm-square composite laminate plate was instrumented with two piezoelectric sensor/actuators, as shown in Figure 12.24. Lamb wave signals travelling between this sensor–actuator pair were recorded every minute with the plate located in an environmental chamber. In the first phase of the test, the environmental chamber was held at a temperature of 25 °C for the first 1355 measurements. In the second phase, the temperature was cycled three times between 10 °C and 30 °C (measurements 1356 to 2482) and the temperature in the chamber was recorded. In the third and final phase, a hole was drilled in the plate and the temperature was again cycled between 10 °C and 30 °C (measurements 2483 to 2944). The temperature profile imposed during phase 2 is shown in Figure 12.25. The features for damage detection were obtained by Fourier transforming the time history of the received Lamb wave signal and then selecting 50 spectral lines around the peak in the spectrum. This process yielded 50-dimensional features, as illustrated in Figure 12.26.

Features extracted from alternate measurements from the 25 °C data (phase 1) and two 10–30 °C cycles of the phase 2 data were used as training data to establish the baseline condition defined in terms of the mean and covariance matrix for these 50-dimensional feature vectors. A Mahalanobis squared-distance (MSD) damage index as described in Section 12.7.3 was then used in a novelty detection analysis.

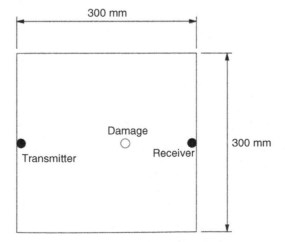

Figure 12.24   Three-mm-thick composite laminate plate with piezoelectric sensor and actuator.

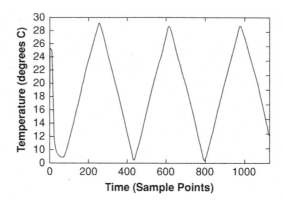

Figure 12.25   Temperature profile from the environmental chamber during phase 2 of the test.

Figure 12.27 summarises the results of this novelty detection. The horizontal dashed line in this figure corresponds to a threshold based on a 1% test of discordancy and the vertical dashed lines separate the three testing regions (from left to right: undamaged at 25 °C, undamaged cycling between 10 °C and 30 °C, and damaged cycling between 10 °C and 30 °C). Once the thermal cycling occurs, the features from the undamaged and damaged cases exceed the threshold that was based on the training data, implying that with these features many false-positive indications of damage will occur because of the EOVs.

Next, the 50-dimensional spectral feature vectors were projected on to the ten smallest minor components yielding new 10-dimensional feature vectors. An outlier analysis was then performed using the same MSD damage index. Training and testing data sets were the same as those used with the 50-dimensional feature vectors. The results of this outlier analysis are shown in Figure 12.28. In this figure it can be seen that all the features from the undamaged data, both constant temperature and temperature fluctuations, fall below the threshold. Features extracted from data corresponding to all the damaged cases fall significantly above the threshold, indicating that new features have been identified that are insensitive to the thermal cycling and that retain their sensitivity to damage.

One must keep in mind that this study was conducted with the damage data available. In general, one can determine that invariance to EOVs has been obtained with only features corresponding to the undamaged condition, but it is difficult to also verify that sensitivity to damage has been maintained by the process if no damage data are available.

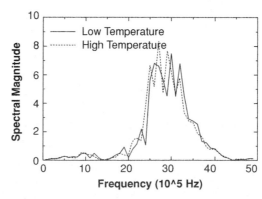

Figure 12.26   The 50-dimensional spectral feature used in the novelty detection analysis.

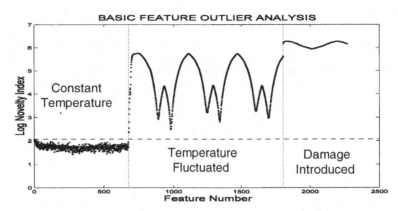

Figure 12.27   Outlier detection using the Mahalanobis squared-distance damage index applied to the 50-dimensional spectral data feature.

12.9   Cointegration

The method described in the last section is a projection method. Features containing evidence of EOVs are projected on to a lower-dimensional subspace in which the variations are absent. This process allows detection of damage as long as the evidence of damage is not also projected out. The criterion for the PCA projection is to preserve as much signal variance (variance in signal power in this case) as possible and, although the method described in the last section worked very well, this criterion is arguably not perfectly matched to the needs of the SHM analysis. Damage sensitivity may be lost as there is no compelling reason why sufficient evidence of damage should manifest itself in different principal components to the EOV variations when these components are fixed on the basis of signal power. The subject of the current section is another projection method based on a different criterion that may have a better match with SHM needs. The idea of *cointegration* comes from econometrics and is used to project out components of data that correspond to long-term trends, that is nonstationarity. As EOVs usually manifest themselves on longer time scales than the dynamics of the structure that is sensitive to damage, the method appears to be well matched to SHM needs.

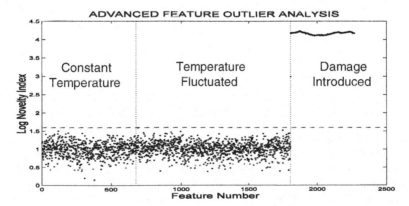

Figure 12.28   Outlier detection using the Mahalanobis squared-distance damage index applied to the 50-dimension spectral data features projected on to the 10 smallest minor components.

12.9.1 Theory

As discussed above, *cointegration* is a property of nonstationary time series (Anderson, 1971; Perman, 1993; Cross, Worden and Chen, 2011). For the purposes of this brief discussion, two or more nonstationary time series will be said to be *cointegrated* if some linear combination of them is stationary. Symbolically, a multivariate nonstationary time series $\{y\}_i$ is cointegrated if a vector $\{\beta\}$ exists such that z_i is stationary, where

$$z_i = \{\beta\}^T \{y\}_i \qquad (12.26)$$

If this is the case, $\{\beta\}^T$ is called a *cointegrating vector*. If $\{y\}_i$ is *n*-dimensional, there may be as many as $n - 1$ linearly independent cointegrating vectors.

To allow the most general definition of cointegration, the concept of an *order of integration* must also be introduced; this is the number of times one must difference a nonstationary time series before it becomes stationary. For engineering applications most variables of interest will be integrated of order 1 (denoted $I(1)$), which implies that their first differences will be stationary. In general a set of time series are cointegrated if they share a common order of integration and a linear combination of the variables exists with a *lower* order of integration.

As the order of integration must be the same for cointegrated variables, the first step in a cointegration analysis will often be to ascertain the order of integration of each of the variables to be included in the analysis. This assessment is commonly achieved in econometrics by testing each variable for a unit root; if a unit root is present in the characteristic equation that defines some time series, then that time series will be inherently nonstationary. The unit root test that will be used here is called the augmented Dickey Fuller (ADF) test and the steps needed to implement it will be described here briefly, but readers should refer to Dickey and Fuller (1979, 1981) or Cross, Worden and Chen (2011) (tutorial) for more details and background theory.

The ADF test involves fitting each variable to a model type of the following form:

$$\Delta y_i = \rho y_{i-1} + \sum_{j=1}^{p-1} b_j \Delta y_{i-j} + \varepsilon_i \qquad (12.27)$$

where the difference operator Δ is defined as $\Delta y_{i-j} = y_{i-j} - y_{i-j-1}$. A suitable number of lags p should be included to ensure that ε_i becomes a white noise process (Anderson, 1971). In this form, the stability (and therefore stationarity) of the model in Equation (12.27) is determined by the value of ρ; if it is statistically close to zero the process will be nonstationary and integrated order 1, $I(1)$. The idea of the ADF is therefore to test the null hypothesis of $\rho = 0$ by comparing the test statistic

$$t_\rho = \frac{\hat{\rho}}{\sigma_\rho} \qquad (12.28)$$

Where $\hat{\rho}$ is the least-squares estimate of ρ and σ_ρ is the variance of the parameter, with critical values that can be found in Fuller (1996), in much the same way one would when conducting a *t*-test. The hypothesis is rejected at level α if $t_\rho < t_\alpha$. If the hypothesis is accepted, the time series has a unit root and is $I(1)$. If the hypothesis is rejected, the test should be repeated for Δy_i; if the hypothesis is then accepted y_i is an $I(2)$ nonstationary sequence. This process can be continued until the integrated order of the time series is found. Additional hypotheses and test statistics are needed if the model form needs to be extended to include shifts or deterministic trends (or both) (Dickey and Fuller, 1979, 1981).

Upon ascertaining the order of integration of each of the variables of interest, those that are integrated of the same order can then be included in a cointegration analysis. The Johansen procedure (Johansen, 1995) is commonly used to find the 'most stationary' linear combination of variables possible; that is, it is a way to find the cointegrating vectors for some set of nonstationary variables. This procedure is especially used with $I(1)$ variables and achieves estimation of the cointegrating vectors through a maximum likelihood argument. The theory behind the Johansen procedure is complex and so will not be included here (see Anderson, 1971, or Johansen, 1995, instead); however, as before, the necessary steps to implement the Johansen procedure will be provided without justification.

The first step of the Johansen procedure is to fit the variables in question to a vector autoregressive (VAR) model, which takes the form

$$\{y\}_i = [A_1]\{y\}_{i-1} + [A_2]\{y\}_{i-2} + \cdots + [A_p]\{y\}_{i-p} + \{\varepsilon\}_i \tag{12.29}$$

where the most suitable model order p has been determined by an Akaike information criterion or similar (see Anderson, 1971, for example).

The most stationary linear combinations of the variables, or cointegrating vectors, are found in the matrix $[\beta]$ in the vector error-correction model (VECM) of the variable set, which takes the form

$$\{z_0\}_i = [\alpha][\beta]^T\{z_1\}_i + [\Psi]\{z_2\}_i + \{\varepsilon\}_i \tag{12.30}$$

where $\{z_0\}_i = \{\Delta y\}_i, \{z_1\}_i = \{y\}_{i-1}, \{z_2\}_i = \{\{\Delta y\}_{i-1}^T, \{\Delta y\}_{i-2}^T, \ldots, \{\Delta y\}_{i-p}^T\}^T$ and p is the model order ascertained previously in Equation (12.27).

To find $[\beta]$, one must first establish the residuals $\{R_0\}_i$ and $\{R_1\}_i$ of the following regressions:

$$\{z_0\}_i = [C_1]\{z_2\}_i + \{R_0\}_i \tag{12.31}$$

$$\{z_1\}_i = [C_1]\{z_2\}_i + \{R_1\}_i \tag{12.32}$$

From these residuals, the following product moment matrices can be defined:

$$[S_{mn}] = \frac{1}{N} \sum_{i=1}^{N} \{R_m\}_i\{R_n\}_i^T, \quad m, n = 0, 1 \tag{12.33}$$

Finally, using the moment matrices, the cointegrating vectors are found as the eigenvectors of the generalised eigenvalue problem

$$(\lambda_i[S_{11}] - [S_{10}][S_{00}]^{-1}[S_{01}])\{v\}_i = 0 \tag{12.34}$$

The cointegrating vector that will result in the most stationary combination of the original variables will be the eigenvector with the corresponding largest eigenvalue. Again readers are referred to Anderson (1971), Johansen (1995) and Cross, Worden and Chen (2011) for more details of the theory behind these steps.

From a practical SHM point of view, the cointegrating vectors of a set of variables should be established using data from some training period from the undamaged structure that encompasses the anticipated EOVs. Upon projecting new data on to a cointegrating vector, the combination will remain stationary all the time the structure continues to act in its normal condition, but should become nonstationary on the introduction of damage. This result will be explored here in the context of a case study.

12.9.2 Illustration

The uses of cointegration are demonstrated here in the context of the experimental data discussed in the last section, that is the Lamb wave data from a composite panel subjected to temperature variations in an

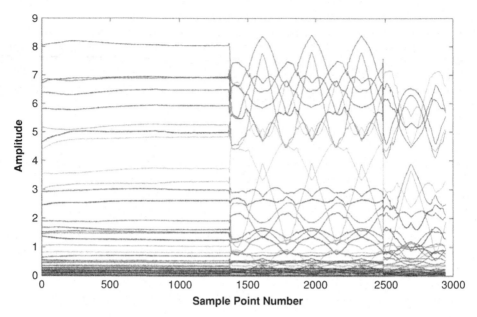

Figure 12.29   Spectral features (lines 46 to 55) over the whole period (three phases) of the test.

environmental chamber (the test setup is illustrated in Figure 12.24). The description of the experiment was given earlier; the features used for analysis here are the same 50-dimensional spectral features as previously discussed in the context of PCA projection. Because cointegration is based on the analysis of time series, it is useful to see how each of the spectral line features evolves with time, which is shown in Figure 12.29.

The data described above provide an ideal platform for exploring the uses of cointegration for SHM; the data includes variations in the recorded signals induced by temperature and in the final stage of the test the effects of damage. If the spectral lines under investigation are cointegrated, given a suitable training period from the data, the Johansen procedure should provide a feature purged of temperature dependence, which would, therefore, be stationary over the first two phases of the Lamb wave test. On the introduction of damage the feature should become nonstationary.

For the training set, 1000 data points were used that incorporated 355 points of data from the first part of the test where the temperature of the plate was held constant and 645 data points from the second period of the test where the temperature was cycled between 10 °C and 30 °C. Figure 12.30 shows the linear combination of all 50 features for the training period chosen. The dashed horizontal lines indicate plus and minus three standard deviations of the training data and are added to act essentially as statistical process control X-chart limits (Manson, 2002a, and see Chapter 10); if a data point is outside these thresholds it can be considered as abnormal. Studying Figure 12.30 one can see that the Johansen procedure has successfully found a linear combination of the 50 features in question that is stationary over the training period, with the exception of a few points occurring around the time when the plate began to undergo its temperature cycles, between the first and second phases of the testing. This anomaly indicates that at the time of switching between the two test phases some more complex relationships between the environmental conditions and the recorded signals existed; happily, after the transition period the feature returns to an equilibrium quickly and is still valid as an anomaly detector.

As the Johansen procedure has successfully created a stationary combination of the variables from a training set it remains to project all of the rest of the data on to this combination and study what happens

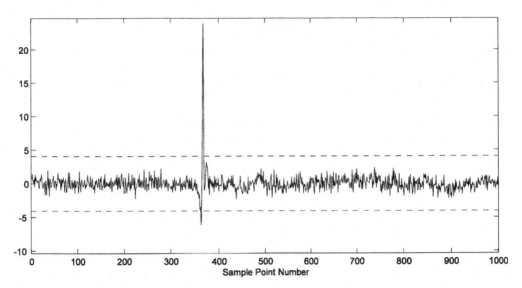

Figure 12.30   Cointegrated signal of the training period (linear combination of 50 spectral lines).

when damage is introduced. The results are shown in Figure 12.31, where the vertical lines indicate the beginning of the temperature cycling period (left vertical line) and the point of the introduction of damage (right vertical line). A clear indication of damage is apparent when the residual becomes nonstationary and deviates significantly outside the control chart boundaries (at plus and minus three standard deviations of the training residual). Cointegration looks to be a very promising approach for the data normalisation problem.

Despite the successful indication of damage by the cointegrated residual, the anomaly in the combination of the training data (in Figure 12.30) occurring after the introduction of the temperature gradient,

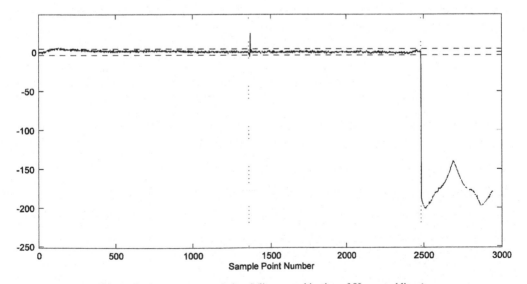

Figure 12.31   Cointegrated signal (linear combination of 50 spectral lines).

although understandable, is a little unsatisfactory. In this instance the problem can be remedied by analysing a smaller subset of features (Montgomery, 2009).

12.10   Summary

In this chapter *data normalisation* has been defined as the procedure used to separate changes in the damage-sensitive features caused by EOVs from the changes caused by damage. First, it is noted that a wide variety of sensors may be available to measure the EOV parameters as well as the structure's dynamic response characteristics. In fact, many deployed structures are monitored with sensors such as thermocouples, anemometers, speedometers, tachometers, strain gauges and accelerometers. The first approach to data normalisation discussed in this chapter makes use of such measurements and attempts to generate parametric models that relate the measured EOV parameters to the system response over a wide range of EOVs when the structure is in an undamaged system state. Various forms of regression analysis are used to develop these models. With such a model in hand, subsequent features can be more directly compared to those obtained from the undamaged structure under similar EOVs.

Next, several methods were presented that allow one to develop a data normalisation procedure when direct measures of the EOV parameters are not available. These methods include the look-up table approach and various machine learning approaches. A basic assumption with these methods is that the damage produces changes in the features that are in some way orthogonal to the changes caused by the EOV. Furthermore, the success of these methods will be dependent on obtaining training data that captures the full range of EOVs. Then a procedure to choose features that are insensitive to the EOVs, but that retain their sensitivity to damage, was presented. The method is based on projection on to a subset of the feature space orthogonal to that containing the EOVs and exploited minor component analysis. However, it was pointed out that it is difficult to verify that the right choice of features has been made without examples of data from the damaged system. This chapter concluded with a recent development in data normalisation referred to as cointegration, which is another projection method for removing EOVs from feature data, but is arguably better matched to the needs of the SHM process than the minor component analysis.

Although studies related to SHM data normalisation issues have been reported in the literature for quite some time (Loland and Dodds, 1976; Cawley, 1997) and review articles on this topic have been published (Sohn, 2004), data normalisation has generally not received as much attention from the research community as other parts of the SHM process. It is speculated that this situation is related to the fact that many SHM research studies have been conducted in well-controlled laboratory settings where such sources of variability are minimal. Furthermore, the approaches reported herein have only investigated the effects of one or two EOV parameters on the damage detection process. In most real-world situations, there will be numerous sources of EOV variability that will influence the damage-sensitive features and further research is needed to verify that the methods presented in this chapter can be extended to such situations. Finally, it is reiterated that data normalisation is, in the authors' opinion, one of the biggest challenges for transitioning SHM technology from research to practice on *in situ* structures.

References

Alampalli, S. (2000) Effects of testing, analysis, damage, and environment on modal parameters. *Mechanical Systems and Signal Processing*, **14**(1), 63–74.

Anderson, T. (1971) *The Statistical Analysis of Time Series*, John Wiley & Sons, Inc., New York.

Askegaard, V. and Mossing, P. (1988) Long term observation of RC-bridge using changes in natural frequencies. *Nordic Concrete Research*, **7**, 20–27.

Brady, T.F. (2004) Global Structural Response Measurement of Swift (HSV-2) from JLOTS and Blue Game Rough Water Trials. Report NSWCCD-65-TR-2004/32, Naval Surface Warfare Center, Carderock Division, Bethesda, MD.

Cawley, P. (1997) Long range inspection of structures using low frequency ultrasound. Proceedings of DAMAS 97: Structural Damage Assessment Using Advanced Signal Processing Procedures, Sheffield, UK.

Cross, E.J., Worden, K. and Chen, Q. (2011) Cointegration; a novel approach for the removal of environmental trends in structural health monitoring data. *Proceedings of the Royal Society, Series A*, **467**, 2712–2732.

Cross, E.J., Koo, K.Y., Brownjohn, J.M.W. and Worden, K. (2010) Long-term monitoring and data analysis of the Tamar Bridge. Proceedings of ISMA, International Conference on Noise and Vibration Engineering.

Dickey, D. and Fuller, W. (1979) Distribution of the estimators for autoregressive time series with a unit root. *Journal of the American Statistical Association*, **74**(366), 427–431.

Dickey, D. and Fuller, W. (1981) Likelihood ratio statistics for autoregressive time series with a unit root. *Econometrica: Journal of the Econometric Society*, **49**(4), 1057–1072.

Farrar, C.R., Cornwell, P.J., Doebling, S.W. and Prime, M.B. (2000) Structural Health Monitoring Studies of the Alamosa Canyon and I-40 Bridges. Los Alamos National Laboratory Report LA-13635-MS.

Figueiredo, E. (2010) *Damage Detection in Civil Infrastructure under Varying Operational and Environmental Conditions*. Department of Civil Engineeering, University of Porto, Porto, Portugal.

Figueiredo, E., Park, G., Farrar, C.R. *et al.* (2011) Machine learning algorithms for damage detection under operational and environmental variability. *International Journal of Structural Health Monitoring*, **10**(6), 559–572.

Fritzen, C.-P., Mengelkamp, G. and Guemes, A. (2003) Elimination of temperature effects on damage detection within a smart structure concept. Proceedings of the 4th International Workshop on SHM, Palo Alto, CA.

Fuller, W. (1996) *Introduction to Statistical Time Series*, Wiley-Interscience.

Helmicki, A., Hunt, V., Shell, M. *et al.* (1999) Multidimensional performance monitoring of a recently constructed steel-stringer bridge. Proceedings of the 2nd International Workshop on SHM, Palo Alto, CA.

Johansen, S. (1995) *Likelihood-Based Inference in Cointegrated Vector Autoregressive Models*, Oxford University Press.

Joreskog, K.G. (1967) Some contributions to maximum likelihood factor analysis. *Psychometrika*, **32**(4), 443–482.

Kim, C.-Y., Jung, D.-S., Kim, N.-S. *et al.* (2003) Effect of vehicle weight on natural frequencies of bridges measured from traffic-induced vibration. *Earthquake Engineering and Engineering Vibration*, **2**(1), 109–115.

Ko, J.M. and Ni, Y.Q. (2005) Technology developments in structural health monitoring of large-scale bridges. *Engineering Structures*, **27**, 1715–1725.

Kramer, M.A. (1991) Nonlinear principal component analysis using autoassociative neural networks. *AIChE Journal*, **37**(2), 233–243.

Kullaa, J. (2003) Is temperature measurement essential in SHM? Proceedings of the 4th International Workshop on SHM, Palo Alto, CA.

Loland, O. and Dodds, J.C. (1976) Experience in developing and operating integrity monitoring system in North Sea. Proceedings of the Offshore Technology Conference, Houston, TX.

Manson, G. (2002a) Identifying damage sensitive, environment insensitive features for damage detection. Proceedings of the IES Conference.

Manson, G. (2002b) Identifying damage sensitive, environmental insensitive features for damage detection. 3rd International Conference on Identification in Engineering Systems, Swansea, UK.

Montgomery, D.C. (2009) *Introduction to Statistical Quality Control*, John Wiley & Sons.

Ni, Y.Q., Zhou, H.F. and Ko, J.M. (2009) Generalization capability of neural network models for temperature–frequency correlation using monitoring data. *Journal of Structural Engineering*, **135**(10), 1290–1300.

Ni, Y., Hua, X., Fan, K. and Ko, J. (2005) Correlating modal properties with temperature using long-term monitoring data and support vector machine technique. *Engineering Structures*, **27**(12), 1762–1773.

Pardo de Vera, C. and Guemes, J. (1997) Embedded self sensing piezoelectrics for damage detection. Proceedings of the International Workshop on SHM, Palo Alto, CA.

Park, G., Cudney, H. and Inman, D. (1999) Impedance-based health monitoring technique for massive structures and high-temperature structures. Proceedings of the Conference on Smart Structures and Materials 1999: Sensory Phenomena and Measurement Instrumentation for Smart Structures and Materials (eds R.O. Claus, W.B. Spillman Jr), vol. 3670, SPIE, pp. 461–469.

Peeters, B. and De Roeck, G. (2001) One-year monitoring of the Z24-bridge: environmental effects versus damage events. *Earthquake Engineering and Structural Dynamics*, **30**, 149–171.

Perman, R. (1993) Cointegration: an introduction to the literature. *Journal of Economic Studies*, **18**(3).

Roberts, G.P. and Pearson, A.J. (1998) Health monitoring of structures – towards a stethoscope for bridge. Proceedings of ISMA 23, the International Conference on Noise and Vibration, Leuven, Belgium.

Rohrmann, R.G., Baessler, M., Said, S. *et al.* (1999) Structural causes of temperature affected modal data of civil structures obtained by long time monitoring. Proceedings of the XVII International Modal Analysis Conference, Kissimmee, FL.

Ruotolo, R. and Surace, C. (1997) Damage detection using singular value decomposition. Proceedings of DAMAS 97: Structural Damage Assessment Using Advanced Signal Processing Procedures, Sheffield, UK.

Sharma, S. (1996) *Applied Multivariate Techniques*, John Wiley & Sons, Inc., New York.

Sohn, H. (2004) Effects of environmental and operational variability on structural health monitoring. *Philosophical Transactions of the Royal Society, Series A*, **365**(1851), 539–560.

Sohn, H., Farrar, C. and Worden, K. (2001) Applying the LANL Statistical Pattern Recognition Paradigm for Structural Health Monitoring to Data from a Surface-Effect Fast Patrol Boat. Los Alamos National Laboratory Report LA-13761-MS, January 2001, available online from Los Alamos National Laboratory Library, http://library.lanl.gov/.

Sohn, H., Worden, K. and Farrar, C.R. (2003) Statistical damage classification under changing environmental and operational conditions. *Journal of Intelligent Materials Systems and Structures*, **13**(9), 561–574.

Sohn, H., Dzwonczyk, M., Straser, E.G. *et al.* (1998) Adaptive modeling of environmental effects in modal parameters for damage detection in civil structures. Proceedings of the Conference on Smart Structures and Materials 1998: Smart Systems for Bridges, Structures, and Highways (ed. S.-C. Liu), SPIE, pp. 127–138.

Soyoz, S. and Feng, M.Q. (2009) Long-term monitoring and identification of bridge structural parameters. *Computer-Aided Civil and Infrastructure Engineering*, **24**, 82–92.

Williams, E.J. and Messina, A. (1999) Applications of the multiple damage location assurance criterion. Proceedings of DAMAS 99: International Conference on Damage Assessment of Structures, Dublin, Ireland.

Worden, K., Sohn, H. and Farrar, C.R. (2002) Novelty detection in a changing environment: regression and interpolation approaches. *Journal of Sound and Vibration*, **258**(4), 741–761.

Xia, Y., Hao, H., Zanardo, G. and Deeks, A. (2006) Long term vibration monitoring of an RC slab: temperature and humidity effect. *Engineering Structures*, **28**, 441–452.

13

Fundamental Axioms of Structural Health Monitoring

13.1 Introduction

As mentioned in the first chapter of this book, the authors feel that the SHM field has matured to the point where several fundamental axioms, or accepted general principles, have emerged (Worden *et al.*, 2007). The subsequent chapters as well as information published in the extensive literature on SHM over the last twenty years (see the surveys by Sohn *et al.*, 2004, and Doebling *et al.*, 1996) further support this notion of a set of fundamental axioms for SHM. Note that the word 'axiom' is being used here in a slightly different way to that which is understood in the literature of mathematics and philosophy. In mathematics, the axioms are sufficient to generate the whole theory. To take arithmetic as an example, the axioms for the field of real numbers not only specify the properties of the numbers but also how the usual arithmetical operators act on them. As a consequence nothing else is needed in order to derive the whole arithmetic of the real numbers. The word 'axiom' here is used to represent a fundamental truth at the root of any SHM methodology. However, the axioms are not sufficient to generate a given methodology. First of all, the authors are not suggesting that the axioms proposed here form a complete set; it is possible that there are other fundamental truths that have yet to be defined. Second, the axioms here do not specify the 'operators' for SHM; in order to generate a methodology from these axioms, it is necessary to add a group of algorithms that will carry the SHM practitioner from the data to a decision. It is the belief of the authors, enunciated throughout this book, that these algorithms should be drawn from the discipline of statistical pattern recognition and implemented in many cases with machine learning methods. Because of this belief, algorithms of this type will be used in the illustrations throughout this chapter. The intention of this chapter is to explicitly state and justify a set of axioms for SHM for which there is strong evidence from studies reported to date, all the while keeping in mind that this set of axioms may be incomplete. Note that many of the examples used to justify these axioms have already been presented in the preceding chapters and the reader will be referred to these examples in several of the subsequent sections. The axioms that will be addressed are:

Axiom I. All materials have inherent flaws or defects.

Axiom II. The damage assessment requires a comparison between two system states.

Axiom III. Identifying the existence and location of damage can be done in an unsupervised learning mode, but identifying the type of damage present and the damage severity can generally only be done in a supervised learning mode.

Structural Health Monitoring: A Machine Learning Perspective, First Edition. Charles R. Farrar and Keith Worden.
© 2013 John Wiley & Sons, Ltd. Published 2013 by John Wiley & Sons, Ltd.

Axiom IVa. Sensors cannot measure damage. Feature extraction through signal processing and statistical classification are necessary to convert sensor data into damage information.

Axiom IVb. Without intelligent feature extraction, the more sensitive a measurement is to damage, the more sensitive it is to changing operational and environmental conditions.

Axiom V. The length and time scales associated with damage initiation and evolution dictate the required properties of the SHM sensing system.

Axiom VI. There is a trade-off between the sensitivity to damage of an algorithm and its noise rejection capability.

Axiom VII. The size of damage that can be detected from changes in system dynamics is inversely proportional to the frequency range of excitation.

Axiom VIII. Damage increases the complexity of a structure.

13.2   Axiom I. All Materials Have Inherent Flaws or Defects

It is easy to imagine a perfect material. Suppose one considers aluminium for example; a perfect sample would comprise a totally regular lattice of aluminium atoms, a perfect periodic structure. It is well known that if one were able to compute the material properties of such a perfect crystal, properties such as the strength would be far higher than those observed experimentally when actual samples of aluminium are tested. The reason for this discrepancy is of course that all materials, and hence all structures, contain defects at the nano-/microstructural level, such as vacancies, inclusions and impurities. Examples of such defects can be seen in Chapter 1, Figure 1.1(a), where inclusions are evident at the grain boundaries in an as-fabricated sample from a U-6Nb plate. Figure 1.1(b) shows cracks forming along the inclusion lines in a similar specimen after it has been subjected to shock loading.

Metals are never perfect single crystals with a perfect periodic lattice. Broberg (1999) provides an excellent account of the inception and growth of microcracks and voids in the *process region* of a metal. In fibre-reinforced plastics (FRPs) defects can also occur at the macrostructural level because of voids produced in manufacturing. These defects compromise the strength of the material, as coalescence of defects in extreme loading regimes will lead to macroscopic failure at the component level and subsequently at the system level. However, engineers have learned to overcome the design problems imposed by the inevitability of material imperfections. For example, in any composite materials text one can find properties of specific fibre/resin systems, virtually always provided as a range of values. These values are totally dependent on the manufacturing process used and any minor variation in the process will cause a departure from nominal values. Therefore, for composites, a basic material evaluation programme is often required at the design stage. The engineer will design a structure using failure criteria based on material property values from the lowest end of the range derived from the testing programme.

In many engineering materials the effects of nano-/microstructural-level defects can be subsumed into the average or bulk material properties, such as yield stress or fatigue limit, and are not typically considered as 'damage'. However, in other circumstances, like in composite materials, this may not be the case and the void content of the material should be regarded as initial 'damage'. There is no way (under dynamic load) of preventing damage evolution from voids and the associated degradation of the material properties.

As all materials contain imperfections or defects, the difficulty is to decide when a structure is 'damaged'. Here the definitions of defects, damage and failure given in Chapter 1 are reiterated because it is important to have a clear taxonomy for further discussion that will be based on a hierarchical relationship between these conditions.

A *defect* is inherent in the material and statistically all materials will contain some amount of defects. *Damage* is present when the structure is no longer operating in its ideal condition, although it can still function satisfactorily, but possibly in a suboptimal manner. *Failure* occurs when the structure can no longer operate satisfactorily. If one defines the *quality* of a structure or system as its fitness for purpose or its ability to meet customer or user requirements, it suffices to define failure as a change in the system that produces an unacceptable reduction in quality.

The above taxonomy results in the notion that defects (which are inevitable in real materials) lead to damage and damage leads to failure. Using this idea it is possible to go beyond the conservative safe-life design philosophy where the structure is designed to reach its operational lifetime without experiencing damage, to design a damage-tolerant structure where damage may be expected at some point in its operational life (Reifsnider and Case, 2002; Grandt, 2003). However, in order to obtain a damage-tolerant structure, it is necessary to introduce monitoring systems so that one can decide when the structure is no longer operating in a satisfactory manner. This requirement means that failure has to have a strict definition; for example, the stiffness of the structure has deteriorated beyond a certain level. The strict definition of failure coupled with a monitoring system allows one to consider the concept of prognosis, which will be discussed in more detail in Chapter 14.

13.3   Axiom II. Damage Assessment Requires a Comparison between Two System States

This axiom is possibly the most basic of the proposed axioms. It is necessary to state it explicitly as it is sometimes stated in the SHM literature that some approach or other 'does not require a baseline'. It is argued here that this statement is simply never true, and any misunderstanding lies with the assumed meaning of 'baseline'. In the pattern recognition approaches to SHM, a training set is always required. In the case of damage detection where novelty detection approaches can be applied, the training set is composed of samples of features that are representative of the normal condition of the system or structure of interest. For higher levels of diagnosis requiring estimates of damage type, location or severity, the training data must still contain samples of normal condition data, but also must be augmented with data samples corresponding to the various damage conditions. In this case, there is no argument that the normal condition data constitutes the 'baseline'. In order to illustrate how this approach is used, it is necessary to fix on a particular algorithm and *outlier analysis* where a discordance test for multivariate data based on the Mahalanobis squared-distance measure will be used as discussed in Chapter 10.

The approach will be illustrated here on a problem of damage detection in a carbon fibre reinforced polymer (composite) (CFRP) plate using a guided-wave-based SHM. The plate was fabricated and instrumented with piezoelectric actuator/sensors by colleagues at INSA, France, as part of the European Union project DAMASCOS (Pierce *et al.*, 2001). A schematic for the plate is shown in Figure 13.1; the piezoelectric patches used as actuators are labelled E (for emitter) and the patches used as sensors are labelled R (for receiver). The emitters were used to launch Lamb waves that propagated across the plate and were measured at the receivers. Damage was initiated in the form of a drilled hole at the geometrical centre of the plate, as shown in Figure 13.1. This damage location means that the damage was on the axis between E1 and R1 and between E3 and R3. R2 was placed off-centre on an edge in order to establish if damage could be detected off-axis.

For the situation discussed here, a Gaussian modulated sine wave was launched from the actuator E1 and the response was measured at sensor R2 (off-axis). A window of 50 time points corresponding to the arrival of the first wave packet was used as the damage-sensitive feature in the outlier analysis (see Figure 13.2a). One hundred samples of this feature vector were used as the training set to define the normal condition of the plate with no damage. Subsequently, the hole was drilled with ten diameters from 1 mm to 10 mm and for each damage severity the wave features were again measured. When the Mahalanobis squared-distance was computed for each of the feature vectors associated with the damaged condition, the results shown in Figure 13.2(b) were obtained. All damage cases equal to or greater than 2 mm in diameter were detected.

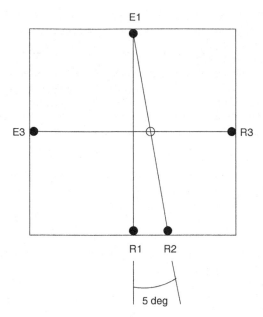

Figure 13.1   Schematic of a composite plate with transducers used for Lamb wave damage detection.

In order to visualise the data better, a Sammon map was used. This procedure is a nonlinear generali-sation of principal component analysis (see Chapters 6 and 9), which was used to reduce the dimension of the feature vector from 50 to 2 while retaining the predominant information contained in the data. The technique is described in some detail in Worden and Manson (1999). The Sammon map is shown in Figure 13.2(c). The normal condition features are the 100 points represented by solid circles. The features corresponding to the damage condition, which are represented by various symbols, are clearly distinct from normal condition features. As the damage progresses, the data move further away from the normal points. This property is reflected in the increasing Mahalanobis squared-distance. Essentially, the damage metric is simply a weighted distance from the mean of the normal cluster. The interpretation of the normal data as a baseline is obvious here. It is equally obvious that the mechanism of detection is simply a comparison between the damage features and those from the normal condition. All novelty detectors in current use rely on the acquisition of a normal condition training set (see Chapter 10); these include auto-associative neural networks (Worden, 1997), probability density estimators (Bishop, 1994) and negative selection algorithms (Dong, Sun and Jia, 2005).

The necessity of Axiom II is not confined to damage detection methods based on pattern recognition; it is also a requirement of the large class of algorithms based on linear algebraic methods. It is perhaps most obviously needed in the case of finite element (FE) updating methods, a class of algorithms that have proved successful at the higher levels of damage identification, that is in the location and quantification of damage (see Chapter 7). The FE updating methodology is based on the construction of a high-fidelity FE model of a structure in its normal condition. To ensure that the model provides an accurate description of the initial-state system, it is usually updated on the basis of experimental data; that is, the parameters of the FE model (e.g. stiffness indices) are adjusted to bring it into closer correspondence with the features derived from experimental observations (e.g. mode shapes and resonance frequencies). The process of damage identification then proceeds by further updating the model on the basis of subsequent data. Clearly, any further need for parameter adjustment will be because the system has changed and this change is assumed to be caused by damage (assuming that benign environmental and operational changes

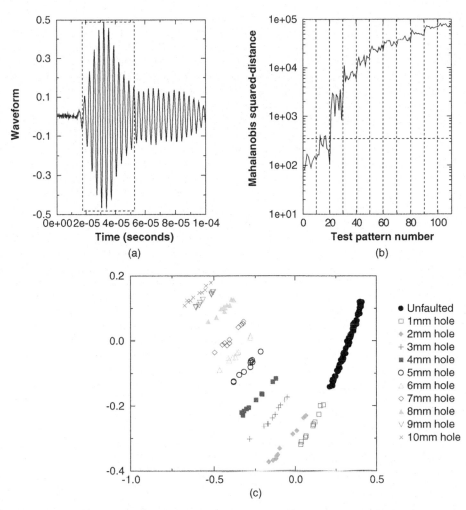

Figure 13.2 (a) Windowed time signal used for novelty detection, (b) novelty index for test patterns including all damage samples, (c) two-dimensional Sammon map of feature data used.

have been ruled out – see Chapter 12). The particular elements adjusted will pinpoint the location of the damage and the size of the adjustment provides an estimate of the severity of the damage.

Several approaches claim to operate without baseline data, which might be interpreted that they do not require a comparison of system states. It is argued here that this is just a matter of terminology. One such method is the strain energy method of Stubbs, Kim and Farrar (1995) discussed in Chapter 7. This approach appears only to operate on data from the damaged structure. Roughly speaking, an estimate of the modal curvature is used to locate and size the damage. In fact, one might argue that there is an implicit baseline or model for these data in the assumption that the undamaged structure behaves as a Euler–Bernoulli beam. Also, the feature computed – the curvature – cannot be used without a threshold of significance, which is computed on the understanding that most of the estimated curvature data comes from the rest of the structure that is undamaged.

Another method based on time-reversal acoustics (Sohn *et al.*, 2005) also makes the claim that no baseline data are needed. However, this method does make the assumption that the undamaged structure responds in a linear, elastic manner and will exhibit the time-reversal property even though this assumption may not be experimentally verified. The baseline here is the assumption of behaviour consistent with that of an ideal elastic solid. In fact, many methods that claim they do not require baseline data in actuality use a numerical model to generate the equivalent of baseline data. Stubbs and Kim (1996) and Sikorsky (1999) use the assumption that the undamaged structure exhibits linear response characteristics or assume that the damage-sensitive features associated with the baseline structure are time-invariant and are not affected by operational and environmental variability.

Another approach that appears not to have a baseline, but explicitly requires a comparison between system states, is the nonlinear acoustic approach described in Section 8.2.11 (Donskoy *et al.*, 2006; Ryles *et al.*, 2006). To recap, the idea is to instrument a plate, for example, with two actuators and a receiver. In a first test, actuator 1 is used to launch an ultrasonic (transient) Lamb wave at a high frequency, f_h, and record the response at the receiver. This response is then converted to a spectrum and can of course be formed by averaging. Now, if the system is linear, the spectrum contains a single line at f_h, as shown in Figure 8.32(a). If the system is nonlinear, as a result of damage, the spectrum as shown may still appear to have a single line, as the other response components will be at the frequencies $2f_h, 3f_h, \ldots$, which are located in the higher-frequency portion of the spectrum. Sampling parameters and anti-aliasing filters may not allow this portion of the spectrum to be observed. If the exercise is now repeated with the second actuator exciting with a low-frequency harmonic signal at f_l and the exercise of forming the spectrum is repeated, the nonlinearity will result in the appearance of sidebands at the frequencies $f_h \pm f_l$, as shown in Figure 8.32(b). The appearance of the sidebands in the second test are the indicator of nonlinearity and, hence, damage. An index of damage extent can then be formed by recording the height of the sidebands or the spread of the sidebands. Because the harmonics in Figure 8.32(a) are not observed, the measurement is a surrogate for the linear system, and hence can be regarded as the baseline. Note that in principle one could infer the presence of nonlinearity from the observation of harmonics in Figure 8.32(b). However, such observations are usually less practical because they would require confidence that the harmonics are not being generated by the instrumentation, which may then require higher-cost data acquisition equipment.

The fact that damage detection algorithms require a comparison of system states is at the root of one of the main problems in SHM. If the normal condition or baseline state changes as a result of environmental or operational variations, the application of a novelty detection algorithm may yield a false-positive indication of damage. This issue will be discussed in more detail later when Axiom IV is considered and is covered in considerably more detail in Chapter 12 of this book.

13.4 Axiom III. Identifying the Existence and Location of Damage Can Be Done in an Unsupervised Learning Mode, but Identifying the Type of Damage Present and the Damage Severity Can Generally Only Be Done in a Supervised Learning Mode

The statistical classification portion of SHM is concerned with the implementation of the algorithms that operate on the extracted damage-sensitive features to quantify the damage state of the structure. As discussed in Chapters 1 and 9, the algorithms used in statistical classification fall into three categories. When data are available from both the undamaged and damaged structures, the statistical pattern recognition algorithms fall under the general classification referred to as *supervised learning* (see Chapter 11). *Group classification* and *regression analysis* are categories of supervised learning algorithms and are generally associated with either discrete or continuous classification, respectively. *Unsupervised learning* refers to algorithms that are applied to data not containing examples from the damaged structure (see Chapter 10). *Novelty detection* is the primary class of algorithms applied in unsupervised learning applications. All of the classification algorithms analyse statistical distributions of the measured or derived features to enhance the damage detection process.

As previously discussed in Chapters 1 and 9, identification of the damage state of a system can be described as a four-step process that answers the following questions. Is there damage in the system (existence)? Where is the damage in the system (location)? What kind of damage is present (type)? How severe is the damage (extent)?

Answers to these questions in the order presented represent increasing knowledge of the damage state. When applied in an unsupervised learning mode, statistical models can typically be used to answer questions regarding the existence and sometimes the location of damage. As an example, if a damage-sensitive feature extracted from measured system response data exceeds some predetermined threshold one can conclude that damage has occurred. This conclusion must also rely on the knowledge that the change in the feature has not been caused by operational or environmental variability. Many approaches to damage detection in rotating machinery, such as those that examine change in the kurtosis values of the acceleration response amplitude to identify bearing damage, are based on such outlier analyses (Worden, Manson and Fieller, 1999). Similarly, changes in features derived from relative information obtained from an array of sensors can be used to locate the damage, as is done in many wave propagation approaches to SHM (Pierce *et al.*, 2001) and impedance measurements (Park *et al.*, 2003). In general, these statistical procedures cannot distinguish between possible damage types or the severity of damage without additional information.

When applied in a supervised learning mode and coupled with analytical models or data obtained from the structure when known types and levels of damage are present, the statistical procedures can, in theory, be used to determine the type of damage and the extent of damage. The previously mentioned finite element model updating approaches to SHM can provide an estimate of damage existence, location and associated stiffness reduction (extent). However, these approaches are typically limited to cases where the structure can be accurately modelled as a linear system before and after damage. Also, these procedures do not usually identify the type of damage present, but instead just assume that the damage produces a local reduction in stiffness at the element level. In the case of rotating machinery, large databases can be obtained from nominally identical pieces of equipment that have been run to some threshold condition level or, in an extreme case, to failure where the type and extent of damage are assessed through some type of equipment autopsy. These data can be used to build group classification and regression models that can assess the type of damage and its extent, respectively.

In the example discussed in association with Axiom II, the novelty index was a monotonic function of the damage size and the reader might suppose therefore that an unsupervised approach may lead to higher levels of damage identification. Unfortunately, this is not the case and counterexamples are not difficult to obtain; the following case study of damage detection in an aircraft wing box will be used to provide a counterexample.

As discussed in Chapter 10, the object of the investigation was to detect damage in an experimentally simulated aircraft wingbox, as shown schematically in Figure 10.10. Damage was induced in the form of a sawcut in one of the stringers. The sawcut was increased from an initial depth of 0.25 mm to 2.25 cm in steps of 0.25 mm, giving nine damage severities. The details of the experiment can be found in Worden, Manson and Allman (2003). Essentially the structure was excited with a Gaussian random noise at a point on the undersurface of the bottom skin and the acceleration responses were recorded from accelerometers on the top surface above the two ends of the damaged stringer (see Figures 10.12 and 10.13). These responses were used to form the transmissibility spectra shown in Figures 10.14 and 10.15, which were then used to construct a feature for novelty detection.

In order to accomplish the novelty detection, a set of fifty spectral lines spanning a single peak in the transmissibility was selected as the damage-sensitive feature. This peak was chosen because it showed systematic variation as the damage severity increased (Figure 10.15). The fifty lines gave a 50-dimensional feature that could be used for outlier detection, as discussed earlier. When the Mahalanobis distance was computed for the undamaged condition and the nine damaged conditions, the results for ten samples corresponding to each damage case were obtained, as shown in Figure 10.18. It is clear from this figure that the novelty index in this case is not a monotonic function of the damage severity and therefore cannot be used to infer severity from the measured features. This result offers direct verification of this axiom.

13.5 Axiom IVa. Sensors Cannot Measure Damage. Feature Extraction through Signal Processing and Statistical Classification Are Necessary to Convert Sensor Data into Damage Information

Sensors measure the response of a system to its operational and environmental input. Therefore, there is nothing surprising about the fact that sensors cannot *directly* measure damage. In a more basic context, it is similarly impossible to measure stress. The solution is to measure a quantity – the strain – from which one can infer the stress. In fact, things are a little more indirect than this simple process of inference. The important point is that the sensor yields a value linearly proportional to the physical quantity of interest. Knowledge of the material and geometric properties then allows one to infer the stress from the strain measure. In other words, the stress σ is a known function of the strain ε, where

$$\sigma = f(\varepsilon) \tag{13.1}$$

In this case the function f is known from observations of basic physics and is particularly simple. The situation is a little more complicated for damage. Suppose for the sake of simplicity that the damage state of a given system is captured by a scalar D; the first objective of damage identification is to measure some quantity, $\{x\}$ (often a multidimensional vector), which is a function of the damage state

$$\{x\} = \{f(D)\} \tag{13.2}$$

and them identify the damage through some inverse modelling approach as discussed in Section 7.8.2. The main difficulty for SHM is that the function $\{f\}$ is generally not known from basic physics and must usually be *learned* from the data. The data in question may often be of high dimensionality such as a computed spectrum or a sampled wave profile.

The main problem associated with pattern recognition techniques implemented through machine learning algorithms to learn the function in Equation (13.2) is their difficulty in dealing with data vectors of high dimensionality. This limitation is sometimes termed the *curse of dimensionality* (see Chapter 6). If one considers methods depending on the availability of *training data*, that is examples of the measurement vectors to be analysed or classified, then the curse can be simply stated as: in order to obtain accurate diagnostics, the amount of training data theoretically grows explosively with the dimension of the feature (Silverman, 1986).

From a pragmatic point of view, there are two solutions to the problem. The first is to obtain adequate training sets. Unfortunately, this will not be possible in some engineering situations because of the limitations on the size and expense of testing programmes. The second approach is to reduce the dimension of the data to a point where the available data are sufficient. However, there is a vital caveat – the reduction or compression of the data must not remove the influence of the damage. This caveat means that the feature extraction must be tailored to the problem. One example of good feature selection for damage identification in a gearbox would be to select only the lines from a spectrum that are at multiples of the meshing frequency, as it is known that these lines are the most strongly affected by damage (Mitchell, 1993). This approach is feature selection on the basis of engineering judgement. More principled approaches to dimension reduction may be pursued, but care should be taken; for example, if one uses principal component analysis (PCA) one certainly obtains a reduced dimension feature. However, this vector is obtained using a criterion that may not preserve the information from damage (as discussed in Chapter 12).

13.6   Axiom IVb. Without Intelligent Feature Extraction, the More Sensitive a Measurement is to Damage, the More Sensitive it is to Changing Operational and Environmental Conditions

This section discusses the concept of *intelligent feature extraction* that was introduced in Section 12.8. The concern being addressed here is that the features derived from measured data will not only depend on the damage state but may also depend on an environmental and/or operational variable. Temperature θ will be used here for illustrative purposes. Equation (13.2) then becomes

$$\underline{x} = \underline{g}(D, \theta) \tag{13.3}$$

The machine learning problem is complicated by the fact that one wishes to learn the dependence on D despite the fact that some of the variation in the measured quantity is likely to be caused by θ varying. The problem of feature extraction is then to find a reduced dimension quantity that depends on the damage, but not the temperature.

This axiom will be illustrated using an example from damage detection based on Lamb wave propagation and outlier analysis. The sample is the 300-mm-square CFRP plate discussed in Section 12.8, where the initial selected feature was the 50-frequency-line portion of the Fourier transform of the wave amplitude time signal, as shown in Figure 12.26. Thermal cycling in an environmental chamber provided a source of controlled environmental variability and a 10-mm hole that was drilled in the plate directly between the two piezoceramic sensor/actuators provided the damage that was to be detected.

Results of the novelty detection based on a Mahalanobis squared-distance damage index showed that the vast majority of the features corresponding to the undamaged constant-temperature condition gave novelty indices below the threshold, as shown in Figure 12.27. However, although all the features from the damage set were substantially above the threshold, so were all the features from the undamaged, temperature-cycled phase. This result is clearly undesirable. However, this effect is not confined to laboratory specimens, as it is shown in Farrar *et al.*, (2002) and Farrar *et al.* (1994) that the natural frequencies of a bridge showed greater variation as a result of the day–night temperature cycle than they do as a result of some substantial local damage.

Note that the statement of this axiom includes the caveat '*without intelligent feature extraction* ... '. It was shown in Section 12.8 that algorithms exist that can project out the environmentally sensitive components of the features while preserving the damage sensitivity. A few techniques have recently emerged for this purpose; the method shown in Section 12.8 involved minor component analysis, but effective procedures exist based on univariate outlier statistics (Manson, 2002), factor analysis (Kullaa, 2001) and cointegration (see Section 12.9 in this book and Cross, Worden and Chen, 2011).

For the features considered in the example from Section 12.8, the principal component decomposition was computed for all the undamaged data including that from the temperature cycling and it emerged that the great majority of the data variance corresponding to the temperature variation was projected on to the first 10 principal components. However, in order to be assured of eliminating these effects, only the last 10 components (the minor components) were extracted for use as an *advanced feature*. When the outlier analysis was carried out on the whole testing set using this advanced feature, including the damage patterns, the results shown in Figure 12.28 were obtained. The results were excellent; the new 10-point feature obtained from projection on to the minor components separated the undamaged and damaged data perfectly, even when the temperature-cycled data were included. In this case the procedure whereby a new set of features were obtained when the initial features were projected on to the minor components can be viewed as the *intelligent feature extraction* process.

An alternative approach to the problem described by Equation (13.3) is to learn both the dependence on damage and the environment (D and θ). Note that this learning problem is a mixed one; supervised learning is used to obtain the temperature dependency but unsupervised learning is used to detect the damage, as discussed in Chapter 12.

13.7   Axiom V. The Length and Time Scales Associated with Damage Initiation and Evolution Dictate the Required Properties of the SHM Sensing System

Axiom I introduced the concept of the length scales associated with damage and pointed out that defects are present in all materials beginning at the atomic length scale and spanning scales, where component- and system-level failures are present. In terms of time scales, it was pointed out in Chapter 1 that damage can accumulate incrementally over periods of time exceeding years and can also result on fraction-of-second time scales from scheduled discrete events and from unscheduled man-made and natural discrete events.

Axiom IVa states that a sensor cannot measure damage. Therefore, the goal of any SHM sensing system is to make the sensor reading as directly correlated with, and as sensitive to, damage as possible. At the same time one also strives to make the sensors as independent as possible from all other sources of environmental and operational variability. To best meet these goals for the SHM sensor and data acquisition system, many system properties must be defined as discussed in Chapter 4, where it was noted that fundamentally there are four factors that control the selection of hardware to address these sensor system design parameters:

1. The length scales on which damage is to be detected.
2. The time scale on which damage evolves.
3. How varying and/or adverse operational and environmental conditions will affect the sensing system.
4. Cost.

The development of a damage detection system for the composite wings of an unmanned aerial vehicle (UAV) (Figure 13.3) is used as an example. In one case damage is assumed to be initiated by foreign object impact on the wing surface. Such damage is often very local in nature and may manifest itself as fibre breakage, matrix cracking, delamination in the wing skin or a debonding between the wing skin

Figure 13.3   Unmanned aerial vehicle (courtesy of www.af.mil).

and spar, on the order of 10 cm^2 or less. Accurate characterisation of the impact phenomena occurs on a micro- to millisecond time scale, which requires the data acquisition system to have relatively high sampling rates (greater than 100 kHz). This damage may then grow to a level that produces failure after being subject to numerous fatigue cycles during many hours of subsequent flight. The time scales associated with the damage initiation and evolution will influence sensing system properties such as sensor types, number and locations; required bandwidth, sensitivity and dynamic range; required sampling intervals; data acquisition/telemetry/storage system requirements; processor/memory requirements and power requirements.

Practically speaking, the UAV impact damage will not produce failure unless it significantly affects the operation of the aircraft. One manner in which this type of damage can affect aircraft operation is by changing the coupling of the bending and torsion modes of the wing, which, in turn, changes the flutter characteristics of the aircraft. Identifying and characterising the local damage associated with foreign object impact may require a local, and somewhat dense, active sensing system while characterising the influence of this damage on the UAV's flutter characteristics will require a more global sensing system capable of identifying bending and torsion modes of the entire wing (Sohn *et al.*, 2004). These length scale considerations will influence the same sensing system properties described in the previous paragraph.

In summary, this example clearly demonstrates that the length and time scales associated with damage initiation and evolution drive many of the SHM sensing system design parameters. A priori quantification of these length and time scales will allow the sensing system to be designed in an efficient manner.

13.8 Axiom VI. There is a Trade-off between the Sensitivity to Damage of an Algorithm and Its Noise Rejection Capability

Once again, this axiom is illustrated via a group of computer simulations involving outlier analysis (see Chapter 10). However, for the sake of variety, the features selected this time return to the low-frequency modal regime. The system of interest is the three-mass lumped-parameter system previously discussed in Chapter 10 and Worden (1997) and illustrated in Figure 10.1. The problem is to detect a loss of stiffness between the centre mass and one of the end masses (the masses are arranged in a simple chain). The feature used here is a window of the transmissibility spectrum between the two masses. Figure 13.4 shows the transmissibility vector for the undamaged state together with the data corresponding to stiffness reductions of 10, 20, 30, 40 and 50%. Each pattern constitutes 50 spectral lines.

The object of this exercise is to investigate the threshold for damage detection as a function of the noise-to-signal ratio. This ratio is expressed as a percentage of the maximum of the transmissibility magnitude over the noise-free normal pattern vector. For a given noise ratio, the training set and testing set for the outlier analysis is generated as follows. The training set is composed of 1000 copies of the clean normal pattern each corrupted by an independent 50-dimensional Gaussian noise vector with zero mean and a diagonal covariance matrix with all variances equal to the noise-to-signal ratio. The testing set is simply the clean normal pattern and the five damage patterns of increasing severity. These vectors are then corrupted with noise vectors with the same statistics as those used for the training set. Once the training set has been generated, it is used as the basis for the mean and covariance statistics used to compute the Mahalanobis squared-distance.

Figure 13.5 shows the outlier metric computed over the testing set for noise-to-signal ratios of 0.03, 0.06 and 0.09. It is clear from the figure that higher noise requires a higher level of damage before the outlier metric crosses the 99% threshold for damage. A more complete study involving many noise ratios in the range 0 to 0.1 leads to Figure 13.6, which shows the minimum detectable damage level as a function of the extent of the noise corruption. As the axiom asserts, the detection level increases monotonically (ignoring the local noise that is a result of generating independent random statistical samples). In fact, the function is remarkably linear in this particular case.

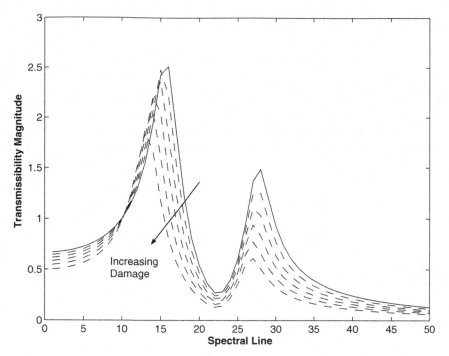

Figure 13.4 Transmissibility feature for normal condition data (solid line) and various damage states (dashed line).

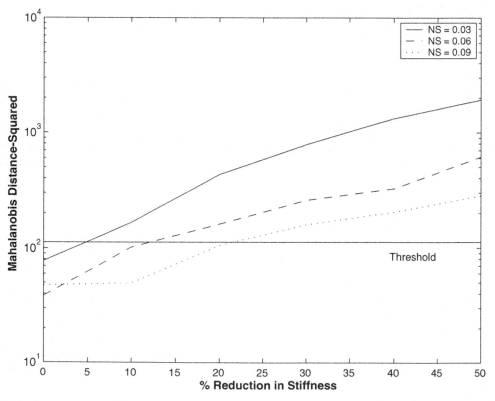

Figure 13.5 Outlier statistic as a function of damage level for various noise-to-signal ratios for the training and testing data.

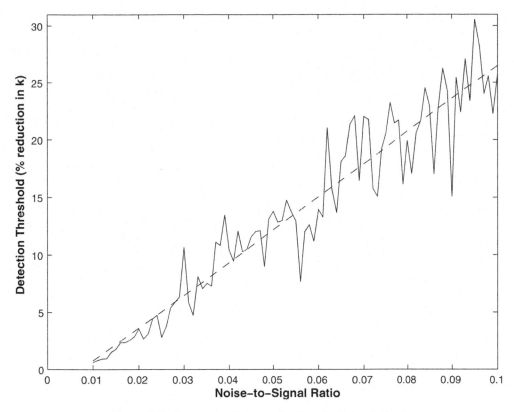

Figure 13.6   Damage sensitivity as a function of pattern noise level.

In summary, this axiom implies that one should attempt to reduce the level of noise in the measured data or the subsequently extracted features as much as possible, which can be accomplished by a variety of hardware and software approaches such as wavelet de-noising, analogue or digital filtering, or even simple averaging (see Appendix A).

13.9   Axiom VII. The Size of Damage that Can Be Detected from Changes in System Dynamics is Inversely Proportional to the Frequency Range of Excitation

In the field of ultrasonic nondestructive testing the diffraction limit is often associated with the minimum size of flaw that can be detected as a function of ultrasonic wavelength. This limit may suggest that flaws of a size comparable with half a wavelength are detectable. The diffraction limit is actually a limit to the resolution of nearby scatterers; that is, if two scatterers are separated by more than a half-wavelength of the incident wave they will be separable. In fact, a flaw will scatter an incident wave for wavelengths below this limit and this amount of scattering decreases with increasing wavelength. This result means that, if instrumentation is available to detect arbitrarily small evidence of scattering, that is arbitrary small reflection coefficients, then arbitrarily small flaws can be detected. However, as described above, scattering is always substantial when the size of the flaw is comparable with the wavelength and so it is advantageous to use small wavelengths in order to detect small flaws.

The wavelength λ is related to the wave phase velocity v and frequency f by

$$\lambda = \frac{v}{f} \tag{13.4}$$

and from this simple relationship it is clear that, for constant velocity, the wavelength will decrease as the frequency increases, which in turn implies that the damage sensitivity will increase.

Note that energy does not necessarily have to be input to the structure at these wavelengths. Non-linear structures can have many frequency up-conversion (and down-conversion) properties that lead to structural response in bandwidths removed from input bandwidths.

Evidence for damage detection well below the diffraction limit for the size of the flaw can be found in numerous places in the literature. Two examples will be cited here. In Alleyne and Cawley (1992), the interaction of Lamb waves with notches in plates was investigated from both an FE simulation viewpoint and by experiment. Among the conclusions of the paper was the fact that Lamb waves could be used to detect notches when the wavelength to notch–depth ratio was of the order of 40. This result was true as long as the notch width was small compared to the wavelength. In support of the axiom, the study found that the sensitivity of given Lamb wave modes to defects increased for incident waves at higher-frequency–thickness products. However, the paper also found that the wavelength of the Lamb wave modes was not the only factor affecting sensitivity. In some regimes, increases in frequency–thickness did not bring correspondingly large improvements in sensitivity. The authors observed that 'appropriate mode selection can sometimes remove the need to go to higher frequencies where the waveform could be more complicated'.

Another interesting study in Valle et al., (2001) concerned a finite element model of circumferential wave propagation in a cylinder with a radial crack and noted that the guided waves could 'detect cracks down to 300 μm – even though the wavelength of these signals is much greater than 300 μm'. The thesis that the damage sensitivity increases with excitation frequency is supported by both of the previously cited papers, but they both also offer the possibility that for detection of a given size flaw; frequencies much lower than those suggested by the diffraction limit may suffice.

The relationship between damage sensitivity and wavelength can be extended to more general types of vibration-based damage detection methods. In these applications the wavelength of the elastic wave travelling through the material is replaced by the 'wavelength' of the standing wave pattern set up in the structure that is interpreted as a mode of vibration. The technical literature is replete with anecdotal evidence that such lower frequency modes are not good indicators of local damage (Doebling et al., 1996; Sohn et al., 2004). The lower-frequency global modes of the structure that have long characteristic wavelengths tend to be insensitive to local damage. For the case of civil engineering infrastructure such as suspension bridges these mode shape wavelengths can be on the order of hundreds of metres (Farrar et al., 1994, 2002) and flaws such as fatigue cracks that must be detected to assure safe operation of the structure are on the order of centimetres in length.

The observations regarding the relationship between the characteristic wavelength and the flaw size has led research to explore other high-frequency active sensing approaches to SHM. These methods are based on Lamb wave propagation (Raghavan and Cesnik, 2007), impedance measurements (Park et al., 2003), high-frequency response functions and time reversal acoustics (Sohn et al., 2005). In these applications the excitation frequency can be as high as several hundred kilohertz and the corresponding wavelengths are on the order of millimetres. However, as the frequency increases and wavelength decreases, scattering effects (e.g. reflection of elastic waves off grain boundaries and other material interfaces) will eventually increase the noise in the measurements and place limits on the sensitivity of the damage detection process. Optimal frequency ranges for damage detection can be determined based on the wavelength of the standing wave pattern and the condition that the damage is located in areas of high curvature associated with the deformation of the wave pattern.

A final example is presented to illustrate how sensitivity to damage increases with increasing frequency. The structure examined was a simple portal-frame structure as shown in Figure 13.7. The structure

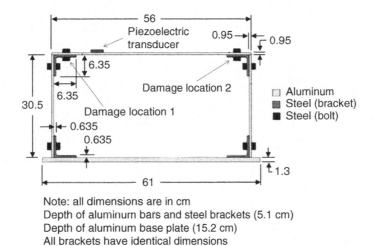

Note: all dimensions are in cm
Depth of aluminum bars and steel brackets (5.1 cm)
Depth of aluminum base plate (15.2 cm)
All brackets have identical dimensions

Figure 13.7 Portal frame test structure.

consists of aluminium bars connected using steel angle brackets and bolts with a simulated rigid base. A piezoelectric transducer was mounted on the top beam of this symmetric structure to measure the impedance signals at frequency ranges of 2–13 kHz and 130–150 kHz. Baseline measurements were first made under the damage-free condition. Two damage states were sequentially introduced at two different locations by loosening both bolts at a particular corner from 18 N m to hand-tight. After implementing the damage, the impedance signals were again recorded from the patch.

The impedance measurements (before and after damage) for Case II are shown in Figures 13.8 and 13.9. These figures show baselines and damaged signals from the structure. The peaks in the impedance measurements correspond to the resonances of the structure. At the frequency range of 130–150 kHz, it is easy to see qualitatively that the damaged signals are quite different, with the appearance of new peaks and shifts at all frequency ranges examined. With increasing levels of the damage, the impedance variation becomes more noticeable. At the frequency range of 2–13 kHz, one clearly observes that there are only relatively small changes, a slight shift of the resonance, seemingly in the range that could be caused by temperature variations. This simple example further demonstrates the varying sensitivity of the SHM techniques at different frequency ranges.

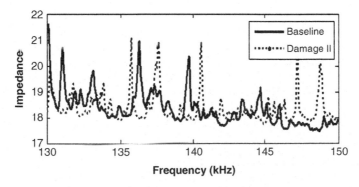

Figure 13.8 High-frequency impedance measurement.

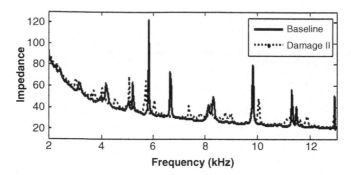

Figure 13.9   Low-frequency impedance measurement.

Finally, it should be noted that there will generally be an increased energy requirement necessary to maintain a comparable amplitude excitation at the higher frequencies. As a result, higher-frequency excitation procedures are typically associated with more local damage detection procedures.

13.10   Axiom VIII. Damage Increases the Complexity of a Structure

If one presents a person with the pictures in Figure 1.3(c) and they are asked to classify this ship as damaged or undamaged, they will almost universally classify it as damaged. There are two reasons for this. The first is based on the fact that the person will usually frame an internalised comparison between all the visual examples of an undamaged structure they have so far experienced and they will note that the images presented deviate from this norm (a qualitative form of pattern recognition and novelty detection). The second reason – and this could apply even if the person has not seen an undamaged exemplar of the structure – relates to the 'appearance' of damage. Most engineered structures are designed to be 'smooth'; there are practical reasons for this, for example minimising drag, and there are aesthetic reasons. In any case, by the latter criterion, the person is likely to be making some qualitative assessment of *complexity*. For the ship in Figure 1.3(c), a person could perceive complexity in the very irregular nature of the ship's bow and note that few engineered structures, particularly ship structures, have such irregular geometries. This concept of complexity can be extended to the case of corrosion, as shown in Figure 13.10. Again, the corrosion adds *complexity* to the image of the girder through the increase in the number of edges or irregularities in this image. Engineers do not like to deal with such qualitative concepts and therefore more quantifiable means of assessing complexity are needed if such a concept is to be used in practice.

A logical first step towards developing more quantifiable measures of complexity is to apply concepts of statistics and signal processing to data from the damaged and undamaged systems. To illustrate this concept, one can examine data from the simulated building structure described in Chapter 5 and reinvestigated at various places in this book. Figure 8.20 shows the acceleration time-history responses to a harmonic base input, measured on the top floor when the bumper used to induce a nonlinear response is not present (undamaged) and when it is present (damaged). The response time history is seen to increase in complexity when damage is present through the irregular nature of the response. Normalised power spectra of these signals (also shown in Figure 8.20) allow for a more quantifiable assessment of this complexity by identifying harmonics of the excitation frequency that are produced as a result of the impacting.

Another means to assess the change in complexity caused by damage is to examine the probability density functions of the measured system response. To quantify complexity through changes in the probability density function more directly, one can again examine the simulated building structure. This time the structure will be subjected to a random base input with and without the bumper mechanism present.

Figure 13.10   Corrosion on a steel bridge girder.

Figure 13.11 shows estimates of the probability density functions for the undamaged and damaged systems obtained with a kernel density estimator. Clear distortions in the density function are evident as a result of the damage. Additionally, statistics such as the skewness and kurtosis have changed significantly.

The ideas of complexity will be further illustrated using synthetic response data acquired by numerical simulation of a simple dynamical system. The system of interest is an SDOF nonlinear (bilinear) oscillator

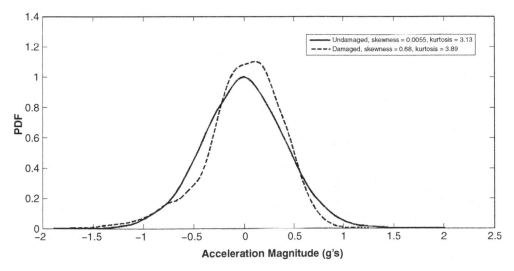

Figure 13.11   Change in the probability density function of the acceleration response measured on the simulated building structure that results from damage.

specified by the equation of motion,

$$m\ddot{y} + c\dot{y} + k(y) = X\sin(\omega t) \tag{13.5}$$

where y is the displacement response to the harmonic excitation, m and c are the usual mass and damping coefficients and $k(y)$ is a nonlinear stiffness force given by

$$
\begin{aligned}
k(y) &= ky & \text{if } k \leq d \\
k(y) &= kd + \alpha k(y - d) & \text{if } k > d
\end{aligned}
\tag{13.6}
$$

The parameter α governs the ratio of stiffness for the two linear regimes and the parameter d specifies the transition point between them. As discussed elsewhere in this book, notably in Chapter 8, this system has some interest in terms of SHM because it is arguably representative of the type of behaviour observed when a breathing crack is present in a structure.

The discussion will be based on a phenomenon already discussed earlier in this chapter, which is that the response of a nonlinear system to a simple harmonic excitation will contain components (harmonics) at multiples of the fundamental forcing frequency. The time data shown in Figure 13.12 were obtained by numerically integrating Equation (13.5). The amplitude of excitation X was taken as 1 unit and the parameter choices $m = 1$, $c = 20$, $k = 10\,000$ and $d = 0.000\,05$ were adopted. Different values of α were used in order to show the growth in complexity of the response as the severity of the nonlinearity increased. Figure 13.12 shows examples of the time-history response for the cases $\alpha = 1$ (linear, undamaged) and $\alpha = 10$ (nonlinear, damaged). Interestingly, there does not appear to be a marked increase in structure or complexity for the more severe nonlinearity. In fact, the increased complexity is more visible in the frequency domain. Figure 13.13(a) and (b) show the Fourier spectra of the two signals from Figure 13.12 and here the presence of the harmonics in the nonlinear case is very marked. In fact, if the nonlinearity becomes even more severe, more structure appears. Figure 13.13(c) shows the spectrum corresponding to the case of $\alpha = 100$. Peaks are beginning to appear in between the harmonics of the forcing frequency, which are the result of subharmonic generation.

This example also allows one to explore a quantitative measure of complexity or, more properly, of information content. It is clear that the more complex pattern will require a greater amount of information in order to specify it than a simpler pattern. One way of encoding the information content of a pattern or

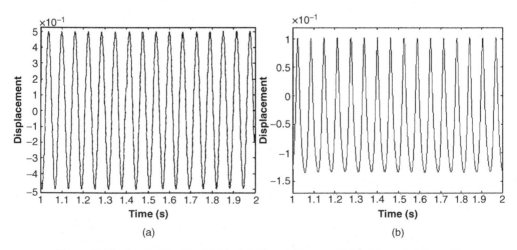

Figure 13.12   Acceleration time history data from a bilinear oscillator: (a) $\alpha = 1$, (b) $\alpha = 10$.

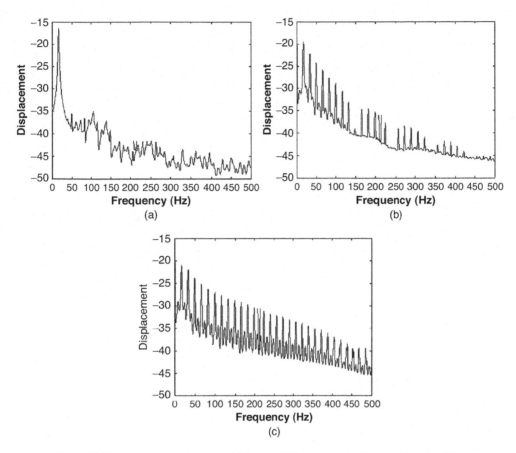

Figure 13.13   Response spectra from a bilinear oscillator: (a) $\alpha = 1$, (b) $\alpha = 10$, (c) $\alpha = 100$.

signal is in terms of Shannon's entropy function (Shannon and Weaver, 1949). Suppose that the state of a system is encoded in a variable x, whose values are governed by a probability density function $p(x)$. Shannon's entropy function or information entropy is given by

$$S = -\int p(x)\log p(x)\mathrm{d}x \qquad (13.7)$$

The function has a minimum if all the probability is concentrated on a single state (total order) and a maximum if it is uniformly distributed over all states. The example considered here is not probabilistic, but it can be cast in a form where a probabilistic analogy is possible. Suppose that one were to normalise the spectrum (not the logarithmic version, but the linear version) by dividing by the total area under the spectrum; with a stretch of the imagination, one might say that the resulting function $p(\omega)$ represents the probability that an 'atom' of 'response energy' would be associated with a frequency ω. Now the situation in Figure 13.12(a) corresponds to 'total order'; all (most of) the energy is associated with the forcing frequency. As the parameter α increases, however, the energy is redistributed across the spectrum and the order reduces. In order to illustrate this, the system in Equation (13.5) was simulated with a range of α values from 1 to 10 and the 'spectral entropy' values were computed using Equation (13.7). The result of this calculation is shown in Figure 13.14.

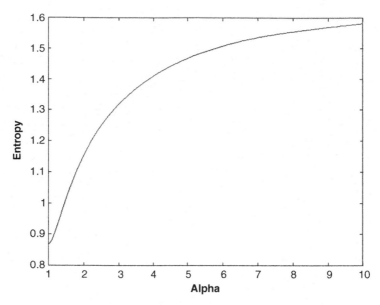

Figure 13.14   Spectral entropy measure as a function of the severity of the bilinear nonlinearity.

Note that the information content or complexity of the spectrum increases monotonically as a function of α, the severity of the nonlinearity. The entropy starts to reach a limiting value as the energy becomes distributed over all the harmonics in a more uniform fashion. Because of this monotonic behaviour, the entropy is revealed as a useful feature for SHM purposes. (Actually, the entropy increases with the nonlinear severity rather than with increased damage here. In the case of a breathing crack, the progressing variable would be the length of the crack rather than the open/closed stiffness ratio.) Features based on, or motivated by, information content have appeared at various points in the development of SHM research (Overbey and Todd, 2009).

Very many of the features proposed in the SHM literature, particularly those based on the assumption that damage will cause an initially linear system to exhibit nonlinear characteristics, are in some way assessing changes in the complexity of a structure. This increase in complexity is assessed through some analysis of the system's measured response. Examples include the previously mentioned change in statistics, identification of harmonic generation, measures of the continuity of a signal by parameters such as the Holder exponent, increase in residual errors associated with time series models and changes in wave propagation characteristics (as discussed in Chapters 7 and 8).

This axiom will impact SHM more directly if one can use it to define better features for a particular SHM application. By defining the type of damage that is of interest and then identifying how such damage increases the complexity of the structure and its associated dynamic response, one can begin to develop a principled approach to feature selection. With further development it is anticipated that this feature-selection process can be directly linked to the definition of a damage observability criteria. Such rigorous procedures for feature selection and damage observability are essential developments that are needed to better transition SHM research to practice on *in situ* structures.

13.11   Summary

In this chapter the authors have attempted to coalesce information that has been reported throughout this book and more generally throughout the technical literature into axioms that form a set of basic principles for SHM. It is reiterated that the term 'axiom' is being used as a 'fundamental truth' as opposed to the mathematical definition where axioms are sufficient to generate the whole theory. In all cases the stated axioms have been supported by examples reported previously in this book and in the more general SHM technical literature. Most of these examples include experimental studies that further support the respective axioms. In cases where one could argue that there are studies that may be interpreted as contradicting a particular axiom, such as Axiom II, it has been argued that often these contradictions are actually related to terminology used in a particular study. When such terminology is put on a common footing, Axiom II is shown to hold true.

Although not backed up by an exhaustive literature review, an attempt has been made to find examples reported in the literature that contradict the reported axioms. To date, none have been found, including those reported in two fairly significant reviews (Doebling *et al.*, 1996; Sohn *et al.*, 2004). Therefore, the authors believe that these axioms do, in fact, represent a starting point for future SHM research, development and applications.

As a closing caveat, the authors make no claim that the axioms presented within this chapter are complete. It is hoped that other researchers in the field will periodically be adding to this set of axioms in an effort to establish further the foundations for structural health monitoring, formalise this multidisciplinary field and better define its relationship to the more traditional and well-established nondestructive evaluation field. Finally, the authors believe that the current axioms and the anticipated future additions will facilitate the transition of SHM from an engineering research topic to accepted engineering practice through the definition and subsequent adoption by the SHM community of these 'fundamental truths'.

References

Alleyne, D. and Cawley, P. (1992) The interaction of Lamb waves with defects. *IEEE Transactions on Ultrasonics, Ferroelectrics and Frequency Control*, **39**, 381–397.

Bishop, C. (1994) Novelty detection and neural network validation. *Novelty IEEE Proceedings on Vision, Image and Signal Processing* **11**(141), 217–222.

Broberg, K. (1999) *Cracks and Fracture*, Cambridge University Press.

Cross, E., Worden, K. and Chen, Q. (2011) Cointegration: a novel approach for the removal of environmental trends in structural health monitoring data. *Proceedings of the Royal Society, Series A*, **467**, 2712–2732.

Doebling, S., Farrar, C.R., Prime, M. and Shevitz, D. (1996) Damage Identification and Health Monitoring of Structural and Mechanical Systems from Changes in their Vibration Characteristics: A Literature Review. Los Alamos National Laboratory Report LA-13070-MS.

Dong, Y., Sun, Z. and Jia, H. (2005) A cosine similarity-based negative selection algorithm for time series novelty detection. *Mechanical Systems and Signal Processing*, **20**(6), 1461–1472.

Donskoy, D., Zagrai, A., Chudnovsky, A. *et al.* (2006) Nonlinear vibro-acoustic modulation technique for life prediction of aging aircraft components. Proceedings of the Third European Workshop on Structural Health Monitoring, Granada, Spain.

Farrar, C.R., Baker, W.E., Bell, T.M. *et al.* (1994) Dynamic Characterization and Damage Detection in the I-40 Bridge over the Rio Grande. Los Alamos National Laboratory Report LA-12767-MS, Los Alamos, NM.

Farrar, C., Cornwell, P., Doebling, S. and Prime, M. (2002) Structural Health Monitoring Studies of the Alamosa Canyon and I-40 Bridges. Los Alamos National Laboratory Report LA-13635-MS, Los Alamos, NM.

Grandt, A. (2003) *Fundamentals of Structural Integrity*, John Wiley & Sons.

Kullaa, J. (2001) Elimination of environmental influences from damage sensitive features in a structural health monitoring system, in *Structural Health Monitoring – The Demands and Challenges*, CRC Press, Palo Alto, CA, pp. 742–749.

Manson, G. (2002) Identifying damage sensitive environment insensitive features for damage detection. Proceedings of International Conference on Identification in Engineering Systems, Swansea, UK, pp. 187–197.

Mitchell, J.S. (1993) *Introduction to Machinery Analysis and Monitoring*, PenWel Books, Tulsa, OK.

Overbey, L. and Todd, M. (2009) Dynamic system change detection using a modification of the transfer entropy. *Journal of Sound and Vibration*, **322**(1–2), 438–453.

Park, G., Sohn, H., Farrar, C. and Inman, D. (2003) Overview of piezoelectric impedance-based health monitoring and path forward *Shock and Vibration Digest*, **35**, 451–463.

Pierce, S., Dong, F., Atherton, K. *et al.* (2001) Damage assessment in smart composite structures: the DAMASCOS programme. Proceedings of Conference on Smart Materials and Structures, 4327, Newport Beach: SPIE, pp. 223–233.

Raghavan, A. and Cesnik, C. (2007) Review of guided-wave structural health monitoring. *The Shock and Vibration Digest*, **39**(2), 91–114.

Reifsnider, K. and Case, S. (2002) *Damage Tolerance and Durability of Material Systems*, Wiley Interscience.

Ryles, M., McDonald, I., Ngau, F. and Staszewski, W. (2006) Ultrasonic wave modulations for damage detection in metallic structures. Proceedings of the Third European Workshop on Structural Health Monitoring, Granada, Spain, pp. 275–282.

Shannon, C.E. and Weaver, W. (1949) *The Mathematical Theory of Information*, University of Illinois Press, Urbana, IL.

Sikorsky, C. (1999) Development of a Health Monitoring System for Civil Structures Using Level IV Non-destructive Evaluation Method. Structural Health Monitoring 2000, pp. 68–81.

Silverman, B.W. (1986) Density Estimation for Statistics and Data Analysis. Report 24. Chapman and Hall Monographs on Statistics and Applied Probability, p. 26.

Sohn, H., Farrar, C., Hemez, F. *et al.* (2004) A Review of Structural Health Monitoring Literature from 1996–2001. Los Alamos National Laboratory Report LA-13976-MS.

Sohn, H., Wait, J., Park, G. and Farrar, C. (2004) Multi-scale structural health monitoring for composite structures. Proceedings of the 2nd European Workshop on Structural Health Monitoring, Munich, Germany.

Sohn, H., Park, H., Law, K. and Farrar, C. (2005) Instantaneous online monitoring of unmanned aerial vehicles without baseline signals. Proceedings of the 23rd International Modal Analysis Conference, Orlando, FL.

Stubbs, N. and Kim, J.-T. (1996) Damage localization without baseline modal parameters. *AIAA Journal*, **34**, 1–6.

Stubbs, N., Kim, J.-T. and Farrar, C. (1995) Field verification of a nondestructive damage localisation and severity estimation algorithm. Proceedings of the 13th International Modal Analysis Conference, pp. 210–218.

Valle, C., Niethammer, M., Qu, J. and Jacobs, L. (2001) Crack characterisation using guided circumferential waves. *Journal of the Acoustical Society of America*, **110**, 1282–1290.

Worden, K. (1997) Structural fault detection using a novelty measure. *Journal of Sound and Vibration*, **201**, 85–101.

Worden, K. and Manson, G. (1999) Visualisation and dimension reduction of high-dimensional data for damage detection. Proceedings of the 17th International Modal Analysis Conference, Orlando, FL.

Worden, K., Manson, G. and Allman, D. (2003) Experimental validation of a structural health monitoring methodology, Part I: novelty detection on a laboratory structure. *Journal of Sound and Vibration*, **259**(2), 323–343.

Worden, K., Manson, G. and Fieller, N. (1999) Damage detection using outlier analysis. *Journal of Sound and Vibration*, **229**, 647–667.

Worden, K., Farrar, C., Manson, G. and Park, G. (2007) The fundamental axioms of structural health monitoring. *Proceedings of the Royal Society A: Mathematical, Physical and Engineering Sciences*, **463**(2082), 1639–1664.

14

Damage Prognosis

14.1 Introduction

As structural health monitoring technology evolves and matures, it will be integrated into a more comprehensive process referred to as *damage prognosis*. Damage prognosis (DP) is defined as the estimate of an engineered system's remaining useful life (Farrar *et al.*, 2003). This estimate is based on a quantified definition of system *failure*, assessments of the structure's current damage state and the output of models that predict the propagation of damage in the system based on some estimate of the system's future loading. Such remaining life predictions will necessarily rely on information from usage monitoring; SHM; past, current and anticipated future environmental and operational conditions; the original design assumptions regarding loading and operational environments, previous component and system level testing; and any available maintenance records. Also, 'softer' information such as user 'feel' for how the system is responding may also be used in the DP process. Stated another way, DP attempts to forecast system performance by assessing the current state of the system (i.e. SHM), estimating the future loading environments for that system and predicting through simulation and past experience the remaining useful life of the system.

To begin the discussion of DP, it is important to distinguish between usage monitoring and SHM. Throughout this book SHM has been defined as the process of identifying the presence of and quantifying the extent of damage in a system based on a statistical pattern recognition approach implemented through machine learning algorithms. *Usage monitoring* is defined as the process of acquiring operational data from a system such as a mechanical response (e.g. strain cycles or accelerations that exceed some threshold) and that preferably also include measures of environmental conditions (e.g. temperature) and operational variables such as payload or speed. Discussions of usage monitoring for applications to aircraft and rotorcraft can be found in Goranson (1997) and Samual and Pines (2005), respectively. In practice both techniques are required for DP as future life predictions are a function of measured and anticipated loading environments and operational conditions as well as the structure's current condition.

Damage prognosis predictions will necessarily be probabilistic in nature unless the future loading and operational parameters can be defined with near certainty, as might be the case with some rotating machinery applications. This chapter will discuss the requisite components of a general DP process followed by some examples and will show how SHM fits into the DP process. It must be emphasised that in most cases DP is still an emerging technology with very few deployed applications on *in situ* systems. In conclusion, this chapter will introduce the concept of cradle-to-grave system state awareness as an engineering 'Grand Challenge' for the 21st century and show how SHM and DP are necessary

Structural Health Monitoring: A Machine Learning Perspective, First Edition. Charles R. Farrar and Keith Worden.
© 2013 John Wiley & Sons, Ltd. Published 2013 by John Wiley & Sons, Ltd.

to realise this capability in an effort to improve the performance of aerospace, civil and mechanical infrastructure while also making these systems more reliable, cost effective, environmentally benign and energy efficient.

14.2 Motivation for Damage Prognosis

As with SHM, the interest in DP solutions is based on this technology's tremendous potential for life safety and/or economic benefits. Damage prognosis has applications to almost all engineered structures and mechanical systems including those associated with all types of defence hardware, civil infrastructure, manufacturing equipment and commercial aerospace systems. As an example, airframe and jet engine manufacturers are moving to a business model where they lease their hardware to the user through so-called *power by the hour* arrangements. Increased profits are then realised by having the ability to assess damage and predict when the damage will reach some critical level that will require corrective action. With such predictions the owners can plan their maintenance schedule better and optimise the amount of time the hardware is available for leasing, which in turn optimises the revenue-generating potential of these assets. In addition, manufacturers of other large capital expenditure equipment such as earth moving equipment for mining operations would like to move to a business model whereby they lease the equipment based on the portion of its life that is used rather than on a time-based leasing arrangement. Such a business model requires the ability to monitor the system's response and predict the damage accumulation during a lease interval as a function of the failure damage level.

A distinction that should be made for aircraft is the differing requirements for structure, power plant and systems. For each category, the support schedule specifies maintenance specific items (MSIs) and safety specific items (SSIs). These are specific components that require regular scheduled inspections to avoid either operational disruption or loss of safety. The selection of monitored components is further complicated by adjacent component failures, an example of which is turbine fan-blade-off (FBO). An occurrence of FBO in flight causes an out of balance in the engine, but, despite powering down the engine, the aerodynamic loads causes the fan assembly to 'windmill'. The outcome of FBO is therefore forced vibration through the whole airframe, so even though the prognostic sensor system may describe the behaviour of the engine adequately, the airframe will have absorbed additional fatigue loading without localised monitoring. Although the focus of prognostic methods has been on power plant and structural components as discrete systems, the FBO example highlights the increasing importance of taking a 'multicomponent' level approach to DP. However, the sophistication and scope of such an approach currently remains a challenging issue. A further consideration is the treatment of 'systems' – rather than structural and power plant components. Aircraft 'systems' may take the form of data networks, printed circuit boards or adaptive software. These examples are typical of the challenges presented for 'systems' prognostics. While the strategy for structural DP has a defined roadmap, many of the issues facing aircraft systems are still to be resolved. The importance of this area of DP should not be underestimated because, over the next 10 years, for civil aircraft the cost of the system elements will rise to 50% of the purchase cost while for military platforms the figure will be closer to two-thirds of the total cost.

For civil infrastructure there is a need for prognosis of structures (e.g. bridges, high-rise buildings) subjected to large-scale discrete events such as earthquakes. As an example, some buildings subjected to the 1995 Kobe earthquake were evaluated for two years before reoccupation. Clearly, there is a need to perform more timely and quantified structural condition assessments and then confidently predict how these structures will respond to future loading, such as the inevitable aftershocks that occur following a major seismic event. For manufacturing facilities, the current slow post-earthquake assessment and reoccupation process can have an extreme economic impact far beyond the reconstruction costs. This economic impact adversely affects the facility owners as well as companies that insure such facilities.

14.3   The Current State of Damage Prognosis

The previous chapters in this book have provided a systematic approach to identifying damage in engineering systems. The most advanced damage detection systems, which have made the transition from research to practice, include the HUMS (see Chapter 2) used for helicopter gearbox monitoring, where the Federal Aviation Administration has already endorsed their effectiveness, and those used to monitor damage accumulation in rotating machinery, such as the Integrated Condition Assessment System deployed on US Navy ships (DiUlio *et al.*, 2003). However, in almost all cases a true prognosis capability still remains elusive. To date, one of the few attempts at integrating DP around a predictive capability is also encountered in the field of rotating machinery. Successful applications of rotating machinery DP exist when extensive data sets are available, some of which include the monitoring of some machines to failure. Also, for this application the damage location and damage types as well as operational and environmental conditions are often well known a priori and often do not vary significantly.

Perhaps the most refined form of combined health and usage monitoring is that found in the helicopter industry. Although it is somewhat short of physical damage-based prognosis, the use of vibration data trending for predictive maintenance can lead to an increased rotor component life of 15% (Silverman, 2005). Of even more significance, the introduction of the HUMS for main rotor and gearbox components on large rotorcraft has been shown to reduce 'the fatal hull loss within the UK to half what could have otherwise been expected had HUMS not been installed' (McColl, 2005). The essential features of this success are that the rotor speed – although not the torque – is maintained typically within 2% of nominal for all flight regimes and that there is a single load path with no redundancy. These constraints provide a basis for a stable vibration spectrum from which a change in measured parameter is attributable to component deterioration. This luxury, consisting of an easily identifiable parametric change coupled with a stable excitation source, makes helicopters particularly amenable to prognosis. Unfortunately, the DP task is complicated when the loading spectrum is constantly varying, as would typically be encountered in automotive, wind energy and military engine applications.

Seismic probabilistic risk assessment (PRA) and seismic margins assessments, as they have been applied to commercial nuclear power plant structures and systems, can to some degree be viewed as forms of DP that have been practised for more than twenty years. The objective of a seismic PRA is to obtain an estimate of the annual probability that failure will occur in a system as a result of some future earthquake. As an example, seismic PRAs may be used to estimate the probability of core melt in a power plant. Alternatively, the PRAs can be used to estimate the consequences of core melt such as the exposure of the neighbouring population to radioactive materials. One input to a seismic PRA is a probabilistic description of the expected failure rates of system-critical components as a function of earthquake ground motion levels. These probabilistic failure rates are based on either analysis, testing or past experience. Additional inputs to the PRA include a probabilistic description of the future earthquake ground motion levels and a fault tree/event tree system model that predicts failure (e.g. core melt) as a result of individual component failures. Historically, the frequency of failure estimates has been used to assist in risk-based maintenance and upgrade of safety class components. For this application, the relative contributions of the various components to system failure are assessed; this information is then used to prioritise equipment for safety upgrades. There has been recent pressure by licensees to move to a risk-based licensing approach and, as a result, the US Nuclear Regulatory Commission has conducted research efforts to establish a risk-based licensing methodology (US Nuclear Regulatory Commission, 1995). It should be noted that in its current state seismic PRA is carried out strictly as an analytical study without experimental verification on a system level (Ellingwood, 1994) and this process does not make use of information from any kind of SHM system. In other words, this process does not account for the inevitable deterioration of the system that will result from normal operations over extended periods of time.

When one looks beyond the examples cited above, few journal papers can be found that discuss DP for other applications. A recently published book (Inman *et al.*, 2005), which is based on a series of papers presented at a 2003 Pan-American Advanced Study Institute focusing on the topic of damage prognosis, is one of the first publications dedicated to this topic. This dearth of publications on DP indicates that this technology is still in the early developmental phase and that there is a need for considerable DP technology development.

14.4   Defining the Damage Prognosis Problem

The definition of the DP problem starts by first asking five general questions:

1. What are the damage cases of concern and how is failure defined for these damage cases?
2. What future loading conditions will the structure experience?
3. What techniques should be used to assess and quantify the damage (e.g. how is SHM implemented)?
4. What type of models will be used to predict the damage propagation in the system?
5. What is the goal of the prognosis?

When developing answers to these questions one will have to consider the length and time scales associated with the damage propagation. As discussed below, the answers to these questions do not have distinct boundaries and for many applications these questions will have to be addressed in an integrated manner.

As with SHM, the success of any DP capability will depend on the ability to define a priori likely damage cases for the structure and the failure criteria associated with these cases. To be effective, quantified definitions must be developed. This portion of the DP process overlaps considerably with the SHM operational evaluation step discussed in Chapter 3.

For each potential failure mode, the loading conditions that cause defects to grow and coalesce into some observable damage and its subsequent evolution to failure fall into three general categories, as previously discussed in Chapter 1. The first category is continuous lower-level loading, where operational and environmental loads are applied over long periods of time, but typically at levels considered to be well below those thought to cause rapid damage propagation. Traffic loading on a bridge structure is one example of continuous lower-level loading. The second category is predictable discrete events. Aircraft landings can be viewed as a predictable discrete loading event that can eventually lead to damage accumulation in the landing gear or airframe. Note that although the number of such landings can be predicted with some accuracy, the severity of each event will be unknown, but most likely will fall within some range that can be defined based on previous experience. Unpredictable discrete events are the third loading category in which unknown and possibly severe loading is applied to the system at unpredictable times. Many natural phenomena hazards such as earthquakes and hurricanes as well as man-made hazards associated with terrorist bomb blasts can produce such unpredictable discrete events. Clearly, many structures will experience all three types of loadings during their lifetime. Regardless of the loading category, a model will be needed that estimates the loading the structure will experience in some future time interval. In most cases, this model will be probabilistic in nature. However, there are cases such as the loading associated with rotating machinery that are operated in a very consistent manner in well-controlled environments where this model may be deterministic.

After identifying the types of damage that are of concern and the loading that will potentially cause the initial defects to evolve into damage, or the current damage to evolve into failure, it is then important to determine which techniques should be used in the damage assessment. This topic has been the focus of the previous chapters in this book and, therefore, only some very general comments will be provided here. The first question that arises concerns whether the assessment should be done on-line, in near real time or off-line at discrete intervals, as this consideration will strongly influence the data acquisition and data processing requirements, as well as set limits on the computational requirements of potential assessment and prognosis techniques. For unpredictable discrete events, the assessment must be done on-line to be of any use, thus limiting the choice of the assessment techniques. However, for gradual lower-level loading, there are cases where the assessment need not be performed in near real time, and hence there is much more flexibility to develop an appropriate assessment technique. It should be noted that DP requires one to not only identify and locate the damage but also to estimate the damage extent. As pointed out previously, such estimates almost always have to be done in a supervised learning mode.

As discussed in Chapters 7 and 8, many of the features used for damage assessment are derived from either physics-based or data-based models (or possibly some combination) that have been trained in some manner with measured system response data. For the damage propagation modelling portion of the DP process, it is anticipated that more emphasis will be placed on physics-based assessments because such approaches are especially useful for predicting a system response to new and possibly unanticipated loading conditions and/or new system configurations (damage states). Also, in many cases it will be difficult to obtain the necessary damaged system data to develop the data-based models for damage propagation predictions. However, one must keep in mind that physics-based assessments are typically more computationally intensive than data-based techniques and necessarily will require extensive validation with experimental data.

Data-based assessment techniques, on the other hand, rely on previous measurements from the system to assess the current damage state, typically by means of the pattern recognition methods discussed throughout this book. These approaches predict future damage propagation through some type of regression analysis assuming data are available from damage states up to failure. However, although data-based assessment techniques may be able to indicate a change in the structure's damage state in the presence of new loading conditions, they will perform poorly when trying to extrapolate beyond the data used in the training process. Typically, the balance between physics-based models and data-based techniques will depend on the amount of relevant training and/or validation data available and the level of confidence in the predictive accuracy of the physics-based models.

A final question to be answered as part of the DP process is to determine the goal for the prognosis. Perhaps the most obvious and desirable type of prognosis estimates how much time remains until maintenance is required or until the system fails. Because predictive models typically have more uncertainty associated with them when the structure responds in a nonlinear manner, as will often be the case when damage accumulates, *an alternate goal might be to estimate how long the system can continue to safely perform in its anticipated environments before one no longer has confidence in the predictive capabilities of the models that are being used to perform the prognosis.* It is critical to recognise that the predictive models themselves may 'fail' at some point; as the eminent statistician George Box famously observed 'Essentially, all models are wrong, but some are useful.'

14.5   The Damage Prognosis Process

The general components of a DP process are depicted in Figure 14.1 where the process has been divided into the portions that are physics based and the portions that are data based. The DP process begins by collecting as much initial system information as possible including testing and analyses that were performed during the system design as well as maintenance and repair information that might be available. Inherent in this information gathering process are the definitions of damage and failure for the system of interest. This information is used to develop initial physics-based numerical models of the system as well as to develop the sensing system that will be used for damage assessment and to define whatever additional sensors are needed to monitor operational and environmental conditions. The physics-based models will also be used to define the necessary sensing system properties (e.g. parameter, location, bandwidth, sensitivity; see Chapter 4). For instance, an understanding of the physics of gear wear may lead to a damage-sensitive feature based on oil conductivity in helicopter gearboxes that results from

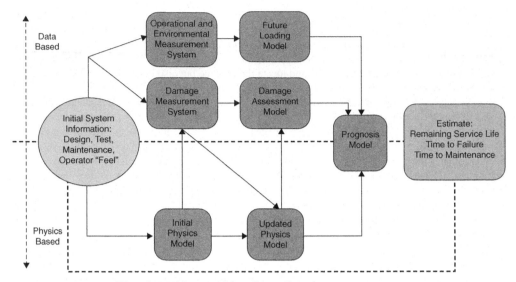

Take action, update system information, continue process

Figure 14.1   A general outline of the damage prognosis process.

metallic contamination after wear and erosion, which in turn helps to define the specific type of sensor
needed to identify this feature.

Because there will be a finite budget for sensing, the physics-based models will often be part of an
optimisation study that will attempt to maximise the observability of damage given constraints on the
sensing system properties. As data become available from the sensing systems, they are used to validate
and update the physics-based models. These data along with output from the physics-based models,
will also be used to perform SHM where ideally the existence, location, type and extent of damage are
quantified. Data from the operational and environmental sensors are used to improve the SHM process
through data normalisation (see Chapter 12) and to develop data-based models, often probabilistic in
nature, that predict the future system loading. The output of the future loading model, the SHM process
and the updated physics-based model will all be input into a predictive tool that estimates damage
propagation and the remaining system life as a function of continued operational time. A key component
of this process is a capability to predict damage propagation (initiation and growth) as a function of
future loading and environmental conditions. If data are available from a system that has experienced
failure, then such damage initiation and growth models can be based on a regression analysis applied to
these data. Alternatively, physics-based models can be developed for such predictions and these models
must necessarily be used in the absence of such run-to-failure data.

Note that the definition of 'remaining life' can take on a variety of meanings depending on the specific
application and this definition will be directly tied to the definition of failure. A key point illustrated in
Figure 14.1 is that various models will have to be employed in the DP process. Also, the data-based and
physics-based portions of the process are not independent. This combination of physics- and data-based
models is the key distinguishing attribute of a prognostic capability from an SHM system: the capacity to
update the remaining life prediction on the basis of new usage data *and* a change in the system's damage
state. This form of DP allows for alternate load path calculations – arising from redundant systems –
and changes in operational parameters that were not anticipated in the original design assumptions.
Finally, the solution process will be iterative, relying on new damage assessments and validation of past
prediction accuracy to improve future predictions.

14.6   Emerging Technologies Impacting the Damage Prognosis Process

In addition to the statistical pattern recognition approaches to SHM discussed in this book, there are many emerging technologies that will have an impact on the development of a damage prognosis capability for various types of engineering systems. Those technologies listed below are not intended to be an exhaustive list, but rather illustrative in nature.

14.6.1   Damage Sensing Systems

Clearly, some of the most rapidly evolving technologies that will impact the ability to perform damage prognosis are associated with low-cost, noninvasive, sensing, processing and telemetry hardware, as discussed in Chapter 4. There are extensive efforts underway at both academic and corporate research centres to develop large-scale, self-organising, embedded sensing networks for a wide variety of applications. These studies focus on developing cost-effective dense sensing arrays and novel approaches to powering the sensing systems that harvest the ambient energy available from the structure's operating environment. Although hardware technologies show every prospect of delivering such systems, their application must be related to the physics of the problem; that is, the damage measurement system must embed sensors that are sensitive to changes in parameters correlated with damage and, most importantly, the sensing system itself must be more reliable than the structure/component it is monitoring.

14.6.2   Prediction Modelling for Future Loading Estimates

A successful DP requires the measurements of the current system state and the prediction of the system deterioration when subjected to future loading. Based on the analysis of previous loading histories, future loading is forecast using various data-driven, time series prediction modelling techniques. For example, metamodelling such as state-space representation (Ljung, 1999) and multivariate time series models (Box, Jenkins and Reinsel, 1994) can be employed to track previous loading and to predict future loading. Then, reliability-based decision analysis provides an appropriate tool to synthesise the damage propagation models with the future loading estimates (Ang and Tang, 1984).

14.6.3   Model Verification and Validation

Because damage prognosis solutions rely on the deployment of a predictive capability, the credibility of numerical simulations must be established. The process of establishing this confidence in the predictive capabilities of the numerical simulations is accomplished through various activities, collectively referred to as verification and validation (V&V). A significant challenge here is to validate nonlinear models and to validate complex system-level models where there will be limited data available from any failure regime. However, the current state-of-the-art (Hemez, Doehling and Anderson, 2004) in this area is still not at the stage where complex dynamic models are routinely validated, particularly for intricate materials such as composites.

14.6.4   Reliability Analysis for Damage Prognosis Decision Making

In reliability analysis, the failure state of a system is represented by a function of the response known as the limit state. The probability of failure is then defined as the integral of the joint probability density function (JPDF) over the unsafe region bounded by the limit state (Haldar and Mahadevan, 2000; Robertson and Hemez, 2005). As an example, the objective of a probabilistic reliability analysis may be to answer the question of how many more fatigue cycles the structure can experience before the damage reaches a critical size. If failure is defined, for example, in terms of a wing flutter condition, then reliability analysis consists of estimating the probability of reaching this limit state (e.g. reduction of the wing's first torsion mode frequency to some critical level) given uncertainties about the model that predicts this frequency reduction as a function of future loading and the anticipated future damage state of the wing. Decision making relies on the estimation of this reliability to decide which course of action should be taken (e.g. perform maintenance and repair, abort mission or continue to operate at some reduced level for a specified time interval).

This analysis begins with identifying the failure modes (such as delamination in a composite material) and the random variables that contribute to these failure modes (such as projectile impact velocity, ply orientation angles, homogenised elasticity parameters, material density). To calculate the probability of failure, the JPDF must be integrated across all random variables that influence the failure region. Because closed-form representations of the failure region are generally not available, integration must be approximated by applying Monte Carlo sampling or approximate expansion methods applied to the previously identified metamodels. Reliability analysis will necessarily be applied to estimate the remaining useful life of the systems under uncertainty.

14.7   A Prognosis Case Study: Crack Propagation in a Titanium Plate

The situation considered will be one where the structure is a finite rectangular metallic plate under uniaxial loading, as shown in Figure 14.2. The damage will take the form of a mode 1 opening crack situated centrally, perpendicular to the axis of loading. In the simple situation of a crack in a metallic plate, a useful approximate damage propagation model is given by the Paris–Erdogan (PE) law. However, it is known that for metallic specimens, the parameters of the PE equation are subject to random variation between samples and even random variation at different points within a sample (Shen, Soboyejo and Soboyejo, 2001; Langoy and Stock, 2001). This variability means that the computed lifetime of a given

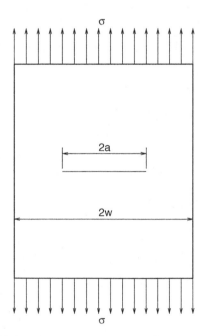

Figure 14.2   Finite plate with a central mode 1 crack under uniaxial loading.

specimen will also be a random variable. Given experimental bounds on the variability of the PE equation parameters, this example will incorporate these sources of variability into predicted lifetime estimates for these samples.

In order to obtain the full statistics of the lifetime estimate, it is necessary to carry out a Monte Carlo analysis; however, this is often computationally intensive. It is made feasible here by the fact that the PE law involves only two parameters. For a general material or structure, it may well be that there are many parameters needed in order to characterise the damage propagation and these parameters may be difficult to characterise experimentally, such as those used in a cohesive zone model for composite materials (Tippetts, Beyerlein and Williams, 2002). It will also generally be the case that the damage propagation cannot be summarised in a simple analytical formula; it may be necessary to execute a complex numerical model in order to investigate the damage growth. If the model is computationally complex and it takes a substantial amount of time to complete a single predictive run of the model, it will not generally be feasible to use a Monte Carlo approach.

14.7.1 The Computational Model

As already stated, the situation in this case study is highly idealised. First of all, the structure under consideration is a thin rectangular plate under a uniaxial tension where the stress resulting from the applied load will be denoted σ. The damage under consideration is a central fatigue crack in mode 1 opening, that is with its length running perpendicular to the loading axis. The width of the crack is denoted by $2a$ and the width of the plate by $2w$.

An extremely important quantity governing the evolution of the crack is the stress intensity factor (SIF), K, which governs the magnification factor for the imposed stress σ in the neighbourhood of the crack tip, which is defined by

$$K = Y(\overline{a})\sigma\sqrt{\pi a} \tag{14.1}$$

$Y(\overline{a})$ is a geometrical factor that is needed in order to accommodate the fact that the plate has finite width. This factor is a function of the dimensionless parameter $\overline{a} = a/(2w)$. In fact, for an infinite plate, an analytical solution yields

$$Y(\overline{a}) = 1 \tag{14.2}$$

One of the reasons for choosing the plate geometry assumed here is that there are a number of well-validated approximations for the geometry factor in the SIF for a plate of finite width. Broek (1982) cites examples:

$$Y(\overline{a}) = 1 + 0.256\overline{a} + 1.152\overline{a}^2 + 12.20\overline{a}^3 \tag{14.3}$$

$$Y(\overline{a}) = \sqrt{\sec(\pi\overline{a})} \tag{14.4}$$

$$Y(\overline{a}) = \frac{1}{\sqrt{1 - (2\overline{a})^2}} \tag{14.5}$$

The effect of using these different approximations will be investigated later.

As previously mentioned, the damage propagation model appropriate for a fatigue crack in a metallic plate is the Paris–Erdogan (PE) law (Broek, 1982). This law relates the crack growth rate da/dn to the crack tip SIF range ΔK as follows:

$$\frac{da}{dn} = C(\Delta K)^m \tag{14.6}$$

where C and m are constants of the material called, respectively, the Paris coefficient and the Paris exponent. (The units of C are $MPa^{-m}m^{m/2+1}$ and m is dimensionless.) As observed already, the parameters depend on the microstructure of the material and therefore can vary significantly from sample to sample. In fact, they can vary with position within a single sample. The variation of these coefficients and its effect on the lifetime of the specimen is the primary challenge for damage prognosis in this simple example. It will be assumed that the material of the plate is the titanium alloy Ti–6Al–4V, because this material can be considered as an approximately isotropic metal and because it has been the subject of research to find values of the Paris coefficient and exponent (Shen, Soboyejo and Soboyejo, 2001; Langoy and Stock, 2001). In reality, the α-phase grain size does vary with direction and this property is neglected here.

The SIF range ΔK in the situation considered here is generated simply by the fact that a time-varying stress $\sigma(t)$ will have a corresponding stress range $\Delta\sigma$, that is,

$$\Delta K = Y(\overline{a})\Delta\sigma\sqrt{\pi a} \tag{14.7}$$

In order to completely specify the computation, it is necessary to specify the bounds of the stress range; this is usually done by defining the load ratio R by

$$R = \frac{\sigma_{\min}}{\sigma_{\max}} = \frac{K_{\min}}{K_{\max}} \tag{14.8}$$

Under conditions of cyclic loading the crack length will grow monotonically. One could specify two termination criteria for the computation, that is the point where the plate fails. The first condition is simply that $a = w$; that is, the crack has reached the edge of the plate. This definition is a little unsatisfactory as the approximations in Equations (14.2) to (14.5) begin to lose their applicability as $a \approx w$. A better termination criterion is to take the point when the crack-tip SIF exceeds the fracture toughness, K_{I_c}, of the material. K_{I_c} can be regarded as a critical value of the stress intensity beyond which the material cannot withstand fracture and is typically determined experimentally (ASTM, 2011). The point at which the crack-tip SIF exceeds the fracture toughness will occur at a critical crack length a_c. This value has to be obtained using numerical methods for Equations (14.3) to (14.5). In the case of the infinite-plate approximation Equation (14.2), one can readily show that

$$a_c = \frac{(1 - R^2)K_{I_c}^2}{(\Delta\sigma)^2\pi} \tag{14.9}$$

The calculations that follow are all terminated when $a = a_c$ and failure is then assumed to occur immediately.

The numerical integration scheme used to step forward Equation (14.6) in time is a simple 'rectangle-rule' (more formally a *forward-Euler explicit*) recursion, so that

$$n_i = n_{i-1} + dn \tag{14.10}$$

$$a_i = a_{i-1} + dn\left(\frac{da}{dn}\right)_{i-1} = a_{i-1} + dnC(Y(\overline{a}_{i-1})\Delta\sigma\sqrt{\pi a_{i-1}})^m \tag{14.11}$$

The initial conditions are $n_0 = 0$ and $a_0 = 10$ mm (corresponding to a crack length of 20 mm). In order to ensure accuracy, an appropriately small step size dn is chosen for each integration on the basis of the size of $(da/dn)_0$. During the integration, steps are taken in groups of 10^5 and after each group the step size is multiplied by a factor of 10. Note that da/dn is usually given in mm/cycles, while the SIF is usually given in $MPa/\sqrt{m}$ (note that 'm' here denotes metres and *not* the Paris exponent). These definitions mean that a little care is needed in handling the various units.

The dimensions of the simulated plate were chosen to give correspondence with the experiments conducted by Shen, Soboyejo and Soboyejo (2001) in order to measure the Paris parameters. An appropriate choice of dimensions and load range to give a ΔK of 10 MPa/$\sqrt{m}$ (the order of the quantity in Shen, Soboyejo and Soboyejo, 2001) is width w of 200 mm, thickness t of 2 mm and $\Delta\sigma$ of 40 MPa. The latter corresponds to a maximum applied load of 16 kN.

Note that as the accuracy of the Monte Carlo approach is dominated by a term of the form $1/\sqrt{S}$, where S is the number of samples (Shreider, 1964), for a univariate simulation 100 runs would usually be sufficient to give accuracy to one decimal place and 10 000 runs would be needed for accuracy to two decimal places. The accuracy here is intermediate between these values, the number of runs being chosen for speed of computation as much as accuracy. One thousand runs were carried out for each of the possible SIF forms described by Equations (14.2) to (14.5). When integrating systems with the SIFs specified by Equations (14.4) and (14.5), it was necessary to use protected versions of the functions to avoid the possibility of complex roots. For each run, the lifetime of the sample was defined as the number of cycles needed for a to reach a_c. The exact lifetime was extracted by interpolation over the last time step.

14.7.2   Monte Carlo Simulation

The Monte Carlo simulation here is based on the Paris parameters tabulated in (Langoy and Stock, 2001). The first point to note is that this example considers a very specific version of the alloy Ti–6Al–4V in that the authors use Ti–6Al–4V–0.1Ru (extra-low interstitials (ELI)). The authors extract one set of Paris parameters from each sample, of which 16 are considered. The tests are carried out in air and in both aerated and deaerated sea water; however, the data from all 16 tests are included here. The Paris parameters in Langoy and Stock (2001) are tabulated and there is no suggestion for an underlying probability distribution; the parameter uncertainty will therefore be considered here as specified by an interval. The interval corresponding to $\log C$ is (–15.0, –11.6) and the interval for m is (3.7, 6.2). In order to carry out the Monte Carlo runs, a uniform distribution was assumed for each parameter covering the appropriate interval. For the sake of simplicity, the parameters were assumed to be statistically independent. The geometry for the plate and the loading regime is exactly as described in the previous section. The value of the fracture toughness is taken as 75 MPa$\sqrt{m}$, as this is the (minimum) value quoted in Langoy and Stock (2001). For each expression of the SIF Equations (14.2) to (14.5), 1000 samples of the crack growth curve were obtained, leading to 1000 estimates of the lifetime. Figure 14.3 shows the 1000 crack growth curves corresponding to the infinite plate SIF in Equation (14.2).

The lifetime PDFs for the various SIF expressions as estimated by kernel density estimation (see Chapter 6) are given in Figure 14.4. The PDFs in Figure 14.4 are all very consistent with each other and are all clearly non-Gaussian. The other point to note is that the distribution corresponding to Equation (14.5) appears to be bimodal. This result is probably worth investigating further. The statistics of the Monte Carlo simulation are given in Table 14.1.

The scatter in the lifetime estimates from this set of Paris parameters yields usable estimates for the specimen lifetime. For example, if the minimum lifetime corresponding to Equation (14.2) is used, one can calculate that 11 817 000 fatigue loading cycles are necessary for the crack to propagate from its initial 20-mm length to the point where the specimen fails. At a nominal fatigue test frequency of 20 Hz, for example, this corresponds to approximately 164 hours. In practice, this time to failure would certainly allow for enough time to take corrective action. The scatter on the lifetimes is still about 5 orders of magnitude.

14.7.3   Issues

To recap the previous example, consider the five general questions that were previously identified as forming the basis of a DP problem definition. The first question deals with identifying the damage that

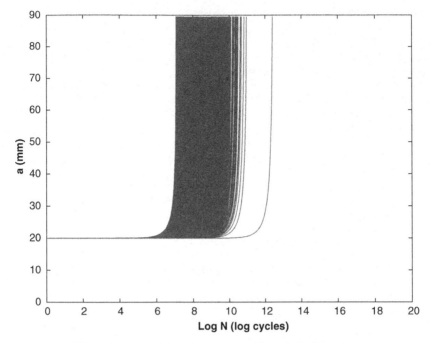

Figure 14.3   Crack growth curves for a Monte Carlo simulation.

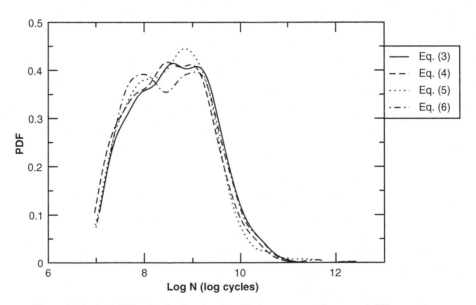

Figure 14.4   Probability density function for a lifetime corresponding to each SIF expression.

Table 14.1 Statistics of Monte Carlo simulations

	Statistics for log N			
SIF expression	Mean	Standard deviation	Minimum	Maximum
Equation (14.3)	8.6618	0.8191	7.0725	12.419
Equation (14.4)	8.5062	0.8158	6.9695	12.308
Equation (14.5)	8.5378	0.8017	6.9870	11.717
Equation (14.6)	8.5691	0.8325	6.9977	11.738

is of concern and the failure criteria for this damage mechanism. Damage is clearly defined as a mode I crack and the failure criteria was defined to be when the crack reaches a size such that the stress intensity factor equals the fracture toughness. Next, there needs to be a model that defines the future loading. In this example that model is a deterministic one where the traction is cycled over 40 MPa. Although the third question deals with the SHM system definition, it was assumed in this example that the type of damage and its initial extent were known exactly. Also, no subsequent estimates of the damage extent were made. Clearly, this question has been neglected as there was no effort made to define an SHM system. To answer the fourth question, a damage propagation model was defined to be the Paris–Erdogan law for crack growth. Finally, because of the idealised nature of this problem, the goal for the prognosis process was not defined. However, it was shown that a time to failure prediction could be made based on the results obtained and such an estimate could very well be a goal for this type of analysis.

It is now worth further exploring exactly how idealised the situation in this example is to keep the reader from thinking that damage prognosis is straightforward and a well-developed technology. There are a number of very important restrictions here.

First, the material under investigation is an isotropic metal. In general, it will be necessary to make lifetime estimates for more complicated materials such as anisotropic composites. It may even be necessary to take into account the anisotropy of some metals. For the material considered here, estimates of the Paris law parameters were available. Even if experimentally obtained Paris law coefficients are available, one must be very careful in considering the nature of the alloy in question. In this paper, the variant of Ti–6Al–4V had markedly different parameters from those in Shen, Soboyejo and Soboyejo (2001). For example, the alloy considered in the latter reference had a mean value for m of 7.3 and this value was not even included in the range of values of m (3.7–6.2) for the alloy considered here.

The structure is simple: a rectangular plate with a central crack and well-defined boundary conditions. This geometry means that experimentally validated expressions for the SIFs needed for the PE law were available. This case will not be true in general. A typical aircraft component may have many geometrical features like ribs and stringers, which complicate matters. Such structural systems will have joints and interfaces and may include a variety of materials. Also, the initial conditions of the structure in this example were known exactly, including the damage state as defined by the parameter a and the assumption that there are no initial stresses in the system prior to applying the tractions at either end of the plate. With current SHM technology one will not, in general, be able to define the damage in such a precise manner and almost all real-world structures will have initial stress conditions that will be difficult to evaluate. Such complications clearly are not considered in this example.

The damage type (a fatigue crack) was the simplest imaginable and a validated damage propagation law – the Paris–Erdogan law – was available. In general, the damage law may well be complex, including phenomena such as strain rate effects, thermal effects and local changes in material properties caused by the diffusion of mobile constituents (e.g. hydrogen) to the crack tip. As such, it may be the case that the prediction of the damage propagation requires a large multiscale, multiphysics calculation. Such laws may be very time-consuming to evaluate, even for the simple geometry used in this example. Also, in composites, for example, which have many different damage scenarios, a damage propagation law will be needed for each type and the interaction of the various damage mechanisms must be considered.

The loading regime is very simple here – the load is uniaxial and harmonic. In general, damage will evolve under multiaxial loading. This situation means that, even in the context of a fatigue crack, it will experience multimode opening. Another vital consideration here is that the future loading was known exactly in this case as a harmonic function – a more realistic situation will demand that the algorithm can deal with uncertain broadband loading. Given these caveats, it is worth stressing that prognosis has *at least* proved possible in the idealised case considered here.

14.8 Damage Prognosis of UAV Structural Components

Next, a reliability framework for DP will be summarised in an effort to show how one might address many of the complications not considered in the previous example. This summary is based on the work presented in Gobbato (2011) and Gobatto *et al.* (2012) and is applied to prognosis of damage to adhesive joints in lightweight unmanned aerial vehicles with particular emphasis on wing skin-to-spar joints. Such joints have been shown to be particularly sensitive to fatigue damage accumulation. This probabilistic framework for DP was adopted from a performance-based seismic analysis framework developed by the earthquake engineering community (Moehle and Deierlein, 2004).

There are multiple damage mechanisms in the UAV composite materials that are of concern. They include debonding, interply delamination, fibre breakage and matrix cracking and in most cases multiple mechanisms will be interacting to produce the observed damage. Such damage can compromise the local structural element strength and it can influence more global aeroelastic performance issues, such as the flutter characteristics of the aircraft. In this example both local and global failure modes are considered. Local failure occurs when the damage propagates to a critical damage size. Global failure occurs when the flutter or limit cycle oscillation velocities are reached.

The framework starts with the assumption that a ground-based and/or in-flight SHM system is available and can provide an estimate of the current damage state. The analysis begins when the outcome of the current SHM assessment (e.g. a damage size and extent vector) is available. Uncertainty in the damage assessment is accounted for through a Bayesian inferencing scheme that is used to update the posterior joint PDF of the damage extent, which is conditioned on material and damage model parameters and previous SHM assessments.

Next, a probabilistic loading analysis is performed to define the joint PDF of the aerodynamic loading in terms of two parameters: the turbulence and flight intensity measures. These parameters are conditioned on the flight profile. A discrete process is employed where the future flight times are divided into q equally spaced time intervals. Within each interval an unknown number of flight segments can occur and information regarding each flight segment is collected, such as altitude, airstream velocity and time of flight, in an effort to define the turbulence and flight intensity measures.

With a probabilistic model of the aerodynamic loading available, a probabilistic structural response analysis can be performed to produce a joint conditional PDF defining the expected size of damage at some future time. This calculation is based on extensive Monte Carlo simulations using finite element calculations with a sophisticated damage accumulation model based on cohesive zone models (Tippetts, Beyerlein and Williams, 2002). The joint PDF describing the system response is conditioned on material and damage model parameters.

In the next step, a probabilistic analysis is used to estimate the joint PDF of the predicted damage size, flutter velocity and limit cycle oscillation velocities at some future time. The flutter velocity is defined as the lowest velocity at which flutter occurs. The limit cycle oscillation velocities define the velocity when the amplitude of wing-tip bending or torsional displacements reach predefined thresholds. With this joint PDF a probability of failure from either local structure modes or global aeroelastic modes is estimated by performing component and system reliability analyses (Ditlevsen and Madsen, 1996).

This approach provides one of the most comprehensive reliability-based damage prognosis method-ologies for composite aircraft structures. Note that there are still limitations to this methodology. First, there is the assumption that fatigue is driving the damage evolution process; other loading conditions like in-flight impact were not considered, but the methodology is sufficiently general that it can accom-modate other loading scenarios. Next, there is the assumption of an on-board SHM system that can give reasonable assessments of the aircraft's damage condition. Finally, surrogate metamodels had to be employed for portions of the structural response analyses and the assessments were limited to only the wings of the aircraft to make the process computationally feasible.

14.9   Concluding Comments on Damage Prognosis

The challenge of damage prognosis is to develop and integrate sensing hardware, data interrogation software and predictive modelling software that will prove more robust than the component- and system-level hardware the DP system is intended to assess. This chapter aims to provide an overview of the issues that must be addressed and technical approaches being used to realise solutions to DP problems. Certainly, considerable technical and cultural challenges remain. The technological aspects are easier to define and anticipate. Already a substantial body of evidence indicates that the individual components of a DP process are realisable, but the integration of all the necessary technologies has been very limited. However, if robust DP solutions are to be adopted, then a change in certification culture must arise that embraces an iterative safety and maintenance process, which may alter significantly from the original design calculations. A 'morphing' safety case based on PRA coupled with a nonlinear iterative model validation will require significant investments in test and analysis correlation studies before proof of robustness is established in each of the constituent technologies. Therefore, it is crucial to apply this technology initially to problems with well-defined damage concerns and where the prognosis system is not relied upon to make decisions impacting life safety. As an example, deployment of a damage prognosis system on an unmanned aerial vehicle may be more appropriate than deployment of such a system on a commercial passenger aircraft. In any case, it is almost certain that new DP approaches will initially be applied in parallel with current system evaluation and maintenance procedures until the DP methodology can be shown to provide a reliable and more cost-effective approach to system operation, assessment and maintenance.

The time scales associated with the development, validation and deployment of DP solutions will vary significantly with specific applications. The previously mentioned examples of DP as it is applied to rotating machinery, helicopter power transmission systems and nuclear power plant seismic hazard studies all are the result of development over time scales on the order of tens of years and with multimillions of dollars in investments. In the case of large multicomponent systems such as nuclear power plants these prognoses have not, in general, been experimentally verified. It is speculated that attempts to apply DP to other classes of structures will require similar development periods and similar monetary investments. Significant variability in the time to realise a DP solution can result from factors such as how consistently the system is operated and the extent that environmental variability can influence operating characteristics. Finally, the time necessary to develop and validate a DP capability can be directly influenced by regulatory authorities. If the DP capabilities are being used to address a life-safety issue, then the time necessary for its development can be on the order of a decade. For such cases significant validation studies will be required before regulatory authorities will allow the DP to dictate operation and maintenance activities.

As this technology evolves, it is anticipated that the DP solutions developed through rigorous validation of each technological component will be used to confirm system-level integrity to normal and extreme loading environments, to estimate the probability of mission completion and personnel survivability, to determine optimal times for preventive maintenance and to develop appropriate design or operation modifications that prevent observed damage propagation. The multidisciplinary and challenging nature of the DP problem, its current embryonic state of development and its tremendous potential for life safety and economic benefits qualify damage prognosis as a 'Grand Challenge' problem for engineers in the 21st century.

14.10   Cradle-to-Grave System State Awareness

To conclude the discussion of SHM and DP, the concept of *cradle-to-grave system state awareness* will now be introduced. Here cradle-to-grave system state awareness is defined as the concept of having complete knowledge of all aspects of an engineered system throughout its lifecycle. This process begins by designing in system functionality and monitoring capabilities at the material and manufacturing level in an effort to minimise the use of raw materials, minimise the generation of scrap and its associated adverse environmental impact, minimise the energy needed to develop and manufacture products and to minimise the time of conception to the finished product. This process necessitates that one starts building 'system intelligence' in at this level through the use of manufacturing process monitoring and multifunctional materials in an effort to increase knowledge of the initial material, component, system and system-of-system states. As shown previously, such knowledge provides the basis for successful SHM and DP.

Once the system is deployed, cradle-to-grave system state awareness dictates that the system is continually monitored to assess its in-service condition in an effort to maximise performance, service life and reliability, minimise maintenance costs, reduce energy demand and minimise environmental impact. SHM is an integral part of this assessment process. This process provides continuous feedback to the material design and manufacturing processes in an effort to improve subsequent system designs and their associated performance.

Finally, cradle-to-grave system state awareness must provide for intelligent system retirement where end-of-life is predicted based on system state assessments and predicted performance degradation rates, anticipated future mission profiles and sound engineering economic analyses. Clearly, SHM and DP will be key components of this intelligent system retirement process. In addition, this part of the process will seek to maximise recyclable components and safely contain and monitor hazardous nonrecyclable components. Again, feedback will be provided to material design and manufacturing processes as well as to the in-service monitoring and assessment process as part of a continual quality improvement procedure.

Cradle-to-grave system state awareness is applicable to all engineered systems ranging from bioengineering of artificial hearts, the next generation of commercial nuclear reactors, the latest civilian and military ships and aircraft, and large-scale scientific infrastructures such as particle accelerators, to name just a few. This brief concluding discussion is meant to show that SHM and DP should not be thought of as independent disciplines, but rather as part of a much more comprehensive life-cycle monitoring and assessment process. In almost all cases cradle-to-grave system state awareness in still just a concept with a tremendous amount of work needed to thoroughly define the requirements of such a process for a given system. However, the engineering community needs to define such high-level concepts that initially appear to be beyond the reach of current technology in an effort to continually motivate and drive innovative research that one day may make such challenging concepts a reality.

References

Ang, A. and Tang, W. (1984) *Probability Concepts in Engineering Planning and Design; Volume II: Decision, Risk, and Reliability*, John Wiley & Sons, Inc., New York.

ASTM (2011) *E1820–11 Standard Test Method for Measurement of Fracture Toughness*, American Society for Testing and Materials, West Conshohocken, PA.

Box, G., Jenkins, G. and Reinsel, L. (1994) *Time Series Analysis – Forecasting and Control*, Prentice Hall, NJ.

Broek, D. (1982) *Elementary Engineering Fracture Mechanics*, Kluwer Academic Publishers.

Ditlevsen, O. and Madsen, H. (1996) *Structural Reliability Methods*, John Wiley & Sons, Ltd, Chichester, West Sussex.

DiUlio, M., Savage, C., Finley, B. and Schneider, E. (2003) Taking the integrated condition assessment system to the year 2010. 13th International Ship Control Systems Symposium, Orlando, FL.

Ellingwood, B. (1994) *Validation of Seismic Probabilistic Risk Assessment of Nuclear Power Plants*, US Nuclear Regulatory Commission.

Farrar, C., Sohn, H., Hemez, F.M. *et al.* (2003) Damage Prognosis: Current Status and Future Needs, Los Alamos National Laboratory Report LA-14051-MS, Los Alamos, NM.

Gobbato, M. (2011) Reliability-based framework for fatigue damage prognosis of bonded structural elements in aerospace composite structures. Doctoral Dissertation, Department of Structural Engineering, University of California, San Diego, La Jolla, CA.

Gobatto, M., Conte, J., Kosmatka, J. and Farrar, C. (2012) Reliability-based framework for fatigue damage prognosis of composite aircraft structures. *Probabilistic Engineering Mechanics*, **29**, 176–199.

Goranson, U. (1997) Jet transport structures performance monitoring, in *Structural Health Monitoring Current Status and Perspectives*, Technomic Publishing Company, Palo Alto, CA, pp. 3–17.

Haldar, A. and Mahadevan, S. (2000) *Probability, Reliability, and Statistical Methods in Engineering Design*, John Wiley & Sons, Inc., New York.

Hemez, F., Doebling, S. and Anderson, M. (2004) A brief tutorial on verification and validation. Proceeding of the 22nd International Modal Analysis Conference, Society of Experimental Mechanics, Dearborn, MI.

Inman, D.J., Farrar, C., Lopes, V.J. and Steffen, V.J. (2005) *Damage Prognosis for Aerospace, Civil and Mechanical Systems*, John Wiley & Sons, Inc., New York.

Langoy, M. and Stock, S. (2001) Fatigue-crack growth in Ti–6Al–4V–0.1Ru in air and seawater: Part I. Design of experiments, assessment, and crack growth rate curves. *Metallurgical and Materials Transactions*, **32**(9), 2297–2314.

Ljung, L. (1999) *System Identification – Theory for the Use*, Prentice Hall, Upper Saddle River, NJ.

McColl, J. (2005) HUMS in the era of CAA, JAA, EASA and ICAO. AIAC Conference, Melbourne.

Moehle, J. and Deierlein, G. (2004) A framework methodology for performance-based earthquake engineering. Proceedings of the 13th Conference on Earthquake Engineering, Vancouver, Canada.

Robertson, A. and Hemez, F. (2005) Reliability methods, in *Damage Prognosis for Aerospace, Civil and Mechanical Systems* (eds D. Inman, C. Farrar, V. Lopes and V. Steffen), John Wiley & Sons, Inc., New York, pp. 221–234.

Samual, P. and Pines, D. (2005) A review of vibration-based techniques for helicopter transmission diagnostics. *Journal of Sound and Vibration*, **282**, 475–508.

Shen, W., Soboyejo, A. and Soboyejo, W. (2001) Probabilistic modeling of fatigue crack growth in Ti–6Al–4V. *International Journal of Fatigue*, **23**, 917–925.

Shreider, Y.A. (1964) *Method of Statistical Testing – Monte Carlo Method*, Elsevier.

Silverman, H. (2005) T HUMS – AH64 lead the fleet (LTF) summary and glimpse at Hermes 450 MT-HUMS. AIAC Conference, Paper 112, Melbourne.

Tippetts, T., Beyerlein, I. and Williams, T. (2002) A multiscale/cohesive zone model for composite laminate impact damage. American Society of Composites 17th Technical Conference, West Lafeyette, IA.

US Nuclear Regulatory Commission (1995) Use of probabilistic risk assessment methods in nuclear regulatory activities; final policy statement. *Federal Register*, **60**(158), 42622–42629.

Appendix A

Signal Processing for SHM

The main aim of this appendix is to make this book as self-contained as possible. For this reason, the background is given here to the great majority of signal processing techniques and algorithms mentioned throughout the book. The appendix also serves to establish the notation used elsewhere in the book. Although a great deal of effort has been made to be comprehensive, it is inevitable that further understanding will be required on some matters. For this purpose, the reader of the current book may well wish to consult more comprehensive treatments and some recommendations for further reading will be made here. A recognised classic in the field is the book by Hamming (1999). The book is excellent on Fourier methods and filters in particular; the explanations are very clear and without heavy mathematics, but the current reader should be aware that the notation is often different from that used here. Another classic is by Stearns and David (1987); there are various versions of this book in existence although not all of them are necessarily available at the moment. The interesting thing is that signal processing routines in various languages are included; certainly Fortran and C are available, and it appears that Matlab is also present in some editions. A slight downside is that the code is not always very transparent, although the theory is generally very well explained. Again, beware of the notation. In terms of digital signal processing instrumentation, which is mentioned often in Chapter 4, but not dealt with explicitly in this appendix, an excellent introductory book is Marven and Ewers (1996), which is very readable; however, this means that the material is sometimes rather superficial.

Before embarking on the matter of signal processing, it is important to be clear on terminology; the next section is designed to provide a discussion of the main terms that will be used throughout the appendix, with particular reference to the time series signals, which are often produced in SHM studies.

A.1 Deterministic and Random Signals

A.1.1 Basic Definitions

It is usually important to be precise with terminology; with this in mind, it is clearly important in the present context to say what one means by the term *signal*. From the Oxford English Dictionary, one has:

> **Signal**: A sign or notice, perceptible by sight or hearing especially for the purpose of conveying warning, direction or information.

This is intended to be very general; some examples fitting this description are given in Table A.1.

Structural Health Monitoring: A Machine Learning Perspective, First Edition. Charles R. Farrar and Keith Worden.
© 2013 John Wiley & Sons, Ltd. Published 2013 by John Wiley & Sons, Ltd.

Table A.1 Examples of signals

Signal	Time dependent?	Physical effect/medium	Encoding	Information
Caution sign	N	Words	Alphabetic	Warning
Smoke signal	Y	Smoke in air	Binary sequence	Versatile
Traffic light	Y	Light	3-bit binary	Warning
Thermometer	Y	Height of column of mercury	Real/analogue	Temperature measurement
Accelerometer	Y	Voltage in circuit	Real/analogue	Acceleration measurement

For most purposes of description and analysis, a signal can be defined simply as a mathematical function,

$$y = f(x) \tag{A.1}$$

where x is the independent variable that specifies the *domain* of the signal.

For example:

$y = \sin(\omega t)$ is a function of a variable in the *time domain* and is thus a time signal;

$X(\omega) = 1/(-m\omega^2 + ic\omega + k)$ is a *frequency-domain* signal (note that throughout this chapter a frequency-domain signal will often be referred to as a *spectrum*, which is common terminology in the field of signal processing);

an image intensity $I(x, y)$ is in the *spatial domain*.

A.1.2   Transducers, Sensors and Calibration

A *transducer* or *instrument* will be defined here as a system or medium – synthetic or natural – that converts one signal into another (Figure A.1). A system may or may not change the domain of the signal.

A *sensor* will be defined as a type of transducer that converts signals that are not necessarily quantifiable or storable on some medium to signals that are. Consider a thermometer (a schematic is given in Figure A.2).

Because the material of this book is within the context of engineering where precision is important, the mathematical definition above will be accepted. Note, though, that the word sensor can be used in a wider context. Consider the human skin; this is sometimes called a sensor as it can generate a signal that responds to pressure or temperature. The only argument with the above definition rests with the term 'quantifiable'. People cannot usually give a precise value of temperature in Centigrade on the basis of how their skin feels. However, if one relaxes the word to 'qualifiable', there is no quarrel; people can usually make a judgement as to whether it is getting hotter or colder.

The most desirable sort of sensor is a *linear* sensor. Staying with the idea of a thermometer, a linear sensor would be defined by

$$\theta(t) \longrightarrow h(t) \Rightarrow \alpha(\theta(t) - \theta_0) \Rightarrow \alpha(h(t) - h_0) \tag{A.2}$$

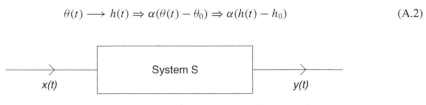

Figure A.1   Schematic of a system.

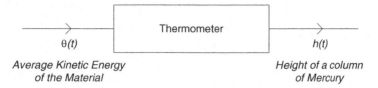

Figure A.2 Thermometer as a basic sensor.

that is, with respect to an appropriate origin (θ_0, h_0), the height of the mercury column doubles when the temperature doubles and so on. Another way of stating this is that

$$h(t) = a\theta(t) + b \tag{A.3}$$

where a and b are constants. If one knows the underlying physics, it may be possible to determine a and b. If the physics is complex, it may be simpler to carry out a *calibration* experiment.

If the sensor is linear, a calibration is simply two pairs of points that lie on the straight line (A.3). Suppose these points are (θ_1, h_1) and (θ_2, h_2). Substituting into Equation (A.3) gives

$$h_1 = a\theta_1 + b, \quad h_2 = a\theta_2 + b \tag{A.4}$$

which are two equations in the two unknowns a and b, which have a unique solution.

Sensors need not be linear. Suppose the physics of the transducer demands that

$$y(t) = ax(t)^2 + bx(t) + c \tag{A.5}$$

Then, in order to fix the three constants a, b and c, one would need three calibration points.

Some common examples of transducers are given in Table A.2, some of which are sensors.

A.1.3 Classification of Deterministic Signals

For most purposes, describing a signal as a mathematical function is too broad, too vague. One can impose a little more order than this by classifying signals into different groups.

Suppose one has a sensor that measures the displacement of a simple pendulum as depicted in Figure A.3 (this could be done, for example, with a laser vibrometer). Basic physics informs one that

$$\theta(t) = \theta_m \cos(\omega t) \tag{A.6}$$

Table A.2 Examples of transducers

Transducer	Input	Physical effect	Output
Thermometer	Temperature	Linear expansivity	Height of column
Litmus paper	pH	Chemical transformation	Colour
Shaker	Voltage	Electromagnetism	Force
Piezoelectric actuator	Voltage	Piezoelectric effect	Force
Piezoelectric accelerometer	Acceleration	Inverse piezoelectric effect	Voltage

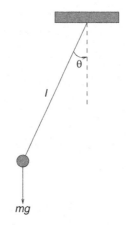

Figure A.3   A simple pendulum.

where θ_m is the peak amplitude of the motion and $\omega = \sqrt{l/g}$ with l the length of the pendulum and g the acceleration due to gravity. (In reality, the simple pendulum is a nonlinear system and things are a little more complicated than this; however, if small amplitude oscillations are assumed, the solution in Equation (A.6) will be a very good approximation to the true behaviour.)

As described above, one would actually see something like a voltage from the laser system making the measurements, but if the system is correctly calibrated, one can assume that one would see Equation (A.6). Now, as the system has a constant amplitude (zero damping is assumed for now), a constant frequency (dictated by physics) and initial conditions (which will be taken as $\theta = \theta_m, \theta = 0$ when $t = 0$), one knows from Equation (A6) the value of $\theta(t)$ for *all* time. Also, it is important to note that two identical pendula released from $\theta = \theta_m$ at $t = 0$ will have the same motions for all time. There is no place for uncertainty here.

If one can uniquely specify the value of θ for all time, that is if the underlying functional relationship between t and θ is known, the motion is *predictable* or *deterministic*. (Actually, the signal is still deterministic even if one does not know the function; it is enough to know that the function exists.)

The opposite situation occurs if one knows all the physics there is to know, but still cannot say what the signal will be at the next time instant – then the signal is *random* or *probabilistic*. This situation will be discussed a little later. This series of observations yields the first level of classification for signals (Figure A.4).

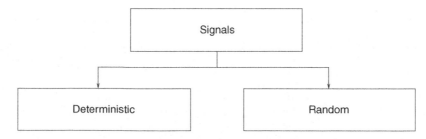

Figure A.4   Classification of signals – first step.

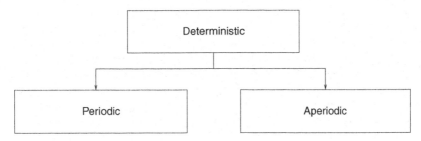

Figure A.5 A deeper level of classification.

If one considers the pendulum further, the motion has a special property. Suppose one defines the *period* τ by $\tau = 2\pi/\omega = 2\pi\sqrt{g/l}$; then for the pendulum,

$$\theta(t) = \theta(t + \tau) \tag{A.7}$$

Any such signals are referred to as *periodic*. Note also that Equation (A.7) implies

$$\theta(t + 2\tau) = \theta([t + \tau] + \tau) = \theta(t + \tau) = \theta(t) \tag{A.8}$$

that is, the signals repeats indefinitely.

Periodic signals arise naturally in systems where rotation is occurring, that is signals for bearings or gearboxes and so on. As discussed later, periodic signals are sometimes a very convenient form of excitation in testing; in that case the response will also be periodic (for linear systems anyway). The concept of periodicity naturally provides a further level of classification (Figure A.5).

Note that periodicity does *not* imply that the signal is sinusoidal; consider the square wave shown in Figure A.6. It will be shown later that the square wave has the functional form (a Fourier series),

$$S(t) = \sum_{i=1}^{\infty} s_i \sin\left(\frac{2\pi i}{\tau}t\right) \tag{A.9}$$

that is, it is made up of an infinite superposition of different sine waves with periods τ, $\tau/2$, $\tau/3$, ... The important point to note is that the component with period $\tau/2$ and so on still repeats after a time τ; that is, $S(t) = S(t + \tau)$ and the square wave in Equation (A.9) is therefore periodic. This all means that the classification tree naturally branches again as in Figure A.7.

If one now concentrates on the aperiodic branch of the classification (Figure A.5), one is led to wonder how such signals might arise? First of all, noting that that periodicity of Equation (A.7) implies that the signal repeats for all time leads to the conclusion that one type of aperiodicity arises if the signal only exists for a *finite* time.

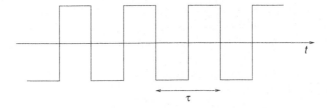

Figure A.6 A square wave; periodic but not sinusoidal.

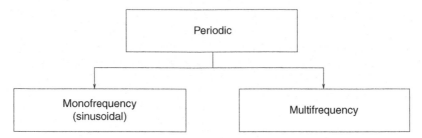

Figure A.7   Another branch in the classification.

Consider the response of a damped pendulum that is released from rest at $\theta = \theta_m$ at $t = 0$. To a good degree of approximation (again assuming small amplitudes),

$$y(t) = \theta_m e^{-ct} \cos(\omega t) \tag{A.10}$$

The signal will only exist for a finite time. (Although $e^{-ct} > 0$ for all time, the real system will be brought to rest by higher-order effects than viscous damping.) The response will be as illustrated in Figure A.8.

Signals existing for only a finite range of time are called *transient*; examples include: an impulse (hammer) excitation response of systems, explosion and shock loading, acoustic emissions, crack propagation (when the crack is the width of the structure, the structure fails).

The opposite of a transient would be some sort of infinite aperiodic function. There, periodicity would have to fail because one or more of the important signal parameters were changing. Suppose the mean of the signal was changing as in Figure A.9, where

$$y(t) = at + b \cos(\omega t) \tag{A.11}$$

Such signals can occur in machines or structures that are about to fail; looking for this type of characteristic is therefore important in condition monitoring and SHM. Also very important are signals with time-varying amplitude, for example as in Figure A.10, where

$$y(t) = at \cos(\omega t) \tag{A.12}$$

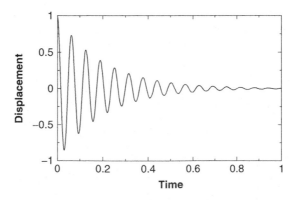

Figure A.8   A transient response.

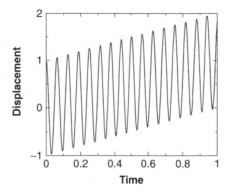

Figure A.9   A signal with time-varying mean.

Signals can also vary in their frequency content, e.g. the so-called chirp signal (as illustrated in Figure A.11)

$$y(t) = a \cos(\omega t^2) \tag{A.13}$$

The preceding discussion suggests a final branch of the classification tree (Figure A.12).

This is as detailed a breakdown as one is likely to need for deterministic signals. To proceed further one returns to the root of the tree to consider random signals.

A.1.4   Classification of Random Signals

As discussed earlier, the opposite of deterministic is *probabilistic* or *random*. For random signals, even if one knows all the physics there is to know one still cannot predict the signal value at the next instant. The most one can learn is the *probability* of the signal taking a certain value. In order to discuss this in detail, one needs to know the basic concepts of probability and statistics; familiarity with Chapter 6 of this book provides a perfectly adequate background. Chapter 6 provides us with the definition of a *random variable* (RV), which is a variable that cannot be predicted, but is governed by probabilistic rules – that

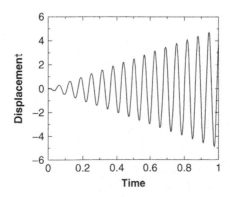

Figure A.10   A signal with time-varying amplitude.

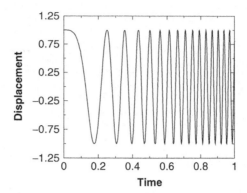

Figure A.11   A signal with time-varying frequency.

is, it will have a given probability distribution. Given the concept of a random variable, it is possible to give a definition of a random signal. What is actually needed is the concept of a *stochastic process*.

A stochastic process is essentially a family of RVs, with X_t parametrised by t (which could be, and usually is, time). The set of times can be continuous $t \in [0, T]$ or discrete $t \in \{t_1, \ldots t_N\}$. Each RV takes values from a probability distribution $p_t(x)$. In complete generality, the RV at t could depend on the values at previous times, that is on $x(t)$ observed for t less than the current value; however, it is often the case that X_t is assumed independent of previous values. The process is then called *independent*. Finally, in the general case, each of the $p_t(x)$ could be a different distribution. An important case occurs when they are all identical. The sequence is then called *identically distributed*. The most important stochastic processes are arguably those that are independent *and* identically distributed (i.i.d.). For such processes, one need only specify one probability density function $p(x)$. This group of observations gives the first branch of the classification tree for random signals (Figure A.13).

The most important i.i.d. processes arise when the underlying probability distribution is Gaussian, as in Equation (6.38); the classification tree immediately branches again as in Figure A.14. Figure A.15 gives an example of a Gaussian random process.

Having established the basic definitions for a random signal, the important question arises as to how to determine the underlying distributions or densities $p_t(x)$ or $p(x)$ for a random signal. If the distributions are not specified by the physics, this can be a very difficult problem (density estimation is discussed in some detail in Section 6.11). Fortunately, in practice, one mainly wants to know the lower-order statistics, that is the mean and variance.

Consider the mean for the moment – how could one estimate $E[X_t]$? The only way to do this in the general case would be to somehow generate several examples of the process starting from the same

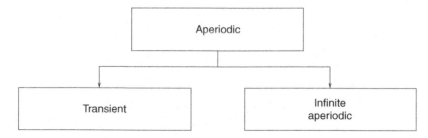

Figure A.12   The final branch for the classification tree for deterministic signals.

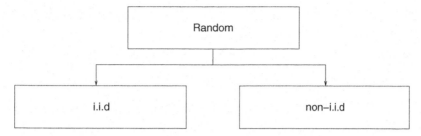

Figure A.13 The first branch of the classification tree for random signals.

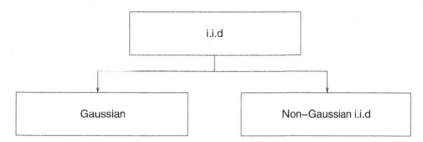

Figure A.14 A further branch in the classification.

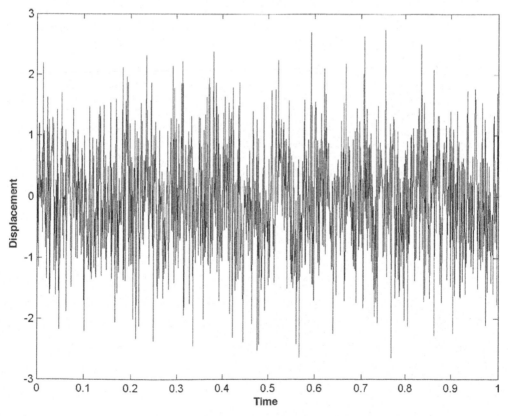

Figure A.15 Gaussian random time series.

initial conditions $x^{(i)}(t)$, $i = 1, \ldots, N_p$, perhaps from a set of N_p identical experiments. Then the mean of the RV X_t is given by

$$E[X_t] = \frac{1}{N_p} \sum_{i=1}^{N_p} x^{(i)}(t) \tag{A.14}$$

This process is called *ensemble averaging*. The individual time signals $x^{(i)}(t)$ are called *realisations* of the process. This procedure clearly presents a problem as in general one might only have one experimental rig. If the process is i.i.d., then $p_t(x)$ is the same for all t and one can imagine estimating the statistics by averaging over time,

$$E[X_t] = \frac{1}{T} \int_0^T xp(x)\mathrm{d}t \tag{A.15}$$

Or, using the more usual discrete estimator,

$$E[X_t] = \frac{1}{N_t} \sum_{i=1}^{N_t} x_i(t) \tag{A.16}$$

where one has N_t samples of the signal, x_i sampled at constant frequency. In fact the signal does not have to be i.i.d., but is does have to be rather special; a process for which time and ensemble averaging are equivalent is called *ergodic*. A weaker condition is that the statistical moments do not change with time. Such signals are called *stationary*. If only the mean and standard deviation are constant, the signal is called *weakly stationary*.

Under a slightly more informal definition of *ergodic*, such signals are those for which the statistical properties can be estimated given a long enough time record. With this definition, one can see that i.i.d. signals will automatically be ergodic and one can continue the classification for random signals with the branch shown in Figure A.16.

There are numerous ways in which signals can fail to be stationary, obvious examples being where the mean or variance of the signals are changing, as in Figures A.17 and A.18.

There is a direct correspondence with the time-varying cases discussed above for deterministic signals.

One could also consider a random signal nonstationary in its frequency content; however, this is rather difficult to visualise and discussion will be postponed until the wavelet transform is introduced later.

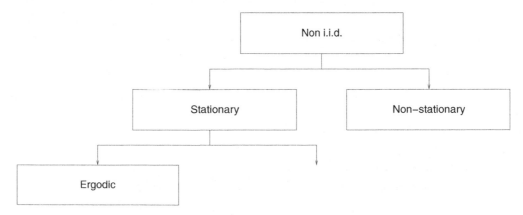

Figure A.16   A further branch in the classification of random signals.

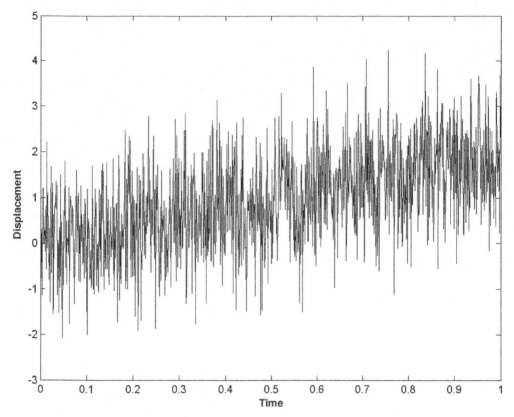

Figure A.17   A random signal nonstationary in the mean.

This completes the classification for random signals as needed for a good understanding of the material of this book.

A.2   Fourier Analysis and Spectra

A.2.1   Fourier Series

The next series of sections will explore the problem of decomposing signals into simple components that are amenable to analysis. Recall from Section A.1 that the square wave could be expressed as a series of sine waves. This is actually the first example here of the *Fourier representation*.

For any function $x(t)$ defined on the interval $[-\tau/2, \tau/2]$, one has a representation of the function on that interval by

$$x(t) = a_0 + \sum_{n=1}^{\infty} a_n \cos\left[\left(\frac{2\pi n}{\tau}\right)t\right] + \sum_{n=1}^{\infty} b_n \sin\left[\left(\frac{2\pi n}{\tau}\right)t\right] \qquad (A.17)$$

Proving that this representation holds and that the series given in Equation (A.17) actually converges is mathematically nontrivial and the reader is accordingly referred to the mathematical literature (Körner, 1989). Fortunately, for the purposes of engineering, one usually only needs to find the coefficients for

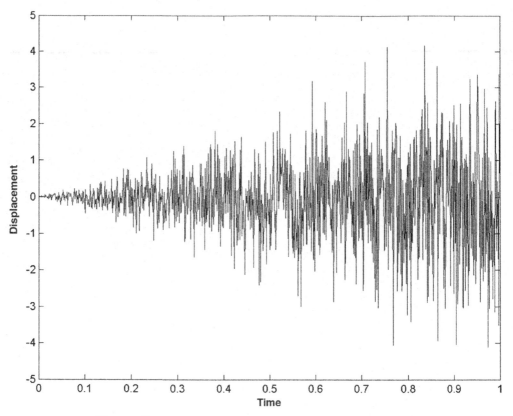

Figure A.18   A random signal nonstationary in the variance (amplitude).

such a representation, and this problem is much more tractable. The coefficents can be obtained as follows. One first considers the integral,

$$I_m = \int_{-\tau/2}^{\tau/2} x(t)\cos\left[\left(\frac{2\pi m}{\tau}\right)t\right] \mathrm{d}t \qquad (A.18)$$

where m is an integer.

When the function $x(t)$ is substituted by its representation from Equation (A.17), the integral in Equation (A18) breaks down into three terms, $I_m^{(1)}$, $I_m^{(2)}$ and $I_m^{(3)}$. The first term is given by

$$
\begin{aligned}
I_m^{(1)} &= \int_{-\tau/2}^{\tau/2} a_0 \cos\left[\left(\frac{2\pi m}{\tau}\right)t\right] \mathrm{d}t \\
&= \left[\frac{a_0\tau}{2\pi m}\sin\left[\left(\frac{2\pi m}{\tau}\right)t\right]\right]_{-\tau/2}^{\tau/2} \\
&= \frac{a_0\tau}{2\pi m}[\sin(\pi m) - \sin(-\pi m)] \\
&= \frac{a_0\tau}{\pi m}\sin(\pi m)
\end{aligned}
$$

but this is zero as $\sin(\pi m) = 0$ for all integer m. The analysis moves to the second integral, which is a little more complicated,

$$I_m^{(2)} = \int_{-\tau/2}^{\tau/2} \sum_{n=1}^{\infty} a_n \cos\left[\left(\frac{2\pi n}{\tau}\right) t\right] \cos\left[\left(\frac{2\pi m}{\tau}\right) t\right] dt \tag{A.19}$$

Assuming that one can change the order of integration and summation (and this is usually justified in cases of interest for engineering), one obtains

$$I_m^{(2)} = \sum_{n=1}^{\infty} a_n \int_{-\tau/2}^{\tau/2} \cos\left[\left(\frac{2\pi n}{\tau}\right) t\right] \cos\left[\left(\frac{2\pi m}{\tau}\right) t\right] dt$$

Now, using the well-known identity

$$\cos A \sin B = \frac{\cos(A + B) + \cos(A - B)}{2}$$

one finds

$$I_m^{(2)} = \sum_{n=1}^{\infty} \frac{a_n}{2} \int_{-\tau/2}^{\tau/2} \left(\cos\left[\left(\frac{2\pi(n + m)}{\tau}\right) t\right] + \cos\left[\left(\frac{2\pi(n - m)}{\tau}\right) t\right]\right) dt$$

$$= \sum_{n=1}^{\infty} \frac{a_n}{2} \left[\frac{\tau}{2\pi(n + m)} \sin\left[\left(\frac{2\pi(n + m)}{\tau}\right) t\right] + \frac{\tau}{2\pi(n - m)} \sin\left[\left(\frac{2\pi(n - m)}{\tau}\right) t\right]\right]_{-\tau/2}^{\tau/2}$$

$$= \sum_{n=1}^{\infty} \frac{a_n}{2} \left[\frac{\tau}{\pi(n + m)} \sin\left[\pi(n + m)\right] + \frac{\tau}{\pi(n - m)} \sin\left[\pi(n - m)\right]\right]$$

Now, if n and m are different integers, then $n + m$ and $n - m$ are both nonzero integers and the sine terms in the last expression vanish. If $n = m$, one has a problem with the second term above. In this case, one could use a limit argument, but it is simpler to go back to Equation (A.19) and set $n = m$. This yields

$$I_m^{(2)} = \frac{a_m}{2} \int_{-\tau/2}^{\tau/2} \left(\cos\left[\left(\frac{4\pi m}{\tau}\right) t\right] + 1\right) dt$$

$$= a_m \left[\frac{\tau}{8\pi m} \sin\left[\left(\frac{4\pi m}{\tau}\right) t\right] + \frac{t}{2}\right]_{-\tau/2}^{\tau/2}$$

and this breaks down fairly easily to

$$I_m^{(2)} = a_m \frac{\tau}{2}$$

The foregoing analysis is essentially a proof of an *orthogonality relation*,

$$\int_{-\tau/2}^{\tau/2} \cos\left[\left(\frac{2\pi n}{\tau}\right) t\right] \cos\left[\left(\frac{2\pi m}{\tau}\right) t\right] dt = \frac{\tau}{2} \delta_{mn} \tag{A.20}$$

where the object δ_{mn} is called the *Kronecker delta* and has the properties

$$\delta_{mn} = \begin{cases} 1 & \text{if } n = m \\ 0 & \text{if } n \neq m \end{cases} \tag{A.21}$$

A similar analysis needs to be carried out for the third integral defined earlier,

$$I_m^{(3)} = \int_{-\tau/2}^{\tau/2} \left[\sum_{n=1}^{\infty} b_n \sin\left[\left(\frac{2\pi n}{\tau}\right) t \right] \right] \cos\left[\left(\frac{2\pi m}{\tau}\right) t \right] dt$$

and this shows that

$$I_m^{(3)} = 0 \tag{A.22}$$

and establishes along the way another orthogonality relation,

$$\int_{-\tau/2}^{\tau/2} \sin\left[\left(\frac{2\pi n}{\tau}\right) t \right] \cos\left[\left(\frac{2\pi m}{\tau}\right) t \right] dt = 0 \tag{A.23}$$

Now as the original integral of interest in Equation (A.18) is $I_m = I_m^{(1)} + I_m^{(2)} + I_m^{(3)}$, one finally has

$$I_m = \frac{\tau}{2} a_m$$

Or, in terms of the original expression,

$$a_m = \frac{2}{\tau} \int_{-\tau/2}^{\tau/2} x(t) \cos\left[\left(\frac{2\pi m}{\tau}\right) t \right] dt \tag{A.24}$$

A completely parallel analysis gives

$$b_m = \frac{2}{\tau} \int_{-\tau/2}^{\tau/2} x(t) \sin\left[\left(\frac{2\pi m}{\tau}\right) t \right] dt \tag{A.25}$$

via the orthogonality relation

$$\int_{-\tau/2}^{\tau/2} \sin\left[\left(\frac{2\pi n}{\tau}\right) t \right] \sin\left[\left(\frac{2\pi m}{\tau}\right) t \right] dt = \frac{\tau}{2} \delta_{mn} \tag{A.26}$$

Finally, the last coefficient in the Fourier expansion is obtained via the expression

$$a_0 = \frac{1}{\tau} \int_{-\tau/2}^{\tau/2} x(t) dt \tag{A.27}$$

Note that a_0 is therefore the mean value of the signal $x(t)$, so for a zero-mean signal one can immediately assume $a_0 = 0$.

A.2.2   The Square Wave Revisited

Consider the square wave depicted in Figure A.6. An appropriate mathematical expression for the signal can be taken as

$$
s(t) = \begin{cases} 1 & \text{if } t \in \left[0, \dfrac{\tau}{2}\right] \\[2mm] -1 & \text{if } t \in \left[-\dfrac{\tau}{2}, 0\right] \end{cases}
\tag{A.28}
$$

Now, noting that the function, as defined, is *odd*, that is

$$
s(-t) = -s(t)
\tag{A.29}
$$

it follows that the Fourier series for $s(t)$ *cannot* include any cosines, as cosines are *even* functions, that is $\cos(-t) = \cos(t)$.

The full Fourier expansion for the square wave is therefore simply

$$
s(t) = \sum_{n=1}^{\infty} b_n \sin\left[\left(\frac{2\pi n}{\tau}\right) t\right]
\tag{A.30}
$$

(This latter equation also assumes that $a_0 = 0$; this follows from the fact that the square wave is zero-mean.)

Recalling Equation (A.25), one has

$$
b_n = \frac{2}{\tau} \int_{-\tau/2}^{\tau/2} s(t) \sin\left[\left(\frac{2\pi n}{\tau}\right) t\right] dt
$$

Now, on substituting the expression for the square wave from Equation (A.28), the coefficient integral splits naturally into two parts,

$$
\begin{aligned}
b_n &= \frac{2}{\tau} \left\{ \int_{0}^{\tau/2} \sin\left[\left(\frac{2\pi n}{\tau}\right) t\right] dt - \int_{-\tau/2}^{0} \sin\left[\left(\frac{2\pi n}{\tau}\right) t\right] dt \right\} \\[2mm]
&= \frac{2}{\tau} \left\{ \left[\frac{-\tau}{2\pi n}\cos\left[\left(\frac{2\pi n}{\tau}\right) t\right]\right]_{0}^{\tau/2} - \left[\frac{-\tau}{2\pi n}\cos\left[\left(\frac{2\pi n}{\tau}\right) t\right]\right]_{-\tau/2}^{0} \right\} \\[2mm]
&= \frac{2}{\tau} \left\{ \frac{-\tau}{2\pi n}[\cos(\pi n) - 1] - \frac{-\tau}{2\pi n}[1 - \cos(-\pi n)] \right\} \\[2mm]
&= \frac{2}{\pi n}[1 - \cos(\pi n)]
\end{aligned}
$$

This latter expression is zero if $n = 0, 2, \ldots, 2k$, as then the term $\cos(\pi n) = 1$. If $n = 1, 3, \ldots, 2k - 1$, then $\cos(\pi n) = -1$ and one has

$$
b_n = \frac{4}{\pi n}
\tag{A.31}
$$

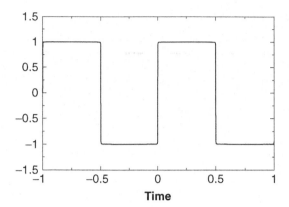

Figure A.19   Fourier representation of the square wave – the first 1000 terms.

So (after a little rearrangement of the formula), the square wave is seen to be represented by

$$s(t) = \sum_{n=1}^{\infty} \frac{4}{\pi(2n-1)} \sin\left[\left(\frac{2\pi(2n-1)}{\tau}\right)t\right] \tag{A.32}$$

Figure A.19 shows the sum of the Fourier series for the square wave up to 1000 terms. In order to illustrate the effects of truncation of the series at a lower order, Figure A.20 shows the summed series with 1, 10 and 100 terms (sine waves).

The 'ringing' or oscillation at the points where the discontinuities occur in the original signal is called the *Gibbs phenomenon*. It occurs because discontinuities in a time signal are very short time-scale events and therefore need the higher frequencies in the Fourier series in order to model them accurately.

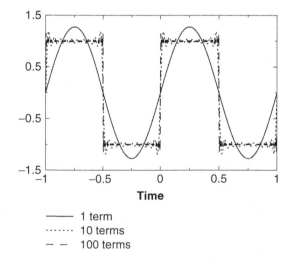

Figure A.20   The Fourier series for the square wave: 1 to 100 terms.

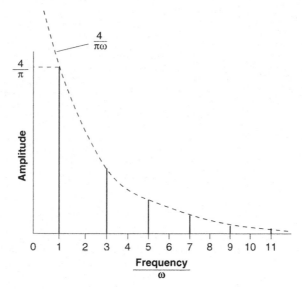

Figure A.21   Spectral representation of the square wave.

A.2.3   A First Look at Spectra

Suppose one were to write expression (A.32) for the square wave in terms of the base (fundamental) frequency ω instead of the period τ; then one finds

$$s(t) = \sum_{n=1}^{\infty} \frac{4}{\pi(2n-1)} \sin\left[(2n-1)\omega t\right] \tag{A.33}$$

This representation has a simple graphical description; one simply plots the amplitude of each component at the corresponding frequency as in Figure A.21, The figure shows a rudimentary form of a *spectrum*.

Now, recall that the analysis for the square wave was simplified enormously by putting the discontinuity at the origin. Suppose now that the discontinuity (up-crossing) was at $t = t_d > 0$, as shown in Figure A.22.

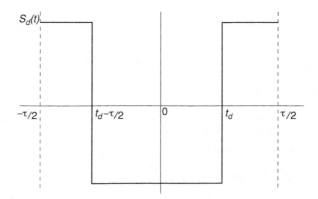

Figure A.22   An 'offset' form of the square wave.

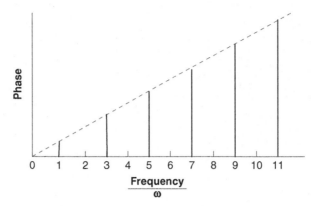

Figure A.23   Phase spectrum of the 'offset' square wave.

The basic Fourier analysis would now seem to require the inclusion of the cosine terms as the signal is neither odd nor even.

However, note that

$$s_d(t) = s(t - t_d)$$

so from Equation (A.33) it follows that

$$s_d(t) = \sum_{n=1}^{\infty} \frac{4}{\pi(2n-1)} \sin\left[(2n-1)\omega(t - t_d)\right]$$

$$= \sum_{n=1}^{\infty} b_{2n-1} \sin\left[(2n-1)\omega(t - \varphi_{2n-1})\right] \tag{A.34}$$

and the amplitudes $b_n = 4/(n\pi)$ are now joined by *phases* φ_n, where

$$\varphi_n = n\omega t_d \tag{A.35}$$

This analysis means that if one wishes to form the spectrum of the 'offset' square wave, one has to add to the amplitude plot in Figure A.21 a phase plot as shown in Figure A.23. In fact Equation (A.34) is an example of the *phase form* of the Fourier series,

$$x(t) = a_0 + \sum_{n=1}^{\infty} b_n \sin\left[n\omega t - \varphi_n\right] \tag{A.36}$$

which is of course quite general. Returning to the phase plot (Figure A.23), one observes that the gradient of the straight line joining the phase values is t_d; it is a general result that if the Fourier series of a signal has its phase points lying on a line of gradient c, the signal $x(t + c)$ will have zero-phase for all frequencies.

A.2.4   The Exponential Form of the Fourier Series

Recall the standard Fourier series (this time in terms of $\omega = 2\pi/\tau$),

$$x(t) = a_0 + \sum_{n=1}^{\infty} a_n \cos(n\omega t) + \sum_{n=1}^{\infty} b_n \sin(n\omega t) \tag{A.37}$$

It is straightforward here to apply de Moivre's theorem in the form

$$\cos(n\omega t) = \frac{1}{2}(e^{in\omega t} + e^{-in\omega t})$$

$$\sin(n\omega t) = \frac{1}{2i}(e^{in\omega t} - e^{-in\omega t})$$

The nth harmonic terms in Equation (A.37) combine so that

$$\frac{a_n}{2}(e^{in\omega t} + e^{-in\omega t}) + \frac{b_n}{2i}(e^{in\omega t} - e^{-in\omega t})$$
$$= \frac{1}{2}(a_n - ib_n)e^{in\omega t} + \frac{1}{2}(a_n + ib_n)e^{-in\omega t} \tag{A.38}$$

and this allows one to write Equation (A.37) in the form

$$x(t) = \sum_{n=-\infty}^{\infty} c_n e^{in\omega t} \tag{A.39}$$

where

$$\begin{aligned}
c_n &= (a_n - ib_n)/2, & n &> 0 \\
c_n &= (a_n + ib_n)/2 & = c_{-n}^*, & n &< 0 \\
c_0 &= a_0, & n &= 0
\end{aligned} \tag{A.40}$$

Using the orthogonality relation,

$$\int_{-\tau/2}^{\tau/2} e^{in\omega t} e^{-im\omega t} \, dt = \tau \delta_{mn} \tag{A.41}$$

one finds

$$c_n = \frac{1}{\tau} \int_{-\tau/2}^{\tau/2} x(t) e^{-in\omega t} \, dt \tag{A.42}$$

which, together with Equation (A.39), represents the exponential form of the Fourier series.

A.3   The Fourier Transform

A.3.1   Basic Transform Theory

In the last section, the idea of the Fourier series led naturally to the concept of the spectrum, which turned out to be a natural way of visualising the frequency content of a given signal. In many situations, particularly if the signals of interest contain substantial random components, a visual inspection of the original time series will not give any particular insight as a result of the complexity of the signal; however, the spectrum will often yield the desired insight. One problem with the analysis shown so far is that it is not sufficiently general for many cases of interest in engineering, or specifically for SHM or condition monitoring. One of the assumptions implicit in the Fourier analysis conducted so far is that the signal of interest is periodic (with period τ). If one wishes to consider the spectral content of nonperiodic signals, one has to let $\tau \to \infty$ as all of the signal over the interval $t \in [-\infty, \infty]$ may contain important information.

Recall the exponential form of the Fourier series, as summarised by Equations (A.39) and (A.42); combining these two equations gives

$$x(t) = \sum_{n=-\infty}^{\infty} \left\{ \frac{1}{\tau} \int_{-\tau/2}^{\tau/2} x(t')e^{-in\omega t'} dt' \right\} e^{in\omega t} \tag{A.43}$$

In order to relax the condition of periodicity, one has to let $\tau \to \infty$. Recall that the spacing between the frequency lines is

$$\omega = \Delta\omega = \frac{2\pi}{\tau} \tag{A.44}$$

so that the kth spectral line is at $\omega_k = k\Delta\omega$. From Equation (A.44), one sees that

$$\frac{1}{\tau} = \frac{\Delta\omega}{2\pi}$$

and substituting this expression into Equation (A.43) yields

$$x(t) = \frac{1}{2\pi} \sum_{n=-\infty}^{\infty} \left\{ \int_{-\tau/2}^{\tau/2} x(t')e^{-i\omega_n t'} dt' \right\} e^{i\omega_n t} \Delta\omega \tag{A.45}$$

Now, in the limit of interest, as $\tau \to \infty$, the ω_n become closer and closer together and the summation 'turns into' an integral with $\Delta\omega = d\omega$ (assuming that $x(t)$ is appropriately well behaved). In that limit,

$$x(t) = \frac{1}{2\pi} \int_{-\infty}^{\infty} \left\{ \int_{-\infty}^{\infty} x(t')e^{-i\omega t'} dt' \right\} e^{i\omega t} d\omega \tag{A.46}$$

This important, but seemingly tautologous, statement is called the *Fourier inversion theorem*. It then follows that if one defines

$$F[x(t)] = X(\omega) = \int_{-\infty}^{\infty} x(t)e^{-i\omega t} dt \tag{A.47}$$

where F denotes the *Fourier transform*, then Equation (A.46) implies that

$$F^{-1}[X(\omega)] = x(t) = \frac{1}{2\pi} \int_{-\infty}^{\infty} X(\omega)e^{i\omega t} d\omega \tag{A.48}$$

and this is the inverse Fourier transform.

Note the formal similarity of the Fourier transform to the Laplace transform; in fact, one can formally obtain the Fourier transform simply by letting $s = i\omega$ in the Laplace transform. The main difference between the two is the comparative simplicity of the inverse Fourier transform – the inverse Laplace transform is a complex contour integral.

The pair $\{x(t), X(\omega)\}$ are a *Fourier transform pair*. As they are uniquely constructable from each other, they must both encode the same information, but in different domains. $X(\omega)$ expresses the frequency content of $x(t)$ (note that the product ωt in Equation (A.47) has to be dimensionless and this forces ω to have the dimensions of frequency) and is another form of *spectrum*. However, note that it has to be a continuous function of ω in order to represent nonperiodic functions. Note also that a family of Fourier methods is emerging here based on the nature of the signals concerned. The Fourier series mapped a continuous-time signal into a discrete frequency signal; the Fourier transform maps a continuous-time signal into a continuous frequency-domain spectrum.

A.3.2 An Interesting Function that is not a Function

It will prove useful here to introduce a new mathematical object. Recall the Kronecker delta δ_{ij} as defined in Equation (A.21); as a result of its definition, it obeys a *projection property*,

$$\sum_{i=-\infty}^{\infty} a_i \delta_{ij} = a_j \tag{A.49}$$

It will prove very useful later to have an analogous property for integrals, that is

$$\int_{-\infty}^{\infty} f(x)\delta(x, y)\mathrm{d}x = f(y) \tag{A.50}$$

A fairly informal argument suggests a means of defining the required function $\delta(x, y)$; suppose one were to reorder Equation (A.46) (assuming as always that this is allowed); then

$$x(t) = \int_{-\infty}^{\infty} \left\{ \frac{1}{2\pi} \int_{-\infty}^{\infty} e^{i\omega(t-t')}\mathrm{d}\omega \right\} x(t')\mathrm{d}t' \tag{A.51}$$

Comparison with Equation (A.50) shows that the object,

$$\frac{1}{2\pi} \int_{-\infty}^{\infty} e^{i\omega(t-t')}\mathrm{d}\omega$$

has the right properties for the desired *delta function*. In fact, as the integral only depends on $t - t'$, one defines

$$\delta(t, t') = \delta(t - t') = \frac{1}{2\pi} \int_{-\infty}^{\infty} e^{i\omega(t-t')}\mathrm{d}\omega \tag{A.52}$$

Now, the question arises: what does $\delta(t - t')$ look like? Suppose first that $t \neq t'$; in this case the term $e^{i\omega(t-t')}$ would oscillate infinitely often as t' varied from $-\infty$ to ∞. This means that it would be negative as often as it was positive and the integral would average to zero. (This is a very sloppy argument indeed!) Now consider the case of $t - t'$; then

$$\delta(0) = \int_{-\infty}^{\infty} \mathrm{d}\omega = \infty$$

Therefore the delta function as defined appears to be zero unless its argument is zero and in that case, it is infinite! An attempt to depict the function is shown in Figure A.24. This behaviour is so strange that $\delta(x)$ is not really entitled to be called a function at all. In order to deal with it rigorously, mathematicians (notably Laurent Schwarz) had to invent a new type of 'function' called a *distribution* (Friedlander and Joshi, 1999). However, any subtleties that arise will be ignored here and the discussion will treat $\delta(x)$ as if it were a function. (The delta function was used extensively by the physicist Paul Dirac in his

Figure A.24   The delta function.

formulation of quantum mechanics and because of this, it is sometimes referred to as the *Dirac delta function.*)

This $\delta(x)$ also arises in other contexts; consider the integral

$$I(y) = \int_{-\infty}^{y} \delta(x)\mathrm{d}x$$

If $y < 0$ then the integrand is zero, as $\delta(x) = 0$ for $x \in [-\infty, y]$. Now, if $y > 0$, one can replace the integral by

$$\int_{-\infty}^{\infty} \delta(x)\mathrm{d}x$$

as $\delta(x) = 0$ for $y < x < \infty$; however, it is clear that the latter integral is unity. Therefore the indefinite (and improper) integral of $\delta(x)$ is a function that is zero for $y < 0$ and unity for $y \geq 0$. This is the definition of the *step function* or *Heaviside function* $H(y)$ (Figure A.25).

This means that

$$\frac{d}{dx}H(x) = \delta(x) \tag{A.53}$$

and a slightly different argument gives

$$\frac{\mathrm{d}}{\mathrm{d}x}H(x - a) = \delta(x - a) \tag{A.54}$$

so the delta function can naturally be regarded as the derivative of the step function.

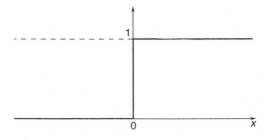

Figure A.25   The Heaviside or step function.

A.3.3   The Fourier Transform of a Periodic Function

One might reasonably ask now, having defined the Fourier transform specifically for nonperiodic functions, what does the transform of a periodic function look like? Suppose one considers the simplest possible case,

$$x(t) = \sin(\Omega t)$$

The associated Fourier transform is

$$X(\omega) = \int_{-\infty}^{\infty} \sin(\Omega t)e^{-i\omega t}\,dt = \int_{-\infty}^{\infty} \frac{1}{2i}(e^{i\Omega t} - e^{-i\Omega t})e^{-i\omega t}\,dt$$

$$= \frac{1}{2i}\int_{-\infty}^{\infty} e^{i(\Omega-\omega)t}\,dt - \frac{1}{2i}\int_{-\infty}^{\infty} e^{i(-\Omega-\omega)t}\,dt$$

so finally one obtains

$$F[\sin(\Omega t)] = \frac{\pi}{i}\delta(\omega - \Omega) - \frac{\pi}{i}\delta(\omega + \Omega) \tag{A.55}$$

Thus the spectrum of the sine wave has only two nonzero components at $\omega = \pm\Omega$ and can be represented graphically as in Figure A.26.

Note that the amplitude of each component is π/i so one needs an amplitude and phase in reality. $|X(\pm\Omega)| = \pi$ and as $1/i = -i = \cos(3\pi/2) + i\sin(3\pi/2)$, then $\angle X(\pm\Omega) = 3\pi/2$ (or $-\pi/2$). The corresponding spectrum for

$$x(t) = \cos(\Omega t)$$

is illustrated in Figure A.27.

Finally, consider the general periodic function with period $\tau = 2\pi/\Omega$,

$$x(t) = a_0 + \sum_{n=1}^{\infty} a_n\cos(n\Omega t) + \sum_{n=1}^{\infty} b_n\sin(n\Omega t)$$

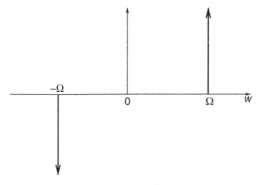

Figure A.26   The spectrum of the sine function.

Figure A.27   The spectrum of the cosine function.

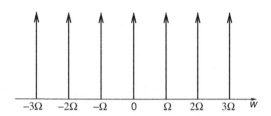

Figure A.28   The spectrum of a general periodic function/signal.

The associated Fourier transform is

$$X(\omega) = 2\pi a_0\delta(0) + \sum_{n=1}^{\infty} \frac{a_n}{2}[\delta(\omega - n\Omega) + \delta(\omega + n\Omega)] + \sum_{n=1}^{\infty} \frac{b_n}{2i}[\delta(\omega - n\Omega) - \delta(\omega + n\Omega)]$$

$$= 2\pi a_0\delta(0) + \sum_{n=1}^{\infty} \frac{1}{2}(a_n - ib_n)\delta(\omega - n\Omega) + \sum_{n=1}^{\infty} \frac{1}{2}(a_n + ib_n)\delta(\omega + n\Omega)$$

Therefore the Fourier transform of a general periodic function can be regarded as an infinite train of delta functions (Figure A.28).

A.3.4   The Fourier Transform of a Pulse/Impulse

An example of practical interest will be the computation of the spectrum of a rectangular pulse of duration τ, as depicted in Figure A.29.

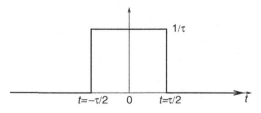

Figure A.29   The rectangular pulse.

An appropriate mathematical definition of the pulse function is

$$P(t) = \begin{cases} \dfrac{1}{\tau} & \text{if} \quad |t| < \dfrac{\tau}{2} \\[2mm] 0 & \text{if} \quad |t| \geq \dfrac{\tau}{2} \end{cases}$$

The corresponding Fourier transform is easily seen to be

$$X(\omega) = \frac{1}{\tau} \int_{-\tau/2}^{\tau/2} e^{-i\omega t} \, dt$$

Now, taking $e^{-i\omega t} = \cos(\omega t) - i\sin(\omega t)$, the integral over the second term is zero as $\sin(\omega t)$ is an odd function; this means that

$$X(\omega) = \frac{1}{\tau} \int_{-\tau/2}^{\tau/2} \cos(\omega t) \, dt$$

$$= \left[\frac{\sin(\omega t)}{\omega \tau} \right]_{-\tau/2}^{\tau/2} = \frac{\sin(\omega \tau/2)}{\omega \tau} - \frac{\sin(-\omega \tau/2)}{\omega \tau} = \frac{\sin(\omega \tau/2)}{(\omega \tau/2)}$$

which has the form shown in Figure A.30 (and is sometimes called the *sinc* function). The *sinc* function has zeros at

$$\omega = \frac{2n\pi}{\tau}$$

Now suppose one approximates matters by saying that the main energy in this spectrum is concentrated between $-2\pi/\tau$ and $2\pi/\tau$. This means that if $\tau \to 0$ – that is, the pulse becomes shorter – the energy content of the spectrum is distributed over a broader range of frequencies. In the limit, one sees a delta function pulse, that is infinitesimal duration, finite energy and

$$X(\omega) = \int_{-\infty}^{\infty} \delta(t) \, dt = 1$$

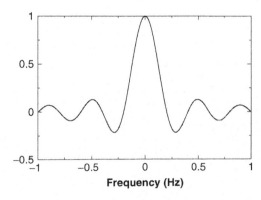

Figure A.30   The spectrum of the rectangular pulse.

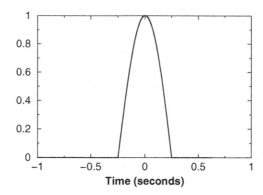

Figure A.31   A half-cosine pulse.

and this has a flat spectrum; that is, all frequencies are represented evenly. This is why the impulse is an excellent excitation force for dynamic testing; a potentially broad range of frequencies is excited.

Given the physical constraints on applying an impact to a system, a more realistic mathematical form of pulse is given by the half-cosine (as illustrated in Figure A.31)

$$P(t) = \begin{cases} \cos\left(\dfrac{\pi t}{\tau}\right) & \text{if} \quad |t| < \dfrac{\tau}{2} \\[2ex] 0 & \text{if} \quad |t| \geq \dfrac{\tau}{2} \end{cases}$$

which has the spectrum

$$X(\omega) = \int_{-\tau/2}^{\tau/2} \cos(\alpha t)\cos(\omega t)\,\mathrm{d}t$$

on setting $\alpha = \pi/\tau$. If one applies the simple trigonometrical identity used earlier, one finds

$$X(\omega) = \frac{1}{2} \int_{-\tau/2}^{\tau/2} \{\cos([\alpha + \omega]t) + \cos([\alpha - \omega]t)\}\,\mathrm{d}t$$

$$= \frac{1}{2} \frac{\sin\left(\frac{(\alpha+\omega)\tau}{2}\right)}{\left(\frac{(\alpha+\omega)\tau}{2}\right)} + \frac{1}{2} \frac{\sin\left(\frac{(\alpha-\omega)\tau}{2}\right)}{\left(\frac{(\alpha-\omega)\tau}{2}\right)}$$

In the limit as $\tau \to 0$, then $\alpha \to 0$ and one recovers the spectrum of the rectangular pulse and all the frequencies are represented again.

A.3.5   The Convolution Theorem

This section is concerned with the effect of certain mathematical operations on a signal or its Fourier transform. First, one observes that the Fourier transform is a linear operator and therefore has a particularly simple operation on a weighted sum of two (or more) functions,

$$F\left[ay_1(t) + by_2(t)\right] = aF\left[y_1(t)\right] + b\left[y_2(t)\right]$$

$$= aY_1(\omega) + bY_2(\omega) \tag{A.56}$$

where a and b are arbitrary complex scalars and $y_1(t)$ and $y_1(t)$ are arbitrary functions of time. Next, one can consider the product of two time signals – what does the Fourier transform look like?

$$F[y_1(t)y_2(t)] = \int_{-\infty}^{\infty} y_1(t)y_2(t)e^{-i\omega t}\,dt \tag{A.57}$$

First, one writes the $y_i(t)$ as the inverse transforms of their corresponding spectra,

$$y_i(t) = \frac{1}{2\pi} \int_{-\infty}^{\infty} Y_i(\omega)e^{i\omega t}\,d\omega$$

and then substitutes the expressions into (A.57), giving

$$F[y_1(t)y_2(t)] = \int_{-\infty}^{\infty} \left\{ \frac{1}{2\pi} \int_{-\infty}^{\infty} Y_1(\alpha)e^{i\alpha t}\,d\alpha \right\} \left\{ \frac{1}{2\pi} \int_{-\infty}^{\infty} Y_2(\Omega)e^{i\Omega t}\,d\Omega \right\} e^{-i\omega t}\,dt$$

Now, assuming that one can interchange the order of the integrals, the overall expression becomes

$$\frac{1}{2\pi} \int_{-\infty}^{\infty} \int_{-\infty}^{\infty} \left\{ \frac{1}{2\pi} \int_{\infty}^{\infty} e^{i(\Omega+\alpha-\omega)t}\,dt \right\} Y_1(\alpha)Y_2(\Omega)\,d\alpha\,d\Omega$$

and the bracketed quantity in the centre is identified as a delta function Equation (A.52). This means that

$$F[y_1(t)y_2(t)] = \frac{1}{2\pi} \int_{\infty}^{\infty} \int_{-\infty}^{\infty} \delta(\Omega+\alpha-\omega)Y_1(\alpha)Y_2(\Omega)\,d\alpha\,d\Omega$$

or

$$F[y_1(t)y_2(t)] = \frac{1}{2\pi} \int_{-\infty}^{\infty} \left\{ \int_{-\infty}^{\infty} \delta(\Omega+\alpha-\omega)Y_1(\alpha)\,d\alpha \right\} Y_2(\Omega)\,d\Omega \tag{A.58}$$

Finally, on applying the projection property of the delta function (A.50), the last equation becomes

$$F[y_1(t)y_2(t)] = \frac{1}{2\pi} \int_{-\infty}^{\infty} Y_1(\omega-\Omega)Y_2(\Omega)\,d\Omega \tag{A.59}$$

The quantity on the right of Equation (A.59) is called the *convolution* or *convolution product* of the two functions $Y_1(\omega)$ and $Y_2(\omega)$ and is often written as

$$(Y_1 * Y_2)(\omega) = \frac{1}{2\pi} \int_{-\infty}^{\infty} Y_1(\omega-\Omega)Y_2(\Omega)\,d\Omega \tag{A.60}$$

Because one could just as easily have projected out the variable Ω in Equation (A.58), one could equally well have

$$F[y_1(t)y_2(t)] = \frac{1}{2\pi} \int_{-\infty}^{\infty} Y_1(\alpha)Y_2(\omega-\alpha)\,d\alpha \tag{A.61}$$

and this, or Equation (A.60), is referred to as the *convolution theorem*. A very similar analysis shows that, for the inverse Fourier transform,

$$F^{-1}[Y_1(\omega)Y_2(\omega)] = \int_{-\infty}^{\infty} y_1(t-\tau)y_2(\tau)\mathrm{d}\tau$$

$$= \int_{-\infty}^{\infty} y_1(\tau)y_2(t-\tau)\mathrm{d}\tau \tag{A.62}$$

A.3.6   Parseval's Theorem

Having established the convolution theorem, one can consider the convolution of a frequency function with itself. One finds from Equation (A.60) that

$$(Y * Y)(\omega) = \frac{1}{2\pi} \int_{-\infty}^{\infty} Y(\omega - \Omega)Y(\Omega)\mathrm{d}\Omega$$

$$= F[y(t)y(t)] = \int_{-\infty}^{\infty} y(t)^2 e^{-i\omega t}\mathrm{d}t$$

Now, letting $\omega = 0$ gives

$$\frac{1}{2\pi} \int_{-\infty}^{\infty} Y(-\Omega)Y(\Omega)\mathrm{d}\Omega = \int_{-\infty}^{\infty} y(t)^2\mathrm{d}t \tag{A.63}$$

It is then simple to observe that

$$Y(-\Omega) = \int_{-\infty}^{\infty} y(t)e^{i\Omega t}\mathrm{d}t = \left(\int_{-\infty}^{\infty} y(t)e^{-i\Omega t}\mathrm{d}t\right)^* = Y(\Omega)^*$$

and substituting this result into Equation (A.63) gives

$$\frac{1}{2\pi} \int_{-\infty}^{\infty} Y(\Omega)^* Y(\Omega)\mathrm{d}\Omega = \int_{-\infty}^{\infty} y(t)^2\mathrm{d}t$$

or, finally,

$$\frac{1}{2\pi} \int_{-\infty}^{\infty} |Y(\Omega)|^2\mathrm{d}\Omega = \int_{-\infty}^{\infty} y(t)^2\mathrm{d}t \tag{A.64}$$

This is *Parseval's theorem*. It shows that if one defines the power of a function $z(.)$ as

$$P[z(.)] = \int_{-\infty}^{\infty} z(.)^2\mathrm{d}.$$

then power is (essentially, up to a global multiple) preserved by the Fourier transform.

A.3.7   The Effect of a Finite Time Window

An important practical observation now is that, in reality, signals are clearly never measured for times from $t = -\infty$ to $t = \infty$. One is forced to question the effect on the Fourier transform of only recording

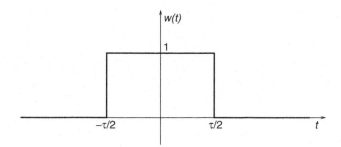

Figure A.32 A rectangular window function.

a finite segment τ of a signal. First of all, by shifting the origin of the time coordinate, one can assume that the signal is measured from $t = -\tau/2$ to $t = \tau/2$. As one cannot say what the signal is outside this interval, one can set it to zero and the Fourier integral becomes

$$X(\omega) = \int_{-\tau/2}^{\tau/2} x(t)\mathrm{e}^{-i\omega t}\,\mathrm{d}t \tag{A.65}$$

An alternate way of thinking about this integral is as

$$X(\omega) = \int_{-\infty}^{\infty} w(t)x(t)\mathrm{e}^{-i\omega t}\,\mathrm{d}t \tag{A.66}$$

where $w(t)$ is a so-called *window function*; in this case it is rectangular, as illustrated in Figure A.32.

Equation (A.66) is the Fourier transform of a product, so by the convolution theorem,

$$F[w(t)x(t)] = \frac{1}{2\pi}\int_{-\infty}^{\infty} W(\Omega)X(\omega - \Omega)\,\mathrm{d}\Omega \tag{A.67}$$

Now, the Fourier transform of $w(t)$ was computed (up to a constant scale) in Section A.3.4,

$$W(\Omega) = \frac{\sin\left(\frac{\Omega\tau}{2}\right)}{\frac{\Omega}{2}} \tag{A.68}$$

so

$$F[w(t)x(t)] = \frac{1}{2\pi}\int_{-\infty}^{\infty} \frac{\sin\left(\frac{\Omega\tau}{2}\right)}{\frac{\Omega}{2}}X(\omega - \Omega)\,\mathrm{d}\Omega \tag{A.69}$$

This is a general formula. In order to see the effect of the window, let $x(t)$ be a sine wave of frequency α,

$$x(t) = \sin(\alpha t)$$

with corresponding spectrum,

$$X(\omega) = \frac{\pi}{i}[\delta(\omega - \alpha) - \delta(\omega + \alpha)] \tag{A.70}$$

as shown graphically in Figure A.33.

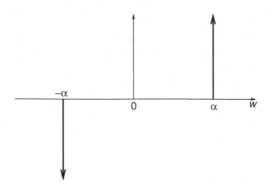

Figure A.33   The spectrum of the sine wave.

Equation (A.69) gives, in this case,

$$F[w(t)\sin(\alpha t)] = \frac{1}{2\pi} \int_{-\infty}^{\infty} \frac{\sin\left(\frac{\Omega\tau}{2}\right)}{\frac{\Omega}{2}} \frac{\pi}{i} [\delta(\omega - \Omega - \alpha) - \delta(\omega - \Omega + \alpha)] d\Omega \qquad (A.71)$$

The first integral in Equation (A.71) is

$$\frac{1}{2i} \int_{-\infty}^{\infty} \frac{\sin\left(\frac{\Omega\tau}{2}\right)}{\frac{\Omega}{2}} \delta(\omega - \Omega - \alpha) d\Omega = \frac{1}{2i} \frac{\sin\left(\frac{(\omega-\alpha)\tau}{2}\right)}{\frac{(\omega-\alpha)}{2}}$$

A similar result for the second integral in Equation (A.71) gives the final expression,

$$F[w(t)\sin(\alpha t)] = \frac{1}{2i} \left\{ \frac{\sin\left(\frac{(\omega-\alpha)\tau}{2}\right)}{\frac{(\omega-\alpha)}{2}} - \frac{\sin\left(\frac{(\omega+\alpha)\tau}{2}\right)}{\frac{(\omega+\alpha)}{2}} \right\} \qquad (A.72)$$

and, graphically, the spectrum is as shown in Figure A.34.

The peaks of the spectrum have width $4\pi/\tau$, which increases as τ gets smaller. This means that the effect of the rectangular window is to smear out in frequency the infinitesimally sharp peaks in the sine wave spectrum. In the limit as $\tau \to \infty$, one recovers the sharp characterisation of the frequencies.

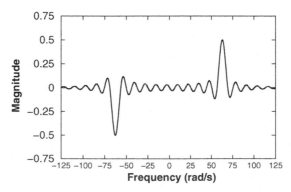

Figure A.34   The spectrum of the sine wave convolved with a rectangular window.

A.3.8   The Effect of Differentiation and Integration

It is useful to consider now the Fourier transform of the time derivative of a signal,

$$\dot{x}(t) = \frac{dx}{dt} \qquad (A.73)$$

In order to do this, the formula for integration by parts will be required,

$$\int u(t)v(t)dt = v(t)\left\{\int u(t)dt\right\} - \int \left\{\int u(t)dt\right\} \frac{dv}{dt}dt$$

Applying this formula to Equation (A.73) with $u(t) = dx/dt$ and $v(t) = e^{-i\omega t}$ gives

$$\int_{-\infty}^{\infty} \frac{dx}{dt}e^{-i\omega t}dt = [e^{-i\omega t}x(t)]_{-\infty}^{\infty} - \int_{-\infty}^{\infty} x(t)(-i\omega)e^{-i\omega t}dt$$
$$= [e^{-i\omega t}x(t)]_{-\infty}^{\infty} + (i\omega)X(\omega) \qquad (A.74)$$

It is, for the first time, now necessary to explain a restriction on the use of the Fourier transform. In order for a signal $x(t)$ to have a 'sensible' Fourier transform, it should have finite power, that is

$$\int_{-\infty}^{\infty} x(t)^2 dt = C < \infty \qquad (A.75)$$

For any real, physical signals, this will clearly be the case (actually, this constraint is not true for the sine wave, but in a sense there is no contradiction, because the Fourier transform of a sine wave is a delta function, which is not a 'sensible' function). Anyway, if one insists on Equation (A.75), it is necessary that

$$\lim_{t \to \pm\infty} x(t) = 0$$

and if this is true, then the first term in Equation (A.74) is zero. So for appropriately well-behaved functions $x(t)$, one has

$$F[\dot{x}(t)] = (i\omega)X(\omega) \qquad (A.76)$$

Therefore one has the result that differentiation in the time domain is equivalent to multiplication by a factor $i\omega$ in the frequency domain. (There is a quicker route to Equation (A.76); however, the route taken here conveniently fitted in the finite power constraint. The short-cut is left as an exercise for the reader.) It follows from Equation (A.76) that

$$F[x^{(n)}(t)] = F\left[\frac{d^n x}{dt^n}\right] = (i\omega)^n X(\omega) \qquad (A.77)$$

Now, let

$$z(t) = \int_{-\infty}^{t} x(\tau)d\tau$$

so that $\dot{z}(t) = x(t)$. This means, from Equation (A.76), that

$$F[\dot{z}] = X(\omega) = (i\omega)F[z(t)] = (i\omega)F\left[\int_{-\infty}^{t} x(\tau)d\tau\right]$$

or

$$F\left[\int_{-\infty}^{t} x(\tau)d\tau\right] = \frac{X(\omega)}{i\omega} \tag{A.78}$$

Therefore integration in the time domain is equivalent to division by $i\omega$ in the frequency domain.

A.4 Frequency Response Functions and the Impulse Response

A.4.1 Basic Definitions

Consider the single-degree-of-freedom (SDOF) system depicted in Figure A.35, and specified by the equation of motion,

$$m\ddot{y} + c\dot{y} + ky = x(t) \tag{A.79}$$

As one can see, this is specified in the time domain by a second-order differential equation which – depending on $x(t)$ – may be more or less difficult to solve. One asks, what about the frequency domain? Taking the Fourier transform of both sides of Equation (A.79) gives

$$F[m\ddot{y} + c\dot{y} + ky] = F[x(t)] \tag{A.80}$$

Applying the linearity property of the Fourier transform Equation (A.56) to Equation (A.80) gives

$$mF[\ddot{y}] + cF[\dot{y}] + kF[y] = F[x(t)] \tag{A.81}$$

and then applying Equation (A.77), the rule for differentiation of signals, yields

$$m(i\omega)^2 Y(\omega) + c(i\omega)Y(\omega) + kY(\omega) = X(\omega)$$

or

$$(-m\omega^2 + ic\omega + k)Y(\omega) = X(\omega) \tag{A.82}$$

So if one defines

$$H(\omega) = \frac{1}{-m\omega^2 + ic\omega + k} \tag{A.83}$$

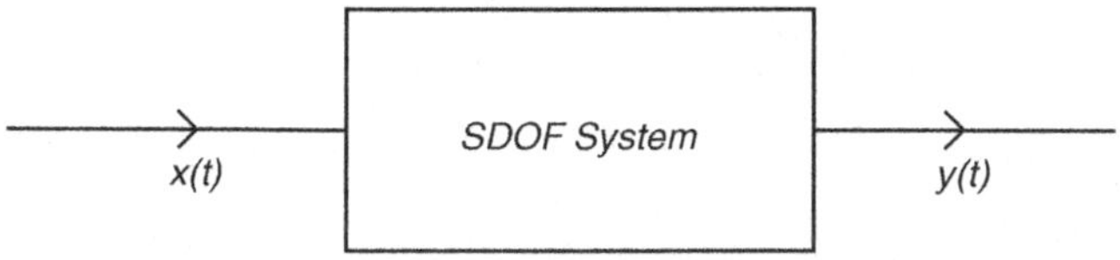

Figure A.35 Single-input–single-output SDOF system.

one finally obtains

$$Y(\omega) = H(\omega)X(\omega) \tag{A.84}$$

and this is a very painless way of defining the *frequency response function* (FRF) $H(\omega)$. This is a remarkable result; a time-domain system, specified by a differential equation in the time domain, only needs the multiplication operator for solution in the frequency domain.

Consider Equation (A.84). If one applies the inverse Fourier transform, one obtains, by the convolution theorem,

$$y(t) = \int_{-\infty}^{\infty} h(\tau)x(t - \tau)\mathrm{d}\tau \tag{A.85}$$

where

$$h(t) = F^{-1}[H(\omega)] \tag{A.86}$$

The obvious question which arises now is – does the time-domain function $h(t)$ have any special meaning? In order to explore this, let $x(t) = \delta(t)$; then, from Equation (A.85),

$$y(t) = \int_{-\infty}^{\infty} h(\tau)\delta(t - \tau)\mathrm{d}\tau = h(t)$$

so $h(t)$ is the solution of

$$m\ddot{h} + c\dot{h} + kh = \delta(t) \tag{A.87}$$

and for this reason is called the *impulse response*.

A.4.2   Harmonic Probing

At this point, it is timely to derive another relationship that will be useful later.

Suppose the input to the SDOF system described above was $x(t) = \mathrm{e}^{i\omega t}$. (Of course, this is impossible in practice as $x(t)$ is complex.) The response of the linear system is

$$
\begin{aligned}
y(t) &= \int_{-\infty}^{\infty} h(\tau)x(t - \tau)\mathrm{d}\tau = \int_{-\infty}^{\infty} h(\tau)\mathrm{e}^{i\omega(t-\tau)}\mathrm{d}\tau \\
&= \mathrm{e}^{i\omega t} \int_{-\infty}^{\infty} h(\tau)\mathrm{e}^{i\omega\tau}\mathrm{d}\tau = H(\omega)\mathrm{e}^{i\omega t}
\end{aligned} \tag{A.88}
$$

Therefore if the input to a linear system is the *harmonic probe* $\mathrm{e}^{i\omega t}$, the output is simply $\mathrm{e}^{i\omega t}$ multiplied by the FRF evaluated at ω.

In fact, the important relations above, Equations (A.84), (A.85) and (A.88), hold for any linear input–output system – SDOF or not. The proofs are not difficult but they will not be presented here.

A.5   The Discrete Fourier Transform

A.5.1   Basic Definitions

The Fourier representations for periodic and aperiodic continuous functions of time have been constructed in various of the previous sections. There has also been an investigation as to what happens when a signal has a finite duration. In reality, not only will the signal be of finite duration, it will be *sampled*. It is not possible to store a continuous function on a computer (although one can store continuous functions in other ways – vinyl records, compact cassettes, etc.) as it would require infinite memory. What one usually does is take measurements from the signal at regular intervals – say Δt seconds apart – so the signal for manipulation takes the form of a finite vector of N samples,

$$\{x_0, \ldots, x_{N-1}\} \quad \text{where } x_i = x(t_i) = x(t_0 + i \, \Delta t) \tag{A.89}$$

where t_0 is a reference time. If one takes $t_0 = 0$ from now on, one will have $t_i = i \, \Delta t$. The question is now: how do we compute the spectrum of such a signal? It is necessary to establish the *discrete Fourier transform* (DFT).

Recall the exponential form of the Fourier series. The spectral coefficients are

$$c_n = \frac{1}{\tau} \int_{-\tau/2}^{\tau/2} x(t) e^{-\frac{2\pi i n}{\tau} t} dt$$

In keeping with the notation for the Fourier transform established earlier, it will be convenient to relabel c_n by X_n from now on. Also, the equation above is not in the most convenient form for the analysis, so it will be modified slightly.

Recall that in the case of the Fourier series, $x(t)$ is assumed periodic with period τ. If one considers the integral

$$\frac{1}{\tau} \int_{-\tau/2}^{0} x(t) e^{-\frac{2\pi i n}{\tau} t} dt$$

and lets $t' = t + \tau$, the integral becomes

$$\frac{1}{\tau} \int_{\tau/2}^{\tau_0} x(t' - \tau) e^{-\frac{2\pi i n}{\tau} (t' - \tau)} dt$$

Now, $x(t') = x(t' - \tau)$ by periodicity; also,

$$e^{-\frac{2\pi i n}{\tau} (t' - \tau)} = e^{-\frac{2\pi i n}{\tau} t} e^{-2\pi i n} = e^{-\frac{2\pi i n}{\tau} t}$$

so

$$\frac{1}{\tau} \int_{-\tau/2}^{0} x(t) e^{-\frac{2\pi i n}{\tau} t} dt = \frac{1}{\tau} \int_{\tau/2}^{\tau} x(t) e^{-\frac{2\pi i n}{\tau} t} dt$$

Now as

$$X_n = \frac{1}{\tau} \int_{-\tau/2}^{0} x(t) e^{-\frac{2\pi i n}{\tau} t} dt + \frac{1}{\tau} \int_{0}^{\tau/2} x(t) e^{-\frac{2\pi i n}{\tau} t} dt$$

$$= \frac{1}{\tau} \int_{\tau/2}^{\tau'} x(t) e^{-\frac{2\pi i n}{\tau} t} dt + \frac{1}{\tau} \int_{0}^{\tau/2} x(t) e^{-\frac{2\pi i n}{\tau} t} dt$$

one finally arrives at

$$X_n = \frac{1}{\tau} \int_0^\tau x(t) e^{-\frac{2\pi i n}{\tau} t} dt \tag{A.90}$$

As one now only has the $x(t)$ sampled at the discrete times $t_r = r \Delta t$, one is forced to approximate the integral (A.90) for the spectrum by a rectangular sum,

$$X_n = \frac{1}{\tau} \sum_{r=0}^{N-1} x(t_r) e^{-i \frac{2\pi n}{\tau} t_r} \Delta t = \frac{1}{\tau} \sum_{r=0}^{N-1} x_r e^{-i \frac{2\pi n}{\tau} r \Delta t} \Delta t \tag{A.91}$$

and as $\tau = N \Delta t$, this becomes

$$X_n = \frac{1}{N} \sum_{r=0}^{N-1} x_r e^{-i \frac{2\pi n}{N} r} \tag{A.92}$$

As the process began with only N independent quantities x_r, it should be clear that one can only derive N independent spectral lines at most. This means that there must be relations between the X_n. The simplest one to demonstrate is periodicity. Consider

$$X_{n+N} = \frac{1}{N} \sum_{r=0}^{N-1} x_r e^{-i \frac{2\pi(n+N)}{N} r}$$

$$= \frac{1}{N} \sum_{r=0}^{N-1} x_r e^{-i \frac{2\pi n}{N} r} e^{-i 2\pi r} = \frac{1}{N} \sum_{r=0}^{N-1} x_r e^{-i \frac{2\pi n}{N} r} = X_n$$

In summary,

$$X_{n+N} = X_n \tag{A.93}$$

and it is confirmed that there are at most N independent spectral lines. In fact, there must actually be less than this as the X_n are complex quantities. Given N real numbers, we can only specify $N/2$ independent complex numbers.

Consider now the exponent in Equation (A.92) $- i(2\pi n / \tau)t$. If this is to be identified with the exponent $i\omega_n t$ of the Fourier transform, one must have

$$\omega_n = n \Delta\omega = \frac{2\pi n}{\tau}$$

or

$$\Delta\omega = \frac{2\pi}{N \Delta t} \tag{A.94}$$

Alternatively, if one specifies the frequency spacing in Hz, one sees

$$\Delta f = \frac{1}{N \Delta t} \tag{A.95}$$

When $n = 0$, the relevant spectral line is given by

$$X_0 = \frac{1}{N} \sum_{r=0}^{N-1} x_r \qquad (A.96)$$

which can be identified as the arithmetic mean or DC component of the signal. Therefore X_0 corresponds to the frequency $\omega = 0$ as one might expect. Given the fixed spacing of the spectral lines from Equation (A.95), this means that the highest frequency one can represent at all is

$$\frac{N}{2} \Delta\omega = \frac{1}{2\Delta t} = \frac{f_s}{2}$$

where f_s is the sampling frequency. This frequency is very important in signal processing and is called the *Nyquist frequency*. This argument says that only the first half of the spectral lines are independent – so what is the meaning of the second half?

Consider

$$X_{N-1} = \frac{1}{N} \sum_{r=0}^{N-1} x_r \mathrm{e}^{-i\frac{2\pi(N-1)}{N}r} = \frac{1}{N} \sum_{r=0}^{N-1} x_r \mathrm{e}^{-i2\pi r} \mathrm{e}^{+i\frac{2\pi}{N}r}$$

However, r is an integer, so $\mathrm{e}^{-i2\pi r} = 1$, which means that

$$X_{N-1} = \frac{1}{N} \sum_{r=0}^{N-1} \mathrm{e}^{+i\frac{2\pi}{N}r} = \left(\frac{1}{N} \sum_{r=0}^{N-1} \mathrm{e}^{-i\frac{2\pi}{N}r} \right)^* = X_1^*$$

A similar argument shows that, generally,

$$X_{N-k} = X_k^* \qquad (A.97)$$

Now, if one recalls that it is a property of the Fourier transform that

$$X(\omega)^* = X(-\omega)$$

This means that the spectral coefficient X_{N-1} corresponds to the frequency $-\Delta\omega$ or, more generally, X_{N-k} corresponds to the frequency $-k\Delta\omega$. Therefore the array X_n stores the spectral/frequency representation of the signal x_r in the manner shown in Figure A.36.

A question that occurs now is: what happens if there is a value of k for which $N - k = k$? The answer is that this implies that $N = 2k$ so the number of sample points is even. Actually, this turns out to be the most usual situation; in fact it is a requirement of the *fast Fourier transform*, which will be discussed a little later. In the event of an even number of points, one has, by Equation (A.97),

$$X_{N/2}^* = X_{N/2} \qquad (A.98)$$

Figure A.36   Storage of spectral lines in a DFT array.

so that the spectral line is real. This finally justifies the earlier assertion that the maximum frequency represented in the spectrum is $N\Delta\omega/2$. The zero line is real and that uses up one degree of freedom. The next $N/2 - 1$ lines are independent and use up $N - 2$ degrees of freedom as they are complex. Finally, the line at $N/2$ is independent but real. The total number of independent reals is therefore N, as required, and the maximum frequency represented (but incompletely) is $N\Delta\omega/2$.

Thus the discrete Fourier transform is defined and there is understanding as to which frequencies correspond to the spectral components. Although initially the transform came from an approximation, it would be useful if it had an exact inverse and one could thus reconstruct the original signal $x_r, r = 1, \ldots, N$ from the spectra X_n.

This turns out to be possible, but one needs another orthogonality relation,

$$\sum_{r=0}^{N-1} e^{i\frac{2\pi r}{N}p} e^{-i\frac{2\pi r}{N}q} = N\delta_{pq} \tag{A.99}$$

which is proved as follows. The left-hand side of Equation (A.99) can be rewritten as follows:

$$\sum_{r=0}^{N-1} e^{i\frac{2\pi r}{N}(p-q)} \tag{A.100}$$

and this is a geometric progression with the first term $a = 0$ and ratio

$$\alpha = e^{i\frac{2\pi}{N}(p-q)}$$

It is basic mathematics that the sum of this progression is

$$a\frac{1-\alpha^N}{1-\alpha} = \frac{1-e^{i2\pi(p-q)}}{1-e^{\frac{i2\pi(p-q)}{N}}}$$

Now if p and q are distinct integers, then

$$e^{i2\pi(p-q)} = 1 \quad \text{and} \quad e^{\frac{i2\pi(p-q)}{N}} \neq 1$$

and the sum in Equation (A.100) is zero. If $p = q$, matters are a little more subtle and one has to use a limiting argument. Let $s = p - q$ be small and expand the sum of the progression as a Taylor series with $e^x = 1 + x + O(x^2)$,

$$\frac{1-e^{i2\pi s}}{1-e^{\frac{i2\pi s}{N}}} = \frac{1-[1+i2\pi s+O(s^2)]}{1-[1+i2\pi s/N+O(s^2)]}$$

$$= \frac{i2\pi s}{i2\pi s/N} + O(s^2) = N + O(s^2)$$

so in the limit $p \to q$ or $s \to 0$, the sum of the progression is N and this is enough to prove Equation (A.99). The required inverse is now almost in sight; one needs the discrete form of the Fourier inversion theorem,

$$x_n = \sum_{r=0}^{N-1}\left\{\frac{1}{N}\sum_{p=0}^{N-1} x_p e^{-i\left(\frac{2\pi r p}{N}\right)}\right\} e^{i\left(\frac{2\pi r n}{N}\right)} \tag{A.101}$$

Consider the right-hand side of Equation (A.101). Assuming that one is allowed to change the order of summations, one obtains

$$\frac{1}{N} \sum_{p=0}^{N-1} x_p \left\{ \sum_{r=0}^{N-1} e^{-i\left(\frac{2\pi pr}{N}\right)} e^{i\left(\frac{2\pi rn}{N}\right)} \right\}$$

but, by the orthogonality relation (A.99), this is

$$\frac{1}{N} \sum_{p=0}^{N-1} x_p \left\{ N \delta_{pn} \right\} = x_n$$

which establishes the required form of the inversion theorem (A.101). This means that if one takes

$$X_n = \frac{1}{N} \sum_{r=0}^{N-1} x_r e^{-i\frac{2\pi nr}{N}}$$

as the definition of the forward discrete Fourier transform (setting aside the fact that this is an approximation to the continuous Fourier transform), then an *exact* inverse is

$$x_r = \sum_{n=0}^{N-1} X_n e^{i\frac{2\pi rn}{N}}$$

In practice, it is rather more convenient to have the factor $1/N$ on the inverse transform (c.f. the 2π in the continuous FT); so the usual definitions of the discrete Fourier transform and inverse are

$$X_n = \sum_{r=0}^{N-1} x_r e^{-i\frac{2\pi nr}{N}} \tag{A.102}$$

$$x_r = \frac{1}{N} \sum_{n=0}^{N-1} X_n e^{i\frac{2\pi rn}{N}} \tag{A.103}$$

A.5.2 More About Sampling

It was established in the last section that the highest frequency one can represent from a signal is the *Nyquist frequency* $f_s/2$. A critical question now is: what happens if the original signal actually has a component above this frequency? In order to answer this, consider the operation of sampling. One might regard this as a switch closing to collect signal energy for ε seconds every Δt seconds, as depicted in Figure A.37.

In the limit as $\varepsilon \to 0$, one obtains a train of impulses; in that case one can write the sampling function as a sequence of delta-function windows,

$$\delta_\tau(t) = \sum_{n=-\infty}^{\infty} \delta(t - n\tau) \tag{A.104}$$

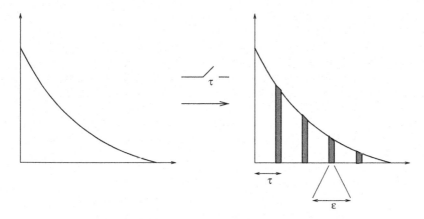

Figure A.37 A model of a sampling process.

and the corresponding sampled signal is therefore

$$x_s(t) = x(t) \sum_{n=\infty}^{\infty} \delta(t - n\tau)$$

or

$$x_s(t) = \sum_{n=-\infty}^{\infty} x_n \delta(t - n\tau) \qquad (A.105)$$

One proceeds by expanding $\delta_\tau(t)$ as a Fourier series,

$$\delta_\tau(t) = \sum_{n=-\infty}^{\infty} c_n e^{in\omega_s t}$$

where

$$c_n = \frac{1}{\tau} \int_0^\tau \delta_\tau(t) e^{-in\omega_s t} \, dt$$

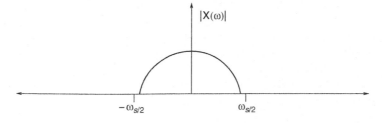

Figure A.38 Spectrum of a continuous signal $x(t)$.

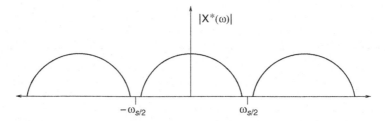

Figure A.39   Spectrum of a sampled signal $x_s(t)$.

and ω_s is the sampling frequency in radians. Substituting (A.104) into the integral for the coefficients gives

$$c_n = \frac{1}{\tau} \int_0^\tau \left\{ \sum_{k=-\infty}^\infty \delta(t - k\tau) \right\} e^{-in\omega_s t} \, dt$$

and, assuming one can change the order of the integration and the summation, one obtains

$$c_n = \frac{1}{\tau} \sum_{k=-\infty}^\infty \left\{ \int_0^\tau \delta(t - k\tau) e^{-in\omega_s t} \, dt \right\}$$

Now, only one of the integrals ever has the argument of the delta function inside the range of integration, so only one integral ever contributes to the sum. Therefore

$$c_n = \frac{1}{\tau}$$

This means that, finally, one obtains

$$\delta_\tau(t) = \frac{1}{\tau} \sum_{k=-\infty}^\infty e^{in\omega_s t} \tag{A.106}$$

and so the sampled signal $x_s(t)$ can be written as

$$x_s(t) = \frac{1}{\tau} \sum_{k=-\infty}^\infty x(t) e^{in\omega_s t} \tag{A.107}$$

Now to get the spectrum of the sampled signal, one takes the Fourier transform,

$$X_s(\omega) = \int_{-\infty}^\infty \left\{ \frac{1}{\tau} \sum_{k=-\infty}^\infty x(t) e^{in\omega_s t} \right\} e^{-i\omega t} \, dt$$

Again assuming one can interchange the sums and integrals, one finds

$$X_s(\omega) = \frac{1}{\tau} \sum_{k=-\infty}^\infty \left\{ \int_{-\infty}^\infty x(t) e^{-i(\omega - n\omega_s)t} \, dt \right\}$$

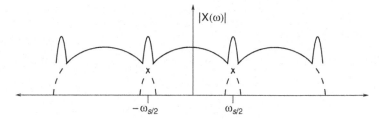

Figure A.40 Spectrum of sampled signal $x_s(t)$ with overlap – aliasing.

Or, finally,

$$X_s(\omega) = \frac{1}{\tau} \sum_{k=-\infty}^{\infty} X(\omega - n\omega_s) \qquad (A.108)$$

This means that the spectrum of the underlying continuous signal is repeated indefinitely. This is partly the periodicity that has already been established, but with this analysis one sees another phenomena. Consider Figures A.38 and A.39, which show fictitious examples of the spectra for corresponding continuous and sampled signals. If the spectrum of the continuous signal extends beyond the Nyquist frequency, one sees overlap as in Figure A.40. If any energy in the original signal extends beyond the Nyquist frequency, it is folded back into the Nyquist interval in the spectrum of the sampled signal. This folding is called *aliasing* (Hamming, 1999).

A.5.3 The Fast Fourier Transform

Recall the definition of the discrete Fourier transform from the last section,

$$X_n = \sum_{r=0}^{N-1} x_r e^{-i \frac{2\pi n}{N} r} \qquad (A.109)$$

For many years, the direct use of this transform was restricted by the expense of the computation. For each spectral line X_n, one has to carry out N function evaluations (the exponential) and N multiplications. In order to obtain the whole spectrum, N^2 multiplications and N^2 function evaluations are required. Concentrating on the multiplications for now, the algorithm is said to be of order N^2 or $O(N^2)$. The situation now is that multiplications are little, if any, more expensive than additions. Historically, however, the need to reduce the number of multiplications was what motivated the development of the *fast Fourier transform* (FFT) (see, for example, Press *et al.*, 2007).

In order to explain the FFT algorithm it is convenient to rewrite Equation (A.109) a little, so

$$X_n = \sum_{r=0}^{N-1} x_r w^{rn} \qquad (A.110)$$

where

$$w = e^{-\frac{2i\pi}{N}} \qquad (A.111)$$

Now, assume that N is divisible by two; in that case, the summation in Equation (A.110) can be divided into two sums: one over the even indices of x_r and the other over the odd,

$$X_n = \sum_{r=0}^{N/2-1} x_{2r} w^{2rn} + \sum_{r=0}^{N/2-1} x_{2r+1} w^{(2r+1)n}$$

or

$$X_n = \sum_{r=0}^{N/2-1} x_{2r} w^{2rn} + w^n \sum_{r=0}^{N/2-1} x_{2r+1} w^{2rn} \tag{A.112}$$

The first summation in Equation (A.112) is simply the DFT of the even-indexed x_r, that is the set $\{x_0, x_2, \ldots, x_{N-2}\}$:

$$E_n = \sum_{r=0}^{N/2-1} x_{2r} w^{2rn} \tag{A.113}$$

The second summation in Equation (A.112) contains the DFT of the odd-indexed x_r, that is the set $\{x_1, x_3, \ldots, x_{N-1}\}$:

$$O_n = \sum_{r=0}^{N/2-1} x_{2r+1} w^{2rn} \tag{A.114}$$

The DFT of the N-point sequence is thus expressible in terms of the DFTs of two $N/2$-point sequences by

$$X_n = E_n + w^n O_n \tag{A.115}$$

Note that as E_n and O_n are $N/2$-point DFTs, the periodicity relation for the DFT shows that

$$E_{n+N/2} = E_n \quad \text{and} \quad O_{n+N/2} = O_n \tag{A.116}$$

In order to evaluate a given spectral line, one now has two $N/2$-point DFTs, each contributing $N^2/4$ multiplications. Adding in the multiplication by w^n, the operation (multiplication) count for the spectrum is $N^2/2 + N$. Neglecting the N compared to the N^2, the new operation count is $O(N^2/2)$; the number of multiplications has halved!

Now, further assume that $N/2$ is itself divisible by two. One can repeat the previous operation and express each of the two $N/2$-point DFTs in terms of two (each) $N/4$-point DFTs and further reduce the number of multiplications by a factor of two. One is drawn to a logical conclusion: if $N = 2^M$ (M is called the *radix*), this procedure can be repeated until one finally obtains N one-point DFTs, which are trivial to estimate, that is

$$X_0 = x_0$$

and there are no multiplications.

What then is the total computational effort? Well, there are M stages, where $M = \log_2 N$. At each stage, one is required to carry out N multiplications. The total cost of the algorithm is therefore $N \log_2 N$. The saving in moving from the DFT to the FFT is

$$\frac{N^2}{N \log_2 N} = \frac{N}{M}$$

To give an example, suppose that $N = 1024$. For the FFT, one achieves a saving of computational effort of the order $100 : 1$, and this saving increases with N. This saving makes the FFT one of the cleverest numerical algorithms of the 20th century.

A.5.3.1 An Example

In order to see how the FFT algorithm works, it will be ueseful to work through an example in some detail. Consider the eight-point FFT of the sampled time sequence $\{x_0, x_1, \ldots, x_7\}$:

$$X_n = \sum_{r=0}^{7} x_r w^{rn} \tag{A.117}$$

where, in this case,

$$w = e^{-\frac{i\pi}{4}} \tag{A.118}$$

Equation (A.117) is divided according to Equation (A.115), such that

$$E_n = \sum_{r=0}^{3} x_{2r} w^{2rn} \quad \text{and} \quad O_n = \sum_{r=0}^{3} x_{2r+1} w^{2rn} \tag{A.119}$$

Now, taking note of the periodicity (expressed most recently in Equation A.116), one has explicitly

$$
\begin{aligned}
X_0 &= E_0 + w^0 O_0 \\
X_1 &= E_1 + w^1 O_1 \\
X_2 &= E_2 + w^2 O_2 \\
X_3 &= E_3 + w^3 O_3 \\
X_4 &= E_4 + w^4 O_4 = E_0 + w^4 O_0 \\
X_5 &= E_5 + w^5 O_5 = E_1 + w^5 O_1 \\
X_6 &= E_6 + w^6 O_6 = E_2 + w^6 O_2 \\
X_7 &= E_7 + w^7 O_7 = E_3 + w^7 O_3
\end{aligned}
\tag{A.120}
$$

The information summarised in the last equation can be shown nicely in a diagram as shown in Figure A.41, sometimes called an *FFT butterfly*.

For the second phase, it will be sufficient to concentrate on the decomposition of the E_n. In this case, the 'even' and 'odd' parts of the sequence are $\{x_0, x_4\}$ and $\{x_2, x_6\}$ respectively. Now let

$$E_n = EE_n + w^{2n} OE_n \tag{A.121}$$

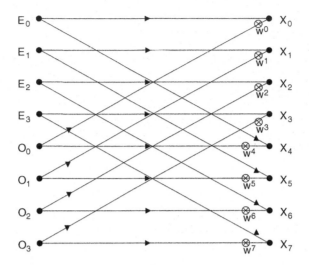

Figure A.41 First phase of an FFT 'butterfly'.

where

$$EE_n = \sum_{r=0}^{1} x_{4r} w^{4rn} \quad \text{and} \quad OE_n = \sum_{r=0}^{1} x_{4r+2} w^{4rn} \tag{A.122}$$

and the detailed breakdown for this stage (using periodicity with period $N/4 = 2$) is

$$\begin{aligned}
E_0 &= EE_0 + w^0 OE_0 \\
E_1 &= EE_1 + w^2 OE_1 \\
E_2 &= EE_2 + w^4 OE_2 = EE_0 + w^4 OE_0 \\
E_3 &= EE_3 + w^6 OE_3 = EE_1 + w^6 OE_1
\end{aligned} \tag{A.123}$$

The corresponding part of the butterfly is shown in Figure A.42.

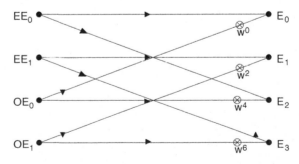

Figure A.42 Part of the second phase of the FFT butterfly.

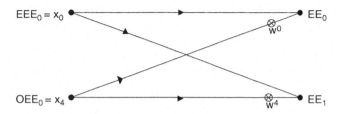

Figure A.43   One of the final parts/stages of the FFT butterfly.

For this particular size of FFT there is only one more stage now ($\log_2 N = 3$). Consider EE_0; this is divided in two DFTs of length $N/8 = 1$ such that

$$EE_n = EEE_n + w^{4n} OEE_n \tag{A.124}$$

so that

$$EE_0 = EEE_0 + w^0 OEE_0 = x_0 + x_4$$
$$EE_1 = EEE_1 + w^4 OEE_1 = EEE_0 + w^4 OEE_0 = x_0 + w^4 x_4$$

and the butterfly reaches its lowest level, as expressed by Figure A.43.

The full butterfly will not be shown here (its construction would make an excellent exercise for the reader); if one did, one would obtain the following final correspondence between time points and spectral lines:

$$
\begin{aligned}
EEE_0 &= x_0 \\
OEE_0 &= x_4 \\
EOE_0 &= x_2 \\
OOE_0 &= x_6 \\
EEO_0 &= x_1 \\
OEO_0 &= x_5 \\
EOO_0 &= x_3 \\
OOO_0 &= x_7
\end{aligned}
\tag{A.125}
$$

This is the order that the original sequence has to be shuffled into to get the most out of the algorithm. It would clearly be nice if there was some systematic way of doing the shuffling, and in fact there is. Suppose one takes a given symbol sequence, say OEO, and interprets it as a binary number with the O's as 1's and the E's as 0's. The sequence OEO above becomes 101 and when this is converted to decimal, it becomes 5, which is the time index corresponding to OEO in the above table of (A.125).

After a little thought this makes sense. At the first stage of decomposition, if a number was even and therefore became part of E_n it would end with 0 as the least significant bit. If the number were odd, the binary sequence would end with 1 and the element would become part of O_n. If the next significant bit ends with 0, the element is assigned to EE_n and so on. In the full FFT butterfly, the correspondence between the leftmost and rightmost columns for the eight-point example would be

$$
\begin{aligned}
x_0 &\rightarrow X_0 \\
x_4 &\rightarrow X_1 \\
x_2 &\rightarrow X_2 \\
x_6 &\rightarrow X_3 \\
x_1 &\rightarrow X_4 \\
x_5 &\rightarrow X_5 \\
x_3 &\rightarrow X_6 \\
x_7 &\rightarrow X_7
\end{aligned}
\tag{A.126}
$$

This correspondence is obtained by a process called *bit reversal*. In order to find which x_r corresponds to which X_n, one can use the following procedure. Suppose one wishes to find the sixth element in the shuffled sequence; first convert the 6 to binary, that is 110; reverse the bits in the binary number, in this case one obtains 011; finally, convert back to decimal – 3. Sure enough, as one can see above, X_6 corresponds to x_3.

It is instructive to consider a computer implementation of the FFT algorithm in order to see the complete elegance of the approach; an extremely compact representation can be found in Press *et al.* (2007), for example.

A.5.4   The DFT of a Sinusoid

So far, the discussion has established a discrete version of the Fourier transform and also described an efficient way of computing it. Now it would be interesting and instructive to see what the DFT of a specific signal looks like. (Of course, there will be no difference to the FFT; the latter is simply an effective means of computing the former.) As always, it is an excellent idea to start with something simple – a sinusoid of frequency f Hz will be considered,

$$
x(t) = A\sin(2\pi f t)
\tag{A.127}
$$

The corresponding sampled data will be labelled as

$$
x_j = A\sin(2\pi f t_j) = A\sin(2\pi f j \Delta t)
\tag{A.128}
$$

In this case the DFT is

$$
X_n = \sum_{r=0}^{N-1} A\sin(2\pi f r \Delta t) e^{-i\frac{2\pi n}{N}r}
\tag{A.129}
$$

or, in terms of exponentials,

$$
\begin{aligned}
X_n &= \frac{A}{2i} \sum_{r=0}^{N-1} (e^{2\pi i f r \Delta t} - e^{-2\pi i f r \Delta t}) e^{-i\frac{2\pi n}{N}r} \\
&= \frac{A}{2i} \left\{ \sum_{r=0}^{N-1} e^{2\pi i r(f\Delta t - \frac{n}{N})} - \sum_{r=0}^{N-1} e^{2\pi i r(-f\Delta t - \frac{n}{N})} \right\}
\end{aligned}
$$

Now, one notes that

$$f \Delta t = \frac{f}{\Delta f N} = \frac{\alpha}{N}$$

where $\alpha = f/\Delta f$; this is an integer *only* if the frequency of the sine wave coincides with one of the spectral lines (this is an important observation). In terms of α one has

$$X_n = \frac{A}{2i} \left\{ \sum_{r=0}^{N-1} e^{i \frac{2\pi(\alpha-n)}{N} r} - \sum_{r=0}^{N-1} e^{i \frac{2\pi(-\alpha-n)}{N} r} \right\}$$

Now, it was established earlier (essentially in Equation (A.99)) that

$$\sum_{r=0}^{N-1} e^{i \frac{2\pi(\alpha-n)}{N} r} = N \delta_{\alpha n}$$

if α is an integer between 0 and $N/2$. In this case the spectrum is simply

$$X_n = \frac{AN}{2i} (\delta_{\alpha,n} - \delta_{-\alpha,n}) \qquad \text{(A.130)}$$

(where commas have been added to the Kronecker deltas for clarification). Now, if α is an integer, there are only two nonzero spectral lines at $\pm f$. If α is *not* an integer, then the summations are not delta functions and the spectrum is nonzero everywhere – but does turn out to be peaked in the vicinity of $\pm f$. If the frequency of the signal does not coincide with one of the spectral lines in the DFT, the energy of the signal is distributed over several lines; the spectrum is not sharp with well-defined lines.

Another way of looking at this is that if α is an integer between 0 and $N/2$, it actually follows that the sine wave is periodic over the window of the DFT and the DFT will correctly resolve that only one frequency is present. If α is not an integer, then the sine wave is not periodic over the window and other frequencies will appear. This phenomenon of energy spreading over several spectral lines if the signal is not periodic over the window is termed *leakage*.

A.6 Practical Matters: Windows and Averaging

A.6.1 Windows

The last section argued that if a sine wave is not periodic over the Fourier window (the overall time interval taken for the transform), the energy in the spectrum will be spread over several lines, this being the phenomenon of *leakage*. This can also be considered in a slightly different way, which conveys further information. Suppose that one is required to transform a periodic signal, but the period of the signal means that it does not repeat perfectly over the given sampling window; the situation is shown schematically in Figure A.44.

The assumption inherent in the DFT is that the signal repeats outside the window with period τ and therefore the DFT will provide the spectral representation of the signal as shown in Figure A.45. Periodicity is *enforced*. However, the new signal has other frequency components. In particular, the spectrum computed will have high-frequency components associated with the sharp transition between the signal segments. Such high-frequency components are clearly spurious in terms of the original signal of interest.

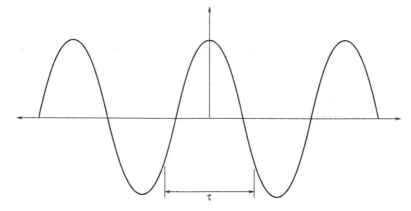

Figure A.44   Period signal not periodic over the Fourier window.

One can think of the spectrum as being the convolution of the true sine wave spectrum, with the spectrum of a rectangular window – as in Section A.3.6. The convolution function has been established earlier as

$$W(\omega) = \frac{\sin\left(\frac{\omega\tau}{2}\right)}{\frac{\omega}{2}} \qquad (A.131)$$

The smearing of the frequencies produced by the rectangular window is illustrated in Figure A.34. A good question at this point is: are there better windows than the rectangular, that is with a narrower convolution function, so that one gets better frequency definition or resolution in the frequency domain? One possibility is to use a window that is zero at both ends and thus forces periodicity; one of the most well-known and often used possibilities is the *Hanning window*,

$$w(t) = 1 - \cos\left(\frac{2\pi t}{\tau}\right) \qquad (A.132)$$

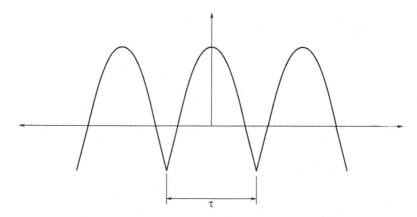

Figure A.45   Perceived periodic signal as enforced by the sampling window.

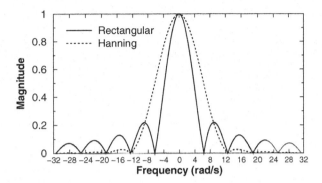

Figure A.46   Comparison of convolution functions for the rectangular and Hanning windows (linear scale).

The spectrum of this window is a little more complicated than that for the rectangular window; one finds

$$W(\omega) = \frac{i(e^{-i\omega t} - 1)}{\omega} - \frac{i(e^{-i(\omega - \omega_\tau)t} - 1)}{2(\omega - \omega_t)} - \frac{i(e^{-i(\omega + \omega_\tau)t} - 1)}{2(\omega + \omega_t)} \tag{A.133}$$

where $\omega_\tau = 2\pi/\tau$. A comparison of the Hanning window spectrum with that of the rectangular window is given in Figure A.46 or on a logarithmic scale in Figure A.47.

There are two major differences. First, the width of the central peak for the rectangular window is a little smaller; in fact it can be shown that there is no window with a narrower main lobe. The other significant difference is that the rectangular window has large side lobes. The question is: how does this affect any signal processing?

A.6.2   The Harris Test

Consider the problem of separating two frequency components, one small in amplitude; that is, suppose one has a signal like

$$x(t) = \sin\left(2\pi\left[\frac{10 f_s}{N}\right]\right) + 0.01\sin\left(2\pi\left[\frac{16 f_s}{N}\right]\right) \tag{A.134}$$

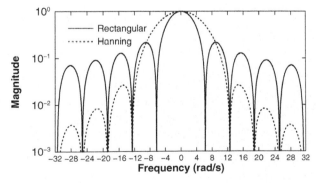

Figure A.47   Comparison of convolution functions for the rectangular and Hanning windows (logarithmic scale).

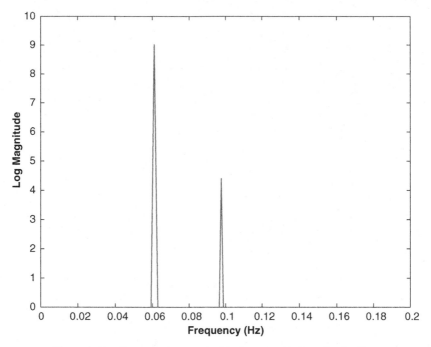

Figure A.48   Spectrum (log magnitude) of the signal in Equation (A.134).

which has been designed here specifically so that the two frequencies will both coincide with spectral lines under the DFT. For the simulation below, the parameters $N = 16\,384$ and $f_s = 1000$ have been specified; Figure A.48 shows the FFT of the signal with no leakage. No window has been used here; that is, the window is basically a rectangular one.

Figure A.49 shows the results of exactly the same calculation except that the factor 10 in the first sine function in Equation (A.134) is changed to 10.5, thus placing its frequency mid-way between two frequency lines. Again, the window is rectangular by default. The small change is extremely destructive to the computed spectrum, and the smaller peak can no longer be resolved.

The third figure (Figure A.50) in this sequence shows the FFT of the leaky signal when a Hanning window is first applied to the signal. Note that the Hanning window allows the resolution of the small component. This shows that the problem here was not the width of the main lobe of the rectangular window, but the magnitude of the sidebands.

There are dozens of different windows in use, each with its own domain of applicability. The important thing here is to recognise that they are available.

A.6.3   Averaging and Power Spectral Density

Suppose that now one wishes to estimate the spectrum of a random signal. In the limit as $\tau \to \infty$ one would get an accurate spectrum, but for τ finite, one has a problem. Any finite realisation of a random process will not exactly represent the long-term frequency content, precisely because it is random. Fortunately, there is a (partial) solution to this problem.

Assuming no problems with aliasing, one will find

$$F[x_\tau(t)] = X(\omega) + \epsilon_\tau(\omega) \tag{A.135}$$

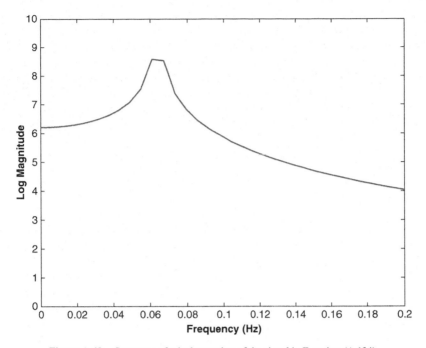

Figure A.49   Spectrum of a leaky version of the signal in Equation (A.134).

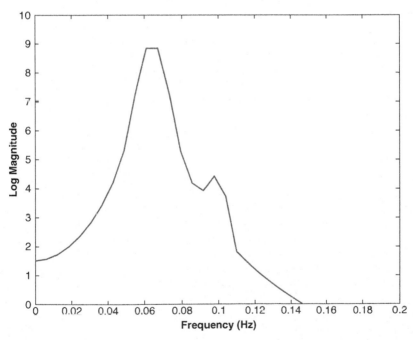

Figure A.50   Spectrum of a leaky version of the signal in Equation (A.134); Hanning window applied in the time domain.

where $X(\omega)$ is the true spectrum and $\epsilon_\tau (\omega)$ is an error term associated with the finite sample size. Now, for each spectral line $\epsilon_\tau (\omega)$ is a random variable and is just as likely to cause an underestimate as an overestimate. This means that one can potentially remove it by averaging, that is

$$E[F[x_\tau(t)]] = E[X(\omega)] + E[\epsilon_\tau (\omega)] = X(\omega) \qquad (A.136)$$

The averaging can be implemented straightforwardly by taking N segments of time data $x_i(t)$ and transforming to $X_i(\omega) = X(\omega) + \epsilon_{\tau i} (\omega)$; then

$$X(\omega) \approx E[X_i(\omega)] = \frac{1}{N} \sum_{i=1}^{N} X_i(\omega) \qquad (A.137)$$

The following figures show the effect of averaging on the spectral estimate for a signal:

$$x(t) = \sin(\omega t) + 0.25N(0, 1) \qquad (A.138)$$

(where the frequency of the sine wave has been chosen so that it is periodic over the window, so there is no need to worry about leakage from the sine wave). The first of the figures (Figure A.51) is one average – a one-shot measurement (for simplicity, the figures show only the magnitude, but one should always remember that the spectra are actually complex).

Although the sine wave (at 10.24 Hz) is visible, there is a lot of background noise from the single average as a result of the random component. Figure A.52 shows the result of taking 10 averages. Finally, Figure A.53 shows the result of 100 averages.

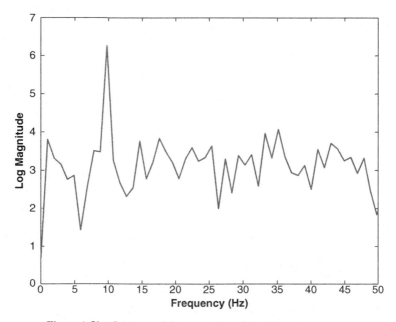

Figure A.51 Spectrum of sine wave + random process: one average.

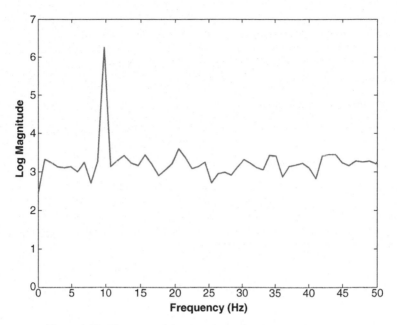

Figure A.52   Spectrum of sine wave + random process: 10 averages.

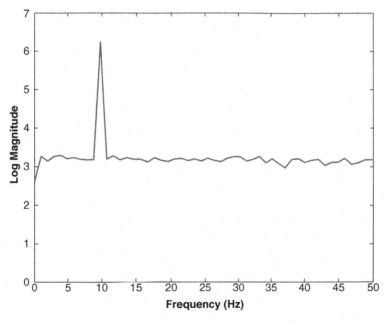

Figure A.53   Spectrum of sine wave + random process: 100 averages.

The smoothing effect of the averaging is very marked. Often, one chooses to plot $E[|X|^2]$ as this is proportional to power. It is called the *power spectral density* (PSD) and is denoted

$$S_{xx}(\omega) = E[|X(\omega)|^2] = E[X(\omega)X(\omega)^*] \tag{A.139}$$

A.7 Correlations and Spectra

The *autocorrelation function* $\varphi_{xx}(\tau)$ of a random process is defined by

$$\varphi_{xx}(\tau) = E[x(t + \tau)x(t)] \tag{A.140}$$

It is essentially a measure of how much a signal looks like itself when shifted in time by an amount τ. It is often used to find regularities in data. Suppose that $x(t) = \sin(2\pi t/\tau')$; then clearly one would expect to see regular peaks in $\varphi_{xx}(\tau)$ when $\tau = n\tau'$. So the autocorrelation function can also be used as a means of detecting periodicities. Equation (A.140) implicitly assumes that the data is stationary; otherwise the autocorrelation function would depend on t as well as τ.

Note that if $x(t)$ is zero-mean, then

$$\varphi_{xx}(0) = E[x(t)^2] = \sigma_x^2 \tag{A.141}$$

and if x is not zero-mean, $\varphi_{xx}(0)$ is the *mean square* of the process. As $x(t)$ is stationary, one can change the origin of time by moving t to $t - \tau$ without changing the autocorrelation function, that is

$$\varphi_{xx}(\tau) = E[x(t + \tau)x(t)] = E[x(t)x(t - \tau)] = E[x(t - \tau)x(t)] = \varphi_{xx}(-\tau) \tag{A.142}$$

so $\varphi_{xx}(\tau)$ must be an even function of τ.

If one has two different stationary random processes $x(t)$ and $y(t)$ (usually the input and output of a system), one can define the *cross-correlation functions*,

$$\varphi_{yx}(\tau) = E[y(t + \tau)x(t)] \tag{A.143}$$

and

$$\varphi_{xy}(\tau) = E[x(t + \tau)y(t)] \tag{A.144}$$

As one might expect, these two functions are related. In fact, because of the stationarity conditions,

$$\varphi_{yx}(\tau) = \varphi_{xy}(-\tau), \qquad \varphi_{xy}(\tau) = \varphi_{yx}(-\tau) \tag{A.145}$$

In a sense $\varphi_{yx}(\tau)$ detects causal relationships between signals – often these are the input and output to some system under investigation or test.

Consider the Fourier transform of $\varphi_{xx}(\tau)$. For simplicity the expectation will be defined as

$$\varphi_{xx}(\tau) = \int_{\infty}^{\infty} x(t + \tau)x(t)\mathrm{d}t \tag{A.146}$$

In fact, this should really be defined as

$$\varphi_{xx}(\tau) = \lim_{T \to \infty} \frac{1}{T} \int_{-T}^{T} x(t + \tau)x(t)\mathrm{d}t \tag{A.147}$$

However, one can adopt Equation (A.146) and just consider it a convenient choice of normalisation. The definition of $\varphi_{xx}(\tau)$ looks very much like a convolution; in fact the same sort of analysis as in Section A.3.6 shows that

$$F[\varphi_{xx}(\tau)] = X(\omega)X(\omega)^* = |X(\omega)|^2 = S_{xx}(\omega) \tag{A.148}$$

Note that as

$$\varphi_{xx}(\tau) = \frac{1}{2\pi} \int_{-\infty}^{\infty} S_{xx}(\omega)e^{i\omega\tau}d\omega \tag{A.149}$$

then one has

$$\varphi_{xx}(0) = \frac{1}{2\pi} \int_{-\infty}^{\infty} S_{xx}(\omega) = \sigma_x^2 \tag{A.150}$$

which shows very clearly why one can interpret $S_{xx}(\omega)$ as a power density.

Suppose now that $S_{xx}(\omega) = P$ – a constant. This means the signal weights all frequencies equally; this is often called *white noise* by analogy with the visible spectrum of light. Then

$$\varphi_{xx}(\tau) = \frac{1}{2\pi} \int_{-\infty}^{\infty} Pe^{i\omega\tau}d\omega = P\delta(\tau) \tag{A.151}$$

(using the definition of the delta function); that is, x is only ever correlated with itself at zero-lags $\tau = 0$ – this is a very clear indication of unpredictability and randomness.

If one now considers two signals $y(t)$ and $x(t)$ again (as before, one might regard these as the output and input of some process), a similar analysis to that earlier shows that

$$F[\varphi_{yx}(\tau)] = S_{yx}(\omega) = Y(\omega)X(\omega)^* \tag{A.152}$$

where $S_{yx}(\omega)$ is called the *cross-spectral density* (CSD), by analogy with $S_{xx}(\omega)$, which is sometimes called the *autospectral density*.

It is now possible to establish some relationships between the quantities defined above when x and y are indeed the input and output of some (linear) system.

Consider the impulse response relationship for a linear system, as discussed earlier – sometimes called *Duhamel's integral*,

$$y(t) = \int_{-\infty}^{\infty} h(\tau)x(t - \tau)d\tau \tag{A.153}$$

Taking the expectation of both sides of the equation gives

$$E[y(t)] = \overline{y} = E\left[\int_{-\infty}^{\infty} h(\tau)x(t - \tau)d\tau\right]$$

However, the only stochastic (random) part of the right-hand side is the term $x(t - \tau)$, so

$$\overline{y} = \int_{-\infty}^{\infty} h(\tau)E[x(t - \tau)]d\tau$$

Furthermore, if one assumes that $x(t)$ is stationary, then $E[x(t - \tau)] = E[x(t)] = \overline{x}$, which is a constant. Finally,

$$\overline{y} = \overline{x} \int_{-\infty}^{\infty} h(\tau) d\tau = H\overline{x} \qquad (A.154)$$

where H here is, in a way, the mean impulse response. An immediate consequence of Equation (A.154) is that, if $x(t)$ – the input to a linear system – is zero-mean, then so is the output $y(t)$.

If one takes a minor variation on Duhamel's integral,

$$y(t + \alpha) = \int_{-\infty}^{\infty} h(\tau) x(t + \alpha - \tau) d\tau$$

then one can form

$$\varphi_{yx}(\alpha) = E[y(t + \alpha)x(t)] = E\left[\int_{-\infty}^{\infty} h(\tau)x(t + \alpha - \tau)d\tau x(t)\right]$$

$$= E\left[\int_{-\infty}^{\infty} h(\tau)x(t + \alpha - \tau)x(t)d\tau\right]$$

$$= \int_{-\infty}^{\infty} h(\tau)E[x(t + \alpha - \tau)x(t)]d\tau$$

so that, finally,

$$\varphi_{yx}(\alpha) = \int_{-\infty}^{\infty} h(\tau)\varphi_{xx}(\alpha - \tau)d\tau \qquad (A.155)$$

For the final derivation in this context, one can multiply (A.153) by its shifted version and take the expectation,

$$\varphi_{yy}(\tau) = E[y(t + \tau)y(t)] = E\left[\int_{-\infty}^{\infty} h(\alpha)x(t + \tau - \alpha)d\alpha \int_{-\infty}^{\infty} h(\beta)x(t - \beta)d\beta\right]$$

$$= E\left[\int_{-\infty}^{\infty}\int_{-\infty}^{\infty} h(\alpha)h(\beta)x(t + \tau - \alpha)x(t - \beta)d\alpha d\beta\right]$$

$$= \int_{-\infty}^{\infty}\int_{-\infty}^{\infty} h(\alpha)h(\beta)E[x(t + \tau - \alpha)x(t - \beta)]d\alpha d\beta$$

$$= \int_{-\infty}^{\infty}\int_{-\infty}^{\infty} h(\alpha)h(\beta)E[x(t + \tau - \alpha + \beta)x(t)]d\alpha d\beta$$

so that, finally,

$$\varphi_{yy}(\tau) = \int_{-\infty}^{\infty}\int_{-\infty}^{\infty} h(\alpha)h(\beta)\varphi_{xx}(\tau - \alpha + \beta)d\alpha d\beta \qquad (A.156)$$

A.8   FRF Estimation and Coherence

A.8.1   FRF Estimation I

Recall Equation (A.155) from the previous section,

$$\varphi_{yx}(\tau) = \int_{-\infty}^{\infty} h(\alpha)\varphi_{xx}(\tau - \alpha)d\alpha$$

Taking the Fourier transform of both sides and using the convolution theorem gives

$$S_{yx}(\omega) = H(\omega)S_{xx}(\omega) \tag{A.157}$$

which leads to a new estimate of the FRF $H(\omega)$,

$$H(\omega) = \frac{S_{yx}(\omega)}{S_{xx}(\omega)} \tag{A.158}$$

where, for finite time signals, the cross-spectrum and auto-spectrum are estimated from an ensemble average, that is

$$S_{yx}(\omega) = E[Y(\omega)X(\omega)^*] \approx \frac{1}{N} \sum_{i=1}^{N} Y_i(\omega)X_i(\omega)^* \tag{A.159}$$

and

$$S_{xx}(\omega) = E[X(\omega)X(\omega)^*] \approx \frac{1}{N} \sum_{i=1}^{N} X_i(\omega)X_i(\omega)^* \tag{A.160}$$

Note that if $x(t)$ and $y(t)$ were samples of infinite duration, one could say

$$\frac{S_{yx}(\omega)}{S_{xx}(\omega)} = \frac{Y(\omega)X(\omega)^*}{X(\omega)X(\omega)^*} = \frac{Y(\omega)}{X(\omega)} = H(\omega) \tag{A.161}$$

Taking the Fourier transform of Equation (A.155) from the last section gives a similar relation; however, one can use a more formal argument. If $x(t)$ and $y(t)$ are samples of infinite duration,

$$\frac{S_{yy}(\omega)}{S_{xx}(\omega)} = \frac{Y(\omega)Y(\omega)^*}{X(\omega)X(\omega)^*} = \left(\frac{Y(\omega)}{X(\omega)}\right)\left(\frac{Y(\omega)^*}{X(\omega)^*}\right) = H(\omega)H(\omega)^*$$

Or, finally,

$$S_{yy}(\omega) = |H(\omega)|^2 S_{xx}(\omega) \tag{A.162}$$

This last equation means that, if one estimates the autocorrelation functions of x and y, one can only recover the *magnitude* of the FRF. In order to obtain the phase, one needs the appropriate cross-correlations or cross-spectra.

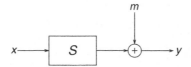

Figure A.54   A linear system with noise polluting the output.

A.8.2   The Coherence Function

The coherence function is essentially a spectrum and is usually used with random or impulse excitation. It can provide a quick visual inspection of the quality of an FRF and, in many cases, is a rapid indicator of noise or nonlinearity in specific frequency bands or resonance regions.

Before discussing nonlinearity, the coherence function will be derived for *linear* systems subject to measurement noise on the output as in Figure A.54. 'Noise' is interpreted here as any component of the response that does not arise from the input or otherwise from any measurable signal.

Systems like the one depicted in Figure A.54 have time-domain equations of motion of the form

$$y(t) = S[x(t)] + m(t) \tag{A.163}$$

where $m(t)$ denotes the measurement noise. In the frequency domain one has the dual representation

$$Y(\omega) = H(\omega)X(\omega) + M(\omega) \tag{A.164}$$

Multiplying the latter equation by its complex conjugate yields

$$YY^* = HXH^*X^* + HXM^* + H^*X^*M + MM^* \tag{A.165}$$

and then taking expectations gives

$$S_{yy}(\omega) = |H(\omega)|^2 S_{xx}(\omega) + H(\omega)S_{xm}(\omega) + H(\omega)^* S_{mx}(\omega) + S_{mm}(\omega) \tag{A.166}$$

Now, if x and m are uncorrelated signals (unpredictable from each other), then $S_{mx}(\omega) = S_{xm}(\omega) = 0$ and Equation (A.166) reduces to

$$S_{yy}(\omega) = |H(\omega)|^2 S_{xx}(\omega) + S_{mm}(\omega) \tag{A.167}$$

A simple rearrangement of this gives

$$\frac{|H(\omega)|^2 S_{xx}(\omega)}{S_{yy}(\omega)} = 1 - \frac{S_{mm}(\omega)}{S_{yy}(\omega)} \tag{A.168}$$

The quantity on the right-hand side of this last equation is simply the fraction of the output power that can be linearly correlated with the input. It is called the *coherence* function and is denoted by $\gamma^2(\omega)$. Now, as $\gamma^2(\omega)$ and $S_{mm}(\omega)/S_{yy}(\omega)$ are both positive quantities, if follows immediately that

$$0 \le \gamma^2 \le 1 \tag{A.169}$$

with $\gamma^2 = 1$ only if $S_{mm}(\omega) = 0$, that is if there is no measurement noise. The coherence function is therefore a means of detecting if there is noise in the output.

In fact, it will be shown later that $\gamma^2 < 1$ if there is noise anywhere in the measurement chain. If the coherence is plotted as a function of ω, any departures from unity will be readily identifiable. The coherence is more usually expressed as

$$\gamma^2(\omega) = \frac{|S_{yx}(\omega)|^2}{S_{yy}(\omega)S_{xx}(\omega)} \tag{A.170}$$

Note that all these quantities are easily computed by commercial spectrum analysers designed to estimate $H(\omega)$, which is why coherence facilities are so readily available in standard instrumentation.

The coherence function also detects nonlinearity as promised above. The relationship between input and output spectra for nonlinear systems is

$$Y(\omega) = H(\omega)X(\omega) + F[X(\omega)] \tag{A.171}$$

where F will be a rather complicated function in general, dependent on the type of nonlinearity present in the system. Multiplying by $\overline{Y}$ in this case and taking expectations gives

$$S_{yy}(\omega) = |H(\omega)|^2 S_{xx}(\omega) + H(\omega)S_{xf}(\omega) + H(\omega)^* S_{fx}(\omega) + S_{ff}(\omega) \tag{A.172}$$

where this time the cross-spectra S_{fx} and S_{xf} will not necessarily vanish. In terms of the coherence,

$$\gamma^2(\omega) = 1 - 2\Re\left(H(\omega)\frac{S_{xf}(\omega)}{S_{yy}(\omega)}\right) - \frac{S_{ff}(\omega)}{S_{yy}(\omega)} \tag{A.173}$$

and the coherence will generally only be unity if $f = 0$; that is, the system is linear.

Consider the Duffing oscillator – arguably the simplest nonlinear structural system – with the equation of motion,

$$m\ddot{y} + c\dot{y} + ky + k_3 y^3 = x(t) \tag{A.174}$$

If the level of excitation is low, the response y will be small and y^3 will be negligible in comparison. In this regime, the system will behave as a linear system and the coherence function for input and output will be unity (Figure A.55). If the excitation is increased, the nonlinear terms will begin to play a part and the coherence will drop as in Figure A.56.

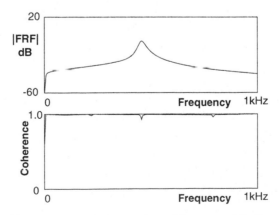

Figure A.55 FRF (top) and coherence (bottom) for an essentially linear system.

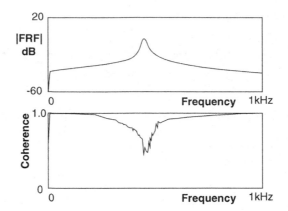

Figure A.56   FRF (top) and coherence (bottom) for a nonlinear system.

Note that the coherence is *only* meaningful if averages are taken. For a single measurement, a value of unity would always occur, that is

$$\gamma^2 = \frac{YX^*XY^*}{YY^*XX^*} = 1 \tag{A.175}$$

A reduction in the level of coherence can be caused by a range of problems, such as extraneous unmeasured inputs and noise on the output and/or input signals, which may in turn be due to incorrect gain settings on amplifiers. Such obvious causes should be checked before structural nonlinearity is suspected.

A.8.3   FRF Estimators II

It was shown in the last section that the coherence function dips below unity if noise (or nonlinearity) is present. This is unfortunately not the only consequence of noise; noise also leads to erroneous or *biased* estimates of the FRF when random excitation is used via Equation (A.161). In order to carry out the analysis this time, a general system will be assumed that has noise on both the input and output (Figure A.57),

The (unknown) clean input is denoted by $u(t)$ and, after the addition of (unknown) noise $n(t)$, gives the measured input $x(t)$. Similarly, the unknown clean output $v(t)$ is corrupted by noise $m(t)$ to give the measured output $y(t)$. It is assumed that $m(t)$, $n(t)$ and $x(t)$ are pairwise uncorrelated. The basic equations in the frequency domain are

$$X(\omega) = U(\omega) + N(\omega) \tag{A.176}$$

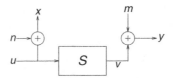

Figure A.57   System with general noise corruption (on both the input and output signals).

and

$$Y(\omega) = H(\omega)U(\omega) + M(\omega) \tag{A.177}$$

Multiplying Equation (A.176) by X^* and taking expectations gives

$$S_{xx}(\omega) = S_{uu}(\omega) + S_{nn}(\omega) \tag{A.178}$$

Similarly, multiplying Equation (A.177) by X^* and taking expectations gives

$$S_{yx}(\omega) = H(\omega)S_{uu}(\omega) \tag{A.179}$$

as $S_{mx}(\omega) = 0$. Taking the ratio of Equations (A.179) and (A.178) yields

$$\frac{S_{yx}(\omega)}{S_{xx}(\omega)} = \frac{H(\omega)S_{uu}(\omega)}{S_{uu}(\omega) + S_{nn}(\omega)} = \frac{H(\omega)}{\left(1 + \frac{S_{nn}(\omega)}{S_{uu}(\omega)}\right)} \tag{A.180}$$

This last equation means that the estimator S_{yx}/S_{xx} – denoted $H_1(\omega)$ – is only equal to the correct FRF $H(\omega)$ if there is no noise on the input ($S_{nn} = 0$). Further, as $S_{nn}/S_{uu} > 0$, the estimator is always an underestimate; that is, $H_1(\omega) < H(\omega)$ if input noise is present. Note that the estimator is completely insensitive to noise on the output.

Now, if one multiplies Equation (A.177) by Y^* and takes expectations, the result is

$$S_{yy}(\omega) = |H(\omega)|^2 S_{uu}(\omega) + S_{mm}(\omega) \tag{A.181}$$

Furthermore, multiplying Equation (A.176) by Y^* and averaging yields

$$S_{xy}(\omega) = H(\omega)^* S_{uu}(\omega) \tag{A.182}$$

And, finally, taking the ratio of Equations (A.181) and (A.182) gives

$$\frac{S_{yy}(\omega)}{S_{xy}(\omega)} = H(\omega)\left(1 + \frac{S_{mm}(\omega)}{S_{uu}(\omega)}\right) \tag{A.183}$$

This means that the estimator S_{yy}/S_{xy} – denoted by $H_2(\omega)$ – is only equal to the true $H(\omega)$ if there is no noise on the output ($S_{mm} = 0$). Also, as $S_{mm}/S_{uu} > 0$, the estimator is always an overestimate; that is, $H_2(\omega) > H(\omega)$ if output noise is present. The estimator is insensitive to noise on the input.

There is a practical consequence of this analysis. If there is noise on the input only, one should always use the FRF estimator H_2; if there is noise only on the output, one should use the estimator H_1. If there is noise on both signals a compromise is clearly needed. As H_1 is known to be an underestimate and H_2 is an overestimate, the sensible estimator would be somewhere in between. With this in mind, a new estimator H_3 can be defined by taking the geometric mean of H_1 and H_2,

$$H_3(\omega) = \sqrt{H_1(\omega)H_2(\omega)} = H(\omega)\sqrt{\frac{S_{mm}(\omega) + S_{uu}(\omega)}{S_{nn}(\omega) + S_{uu}(\omega)}} \tag{A.184}$$

and this is the FRF estimator of choice if both the input and output signals are corrupted by noise.

Note that a useful by-product of this analysis is a general expression for the coherence function,

$$\gamma^2(\omega) = \frac{|S_{yx}(\omega)|^2}{S_{yy}(\omega)S_{xx}(\omega)} = \frac{1}{\left(1 + \frac{S_{mm}(\omega)}{S_{vv}(\omega)}\right)\left(1 + \frac{S_{nn}(\omega)}{S_{vv}(\omega)}\right)} \tag{A.185}$$

from which it follows that $\gamma^2 < 1$ if *either* input or output noise is present. It also follows from the equations above that $\gamma^2 = H_1/H_2$ or

$$H_2(\omega) = \frac{H_1(\omega)}{\gamma^2(\omega)} \tag{A.186}$$

so the three quantities are not independent.

A.9   Wavelets

While there are a number of established classic texts and monographs on wavelets for mathematicians, there are very few books on wavelets that are good for engineers, as most of them have also been written by mathematicians. The main exception to this rule is Hubbard (1998). This is very readable and light on the mathematics. If you the reader have a strong stomach (mathematically speaking) and are curious about the orthogonal wavelets discussed a little later, you may care to try Mallat (1999), which is a very comprehensive book.

A.9.1   Introduction and Continuous Wavelets

The Fourier series and transform have proved to be a powerful means of characterising and visualising the frequency content of signals; however, they are subject to an important limitation – they are designed for use with stationary signals only. This is because they are expansions in terms of sine and cosine waves, which extend in time forever; if the representation has a certain frequency content at one time, it is constrained to have the same content for all time.

For nonstationary signals – as will be recalled from Section A.1 – this is too restricted. Consider the two signals below. These are both formed from two components $f_1(t)$ and $f_2(t)$, defined as follows:

$$f_1(t) = \sin\left(2\pi \times 10\frac{f_s}{N}t\right) \tag{A.187}$$

and

$$f_2(t) = \sin\left(2\pi \times 80\frac{f_s}{N}t\right) \tag{A.188}$$

where f_s is the sampling frequency and N is the number of points. Figure A.58 is made by simply adding f_1 and f_2 and is clearly stationary. In contrast, Figure A.59 is made by concatenating f_1 and f_2 and is nonstationary in amplitude and frequency. The corresponding spectra are given in Figures A.60 and A.61. The two spectra are almost indistinguishable ($S_2(f)$ has a little frequency smearing as a result of the discontinuity in the middle of the record).

The clear problem is that the Fourier transform cannot tell that the second signal has a finite end point for the 10 Hz component and start point for the 80 Hz component. The reason for this is that the *basis functions* of the Fourier transform are unbounded in extent (in time). It is impossible, from the spectrum, to localise a given time event, like the change from 10 to 80 Hz.

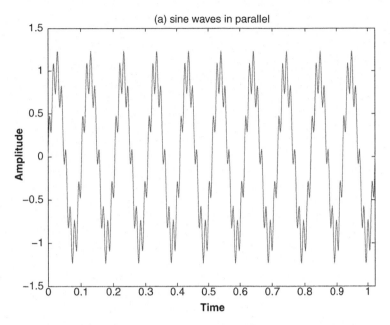

Figure A.58   Stationary signal formed by adding $f_1(t)$ and $f_2(t)$.

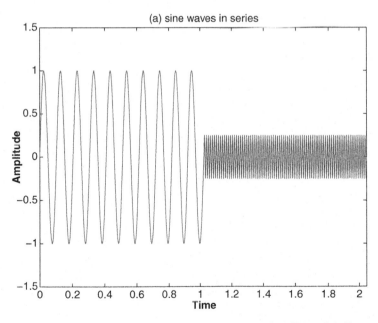

Figure A.59   Nonstationary signal formed by concatenating $f_1(t)$ and $f_2(t)$.

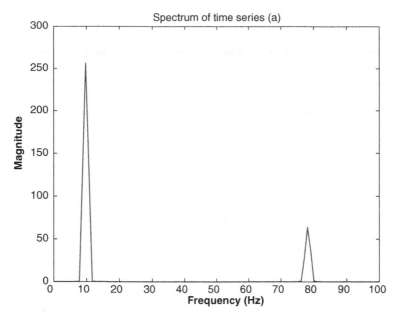

Figure A.60   Fourier spectrum $S_1(f)$ of the stationary signal from Figure A.58.

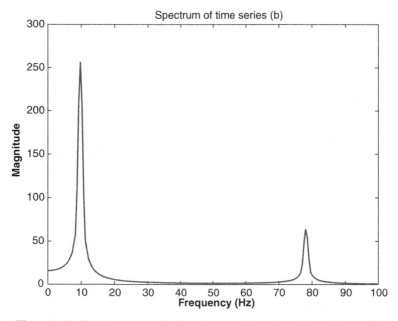

Figure A.61   Fourier spectrum $S_2(f)$ of the nonstationary signal from Figure A.59.

As the problem here has been traced to the use of the infinitely long sine and cosine functions, a solution may well be obtained by using basis functions of finite time duration. One way of accomplishing the latter would be to introduce a localised time window (one would still need to assume stationarity over the duration of the window). With this in mind, one is led to consider

$$X(\tau, \omega) = \int_{-\infty}^{\infty} w(t - \tau)e^{i\omega t}x(t)dt \qquad (A.189)$$

so that now the basis functions for the expansion are the time–frequency 'atoms',

$$W_{\tau,\omega} = w(t - \tau)e^{i\omega t} \qquad (A.190)$$

and the windowed harmonic has a finite extent, say T, to give time localisation. This transform is called the *short-time Fourier transform* (STFT).

Suppose the window or kernel is specified by

$$w(t - \tau) = \exp\left[-\frac{1}{2}\left(\frac{t - \tau}{\sigma}\right)^2\right] \qquad (A.191)$$

that is, a Gaussian function. Although this function is actually nonzero everywhere, because it decays so rapidly away from the centre, it is nominally nonzero only in the region $(\tau - 3\sigma, \tau + 3\sigma)$ and looks like the function depicted in Figure A.62 with $\sigma = 1$.

Now, if one recalls the relation, $\Delta\omega = 1/N\,\Delta t$, or

$$\Delta\omega T = 1 \qquad (A.192)$$

this clearly means that one can have good time localisation (T is small) or good frequency localisation ($\Delta\omega$ is small), but not both. For the STFT, the window width T is independent of $\Delta\omega$ in the frequency domain. This can be visualised using *time–frequency cells* as in Figure A.63. The diagram shows in a pictorial manner that the time localisation ($= T$) is indeed independent of time and frequency and that the frequency localisation ($= \Delta\omega$) is similarly independent of time and frequency. Furthermore, it is clear that $\Delta\omega T = 1$; that is, the area of the cells is constant.

For nonstationary signals, wide time windows T are not a good idea, as the signal frequency *content* may then vary substantially over the duration of the window, although the frequency resolution will be

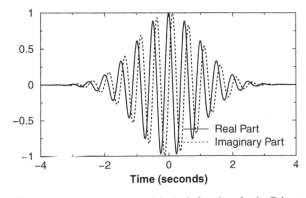

Figure A.62   Real and imaginary parts of the basis functions for the Gabor transform.

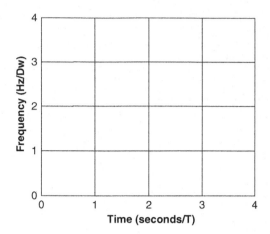

Figure A.63 Time–frequency cell structure for the STFT.

good within the window. Alternatively, if T is kept small in order to allow one to track changes in the frequency content, the frequency resolution within the window will be poor. Another serious problem with the STFT is that there is no obvious analytical inverse; that is, the original signal cannot easily be reconstructed from the time–frequency map or *spectrogram*. The *wavelet transform* (WT) solves both problems.

The first problem is solved (in a similar manner to the STFT) by using basis functions with variable window sizes; one expands as a series of *little waves*. If the basis function at a given frequency ω looks like the form in Figure A.64, then to get the basis function at $\omega/2$, one simply stretches or *dilates* the original basis function as in Figure A.65.

The difference between this idea and the STFT is that the STFT would have kept the window size constant and therefore the number of cycles of the basis function would have changed. The question now arises as to how the dilation might be implemented; consider the Fourier transform in the phase form,

$$X(\omega) = \int_{-\infty}^{\infty} x(t)\sin(2\pi f t - \varphi(f))\mathrm{d}t \tag{A.193}$$

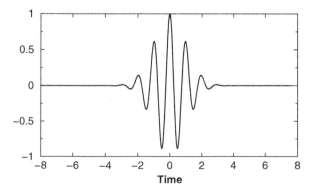

Figure A.64 Wavelet basis function corresponding to frequency ω.

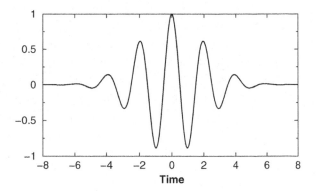

Figure A.65 Wavelet basis function corresponding to frequency $\omega/2$.

If one introduces a little new notation: $a_f = 1/(2\pi f)$ and $b_f = \varphi_f/(2\pi f)$, then the transform becomes

$$X(\omega) = \int_{-\infty}^{\infty} x(t)\sin\left(\frac{t - b_f}{a_f}\right) dt \qquad (A.194)$$

where the parameter a_f controls the frequency by dilating or scaling the time variable. The latter variable will be called the *scale parameter*; note that a_f is inversely proportional to f. The parameter b_f is seen here to move or *translate* the basis sine wave up and down the t axis – it will be called the *translation parameter*. One must note that in this situation the translation parameter has a very limited effect as the sine wave is defined over an infinite time interval; it can, however, change the sine function to a cosine function as is required for a full Fourier expansion.

Suppose now that one now replaces the sine wave in Equation (A.194) by any localised oscillatory function – such as a wavelet ψ. One obtains the *wavelet transform* (WT),

$$W_\psi(a, b) = \int_{-\infty}^{\infty} x(t)\psi\left(\frac{t - b}{a}\right) dt \qquad (A.195)$$

where the dependence of a and b on f has been removed and they have been promoted to parameters in their own right. In practice, for technical reasons, one normally uses ψ^* – the complex conjugate of ψ – and scales the overall transform by $1/\sqrt{a}$ (this ensures equal energy of the ψ at all time scales). The usual definition of the WT is therefore

$$W_\psi(a, b) = \frac{1}{\sqrt{a}} \int_{-\infty}^{\infty} x(t)\psi^*\left(\frac{t - b}{a}\right) dt \qquad (A.196)$$

The function $\psi(t)$ from which all the basis functions are derived is commonly called the *mother wavelet*.

This definition of the transform solves the resolution problem discussed above as the window length is high for low frequencies. Therefore frequency resolution is good for low frequencies (long scales) and the time resolution is good at high frequencies (short scales). The time–frequency resolution looks like the representation given in Figure A.66.

It is important to note that not just any function is allowed as a mother wavelet; the mother wavelet must satisfy certain properties. The first of these is

$$\int_{-\infty}^{\infty} \psi(t)dt = 0 \qquad (A.197)$$

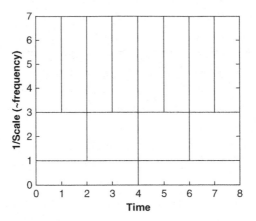

Figure A.66 Time–frequency cell structure for WT.

This ensures that the mother wavelet is positive as often as it is negative, a necessary condition if the mother wavelet is to be *oscillatory*.

The second important property is a usual one required of a time signal; it is simply that the mother wavelet should have *finite power*, that is

$$\int_{-\infty}^{\infty} |\psi(t)|^2 \mathrm{d}t = 0 \tag{A.198}$$

The final condition is called the *admissability condition*,

$$\int_{-\infty}^{\infty} \frac{|\Psi(\omega)|^2}{|\omega|} \mathrm{d}\omega = C_\psi < \infty \tag{A.199}$$

where $\Psi(\omega)$ is the Fourier transform of the mother wavelet. This condition has no simple physical interpretation. It is a technical condition that is necessary for the WT to have an inverse; that inverse is found to be

$$x(t) = \frac{1}{C_\psi \sqrt{a}} \int_{-\infty}^{\infty} W_\psi(a, b)\psi\left(\frac{t-b}{a}\right)\frac{\mathrm{d}a\mathrm{d}b}{a^2} \tag{A.200}$$

Because this transform operates on continuous-time signals, it is called the *continuous wavelet transform* (CWT). There are discrete variants of the transform in much the same way that the Fourier transform has a discrete variant in addition to its continuous form, and these are discussed a little later.

One of the most often used mother wavelets for the CWT is the *Morlet* wavelet, given by

$$\psi(t) = \mathrm{e}^{i\omega_0 t}\mathrm{e}^{-\frac{t^2}{2}} \tag{A.201}$$

which is a complex wavelet and is shown in Figure A.67. The Fourier transform of this mother wavelet is given by

$$\Psi(\omega) = \sqrt{2}\pi\mathrm{e}^{-\frac{(a\omega-\omega_0)^2}{2}} \tag{A.202}$$

which is a Gaussian function centred on the frequency ω_0.

Note that if t is dilated and goes to t/a, then the centre frequency of the associated wavelet goes from ω_0 to ω_0/a. Dilation moves the frequency of the wavelet, but also changes the bandwidth in the

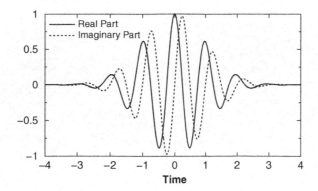

Figure A.67   Complex Morlet mother wavelet.

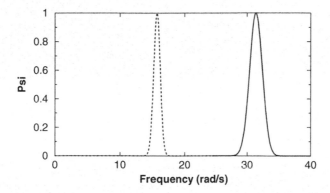

Figure A.68   Effect of dilation on the frequency content of a wavelet basis function.

frequency domain. Figure A.68 shows the transforms associated with $\omega_0 = 20\pi$ with dilations of $a = 1.0$ and $a = 2.0$.

Figure A.68 shows clearly that the centre of the wavelet spectrum shifts as a function of dilation. If a increases (i.e. one moves to a longer scale/lower frequency), the spectrum shifts down and narrows; that is, one obtains good frequency resolution at low frequencies.

To show how well the wavelet transformation represents nonstationary signals, one can consider how it performs on the signals shown at the beginning of this chapter. First the sum signal (Figure A.69) and second, the signal are formed by concatenating $f_1(t)$ and $f_2(t)$ (Figure A.70). In contrast to the situation with the Fourier spectra, the differences between the signals are shown very clearly. The major component is the high-amplitude high-scale (low-frequency) one. In the first CWT, this persists throughout the signal with the low-scale (high-frequency) component hardly visible. In the WT of the second signal, the dominant component disappears halfway along, exactly as it should with the minor component appearing.

As a more practical illustration of the use of the wavelet transform, Figure A.71 shows the CWT of the response of a model building structure on a shake table to a recording of the El Centro earthquake.[1] The figure shows very clearly that as the frequency content of the earthquake changed as it progressed,

---

[1] US National Geophysical Data Centre; http://www.ngdc.noaa.gov/hazard/hazards.shtml.

Absolute Values of Ca,b Coefficients for a = 1 2 3 4 5 ...

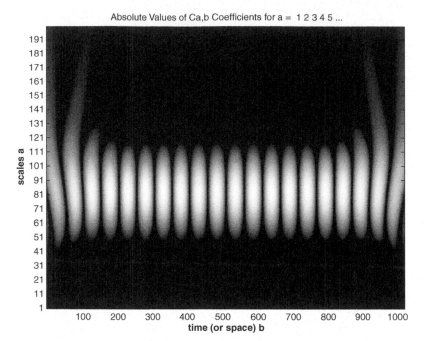

Figure A.69   Continuous wavelet transform of a signal from Figure A.58.

Absolute Values of Ca,b Coefficients for a = 1 2 3 4 5 ...

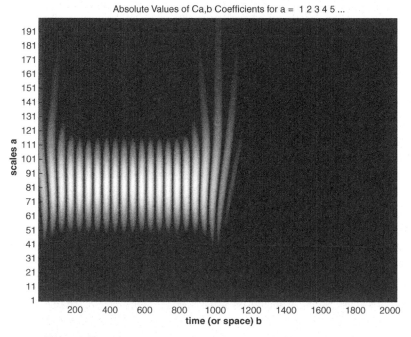

Figure A.70   Continuous wavelet transform of a signal from Figure A.59.

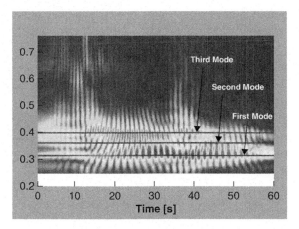

Figure A.71 CWT of a model building response to an earthquake.

because it was a nonstationary event in terms of frequency, different vibration modes of the building were excited as time went by.

A.9.2   Discrete and Orthogonal Wavelets

As in the case of the Fourier transform, if one wishes to conduct analysis on sampled data, there is a more convenient variant of the wavelet transform than the continuous version described in the last section. However, the way in which the transform is discretised is a little more subtle than in the Fourier case.

A.9.2.1   Discretisation – Fourier Analysis

As discussed in Section A.3, the Fourier transform maps from the time domain to the frequency domain via the integral

$$X(\omega) = \int_{-\infty}^{\infty} e^{-i\omega t} x(t) dt$$

when the time signals are *sampled* at discrete intervals. The available measurements are also finite in number,

$$x(t_0), x(t_0 + \Delta t), x(t_0 + 2\Delta t), \ldots, x(t_0 + (N-1)\Delta t)$$

and one passes to the frequency domain by replacing the Fourier integral with a discrete sum,

$$X(\omega) = \Delta t \sum_{j=0}^{N-1} x_j e^{-i(j\Delta t)\omega}$$

where $x_j = x(t_0 + j\Delta t)$.

This appears to go from N real numbers to a continuum of complex numbers. There must therefore be some *redundancy* in the frequency-domain representation; that is, there must be relationships between the transformed quantities. As before, one can think about the inverse transform and consider the quantities of information involved. In order to reconstruct the time signal, one only needs N frequency lines (actually $N/2$, but this will be forgotten for now.).

If one defines the discrete spectrum on the points,

$$\omega_0, \omega_0 + \Delta\omega, \omega_0 + 2\Delta\omega, \dots, \omega_0 + (N-1)\Delta\omega$$

then, with the right choices of ω_0 and $\Delta\omega$, which turn out to be

$$\omega_0 = -\frac{1}{2\Delta t}, \qquad \Delta\omega = \frac{1}{N\Delta t}$$

one finds that the inverse is

$$x_k = \frac{\Delta\omega}{2\pi} \sum_{j=0}^{N-1} X_j e^{i(j\Delta\omega)(k\Delta t)}$$

where $X_j = X(\omega_0 + j\Delta\omega)$.

A.9.2.2   Discretisation – Wavelet Analysis

As indicated above, discretisation for wavelet analysis is a little more involved.

Consider the continuous transform,

$$W_\psi(a, b) = \frac{1}{\sqrt{a}} \int_{-\infty}^{\infty} x(t)\psi^*\left(\frac{t-b}{a}\right) dt$$

Given discrete time data, the transform is approximated by a rectangular sum as

$$W_\psi(a, b) = \frac{\Delta t}{\sqrt{a}} \sum_{j=0}^{N-1} x_j \, \psi^*\left(\frac{t_j - b}{a}\right) \tag{A.203}$$

In this case, as the transform is defined on the whole (a, b) plane, there is considerably more redundancy than for the Fourier transform. If one wishes to arrive at a unique inverse as in the Fourier case, how should one choose N points from the (a, b) plane to reconstruct the x_j data and thus eliminate redundancy?

To solve the problem, first of all, one assumes a discrete set of scales a_i, $i = 0, \dots L - 1$. Having done this, the next problem is to set the sampling rate at each scale, that is to specify how many b values one needs in order to represent the data. Suppose that a_i is chosen as half a_{i-1}. It follows that to have *constant relative scale (frequency) resolution*, there will need to be twice as many b samples at scale i as there are at scale $i - 1$. It transpires that halving the scale at each level is convenient and leads to very elegant mathematics (Mallat, 1999).

After some thought, this gives a discretisation,

$$W_\psi(a_m, b_{m,n}) = \frac{1}{\sqrt{a_m}} \sum_{j=0}^{N-1} x_j \, \psi^*\left(\frac{t_j - b_{m,n}}{a_m}\right) \tag{A.204}$$

where (the time scale has been set so that $\Delta t = 1$)

$$a_m = 2^{-m}, \qquad b_{m,n} = n2^{-m} \tag{A.205}$$

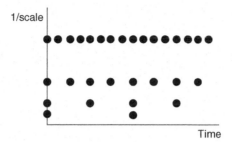

Figure A.72   Dyadic sampling grid for the discrete wavelet transform.

where the index n runs over the number of samples needed at scale level m. Defining

$$\psi_{m,n}(t) = 2^{m/2}\psi(2^m t - n) \tag{A.206}$$

the transform becomes

$$W_\psi(m, n) = \sum_{j=0}^{N-1} x_j\, \psi_{m,n}^*(t_j) \tag{A.207}$$

and the time-scale grid that supports the sum is as shown in Figure A.72. This type of grid is termed *dyadic*; the immediate question that arises here is of how many scales are allowed.

Suppose half the points are assigned to the finest scale, a quarter are assigned to the next finest and so on; a little thought suffices to show that this gives $L = \log_2 N$ scales. In order that L is an integer, it is necessary to ensure that

$$N = 2^L$$

However, this constraint is already completely familiar from discrete Fourier analysis. Having established a discretisation for the forward CWT, the question of how to compute an exact inverse naturally arises.

A.9.2.3   Inversion

Having eliminated the redundancy in the forward CWT by going to a dyadic grid, one also desires, by analogy with what has already been said for the Fourier transform, that the transform should have a unique inverse. Recall that (Equation A.207)

$$W_\psi(m, n) = \sum_{j=0}^{N-1} x_j\, \psi_{m,n}^*(t_j)$$

If the $W_\psi(m, n)$ are known, one immediately sees that this constitutes a system of linear simultaneous equations, which one can in principle solve uniquely for the x_j. However, to simplify the analysis, consider the *ansatz*,

$$x_j = \sum_k \sum_l W_\psi(k, l)\psi_{k,l}(t_j) \tag{A.208}$$

which, substituting into Equation (A.207), gives directly

$$W_\psi(m, n) = \sum_{j=0}^{N-1} \sum_k \sum_l W_\psi(k, l) \psi_{k,l}(t_j) \psi_{m,n}^*(t_j) \tag{A.209}$$

Rearranging this expression, one finds

$$W_\psi(m, n) = \sum_k \sum_l W_\psi(k, l) \left(\sum_{j=0}^{N-1} \psi_{k,l}(t_j) \psi_{m,n}^*(t_j) \right) \tag{A.210}$$

and this is actually identically satisfied if

$$\sum_{j=0}^{N-1} \psi_{k,l}(t_j) \psi_{m,n}^*(t_j) = \delta_{km} \delta_{ln} \tag{A.211}$$

In words, the sum is 0 unless $k = m$ and $l = n$ and in that case the sum is unity.

Thus, if the basis vectors $\psi_{k,l}(t_j)$ are *orthogonal* to each other, the inverse is simply

$$x_j = \sum_k \sum_l W_\psi(k, l) \psi_{k,l}(t_j) \tag{A.212}$$

In order to make progress with the analysis now, it is useful to make a digression on the subject of *orthogonality*, in a little more general terms than previously. Consider a general point in the plane; one can represent it using any basis of vectors one likes as long as they span the space; that is, multiples of the basis set can point to any point in the plane with an appropriate linear combination (Figure A.73). Essentially, the requirement for the basis vectors is that some have a horizontal component and some have a vertical component; they cannot all be parallel. However, the representation described above

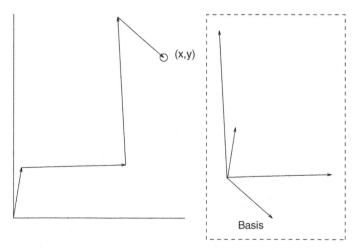

Figure A.73   Redundant or dependent basis of vectors for the plane.

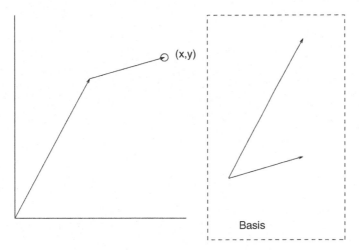

Figure A.74 Independent basis of vectors for the plane.

is *redundant*; one actually only needs two basis vectors as long as they are not parallel to each other (Figure A.74).

Note that if one demands that the basis vectors are all orthogonal to each other (and not zero length), one ensures that they are not parallel to each other and also arrives at a minimal basis – we cannot have more than two – which is the dimension of the space (the plane). Orthogonality therefore means *no redundancy* (Figure A.75).

To recap, in the wavelet analysis so far, the required basis functions established on the dyadic grid have the form

$$\psi_{m,n}(t) = 2^{m/2}\psi(2^m t - n)$$

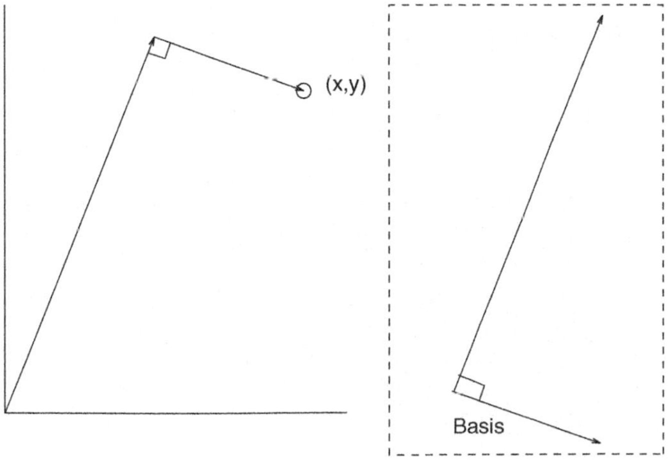

Figure A.75 Orthogonal basis of vectors for the plane.

and orthogonality – no redundancy – demands,

$$\sum_{j=0}^{N-1} \psi_{k,l}(t_j)\psi_{m,n}^*(t_j) = \delta_{km}\delta_{ln}$$

with ψ some mother wavelet. It is important to note that restricting a transform with a given continuous mother wavelet to a dyadic grid does *not* automatically give an orthogonal wavelet. They are more difficult to construct.

A.9.2.4 Scaling Functions and Mother Wavelets

The most often used orthogonal mother wavelets are those of *Daubechies*. The mathematics of their construction is deep and complex (Mallat, 1999). What one finds, surprisingly, is that there is no explicit formula for any of the family. Instead they are constructed from *scaling functions* $\varphi(t)$ defined by a *dilation equation*,

$$\varphi(t) = \sum_{n=0}^{N} c_n \varphi(2t - n) \tag{A.213}$$

The c_n can be regarded as filter coefficients; different sets result in different mother wavelets. The different wavelets are referred to as a *family*. Once the scaling function has been obtained, the corresponding mother wavelet is defined by

$$\psi(t) = \sum_{n} (-1)^n c_n \varphi(2t + n - N + 1) \tag{A.214}$$

The scaling function is chosen to preserve its area under each iteration and this forces a condition,

$$\int_{-\infty}^{\infty} dt\ \varphi(t) = 1 \tag{A.215}$$

Integrating the dilation Equation (A.213) and using this condition gives

$$\sum_{n=0}^{N-1} c_n = 2 \tag{A.216}$$

Therefore stability of the iteration forces a condition on the allowed coefficients c_n.

If a finite sum of wavelets is to represent the signal as accurately as possible, another set of conditions on the coefficients is obtained (Mallat, 1999),

$$\sum_{n=0}^{N-1} (-1)^n n^m c_n = 2 \tag{A.217}$$

where $m = 0, 1, 2, \ldots, N/2 - 1$. Furthermore, requiring orthogonality forces the condition

$$\sum_{n=0}^{N-1} c_n c_{2n+2m} = 0 \tag{A.218}$$

where $m = 0, 1, 2, \ldots, N/2 - 1$. Finally, if the scaling function is required to be orthogonal one finds

$$\sum_{n=0}^{N-1} c_n^2 = 2 \tag{A.219}$$

A set of filter coefficients satisfying all these conditions leads to an allowed mother wavelet via the dilation equation. The simplest such mother wavelet is the *Haar wavelet*.

Consider the necessary constraints on the c_n for $N = 2$. The stability condition enforces

$$c_0 + c_1 = 2$$

The accuracy condition implies

$$c_0 - c_1 = 0$$

and the orthogonality of $\varphi(t)$ gives

$$c_0^2 + c_1^2 = 2$$

The unique solution is $c_0 = c_1 = 1$ and so the dilation equation for $N = 2$ is

$$\varphi(t) = \varphi(2t - 1) + \varphi(2t) \tag{A.220}$$

It transpires that the dilation equation in this case is satisfied by a step function (Figure A.76 shows how such a function satisfies the recursion).

The mother wavelet is then given by (Figure A.77)

$$\psi(t) = \varphi(2t - 1) - \varphi(2t) \tag{A.221}$$

This is known from classical approximation theory and is referred to in modern terms as the *Haar wavelet*.

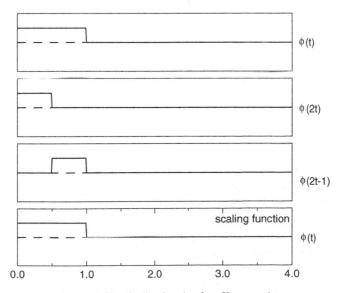

Figure A.76   Scaling function for a Haar wavelet.

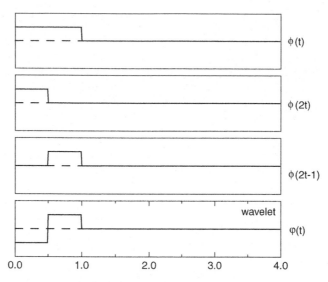

Figure A.77   Haar mother wavelet.

If $N = 4$, a mother wavelet not known from 'classical' theory is found; in fact, if $N \neq 2$, the wavelets are all new – they constitute the *Daubechies family*. The $N = 4$ equations for the filter coefficients are

$$c_0 + c_1 + c_2 + c_3 = 2$$
$$c_0 - c_1 + c_2 - c_3 = 0$$
$$-c_1 + 2c_2 - 3c_3 = 0$$
$$c_0 c_2 + c_1 c_3 = 0$$
$$c_0^2 + c_1^2 + c_2^2 + c_3^2 = 2 \qquad\qquad\qquad (A.222)$$

The solutions are

$$c_0 = \frac{1 + \sqrt{3}}{4}, \quad c_1 = \frac{3 + \sqrt{3}}{4}$$
$$c_2 = \frac{3 - \sqrt{3}}{4}, \quad c_3 = \frac{1 - \sqrt{3}}{4}$$

The scaling function can be constructed by an iteration starting from a step function, as illustrated in Figure A.78.

Note that the scaling function shown in Figure A.78 – and hence the derived mother wavelet – is not smooth. In fact the following properties can be proved for the whole Daubechies family, labelled D_N, with the associated mother wavelet $\psi_N(t)$.

- The support for D_N is on the interval $[0, 2N - 1]$.
- $\psi_N(t)$ has N vanishing moments. (This is a condition on smoothness.)
- $\psi_N(t)$ has γN continuous derivatives (γ is about 0.2).

D_4 has approximately 0.8 continuous derivatives – not very smooth. One can contrast the smoothness of the different Daubechies wavelets; Figure A.79 illustrates this with: (a) D_5 and (b) D_{16}. More mother wavelets of different families (different defining relations) are shown in Figure A.80.

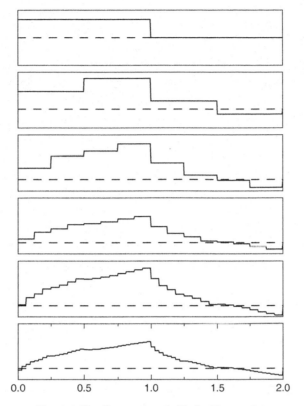

Figure A.78 Construction of a Daubechies wavelet.

A.9.2.5 Wavelet Levels

Recall that the definition of the discrete wavelet transform expressed functions as summations of terms at common dilation scales,

$$x_j = \sum_m \sum_n W_\psi(m, n) \psi_{m,n}(t_j)$$

if one defines x_{mj} by

$$x_{mj} = \sum_n W_\psi(m, n) \psi_{m,n}(t_j) \tag{A.223}$$

This quantity is the time series component of the original signal associated with the scale m. It is called the mth level. It follows that

$$x_j = \sum_m x_{mj} \tag{A.224}$$

As the level is composed of basis functions at the same scale, it represents a component of the original signal associated with a certain frequency band. The lower levels are the low-frequency component, while the higher levels represent the high-frequency component. This decomposition of the signal can prove very useful in a number of contexts, SHM being one. Figure A.81 shows the levels associated with a Lamb wave signal recorded in the context of a wave-propagation-based SHM experiment. Figure A.82 shows

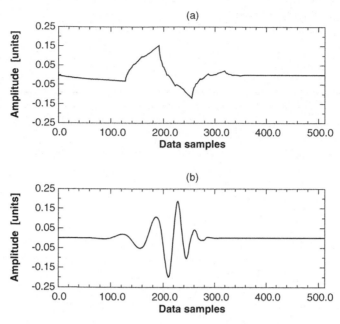

Figure A.79   Daubechies wavelets of order (a) 5, (b) 16.

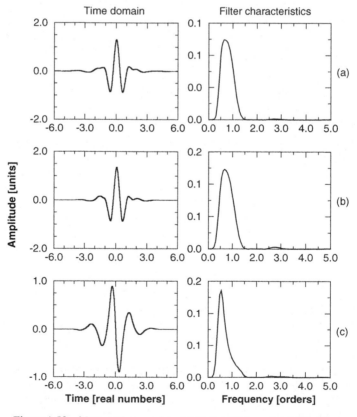

Figure A.80   More mother wavelets: (a) Coiflet, (b) Symmlet, (c) B-spline.

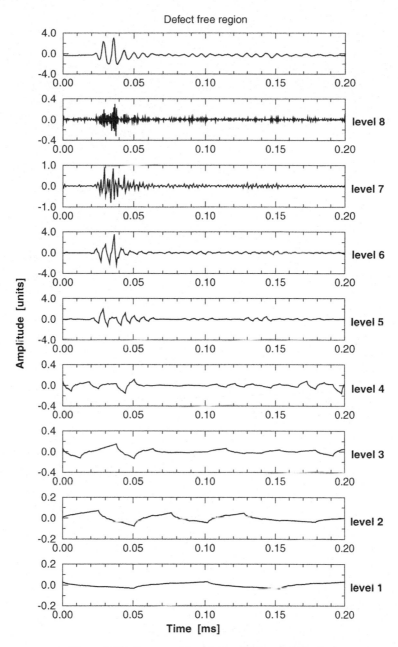

Figure A.81 Decomposition of a signal into wavelet levels.

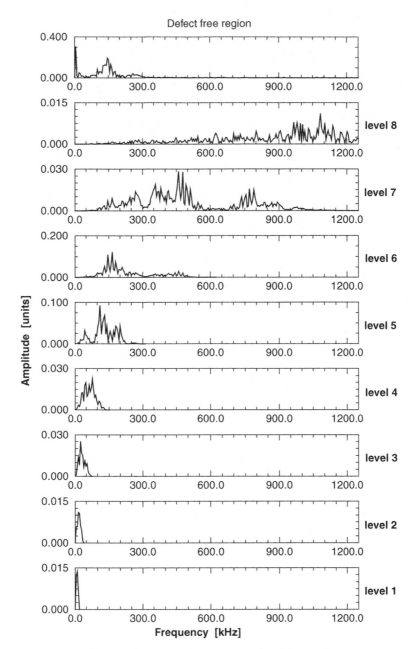

Figure A.82   Spectra of wavelet levels from Figure A.81.

the spectra of the associated level time series from Figure A.81. Often when the signal is decomposed as described, the higher levels are associated with noise rather than a true signal; in this case, if one reconstructs the signal from only the lower levels, one arrives at a denoised or filtered version of the original signal.

A.9.2.6 Computation

As a brief summary of the cost of computing the Fourier and wavelet transforms, one finds:

Discrete Fourier transform (DFT) : $O(N^2)$ multiplications.

Fast Fourier transform (FFT) (Cooley–Tukey algorithm): $O(N \log N)$ multiplications.

Fast wavelet transform (FWT) (Mallat's pyramidal algorithm): $O(N)$ multiplications.

FWT is *faster* than FFT!

A.9.2.6.1   An Application of the DWT – Detection of Discontinuities

Recall from Section A.2 that the Fourier series for the square wave had difficulties representing the discontinuity in the function (as shown in Figure A.83). The discontinuities required high frequencies for proper modelling and led to the Gibbs phenomenon when not enough terms were included in the series.

The type of discontinuity found in the square-wave function is the most severe possible, but it is not the only type possible. More subtle discontinuities can occur as shown in Figure A.84; this function has a discontinuity in the first derivative or gradient.

These discontinuities are very simple to detect. However, one may be interested in even more subtle discontinuities. Suppose one is given a displacement signal and one wishes to detect a discontinuity in the acceleration. The wavelet decomposition can be used to detect discontinuities as they will show up as high-frequency components in the short-scale levels; Figure A.85 shows (essentially) part of the level decomposition for a function with a discontinuity in the second derivative. As one can see from the 'original' function in Figure A.85, such subtle discontinuities are not readily detectable with the naked eye.

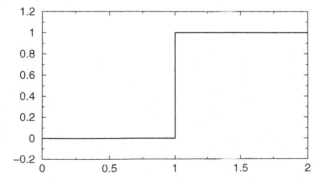

Figure A.83   Discontinuity in a simple function.

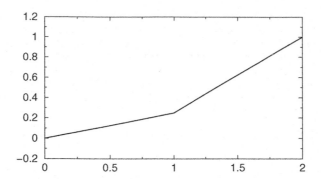

Figure A.84   Simple function with a discontinuity in the first derivative.

A.9.2.7   An Application of the DWT – Compression

The wavelet decomposition can also be used to reduce storage requirements for signals. Suppose a signal is decomposed into a sequence of wavelet coefficients; one could select only the largest components and still regenerate most of the signal. This is a more general aspect of the denoising properties mentioned earlier, for if the noise in the signal is associated with the smaller components, the signal is also denoised by this process.

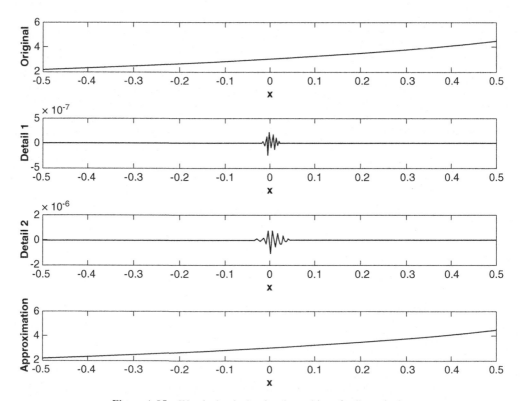

Figure A.85   Wavelet levels showing the position of a discontinuity.

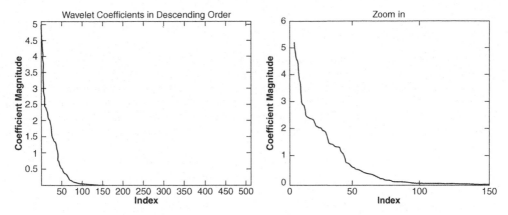

Figure A.86 Wavelet coefficients for a signal in order of magnitude.

Figure A.86 shows the wavelet coefficients for a chirp signal ranked in order of magnitude. If one were to reconstruct the signal only from those with the largest relative values, one would find that the main features of the original signal were recovered, the remainder being unimportant. It is clear that if only a small number of wavelets contribute to reconstruction then the coefficients form a highly compressed representation of the original signal.

Figure A.87 illustrates how a signal containing 512 sampled points can be reconstructed with great accuracy from only 128, or even 64, coefficients.

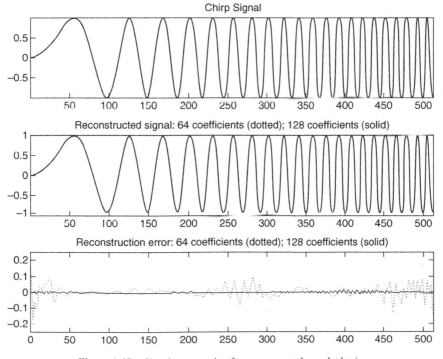

Figure A.87 Signal regeneration from compressed wavelet basis.

A.10   Filters

A.10.1   Introduction to Filters

So far in this appendix, all of the signal processing has been *passive* in the sense that it has been concerned with visualising signals without changing them in any substantial sense. This section will adopt an active approach to signal processing in the sense that one can use the techniques discussed here to actually change the nature of the signals that have been acquired; this process can be thought of as a form of data cleansing (which is an essential part of the SHM process described in Chapter 1). There are a number of reasons why one might wish to do this; one very important reason will be given first as motivation for all that follows.

Suppose that one is trying to develop a Fourier transforming device that can give the spectrum of a given time signal. Suppose further that limitations on the instrumentation mean that one has a maximum sampling frequency of 1000 Hz, that is a Nyquist frequency of 500 Hz. Now imagine that the time signal has a broadband spectrum that is flat up to 750 Hz. What will the estimated spectrum look like? Figure A.88 provides the answer to this question.

Thus energy is aliased into the range 250–500 Hz from the range 500–750 Hz and one obtains a completely fictitious spectrum. This is a serious problem and demands a resolution: how can we help this? Suppose one had a device that removed the part of the signal at frequencies between 500 and 750 Hz. Then one would have changed the signal admittedly, but the FFT would at least give an accurate spectrum all the way up to 500 Hz. Such a device, which passes parts of a signal's frequency content and suppresses others, is called a *filter*. The particular filter described above is called an *anti-aliasing filter* for obvious reasons.

In practical terms, one can implement a filter in essentially two ways. If the filter is a mechanical or electrical device that operates on the continuous-time physical signal, it is called an *analogue filter*. If the filter is a numerical algorithm or mechanical device that operates on sampled data, it is called a *digital filter*.

It is convenient to start the discussion here with an electrical analogue filter. Consider the circuit shown in Figure A.89 with an alternating voltage input,

$$V_i(t) = V_i\cos(\omega t) \tag{A.225}$$

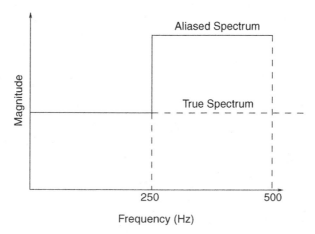

Figure A.88   Effect of aliasing on a broadband spectrum.

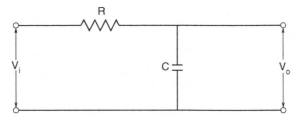

Figure A.89   Low-pass filter electrical circuit.

Elementary circuit theory gives the output voltage $V_o(t)$ as the solution of the differential equation,

$$RC\frac{dV_o}{dt} + V_o = V_i(t) \tag{A.226}$$

Passing to the frequency domain via the Fourier transform gives

$$iRC\omega V_o(\omega) + V_o(\omega) = V_i(\omega)$$

so that

$$V_o(\omega) = H(\omega)V_i(\omega) \tag{A.227}$$

In terms of an FRF where

$$H(\omega) = \frac{1}{1 + iRC\omega} \tag{A.228}$$

the gain of this system is

$$|H(\omega)| = \frac{1}{\sqrt{1 + R^2C^2\omega^2}} \tag{A.229}$$

and the corresponding phase is

$$\angle H(\omega) = -\tan^{-1}(RC\omega) \tag{A.230}$$

The gain of the circuit is illustrated in Figure A.90. When $RC\omega = 0.1$, $|H(\omega)| = 0.995$; when $RC\omega = 10.0$, $|H(\omega)| = 0.0995$.

As this circuit is essentially a filter that attenuates high frequencies and passes low frequencies, it is called a *low-pass filter*. This filter is an analogue filter as defined above. The next section will discuss how to implement a low-pass filter in digital form. It is important to note, though, that one cannot use a digital filter as an anti-aliasing filter; once the signal has been sampled, any damage has already been done.

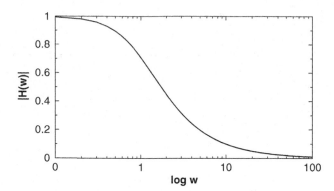

Figure A.90   Gain of electrical low-pass filter ($RC = 1$).

A.10.2   A Digital Low-Pass Filter

Now the question arises: how can we implement a filter of the sort described above on sampled data? First, one can adopt a more general notation. Let $x(t)$ or x_i be the input to the filter and $y(t)$ or y_i be the output. The differential equation of the electrical analogue filter is then

$$RC\frac{\mathrm{d}y}{\mathrm{d}t} + y = x(t) \tag{A.231}$$

Now, further suppose that x and y are sampled with an interval Δt, so that $x(t) \longrightarrow x_i = x(t_i) = x(i\Delta t)$ and $y(t) \longrightarrow y_i = y(t_i) = y(i\Delta t)$. The derivative in Equation (A.231) can be approximated by

$$\frac{\mathrm{d}y_i}{\mathrm{d}t} = \frac{y_i - y_{i-1}}{\Delta t} \tag{A.232}$$

and making this approximation results in a *difference equation*,

$$\frac{RC}{\Delta t}(y_i - y_{i-1}) + y_i = x_i$$

or, with a little straightforward rearrangement,

$$y_i = \left(\frac{RC}{\frac{RC}{\Delta t} + 1}\right)y_{i-1} + y_i = \left(\frac{1}{\frac{RC}{\Delta t} + 1}\right)x_i$$

or

$$y_i = a_1 y_{i-1} + b_0 x_i \tag{A.233}$$

with appropriate definitions for a_1 and b_0.

   In order to illustrate the action of a filter like the one in Equation (A.233), which is one of the simplest conceivable digital filters, consider the signal

$$x(t) = \sin(2\pi \times 5t) + 0.1\sin(2\pi \times 50t)$$

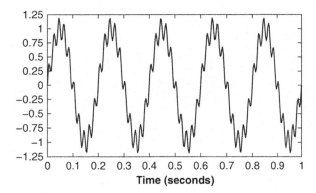

Figure A.91 'Noisy' sine wave.

(sampled with $\Delta t = 0.003$) as depicted in Figure A.91. This example is not without meaning (well in the UK anyway); it could represent a 5 Hz sine wave polluted by a small 50 Hz component picked up by the instrumentation from the electrical main.

After one pass through the digital filter,

$$y_i = 0.6655 y_{i-1} + 0.3345 x_i$$

The resulting signal is shown in Figure A.92. (The constants in this filter were obtained in an ad hoc manner; systematic means of determining filter coefficients will be given later.) The results are by no means perfect; the sine wave still carries some of the high-frequency component. The reason for this is that the simple low-pass filter above has a long 'tail' in the frequency domain and there is still some positive gain at 50 Hz that allows some of the noise signal through. The means of designing better filters will be covered a little later; in the meantime, one way of getting better performance is to apply the filter multiple times. For example, the resulting signal after five passes through the filter is shown in Figure A.93.

This example shows that the filter is able to remove the high-frequency sine wave effectively. However, it is very important to note that the amplitude of the low-frequency carrier signal has also been attenuated. Also, importantly, the phase of the signal has been altered.

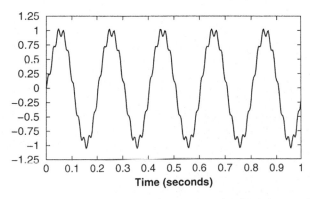

Figure A.92 'Noisy' sine wave after one pass through a simple low-pass filter.

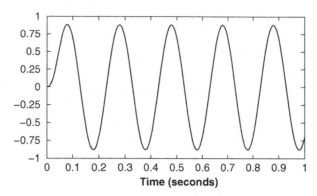

Figure A.93 'Noisy' sine wave after five passes through a simple low-pass filter.

The next question one should ask is – what is the frequency response of the new digital filter derived from the original analogue filter? In order to answer this question, one uses a theorem from Section A.4.2, which says that if the input to a system is $x(t) = \mathrm{e}^{i\omega t}$, then the response will be $y(t) = H(\omega)\mathrm{e}^{i\omega t}$; this is the *harmonic probing* algorithm.

As an aid to notation one can define the backward shift operator Z^{-1} by

$$Z^{-1}y_i = y_{i-1} \tag{A.234}$$

As one might imagine, this is related to the Z-transform, which is often used in the analysis of sampled data systems; however, the full potential of that approach will not be pursued here. The shift operator allows one to write Equation (A.233) as

$$(1 - a_1 Z^{-1})y_i = b_0 x_i \tag{A.235}$$

Let $x_i = \mathrm{e}^{i\omega t_i}$, so that then $y_i = H(\omega)\mathrm{e}^{i\omega t_i}$, and one observes that

$$Z^{-1}\mathrm{e}^{i\omega t_i} = \mathrm{e}^{i\omega t_{i-1}} = \mathrm{e}^{i\omega(t_i - \Delta t)} = \mathrm{e}^{-i\omega \Delta t}\mathrm{e}^{i\omega t_i}$$

Then Equation (A.235) becomes

$$(1 - a_1 \mathrm{e}^{-i\omega \Delta t})y_i = b_0 x_i$$

and it then follows that the FRF for the input–output process (and therefore the filter) is

$$H(\omega) = \frac{b_0}{1 - a_1 \mathrm{e}^{-i\omega \Delta t}} \tag{A.236}$$

Figure A.94 (in schematic form) compares the FRF gain of the digital filter with the FRF gain of the original analogue filter, and thus exposes an important difference.

Because of the presence of the complex exponential in the digital filter FRF, it is actually periodic in frequency. This is a reflection of the fact that the FRF cannot give any information about the system (filter) behaviour above the Nyquist frequency. It is also important to note that the final digital filter does not coincide with the analogue one even over the Nyquist interval. When the means of deriving general digital filters is discussed later, it will be critical to take account of this.

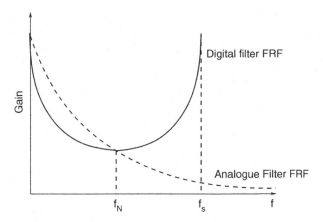

Figure A.94   FRF of an analogue low-pass filter and associated digital filter.

In terms of the FRF derived above, one has enough now to obtain a general result. A general (linear) digital filter would have the form

$$y_i = \sum_{j=1}^{n_y} a_j y_{i-j} + \sum_{j=0}^{n_x} b_j x_{i-j} \tag{A.237}$$

This differs from the simple low-pass filters discussed so far only in terms of how many past (or *lagged*) values of the signals are used. In terms of the backward shift operator one has

$$\left(1 - \sum_{j=1}^{n_y} a_j Z^{-j}\right) y_i = \left(\sum_{j=0}^{n_x} b_j Z^{-j}\right) x_i$$

so that symbolically one can write

$$\frac{y_i}{x_l} = \frac{\sum_{j=0}^{n_x} b_j Z^{-j}}{1 - \sum_{j=1}^{n_y} a_j Z^{-j}} = G(Z^{-1}) \tag{A.238}$$

where $G(Z^{-1})$ is commonly called the *pulse transfer function*. The standard FRF follows from substituting $Z^{-1} = e^{-i\omega \Delta t}$,

$$H(\omega) = \frac{\sum_{j=0}^{n_x} b_j e^{-ij\omega\Delta t}}{1 - \sum_{j=1}^{n_y} a_j e^{-ij\omega\Delta t}} \tag{A.239}$$

A.10.3   A High-Pass Filter

Recall the earlier formula for the low-pass filter FRF,

$$H(\omega) = \frac{1}{1 + iRC\omega} \tag{A.240}$$

This clearly encodes the essential properties of a low-pass filter, which are that

$$|H(\omega)| \longrightarrow 1 \quad \text{as} \quad \omega \longrightarrow 0$$

$$|H(\omega)| \longrightarrow 0 \quad \text{as} \quad \omega \longrightarrow \infty$$

By analogy, the required characteristics for a *high-pass filter* would be

$$|H(\omega)| \longrightarrow 0 \quad \text{as} \quad \omega \longrightarrow 0$$

$$|H(\omega)| \longrightarrow 1 \quad \text{as} \quad \omega \longrightarrow \infty$$

One observes that it is possible to obtain this characteristic by making a simple transformation on Equation (A.240), namely

$$i\omega \longrightarrow \frac{1}{i\omega} \quad \text{or} \quad \omega \longrightarrow -\frac{1}{\omega} \tag{A.241}$$

The FRF of the low-pass filter is transformed into

$$H(\omega) = \frac{1}{\frac{RC}{i\omega} + 1} = \frac{\frac{i\omega}{RC}}{1 + \frac{i\omega}{RC}}$$

and the resulting gain is

$$|H(\omega)| = \frac{\frac{\omega}{RC}}{\sqrt{1 + \frac{\omega^2}{R^2C^2}}}$$

which is high-pass as required. When $\omega/RC = 0.1$, $|H(\omega)| = 0.0995$; when $\omega/RC = 10.0$, $|H(\omega)| = 0.995$. The gain of this simple high-pass filter is illustrated in Figure A.95.

A.10.4   A Simple Classification of Filters

The following diagram (Figure A.96) shows the important features in the FRF of a low-pass filter. The figure shows a feature that has not appeared in the discussion so far – the passband ripple. It is sometimes

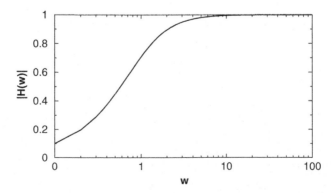

Figure A.95   Gain of a simple high-pass (analogue) filter ($RC = 1$).

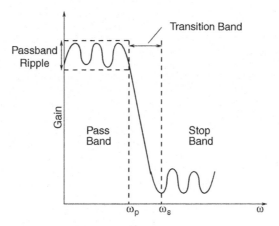

Figure A.96   Gain of a generic low-pass filter.

possible (and desirable) to trade off steepness of the filter cut-off against the smoothness of the gain curve; this depends on the family of filters under consideration. The diagram corresponding to Figure A.96 for a high-pass filter is given in Figure A.97.

In general, one can define two other main classes of filter: the *band-pass filter* (Figure A.98) and the *band-stop filter* (Figure A.99).

A.10.5   Filter Design

So far the only means of establishing a digital filter has been through the intermediate 'design' of an electrical analogue filter. It is important to have more flexibility in designing filters; for example, one would like to specify characteristics like the rate of fall-off of the response after the cut-off frequency. As alluded to above, the first stage in the design of a given filter is to choose which of the available families of filters is appropriate. One always starts by considering analogue filters and one of the most useful families of analogue filters is that of *Butterworth filters*. This family is one of the most basic,

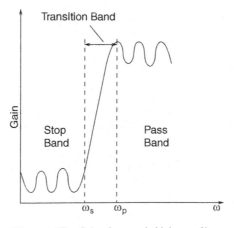

Figure A.97   Gain of a generic high-pass filter.

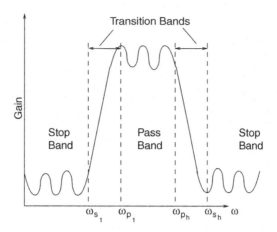

Figure A.98   Gain of a generic band-pass filter.

but is nonetheless very often used. Among the general characteristics of the Butterworth filters is that they have smooth gain curves; as described above, one can adopt a family that allows passband ripple – like the Chebyshev family – and this would potentially allow steeper cut-off characteristics for the same 'size' (number of lags) of filter.

The Butterworth filters are controlled by two parameters for the low-pass filter case. (Only low-pass filters will be considered here; other filters like those in the last section can be obtained by transformation of low-pass filters, so, in a sense, low-pass filters are the building blocks of the filter world.) The FRF gain of the Butterworth family is specified by

$$|H(\omega)|^2 = \frac{1}{1 + \left(\frac{\omega}{\omega_c}\right)^{2n}} = \frac{1}{1 + \left(\frac{f}{f_c}\right)^{2n}} \qquad (A.242)$$

where ω_c is the cut-off frequency and n is a sort of steepness factor that specifies how fast the signal should die away after the cut-off frequency.

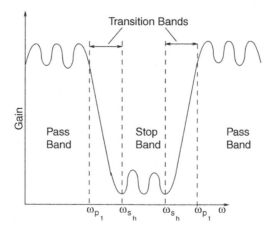

Figure A.99   Gain of a generic band-stop filter.

Suppose n is very large. If $\omega < \omega_c$, then $\frac{\omega}{\omega_c} < 1$ and $\left(\frac{\omega}{\omega_c}\right)^{2n} \approx 0$. Therefore $H(\omega) \approx 1$ for $\omega < \omega_c$. If $\omega > \omega_c$, then $\frac{\omega}{\omega_c} > 1$ and $\left(\frac{\omega}{\omega_c}\right)^{2n} \gg 1$. Therefore $H(\omega) \approx 0$ for $\omega > \omega_c$.

The critical question is: how can one specify the requirements for a digital filter and translate them into design parameters based on Equation (A.242)? One might require a cut-off frequency of $f_c = 2\pi\omega_c = 20$ Hz at a sampling frequency of 250 Hz. (It will become clear that all frequencies are considered relative to the sampling frequency and that the sampling frequency is therefore a crucial ingredient in the filter specification.) Furthermore, in order to specify the sharpness of the cut-off, one may require a 20 dB fall-off by 50 Hz.

The decibel is often used as a means of stating filter design requirements, so it is useful to recall that the definition of the decibel, or dB, which is a way of stating the difference between two signals levels s_1 and s_2, is

$$dB(s_2, s_1) = 20 \log_{10} \frac{s_2}{s_1}$$

and a difference (decrease) of 20 dB therefore corresponds to the following:

$$-20 = 20 \log_{10} \frac{s_2}{s_1}$$

$$-1 = \log_{10} \frac{s_2}{s_1}$$

$$10^{-1} = \frac{s_2}{s_1}$$

$$s_2 = \frac{s_1}{10}$$

so a -20 db fall-off corresponds to a gain reduction by a factor of 10. This summarises the correspondence between ratios of signal levels and the dB difference.

Returning to the filter design, the above specification means that the filter should have a unit gain (passes all signals) as far as possible up to 20 Hz, but should only be passing a tenth of the signal by 50 Hz. (One must remember that all of this is relative to a sampling frequency of 250 Hz.) However, care is needed here; these specifications refer to the digital filter that one ultimately wishes to arrive at, but Equation (A.242) is an analogue filter. One should recall from Figure A.94 and the surrounding discussion that the filter characteristics change in passing from analogue to digital and this effect has to be corrected for. One needs a means of translating from 'digital' frequencies to 'analogue' frequencies; this is the idea of *frequency warping*.

A.10.6 The Bilinear Transformation

Consider the simple filter with FRF,

$$H(\omega) = \frac{b}{i\omega + a} \tag{A.243}$$

If one wishes to reconstruct the input–output system corresponding to this FRF, one can reorganise the equation so that

$$Y(\omega) = H(\omega)X(\omega)$$

where $X(\omega)$ and $Y(\omega)$ are the spectra corresponding to $x(t)$ and $y(t)$, the proposed filter input and output. Substituting Equation (A.243) into the last equation gives

$$(i\omega + a)Y(\omega) = bX(\omega)$$

and taking the inverse Fourier transform (using the rules established earlier for derivatives) yields

$$\dot{y}(t) + ay(t) = bx(t) \qquad\qquad (A.244)$$

Now, according to the trapezium rule for integration (of a sampled signal), one has

$$y_i = y_{i-1} = \frac{\Delta t}{2}(\dot{y}_i + \dot{y}_{i-1}) \qquad\qquad (A.245)$$

and from Equation (A.244) one has

$$\dot{y}_i = -ay_i + bx_i$$
$$\dot{y}_{i-1} = -ay_{i-1} + bx_{i-1}$$

Substituting these relations into Equation (A.245) gives

$$y_i = y_{i-1} = \frac{\Delta t}{2}[-a(y_i + y_{i-1}) + b(x_i + x_{i-1})]$$

Rearranging the latter expression gives

$$\left(1 + a\frac{\Delta t}{2}\right) y_i - \left(1 + a\frac{\Delta t}{2}\right) y_{i-1} = b\frac{\Delta t}{2}(x_i + x_{i-1})$$

or

$$\left[\left(1 + a\frac{\Delta t}{2}\right) - \left(1 + a\frac{\Delta t}{2}\right) Z^{-1}\right] y_{i-1} = b\frac{\Delta t}{2}(1 + Z^{-1})x_i$$

so the pulse transfer function of this system is given by

$$G(Z^{-1}) = \frac{b\frac{\Delta t}{2}(1 + Z^{-1})}{(1 + a\frac{\Delta t}{2}) - (1 + a\frac{\Delta t}{2})Z^{-1}}$$

$$= \frac{b\frac{\Delta t}{2}(1 + Z^{-1})}{a\frac{\Delta t}{2}(1 + Z^{-1}) + 1 - Z^{-1}}$$

$$\frac{b}{\frac{2}{\Delta t}\left(\frac{1 - Z^{-1}}{1 + Z^{-1}}\right) + a} \qquad\qquad (A.246)$$

Now, in terms of the 'digital' frequencies ω_d, one recalls from earlier that

$$Z^{-1} = e^{-i\omega_d \Delta t}$$

so the 'digital' FRF is

$$\frac{b}{\frac{2}{\Delta t}\left(\frac{1-e^{-i\omega_d\Delta t}}{1+e^{-i\omega_d\Delta t}}\right)+a} \tag{A.247}$$

If this is to coincide with the 'analogue' FRF defined by Equation (A.243) there has to be a correspondence between 'digital' frequencies and 'analogue' frequencies (which will be denoted ω_a), given by

$$i\omega_a = \frac{2}{\Delta t}\left(\frac{1-e^{-i\omega_d\Delta t}}{1+e^{-i\omega_d\Delta t}}\right) \tag{A.248}$$

This is the promised *frequency–warping* relationship, which is needed in order to make the digital and analogue FRFs coincide.

The relationship in Equation (A.248) can be simplified to an extent. Suppose one multiplies the top and bottom of Equation (A.248) by $\exp(i\omega_d\Delta t/2)$; one obtains

$$i\omega_a = \frac{2}{\Delta t}\left(\frac{e^{i\omega_d\frac{\Delta t}{2}}-e^{-i\omega_d\frac{\Delta t}{2}}}{e^{i\omega_d\frac{\Delta t}{2}}+e^{-i\omega_d\frac{\Delta t}{2}}}\right)$$

or

$$\omega_a = \frac{2}{\Delta t}\left(\frac{\frac{1}{2i}(e^{i\omega_d\frac{\Delta t}{2}}-e^{-i\omega_d\frac{\Delta t}{2}})}{\frac{1}{2}(e^{i\omega_d\frac{\Delta t}{2}}+e^{-i\omega_d\frac{\Delta t}{2}})}\right)$$

$$\frac{2}{\Delta t}\frac{\sin\left(\omega_d\frac{\Delta t}{2}\right)}{\cos\left(\omega_d\frac{\Delta t}{2}\right)}$$

or, finally,

$$\omega_a = \frac{2}{\Delta t}\tan\left(\omega_d\frac{\Delta t}{2}\right) \tag{A.249}$$

A little further rearrangement gives the very symmetrical form,

$$\pi\frac{f_a}{f_s} = \tan\left(\pi\frac{f_d}{f_s}\right) \tag{A.250}$$

where f_s is the sampling frequency. This last equation makes it very clear that the absolute frequencies are essentially irrelevant; only the values relative to the sampling frequency are significant. Note that the interval $[0,\infty]$ for the analogue frequency is mapped uniquely to and from the interval $[0, f_N]$ for the digital frequency. This means that the warping function is squashing the whole analogue frequency range on to the Nyquist interval for the digital frequency.

What does Equation (A.249) or (A.250) mean in practical terms? Suppose one requires a cut-off frequency of ω_{cd} for the cut-off of the digital filter; this is a 'digital frequency' in the terms discussed above, and one now knows that this should not be used directly in the analogue FRF specification, as it is known that the analogue and digital FRFs do not correspond. One therefore translates the frequency of interest here into the appropriate analogue cut-off $\omega_{ca} = 2/\Delta t\tan(\omega_{cd}\Delta t/2)$. With this specification, when the analogue filter is discretised, the digital cut-off is in the desired place.

A.10.7   An Example of Digital Filter Design

Consider the filter specification discussed earlier, that is a digital cut-off at 20 Hz, signal 20 dB down at 50 Hz. As discussed previously, the first thing to do in the construction of the filter is to convert to an analogue specification for the frequencies. First, the cut-off frequency,

$$f_{ca} = \frac{1}{\pi \Delta t} \tan \left(\frac{20\pi}{250} \right) = \frac{0.257}{\pi \Delta t}$$

and then the frequency of the -20 dB reference point,

$$f_{(-20\mathrm{dB})a} = \frac{1}{\pi \Delta t} \tan \left(\frac{50\pi}{250} \right) = \frac{0.727}{\pi \Delta t}$$

Then

$$\frac{f_{(-20\mathrm{dB})a}}{f_{ca}} = 2.83$$

so the gain condition for the filter at the -20 dB point is

$$|H(2\pi f_{(-20\mathrm{dB})a})|^2 = \frac{1}{1 + 2.83^{2n}} \le 10^{-2}$$

Or, alternatively,

$$1 + 2.83^{2n} \ge 10^2$$

This finally implies

$$n \ge 2.2$$

but n must be an integer (this will become clear later), so one has to take $n = 3$ in order to get the required speed of fall-off. This means that, in order to obtain a digital filter with the required specification, one has to begin with an analogue filter satisfying

$$|H(\omega)|^2 = \frac{1}{\left(\frac{\pi \Delta t f}{0.257} \right)^6 + 1} = \frac{1}{\left(\frac{\Delta t \omega}{0.514} \right)^6 + 1} = \frac{1}{\left(\frac{\omega}{\omega_{ca}} \right)^6 + 1} \qquad (A.251)$$

and a simple rearrangement gives

$$|H(\omega)|^2 = \frac{\omega_{ca}^6}{\omega^6 + \omega_{ca}^6} \qquad (A.252)$$

Now, the poles of the above expression (which will prove critical in the analysis) are the roots of the equation

$$\omega^6 + \omega_{ca}^6 = 0$$

Furthermore, as the coefficients in the latter polynomial equation are real, the roots will appear in complex conjugate pairs; suppose one labels them ω_1, ω_1^*, ω_2, ω_2^*, ω_3 and ω_3^*. It is known from complex analysis

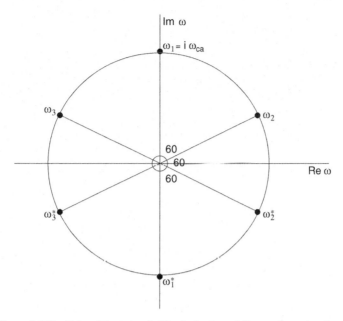

Figure A.100 Poles of the example filter in the Argand diagram (complex plane).

that the zeros of interest here, and hence the poles of the filter, will be spaced equidistantly on a circle of radius ω_{ca} in the Argand diagram (complex plane) as in Figure A.100. It follows from the diagram that one can choose the poles

$$\omega_1 = i\omega_{ca}$$

$$\omega_2 = \omega_{ca}e^{i\pi/6} = \omega_{ca}\left(\cos\left(\frac{\pi}{6}\right) + i\sin\left(\frac{\pi}{6}\right)\right) = \omega_{ca}\left(\frac{\sqrt{3}}{2} + \frac{i}{2}\right)$$

$$\omega_3 = \omega_{ca}e^{i5\pi/6} = \omega_{ca}\left(-\frac{\sqrt{3}}{2} + \frac{i}{2}\right)$$

to be in the upper complex plane so that ω_1^*, ω_2^* and ω_3^* are in the lower plane.

Now, factorising the denominator of (A.252) gives

$$|H(\omega)|^2 = \frac{\omega_{ca}^6}{(\omega - \omega_1)(\omega - \omega_2)(\omega - \omega_3)(\omega - \omega_1^*)(\omega - \omega_2^*)(\omega - \omega_3^*)}$$

and one of the means by which this can be satisfied is if

$$H(\omega) = \frac{i\omega_{ca}^3}{(\omega - \omega_1)(\omega - \omega_2)(\omega - \omega_3)} \tag{A.253}$$

where one has specifically chosen the stable poles of the filter (i.e. those in the upper half of the ω plane).

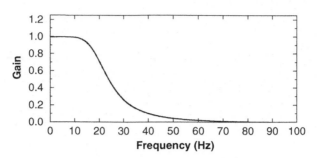

Figure A.101 Gain of a third-order low-pass Butterworth filter.

However, ω_1, ω_2 and ω_3 are all known from the above, so one can expand the polynomial, and with a certain amount of algebraic effort one obtains

$$H(\omega) = \frac{i\omega_{ca}^3}{(i\omega + \omega_{ca})((i\omega)^2 + \omega_{ca}(i\omega) + \omega_{ca})}$$

It is convenient to group the imaginary i's with the ω's because one now substitutes

$$i\omega = \frac{2}{\Delta t}\left(\frac{1 - Z^{-1}}{1 + Z^{-1}}\right) \quad \text{and} \quad \omega_{ca} = \frac{0.514}{\Delta t}$$

and thus obtains

$$\frac{0.514^3}{\left(2\left(\frac{1-Z^{-1}}{1+Z^{-1}}\right) + 0.514\right)\left(4\left(\frac{1-Z^{-1}}{1+Z^{-1}}\right)^2 + 1.028\left(\frac{1-Z^{-1}}{1+Z^{-1}}\right) + 0.2642\right)}$$

This is finally the pulse transfer function of the required digital filter. A certain amount of tedious arithmetic occurs at this point, but eventually one finds

$$G(Z^{-1}) = \frac{0.0102(1 + 3Z^{-1} + 3Z^{-2} + Z^{-3})}{1 - 2.003Z^{-1} + 1.446Z^{-2} - 0.361Z^{-3}}$$

and (using the example leading to Equation (A.146) as a guide) one can then read off the required digital filter as

$$y_i = 2.081y_{i-1} - 1.446y_{i-2} + 0.361y_{i-3} + 0.0102(x_i + 3x_{i-1} + 3x_{i-2} + x_{i-3})$$

The gain of this filter is shown in Figure A.101; note that the response below the cut-off is much flatter than for the single-pole filter shown in the earlier sections.

It is now possible to give the general design procedure for a low-pass Butterworth filter; however a short digression is needed first.

A.10.8   Combining Filters

A *cascade* of n filters with individual FRFs $H_i(\omega)$, $i = 1, \ldots, n$, has FRF,

$$H_c(\omega) = H_1(\omega)H_2(\omega)\cdots H_n(\omega) \tag{A.254}$$

and has the diagrammatic form shown in Figure A.102.

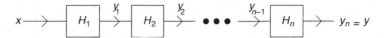

Figure A.102   Cascade combination of filters.

A *parallel* sum of filters has

$$H_p(\omega) = H_1(\omega) + H_2(\omega) + \cdots + H_n(\omega)$$

(A.255)

and has the diagrammatic form shown in Figure A.103.

A.10.9   General Butterworth Filters

Everything is now in place that is needed in order to construct a general Butterworth filter.

First suppose that the specification of the cut-off and so on has determined the Butterworth parameters ω_c and n. In order to simplify matters a little, the following discussion will work in the s plane or Laplace domain, where

$$s = i\omega$$

(A.256)

In this case, the Butterworth low-pass filter has the characteristic

$$|H(s)|^2 = \frac{1}{1 + \left(\frac{s}{i\omega_c}\right)^{2n}} = \frac{(i\omega_c)^{2n}}{s^{2n} + (i\omega_c)^{2n}}$$

(A.257)

and the poles of the filter are therefore the solutions of the polynomial equation

$$s^{2n} = -i^{2n}\omega_c^{2n}$$

Just as in the example case earlier (as in Figure A.100) the poles will be equispaced in the s plane (Argand diagram) on a circle of radius ω_c. As $i = e^{i\pi/2}$ and $-1 = e^{i\pi}$, one has

$$s^{2n} = e^{i\pi}e^{in\pi}\omega_c^{2n} = e^{i\pi(1+n)}\omega_c^{2n}$$

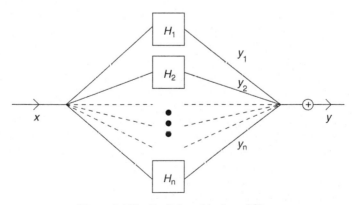

Figure A.103   Parallel combination of filters.

It follows that one solution is

$$s = e^{i\frac{\pi}{2}(1+\frac{1}{n})}\omega_c$$

But, because $e^{i2\pi} = 1$, it is also true that

$$s^{2n} = e^{i\pi}e^{in\pi}\omega_c^{2n} = e^{i\pi(1+n+2k)}\omega_c$$

where $k = 0, \ldots, 2n - 1$. This gives the full set of solutions,

$$s_k = \omega_c \exp\left(i\pi\frac{(2k+n+1)}{2n}\right), \quad k = 0, \ldots, 2n - 1$$

or, if one wishes the index k to run from 1 to $2n$,

$$s_k = \omega_c \exp\left(i\pi\frac{(2k+n-1)}{2n}\right), \quad k = 1, \ldots, 2n \tag{A.258}$$

To give a concrete example, suppose that $n = 3$; the solutions are seen to be

$$s_k = \omega_c \exp\left(i\pi\frac{(k+1)}{3}\right), \quad k = 1, \ldots, 6$$

or

$$s_1 = \omega_c e^{i\pi\frac{2}{3}}, s_2 = \omega_c e^{i\pi}, s_3 = \omega_c e^{i\pi\frac{4}{3}}, s_4 = \omega_c e^{i\pi\frac{5}{3}}, s_5 = \omega_c, s_6 = \omega_c e^{i\pi\frac{7}{3}},$$

as depicted in Figure A.104. Note that with an appropriate labelling,

$$s_1 = s_3^*, \quad s_2 = s_2^*, \quad s_4 = s_6^* = s_0^*, \quad s_5 = s_5^* = s_{-1}^*$$

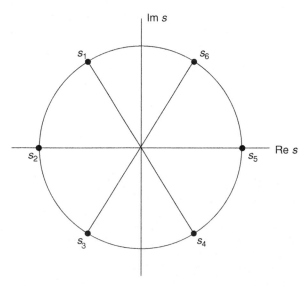

Figure A.104   Poles for a third-order Butterworth low-pass filter (s plane).

or, more generally,

$$s_k = s_{n-k+1}^*$$ (A.259)

where the solutions are periodic in k with period $2n$.

In order to have a stable filter, one only uses the poles in the left-hand side of the s plane and these are the poles $s_1, \ldots, s_n$. Now, as observed earlier, one of the possible filters satisfying (A.257) is simply

$$H(s) = \frac{\omega_c^n}{(s - s_1)(s - s_2) \cdots (s - s_n)}$$ (A.260)

and one should be aware that this choice has made a prescription for the phase of the filter FRF and thus the filter. It follows from Equation (A.260) that

$$H(0) = 1 = \frac{\omega_c^n}{s_1 s_2 \cdots s_n}$$

which implies $s_1 s_2 \ldots s_n = \omega_c^n$ so that Equation (A.260) can be rewritten as

$$H(s) = \frac{s_1 s_2 \cdots s_n}{(s - s_1)(s - s_2) \cdots (s - s_n)}$$ (A.261)

It is a simple matter to implement this filter as a cascade of second-order filters or *sections*,

$$P_k(s) = \frac{s_k s_{n-k+1}}{(s - s_k)(s - s_{n-k+1})}$$

From Equation (A.259) it follows that $s_k s_{n-k+1} = s_k s_k^* = |s_k|^2 = \omega_c^2$, so that, finally,

$$P_k(s) = \frac{\omega_c^2}{s^2 - 2\mathrm{Re}\, s_k s \omega_c + \omega_c^2}$$

and using Equation (A.258),

$$P_k(s) = \frac{\omega_c^2}{s^2 - 2s\omega_c \cos\pi\left(\frac{2k+n-1}{2n}\right) + \omega_c^2}$$ (A.262)

If n is even, the required filter is the cascade

$$H(s) = \prod_{k=1}^{n/2} P_k(s)$$ (A.263)

If n is odd, then $s_{(n+1)/2}$ does not pair up with another pole, but $s_{(n+1)/2}$ in this case is always equal to $\omega_c e^{i\pi} = -\omega_c$; this means that one simply adds a first-order section and the transfer function of the filter is then

$$H(s) = \frac{\omega_c}{s + \omega_c} \prod_{k=1}^{n-1/2} P_k(s)$$ (A.264)

(The final result has been multiplied here by -1, but that will not change the overall gain of the filter, only the phase.)

It is now straightforward to consider the implementation as a digital filter. One recalls from an earlier section that

$$s = i\omega = \frac{2}{\Delta t}\left(\frac{1 - Z^{-1}}{1 + Z^{-1}}\right)$$

is used for conversion to the digital filter. This means each second-order section becomes

$$P_k(Z^{-1}) = \frac{\omega_c^2}{\frac{4}{\Delta t^2}\left(\frac{1-Z^{-1}}{1+Z^{-1}}\right)^2 + q_k\frac{2}{\Delta t}\left(\frac{1-Z^{-1}}{1+Z^{-1}}\right) + \omega_c^2} \tag{A.265}$$

where this last equation incorporates the definition

$$q_k = -\omega_c\cos\left(\frac{2k + n - 1}{2n}\right)$$

After a fairly tedious calculation, Equation (A.243) ultimately leads to

$$P_k(Z^{-1}) = \frac{b^k(1 + 2Z^{-1} + Z^{-2})}{1 - a_1 Z^{-1} - a_2^k Z^{-2}} \tag{A.266}$$

where

$$b_k = \left(\frac{4}{\Delta t^2} + \frac{2q_k}{\Delta t} + \omega_c^2\right)^{-1} \tag{A.267}$$

$$a_1 = -\left(2 - \frac{8}{\Delta t^2}\right)b_k \tag{A.268}$$

and

$$a_2^k = -\left(\frac{4}{\Delta t^2} - \frac{2q_k}{\Delta t} + \omega_c^2\right)b_k \tag{A.269}$$

So, finally, each second-order section implemented as a digital filter takes the form

$$y_i^k = a_1 y_{i-1}^k + a_2^k y_{i-2}^k + b^k\left(y_i^{k-1} + 2y_{i-1}^{k-1} + y_{i-2}^{k-1}\right) \tag{A.270}$$

where y^{k-1} is the input to the kth section and y^k is the corresponding output, with $y_0 = x$, the overall input to the filter and $y^n = y$, the overall output.

If n is odd, the first-order section in Equation (A.264) converts to the digital form

$$\frac{\omega_c}{\frac{2}{\Delta t}\left(\frac{1-Z^{-1}}{1+Z^{-1}}\right) + \omega_c}$$

which can be manipulated into the form

$$\frac{b(1 + Z^{-1})}{1 - a_1 Z^{-1}} \tag{A.271}$$

where

$$b = \left(\omega_c + \frac{2}{\Delta t} \right)^{-1} \qquad (A.272)$$

and

$$a_1 = \left(\omega_c - \frac{2}{\Delta t} \right) b \qquad (A.273)$$

The corresponding implementation on the sampled data is

$$y_i^k = a_1 y_{i-1}^k + b \left(y_i^{k-1} + y_{i-1}^{k-1} \right) \qquad (A.274)$$

(for the appropriate $k = (n+1)/2$).

If n is even, the FRF of the digital filter is

$$H(\omega) = \prod_{k=1}^{n/2} P_k(\omega) \qquad (A.275)$$

where

$$P_k(Z^{-1}) = \frac{b^k(1 + 2e^{-i\omega\Delta t} + e^{-2i\omega\Delta t})}{1 - a_1 e^{-i\omega\Delta t} - a_2^k e^{-2i\omega\Delta t}} \qquad (A.276)$$

with the coefficients defined as above. If n is odd one has

$$H(\omega) = A(\omega) \prod_{k=1}^{(n-1)/2} P_k(\omega) \qquad (A.277)$$

where

$$A(\omega) = \frac{b(1 + e^{-i\omega\Delta t})}{1 - a_1 e^{-i\omega\Delta t}} \qquad (A.278)$$

with the appropriate coefficients.

A.11 System Identification

A.11.1 Introduction

System identification is another activity in which signal processing is used in an *active* sense. The discipline is based on the idea that, given examples of input and output signals to and from a given system, one can, under certain circumstances, estimate the equations of motion of the system. To expand a little on this definition, consider a given physical system that responds in some measurable way $y_s(t)$ when an external stimulus or excitation $x(t)$ is applied. Now, suppose that a mathematical model of the system is required that responds with an identical output $y_m(t)$ when presented with the same stimulus.

The model will, in the most general terms, be some functional $S[.]$ that maps the input $x(t)$ to the output $y_m(t)$. In mathematical terms,

$$y_m(t) = S[x](t) \tag{A.279}$$

If the model changes when the frequency or amplitude characteristics of the excitation change, it is said to be *input-dependent*. Such models are unsatisfactory in that they may have very limited predictive capabilities. This situation commonly arises when one attempts to fit a linear model to a system that is in fact nonlinear.

The problem of system identification is therefore to obtain an appropriate functional $S[]$ for a given system. If a priori information about the system is available, the complexity of the problem can be reduced considerably. For example, suppose that the system is known to be a continuous-time linear SDOF dynamical system; in this case the form of the equation relating the input $x(t)$ and the response $y(t)$ is known to be (the subscripts on y will be omitted where the meaning is clear from the context)

$$m\ddot{y} + c\dot{y} + ky = x(t) \tag{A.280}$$

In this case the implicit structure of the functional $S[]$ is known and the only unknowns are the coefficients or parameters m, c and k; the problem has been reduced to one of *parameter estimation*. In structural dynamics, one is often helped by the fact that Newton's laws of motion generally lead to second-order differential equations and this means that it is natural to assume a continuous-time representation of systems. Such a representation defines the motion at *all* times. In reality, most observations of system behaviour – measurements of input and output signals – will be carried out at discrete intervals. The system data are then a discrete set of measurement values $\{x_i, y_i, i = 1, \dots, N\}$.

For modelling purposes one might therefore ask if there exists a model structure – like a digital filter – that maps the discrete measured inputs x_i directly to the corresponding discrete outputs y_i; such models do exist and in many cases offer advantages over the continuous-time representation.

Consider the general linear SDOF system specified by Equation (A.280) and suppose that one is only interested in the value of the output at a sequence of regularly spaced times t_i where $t_i = (i - 1)\Delta t$ (as usual, Δt is the *sampling interval* and the associated frequency $f_s = 1/\Delta t$ is the *sampling frequency*). Because the equation of motion holds for all times, it is certainly true that at each sampling instant t_i,

$$m\ddot{y}_i + c\dot{y}_i + ky_i = x_i \tag{A.281}$$

where $x_i = x(t_i)$ and so on. As discussed earlier in relation to digital filters, the derivatives $\dot{y}(t_i)$ and $\ddot{y}(t_i)$ can be approximated by the discrete forms

$$\dot{y}_i = \dot{y}(t_i) \approx \frac{y(t_i) - y(t_i - \Delta t)}{\Delta t} = \frac{y_i - y_{i-1}}{\Delta t} \tag{A.282}$$

$$\ddot{y}(t_i) \approx \frac{y_{i+1} - 2y_i + y_{i-1}}{\Delta t^2} \tag{A.283}$$

Substituting these approximations into Equation (A.281) yields, after a little rearrangement,

$$y_i = \left(2 - \frac{c\Delta t}{m} - \frac{k\Delta t^2}{m}\right) y_{i-1} + \left(\frac{c\Delta t}{m} - 1\right) y_{i-2} + \frac{\Delta t^2}{m} x_{i-1} \tag{A.284}$$

or

$$y_i = a_1 y_{i-1} + a_2 y_{i-2} + b_1 x_{i-1} \tag{A.285}$$

where the constants a_1, a_2, b_1 are defined by the previous equation. Equation (A.285) is a *discrete-time* representation of the SDOF system under study.

Note that the motion for all discrete times is fixed by the input sequence x_i together with values for y_1 and y_2. The specification of the first two values of the output sequence is directly equivalent to the specification of initial values for $y(t)$ and $\dot{y}(t)$ in the continuous-time case. An obvious advantage of using a discrete model like Equation (A.285) is that it is much simpler to numerically predict the output in comparison with a differential equation. The price one pays is a loss of generality, because the coefficients in Equation (A.285) are functions of the sampling interval Δt; one could only use this model to predict responses with the same spacing in time.

Although arguably less familiar, the theory for the solution of difference equations is no more difficult than the corresponding theory for differential equations. Almost every fact about a continuous-time system is mirrored in the discrete-time case and many examples of this are given in Section B.4 in Appendix B. This all means that the discrete-time form of the system equations of motion is often the most convenient for system identification or parameter estimation purposes. As in the continuous-time case, it is possible to derive dual frequency-domain representations for the systems; however, before moving on to discuss this, it will be useful to digress slightly in order to discuss the taxonomy of difference equations, particularly as they will feature in later discussions and in the main body of this book.

A.11.1.1 Autoregressive (AR) Models

As suggested by the name, an autoregressive model expresses the present output y_i from a system as a linear combination of past outputs; that is, the variable is regressed on itself. The general expression for such a model is

$$y_i = \sum_{j=1}^{p} a_j y_{i-j} \tag{A.286}$$

and is termed an AR(p) model.

A.11.1.2 Moving-Average (MA) Models

In this case the output is expressed as a linear combination of past inputs. One can think of the output as a weighted average of the inputs over a finite window that moves with time – hence the name. The general form is

$$y_i = \sum_{j=1}^{q} b_j x_{i-j} \tag{A.287}$$

and is called an MA(q) model. All linear continuous-time systems have a canonical representation as a moving-average model as a consequence of the canonical linear input–output relationship

$$y(t_i) = \int_0^{+\infty} d\tau \, h(\tau) x(t_i - \tau)$$

which can be approximated by the discrete sum

$$y_i = \sum_{j=0}^{\infty} \Delta t \, h(j\Delta t)x(t_i - j\Delta t)$$

As $t_i - j\Delta t = t_{i-j}$, one has

$$y_i = \sum_{j=0}^{\infty} b_j x_{i-j}$$

which is an MA(∞) model with $b_j = \Delta t \, h(j\Delta t)$.

A.11.1.3　Autoregressive Moving-Average (ARMA) Models

As the name suggests, these are simply a combination of the two model types discussed above. The general form is the ARMA(p, q) model,

$$y_i = \sum_{j=1}^{p} a_j y_{i-j} + \sum_{j=1}^{q} b_j x_{i-j} \tag{A.288}$$

which is quite general in the sense that any discretisation of a linear differential equation will yield an ARMA model. Equation (A.288) for the discrete version of an SDOF system is an ARMA(2, 1) model, as shown in Equation (A.285).

Note that a given continuous-time system will have in general many discrete-time representations. By virtue of the arguments above, the linear SDOF system can be modelled using either an MA(∞) or an ARMA(2, 1) structure. The advantage of using the ARMA form is that far fewer past values of the variables need to be included to predict with the same accuracy as the MA model.

A.11.2　Discrete-Time Models in the Frequency Domain

The aim of this short section is to give the simple construction of the FRF for a discrete-time system; it is essentially a repetition of what has already been given for digital filters. The discussion of the preceding section shows that the ARMA(p, q) structure is sufficiently general in the linear case; that is, the system of interest is given by Equation (A.288) (and this is, of course, nothing more than the linear digital filter).

Recalling the backward shift operator Z^{-1}, which is defined by its action on signals $Z^{-k} y_i = y_{i-k}$, allows one to rewrite Equation (A.288) as

$$y_i = \left(\sum_{j=1}^{p} a_j Z^{-j} \right) y_i + \left(\sum_{j=1}^{q} b_j Z^{-j} \right) x_i \tag{A.289}$$

or

$$\left(1 - \sum_{j=1}^{p} a_j Z^{-j} \right) y_i = \left(\sum_{j=1}^{q} b_j Z^{-j} \right) x_i \tag{A.290}$$

Now, one simply defines the FRF $H(\omega)$ by the means suggested in Section A.10.2; if the input to the system is $e^{i\omega t}$, the output is $H(\omega)e^{i\omega t}$. At the risk of a little repetition, the action of Z^{-1} on the signals is given by

$$Z^{-m}x_k = Z^{-m}e^{i\omega k\Delta t} = e^{i\omega(k-m)\Delta t} = e^{-im\omega\Delta t}x_k$$

on the input and

$$Z^{-m}y_k = Z^{-m}H(\omega)x_k = H(\omega)Z^{-m}e^{i\omega k\Delta t} = H(\omega)e^{i\omega(k-m)\Delta t} = H(\omega)e^{-im\omega\Delta t}x_k$$

on the output. Substituting these results into Equation (A.290) yields

$$\left(1 - \sum_{j=1}^{p} a_j e^{-ij\omega\Delta t}\right)H(\omega)x_i = \left(\sum_{j=1}^{q} b_j e^{-ij\omega\Delta t}\right)x_i$$

which, on simple rearrangement, gives the required result

$$H(\omega) = \frac{\sum_{j=1}^{q} b_j e^{-ij\omega\Delta t}}{\left(1 - \sum_{j=1}^{p} a_j e^{-ij\omega\Delta t}\right)} \tag{A.291}$$

Note that this expression is periodic in ω.

A.11.3 Least-Squares Parameter Estimation

Having seen that the AR(p,q) form is versatile enough to accommodate the general linear system, the object of the present section is to give a brief description of the least-squares methods that can be used to estimate the model parameters given samples of measured data. To assume a concrete example, suppose a model of the ARMA(2,1) form, as in Equation (A.285), is proposed for a set of measured input and output data $\{x_i, y_i; i = 1, \ldots, N\}$. Taking measurement noise into account, one has at each sampling instant the relation

$$y_i = a_1 y_{i-1} + a_2 y_{i-2} + b_1 x_{i-1} + \zeta_i \tag{A.292}$$

where the residual signal ζ_i is assumed to contain the output noise and an error component due to the fact that the parameter estimates may be incorrect.

The *least-squares* estimator finds the set of parameter estimates that minimise the error function,

$$J = \sum_{i=1}^{N} \zeta_i^2 \tag{A.293}$$

The parameter estimates obtained by minimising the errors will hopefully reduce the residual sequence to measurement noise only.

The problem is best expressed in terms of matrices. Assembling each equation of the form of (A.292) for $i = 3, \ldots, N$ into a matrix equation gives

$$
\begin{pmatrix} y_3 \\ y_4 \\ \vdots \\ y_N \end{pmatrix} = \begin{pmatrix} y_2 & y_1 & x_2 \\ y_3 & y_2 & x_3 \\ \vdots & \vdots & \vdots \\ y_{N-1} & y_{N-2} & x_{N-1} \end{pmatrix} \begin{pmatrix} a_1 \\ a_2 \\ b_1 \end{pmatrix} + \begin{pmatrix} \zeta_3 \\ \zeta_4 \\ \vdots \\ \zeta_N \end{pmatrix}
\tag{A.294}
$$

or

$$
\{Y\} = [A]\{\beta\} + \{\zeta\}
\tag{A.295}
$$

in matrix notation. As usual, matrices shall be denoted by square brackets and column vectors by curly brackets. $[A]$ is called the design matrix, $\{\beta\}$ is the vector of parameters and $\{\zeta\}$ is the residual vector. In this notation the sum of squared errors is given by

$$
J(\{\beta\}) = \{\zeta\}^T\{\zeta\} = (\{Y\}^T - \{\beta\}^T[A]^T)(\{Y\} - [A]\{\beta\})
\tag{A.296}
$$

Minimising this expression with respect to variations of the parameters proceeds as follows. The derivatives of J with respect to the parameters are evaluated and set equal to zero; the resulting linear system of equations yields the parameter estimates. Expanding Equation (A.296) gives

$$
J(\{\beta\}) = \{Y\}^T\{Y\} - \{Y\}^T[A]\{\beta\} - \{\beta\}^T[A]^T\{Y\} + \{\beta\}^T[A]^T[A]\{\beta\}
\tag{A.297}
$$

and differentiating this expression with respect to $\{\beta\}^T$ yields

$$
\frac{\partial J(\{\beta\})}{\partial \{\beta\}^T} = -[A]^T\{Y\} + [A]^T[A]\{\beta\}
\tag{A.298}
$$

Finally, setting the derivative in this last equation to zero gives the well-known *normal equations* for the best parameter estimates $\{\hat{\beta}\}$,

$$
[A]^T[A]\{\hat{\beta}\} = [A]^T\{Y\}
\tag{A.299}
$$

which are trivially solved by

$$
\{\hat{\beta}\} = ([A]^T[A])^{-1}[A]^T\{Y\}
\tag{A.300}
$$

provided that $[A]^T[A]$ is invertible. In practice, it is not necessary to invert this matrix in order to obtain the parameter estimates. However, as shown below, the matrix $([A]^T[A])^{-1}$ contains valuable information. In practice, direct solution of the normal equations via Equation (A.300) is not always recommended as problems can arise if the matrix $[A]^T[A]$ is close to singularity. Suppose that the RHS of Equation (A.300) has a small error $\{\delta Y\}$ (due to roundoff, say); the resulting error in the estimated parameters is given by

$$
\{\delta\beta\} = ([A]^T[A])^{-1}[A]^T\{\delta Y\}
\tag{A.301}
$$

As the elements in the inverted matrix are inversely proportional to the determinant of $[A]^T[A]$, they can be arbitrarily large if $[A]^T[A]$ is close to singularity. As a consequence, parameters with arbitrarily large errors could be obtained. A more stable means of extracting the parameters is via singular value decomposition (SVD).

A.11.4 Parameter Uncertainty

Because of random errors in the measurements, different samples of data will contain different noise components and consequently they will lead to slightly different parameter estimates. The parameter estimates therefore constitute a random sample from a population of possible estimates, this population being characterised by a probability distribution. Clearly, it is desirable that the expected value of this distribution should coincide with the true parameters. If such a condition holds, the parameter estimator is said to be *unbiased*.

Now, given that the unbiased estimates are distributed about the true parameters, knowledge of the variance of the parameter distribution would provide valuable information about the possible scatter in the estimates. This information turns out to be readily available; the *covariance matrix* $[\Sigma]$ for the parameters is defined by

$$[\Sigma](\{\hat{\beta}\}) = E[(\{\hat{\beta}\} - E[\{\hat{\beta}\}])(\{\hat{\beta}\} - E[\{\hat{\beta}\}])^{\mathrm{T}}] \tag{A.302}$$

where the quantities with carets are the estimates and the expectation E is taken over all possible estimates. The diagonal elements of this matrix, σ_{ii}^2, are the variances of the parameter estimates $\hat{\beta}_i$.

Under the assumption that the estimates are unbiased and therefore $E[\{\hat{\beta}\}] = \{\beta\}$, where $\{\beta\}$ denotes the true parameters, one has

$$[\Sigma](\{\hat{\beta}\}) = E[(\{\hat{\beta}\} - \{\beta\})(\{\hat{\beta}\} - \{\beta\})^{\mathrm{T}}] \tag{A.303}$$

Now, substituting Equation (A.295) containing the *true* parameters into Equation (A.300) for the estimates yields

$$\{\hat{\beta}\} = \{\beta\} + ([A]^{\mathrm{T}}[A])^{-1}[A]^{\mathrm{T}}\{\zeta\}$$

or, trivially,

$$\{\hat{\beta}\} - \{\beta\} = ([A]^{\mathrm{T}}[A])^{-1}[A]^{\mathrm{T}}\{\zeta\}$$

which can be immediately substituted into Equation (A.303) to give

$$[\Sigma] = E[([A]^{\mathrm{T}}[A])^{-1}[A]^{\mathrm{T}}\{\zeta\}\{\zeta\}^{\mathrm{T}}[A]([A]^{\mathrm{T}}[A])^{-1}] \tag{A.304}$$

It has been assumed that the only variable that changes from measurement to measurement if the excitation is repeated exactly is $\{\zeta\}$. Further, if $\{\zeta\}$ is independent of $[A]$, that is independent of x_i and y_i and so on, then in this particular case,

$$[\Sigma] = ([A]^{\mathrm{T}}[A])^{-1}[A]^{\mathrm{T}}E[\{\zeta\}\{\zeta\}^{\mathrm{T}}][A]([A]^{\mathrm{T}}[A])^{-1}] \tag{A.305}$$

In order to proceed further, more assumptions must be made. First assume that the noise process $\{\zeta\}$ is zero-mean, that is $E[\{\zeta\}] = 0$. In this case the expectation in Equation (A.305) is the covariance matrix of the noise process, that is

$$E[\{\zeta\}\{\zeta\}^{\mathrm{T}}] = [E[\zeta_i\zeta_j]] \tag{A.306}$$

and further assume that

$$E[\zeta_i \zeta_j] = \sigma_\zeta^2 \delta_{ij} \tag{A.307}$$

where σ_ζ^2 is the variance of the residual sequence ζ_i and δ_{ij} is the Kronecker delta. Under this condition, the expression (A.305) collapses to

$$[\Sigma] = \sigma_\zeta^2 ([A]^T[A])^{-1} \tag{A.308}$$

The standard deviation for each estimated parameter is therefore

$$\sigma_i = \sigma_\zeta \sqrt{([A]^T[A])_{ii}^{-1}} \tag{A.309}$$

Now, if the parameter distributions are Gaussian, standard theory yields a 95% confidence interval of $\{\hat{\beta}\} \pm 1.96\{\sigma\}$; that is, there is a 95% probability that the true parameters fall within this interval.

A.11.5   A Case Study

The data for the simple illustrative case study here were generated by numerically integrating the equation of motion (A.280) with parameter values $m = 1$, $c = 20$ and $k = 10^4$. A simple fourth-order Runge–Kutta scheme was used (Press *et al.* 2007) and the excitation was a Gaussian white noise sequence with zero-mean and unit variance, band-limited onto the interval 0 to 200 Hz. The sampling frequency was taken as 1000 Hz and 1000 points of displacement data were recorded.

The parameters of the model structure (A.292) were estimated for the data using the simple least-squares approach outlined in the previous sections. In order to consider the effects of noise a white Gaussian sequence was added to the $x(t)$ data. Three levels of corruption were investigated, where the noise RMS was 1%, 10% and 50% of the input RMS. A further simulation used the clean data. Note that the way the noise was added makes sure that the estimator was unbiased. The parameter estimates for the various data sets are given in Table A.3.

The parameters for the clean data were used to predict the response $\hat{y}_i$ from the excitation x_i and the result is shown in Figure A.105 as a comparison with the true data. It is clear that the discrete-time model structure (A.292) is appropriate.

Table A.3   Least-squares parameter estimates for a linear system

Parameter	Noise level	Estimate	Standard error
a_1	0.0	1.9718	4.6337×10^{-4}
a_2	0.0	-0.9817	4.6241×10^{-4}
b_1	0.0	9.0520×10^{-7}	4.2389×10^{-9}
a_1	1.0	1.9718	4.6609×10^{-4}
a_2	1.0	-0.9817	4.6513×10^{-4}
b_1	1.0	9.0499×10^{-7}	4.2631×10^{-9}
a_1	10.0	1.9721	7.6544×10^{-4}
a_2	10.0	-0.9820	7.6389×10^{-4}
b_1	10.0	8.9483×10^{-7}	6.9589×10^{-9}
a_1	50.0	1.9751	2.7864×10^{-3}
a_2	50.0	-0.9848	2.7813×10^{-3}
b_1	50.0	7.1624×10^{-7}	2.2644×10^{-8}

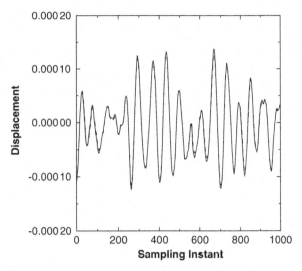

Figure A.105   Model predictions for clean data: true data (solid), prediction (dashed).

Figure A.106 shows that the prediction from the model degrades considerably when the 50% noise situation is considered.

A.12   Summary

All the material in this appendix is included in the hope of making the book as self-contained as possible. All of the concepts discussed here will appear at some point when one is attempting to carry out analysis of data for SHM purposes. This does not mean to say that everything one needs to know is here. However,

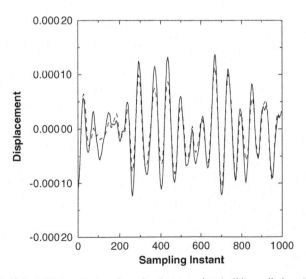

Figure A.106   Model predictions for noisy data: true data (solid), predictions (dashed).

there is a wealth of further information regarding signal processing in the references cited here and in the literature at large, and the reader should not hesitate to consult those references when matters are unclear.

References

Friedlander, F.G. and Joshi, M. (1999) *Introduction to the Theory of Distributions*, 2nd edn, Cambridge University Press.

Hamming, R.W. (1999) *Digital Filters*, revised edn, Dover Publications Inc.

Hubbard, B.B. (1998) *The World According to Wavelets: The Story of a Mathematical Technique in the Making*, 2nd edn, CRC Press.

Körner, T.W. (1989) *Fourier Analysis*, reprint edn, Cambridge University Press.

Mallat, S. (1999) *A Wavelet Tour of Signal Processing (Wavelet Analysis and Its Applications)*, 2nd edn, Academic Press.

Marven, C. and Ewers, G. (1996) *A Simple Approach to Digital Signal Processing*, Wiley-Blackwell.

Press, W.H., Teukolsky, S.A., Vetterling, W.T. and Flannery, B.P. (2007) *Numerical Recipes: The Art of Scientific Computing*, 3rd edn, Cambridge University Press.

Stearns, S.D. and David, R.A. (1987) *Signal Processing Algorithms*, Prentice-Hall.

Appendix B

Essential Linear Structural Dynamics

Almost all of the material of the current book is centred on vibration-based SHM. This means that it presupposes of the reader a certain degree of familiarity with the theory of structural vibrations. In order to make the book more self-contained, this appendix is presented with the aim of providing coverage of the basic notions of linear structural dynamics and of establishing notation. It is not intended to be comprehensive, although parts of the theory that are not commonly covered in elementary textbooks are treated in a little more detail.

For more comprehensive coverage, any book on engineering dynamics or mechanical vibrations will serve as reference for the sections below on continuous-time systems, for example Thompson (1965) or the more modern work by Inman (1994). For the material on discrete-time systems, any recent book on system identification can be consulted; Söderstrom and Stoica (1988) is an excellent example.

B.1 Continuous-Time Systems: The Time Domain

In order to start the discussion at the simplest point possible, it is assumed that the system of interest is a single point particle of mass m moving in one dimension subject to an applied force $x(t)$. The equation of motion for such an object is provided by Newton's second law,

$$\frac{d}{dt}(mv) = x(t) \tag{B.1}$$

where v is the velocity of the particle. The advantage of thinking in terms of point masses for the moment is that the equations of motion are ordinary differential equations. In fact, almost all systems of engineering interest, beams, plates, shells and so on, are governed by partial differential equations. However, dealing with partial differential equations at this point will only complicate matters and potentially obscure the underlying basic concepts of dynamics. In any case, the final results of this appendix will show that a modal approach to vibrations allows one to approximate any system meaningfully as a series of *lumped masses*.

If the mass m is constant in Equation (B.1), the equation becomes a second-order differential equation, as promised earlier,

$$m\ddot{y} = x(t) \tag{B.2}$$

Structural Health Monitoring: A Machine Learning Perspective, First Edition. Charles R. Farrar and Keith Worden.
© 2013 John Wiley & Sons, Ltd. Published 2013 by John Wiley & Sons, Ltd.

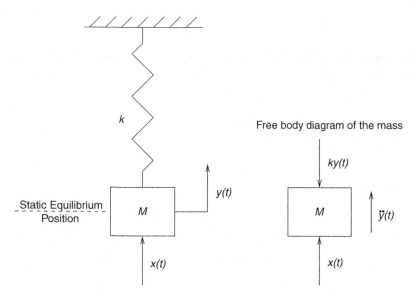

Figure B.1   Single degree-of-freedom (SDOF) mass–spring system.

in the standard notation where overdots denote differentiation with respect to time. In general, throughout this book, input variables to systems or structures will be denoted by x and output variables denoted by y.

Apart from the obvious restrictions inherent in the analysis so far (all real systems have more than one governing variable or *degree of freedom* (DOF)), this equation is unrealistic in that there is no resistance to the motion. Even if $x(t) = 0$, the particle can move with constant velocity for all time. The simplest way of providing resistance to motion is to add an *internal* or *restoring force* $f_r(y)$ that always acts in the opposite direction to the motion:

$$m\ddot{y} = x(t) - f_r(y) \tag{B.3}$$

In the study of vibrations, the paradigm for this type of equation is a mass on a spring (Figure B.1). The form of the restoring force in this case is given by Hooke's law; for a static displacement y of the mass, the restoring force is given by

$$f_r(y) = ky \tag{B.4}$$

Substituting this expression into the equation of motion (B.3) gives

$$m\ddot{y} + ky = x(t) \tag{B.5}$$

Note that as the restoring force vanishes when $y = 0$, this will be the static equilibrium position of the motion, that is, the position of rest when there is no force.

In structural dynamics, it is traditional to use k for the coefficient of y and to refer to it as the *elastic stiffness* or simply *stiffness* of the system.

The solution of Equation (B.5) is elementary and is given in any book on vibrations or differential equations; Simmons (1974) is an excellent example. An interesting special case is where $x(t) = 0$ and one observes the unforced or *free* motion,

$$\ddot{y} + \frac{k}{m} y = 0 \tag{B.6}$$

There is a trivial solution to this equation given by $y(t) = 0$, which results from specifying the initial conditions $y(0) = 0$ and $\dot{y}(0) = 0$. Any point at which the mass remains stationary for all time is termed an *equilibrium* or *fixed point* for the system. It is clear from the equation that the only equilibrium for this system is the origin $y = 0$, that is, the static equilibrium position. This is typical of linear systems but need not be the case for nonlinear systems. A more interesting solution results from specifying different initial conditions, for example $y(0) = A$, $\dot{y} = 0$; in this case the mass is released from rest at $t = 0$ a distance A from the equilibrium. As before, it is straightforward to solve the equation and the result is

$$y(t) = A \cos(\omega_n t) \tag{B.7}$$

This is a periodic oscillation about $y = 0$ with angular frequency $\omega_n = \sqrt{k/m}$ radians per second, frequency $f_n = 1/(2\pi)\sqrt{k/m}$ Hz and period of oscillation $T_n = 2\pi\sqrt{m/k}$ seconds. Because the frequency is of the free oscillations, it is termed the *undamped natural frequency* of the system – hence the subscript n.

The first thing to note here is that the oscillations persist indefinitely without attenuation. This sort of behaviour is forbidden by the second law of thermodynamics (no perpetual motion), so some modification of the model is necessary in order that free oscillations are not allowed to continue forever. If one thinks in terms of a mass on a spring, two mechanisms become apparent by which energy is dissipated or *damped*. First, unless the motion is taking place in a vacuum, there will be resistance to motion by the ambient fluid (air in this case). Second, energy will be dissipated in the material of the spring. Of these mechanisms, only the first is understood to any great extent. Fortunately, experiment shows that it is fairly common. In fact, at low velocities, the fluid offers a resistance proportional to and in opposition to the velocity of the mass. A *damping force* can therefore be represented by $f_d(\dot{y}) = -c\dot{y}$ in the model, where c is a *damping constant*. The equation of motion becomes

$$m\ddot{y} = x(t) - f_d(\dot{y}) - f_r(y) \tag{B.8}$$

or

$$m\ddot{y} + c\dot{y} + ky = x(t) \tag{B.9}$$

This equation still describes a single point mass moving in one dimension; such a system is referred to as a single-degree-of-freedom (SDOF) system. If the point mass were allowed to move in three dimensions, the displacement $y(t)$ would be a vector whose components would be specified by three equations of motion. Such a system is said to have three degrees of freedom and would be referred to as a multi-degree-of-freedom (MDOF) system. An MDOF system would also result from considering the motion of more than one point particle.

Considered as a differential equation, (B.9) is *linear*. An immediate and important consequence of this is that the *principle of superposition* holds. This can be stated as follows: if the response of the system to an arbitrary applied force $x_1(t)$ is $y_1(t)$ and the response to a second independent input $x_2(t)$ is $y_2(t)$, then the response to the superposition $\alpha x_1(t) + \beta x_2(t)$ (with appropriate initial conditions) is $\alpha y_1(t) + \beta y_2(t)$ for *any* values of the constants α, β. This is a fundamental property of linear systems and is the reason why analysis methods based on superpositions of terms, like Fourier series (see Appendix A), can work at all.

Systems whose equations of motion are differential equations are termed *continuous-time systems* and for such systems the evolution of the system from given initial conditions is usually specified for a continuum of times $t \geq 0$.

Returning now to Equation (B.9), elementary theory shows that the solution for the free motion $(x(t) = 0)$ with initial conditions $y(0) = A$, $\dot{y} = 0$, is

$$y_t(t) = Ae^{-\zeta \omega_n t} \cos(\omega_d t) \tag{B.10}$$

where

$$\zeta = \frac{c}{2\sqrt{mk}} \tag{B.11}$$

$$\omega_d = \omega_n (1 - \zeta^2)^{1/2} \tag{B.12}$$

and $\omega_n = \sqrt{k/m}$ is the undamped natural frequency as before. Note that the frequency of free oscillations in this case is $\omega_d \neq \omega_n$ and is termed the *damped natural frequency*; ζ is called the *damping ratio*. The main features of this solution can be summarised as follows:

- The damped natural frequency is always less than the undamped natural frequency, which it approaches in the limit as $c \to 0$ or equivalently as $\zeta \to 0$.
- If $1 > \zeta > 0$ the oscillations decay exponentially with a certain *time-constant* τ_ζ. *This is defined as the time taken for the amplitude to decay from a given value Y to the value Y/e*; $\tau_\zeta = 1/(\zeta \omega_n)$ (Figure B.2). Because of this, the solution (B.12) is termed the *transient* solution – hence the subscript t on the response. If $\zeta < 0$ or equivalently $c < 0$ the oscillations grow exponentially (Figure B.3). In order to ensure that the system is stable (in the sense that a bounded input generates a bounded output), ζ and hence c must be positive.
- If $\zeta = 1$, then $\omega_d = 0$ and the system does not oscillate but simply tends monotonically from $y(0) = A$ to zero as $t \to \infty$ (Figure B.4). The system is said to be *critically damped*. The critical value for the damping constant c is easily seen to be $2\sqrt{mk}$.
- If $\zeta > 1$, the system is said to be *overdamped* and the situation is similar to critical damping; the system is nonoscillatory but gradually returns to its equilibrium when disturbed.

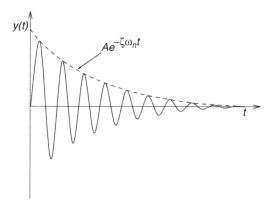

Figure B.2   Transient motion of an SDOF oscillator with positive damping. The envelope of the response is also shown.

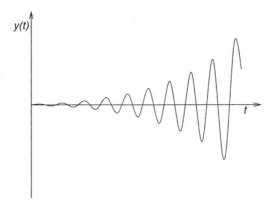

Figure B.3 Unforced motion of an SDOF oscillator with negative damping. The system displays instability.

Analysis of the free motion has proved useful in that it has allowed a physical *positivity* constraint on ζ or c to be derived. However, the most interesting and more generally applicable solutions of the equation will be for forced motion. If attention is focused on deterministic (see Section A.1 in Appendix A) force signals $x(t)$, Fourier analysis allows one to express an arbitrary periodic signal as a linear sum of sinusoids of different frequencies. One can then invoke the principle of superposition, which allows one to concentrate on the solution where $x(t)$ is a single sinusoid, that is

$$m\ddot{y} + c\dot{y} + ky = X\cos(\omega t) \tag{B.13}$$

where $X > 0$ and ω is the *constant* frequency of excitation. Standard differential equation theory (Simmons, 1974) asserts that the general solution of Equation (B.13) is given by

$$y(t) = y_t(t) + y_s(t) \tag{B.14}$$

where the *complementary function* (or transient response according to the above) $y_t(t)$ is the unique solution for the *free* equation of motion and contains arbitrary constants that are fixed by initial conditions. The complementary function $y_t(t)$ for Equation (B.13) has already been established here and is given

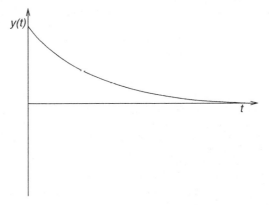

Figure B.4 Transient motion of an SDOF oscillator with critical damping showing that no oscillations occur.

by Equation (B.10). The remaining part of the solution $y_s(t)$, which is termed the *particular integral*, is independent of the initial conditions and persists after the transient $y_t(t)$ has decayed away. For this reason $y_s(t)$ is also called the *steady-state response* of the solution.

For linear systems, the steady-state response to a periodic force is periodic with the same frequency, but not necessarily in phase with the input due to the energy dissipation by the damping term, which causes the output to lag the input. In order to find $y_s(t)$ for Equation (B.13), one can therefore substitute the trial solution

$$y_s(t) = Y \cos(\omega t - \varphi) \tag{B.15}$$

where $Y > 0$, and one obtains

$$-m\omega^2 Y \cos(\omega t - \varphi) + -c\omega Y \sin(\omega t - \varphi) + kY \cos(\omega t - \varphi) = X \cos(\omega t) \tag{B.16}$$

A shift of the time variable $t \to t + \phi/\omega$ yields the simpler expression,

$$-m\omega^2 Y \cos(\omega t) + -c\omega Y \sin(\omega t) + kY \cos(\omega t) = X \cos(\omega t + \varphi)$$
$$= X \cos \omega t \cos \varphi - X \sin \omega t \sin \varphi \tag{B.17}$$

Equating the coefficients of the sine and cosine terms gives

$$-m\omega^2 Y + kY = X \cos \varphi \tag{B.18}$$

$$c\omega Y = X \sin \varphi \tag{B.19}$$

Squaring and adding these equations gives

$$\{(-m\omega^2 + k)^2 + c^2\omega^2\}Y^2 = X^2(\cos^2 \varphi + \sin^2 \varphi) = X^2 \tag{B.20}$$

so that

$$\frac{Y}{X} = \frac{1}{\sqrt{(-m\omega^2 + k)^2 + c^2\omega^2}} \tag{B.21}$$

This is the *gain* of the system at frequency ω, that is, the proportional change in the amplitude of the signal as it passes through the system $x(t) \to y(t)$. Because X and Y are both positive real numbers, so is the gain.

Taking the ratio of Equations (B.19) and (B.18) yields

$$\tan\varphi = \frac{c\omega}{k - m\omega^2} \tag{B.22}$$

The *phase* φ represents the degree by which the output signal $y(t)$ lags the input $x(t)$ as a consequence of passage through the damped system.

One can now examine how the response characteristics vary as the excitation frequency ω is changed. First, one can rewrite Equation (B.21) in terms of the quantities ω_n and ζ as

$$\frac{Y}{X}(\omega) = \frac{1}{m\sqrt{\left(\omega^2 - \omega_n^2\right)^2 + 4\zeta^2\omega_n^2\omega^2}} \tag{B.23}$$

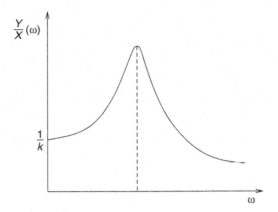

Figure B.5 SDOF system gain as a function of frequency ω.

This function will clearly be a maximum when

$$\left(\omega^2 - \omega_n^2\right)^2 + 4\zeta^2 \omega_n^2 \omega^2 \tag{B.24}$$

is a minimum, that is, when

$$\frac{\mathrm{d}}{\mathrm{d}\omega}\left[\left(\omega^2 - \omega_n^2\right)^2 + 4\zeta^2\omega_n^2\omega^2\right] = 4\omega\left(\omega^2 - \omega_n^2\right) + 8\zeta^2\omega_n^2\omega = 0 \tag{B.25}$$

so that

$$\omega^2 = \omega_n^2(1 - 2\zeta^2) \tag{B.26}$$

This frequency corresponds to the *only* extreme value of the gain (for an SDOF system) and is termed the *resonance*[1] frequency of the system and denoted by ω_r. Note that for the damped system under study $\omega_r \neq \omega_d \neq \omega_n$. It is easy to show that for an undamped system $\omega_r = \omega_d = \omega_n$ and that the gain of the undamped system is infinite for excitation at the resonance frequency. In general if the excitation is at $\omega = \omega_r$, the system is said to be *at resonance*.

Equation (B.21) shows that $Y/X = 1/k$ when $\omega = 0$ and that $Y/X \to 0$ as $\omega \to \infty$. The information accumulated so far is sufficient to define the (qualitative) behaviour of the system gain as a function of the frequency of excitation ω. The resulting graph is plotted in Figure B.5.

The behaviour of the phase $\varphi(\omega)$ is now needed in order to completely specify the system response as a function of frequency. Equation (B.22) gives

$$\tan\varphi(\omega) = \frac{c\omega}{m\left(\omega_n^2 - \omega^2\right)} = \frac{2\zeta\omega_n\omega}{\omega_n^2 - \omega^2} \tag{B.27}$$

As $\omega \to 0$, $\tan\varphi \to 0$ from above, corresponding to $\varphi \to 0$. As $\omega \to \infty$, $\tan\varphi \to 0$ from below, corresponding to $\varphi \to \pi$. At $\omega = \omega_n$ the undamped natural frequency, $\tan\varphi = \infty$, corresponding to $\varphi = \pi/2$. This is sufficient to define φ (qualitatively) as a function of ω. The graph of $\varphi(\omega)$ is given in Figure B.6.

---

[1] The term *resonant frequency* is also often used; however, this is inappropriate as it is the system or structure that is resonant, not the frequency.

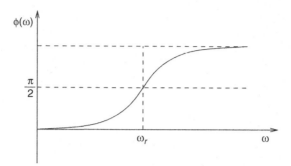

Figure B.6   SDOF system phase as a function of frequency ω.

The graphs of $Y/X(\omega)$ and $\varphi(\omega)$ are usually given together as they specify between them all properties of the system response to a harmonic input. This type of plot is usually called a *Bode plot*. If Y/X and $\varphi(\omega)$ are interpreted as the amplitude and phase of a complex function, this is called the *frequency response function*, or FRF. In practice, FRFs are obtained in dynamic tests from a process of numerical analysis based on the acquisition of time data. The associated analysis is covered in great detail in Sections A.2 to A.9 in this book.

At the risk of a little duplication, an example is given in Figure B.7 for the Bode plot of a specific SDOF system,

$$\ddot{y} + 20\dot{y} + 10^4 y = x(t) \tag{B.28}$$

(The particular routine used to generate this plot actually shows $-\varphi$ in keeping with a common convention.) For this system, the undamped natural frequency is 100 rad/s, the damped natural frequency is 99.5 rad/s, the resonance frequency is 99.0 rad/s and the damping ratio is 0.1 or 10% of critical.

A more direct construction of the system representation in terms of the Bode plot will be given in the following section. Note that the gain and phase in expressions (B.21) and (B.22) are independent of the magnitude of the forcing level X. This means that the FRF is an *invariant* of the amplitude of excitation. In fact, this is only true for linear systems and breakdown in the amplitude invariance of the FRF can sometimes be used as a test for nonlinearity.

B.2   Continuous-Time Systems: The Frequency Domain

As discussed in detail in Appendix A, the input and output time signals $x(t)$ and $y(t)$ for the SDOF system discussed above will both have dual *frequency-domain* representations or *Fourier spectra* $X(\omega) = F\{x(t)\}$ and $Y(\omega) = F\{y(t)\}$ obtained by Fourier transformation, where

$$G(\omega) = F\{g(t)\} = \int_{-\infty}^{+\infty} e^{-i\omega t} g(t) dt \tag{B.29}$$

defines the Fourier transform F and the corresponding inverse transform is given by

$$g(t) = F^{-1}\{G(\omega)\} = \frac{1}{2\pi} \int_{-\infty}^{+\infty} e^{-i\omega t} G(\omega) d\omega \tag{B.30}$$

It is natural to ask now if there is a frequency-domain representation of the system itself that maps the input spectrum $X(\omega)$ directly to the output spectrum $Y(\omega)$. The answer to this is 'yes' and the mapping is remarkably simple, as discussed in Section A.4; one finds

$$Y(\omega) = H(\omega)X(\omega) \tag{B.31}$$

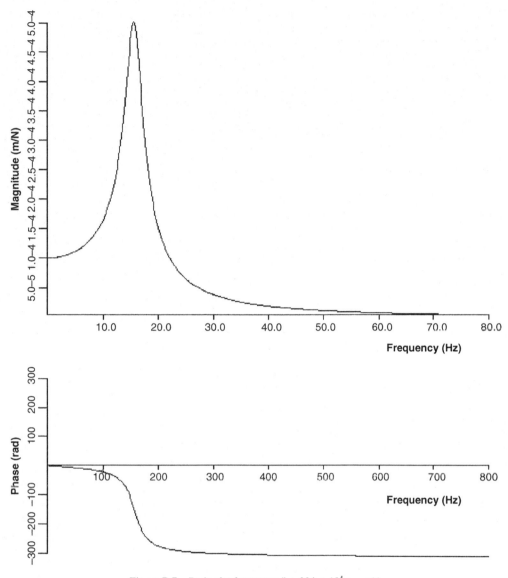

Figure B.7   Bode plot for system $\ddot{y} + 20\dot{y} + 10^4 y = x(t)$.

where the exact form of the *frequency response function* (FRF) $H(\omega)$ depends on the specific system. For the SDOF oscillator of Equation (B.13), one finds

$$H(\omega) = \frac{1}{-m\omega^2 + ic\omega + k} = \frac{1}{k - m\omega^2 + ic\omega} \tag{B.32}$$

Therefore, in the frequency domain, mapping input $X(\omega)$ to output is $Y(\omega)$ is simply a matter of multiplying X by a complex function H. All system information is contained in the FRF; all coefficients from the time domain are present and the number and order of the derivatives in Equation (B.9) is encoded in the powers of $i\omega$ present. It is a simple matter to convince oneself that the relation (B.31) holds in the frequency domain for *any* system whose equation of motion is a linear differential equation, although, of course, the form of the function $H(\omega)$ will depend on the particular system.

As $H(\omega)$ is a complex function, it has a representation in terms of magnitude $|H(\omega)|$ and phase $\angle H(\omega)$:

$$H(\omega) = |H(\omega)|e^{i\angle H(\omega)} \tag{B.33}$$

The $|H(\omega)|$ and $\angle H(\omega)$ so defined correspond exactly to the gain $Y/X(\omega)$ and phase $\varphi(\omega)$ defined in the previous section. This result provides a direct interpretation of the FRF $H(\omega)$ in terms of the gain and phase of the response when the system is presented with a harmonic input.

It is now clear why the Bode plot defined in the previous section suffices to characterise the system. An alternative means of presenting the information in $H(\omega)$ is the commonly used *Nyquist plot*, which graphs the locus of $H(\omega)$ in the complex plane or Argand diagram as $\omega \rightarrow \infty$ (or $\omega \rightarrow$ the limit of measurable ω). The Nyquist plot corresponding to the system in Equation (B.28) is given in Figure B.8.

The FRF for the system given above is for the process $x(t) \rightarrow y(t)$. It is called the *receptance* form and is sometimes denoted $H_R(\omega)$. One is often interested in the FRFs for the processes $x(t) \rightarrow \dot{y}(t)$ and

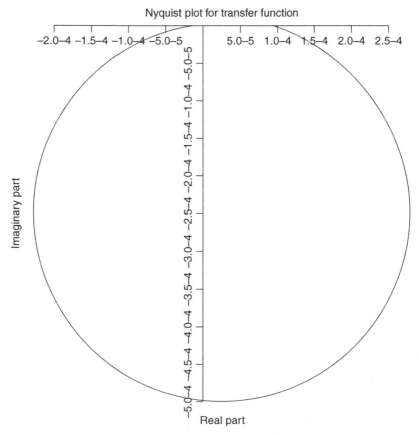

Figure B.8   Nyquist plot for system $\ddot{y} + 20\dot{y} + 10^4 y = x(t)$ – receptance.

$x(t) \to \ddot{y}(t)$ as it may well be that velocity or acceleration is measured rather than displacement; these are easily shown to be

$$H_M(\omega) = \frac{i\omega}{-m\omega^2 + ic\omega + k} \tag{B.34}$$

and

$$H_I(\omega) = \frac{-\omega^2}{-m\omega^2 + ic\omega + k} \tag{B.35}$$

They are respectively referred to as the *mobility* form and *accelerance* or *inertance* form. The Nyquist plots for these forms of the FRF are very similar to that for receptance in the sense that they are circles or ellipses in the complex plane, the major difference being that the ellipses are rotated in the plane relative to the receptance form in Figure B.8.

B.3 The Impulse Response

Given the *general* frequency-domain relationship (B.31) for linear systems, as discussed in Section A.4, one can now pass back to the time domain and obtain a parallel relationship,

$$y(t) = \int_{-\infty}^{+\infty} h(\tau)x(t - \tau)d\tau \tag{B.36}$$

These equations provide another time-domain version of the system input–output relationship. All system information is encoded in the function $h(t)$. One can now ask if $h(t)$ has a physical interpretation. The answer is 'yes', as given in Appendix A; however, it is useful to expand on the argument here.

Suppose one wishes to know the response of a system to a transient input, that is $x(t)$ where $x(t) = 0$ if $|t| > \varepsilon$, say (Figure B.9). All the energy is communicated to the system in time 2ε, after which

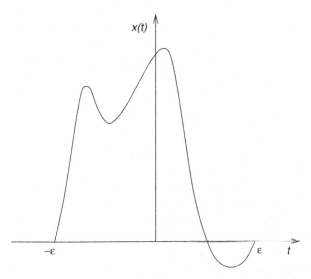

Figure B.9 Example of a transient excitation whose duration is 2ε.

the system follows the unforced equations of motion. An ideal transient excitation or *impulse* would communicate all energy in an instant. No such *physical* signal exists for obvious reasons. However, there is a mathematical object, the Dirac delta function $\delta(t)$ (see Section A.3.2), which has the properties of an ideal impulse,

$$\text{Infinitesimal duration:} \quad \delta(t) = 0 \quad t \neq 0 \tag{B.37}$$

$$\text{Finite power:} \quad \int_{-\infty}^{+\infty} |x(t)|^2 dt = 1 \tag{B.38}$$

And, at the risk of repetition, the defining relationship for the delta function is

$$\int_{-\infty}^{+\infty} f(t)\delta(t-a)dt = f(a) \quad \text{for any } f(t) \tag{B.39}$$

Now, according to Equation (B.36), the system response to a delta function input $y_\delta(t)$ is given by

$$y_\delta(t) = \int_{-\infty}^{+\infty} h(\tau)\delta(t-\tau)d\tau \tag{B.40}$$

so applying the relation (B.39) immediately gives

$$y_\delta(t) = h(t) \tag{B.41}$$

which provides the required interpretation of $h(t)$ – it is the *impulse response* of the system, that is the solution of the equation

$$m\ddot{h}(t) + c\dot{h}(t) + kh(t) = \delta(t) \tag{B.42}$$

It is not an entirely straightforward matter to evaluate $h(t)$ for the general SDOF system; integration in the complex plane is needed. However, as a means of bypassing the rigorous analysis, a more formal argument is provided.

The impulse response is the solution of Equation (B.42) and must therefore have the general form

$$y(t) = e^{-\zeta\omega_n t}[A\cos(\omega_d t) + B\sin(\omega_d t)] \tag{B.43}$$

where A and B are fixed by the initial conditions.

The initial displacement $y(0)$ is assumed zero and the initial velocity is assumed to follow from the initial momentum coming from the impulsive force $I(t) = \delta(t)$,

$$m\ddot{y}(0) = \int I(t)dt = \int \delta(t)dt = 1 \tag{B.44}$$

from Equation (B.41); so it follows that $\dot{y}(0) = 1/m$. Substituting these initial conditions into Equation (B.45) yields $A = 0$ and $B = 1/(m\omega_d)$, and the impulse response is therefore

$$h(t) = \frac{1}{m\omega_d}e^{-\zeta\omega_d t}\sin(\omega_n t) \tag{B.45}$$

for $t > 0$. The impulse response is therefore a decaying harmonic motion at the damped natural frequency. Note that $h(t)$ is zero before $t = 0$, the time at which the impulse is applied. This is an expression of the principle of causality, that is that effect cannot precede cause. A consequence of $h(t)$ vanishing for negative times is that one can change the lower limit of the integral in Equation (B.36) from $-\infty$ to 0 with no effect.

B.4 Discrete-Time Models: Time Domain

The fact that Newton's laws of motion are differential equations leads directly to the continuous-time representation of systems described above. This representation defines the motion at *all* times. In reality, most observations of system behaviour – measurements of input and output signals – will be carried out at discrete intervals. The system data are then a discrete set of values $\{x_i, y_i, i = 1, \ldots, N\}$. For modelling purposes one might therefore ask if there exists a model structure that maps the discrete inputs x_i directly to the discrete outputs y_i. Such models do exist and in many cases offer advantages over the continuous-time representation. Consider the general linear SDOF system,

$$m\ddot{y} + c\dot{y} + ky = x(t) \tag{B.46}$$

Suppose that one is only interested in the value of the output at a sequence of regularly spaced times t_i where $t_i = (i - 1)\Delta t$ (Δt is called the *sampling interval* and the associated frequency $f_s = 1/\Delta t$ is called the *sampling frequency*). By approximating the derivatives in Equation (B.46) and following the analysis described in Section A.11.1, one arrives at a model of the form

$$y_i = a_1 y_{i-1} + a_2 y_{i-2} + b_1 x_{i-1} \tag{B.47}$$

where the constants a_1, a_2, b_1 are defined by Equation (A.284). Equation (B.47) is a *discrete-time* representation of the SDOF system under study; other forms can arise by taking different discrete approximations to the derivatives. The terminology of discrete-time models is discussed in detail in Section A.11.1; the model in Equation (B.47) is an example of an ARMA(2, 1) model.

The motion for all discrete times is fixed by the input sequence x_i together with values for y_1 and y_2. The specification of the first two values of the output sequence is directly equivalent to the specification of initial values for $y(t)$ and $\dot{y}(t)$ in the continuous-time case. An obvious advantage of using a discrete model like Equation (B.47) is that it is much simpler numerically to predict the output in comparison with a differential equation. The price one pays is a loss of generality, because the coefficients in Equation (B.47) are functions of the sampling interval Δt; one could only use this model to predict responses with the same spacing in time.

Although arguably less familiar, the theory for the solution of difference equations is no more difficult than the corresponding theory for differential equations. The basic principles will be given here. Consider the free motion for the system in Equation (B.47); this is specified by

$$y_i = a_1 y_{i-1} + a_2 y_{i-2} \tag{B.48}$$

Substituting a trial solution $y_i = \alpha^i$ with α constant yields

$$\alpha^{i-2}(\alpha^2 - a_1\alpha - a_2) = 0 \tag{B.49}$$

which has nontrivial solutions,

$$\alpha_{\pm} = \frac{a_1}{2} \pm \frac{1}{2}\sqrt{4a_2 + a_1^2} \tag{B.50}$$

The general solution of Equation (B.48) is therefore

$$y_i = A\alpha_+^i + B\alpha_-^i \tag{B.51}$$

where A and B are arbitrary constants, which can be fixed in terms of the initial values y_1 and y_2 as follows. According to the solution above, $y_1 = A\alpha_+ + B\alpha_-$ and $y_2 = A\alpha_+^2 + B\alpha_-^2$, which can be regarded as simultaneous equations for A and B, the solution being

$$A = \frac{y_2 - \alpha_- y_1}{\alpha_+(\alpha_+ - \alpha_-)} \tag{B.52}$$

$$B = \frac{\alpha_+ y_1 - y_2}{\alpha_-(\alpha_+ - \alpha_-)} \tag{B.53}$$

Analysis of the stability of this system is straightforward. If either $|\alpha_+| > 1$ or $|\alpha_-| > 1$ the solution grows exponentially; otherwise the solution decays exponentially. More precisely, if the *magnitudes* of the *alpha* are greater than 1 – as they may be complex – the solutions are unstable. In the differential equation case the stability condition was simply $c > 0$. The stability condition in terms of the difference equation parameters is the slightly more complicated expression

$$\left| \frac{a_1}{2} \pm \frac{1}{2}\sqrt{4a_2 + a_1^2} \right| < 1 \tag{B.54}$$

Note that each difference equation property parallels a differential equation property; it is this that allows either representation when modelling a system.

As for the differential equation, the principle of superposition holds for linear difference equations so it is sufficient to consider a harmonic excitation $x_i = X \cos(\omega t_i)$ in order to explore the characteristics of the forced equation. As in the continuous-time case, the general solution of the forced equation,

$$y_i - a_1 y_{i-1} - a_2 y_{i-2} = X \cos(\omega t_{i-1}) \tag{B.55}$$

will comprise a transient part, specified in Equation (B.51), and a steady-state part independent of the initial conditions. In order to find the steady-state solution one can assume that the response will be a harmonic at the forcing frequency. This provides the form of the trial solution,

$$y_i = Y \cos(\omega t_i + \varphi) \tag{B.56}$$

Substituting this expression into Equation (B.47) and shifting the time $t \to t + \Delta t - \phi/\omega$ yields

$$Y(\cos(\omega t_i + \omega \Delta t) - a_1 \cos(\omega t_i) - a_2 \cos(\omega t_i - \omega \Delta t)) = X \cos(\omega t_i - \varphi) \tag{B.57}$$

Expanding and comparing the coefficients for the sine and cosine terms in the result yields the two equations,

$$Y(-a_1 + (1 - a_2)C) = X \cos \varphi \tag{B.58}$$

$$Y(-(1 + a_2)S) = X \sin \varphi \tag{B.59}$$

where $C = \cos(\omega \Delta t)$ and $S = \sin(\omega \Delta t)$. It is a now a simple matter to obtain the expressions for the system gain and phase,

$$\frac{Y}{X} = \frac{1}{\sqrt{a_1^2 - 2a_1(1 - a_2)C + (1 - a_2)^2 C^2 + (1 + a_2)^2 S^2}} \tag{B.60}$$

$$\tan\varphi = \frac{(1 + a_2)S}{a_1 + (a_2 - 1)C} \tag{B.61}$$

One point about these equations is worth noting. The expressions for gain and phase are functions of frequency ω through the variables C and S. However, these variables are periodic with period $1/\Delta t = f_s$. As a consequence, the gain and phase formulae simply repeat indefinitely as $\omega \to \infty$. This means that knowledge of the response functions in the interval $[-f_s/2, f_s/2]$ is sufficient to specify them for all frequencies. An important consequence of this is that a discrete representation of a system can be accurate in the frequency domain *only* on a finite interval. As discussed in Section A.5, the frequency $f_s/2$ which prescribes this interval is called the *Nyquist frequency*.

The classification/taxonomy for different forms of difference equation models and their frequency-domain representations can be found in Section A.11.1.

B.5   Multi-Degree-of-Freedom (MDOF) Systems

The discussion so far has been restricted to the case of a single mass point. This has proved useful in that it has allowed the development of most of the basic theory used in modelling systems. However, the assumption of single-degree-of-freedom behaviour for all systems is clearly unrealistic. In general, one will have to account for the motion of several mass points or even a continuum. To see this, consider the transverse vibrations of a simply supported beam (Figure B.10).

A basic analysis of the statics of the situation, shows that an applied force F at the centre of the beam produces a displacement y given by,

$$F = ky, \quad k = \frac{48EI}{L^3} \tag{B.62}$$

where L is the length of the beam. k is called the *flexural stiffness*.

If it is now assumed that the mass is concentrated at the centre (Figure B.11); by considering the kinetic energy of the beam vibrating with a maximum displacement at the centre, it can be shown that the point mass is equal to half the total mass of the beam $M/2$ (Thompson, 1965). The appropriate equation of motion is

$$M\ddot{y} + ky = x(t) \tag{B.63}$$

for the displacement of the centre point, under a time-dependent excitation $x(t)$.

Figure B.10   A uniform simply supported beam under transverse vibration.

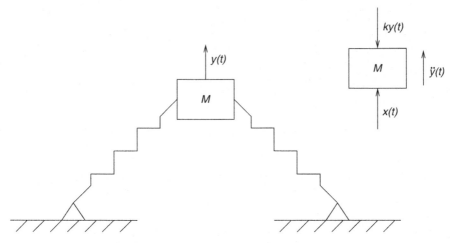

Figure B.11 Central point mass approximation for the beam of Figure B.10.

Damping effects are neglected for the present. If $x(t)$ is assumed harmonic, the theory developed in previous sections shows that the response will be harmonic at the same frequency. Unfortunately, as the beam has been replaced by a mass point in this approximation, one cannot obtain any information about the profile of the beam while vibrating. If the free equation of motion is considered, a natural frequency of $\omega_n = \sqrt{k/M}$ follows. Extrapolation from the static case suggests that the profile of the beam at this frequency will show its maximum displacement in the centre; the displacement of other points will fall monotonically as they approach the ends of the beam. No points except the end points will have zero displacement for all time. This mode of vibration is termed the *fundamental* mode. The word *mode* has acquired a technical sense here and refers to the shape of the beam vibrating at its natural frequency.

In order to obtain more information about the profile of the beam, the mass can be assumed to be concentrated at two points spaced evenly on the beam (Figure B.12). This time an energy analysis shows that one-third of the beam mass should be concentrated at each point. The equations of motion for this system are

$$\frac{M}{3}\ddot{y}_1 + k_{11}^f y_1 + k_{12}^f (y_1 - y_2) = x_1(t) \tag{B.64}$$

$$\frac{M}{3}\ddot{y}_2 + k_{22}^f y_2 + k_{12}^f (y_2 - y_1) = x_2(t) \tag{B.65}$$

These equations can be placed in a compact matrix form,

$$[m]\{\ddot{y}\} + [k]\{y\} = \{x\} \tag{B.66}$$

where curved braces denote vectors and square braces denote matrices:

$$[m] = \begin{pmatrix} m_{11} & m_{12} \\ m_{21} & m_{22} \end{pmatrix} = \begin{pmatrix} \dfrac{M}{3} & 0 \\ 0 & \dfrac{M}{3} \end{pmatrix}, \quad [k] = \begin{pmatrix} k_{11} & k_{12} \\ k_{21} & k_{22} \end{pmatrix} = \begin{pmatrix} k_{11}^f + k_{12}^f & -k_{12}^f \\ -k_{12}^f & k_{12}^f + k_{22}^f \end{pmatrix} \tag{B.67}$$

$$\{y\} = \begin{pmatrix} y_1 \\ y_2 \end{pmatrix}, \quad \{x\} = \begin{pmatrix} x_1 \\ x_2 \end{pmatrix} \tag{B.68}$$

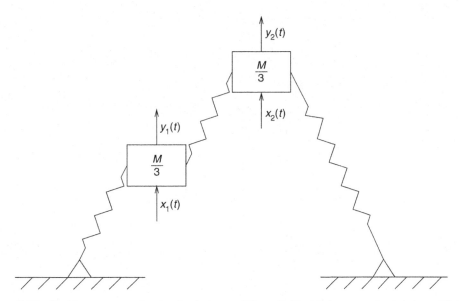

Figure B.12 Double mass approximation for the beam of Figure B.10 with the masses located at one-third and two-thirds of the length.

Here $[m]$ and $[k]$ are called the mass and stiffness matrices respectively. Note that $k_{12} = k_{21}$; this is an expression of a general principle, that of *reciprocity*. In general, reciprocity means that the mass and stiffness matrices will be symmetric. (Note that reciprocity is a property that only holds for linear systems. Violations of reciprocity can be used to indicate the presence of nonlinearity.)

In order to find the natural frequencies (it will turn out that there are more than one), consider the unforced equation of motion,

$$[m]\{\ddot{y}\} + [k]\{y\} = \{0\} \qquad\qquad (B.69)$$

To solve these equations, one can make use of a result of linear algebra theory, which asserts that there exists a matrix $[\psi]$ that simultaneously diagonalises $[m]$ and $[k]$, that is

$$[\psi]^{\mathrm{T}}[m][\psi] = [M] = \begin{pmatrix} m_1 & 0 \\ 0 & m_2 \end{pmatrix} \qquad\qquad (B.70)$$

$$[\psi]^{\mathrm{T}}[k][\psi] = [K] = \begin{pmatrix} k_1 & 0 \\ 0 & k_2 \end{pmatrix} \qquad\qquad (B.71)$$

Now, make the linear change of coordinates from $\{y\}$ to $\{z\}$, where $\{y\} = [\psi]\{z\}$, that is

$$\begin{aligned} y_1 &= \psi_{11}z_1 + \psi_{12}z_2 \\ y_2 &= \psi_{21}z_1 + \psi_{22}z_2 \end{aligned} \qquad\qquad (B.72)$$

Equation (B.69) becomes

$$[m][\psi]\{\ddot{z}\} + [k][\psi]\{z\} = \{0\} \qquad\qquad (B.73)$$

and on premultiplying this expression by $[\psi]^T$, one obtains

$$[M]\{\ddot{z}\} + [K]\{z\} = \{0\} \tag{B.74}$$

which represents the following scalar equations:

$$
\begin{aligned}
m_1 \ddot{z}_1 + k_1 z_1 &= 0 \\
m_2 \ddot{z}_2 + k_2 z_2 &= 0
\end{aligned}
\tag{B.75}
$$

which represent two uncoupled SDOF systems. The solutions are

$$
\begin{aligned}
z_1(t) &= A_1 \cos(\omega_1 t) + B_1 \cos(\omega_1 t) \\
z_2(t) &= A_2 \cos(\omega_2 t) + B_2 \cos(\omega_2 t)
\end{aligned}
\tag{B.76}
$$

(For simplicity, the sine terms, which are parts of a completely general solution, are ignored. This can be arranged by setting the initial conditions appropriately.) One then has

$$
\begin{aligned}
z_1(t) &= A_1 \cos(\omega_1 t) \\
z_2(t) &= A_2 \cos(\omega_2 t)
\end{aligned}
\tag{B.77}
$$

The *two* natural frequencies are $\omega_1 = \sqrt{k_1/m_1}$ and $\omega_2 = \sqrt{k_2/m_2}$. Each of the z coordinates is associated with a distinct frequency and, as will be shown later, a distinct mode of vibration. For this reason the z coordinates are referred to as *modal coordinates*. The elements of the diagonal mass and stiffness matrices are referred to as the *modal masses* and *modal stiffnesses* respectively.

On transforming back to the physical y-coordinate system using Equation (B.72), one obtains

$$
\begin{aligned}
y_1 &= \psi_{11} A_1 \cos(\omega_1 t) + \psi_{12} A_2 \cos(\omega_2 t) \\
y_2 &= \psi_{21} A_1 \cos(\omega_1 t) + \psi_{22} A_2 \cos(\omega_2 t)
\end{aligned}
\tag{B.78}
$$

One observes that both natural frequencies are present in the solution for the physical coordinates.

The solution above is unrealistic in that the motion is undamped and therefore persists indefinitely; some damping mechanism is required. The equations of motion of the two-mass system should be modified to give

$$[m]\{\ddot{y}\} + [c]\{\dot{y}\} + [k]\{y\} = \{0\} \tag{B.79}$$

where $[c]$ is called the *damping matrix*. A problem arises now if one tries to repeat the analysis above for the damped system. Generally, there is no matrix $[\psi]$ that will simultaneously diagonalise *three* matrices $[m]$, $[c]$ and $[k]$. Consequently, no transformation exists that uncouples the equations of motion. The simplest means of circumnavigating this problem is to assume *proportional* or *Rayleigh damping*. This means that one assumes

$$[c] = \alpha[m] + \beta[k] \tag{B.80}$$

where α and β are constants. This is a fairly restrictive assumption and, in many cases, it does not hold. In particular, if the damping is nonlinear, one cannot apply this assumption. However, with this form of damping, one finds that the diagonalising matrix $[\psi]$ for the undamped motion also suffices for the damped motion. In fact,

$$[\psi]^T[c][\psi] = [C] = \alpha[M] + \beta[K] \tag{B.81}$$

with diagonal entries the *modal dampings*, given by

$$c_i = \alpha m_i + \beta k_i \tag{B.82}$$

For this type of damping, the equations of motion uncouple as before on transforming to modal coordinates, so that

$$m_1 \ddot{z}_1 + c_1 \dot{z}_1 + k_1 z_1 = 0$$
$$m_2 \ddot{z}_2 + c_2 \dot{z}_2 + k_2 z_2 = 0 \tag{B.83}$$

The solutions are

$$z_1 = A_1 e^{-\zeta_1 \omega_1 t} \cos(\omega_{d1} t)$$
$$z_2 = A_2 e^{-\zeta_2 \omega_2 t} \cos(\omega_{d2} t) \tag{B.84}$$

where the damped natural frequencies and *modal* damping ratios are specified by

$$\zeta_i = \frac{c_i}{2\sqrt{m_i k_i}}, \quad \omega_{di}^2 = \omega_i^2 \left(1 - \zeta_i^2\right) \tag{B.85}$$

On transforming back to the physical coordinates, one obtains

$$y_1 = \psi_{11} A_1 e^{-\zeta_1 \omega_1 t} \cos(\omega_{d1} t) + \psi_{12} A_2 e^{-\zeta_2 \omega_2 t} \cos(\omega_{d2} t)$$
$$y_2 = \psi_{21} A_1 e^{-\zeta_1 \omega_1 t} \cos(\omega_{d1} t) + \psi_{22} A_2 e^{-\zeta_2 \omega_2 t} \cos(\omega_{d2} t) \tag{B.86}$$

and the free motion is a sum of damped harmonics at the damped natural frequencies. Note that the rates of decay are different for each frequency component.

The forced response of the system can be obtained in much the same manner as for the SDOF system. In order to simplify matters slightly, the excitation vector is assumed to have the form

$$\{r\} = \begin{pmatrix} x_1(t) \\ 0 \end{pmatrix} \tag{B.87}$$

On transforming the forced equation to modal coordinates, one obtains

$$[M]\{\ddot{z}\} + [C]\{\dot{z}\} + [K]\{z\} = \{p\} = [\psi]^T\{x\} \tag{B.88}$$

where

$$\{p\} = \begin{pmatrix} p_1 \\ p_2 \end{pmatrix} = \begin{pmatrix} \psi_{11} x_1 \\ \psi_{12} x_1 \end{pmatrix} \tag{B.89}$$

so that

$$m_1 \ddot{z}_1 + c_1 \dot{z}_1 + k_1 z_1 = p_1$$
$$m_2 \ddot{z}_2 + c_2 \dot{z}_2 + k_2 z_2 = p_2$$
(B.90)

For a harmonic input $x_1(t)$ these SDOF equations can be solved directly as in Section B.1.

The representation of the system in the frequency domain is obtained by Fourier transforming the Equations (B.89) and (B.90). The results are

$$Z_1(\omega) = \frac{\psi_{11}}{-m_1\omega^2 + ic_1\omega + k_1} X_1(\omega)$$
(B.91)

$$Z_2(\omega) = \frac{\psi_{12}}{-m_2\omega^2 + ic_2\omega + k_2} X_1(\omega)$$
(B.92)

and linearity of the Fourier transform implies, from Equation (B.72),

$$Y_1(\omega) = \psi_{11} Z_1(\omega) + \psi_{12} Z_2(\omega) = \left\{ \frac{\psi_{11}^2}{-m_1\omega^2 + ic_1\omega + k_1} + \frac{\psi_{12}^2}{-m_2\omega^2 + ic_2\omega + k_2} \right\} X_1(\omega)$$
(B.93)

$$Y_2(\omega) = \psi_{21} Z_1(\omega) + \psi_{22} Z_2(\omega) = \left\{ \frac{\psi_{21}\psi_{11}}{-m_1\omega^2 + ic_1\omega + k_1} + \frac{\psi_{12}\psi_{22}}{-m_2\omega^2 + ic_2\omega + k_2} \right\} X_1(\omega)$$
(B.94)

Recalling that $Y(\omega) = H(\omega)X(\omega)$, the overall FRFs for the processes $x_1(t) \to y_1(t)$ and $x_1(t) \to y_2(t)$ are therefore given by

$$H_{11}(\omega) = \frac{Y_1(\omega)}{X_1(\omega)} = \frac{\psi_{11}^2}{-m_1\omega^2 + ic_1\omega + k_1} + \frac{\psi_{12}^2}{-m_2\omega^2 + ic_2\omega + k_2}$$
(B.95)

$$H_{12}(\omega) = \frac{Y_2(\omega)}{X_1(\omega)} = \frac{\psi_{21}\psi_{11}}{-m_1\omega^2 + ic_1\omega + k_1} + \frac{\psi_{12}\psi_{22}}{-m_2\omega^2 + ic_2\omega + k_2}$$
(B.96)

On referring back to the formula for the resonant frequency of an SDOF system, it is clear from these expressions that the Bode plot for each of these expressions will show two peaks or resonances at the frequencies

$$\omega_{r1} = \omega_1 \sqrt{1 - 2\zeta_1^2}$$
$$\omega_{r2} = \omega_2 \sqrt{1 - 2\zeta_2^2}$$
(B.97)

As an example, the Bode plot for the first coordinate y_1 of the system

$$\begin{pmatrix} 1 & 0 \\ 0 & 1 \end{pmatrix} \begin{pmatrix} \ddot{y}_1 \\ \ddot{y}_2 \end{pmatrix} + 20 \begin{pmatrix} 1 & 0 \\ 0 & 1 \end{pmatrix} \begin{pmatrix} \dot{y}_1 \\ \dot{y}_2 \end{pmatrix} + 10^4 \begin{pmatrix} 2 & -1 \\ -1 & 2 \end{pmatrix} \begin{pmatrix} y_1 \\ y_2 \end{pmatrix} = \begin{pmatrix} x_1 \\ 0 \end{pmatrix}$$
(B.98)

is given in Figure B.13.

It has proved useful to consider a 2-DOF system to discuss how natural frequencies and so on generalise to MDOF systems. However, as one might expect, it is possible to deal with linear systems with arbitrary numbers of DOF at the expense of a little more abstraction. This is the subject of the last section.

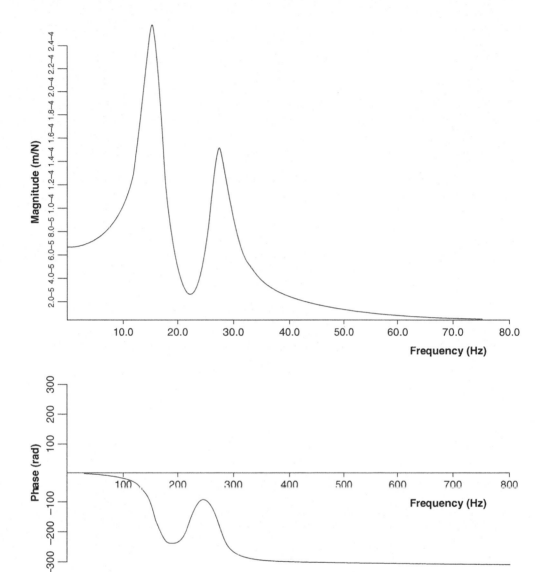

Figure B.13 H_{11} Bode plot for a two-degree-of-freedom system of Equation (B.98).

B.6   Modal Analysis

B.6.1   Free, Undamped Motion

The object of this section is to formalise the arguments given previously for MDOF systems and state them in their full generality. As above, the theory will be provided in stages, starting with the simplest case, that of an undamped unforced system. The equation of motion for such a linear system is

$$[m]\{\ddot{y}\} + [k]\{y\} = 0 \tag{B.99}$$

where $\{y\}$ is now an $n \times 1$ column vector and $[m]$ and $[k]$ are $n \times n$ matrices. As always, the excitation is assumed to be harmonic, so the solution is assumed to have the form

$$\{y(t)\} = \{\psi\}e^{i\omega t} \tag{B.100}$$

where $\{\psi\}$ is a constant $n \times 1$ vector. This ansatz basically assumes that all points on the structure move in phase with the same frequency. Substituting into Equation (B.99) yields

$$-\omega^2[m]\{\psi\} + [k]\{\psi\} = 0 \tag{B.101}$$

which is a standard linear eigenvalue problem with n solutions ω_{ni} and $\{\psi_i\}$. (The eigenvalue problem is of the most fundamental in all linear algebra or matrix algebra. There are any number of excellent texts the reader could consult for details; Fox (1964) can be considered somewhat of a classic.) These are the undamped natural frequencies and the mode shapes. The interpretation is well known: if the system is excited at a frequency ω_{ni}, all points will move in phase with a profile given by $\{\psi_i\}$.

If it is assumed that $[m]$ is invertible (and this is usually true), it is a simple matter to rewrite Equation (B.101) in the more usual form for an eigenvalue problem,

$$[m]^{-1}[k]\{\psi_i\} - \frac{1}{\omega_{ni}^2}\{\psi_i\} = [D]\{\psi_i\} - \lambda_i\{\psi_i\} = 0 \tag{B.102}$$

with a little notation added. Note that the normalisation of $\{\psi_i\}$ is arbitrary; that is, if $\{\psi_i\}$ is a solution of Equation (B.102), then so is $\alpha\{\psi_i\}$ for any real number α. Common normalisations for mode shapes include setting the largest element to unity or setting the length of the vector to unity, that is $\{\psi_i\}^T\{\psi_i\} = 1$.

Nontrivial solutions of Equation (B.102) must have $\{\psi_i\} \neq \{0\}$. This forces the *characteristic equation*,

$$\det([D] - \lambda_i[1]) = 0 \tag{B.103}$$

which has n solutions for the λ_i as required.

This apparently flexible system of equations turns out to have rather constrained solutions for the mode shapes. The reason is that $[m]$ and $[k]$ can almost always be assumed symmetric. This is a consequence of the property of reciprocity mentioned earlier.

Suppose that ω_{ni}^2 and ω_{nj}^2 are *distinct* eigenvalues of Equation (B.102); then

$$\begin{aligned} -\omega_{ni}^2[m]\{\psi_i\} &= [k]\{\psi_i\} \\ -\omega_{nj}^2[m]\{\psi_j\} &= [k]\{\psi_j\} \end{aligned} \tag{B.104}$$

Now, premultiplying the first of these expressions by $\{\psi_j\}^T$ and the second by $\{\psi_i\}^T$ gives

$$\begin{aligned} -\omega_{ni}^2\{\psi_j\}^T[m]\{\psi_i\} &= \{\psi_j\}^T[k]\{\psi_i\} \\ -\omega_{nj}^2\{\psi_i\}^T[m]\{\psi_j\} &= \{\psi_i\}^T[k]\{\psi_j\} \end{aligned} \tag{B.105}$$

and as $[m]$ and $[k]$ are symmetric, it follows that

$$\begin{aligned} (\{\psi_j\}^T[m]\{\psi_i\})^T &= \{\psi_i\}^T[m]\{\psi_j\} \\ (\{\psi_j\}^T[k]\{\psi_i\})^T &= \{\psi_i\}^T[k]\{\psi_j\} \end{aligned} \tag{B.106}$$

Transposing the first expression in Equation (B.105) and subtracting from the second expression yields

$$\left(\omega_{ni}^2 - \omega_{nj}^2\right)\{\psi_i\}^{\mathrm{T}}[m]\{\psi_j\} = 0 \qquad (B.107)$$

and, as $\omega_{ni} \neq \omega_{nj}$, it follows that

$$\{\psi_i\}^{\mathrm{T}}[m]\{\psi_j\} = 0 \qquad (B.108)$$

From Equation (B.105) it follows that

$$\{\psi_i\}^{\mathrm{T}}[k]\{\psi_j\} = 0 \qquad (B.109)$$

Therefore the mode shapes belonging to distinct eigenvalues are *orthogonal with respect to the mass and stiffness matrices*. This is referred to as *weighted orthogonality*. The situation where the eigenvalues are not distinct is a little more complicated and will not be discussed here; the reader can refer to Ewins (1984). Note that unless the mass or stiffness matrix is the unit matrix, the eigenvectors or mode shapes are *not* orthogonal in the usual sense, that is $\{\psi_i\}^{\mathrm{T}}\{\psi_j\} \neq 0$. Assuming n distinct eigenvalues, one can form the *modal matrix* $[\Psi]$ by taking an array of the mode shapes,

$$[\Psi] = \{\{\psi_1\}, \{\psi_2\}, \ldots, \{\psi_n\}\} \qquad (B.110)$$

Consider the matrix

$$[M] = [\Psi]^{\mathrm{T}}[m][\Psi] \qquad (B.111)$$

A little algebra shows that the elements are

$$M_{ij} = \{\psi_i\}^{\mathrm{T}}[m]\{\psi_j\} \qquad (B.112)$$

and these are zero if $i \neq j$ by the weighted orthogonality (B.108). This means that $[M]$ is diagonal. The diagonal elements $m_1, m_2, \ldots, m_n$ are referred to as the *generalised masses* or modal masses, as discussed in the last section. By a similar argument, the matrix

$$[K] = [\Psi]^{\mathrm{T}}[k][\Psi] \qquad (B.113)$$

is diagonal with elements $k_1, k_2, \ldots, k_n$, which are termed the *generalised* or modal stiffnesses. The implications for the equations of motion (B.99) are important. Consider the change of coordinates,

$$[\Psi]\{u\} = \{y\} \qquad (B.114)$$

Equation (B.99) becomes

$$[m][\Psi]\{\ddot{u}\} + [k][\Psi]\{u\} = 0 \qquad (B.115)$$

and premultiplying by $[\Psi]^{\mathrm{T}}$ gives

$$[\Psi]^{\mathrm{T}}[m][\Psi]\{\ddot{u}\} + [\Psi]^{\mathrm{T}}[k][\Psi]\{u\} = 0 \qquad (B.116)$$

or

$$[M]\{\ddot{u}\} + [K]\{u\} = 0 \tag{B.117}$$

by virtue of Equations (B.111) and (B.113). The system has been decoupled into n SDOF equations of motion of the form

$$m_i \ddot{u}_i + k_i u_i = 0, \quad i = 1, \ldots, n \tag{B.118}$$

and it follows, by premultiplying the first equation of (B.104) by $\{\psi_i\}$, that

$$\omega_{ni}^2 = \frac{k_i}{m_i} \tag{B.119}$$

and Equation (B.118) becomes

$$\ddot{u}_i + \omega_{ni}^2 u_i = 0 \tag{B.120}$$

which is the equation of an undamped SDOF oscillator with undamped natural frequency ω_{ni}. The coordinates u_i are termed *generalised*, *modal* or *normal* coordinates. Now, following the SDOF theory developed in the course of this appendix, the solution of Equation (B.120) is simply

$$u_i = U_i \cos(\omega_{ni} t) \tag{B.121}$$

and in the original physical coordinates, the response can contain components at *all* natural frequencies,

$$y_i = \sum_{j=1}^{n} \Psi_{ij} U_j \cos(\omega_{nj} t) \tag{B.122}$$

Before passing to the damped case, it is worthwhile to return to the question of normalisation. Different normalisations lead to different modal masses and stiffnesses; however, they are always constrained to satisfy $k_i/m_i = \omega_{ni}^2$. A common approach is to use *mass normalisation* as follows. Suppose a modal matrix $[\Psi]$ is specified such that the modal mass matrix is $[M]$; if one defines $[\Phi]$ by

$$[\Phi] = [\Psi][M]^{-1/2} \tag{B.123}$$

it follows that

$$[\Phi]^{\mathrm{T}}[m][\Phi] = [1] \tag{B.124}$$

$$[\Phi]^{\mathrm{T}}[k][\Phi] = [\Lambda]^2 \tag{B.125}$$

where

$$[\Lambda] = \mathrm{diag}(\omega_{n1}, \omega_{n2}, \ldots, \omega_{nn}) \tag{B.126}$$

and this representation is unique. Equation (B.131) amounts to choosing

$$\{\varphi_i\} = \frac{1}{\sqrt{m_i}}\{\psi_i\} \tag{B.127}$$

B.6.2   Free, Damped Motion

It is a simple matter to generalise Equation (B.99) to the damped case. The relevant equation is

$$[m]\{\ddot{y}\} + [c]\{\dot{y}\} + [k]\{y\} = 0 \tag{B.128}$$

with $[c]$ termed the (viscous) *damping matrix*. (In many cases, it will be desirable to consider structural damping. The reader is referred to Nashif et al. (1985), which is a standard work on damping in structural dynamics.) The desired result is to decouple Equation (B.128) into SDOF oscillators in much the same way as for the undamped case. Unfortunately, this is generally impossible, as observed in the last section. Rather than give up, the usual recourse is to assume the MDOF form of Rayleigh or proportional damping (see Equation (B.80)),

$$[c] = \alpha[m] + \beta[m] \tag{B.129}$$

and $[\psi]$ will then diagonalise it also, that is

$$[\Psi]^{\mathrm{T}}[c][\Psi] = [C] = \mathrm{diag}(c_1, \ldots, c_n) \tag{B.130}$$

with

$$c_i = \alpha m_i + \beta k_i \tag{B.131}$$

With this assumption, the modal matrix decouples the system (B.128) into n SDOF systems in much the same way as for the undamped case. The relevant equations are (after the transformation corresponding to Equation (B.116))

$$m_i \ddot{u}_i + c_i \dot{u}_i + k_i u_i = 0, \quad i = 1, \ldots, n \tag{B.132}$$

These have solutions

$$u_i = A_i e^{-\zeta_i \omega_{ni} t} \sin(\omega_{di} t - \theta_i) \tag{B.133}$$

where A_i and θ_i are fixed by the initial conditions and

$$\zeta_i = \frac{c_i}{2\sqrt{m_i k_i}} \tag{B.134}$$

is the ith modal damping ratio and

$$\omega_{di}^2 = \omega_{ni}^2 \left(1 - \zeta_i^2\right) \tag{B.135}$$

is the ith damped natural frequency. Transforming back to physical coordinates using Equation (B.114) yields

$$y_i = \sum_{j=1}^{n} \Psi_{ij} A_j e^{-\zeta_i \omega_{ni} t} \sin(\omega_{di} t - \theta_i) \tag{B.136}$$

B.6.3   Forced, Damped Motion

The general forced linear MDOF system is

$$[m]\{\ddot{y}\} + [c]\{\dot{y}\} + [k]\{y\} = \{x(t)\} \tag{B.137}$$

where $\{x(t)\}$ is an $n \times 1$ vector of time-dependent excitations. As in the free, damped case, one can change to modal coordinates; the result is

$$[M]\{\ddot{u}\} + [C]\{\dot{u}\} + [K]\{u\} = [\Psi]^T\{x(t)\} = \{p\} \tag{B.138}$$

which serves to define $\{p\}$, the vector of *generalised forces*. As before (under the assumption of proportional damping), the equations decouple into n SDOF systems,

$$m_i \ddot{u}_i + c_i \dot{u}_i + k_i u_i = p_i, \quad i = 1, \dots, n \tag{B.139}$$

and all of the analysis relevant to SDOF systems developed previously applies.

It is instructive to develop the theory in the frequency domain. Suppose the excitations p_i are broadband random; it is sensible to think in terms of FRFs. The ith modal FRF (i.e. the FRF associated with the process $p_i \longrightarrow u_i$) is

$$G_i(\omega) = \frac{S_{u_i p_i}(\omega)}{S_{u_i u_i}(\omega)} = \frac{1}{-m_i \omega^2 + i c_i \omega + k_i} \tag{B.140}$$

In order to allow a simple derivation of the FRFs in physical coordinates, one makes the formal definition,

$$\{Y(\omega)\} = [H(\omega)]\{X(\omega)\} \tag{B.141}$$

of $[H(\omega)]$ the FRF matrix. According to Equation (A.138), the corresponding relation in modal coordinates is

$$\{U(\omega)\} = [G(\omega)]\{P(\omega)\} \tag{B.142}$$

with $[G(\omega)] = \mathrm{diag}(G_1(\omega), \dots, G_n(\omega))$ diagonal. Substituting for $\{U\}$ and $\{P\}$ in the last expression gives

$$[\Psi]^{-1}\{Y(\omega)\} = [G(\omega)][\Psi]^T\{X(\omega)\} \tag{B.143}$$

or

$$\{Y(\omega)\} = [\Psi][G(\omega)][\Psi]^T\{X(\omega)\} \tag{B.144}$$

which identifies

$$[H(\omega)] = [\Psi][G(\omega)][\Psi]^T \tag{B.145}$$

In terms of the individual elements of $[H]$, Equation (B.145) yields

$$H_{ij}(\omega) = \sum_{l=1}^{n} \sum_{k=1}^{n} \psi_{il} G(\omega)_{lk} \psi_{kj}^T = \sum_{k=1}^{n} \psi_{ik} G_{kk}(\omega) \psi_{jk} \tag{B.146}$$

and, finally,

$$H_{ij}(\omega) = \sum_{k=1}^{n} \frac{\psi_{ik}\psi_{jk}}{-m_i\omega^2 + ic_i\omega + k_i} \tag{B.147}$$

or

$$H_{ij}(\omega) = \sum_{k=1}^{n} \frac{{}_k A_{ij}}{-\left(\omega^2 - \omega_{nk}^2\right) + 2i\zeta_k\omega_{nk}\omega} \tag{B.148}$$

where

$$_k A_{ij} = \frac{\psi_{ik}\psi_{jk}}{m_k} = \varphi_{ik}\varphi_{jk} \tag{B.149}$$

are the *residues* or modal constants.

It follows from the equations above that the FRF for any process $x_i \longrightarrow y_j$ of an MDOF linear system is the sum of n SDOF FRFs, one for each natural frequency. It is straightforward to show that each individual mode has a resonance frequency,

$$\omega_{ri} = \omega_{ni}\sqrt{1 - 2\zeta_i^2} \tag{B.150}$$

Taking the inverse Fourier transform of the expression (B.142) gives the general form of the impulse response for an MDOF system,

$$h_{ij}(t) = \sum_{k=1}^{n} \frac{{}_k A_{ij}}{\omega_{dk}} e^{-\zeta_k\omega_k t} \cos(\omega_{dk}t - \theta_k) \tag{B.151}$$

The response of a general MDOF system to a transient is a sum of decaying harmonics with individual decay rates and frequencies.

A final remark is required about the proportionality assumption for the damping. For a little more effort than that expended above, one can obtain the system FRFs for an arbitrarily damped linear system (Ewins, 1984). The only change in the final form (B.148) is that the constants ${}_k A_{ij}$ become complex.

All the expressions above are given in receptance form. Parallel mobility and accelerance forms exist and are obtained by multiplying the receptance form by $i\omega$ and $-\omega^2$, respectively.

There are well-established signal processing techniques that allow one to determine experimentally the FRFs of a system as discussed in Appendix A in considerable detail. It is found for linear structural systems that the representation as a sum of resonances given in Equation (B.148) is remarkably accurate.

So far in this appendix there have only been very simple examples of FRFs; only the results from true lumped-mass systems with one or two DOFs have been shown. As an example of how much complexity a real system FRF can display, a small case study is given. The structure of interest is a composite laminate plate with a laminate stiffener with a foam core across the underside. The dimensions of the top plate are 600 mm by 150 mm and it has a thickness of 3 mm built up from 12 layers of woven carbon fibre reinforced polymer; the exact and detailed dimensions are not particularly important here. The sample is shown in Figure B.14 in the test rig used for modal analysis.

The sample was suspended by soft springs and fishing line from a metal framework in order to simulate free boundary conditions. Excitation was provided by an instrumented hammer and the response was sensed with a small piezoelectric accelerometer. Data were sampled using an acquisition system under the control of a PC.

Figure B.14 Composite sample under modal test.

Each individual FRF was obtained by sampling 8192 time points and thus yielded 4096 spectral lines (labelled 'points' in Figure B.15). The frequency range covered in the experiment was 0 to 10 240 Hz, giving a frequency resolution of 2.5 Hz. The final FRF was formed from 180 impacts/averages, which accounts for its smoothness. The magnitude of the averaged FRF is shown in Figure B.15.

On the order of 50 resonances are visible in the FRF. This is by no means unusual for a modal test but perhaps more typical of mechanical structures rather than civil, as it can sometimes prove difficult to excite many modes of very large structures.

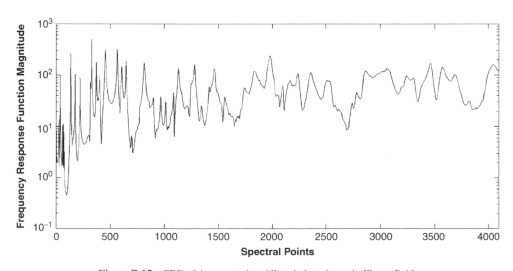

Figure B.15 FRF of the composite stiffened plate shown in Figure B.18.

After obtaining an experimental curve for some FRF $H(\omega)$ the data can be curve-fitted to the form in Equation (B.148) and the best-fit values for the parameters ω_{ni}, ζ_{ni}, $_kA_{ij}$, $i, j, k = 1, \ldots, N$, can be obtained. The resulting model is called a *modal model* of the system. The body of theory that covers the estimation of FRFs and so on, together with the algorithms for extracting parameters, is called *modal analysis*. Modal analysis is now a mature subject and the academic and industrial standard for the analysis of linear structures; excellent references are Ewins (1984) (and the later edition) and Maia *et al.* (1997). To discuss curve-fitting algorithms in any detail here would be to stray too far from the core material of this book; however, there is a very large literature that the curious reader can consult, beginning with the two references just mentioned. Many of the more recent developments in modal analysis have centred around the development of a very powerful class of algorithms called *subspace algorithms* (van Overshee and Moor, 1996); these are based on a state-space formulation of the equations of motion. The subspace algorithms can also be applied very effectively in the difficult case when the input excitation to a system is not known. This case is very important in SHM, particularly of civil infrastructure, as the systems of interest will actually be forced by ambient excitation, which cannot be measured. A good example of this situation is wind or traffic excitation of bridges.

References

Ewins, D.J. (1984) *Modal Testing: Theory and Practice*, Research Studies Press.

Fox, L. (1964) *An Introduction to Numerical Linear Algebra*, Monographs on Numerical Analysis, Clarendon Press, Oxford.

Inman, D.J. (1994) *Engineering Vibration*, Prentice Hall.

Maia, N.M.M. *et al.* (1997) *Theoretical and Experimental Modal Analysis*, Mechanical Engineering Research Studies: Engineering Dynamics Series, Wiley-Blackwell.

Nashif, A.D., Jones, D.I.G. and Henderson, J.P. (1985) *Vibration Damping*, John Wiley & Sons, Ltd.

Simmons, G.F. (1974) *Differential Equations*, McGraw-Hill.

Söderstrom, T. and Stoica, P. (1988) *System Identification*, Prentice-Hall, London.

Thompson, W.T. (1965) *Mechanical Vibrations with Applications*, George Allen and Unwin.

van Overschee, P. and Moor, B. (1996) *Subspace Identification for Linear Systems: Theory-Implementation-Applications*, Kluwer Academic.

Index

Structural Health Monitoring: A Machine Learning Perspective, First Edition. Charles R. Farrar and Keith Worden.
© 2013 John Wiley & Sons, Ltd. Published 2013 by John Wiley & Sons, Ltd.